THE MECHANICS AND THERMODYNAMICS OF CONTINUA

The Mechanics and Thermodynamics of Continua presents a unified treatment of continuum mechanics and thermodynamics that emphasizes the universal status of the basic balances and the entropy imbalance. These laws are viewed as fundamental building blocks on which to frame theories of material behavior. As a valuable reference source, this book presents a detailed and complete treatment of continuum mechanics and thermodynamics for graduates and advanced undergraduates in engineering, physics, and mathematics. The chapters on plasticity discuss the standard isotropic theories and crystal plasticity and gradient plasticity.

Morton E. Gurtin is the Alumni Professor Emeritus of Mathematics at Carnegie Mellon University. His research concerns nonlinear continuum mechanics and thermodynamics, with recent emphasis on applications to problems in materials science. Among his many awards are the 2004 Timoshenko Medal of the American Society of Mechanical Engineers (ASME) "in recognition of distinguished contributions to the field of applied mechanics"; the Agostinelli Prize (an annual prize in pure and applied mathematics and mathematical physics); Accademia Nazionale dei Lincei, Italy; Dottore Honoris Causa, Civil Engineering, University of Rome; Distinguished Graduate School Alumnus Award, Brown University; and the Richard Moore Education Award, Carnegie Mellon University. In addition to his numerous archival research publications, Professor Gurtin is the author of *Configurational Forces as Basic Concepts in Continuum Physics*, *An Introduction to Continuum Mechanics*, *Thermomechanics of Evolving Phase Boundaries in the Plane*, *Topics in Finite Elasticity*, *The Linear Theory of Elasticity*, *Handbuch der Physik*, Volume VIa/2, and *Wave Propagation in Dissipative Materials* (with B. D. Coleman, I. Herrera, and C. Truesdell).

Eliot Fried is a Professor of Mechanical Engineering at McGill University, where he holds the Tier I Canada Research Chair in Interfacial and Defect Mechanics. His research focuses on the mechanics and thermodynamics of novel materials, including liquid crystals, surfactant solutions, hydrogels, granular materials, biovesicles, and nanocrystalline alloys. He is the recipient of an NSF Mathematical Sciences Postdoctoral Fellowship, a Japan Society for the Promotion of Science Postdoctoral Research Fellowship, and an NSF Research Initiation Award. Prior to joining McGill, he held tenured faculty positions in the Department of Theoretical and Applied Mechanics at the University of Illinois at Urbana-Champaign and in the Department of Mechanical, Aerospace and Structural Engineering at Washington University in St. Louis. At Illinois, he was a Fellow of the Center of Advanced Study and was awarded a Critical Research Initiative Grant. His current research is funded by the National Science Foundation, the Department of Energy, the National Institute of Health, the Natural Sciences and Engineering Research Council of Canada, the Canada Research Chairs Program, and the Canada Foundation for Innovation.

Lallit Anand is the Rohsenow Professor of Mechanical Engineering at MIT. He has had more than twenty-five years of experience conducting research and teaching at MIT, as well as substantial research experience in industry. In 1975 he began his career as a Research Scientist in the Mechanical Sciences Division of the Fundamental Research Laboratory of U.S. Steel Corporation, and he joined the MIT faculty in 1982. His research concerns continuum mechanics of solids, with focus on inelastic deformation and failure of engineering materials. In 1992, Anand was awarded the Eric Reissner Medal for "outstanding contributions to the field of mechanics of materials" from the International Society for Computational Engineering Science. In 2007 he received the Khan International Medal for "outstanding lifelong contributions to the field of plasticity" from the *International Journal of Plasticity*. He is also a Fellow of the ASME.

The Mechanics and Thermodynamics of Continua

Morton E. Gurtin
Carnegie Mellon University

Eliot Fried
McGill University

Lallit Anand
Massachusetts Institute of Technology

CAMBRIDGE
UNIVERSITY PRESS

CAMBRIDGE UNIVERSITY PRESS
Cambridge, New York, Melbourne, Madrid, Cape Town,
Singapore, São Paulo, Delhi, Mexico City

Cambridge University Press
32 Avenue of the Americas, New York NY 10013-2473, USA

www.cambridge.org
Information on this title: www.cambridge.org/9781107617063

First published 2010
Reprinted 2011
First paperback edition 2013

A catalogue record for this publication is available from the British Library

Library of Congress Cataloguing in Publication Data

Gurtin, Morton E.
The mechanics and thermodynamics of continua / Morton E. Gurtin, Eliot Fried,
and Lallit Anand.
p. cm.
Includes bibliographical references and index.
ISBN 978-0-521-40598-0 (hardback)
1. Continuum mechanics – Mathematics. 2. Thermodynamics – Mathematics.
I. Fried, Eliot. II. Anand, Lallit. III. Title.
QA808.2.G863 2009
531–dc22 2009010668

ISBN 978-0-521-40598-0 Hardback
ISBN 978-1-107-61706-3 Paperback

Contents

Preface

The Central Thrust of This Book

A large class of theories in continuum physics takes as its starting point the balance laws for mass, for linear and angular momenta, and for energy, together with an entropy imbalance that represents the second law of thermodynamics. Unfortunately, most engineering curricula teach the momentum balance laws for an array of materials, often without informing students that these laws are actually independent of those materials. Further, while courses do discuss balance of energy, they often fail to mention the second law of thermodynamics, even though its place as a basic law for continua was carefully set forth by Truesdell and Toupin[1] almost half a century ago.

This book presents a unified treatment of continuum mechanics and thermodynamics that emphasizes the universal status of the basic balances and the entropy imbalance. These laws and an hypothesis – the principle of frame-indifference, which asserts that physical theories be independent of the observer (i.e., frame of reference) – are viewed as fundamental building blocks upon which to frame theories of material behavior.

The basic laws and the frame-indifference hypothesis – being independent of material – are common to all bodies that we discuss. On the other hand, particular materials are defined by additional equations in the form of constitutive relations (such as Fourier's law) and constraints (such as incompressibility). Trivially, such constitutive assumptions reflect the fact that two bodies, one made of steel and the other of wood, generally behave differently when subject to prescribed forces – even though the two bodies obey the same basic laws.

Our general discussion of constitutive equations is based on:

(i) the principle of frame-indifference;
(ii) the use of thermodynamics to restrict constitutive equations via a paradigm generally referred to as the *Coleman–Noll procedure.*

[1] Truesdell & Toupin (1960, p. 644). In the 1960s and early 1970s this form of the second law, generally referred to as the Clausius–Duhem inequality (cf. footnote 152), was considered to be controversial because – as the argument went – the notions of entropy and temperature make no sense outside of equilibrium, an argument that stands in stark contrast to the fact that temperatures are routinely measured at shock waves. The religious nature of this argument together with the observation that most conventional theories are consistent with this form of the second law gradually led to its general acceptance – and its overall power in describing new and more general theories gave additional credence to its place as a basic law of continuum physics.

Because frame-indifference and the Coleman–Noll procedure represent powerful tools for developing physically reasonable constitutive equations, we begin our discussion by developing such equations for:

(I) the conduction of heat in a rigid medium, as this represents an excellent vehicle for demonstrating the power of the Coleman–Noll procedure;
(II) the mechanical theories of both compressible and incompressible, linearly viscous fluids, where frame-indifference applied within a very general constitutive framework demonstrates the veracity of conventional constitutive relations for fluids.

Based on frame-indifference and using the Coleman–Noll procedure, we discuss the following topics: elastic solids under isothermal and nonisothermal conditions; coupled elastic deformation and species transport, where the species in question may be ionic, atomic, molecular, or chemical; both isotropic and crystalline plastic solids; and viscoplastic solids. In our treatment of these subjects, we consider general large-deformation theories as well as corresponding small-deformation theories.

Our discussion of rate-independent and rate-dependent plasticity is *not* traditional. Unlike – but compatible with – conventional treatments, we consider flow rules that give the deviatoric stress as a function of the plastic strain-rate (and an internal variable that represents hardening).[2] We also provide a parallel description of the conventional theory based on the *principle of virtual power*. We do this because: (i) it allows us to account separately for the stretching of the microscopic structure and the flow of dislocations through that structure as described, respectively, by the elastic and plastic strain-rates; (ii) it allows for a precise discussion of material stability; and (iii) it provides a basic structure within which one can formulate more general theories. In this last regard, conventional plasticity cannot characterize recent experimental results exhibiting size effects. To model size-dependent phenomena requires a theory of plasticity with one or more material length-scales. A number of recent theories – referred to as gradient theories – accomplish this by allowing for constitutive dependencies on gradients of plastic strain and/or its rate. Such dependencies generally lead to nonlocal flow rules in the form of partial differential equations with concomitant boundary conditions. For that reason, we find it most useful to develop gradient theories via the principle of virtual power, a paradigm that automatically delivers the partial-differential equations and boundary conditions from natural assumptions regarding the expenditure of power.

Requirements of space and pedagogy led us to omit several important topics such as liquid crystals, non-Newtonian fluids, configurational forces, relativistic continuum mechanics, computational mechanics, classical viscoelasticity, and couple-stress theory.

For Whom Is This Book Meant?

Our goal is a book suitable for engineers, physicists, and mathematicians. Moreover, with the intention of providing a valuable reference source, we have tried to present a fairly detailed and complete treatment of continuum mechanics and thermodynamics. Such an ambitious scope requires a willingness to bore some when discussing issues not familiar to others. We have used parts of this book with good

---

[2] We do this for consistency with the remainder of the book, which is based on the requirement that "the stress in a body is determined by the history of the motion of that body"; cf. TRUESDELL & NOLL (1965, p. 56). When discussing crystalline bodies, the flow rules express the resolved shear on the individual slip systems in terms of corresponding slip rates.

success in teaching graduates and advanced undergraduates in engineering, physics, and mathematics.

Direct Notation

For the most part, we use direct – as opposed to component (i.e., index) – notation. While some engineers and physicists might find this difficult, at least at first, we believe that the gain in clarity and insight more than compensates for the initial effort required. For those not familiar with direct notation, we have included helpful sections on vector and tensor algebra and analysis, and we present the most important results in both direct and component form.

Rigor

We present careful proofs of the basic theorems of the subject. However, when the proofs are complicated or lengthy they generally appear in petite at the end of the section in question. We also do not normally state smoothness hypotheses. Indeed, standard differentiability assumptions sufficient to make an argument rigorous are generally obvious to mathematicians and of little interest to engineers and physicists.

Attributions and Historical Issues

Our emphasis is on basic concepts and central results, not on the history of our subject. For correct references before 1965, we refer the reader to the great encyclopedic handbook articles of TRUESDELL & TOUPIN (1960) and TRUESDELL & NOLL (1965). These articles do not discuss plasticity; for the early history of that subject we refer the reader to the books of HILL (1950) and MALVERN (1969). For more recent work, we attempted to cite the contributions most central to our presentation, and we apologize in advance if we have not done so faultlessly.

Our Debt

We owe much to the chief cultivators of continuum mechanics and thermodynamics whose great work during the years 1947–1965 led to a rennaisance of the field. Their names, listed chronologically with respect to their earliest published contributions, are Ronald Rivlin, Clifford Truesdell, Jerald Ericksen, Richard Toupin, Walter Noll, and Bernard Coleman. With the exception of plasticity theory, much of this book stems from the work of these scholars – work central to the development of a unified treatment of continuum mechanics and thermodynamics based on (a) a precise statement of the balance laws for mass, linear and angular momentum, and energy, together with an entropy imbalance (the Clausius–Duhem inequality) that represents the second law of thermodynamics; (b) the unambiguous distinction between these basis laws and the notion of constitutive assumptions; and (c) a clear and compelling statement of material frame-indifference.

We are grateful to Paolo Podio-Guidugli, Guy Genin, and Giuseppe Tomassetti for their many valuable comments concerning the section on plasticity; to B. Daya Reddy for his help in developing material on variational inequalities for plasticity; and to Ian Murdoch for extensive discussions that expanded our understanding of the frame-indifference principle. Others who have contributed to this work are Paolo Cermelli, Xuemei Chen, Shaun Sellers, and Oleg Shklyaev.

PART I

VECTOR AND TENSOR ALGEBRA

Throughout this book:

(i) *Lightface* Latin and Greek letters generally denote **scalars**.
(ii) *Boldface lowercase* Latin and Greek letters generally denote **vectors**, but the letters **o**, **x**, **y**, and **z** are reserved for **points**.
(iii) *Boldface uppercase* Latin and Greek letters generally denote **tensors**, but the letters **X**, **Y**, and **Z** are reserved for **points**.

1 Vector Algebra

We assume that the reader has had a basic course in vector algebra and, therefore, in introducing this subject, we take a relaxed approach that does not begin with formal definitions of point and vector spaces.

Roughly speaking, a point **x** is a dot in space and a vector **v** is an arrow that may be placed anywhere in space. As an example from everyday life, on a street map of a city

$$\mathbf{x} \stackrel{\text{def}}{=} \text{the corner of 4th Street and 5th Avenue}$$

might describe a point on the map, and to describe a second point **y** one-quarter of a mile northeast of **x** one might let

$$\mathbf{v} \stackrel{\text{def}}{=} \text{the vector whose direction is northeast and whose length is one-quarter mile}$$

and write $\mathbf{y} = \mathbf{x} + \mathbf{v}$. Thus, it would seem that a reasonable definition of a point space would require two basic notions: that of a point and that of a vector. Of course, granted a choice **o** of origin, one can identify all points with their vectors from **o**; but such an identification is artificial, since there is no intrinsic way of defining **o**. (What point on a street map would you call the origin?)

The space under consideration will always be a **three-dimensional Euclidean point space** $\mathcal{E}$. The term **point** will be reserved for elements of $\mathcal{E}$ and the term **vector** for elements of the associated vector space $\mathcal{V}$. Then:

(i) The difference $\mathbf{v} = \mathbf{y} - \mathbf{x}$ between the points **y** and **x** is a vector.
(ii) The sum $\mathbf{y} = \mathbf{x} + \mathbf{v}$ of a point **x** and a vector **v** is a point.
(iii) The zero vector is denoted by **0**: for **v** any vector, $\mathbf{v} = \mathbf{v} + \mathbf{0}$.
(iv) Unlike the sum of two vectors, the sum of two points has *no* meaning.

1.1 Inner Product. Cross Product

Our assumption that the point space $\mathcal{E}$ be Euclidean automatically endows the associated vector space $\mathcal{V}$ with an inner product.[3] We use the standard notation of vector analysis. In particular,

- The **inner product** (a scalar) and **cross product** (a vector)[4] of vectors **u** and **v** are respectively designated by

$$\mathbf{u} \cdot \mathbf{v} \qquad \text{and} \qquad \mathbf{u} \times \mathbf{v}.$$

---

[3] The inner product is often referred to as the dot product.

[4] We assume that the reader has some familiarity with these notions. The cross product is ordered in the sense that the cross product $\mathbf{u} \times \mathbf{v}$ of **u** and **v** is not generally equal to the cross product $\mathbf{v} \times \mathbf{u}$ of **v** and **u**.

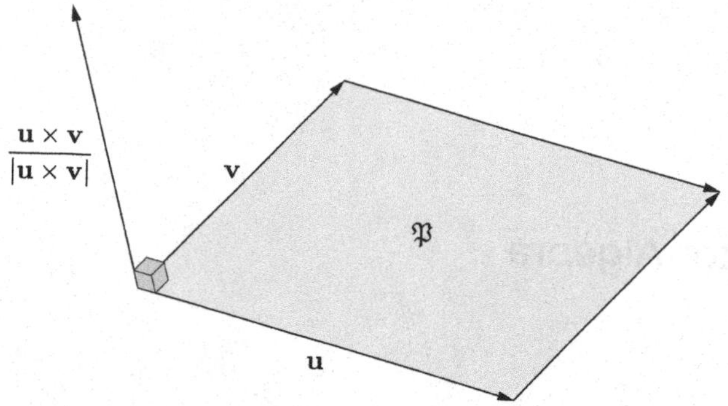

Figure 1.1. The parallelogram $\mathfrak{P}$ defined by the vectors **u** and **v** and the direction of $\mathbf{u}\times\mathbf{v}$ determined by the right-hand screw rule.

The inner product determines the magnitude (or length) of a vector **u** via the relation

$$|\mathbf{u}| = \sqrt{\mathbf{u}\cdot\mathbf{u}};$$

and the **angle**

$$\theta = \angle(\mathbf{u}, \mathbf{v}) \tag{1.1}$$

between nonzero vectors **u** and **v** is defined by

$$\cos\theta = \frac{\mathbf{u}\cdot\mathbf{v}}{|\mathbf{u}||\mathbf{v}|} \qquad (0 \le \theta \le \pi). \tag{1.2}$$

Since $-|\mathbf{u}||\mathbf{v}| \le \mathbf{u}\cdot\mathbf{v} \le |\mathbf{u}||\mathbf{v}|$, this definition assigns exactly one angle θ to each pair of nonzero vectors **u** and **v**. Trivially,

$$\mathbf{u}\cdot\mathbf{v} = |\mathbf{u}||\mathbf{v}|\cos\theta;$$

this relation is often used to define the inner product.

With regard to the cross product, the magnitude

$$|\mathbf{u}\times\mathbf{v}| \tag{1.3}$$

represents the **area spanned** by the vectors **u** and **v**; that is, the *area* of the parallelogram $\mathfrak{P}$ **defined by** these vectors as indicated in Figure 1.1; this area is nonzero if and only if **u** and **v** are *linearly independent.*[5] Further, and what is most important, if **u** and **v** are linearly independent, then:

(i) the magnitude of $\mathbf{u}\times\mathbf{v}$ is given by

$$|\mathbf{u}\times\mathbf{v}| = |\mathbf{u}||\mathbf{v}|\sin\theta \qquad (0 < \theta < \pi);$$

(ii) the vector $\mathbf{u}\times\mathbf{v}$ is *orthogonal* to both **u** and **v** with *direction* given by the right-hand screw rule.[6]

---

[5] Vectors **u**, **v**, and **w** are linearly dependent if, for some choice of the scalars a, b, and c, *not all zero*,

$$a\mathbf{u} + b\mathbf{v} + c\mathbf{w} = \mathbf{0}.$$

A similar definition applies to a pair of vectors **u** and **v**. Finally, a set of vectors is *linearly independent* if it is not linearly dependent.

[6] Asserting formally that a right-hand screw revolved from **u** to **v** will advance in its nut toward $\mathbf{u}\times\mathbf{v}$ (Figure 1.1); cf. Brand (1947, §16) and Jeffrey (2002, §2.3). A simple consequence of this is that $\mathbf{v}\times\mathbf{u} = -\mathbf{u}\times\mathbf{v}$ and thus, in particular, that the cross product of **u** and **v** vanishes whenever **u** and **v** are linearly dependent (Footnotes 4 and 5).

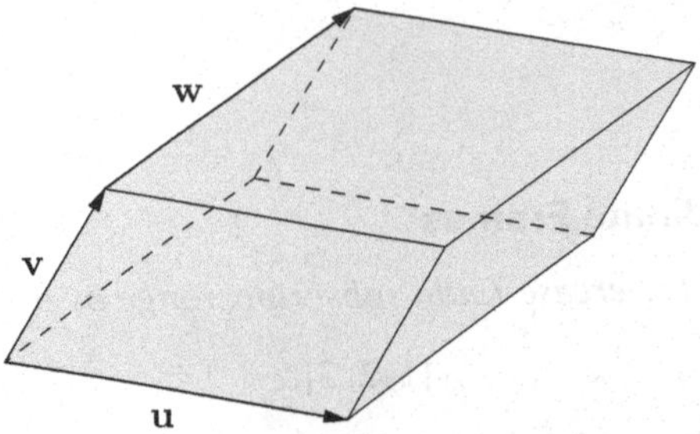

Figure 1.2. The parallelepiped defined by the vectors **u**, **v**, and **w**.

Since $\mathbf{u} \times \mathbf{v}$ gives the area of $\mathfrak{P}$, while $\mathbf{u} \times \mathbf{v}$ is orthogonal to $\mathfrak{P}$, $\mathbf{u} \times \mathbf{v}$ is sometimes referred to as the **area vector** of $\mathfrak{P}$.

Similarly,

$$|\mathbf{u} \cdot (\mathbf{v} \times \mathbf{w})| \tag{1.4}$$

represents the **volume spanned** by the vectors **u**, **v**, and **w**; that is, the volume of the parallelepiped **defined by** these vectors as indicated in Figure 1.2. If this volume is nonzero, then **u**, **v**, and **w** are *linearly independent.*

If **u**, **v**, and **w** are linearly independent, then the **triad** $\{\mathbf{u}, \mathbf{v}, \mathbf{w}\}$ forms a **basis** for $\mathcal{V}$ in the sense that any vector **a** may be uniquely represented in terms of that triad, that is, there are unique scalars α, β, and γ such that

$$\mathbf{a} = \alpha\mathbf{u} + \beta\mathbf{v} + \gamma\mathbf{w}.$$

A basis $\{\mathbf{u}, \mathbf{v}, \mathbf{w}\}$ is **positively oriented**[7] if

$$\mathbf{u} \cdot (\mathbf{v} \times \mathbf{w}) > 0.$$

Two bases $\{\mathbf{a}, \mathbf{b}, \mathbf{c}\}$ and $\{\mathbf{u}, \mathbf{v}, \mathbf{w}\}$ have the **same orientation** if

$$\mathbf{a} \cdot (\mathbf{b} \times \mathbf{c}) \text{ and } \mathbf{u} \cdot (\mathbf{v} \times \mathbf{w}) \text{ have the same sign.} \tag{1.5}$$

A basis $\{\mathbf{u}, \mathbf{v}, \mathbf{w}\}$ is **orthonormal** if

$$\mathbf{u} \cdot \mathbf{v} = \mathbf{v} \cdot \mathbf{w} = \mathbf{w} \cdot \mathbf{u} = 0 \qquad \text{and} \qquad |\mathbf{u}| = |\mathbf{v}| = |\mathbf{w}| = 1, \tag{1.6}$$

so that **u**, **v**, and **w** are mutually orthogonal and of unit length.

A standard method of showing that two vectors **a** and **b** are equal uses the following result:

$$\mathbf{a} \cdot \mathbf{v} = \mathbf{b} \cdot \mathbf{v} \text{ for all vectors } \mathbf{v} \text{ if and only if } \mathbf{a} = \mathbf{b}. \tag{1.7}$$

The verification of this result is not difficult. Assume that $\mathbf{a} \cdot \mathbf{v} = \mathbf{b} \cdot \mathbf{v}$ for all **v**. Then the choice $\mathbf{v} = \mathbf{a} - \mathbf{b}$ yields $|\mathbf{a} - \mathbf{b}|^2 = 0$ and, hence, $\mathbf{a} = \mathbf{b}$. Similarly,

$$\mathbf{a} \times \mathbf{v} = \mathbf{b} \times \mathbf{v} \text{ for all vectors } \mathbf{v} \text{ if and only if } \mathbf{a} = \mathbf{b}, \tag{1.8}$$

which is a result whose proof we leave as an exercise.

A subset $\mathcal{K}$ of vectors is referred to as a **subspace** if, given any vectors **u** and **v** belonging to $\mathcal{K}$ and any scalars α and β, the **linear combination**

$$\alpha\mathbf{u} + \beta\mathbf{v} \text{ belongs to } \mathcal{K}. \tag{1.9}$$

Examples of subspaces of $\mathcal{V}$ are the singleton $\{\mathbf{0}\}$, a line through the origin, a plane through the origin, and $\mathcal{V}$ itself. There are no other examples.

---

[7] We view a positively oriented basis as right-handed, since the definition of the cross product is based on the right-hand screw rule (Footnote 6).

EXERCISE

1. Verify (1.8).

1.2 Cartesian Coordinate Frames

Throughout this book, *lowercase Latin subscripts range over the subset of integers*

$$\{1, 2, 3\}.$$

A **Cartesian coordinate frame** for $\mathcal{E}$ consists of a reference point $\mathbf{o}$ called **the origin** together with a positively oriented orthonormal basis $\{\mathbf{e}_1, \mathbf{e}_2, \mathbf{e}_3\}$ for $\mathcal{V}$. Being positively oriented and orthonormal, the basis vectors obey

$$\mathbf{e}_i \cdot \mathbf{e}_j = \delta_{ij} \qquad \text{and} \qquad \mathbf{e}_i \cdot (\mathbf{e}_j \times \mathbf{e}_k) = \epsilon_{ijk}. \tag{1.10}$$

Here δ_{ij}, the Kronecker delta, is defined by

$$\delta_{ij} = \begin{cases} 1, & \text{if } i = j, \\ 0, & \text{if } i \neq j, \end{cases} \tag{1.11}$$

while ϵ_{ijk}, the alternating symbol, is defined by

$$\epsilon_{ijk} = \begin{cases} 1, & \text{if } \{i, j, k\} = \{1, 2, 3\}, \{2, 3, 1\}, \text{ or } \{3, 1, 2\}, \\ -1, & \text{if } \{i, j, k\} = \{2, 1, 3\}, \{1, 3, 2\}, \text{ or } \{3, 2, 1\}, \\ 0, & \text{if an index is repeated,} \end{cases} \tag{1.12}$$

and, hence, has the value $+1$, -1, or 0 according to whether $\{i, j, k\}$ is an even permutation, an odd permutation, or not a permutation of $\{1, 2, 3\}$.

For brevity,

$$\{\mathbf{e}_i\} \stackrel{\text{def}}{=} \{\mathbf{e}_1, \mathbf{e}_2, \mathbf{e}_3\}$$

denotes a positively oriented orthonormal basis.

1.3 Summation Convention. Components of a Vector and a Point

Throughout this book, we employ the Einstein *summation convention* according to which summation over the range 1, 2, 3 is implied for any index that is repeated twice in any term, so that, for instance,

$$u_i v_i = u_1 v_1 + u_2 v_2 + u_3 v_3,$$

$$S_{ij} u_j = S_{i1} u_1 + S_{i2} u_2 + S_{i3} u_3,$$

$$S_{ik} T_{kj} = S_{i1} T_{1j} + S_{i2} T_{2j} + S_{i3} T_{3j}.$$

In the expression $S_{ij} u_j$, the subscript i is *free*, because it is not summed over, while j is a *dummy* subscript, since

$$S_{ij} u_j = S_{ik} u_k = S_{im} u_m.$$

When an expression in which an index is repeated twice *but summation is not to be performed* we state so explicitly. For example,

$$u_i v_i \qquad \text{(no sum)}$$

signifies that the subscript i is not to be summed over.

Next, if $S_{jk} = S_{kj}$, then

$$\begin{aligned}\epsilon_{ijk}S_{jk} &= \epsilon_{ikj}S_{kj} \\ &= -\epsilon_{ijk}S_{kj} \\ &= -\epsilon_{ijk}S_{jk}\end{aligned}$$

and vice versa; therefore

$$S_{ij} = S_{ji} \qquad \text{if and only if} \qquad \epsilon_{ijk}S_{jk} = 0. \tag{1.13}$$

Because $\{\mathbf{e}_i\}$ is a basis, every vector $\mathbf{u}$ admits the unique expansion

$$\mathbf{u} = u_j\mathbf{e}_j; \tag{1.14}$$

the scalars u_i are called the (Cartesian) **components of u** (relative to this basis). If we take the inner product of (1.14) with $\mathbf{e}_i$, we find that, since $\mathbf{e}_i \cdot \mathbf{e}_j = \delta_{ij}$,

$$u_i = \mathbf{u} \cdot \mathbf{e}_i. \tag{1.15}$$

Guided by this relation, we define the coordinates of a point $\mathbf{x}$ with respect to the origin $\mathbf{o}$ by

$$x_i = (\mathbf{x} - \mathbf{o}) \cdot \mathbf{e}_i. \tag{1.16}$$

In view of (1.14), the inner and cross products of vectors $\mathbf{u}$ and $\mathbf{v}$ may be expressed as

$$\begin{aligned}\mathbf{u} \cdot \mathbf{v} &= (u_i\mathbf{e}_i) \cdot (v_j\mathbf{e}_j) \\ &= u_i v_j \delta_{ij} \\ &= u_i v_i\end{aligned} \tag{1.17}$$

and

$$\begin{aligned}\mathbf{u} \times \mathbf{v} &= (u_j\mathbf{e}_j) \times (v_k\mathbf{e}_k) \\ &= u_j v_k \mathbf{e}_j \times \mathbf{e}_k \\ &= \epsilon_{ijk} u_j v_k \mathbf{e}_i.\end{aligned} \tag{1.18}$$

In particular, (1.18) implies that the vector $\mathbf{u} \times \mathbf{v}$ has the component form

$$(\mathbf{u} \times \mathbf{v})_i = \epsilon_{ijk} u_j v_k. \tag{1.19}$$

When working with the cross product, the epsilon-delta identity

$$\epsilon_{ijk}\epsilon_{ipq} = \delta_{jp}\delta_{kq} - \delta_{jq}\delta_{kp}, \tag{1.20}$$

and its consequences

$$\epsilon_{ijk}\epsilon_{ijl} = 2\delta_{kl} \qquad \text{and} \qquad \epsilon_{ijk}\epsilon_{ijk} = 6, \tag{1.21}$$

are useful. Also useful is the identity

$$\mathbf{e}_i = \tfrac{1}{2}\epsilon_{ijk}\mathbf{e}_j \times \mathbf{e}_k. \tag{1.22}$$

Let **u**, **v**, and **w** be vectors. Useful relations involving the inner and cross products then include

$$\begin{aligned} \mathbf{u}\cdot(\mathbf{v}\times\mathbf{w}) &= \mathbf{v}\cdot(\mathbf{w}\times\mathbf{u}) = \mathbf{w}\cdot(\mathbf{u}\times\mathbf{v}), \\ \mathbf{u}\cdot(\mathbf{u}\times\mathbf{v}) &= 0, \\ \mathbf{u}\times\mathbf{v} &= -\mathbf{v}\times\mathbf{u}, \\ \mathbf{u}\times(\mathbf{v}\times\mathbf{w}) &= (\mathbf{u}\cdot\mathbf{w})\mathbf{v}-(\mathbf{u}\cdot\mathbf{v})\mathbf{w}. \end{aligned} \tag{1.23}$$

EXERCISES

1. Verify (1.22) and (1.23).
2. Establish the following identities for any vectors **u**, **v**, and **w**:

$$(\mathbf{u}\cdot\mathbf{v})^2 + |\mathbf{u}\times\mathbf{v}|^2 = |\mathbf{u}|^2|\mathbf{v}|^2,$$

$$\mathbf{u}\times(\mathbf{v}\times\mathbf{w}) + \mathbf{v}\times(\mathbf{w}\times\mathbf{u}) + \mathbf{w}\times(\mathbf{u}\times\mathbf{v}) = \mathbf{0}.$$

2 Tensor Algebra

2.1 What Is a Tensor?

We use the term **tensor** as a synonym for the phrase "linear transformation from $\mathcal{V}$ into $\mathcal{V}$." A tensor $\mathbf{S}$ is therefore a *linear* mapping of vectors to vectors; that is, given a vector $\mathbf{u}$,

$$\mathbf{v} = \mathbf{S}\mathbf{u} \tag{2.1}$$

is also a vector. One might think of a tensor $\mathbf{S}$ as a machine with an input and an output: if a vector $\mathbf{u}$ is the input, then the vector $\mathbf{v} = \mathbf{S}\mathbf{u}$ is the output (Figure 2.1). The **linearity** of a tensor $\mathbf{S}$ is embodied by the requirements:

$$\mathbf{S}(\mathbf{u} + \mathbf{v}) = \mathbf{S}\mathbf{u} + \mathbf{S}\mathbf{v} \qquad \text{for all vectors } \mathbf{u} \text{ and } \mathbf{v};$$

$$\mathbf{S}(\alpha\mathbf{u}) = \alpha\mathbf{S}\mathbf{u} \qquad \text{for all vectors } \mathbf{u} \text{ and scalars } \alpha.$$

Tensors $\mathbf{S}$ and $\mathbf{T}$ are equal if their outputs are the same whenever their inputs are equal; precisely,

$$\mathbf{S} = \mathbf{T} \text{ if and only if } \mathbf{S}\mathbf{v} = \mathbf{T}\mathbf{v} \text{ for all vectors } \mathbf{v}. \tag{2.2}$$

One way of showing that tensors $\mathbf{S}$ and $\mathbf{T}$ are equal is a conseqence of the following result:

$$\mathbf{a} \cdot \mathbf{S}\mathbf{b} = \mathbf{a} \cdot \mathbf{T}\mathbf{b} \text{ for all vectors } \mathbf{a} \text{ and } \mathbf{b} \text{ if and only if } \mathbf{S} = \mathbf{T}. \tag{2.3}$$

To prove this, we write $\mathbf{a} \cdot \mathbf{S}\mathbf{b} = \mathbf{a} \cdot \mathbf{T}\mathbf{b}$ in the form $\mathbf{a} \cdot (\mathbf{S}\mathbf{b} - \mathbf{T}\mathbf{b})$, which, by (1.7), holds for all $\mathbf{a}$ if and only if $\mathbf{S}\mathbf{b} = \mathbf{T}\mathbf{b}$ and the validity of this for all $\mathbf{b}$ yields, by (2.2), $\mathbf{S} = \mathbf{T}$.

Consistent with (2.2), tensors are generally defined by their actions on arbitrary vectors. For example, the sum $\mathbf{S} + \mathbf{T}$ of tensors $\mathbf{S}$ and $\mathbf{T}$ and the product $\alpha\mathbf{S}$ of a tensor $\mathbf{S}$ and a scalar α are defined as follows:

$$\left.\begin{aligned} (\mathbf{S} + \mathbf{T})\mathbf{v} &= \mathbf{S}\mathbf{v} + \mathbf{T}\mathbf{v}, \\ (\alpha\mathbf{S})\mathbf{v} &= \alpha(\mathbf{S}\mathbf{v}), \end{aligned}\right\} \tag{2.4}$$

for all vectors $\mathbf{v}$. As a consequence of these definitions, the set of all tensors forms a vector space (of dimension 9).

EXERCISE

1. Show that $\mathbf{S} + \mathbf{T}$ and $\alpha\mathbf{S}$ defined by (2.4) are actually *linear* and hence tensors.

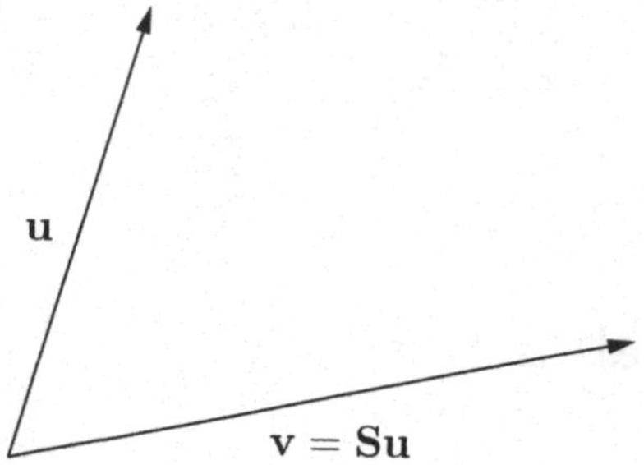

Figure 2.1. A vector **u** and its transformation $\mathbf{v} = \mathbf{S}\mathbf{u}$ by a tensor **S**.

2.2 Zero and Identity Tensors. Tensor Product of Two Vectors. Projection Tensor. Spherical Tensor

Two basic tensors are the **zero tensor 0** and the **identity tensor 1**, defined by

$$\mathbf{0}\mathbf{v} = \mathbf{0} \qquad \text{and} \qquad \mathbf{1}\mathbf{v} = \mathbf{v}$$

for all vectors **v**. Although the zero vector and the zero tensor are denoted by the same symbol, what is meant should always be clear from the context.

Another example of a tensor is the **tensor product** $\mathbf{u} \otimes \mathbf{v}$, of two vectors **u** and **v**, defined by

$$(\mathbf{u} \otimes \mathbf{v})\mathbf{w} = (\mathbf{v} \cdot \mathbf{w})\mathbf{u} \tag{2.5}$$

for all **w**. By (2.5), the tensor $\mathbf{u} \otimes \mathbf{v}$ maps any vector **w** onto a scalar multiple of **u**. Let **e** be a unit vector. Then, since

$$(\mathbf{e} \otimes \mathbf{e})\mathbf{u} = (\mathbf{u} \cdot \mathbf{e})\mathbf{e}$$

for any vector **u**, the tensor $\mathbf{e} \otimes \mathbf{e}$ maps each vector **u** to the projection $(\mathbf{u} \cdot \mathbf{e})\mathbf{e}$ of **u** onto the vector **e**. Similarly,

$$(\mathbf{1} - \mathbf{e} \otimes \mathbf{e})\mathbf{u} = \mathbf{u} - (\mathbf{u} \cdot \mathbf{e})\mathbf{e},$$

so that, for any vector **u**, the tensor $\mathbf{1} - \mathbf{e} \otimes \mathbf{e}$ maps each vector **u** to the projection $\mathbf{u} - (\mathbf{u} \cdot \mathbf{e})\mathbf{e}$ of **u** onto the plane perpendicular to **e**. The tensors

$$\mathbf{e} \otimes \mathbf{e} \qquad \text{and} \qquad \mathbf{1} - \mathbf{e} \otimes \mathbf{e} \tag{2.6}$$

therefore define **projections** onto **e** and onto the plane perpendicular to **e** (Figure 2.2).

Next, note that

$$\begin{aligned} \mathbf{1}\mathbf{v} &= \mathbf{v} \\ &= (\mathbf{v} \cdot \mathbf{e}_i)\mathbf{e}_i \\ &= (\mathbf{e}_i \otimes \mathbf{e}_i)\mathbf{v} \end{aligned}$$

for all **v**; thus, by (2.2), we have the useful identity

$$\mathbf{1} = \mathbf{e}_i \otimes \mathbf{e}_i. \tag{2.7}$$

Finally, a tensor **S** of the form

$$\mathbf{S} = \alpha\mathbf{1}, \tag{2.8}$$

with α a scalar, is called a **spherical tensor**.

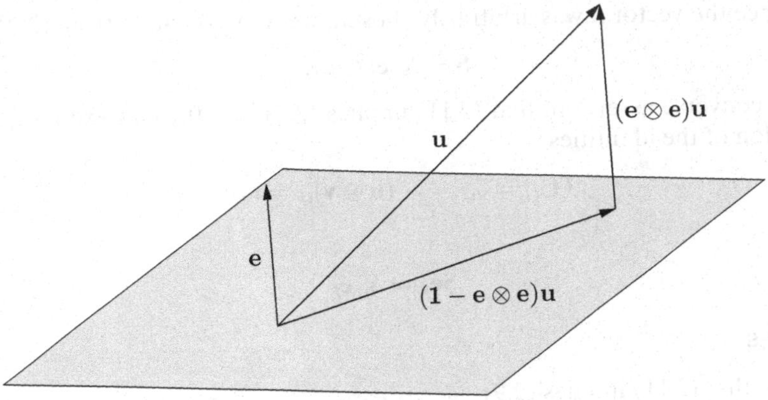

Figure 2.2. The projections $(\mathbf{e} \otimes \mathbf{e})\mathbf{u} = (\mathbf{u} \cdot \mathbf{e})\mathbf{e}$ and $(\mathbf{1} - \mathbf{e} \otimes \mathbf{e})\mathbf{u} = \mathbf{u} - (\mathbf{u} \cdot \mathbf{e})\mathbf{e}$ of vector $\mathbf{u}$ onto a unit vector $\mathbf{e}$ and onto the plane perpendicular to that unit vector.

EXERCISE

1. Show that, for **a** and **b** nonzero,

$$\mathbf{a} \otimes \mathbf{b} = \mathbf{c} \otimes \mathbf{d} \qquad \text{if and only if } \mathbf{c} = \alpha\mathbf{a} \text{ and } \mathbf{d} = \beta\mathbf{b} \text{ with } \alpha\beta = 1.$$

2.3 Components of a Tensor

Given a tensor **S**, choose an arbitrary vector **u** and let

$$\mathbf{v} = \mathbf{S}\mathbf{u}.$$

Then,

$$\begin{aligned} v_k \mathbf{e}_k &= \mathbf{S}(u_j \mathbf{e}_j) \\ &= u_j \mathbf{S}\mathbf{e}_j \end{aligned}$$

and, by (1.10), taking the inner product of this relation with $\mathbf{e}_i$ yields

$$v_i = (\mathbf{e}_i \cdot \mathbf{S}\mathbf{e}_j) u_j;$$

thus, defining the **components** of **S** by

$$\begin{aligned} S_{ij} &= (\mathbf{S})_{ij} \\ &= \mathbf{e}_i \cdot \mathbf{S}\mathbf{e}_j, \end{aligned} \tag{2.9}$$

we see that the component form of the relation $\mathbf{v} = \mathbf{S}\mathbf{u}$ is

$$v_i = S_{ij} u_j. \tag{2.10}$$

Further, this relation implies that

$$\begin{aligned} \mathbf{v} &= v_i \mathbf{e}_i \\ &= (S_{ij} u_j)\mathbf{e}_i \\ &= (S_{ij} \mathbf{e}_j \cdot \mathbf{u})\mathbf{e}_i \\ &= (S_{ij} \mathbf{e}_i \otimes \mathbf{e}_j)\mathbf{u}, \end{aligned}$$

and since $\mathbf{v} = \mathbf{S}\mathbf{u}$ we must have

$$\mathbf{S}\mathbf{u} = (S_{ij} \mathbf{e}_i \otimes \mathbf{e}_j)\mathbf{u}.$$

Thus, since the vector **u** was arbitrarily chosen, we may conclude from (2.2) that

$$\mathbf{S} = S_{ij}\mathbf{e}_i \otimes \mathbf{e}_j. \tag{2.11}$$

The converse assertion, that (2.11) implies (2.9), is left as an exercise, as is the verification of the identities

$$(\mathbf{1})_{ij} = \delta_{ij}, \qquad (\mathbf{u}\otimes\mathbf{v})_{ij} = u_i v_j, \tag{2.12}$$

and

$$\mathbf{S}\mathbf{e}_j = S_{ij}\mathbf{e}_i. \tag{2.13}$$

EXERCISES

1. Show that (2.11) implies (2.9).
2. Verify (2.12) and (2.13).
3. Let **e** and **f** be orthogonal unit vectors. Describe the geometric nature of the tensor
$$\mathbf{e}\otimes\mathbf{e}+\mathbf{f}\otimes\mathbf{f}.$$
4. Show that $(\mathbf{S}\mathbf{e}_i)\otimes\mathbf{e}_i = \mathbf{S}$.
5. Bearing in mind $(1.23)_4$, show that
$$\mathbf{u}\times(\mathbf{v}\times\mathbf{w}) = [(\mathbf{w}\cdot\mathbf{u})\mathbf{1} - \mathbf{w}\otimes\mathbf{u}]\mathbf{v}.$$
6. Given a unit vector **e**, show that
$$(\mathbf{1}-\mathbf{e}\otimes\mathbf{e})\mathbf{u} = -\mathbf{e}\times(\mathbf{e}\times\mathbf{u}).$$

2.4 Transpose of a Tensor. Symmetric and Skew Tensors

The **transpose** $\mathbf{S}^{\top}$ of a tensor **S** is the unique tensor with the property that

$$\boxed{\mathbf{u}\cdot\mathbf{S}\mathbf{v} = \mathbf{v}\cdot\mathbf{S}^{\top}\mathbf{u}} \tag{2.14}$$

for all vectors **u** and **v**. To establish the uniqueness of the transpose, assume that **S** has a second "transpose" $\mathbf{S}^{\flat}$. Then,

$$\mathbf{v}\cdot\mathbf{S}^{\top}\mathbf{u} = \mathbf{v}\cdot\mathbf{S}^{\flat}\mathbf{u}$$

for all **u** and **v**, so that, by (2.3), $\mathbf{S}^{\flat} = \mathbf{S}^{\top}$. Next, if **S** has a transpose, then

$$\mathbf{e}_i\cdot\mathbf{S}^{\top}\mathbf{e}_j = \mathbf{e}_j\cdot\mathbf{S}\mathbf{e}_i = S_{ji},$$

so that

$$(\mathbf{S}^{\top})_{ij} = S_{ji}. \tag{2.15}$$

This calculation shows that every tensor **S** has a transpose:

$$\mathbf{S}^{\top} = S_{ji}\mathbf{e}_i\otimes\mathbf{e}_j. \tag{2.16}$$

Consequences of (2.14) are

$$\left.\begin{aligned}\mathbf{1}^{\top} &= \mathbf{1},\\ (\mathbf{S}+\mathbf{T})^{\top} &= \mathbf{S}^{\top}+\mathbf{T}^{\top},\\ (\mathbf{S}^{\top})^{\top} &= \mathbf{S}.\end{aligned}\right\} \tag{2.17}$$

A tensor **S** is **symmetric** if

$$\mathbf{S} = \mathbf{S}^{\top} \tag{2.18}$$

or **skew** if

$$\mathbf{S} = -\mathbf{S}^{\top}. \tag{2.19}$$

We write

$$\begin{aligned} \operatorname{sym}\mathbf{S} &= \tfrac{1}{2}(\mathbf{S} + \mathbf{S}^{\top}), \\ \operatorname{skw}\mathbf{S} &= \tfrac{1}{2}(\mathbf{S} - \mathbf{S}^{\top}), \end{aligned} \tag{2.20}$$

and refer to sym **S** as the **symmetric part** of **S** and to skw **S** as the **skew part** of **S**. As a consequence of these definitions,

$$\operatorname{skw}(\operatorname{sym}\mathbf{S}) = \operatorname{sym}(\operatorname{skw}\mathbf{S}) = \mathbf{0} \tag{2.21}$$

for any tensor **S**. Clearly,

$$\mathbf{S} = \tfrac{1}{2}(\mathbf{S} + \mathbf{S}^{\top}) + \tfrac{1}{2}(\mathbf{S} - \mathbf{S}^{\top});$$

thus, every tensor **S** admits the decomposition

$$\mathbf{S} = \operatorname{sym}\mathbf{S} + \operatorname{skw}\mathbf{S} \tag{2.22}$$

into symmetric and skew parts; as a consequence of (2.20), this decomposition is unique. In components, by (2.11), (2.15), and (2.20),

$$\begin{aligned} (\operatorname{sym}\mathbf{S})_{ij} &= \tfrac{1}{2}(S_{ij} + S_{ji}), \\ (\operatorname{skw}\mathbf{S})_{ij} &= \tfrac{1}{2}(S_{ij} - S_{ji}). \end{aligned} \tag{2.23}$$

EXERCISE

1. Using the definition (2.14) of the transpose, establish (2.17).

2.5 Product of Tensors

Given tensors **S** and **T**, the **product ST** is defined by composition; that is, **ST** is defined by

$$(\mathbf{ST})\mathbf{v} = \mathbf{S}(\mathbf{T}\mathbf{v}) \tag{2.24}$$

for all vectors **v**. By (2.11), $\mathbf{T}\mathbf{e}_j = T_{lj}\mathbf{e}_l$ and by (2.16), $\mathbf{S}^{\top}\mathbf{e}_i = S_{ik}\mathbf{e}_k$; thus,

$$\begin{aligned} \mathbf{e}_i \cdot \mathbf{S}\mathbf{T}\mathbf{e}_j &= \mathbf{S}^{\top}\mathbf{e}_i \cdot \mathbf{T}\mathbf{e}_j \\ &= (S_{ik}\mathbf{e}_k) \cdot (T_{lj}\mathbf{e}_l) \\ &= S_{ik}T_{lj}\underbrace{(\mathbf{e}_k \cdot \mathbf{e}_l)}_{\delta_{kl}} \\ &= S_{ik}T_{kj}, \end{aligned}$$

and, hence,

$$(\mathbf{ST})_{ij} = S_{ik}T_{kj}. \tag{2.25}$$

Generally, $\mathbf{ST} \neq \mathbf{TS}$. When $\mathbf{ST} = \mathbf{TS}$, the tensors **S** and **T** are said to **commute**.

Consequences of the definition (2.24), for tensors **R**, **S**, and **T**, are that

$$\begin{aligned} (\mathbf{RS})\mathbf{T} &= \mathbf{R}(\mathbf{ST}) = \mathbf{RST}, \\ (\mathbf{R} + \mathbf{S})\mathbf{T} &= \mathbf{RT} + \mathbf{ST}. \end{aligned} \tag{2.26}$$

We write $\mathbf{S}^2 = \mathbf{SS}$, and so forth; then, for m and n nonnegative integers,

$$\left.\begin{aligned} \mathbf{S}^m\mathbf{S}^n &= \mathbf{S}^{m+n} = \mathbf{S}^n\mathbf{S}^m, \\ (\alpha\mathbf{S})^m &= \alpha^m\mathbf{S}^m, \\ (\mathbf{S}^m)^n &= \mathbf{S}^{mn} = (\mathbf{S}^n)^m. \end{aligned}\right\} \tag{2.27}$$

Given vectors $\mathbf{u}$, $\mathbf{v}$, $\mathbf{a}$, and $\mathbf{b}$, it follows that

$$\begin{aligned} (\mathbf{u}\otimes\mathbf{v})(\mathbf{a}\otimes\mathbf{b})\mathbf{w} &= (\mathbf{b}\cdot\mathbf{w})(\mathbf{u}\otimes\mathbf{v})\mathbf{a} \\ &= (\mathbf{b}\cdot\mathbf{w})(\mathbf{v}\cdot\mathbf{a})\mathbf{u} \\ &= (\mathbf{v}\cdot\mathbf{a})(\mathbf{u}\otimes\mathbf{b})\mathbf{w} \end{aligned}$$

for any vector $\mathbf{w}$ and, thus, that

$$(\mathbf{u}\otimes\mathbf{v})(\mathbf{a}\otimes\mathbf{b}) = (\mathbf{v}\cdot\mathbf{a})\mathbf{u}\otimes\mathbf{b}. \tag{2.28}$$

The identity (2.28) may also be established using components: by (2.12) and (2.25),

$$\begin{aligned} [(\mathbf{u}\otimes\mathbf{v})(\mathbf{a}\otimes\mathbf{b})]_{ij} &= (u_i v_k)(a_k b_j) \\ &= (v_k a_k)(u_i b_j) \\ &= [(\mathbf{v}\cdot\mathbf{a})\mathbf{u}\otimes\mathbf{b}]_{ij}. \end{aligned} \tag{2.29}$$

Further, given a tensor $\mathbf{S}$ and vectors $\mathbf{u}$ and $\mathbf{v}$, we have the useful identities

$$\left.\begin{aligned} \mathbf{S}(\mathbf{u}\otimes\mathbf{v}) &= (\mathbf{Su})\otimes\mathbf{v}, \\ (\mathbf{u}\otimes\mathbf{v})\mathbf{S} &= \mathbf{u}\otimes(\mathbf{S}^{\top}\mathbf{v}), \\ (\mathbf{u}\otimes\mathbf{v})^{\top} &= \mathbf{v}\otimes\mathbf{u}, \end{aligned}\right\} \tag{2.30}$$

whose verification we leave as an exercise.

Given tensors $\mathbf{S}$ and $\mathbf{T}$, it follows from the definition (2.14) of the transpose and (2.24)

$$\begin{aligned} \mathbf{u}\cdot(\mathbf{ST})^{\top}\mathbf{v} &= (\mathbf{ST})\mathbf{u}\cdot\mathbf{v} \\ &= \mathbf{S}(\mathbf{Tu})\cdot\mathbf{v} \\ &= \mathbf{Tu}\cdot\mathbf{S}^{\top}\mathbf{v} \\ &= \mathbf{u}\cdot\mathbf{T}^{\top}\mathbf{S}^{\top}\mathbf{v} \end{aligned}$$

for all vectors $\mathbf{u}$ and $\mathbf{v}$. Thus,

$$(\mathbf{ST})^{\top} = \mathbf{T}^{\top}\mathbf{S}^{\top}. \tag{2.31}$$

EXERCISE

1. Verify (2.30).
2. For $\mathbf{A}$ and $\mathbf{S}$ tensors, show that

$$\operatorname{sym}(\mathbf{A}^{\top}\mathbf{SA}) = \mathbf{A}^{\top}\operatorname{sym}(\mathbf{S})\mathbf{A}. \tag{2.32}$$

2.6 Vector Cross. Axial Vector of a Skew Tensor

The **vector cross** $\mathbf{w}\times$ of a vector $\mathbf{w}$ is the tensor defined via the (natural) requirement that

$$(\mathbf{w}\times)\mathbf{u} = \mathbf{w} \times \mathbf{u} \tag{2.33}$$

for all vectors $\mathbf{u}$. To determine the component representation of $\mathbf{w}\times$, choose a vector $\mathbf{u}$ and let $\mathbf{v} = (\mathbf{w}\times)\mathbf{u} = \mathbf{w} \times \mathbf{u}$, which, in view of (1.19), has the component form

$$v_i = \epsilon_{ikj} w_k u_j.$$

Comparing this relation to (2.10), we find that, since $\mathbf{u}$ is arbitrary,

$$(\mathbf{w}\times)_{ij} = \epsilon_{ikj} w_k. \tag{2.34}$$

Thus, since $\epsilon_{ikj} = -\epsilon_{jki}$,

$$(\mathbf{w}\times)^{\top} = -\mathbf{w}\times . \tag{2.35}$$

Given any *skew* tensor $\mathbf{\Omega}$, there is a unique vector $\boldsymbol{\omega}$ — called the **axial vector** of $\mathbf{\Omega}$ — such that

$$\mathbf{\Omega} = \boldsymbol{\omega}\times, \tag{2.36}$$

and, hence, such that

$$\mathbf{\Omega}\mathbf{u} = \boldsymbol{\omega} \times \mathbf{u} \tag{2.37}$$

for all vectors $\mathbf{u}$. The uniqueness of $\boldsymbol{\omega}$ follows from (1.8). To establish the existence of such a vector, it suffices to show that $\boldsymbol{\omega}$ defined by

$$\omega_i = -\tfrac{1}{2}\epsilon_{ijk}\Omega_{jk} \tag{2.38}$$

satisfies (2.36). By (2.38) and the epsilon-delta identity (1.20),

$$\begin{aligned}
\epsilon_{ipq}\omega_i &= -\tfrac{1}{2}\epsilon_{ipq}\epsilon_{ijk}\Omega_{jk} \\
&= -\tfrac{1}{2}(\delta_{jp}\delta_{kq} - \delta_{jq}\delta_{kp})\Omega_{jk} \\
&= -\tfrac{1}{2}(\Omega_{pq} - \Omega_{qp}) \\
&= \Omega_{qp}.
\end{aligned}$$

Thus, $\Omega_{qp} = \epsilon_{qip}\omega_i$, which, by (2.34), is the component form of (2.36).

EXERCISES

1. Given vectors $\mathbf{u}$, $\mathbf{v}$, and $\mathbf{w}$, establish the identities

$$\left.\begin{aligned}
(\mathbf{u}\times)(\mathbf{v}\otimes\mathbf{w}) &= (\mathbf{u}\times\mathbf{v})\otimes\mathbf{w}, \\
(\mathbf{u}\otimes\mathbf{v})(\mathbf{w}\times) &= \mathbf{u}\otimes(\mathbf{v}\times\mathbf{w}), \\
(\mathbf{u}\times\mathbf{v})\times &= \mathbf{v}\otimes\mathbf{u} - \mathbf{u}\otimes\mathbf{v}, \\
(\mathbf{u}\times)(\mathbf{v}\times) &= \mathbf{v}\otimes\mathbf{u} - (\mathbf{u}\cdot\mathbf{v})\mathbf{1}, \\
(\mathbf{u}\times)(\mathbf{v}\times)(\mathbf{w}\times) &= \mathbf{v}\otimes(\mathbf{u}\times\mathbf{w}) - (\mathbf{u}\cdot\mathbf{v})\mathbf{w}\times.
\end{aligned}\right\} \tag{2.39}$$

2. Show that the axial vector of the skew part $\frac{1}{2}(\mathbf{u}\otimes\mathbf{v} - \mathbf{v}\otimes\mathbf{u})$ of the tensor $\mathbf{u}\otimes\mathbf{v}$ is

$$-\tfrac{1}{2}\mathbf{u}\times\mathbf{v}. \tag{2.40}$$

2.7 Trace of a Tensor. Deviatoric Tensors

The **trace** is the linear operation that assigns to each tensor $\mathbf{S}$ a scalar $\operatorname{tr}\mathbf{S}$ and satisfies

$$\operatorname{tr}(\mathbf{u}\otimes\mathbf{v}) = \mathbf{u}\cdot\mathbf{v} \tag{2.41}$$

for any vectors $\mathbf{u}$ and $\mathbf{v}$. Linearity is the requirement that

$$\operatorname{tr}(\alpha\mathbf{S}+\beta\mathbf{T}) = \alpha\operatorname{tr}(\mathbf{S})+\beta\operatorname{tr}(\mathbf{T})$$

for all tensors $\mathbf{S}$ and $\mathbf{T}$ and all scalars α and β. Thus, by (2.11),

$$\begin{aligned}\operatorname{tr}\mathbf{S} &= \operatorname{tr}(S_{ij}\mathbf{e}_i\otimes\mathbf{e}_j)\\ &= S_{ij}\operatorname{tr}(\mathbf{e}_i\otimes\mathbf{e}_j)\\ &= S_{ij}(\mathbf{e}_i\cdot\mathbf{e}_j)\\ &= S_{ii},\end{aligned} \tag{2.42}$$

and the trace is well-defined. Some useful properties of the trace are

$$\begin{aligned}\operatorname{tr}(\mathbf{Su}\otimes\mathbf{v}) &= \mathbf{v}\cdot\mathbf{Su},\\ \operatorname{tr}(\mathbf{S}^{\top}) &= \operatorname{tr}\mathbf{S},\\ \operatorname{tr}(\mathbf{ST}) &= \operatorname{tr}(\mathbf{TS}),\\ \operatorname{tr}\mathbf{1} &= 3.\end{aligned} \tag{2.43}$$

As a consequence of $(2.43)_2$,

$$\operatorname{tr}\mathbf{S} = 0 \quad \text{whenever } \mathbf{S} \text{ is skew.} \tag{2.44}$$

A tensor $\mathbf{S}$ is **deviatoric** (or traceless) if

$$\operatorname{tr}\mathbf{S} = 0, \tag{2.45}$$

and we refer to

$$\begin{aligned}\mathbf{S}_0 &\equiv \operatorname{dev}\mathbf{S}\\ &\overset{\text{def}}{=} \mathbf{S} - \tfrac{1}{3}(\operatorname{tr}\mathbf{S})\mathbf{1}\end{aligned} \tag{2.46}$$

as the **deviatoric part** of $\mathbf{S}$,[8] since

$$\operatorname{tr}\mathbf{S}_0 = \mathbf{0},$$

and to

$$\tfrac{1}{3}(\operatorname{tr}\mathbf{S})\mathbf{1} \tag{2.47}$$

as the **spherical part** of $\mathbf{S}$. Trivially,

$$\mathbf{S} = \underbrace{\mathbf{S} - \tfrac{1}{3}(\operatorname{tr}\mathbf{S})\mathbf{1}}_{\mathbf{S}_0} + \underbrace{\tfrac{1}{3}(\operatorname{tr}\mathbf{S})\mathbf{1}}_{s\mathbf{1}}.$$

Every tensor $\mathbf{S}$ thus admits the decomposition

$$\mathbf{S} = \mathbf{S}_0 + s\mathbf{1} \tag{2.48}$$

into the sum of a deviatoric tensor and a spherical tensor.

[8] The operation "dev" is useful for denoting the deviatoric part of the product of many tensors, e.g., $\operatorname{dev}(\mathbf{ST}\cdots\mathbf{M})$.

EXERCISES

1. Establish (2.43) and (2.44).
2. Establish the uniqueness of the decomposition (2.48).

2.8 Inner Product of Tensors. Magnitude of a Tensor

The space of all tensors has a natural **inner product**[9]

$$\mathbf{S}:\mathbf{T} = \operatorname{tr}(\mathbf{S}^{\top}\mathbf{T}) = \operatorname{tr}(\mathbf{S}\mathbf{T}^{\top}), \tag{2.49}$$

which, by (2.11) and (2.41), has the component form

$$\begin{aligned} \mathbf{S}:\mathbf{T} &= (\mathbf{S}^{\top}\mathbf{T})_{jj} \\ &= S^{\top}_{ji}T_{ij} \\ &= S_{ij}T_{ij}. \end{aligned} \tag{2.50}$$

By $(2.43)_{2,3}$ and (2.49),

$$\begin{aligned} \mathbf{S}:\mathbf{T} &= \mathbf{T}:\mathbf{S}, \\ \mathbf{1}:\mathbf{S} &= \operatorname{tr}\mathbf{S}. \end{aligned} \tag{2.51}$$

Another useful identity is given by

$$\mathbf{R}:(\mathbf{S}\mathbf{T}) = (\mathbf{S}^{\top}\mathbf{R}):\mathbf{T} \tag{2.52}$$

and should be compared to

$$\mathbf{r}\cdot(\mathbf{S}\mathbf{t}) = (\mathbf{S}^{\top}\mathbf{r})\cdot\mathbf{t},$$

which is the defining identity for the transpose. The verification of (2.52), which relies on $(2.17)_3$ and (2.49), proceeds as follows:

$$\begin{aligned} \mathbf{R}:(\mathbf{S}\mathbf{T}) &= \operatorname{tr}(\mathbf{R}(\mathbf{S}\mathbf{T})^{\top}) \\ &= \operatorname{tr}[\mathbf{R}^{\top}(\mathbf{S}\mathbf{T})] \\ &= \operatorname{tr}[(\mathbf{S}^{\top}\mathbf{R})^{\top}\mathbf{T}] \\ &= (\mathbf{S}^{\top}\mathbf{R}):\mathbf{T}. \end{aligned}$$

Alternatively, working with components and using (2.25) and (2.50),

$$\begin{aligned} \mathbf{R}:(\mathbf{S}\mathbf{T}) &= R_{ij}(\mathbf{S}\mathbf{T})_{ij} \\ &= R_{ij}S_{ik}T_{kj} \\ &= (S_{ik}R_{ij})T_{kj} \\ &= (\mathbf{S}^{\top}\mathbf{R})_{kj}T_{kj} \\ &= (\mathbf{S}^{\top}\mathbf{R}):\mathbf{T}. \end{aligned}$$

Similar steps may be used to verify the following counterpart of (2.52):

$$\mathbf{R}:(\mathbf{S}\mathbf{T}) = (\mathbf{R}\mathbf{T}^{\top}):\mathbf{S}. \tag{2.53}$$

[9] Cf. $(2.43)_{2,3}$.

By (2.50),

$$\mathbf{S}:\mathbf{S} \geq 0 \qquad \text{with } \mathbf{S}:\mathbf{S} = 0 \text{ only when } \mathbf{S} = \mathbf{0}. \tag{2.54}$$

By analogy to the notion of the magnitude $|\mathbf{u}|$ of a vector $\mathbf{u}$, the magnitude $|\mathbf{S}|$ of a tensor $\mathbf{S}$ is defined by

$$|\mathbf{S}| = \sqrt{\mathbf{S}:\mathbf{S}} \geq 0. \tag{2.55}$$

The following assertion is a useful consequence of (2.54)[10] given a tensor $\mathbf{S}$,

$$\text{If } \mathbf{S}:\mathbf{T} = 0 \text{ for all tensors } \mathbf{T}, \text{ then } \mathbf{S} = \mathbf{0}. \tag{2.56}$$

To verify (2.56), we simply take $\mathbf{T} = \mathbf{S}$.

Let $\mathbf{S}$ and $\mathbf{W}$ be symmetric and skew, respectively. Then, by (2.49) and the definitions (2.18) and (2.19) of skew and symmetric tensors,

$$\begin{aligned}\mathbf{S}:\mathbf{W} &= \mathrm{tr}\,(\mathbf{S}\mathbf{W}^{\top}) \\ &= -\mathrm{tr}\,(\mathbf{S}\mathbf{W}) \\ &= -\mathrm{tr}\,(\mathbf{S}^{\top}\mathbf{W}) \\ &= -\mathbf{S}:\mathbf{W};\end{aligned}$$

therefore,

$$\mathbf{S}:\mathbf{W} = 0 \qquad \text{for } \mathbf{S} \text{ symmetric and } \mathbf{W} \text{ skew} \tag{2.57}$$

and *the tensors* $\mathbf{S}$ *and* $\mathbf{W}$ *are orthogonal.* A conseqence of (2.57) is that, for any tensor $\mathbf{T}$,

$$|\mathbf{T}|^2 = |\mathrm{sym}\,\mathbf{T}|^2 + |\mathrm{skw}\,\mathbf{T}|^2. \tag{2.58}$$

Similarly,

$$\mathbf{D}:\mathbf{S} = 0 \qquad \text{for } \mathbf{D} \text{ deviatoric and } \mathbf{S} \text{ spherical}, \tag{2.59}$$

and, for any tensor $\mathbf{T}$,

$$|\mathbf{T}|^2 = |\mathbf{T}_0|^2 + \tfrac{1}{3}(\mathrm{tr}\,\mathbf{T})^2. \tag{2.60}$$

Some other important identities, which hold for arbitrary choices of $\mathbf{T}$, are

$$\left.\begin{aligned}&\mathbf{S}:\mathbf{T} = \mathbf{S}:\mathbf{T}^{\top} = \mathbf{S}:\mathrm{sym}\,\mathbf{T} && \text{for } \mathbf{S} \text{ symmetric}, \\ &\mathbf{W}:\mathbf{T} = -\mathbf{W}:\mathbf{T}^{\top} = \mathbf{W}:\mathrm{skw}\,\mathbf{T} && \text{for } \mathbf{W} \text{ skew}, \\ &\mathbf{S}:\mathbf{T} = \mathbf{S}:\mathbf{T}_0 && \text{for } \mathbf{S} \text{ deviatoric}.\end{aligned}\right\} \tag{2.61}$$

EXERCISES

1. Verify (2.53).
2. Establish (2.58), (2.59), (2.60), and (2.61).
3. Show that for $\boldsymbol{\omega}$ the axial vector of a skew tensor $\boldsymbol{\Omega}$,

$$|\boldsymbol{\Omega}|^2 = 2|\boldsymbol{\omega}|^2. \tag{2.62}$$

4. Show that for $\mathbf{S}$ and $\mathbf{T}$ arbitrary

$$\mathbf{S}:\mathbf{T} = (\mathrm{sym}\,\mathbf{S}):(\mathrm{sym}\,\mathbf{T}) + (\mathrm{skw}\,\mathbf{S}):(\mathrm{skw}\,\mathbf{T}). \tag{2.63}$$

---

[10] Cf. (1.7).

5. Show that for any two vectors $\mathbf{v}$ and $\mathbf{w}$,
$$(\mathbf{v}\times):(\mathbf{w}\times) = 2\mathbf{v}\cdot\mathbf{w}, \tag{2.64}$$
and thus conclude that $|\mathbf{v}\times| = \sqrt{2}|\mathbf{v}|$.
6. Show that the components S_{ij} of a tensor $\mathbf{S}$ as defined in (2.9) are given by
$$S_{ij} = (\mathbf{e}_i \otimes \mathbf{e}_j):\mathbf{S}. \tag{2.65}$$
7. Establish the identity
$$(\mathbf{a}\otimes\mathbf{b}):(\mathbf{c}\otimes\mathbf{d}) = (\mathbf{a}\cdot\mathbf{c})(\mathbf{b}\cdot\mathbf{d}). \tag{2.66}$$

2.9 Invertible Tensors

A tensor $\mathbf{S}$ is **invertible** if there is a tensor $\mathbf{S}^{-1}$, called the **inverse of S**, such that
$$\mathbf{S}\mathbf{S}^{-1} = \mathbf{S}^{-1}\mathbf{S} = \mathbf{1}. \tag{2.67}$$
Conditions equivalent to invertibility are furnished by the next result, whose proof is given at the end of this subsection.

(‡) *Given a tensor* $\mathbf{S}$*, the following four conditions are equivalent*:

- (i) $\mathbf{S}$ *is invertible.*
- (ii) *For any vector* $\mathbf{u}$, $\mathbf{S}\mathbf{u} = \mathbf{0}$ *implies that* $\mathbf{u} = \mathbf{0}$.
- (iii) *For any basis* $\{\mathbf{u}, \mathbf{v}, \mathbf{w}\}$, *the triad* $\{\mathbf{S}\mathbf{u}, \mathbf{S}\mathbf{v}, \mathbf{S}\mathbf{w}\}$ *is a basis.*
- (iv) *For any two vectors* $\mathbf{u}$ and $\mathbf{v}$, $\mathbf{u}\times\mathbf{v} \neq \mathbf{0}$ *implies that* $\mathbf{S}\mathbf{u}\times\mathbf{S}\mathbf{v} \neq \mathbf{0}$.

The product $\mathbf{ST}$ of two invertible tensors is also invertible with
$$(\mathbf{ST})^{-1} = \mathbf{T}^{-1}\mathbf{S}^{-1}, \tag{2.68}$$
since
$$(\mathbf{ST})\mathbf{T}^{-1}\mathbf{S}^{-1} = \mathbf{1},$$
$$\mathbf{T}^{-1}\mathbf{S}^{-1}(\mathbf{ST}) = \mathbf{1}.$$
As a consequence of this result, if $\mathbf{S}$ is invertible, then so also is $\mathbf{S}^m$ for m a nonnegative integer; in particular,
$$(\mathbf{S}^{-1})^m = (\mathbf{S}^m)^{-1}. \tag{2.69}$$

For $\mathbf{S}$ an invertible tensor,
$$\mathbf{S}^{\top}(\mathbf{S}^{-1})^{\top} = (\mathbf{S}^{-1}\mathbf{S})^{\top} = \mathbf{1},$$
and $\mathbf{S}^{\top}$ *is invertible* with
$$(\mathbf{S}^{\top})^{-1} = (\mathbf{S}^{-1})^{\top};$$
we therefore write
$$\mathbf{S}^{-\top} \stackrel{\text{def}}{=} (\mathbf{S}^{-1})^{\top} = (\mathbf{S}^{\top})^{-1}. \tag{2.70}$$

Further, if $\mathbf{S}$ is invertible, then, by $(2.43)_3$,
$$\operatorname{tr}(\mathbf{STS}^{-1}) = \operatorname{tr}(\underbrace{\mathbf{S}^{-1}\mathbf{S}}_{\mathbf{1}}\mathbf{T}),$$
and we have the important identity
$$\operatorname{tr}(\mathbf{STS}^{-1}) = \operatorname{tr}\mathbf{T}. \tag{2.71}$$

Finally, we write

$$S_{ij}^{-1} = (\mathbf{S}^{-1})_{ij} = \mathbf{e}_i \cdot \mathbf{S}^{-1}\mathbf{e}_j$$

for the components of an invertible tensor $\mathbf{S}$, so that

$$S_{ik}S_{kj}^{-1} = S_{ik}^{-1}S_{kj} = \delta_{ij}. \tag{2.72}$$

2.9.1 Proof of (‡)

It suffices to show that (i) and (ii) are equivalent and that (ii) is equivalent to (iii) and also to (iv).

Step 1. (i) *and* (ii) *are equivalent.* Assume that (i) is satisfied. Premultiplying the equation $\mathbf{Su} = \mathbf{0}$ by $\mathbf{S}^{-1}$ then yields $\mathbf{u} = \mathbf{0}$, and (ii) is satisfied. For a proof that (ii) implies (i), cf., for example, Halmos (1958, §36).

Step 2. (ii) *and* (iii) *are equivalent.* Let $\mathbf{u}$, $\mathbf{v}$, and $\mathbf{w}$ be linearly independent. Note that for any choice of the scalars u, v, and w,

$$\mathbf{S}(a\mathbf{u} + b\mathbf{v} + c\mathbf{w}) = a\mathbf{Su} + b\mathbf{Sv} + c\mathbf{Sw}. \tag{2.73}$$

Assume that (ii) is satisfied. We show by contradiction that (iii) is satisfied. Assume that (iii) is not satisfied. Then, $\mathbf{Su}$, $\mathbf{Sv}$, and $\mathbf{Sw}$ must be linearly dependent so that for some choice of the scalars a, b, and c, *not all zero*, the right side of (2.73) vanishes. Thus, by (2.73),

$$\mathbf{S}(a\mathbf{u} + b\mathbf{v} + c\mathbf{w}) = \mathbf{0},$$

so that by (ii), $a\mathbf{u} + b\mathbf{v} + c\mathbf{w} = \mathbf{0}$, which contradicts the assumption that $\mathbf{u}$, $\mathbf{v}$, and $\mathbf{w}$ are linearly independent. Thus, (iii) is valid.

To prove that (iii) implies (ii), assume that (iii) is satisfied but that (ii) is not satisfied. Then, there is a vector $\mathbf{k}$ such that

$$\mathbf{Sk} = \mathbf{0}, \qquad \mathbf{k} \neq \mathbf{0}. \tag{2.74}$$

Then, by (iii), assuming that $\mathbf{u}$, $\mathbf{v}$, and $\mathbf{w}$ are linearly independent (so that $\mathbf{Su}$, $\mathbf{Sv}$, and $\mathbf{Sw}$ are linearly independent), there are scalars u, v, and w, *not all zero*, such that

$$\mathbf{k} = u\mathbf{u} + v\mathbf{v} + w\mathbf{w}.$$

Hence, by (2.73) and (2.74),

$$u\mathbf{Su} + v\mathbf{Sv} + w\mathbf{Sw} = \mathbf{Sk} = \mathbf{0};$$

hence, $\mathbf{Su}$, $\mathbf{Sv}$, and $\mathbf{Sw}$ are linearly dependent, which contradicts (iii). Thus, (iii) implies (ii).

Step 3. (ii) *and* (iv) *are equivalent.* To prove this, we rephrase (iv) in the *equivalent* form:

(v) For any two vectors $\mathbf{u}$ and $\mathbf{v}$, if $\mathbf{u}$ and $\mathbf{v}$ are linearly independent, then so also are $\mathbf{Su}$ and $\mathbf{Sv}$.

The proof that (ii) implies (v) follows exactly the steps showing that (ii) implies (iii). To show that (v) implies (ii), we again use contradiction. Thus, assume that (v) is satisfied but that (ii) is not satisfied, so that (2.74) holds for some vector $\mathbf{v}$. Let $\mathbf{j}$ be any nonzero vector such that $\mathbf{j}$ and $\mathbf{k}$ are linearly independent. Then, by (v), $\mathbf{Sj}$ and $\mathbf{Sk}$ must be linearly independent; but this is not possible, since, by (2.74), $\mathbf{Sk} = \mathbf{0}$.

This completes the proof of (‡).

EXERCISES

1. Verify that for $\mathbf{u} \cdot \mathbf{v} \neq -1$, the tensor $\mathbf{1} + \mathbf{u} \otimes \mathbf{v}$ is invertible with inverse

$$(\mathbf{1} + \mathbf{u} \otimes \mathbf{v})^{-1} = \mathbf{1} - (1 + \mathbf{u} \cdot \mathbf{v})^{-1}\mathbf{u} \otimes \mathbf{v}. \tag{2.75}$$

2. Verify that for any vector $\mathbf{w}$, the tensor $\mathbf{1} + \mathbf{w}\times$ is invertible with inverse

$$(\mathbf{1} + \mathbf{w}\times)^{-1} = \mathbf{1} - (1 + |\mathbf{w}|^2)^{-1}(|\mathbf{w}|^2\mathbf{1} - \mathbf{w} \otimes \mathbf{w} + \mathbf{w}\times). \tag{2.76}$$

3. Let $\mathbf{S}$ be invertible and let $\mathbf{u}$ and $\mathbf{v}$ be vectors chosen so that $\mathbf{v} \cdot \mathbf{S}^{-1}\mathbf{u} \neq -1$. Verify that $\mathbf{T} = \mathbf{S} + \mathbf{u} \otimes \mathbf{v}$ is invertible with inverse

$$\mathbf{T}^{-1} = \mathbf{S}^{-1} - (1 + \mathbf{v} \cdot \mathbf{S}^{-1}\mathbf{u})^{-1}\mathbf{S}^{-1}\mathbf{u} \otimes \mathbf{S}^{-\top}\mathbf{v}. \tag{2.77}$$

Note that (2.75) follows from (2.77) on setting $\mathbf{S} = \mathbf{1}$ (so that $\mathbf{T} = \mathbf{1} + \mathbf{u} \otimes \mathbf{v}$).

4. Show that for $\mathbf{S}$ and $\mathbf{T}$ tensors, with $\mathbf{T}$ invertible,

$$(\mathbf{TST}^{-1})_0 = \operatorname{dev}(\mathbf{TST}^{-1}) = \mathbf{TS}_0\mathbf{T}^{-1}. \tag{2.78}$$

2.10 Determinant of a Tensor

The next result allows us to define the determinant of a tensor in a manner most suitable to our needs.[11]

(†) The ratio

$$\frac{\mathbf{Su}\cdot(\mathbf{Sv}\times\mathbf{Sw})}{\mathbf{u}\cdot(\mathbf{v}\times\mathbf{w})}$$

is the same for all choices of the basis $\{\mathbf{u}, \mathbf{v}, \mathbf{w}\}$.

The **determinant** is an operation that assigns to each tensor $\mathbf{S}$ a scalar $\det\mathbf{S}$ defined by

$$\det\mathbf{S} = \frac{\mathbf{Su}\cdot(\mathbf{Sv}\times\mathbf{Sw})}{\mathbf{u}\cdot(\mathbf{v}\times\mathbf{w})} \tag{2.79}$$

for any basis $\{\mathbf{u}, \mathbf{v}, \mathbf{w}\}$. Thus, $|\det\mathbf{S}|$ is the ratio of the volume of the parallelepiped defined by the vectors $\mathbf{Su}$, $\mathbf{Sv}$, and $\mathbf{Sw}$ to the volume of the parallelepiped defined by the vectors $\mathbf{u}$, $\mathbf{v}$, and $\mathbf{w}$. An immediate consequence of this definition and (1.5) is that

(‡) *two bases* $\{\mathbf{u}, \mathbf{v}, \mathbf{w}\}$ *and* $\{\mathbf{Su}, \mathbf{Sv}, \mathbf{Sw}\}$ *have the same orientation if and only if*

$$\det\mathbf{S} > 0. \tag{2.80}$$

By (2.79), granted that $\{\mathbf{u}, \mathbf{v}, \mathbf{w}\}$ is a basis, $\det\mathbf{S} \neq 0$ if and only if $\mathbf{Su}\cdot(\mathbf{Sv}\times\mathbf{Sw}) \neq 0$ and, hence, in view of the discussion given in the paragraph containing (1.4), if and only if $\{\mathbf{Su}, \mathbf{Sv}, \mathbf{Sw}\}$ is a basis. Thus, appealing to the proposition (‡) on page 19,

$$\boxed{\mathbf{S}\ \text{is invertible}\quad \text{if and only if}\quad \det\mathbf{S} \neq 0.} \tag{2.81}$$

Some useful properties of the determinant, which we state without proof,[12] are

$$\left.\begin{aligned} \det(\mathbf{S}^{\top}) &= \det\mathbf{S}, \\ \det(\mathbf{ST}) = \det(\mathbf{TS}) &= (\det\mathbf{S})(\det\mathbf{T}), \\ \det(\alpha\mathbf{S}) &= \alpha^3\det\mathbf{S}, \end{aligned}\right\} \tag{2.82}$$

and, for $\mathbf{S}$ invertible,

$$\det(\mathbf{S}^{-1}) = (\det\mathbf{S})^{-1}. \tag{2.83}$$

Recalling our agreement that $\{\mathbf{e}_i\}$ is a positively oriented orthonormal basis, we let $\mathbf{u} = \mathbf{e}_p$, $\mathbf{v} = \mathbf{e}_q$, and $\mathbf{w} = \mathbf{e}_r$ (so that no two of the subscripts p, q, and r may coincide). Then, by $(1.10)_2$, (1.14), (2.11), and (2.79),

$$\epsilon_{pqr}\det\mathbf{S} = (\mathbf{Se}_p \times \mathbf{Se}_q)\cdot\mathbf{Se}_r.$$

---

11 Cf., e.g., Nickerson, Spencer & Steenrod (1959, §5.2).

12 The proofs, which are not particularly illuminating, can be obtained using the fact that the determinant of a tensor $\mathbf{S}$ is equal to the determinant of its matrix representation $[\mathbf{S}]$; cf. (2.108). Alternatively, the properties (2.82) of the determinant may be established by direct recourse to the definition (2.79).

Further, this result holds for any choice of the subscripts p, q, and r, since both sides of this equation vanish when two subscripts coincide. Thus,

$$\begin{aligned}\epsilon_{pqr}\det\mathbf{S} &= (S_{ip}\mathbf{e}_i \times S_{jq}\mathbf{e}_j)\cdot(S_{kr}\mathbf{e}_k)\\ &= \epsilon_{ijk}S_{ip}S_{jq}S_{kr},\end{aligned} \tag{2.84}$$

and $(1.21)_2$ yields

$$\det\mathbf{S} = \tfrac{1}{6}\epsilon_{ijk}\epsilon_{pqr}S_{ip}S_{jq}S_{kr}. \tag{2.85}$$

Alternative expressions for $\det\mathbf{S}$ that arise from (2.85) are

$$\det\mathbf{S} = \epsilon_{ijk}S_{i1}S_{j2}S_{k3} = \epsilon_{ijk}S_{1i}S_{2j}S_{3k}. \tag{2.86}$$

EXERCISES

1. Show that
$$\det(\alpha\mathbf{S}) = \alpha^3\det\mathbf{S}$$
and, provided that $\mathbf{S}$ is invertible, that
$$\det(\mathbf{S}^{-1}) = (\det\mathbf{S})^{-1}.$$
2. Show that for $\boldsymbol{\Omega}$ skew,
$$\det(\mathbf{1}+\boldsymbol{\Omega}) = 1 + \tfrac{1}{2}|\boldsymbol{\Omega}|^2$$
and, hence, conclude that any tensor of the form $\mathbf{1}+\boldsymbol{\Omega}$ with $\boldsymbol{\Omega}$ skew is invertible.
3. Show that for $\mathbf{S}$ invertible and any pair of vectors $\mathbf{u}$ and $\mathbf{v}$,
$$\det(\mathbf{S}+\mathbf{u}\otimes\mathbf{v}) = (1+\mathbf{v}\cdot\mathbf{S}^{-1}\mathbf{u})\det\mathbf{S}.$$

2.11 Cofactor of a Tensor

Let $\mathbf{S}$ be an invertible tensor. Further, let $\mathbf{u}$ and $\mathbf{v}$ be linearly independent and consider the *area vector*[13] $\mathbf{u}\times\mathbf{v}$ of the parallelogram defined by $\mathbf{u}$ and $\mathbf{v}$. Then, by (iv) of (‡) on page 19, $\mathbf{Su}\times\mathbf{Sv}\neq\mathbf{0}$; hence,

$$\mathbf{n} \overset{\text{def}}{=} \mathbf{Su}\times\mathbf{Sv}\neq\mathbf{0} \tag{2.87}$$

is the area vector of the parallelogram defined by $\mathbf{Su}$ and $\mathbf{Sv}$. Thus,

$$\mathbf{n}\cdot\mathbf{Su} = 0 \qquad \text{and} \qquad \mathbf{n}\cdot\mathbf{Sv} = 0$$

or, equivalently,

$$\mathbf{S}^{\top}\mathbf{n}\cdot\mathbf{u} = 0 \qquad \text{and} \qquad \mathbf{S}^{\top}\mathbf{n}\cdot\mathbf{v} = 0.$$

$\mathbf{S}^{\top}\mathbf{n}$ is therefore perpendicular to $\mathbf{u}$ and $\mathbf{v}$, and there is then a scalar γ such that

$$\mathbf{S}^{\top}\mathbf{n} = \gamma\mathbf{u}\times\mathbf{v}$$

and, by (2.87),

$$\mathbf{S}^{\top}(\mathbf{Su}\times\mathbf{Sv}) = \gamma\mathbf{u}\times\mathbf{v}. \tag{2.88}$$

Finally, letting $\mathbf{w}=\mathbf{u}\times\mathbf{v}$, so that $\mathbf{w}$, $\mathbf{u}$, and $\mathbf{v}$ are linearly independent, we see that

$$\begin{aligned}\gamma\mathbf{w}\cdot(\mathbf{u}\times\mathbf{v}) &= \mathbf{w}\cdot\mathbf{S}^{\top}(\mathbf{Su}\times\mathbf{Sv})\\ &= \mathbf{Sw}\cdot(\mathbf{Su}\times\mathbf{Sv});\end{aligned}$$

[13] Cf. the paragraph containing (1.3).

and, dividing this equation by $\mathbf{w}\cdot(\mathbf{u}\times\mathbf{v})$, and appealing to the definition (2.79) of the determinant, we find that

$$\gamma = \det\mathbf{S}.$$

Thus, (2.88) becomes

$$\mathbf{S}\mathbf{u}\times\mathbf{S}\mathbf{v} = (\det\mathbf{S})\mathbf{S}^{-\top}(\mathbf{u}\times\mathbf{v}). \tag{2.89}$$

For $\mathbf{S}$ invertible, the tensor $\mathbf{S}^{\mathrm{C}}$ defined by

$$\mathbf{S}^{\mathrm{C}} = (\det\mathbf{S})\mathbf{S}^{-\top} \tag{2.90}$$

is called the **cofactor** of $\mathbf{S}$; upon using this tensor, (2.89) takes the simple form[14]

$$\mathbf{S}^{\mathrm{C}}(\mathbf{u}\times\mathbf{v}) = \mathbf{S}\mathbf{u}\times\mathbf{S}\mathbf{v} \tag{2.91}$$

for all linearly independent vectors $\mathbf{u}$ and $\mathbf{v}$. Thus, $\mathbf{S}^{\mathrm{C}}$ transforms the *area vector* $\mathbf{u}\times\mathbf{v}$ of the parallelogram defined by $\mathbf{u}$ and $\mathbf{v}$ into the *area vector* $\mathbf{S}\mathbf{u}\times\mathbf{S}\mathbf{v}$ of the parallelogram defined by $\mathbf{S}\mathbf{u}$ and $\mathbf{S}\mathbf{v}$.

Given a tensor $\mathbf{S}$, we write

$$\begin{aligned} S^{\mathrm{C}}_{ij} &= (\mathbf{S}^{\mathrm{C}})_{ij} \\ &= \mathbf{e}_i\cdot\mathbf{S}^{\mathrm{C}}\mathbf{e}_j \end{aligned}$$

for the components of its cofactor $\mathbf{S}^{\mathrm{C}}$. Thus, using (1.22) to write $\mathbf{e}_j = \frac{1}{2}\epsilon_{jmn}\mathbf{e}_m\times\mathbf{e}_n$ and invoking (2.91), we find that

$$\begin{aligned} S^{\mathrm{C}}_{ij} &= \mathbf{e}_i\cdot[\mathbf{S}^{\mathrm{C}}(\tfrac{1}{2}\epsilon_{jmn}\mathbf{e}_m\times\mathbf{e}_n)] \\ &= \tfrac{1}{2}\epsilon_{jmn}[\mathbf{e}_i\cdot(\mathbf{S}\mathbf{e}_m\times\mathbf{S}\mathbf{e}_n)] \\ &= \tfrac{1}{2}\epsilon_{jmn}[\epsilon_{ikl}(\mathbf{e}_k\cdot\mathbf{S}\mathbf{e}_m)(\mathbf{e}_l\cdot\mathbf{S}\mathbf{e}_n)] \\ &= \tfrac{1}{2}\epsilon_{ikl}\epsilon_{jmn}S_{km}S_{ln}, \end{aligned} \tag{2.92}$$

whereby the components of $\mathbf{S}^{\mathrm{C}}$ depend quadratically on those of $\mathbf{S}$. Also, by (2.90),[15] for $\mathbf{S}$ invertible, $\mathbf{S}^{-1} = (\det\mathbf{S})^{-1}(\mathbf{S}^{\mathrm{C}})^{\top}$, and it follows from (2.92) that the components of $\mathbf{S}^{-1}$ are given in terms of those of $\mathbf{S}$ by

$$S^{-1}_{ij} = \tfrac{1}{2}(\det\mathbf{S})^{-1}\epsilon_{ikl}\epsilon_{jmn}S_{mk}S_{nl}. \tag{2.93}$$

EXERCISES

1. Verify (2.93).
2. Show that for any invertible tensor $\mathbf{S}$

$$\begin{aligned} \det(\mathbf{S}^{\mathrm{C}}) &= (\det\mathbf{S})^2, \\ (\alpha\mathbf{S})^{\mathrm{C}} &= \alpha^2\mathbf{S}^{\mathrm{C}}, \\ (\mathbf{S}^{-1})^{\mathrm{C}} &= (\det\mathbf{S})^{-1}\mathbf{S}^{\top}, \\ (\mathbf{S}^{\mathrm{C}})^{-1} &= (\det\mathbf{S})^{-1}\mathbf{S}^{\top}, \\ (\mathbf{S}^{\mathrm{C}})^{\mathrm{C}} &= (\det\mathbf{S})\mathbf{S}. \end{aligned}$$

---

14 By (iv) of (‡) on page 19.

15 Which is to be expected on the basis of (2.91).

3. Show that for any invertible tensor $\mathbf{S}$ and any vector $\mathbf{u}$

$$(\mathbf{S}\mathbf{u})\times = \mathbf{S}^{\mathrm{C}}(\mathbf{u}\times)\mathbf{S}^{-1},$$

$$(\mathbf{S}^{\mathrm{C}}\mathbf{u})\times = \mathbf{S}(\mathbf{u}\times)\mathbf{S}^{\top}.$$

4. Let $\boldsymbol{\Omega}$ be skew with axial vector $\boldsymbol{\omega}$. Given vectors $\mathbf{u}$ and $\mathbf{v}$, show that

$$\boldsymbol{\Omega}\mathbf{u} \times \boldsymbol{\Omega}\mathbf{v} = (\boldsymbol{\omega} \otimes \boldsymbol{\omega})(\mathbf{u} \times \mathbf{v})$$

and, hence, conclude that $\boldsymbol{\Omega}$ has a unique cofactor given by

$$\boldsymbol{\Omega}^{\mathrm{C}} = \boldsymbol{\omega} \otimes \boldsymbol{\omega}.$$

5. The cofactor $\mathbf{S}^{\mathrm{C}}$ of a not necessarily invertible tensor $\mathbf{S}$ is defined using as a starting point the requirement that (2.91) hold for all vectors $\mathbf{u}$ and $\mathbf{v}$. This definition yields the representation

$$\mathbf{S}^{\mathrm{C}} = \big[\mathbf{S}^2 - (\operatorname{tr}\mathbf{S})\mathbf{S} + \tfrac{1}{2}[\operatorname{tr}^2(\mathbf{S}) - \operatorname{tr}(\mathbf{S}^2)]\mathbf{1}\big]^{\top}. \tag{2.94}$$

Verify that the component form of (2.94) is consistent with (2.92).

6. Use the representation (2.94) to show that, for any vector $\mathbf{u}$,

$$(\mathbf{u}\times)^{\mathrm{C}} = \mathbf{u} \otimes \mathbf{u},$$

$$(\mathbf{u} \otimes \mathbf{u})^{\mathrm{C}} = \mathbf{0}.$$

7. Using the relation (2.79) defining the determinant of a tensor and the identity (2.91), show that for any pair of tensors $\mathbf{S}$ and $\mathbf{T}$,

$$\det(\mathbf{S} + \mathbf{T}) = \det\mathbf{S} + \operatorname{tr}(\mathbf{S}^{\top}\mathbf{T}^{\mathrm{C}}) + \operatorname{tr}(\mathbf{S}^{\top}\mathbf{T}^{\mathrm{C}}) + \det\mathbf{T}. \tag{2.95}$$

8. Using (2.94), show that for any pair of tensors $\mathbf{S}$ and $\mathbf{T}$,

$$(\mathbf{S}\mathbf{T})^{\mathrm{C}} = \mathbf{S}^{\mathrm{C}}\mathbf{T}^{\mathrm{C}}$$

and

$$\begin{aligned}(\mathbf{S} + \mathbf{T})^{\mathrm{C}} = {} & \mathbf{S}^{\mathrm{C}} + \mathbf{T}^{\mathrm{C}} + \mathbf{T}^{\top}\mathbf{S}^{\top} + \mathbf{S}^{\top}\mathbf{T}^{\top} \\ & - (\operatorname{tr}\mathbf{T})\mathbf{S}^{\top} - (\operatorname{tr}\mathbf{S})\mathbf{T}^{\top} + [(\operatorname{tr}\mathbf{S})(\operatorname{tr}\mathbf{T}) - \operatorname{tr}(\mathbf{S}\mathbf{T})]\mathbf{1}.\end{aligned} \tag{2.96}$$

9. Let $\mathbf{S}$ be an invertible tensor and let $\mathbf{u}$ and $\mathbf{v}$ be vectors. Specialize (2.95) to derive the identities

$$\det(\mathbf{S} + \mathbf{1}) = \det\mathbf{S} + (\det\mathbf{S})\operatorname{tr}\mathbf{S}^{-1} + \operatorname{tr}\mathbf{S} + 1,$$

$$\det(\mathbf{S} + \mathbf{u}\times) = (\det\mathbf{S})(1 + (\mathbf{u}\times)\colon\mathbf{S}^{-1}) + \mathbf{u}\cdot\mathbf{S}\mathbf{u},$$

$$\det(\mathbf{S} + \mathbf{u}\otimes\mathbf{v}) = \det\mathbf{S} + \mathbf{u}\cdot\mathbf{S}^{\mathrm{C}}\mathbf{v}.$$

10. Let $\mathbf{S}$ be a tensor and let $\mathbf{u}$ and $\mathbf{v}$ be vectors. Specialize (2.96) to derive the identities

$$(\mathbf{S} + \mathbf{1})^{\mathrm{C}} = \mathbf{S}^{\mathrm{C}} + \mathbf{1} + (\operatorname{tr}\mathbf{S})\mathbf{1} - \mathbf{S}^{\top},$$

$$(\mathbf{S} + \mathbf{u}\times)^{\mathrm{C}} = \mathbf{S}^{\mathrm{C}} - [\mathbf{S}^{\top} - (\operatorname{tr}\mathbf{S})\mathbf{1}]\mathbf{u}\times - (\mathbf{u}\times)\mathbf{S}^{\top} + [\mathbf{S}\colon(\mathbf{u}\times)]\mathbf{1} + \mathbf{u}\otimes\mathbf{u},$$

$$(\mathbf{S} + \mathbf{u}\otimes\mathbf{v})^{\mathrm{C}} = \mathbf{S}^{\mathrm{C}} - (\mathbf{u}\times)\mathbf{S}(\mathbf{v}\times).$$

2.12 Orthogonal Tensors

A tensor $\mathbf{Q}$ is **orthogonal** if

$$\mathbf{Qu} \cdot \mathbf{Qv} = \mathbf{u} \cdot \mathbf{v} \tag{2.97}$$

for all vectors $\mathbf{u}$ and $\mathbf{v}$. Choosing $\mathbf{v} = \mathbf{u}$ in (2.97),

$$\begin{aligned} |\mathbf{Qu}|^2 &= \mathbf{Qu} \cdot \mathbf{Qu} \\ &= \mathbf{u} \cdot \mathbf{u} \\ &= |\mathbf{u}|^2 \end{aligned}$$

so that

$$|\mathbf{Qu}| = |\mathbf{u}|; \tag{2.98}$$

the action of an orthogonal tensor $\mathbf{Q}$ on a vector $\mathbf{u}$ therefore leaves the length of $\mathbf{u}$ unchanged. Consider next the angle $\theta = \angle(\mathbf{u}, \mathbf{v})$ between nonzero vectors $\mathbf{u}$ and $\mathbf{v}$ and the angle $\beta = \angle(\mathbf{Qu}, \mathbf{Qv})$ between the transformed vectors $\mathbf{Qu}$ and $\mathbf{Qv}$.[16] Then, by (2.97) and (2.98),

$$\begin{aligned} \cos\beta &= \frac{\mathbf{Qu} \cdot \mathbf{Qv}}{|\mathbf{Qu}||\mathbf{Qv}|} \\ &= \frac{\mathbf{u} \cdot \mathbf{v}}{|\mathbf{u}||\mathbf{v}|} \\ &= \cos\theta; \end{aligned} \tag{2.99}$$

the angle between $\mathbf{u}$ and $\mathbf{v}$ is therefore preserved whenever $\mathbf{u}$ and $\mathbf{v}$ are transformed by an orthogonal tensor $\mathbf{Q}$.

The results derived above represent the essential geometrical properties of an orthogonal tensor: the preservation of lengths and angles. The next result gives the basic algebraic property of an orthogonal tensor.

(‡) *A condition both necessary and sufficient for a tensor* $\mathbf{Q}$ *to be orthogonal is that* $\mathbf{Q}$ *be invertible with transpose equal to its inverse*

$$\mathbf{Q}^{\top} = \mathbf{Q}^{-1}. \tag{2.100}$$

The result (‡) is central to our treatment of continuum mechanics; its proof is given at the end of this subsection.

In view of (2.67), an immediate consequence of (‡) is the important relation

$$\boxed{\mathbf{Q}^{\top}\mathbf{Q} = \mathbf{Q}\mathbf{Q}^{\top} = \mathbf{1},} \tag{2.101}$$

which is valid for any orthogonal tensor $\mathbf{Q}$. Further, (2.82) implies that $\det(\mathbf{Q}^{\top}\mathbf{Q}) = (\det\mathbf{Q})^2$; therefore, by (2.101), if $\mathbf{Q}$ is orthogonal, then

$$\det\mathbf{Q} = \pm 1. \tag{2.102}$$

An orthogonal tensor $\mathbf{Q}$ is a **rotation** if $\det\mathbf{Q} = 1$ and a **reflection** otherwise. Every reflection can be expressed as the product of a rotation with $-\mathbf{1}$. For example, on multiplication by $-\mathbf{1}$, the rotation, by π radians, $2\mathbf{e}\otimes\mathbf{e} - \mathbf{1}$ about an axis with unit normal $\mathbf{e}$ becomes the reflection $\mathbf{1} - 2\mathbf{e}\otimes\mathbf{e}$ about that same axis.

16 Cf. (1.1) and (1.2).

2.12.1 Proof of (‡)

Necessity. Let $\mathbf{Q}$ be orthogonal. Then, by (2.98), $\mathbf{Q}$ satisfies condition (ii) of (‡) on page 19; $\mathbf{Q}$ is therefore invertible. Further, by (2.14) and (2.97),

$$\begin{aligned}\mathbf{u}\cdot\mathbf{u} &= \mathbf{Qu}\cdot\mathbf{Qu}\\ &= \mathbf{u}\cdot\mathbf{Q}^{\mathsf{T}}\mathbf{Qu}\end{aligned}$$

and hence,

$$\mathbf{u}\cdot\underbrace{(\mathbf{Q}^{\mathsf{T}}\mathbf{Q}-\mathbf{1})}_{\mathbf{T}}\mathbf{u}=0 \tag{2.103}$$

for all $\mathbf{u}$. On the other hand, since $\mathbf{T}=\mathbf{Q}^{\mathsf{T}}\mathbf{Q}-\mathbf{1}$ is symmetric,

$$2\mathbf{u}\cdot\mathbf{Tv}=(\mathbf{u}+\mathbf{v})\cdot\mathbf{T}(\mathbf{u}+\mathbf{v})-\mathbf{u}\cdot\mathbf{Tu}-\mathbf{v}\cdot\mathbf{Tv}.$$

But by (2.103), the right side of this equation vanishes. Thus, $\mathbf{u}\cdot\mathbf{Tv}=0$ for all $\mathbf{u}$ and $\mathbf{v}$, which, by (2.3), implies that $\mathbf{T}=\mathbf{0}$; hence,

$$\mathbf{Q}^{\mathsf{T}}\mathbf{Q}=\mathbf{1}. \tag{2.104}$$

Finally, since $\mathbf{Q}$ is invertible, if we multiply the right side of (2.104) by $\mathbf{Q}^{-1}$ we see that

$$\mathbf{Q}^{\mathsf{T}}=\mathbf{Q}^{-1}, \tag{2.105}$$

and consequently, by (2.67), that (2.101) is satisfied.

Sufficiency. Assume that (2.101) is satisfied. Then

$$\mathbf{Qu}\cdot\mathbf{Qv}=\mathbf{u}\cdot\mathbf{Q}^{\mathsf{T}}\mathbf{Qv}=\mathbf{u}\cdot\mathbf{v}$$

for all $\mathbf{u}$ and $\mathbf{Q}$ is orthogonal.

EXERCISE

1. Show that given a unit vector $\mathbf{e}$, the tensor

$$\mathbf{Q}(\theta)=(\cos\theta)\mathbf{1}+(1-\cos\theta)\mathbf{e}\otimes\mathbf{e}+(\sin\theta)\mathbf{e}\times \tag{2.106}$$

   is a *rotation* for $0\le\theta\le 2\pi$. Show further that
   a) $\mathbf{Q}(0)=\mathbf{Q}(2\pi)=\mathbf{1}$,
   b) $\mathbf{Q}(\theta+\phi)=\mathbf{Q}(\theta)\mathbf{Q}(\phi)$,
   c) $\mathbf{Q}(\theta)\mathbf{e}=\mathbf{e}$,
   d) $\mathbf{Q}(\theta)(\mathbf{1}-\mathbf{e}\otimes\mathbf{e})=(\cos\theta)(\mathbf{1}-\mathbf{e}\otimes\mathbf{e})+(\sin\theta)\mathbf{e}\times$,
   and, on the basis of these properties, conclude that $\mathbf{Q}(\theta)$ rotates any vector by an angle θ about an axis parallel to $\mathbf{e}$.
2. Using (ii) of Appendix 114, show that $\mathbf{Q}(\theta)$ defined in (2.106) can be represented alternatively as

$$\mathbf{Q}(\theta)=\mathrm{e}^{\theta\mathbf{e}\times}.$$

3. Let $\mathbf{Q}$ be a rotation and show that for any pair of vectors $\mathbf{u}$ and $\mathbf{v}$,

$$\mathbf{Q}(\mathbf{u}\times\mathbf{v})=\mathbf{Qu}\times\mathbf{Qv}.$$

   (Hint: Note from (2.90) and (2.100) that $\mathbf{Q}^{\mathrm{c}}=(\det\mathbf{Q})\mathbf{Q}$ for any orthogonal tensor $\mathbf{Q}$ and use (2.91) and (2.102).)

2.13 Matrix of a Tensor

We now show that the operations of multiplication, transposition of tensors as well as the operators defining the trace and determinant of tensors are in one-to-one correspondence with these same operations and operators applied to matrices.

We write $[\mathbf{u}]$ and $[\mathbf{S}]$ for the matrix representations of a vector $\mathbf{u}$ and a tensor $\mathbf{S}$ with respect to the basis $\{\mathbf{e}_i\}$:

$$[\mathbf{u}]=\begin{bmatrix}u_1\\u_2\\u_3\end{bmatrix},\qquad [\mathbf{S}]=\begin{bmatrix}S_{11}&S_{12}&S_{13}\\S_{21}&S_{22}&S_{23}\\S_{31}&S_{32}&S_{33}\end{bmatrix}.$$

Then,

$$
\begin{aligned}
[\mathbf{S}][\mathbf{u}] &= \begin{bmatrix} S_{11} & S_{12} & S_{13} \\ S_{21} & S_{22} & S_{23} \\ S_{31} & S_{32} & S_{33} \end{bmatrix} \begin{bmatrix} u_1 \\ u_2 \\ u_3 \end{bmatrix} \\
&= \begin{bmatrix} S_{11}u_1 + S_{12}u_2 + S_{13}u_3 \\ S_{21}u_1 + S_{22}u_2 + S_{23}u_3 \\ S_{31}u_1 + S_{32}u_2 + S_{33}u_3 \end{bmatrix} \\
&= \begin{bmatrix} S_{1i}u_i \\ S_{2i}u_i \\ S_{3i}u_i \end{bmatrix} \\
&= [\mathbf{Su}],
\end{aligned}
$$

so that the action of a tensor on a vector is consistent with that of a 3×3 matrix on a 3×1 matrix. Further, bearing in mind (2.15), transposition of the matrix $[\mathbf{S}]$ yields

$$
\begin{aligned}
[\mathbf{S}]^{\scriptscriptstyle\mathsf{T}} &= \begin{bmatrix} S_{11} & S_{21} & S_{31} \\ S_{12} & S_{22} & S_{32} \\ S_{13} & S_{23} & S_{33} \end{bmatrix} \\
&= [\mathbf{S}^{\scriptscriptstyle\mathsf{T}}].
\end{aligned}
$$

In view of (2.42) and $(2.86)_1$, the conventional definitions of the trace and determinant from matrix algebra yield

$$
\begin{aligned}
\operatorname{tr}[\mathbf{S}] &= S_{11} + S_{22} + S_{33} = S_{ii} \\
&= \operatorname{tr}\mathbf{S}
\end{aligned}
\tag{2.107}
$$

and

$$
\begin{aligned}
\det[\mathbf{S}] &= \begin{vmatrix} S_{11} & S_{12} & S_{13} \\ S_{21} & S_{22} & S_{23} \\ S_{31} & S_{32} & S_{33} \end{vmatrix} \\
&= S_{11}(S_{22}S_{33} - S_{23}S_{32}) - S_{12}(S_{21}S_{33} - S_{23}S_{31}) + S_{13}(S_{21}S_{32} - S_{22}S_{31}) \\
&= \epsilon_{ijk}S_{i1}S_{j2}S_{k3} \\
&= \det\mathbf{S}.
\end{aligned}
\tag{2.108}
$$

EXERCISE

1. Use the conventional definition of the cofactor $[\mathbf{S}]^{\mathrm{c}}$ of a matrix $[\mathbf{S}]$ to show that

$$
\begin{aligned}
[\mathbf{S}]^{\mathrm{c}} &= \begin{bmatrix} S_{22}S_{33} - S_{23}S_{32} & S_{23}S_{31} - S_{21}S_{33} & S_{21}S_{32} - S_{22}S_{31} \\ S_{13}S_{32} - S_{33}S_{12} & S_{11}S_{33} - S_{13}S_{31} & S_{12}S_{31} - S_{11}S_{32} \\ S_{12}S_{23} - S_{13}S_{22} & S_{13}S_{21} - S_{11}S_{23} & S_{11}S_{22} - S_{12}S_{21} \end{bmatrix} \\
&= [\mathbf{S}^{\mathrm{c}}]
\end{aligned}
\tag{2.109}
$$

and thus to conclude from (2.92) that the definition (2.91) of the cofactor $\mathbf{S}^{\mathrm{c}}$ of a tensor is consistent with what arises in matrix algebra.

2.14 Eigenvalues and Eigenvectors of a Tensor. Spectral Theorem

A scalar ω is an **eigenvalue** of a tensor $\mathbf{S}$ if there is a unit vector $\mathbf{e}$ such that

$$\mathbf{S}\mathbf{e} = \omega\mathbf{e}, \tag{2.110}$$

in which case $\mathbf{e}$ is an **eigenvector** of $\mathbf{S}$ corresponding to the eigenvalue ω and ω is an eigenvalue of $\mathbf{S}$ corresponding to the eigenvector $\mathbf{e}$. The **characteristic space** for $\mathbf{S}$ corresponding to ω is the *subspace* $\mathcal{U}$ of vectors $\mathbf{v}$ satisfying the equation[17]

$$\mathbf{S}\mathbf{v} = \omega\mathbf{v}. \tag{2.111}$$

Suppose that $\mathbf{S}$ is *symmetric* and that $\omega_1 \neq \omega_2$ are eigenvalues of $\mathbf{S}$ with corresponding eigenvectors $\mathbf{e}_1$ and $\mathbf{e}_2$. Then, since $\mathbf{S}$ is symmetric,

$$\begin{aligned} \omega_1\mathbf{e}_1 \cdot \mathbf{e}_2 &= (\mathbf{S}\mathbf{e}_1) \cdot \mathbf{e}_2 \\ &= (\mathbf{S}\mathbf{e}_2) \cdot \mathbf{e}_1 \\ &= \omega_2\mathbf{e}_2 \cdot \mathbf{e}_1. \end{aligned}$$

Thus,

$$(\omega_1 - \omega_2)\mathbf{e}_1 \cdot \mathbf{e}_2 = 0,$$

and since $\omega_1 \neq \omega_2$, we must have

$$\mathbf{e}_1 \cdot \mathbf{e}_2 = 0.$$

For a symmetric tensor, eigenvectors corresponding to distinct eigenvalues are therefore orthogonal.

The next result, whose proof we omit,[18] is a central theorem of linear algebra and one of substantial value in continuum mechanics.

SPECTRAL THEOREM *Let* $\mathbf{S}$ *be symmetric. Then there is an orthonormal basis* $\{\mathbf{e}_i\}$ *of eigenvectors of* $\mathbf{S}$ *and, what is most important,*

$$\mathbf{S} = \sum_{i=1}^{3} \omega_i \mathbf{e}_i \otimes \mathbf{e}_i, \tag{2.112}$$

where, for each i, ω_i *is an eigenvalue of* $\mathbf{S}$ *and* $\mathbf{e}_i$ *is a corresponding eigenvector.*

The relation (2.112), which is called a **spectral decomposition** of $\mathbf{S}$, gives $\mathbf{S}$ as a linear combination of projections,[19] with each $\mathbf{e}_i \otimes \mathbf{e}_i$ (no sum) a projection onto the eigenvector $\mathbf{e}_i$. Such a decomposition is unique if and only if the eigenvalues of $\mathbf{S}$ are distinct. The geometric nature of the characteristic spaces of $\mathbf{S}$ depend on the number of distinct eigenvalues of $\mathbf{S}$:

(i) If the eigenvalues are distinct, then $\mathbf{S}$ has exactly *three* characteristic spaces: the three mutually perpendicular lines l_i, with each l_i parallel to the corresponding eigenvector $\mathbf{e}_i$.

(ii) If $\mathbf{S}$ has but two distinct eigenvalues, say

$$\omega_1 \neq \omega_2 = \omega_3,$$

then since $\mathbf{1} = \mathbf{e}_i \otimes \mathbf{e}_i$,

$$\omega_2\mathbf{e}_2 \otimes \mathbf{e}_2 + \omega_3\mathbf{e}_3 \otimes \mathbf{e}_3 = \omega_2(\mathbf{1} - \mathbf{e}_1 \otimes \mathbf{e}_1).$$

---

[17] Cf. (1.9). $\mathcal{U}$ is clearly a subspace, since any linear combination of solutions of (2.111) is itself a solution.

[18] Cf., e.g., HALMOS (1958, §79), STEWART (1963, §37), and BOWEN & WANG (1976, §27).

[19] Projections are discussed in the paragraph containing (2.6).

The spectral decomposition (2.112) therefore has a form

$$\mathbf{S} = \omega_1 \mathbf{e}_1 \otimes \mathbf{e}_1 + \omega_2(\mathbf{1} - \mathbf{e}_1 \otimes \mathbf{e}_1) \tag{2.113}$$

giving $\mathbf{S}$ as a linear combination of the projection $\mathbf{e}_1 \otimes \mathbf{e}_1$ onto $\mathbf{e}_1$ and the projection $\mathbf{1} - \mathbf{e}_1 \otimes \mathbf{e}_1$ onto the plane perpendicular to $\mathbf{e}_1$. In this case, $\mathbf{S}$ has exactly *two* characteristic spaces: the line l through $\mathbf{o}$ parallel to $\mathbf{e}_1$ and the plane Π through $\mathbf{0}$ perpendicular to $\mathbf{e}_1$ and hence to l.

(iii) If $\omega_1 = \omega_2 = \omega_3 = s$ then (and only then)

$$\mathbf{S} = s\mathbf{1} \tag{2.114}$$

and is, hence, spherical. In this case, every unit vector is an eigenvector and the *sole characteristic space* of $\mathbf{S}$ is the entire vector space $\mathcal{V}$.

We therefore have the following decompositions of a vector $\mathbf{v}$ into a sum of vectors: *one vector for each characteristic space*:

$$\left.\begin{array}{ll} \text{case (i):} & \mathbf{v} = \underbrace{(\mathbf{v}\cdot\mathbf{e}_i)\mathbf{e}_i}_{\text{a vector in } l_i}, \\[2ex] \text{case (ii):} & \mathbf{v} = \underbrace{(\mathbf{e}_1 \otimes \mathbf{e}_1)\mathbf{v}}_{\text{a vector in } l} + \underbrace{(\mathbf{1} - \mathbf{e}_1 \otimes \mathbf{e}_1)\mathbf{v}}_{\substack{\text{a vector in the plane} \\ \Pi \text{ perpendicular to } l}}, \\[2ex] \text{case (iii):} & \mathbf{v} = \mathbf{v}. \end{array}\right\} \tag{2.115}$$

Thus, if we denote the characteristic spaces of an arbitrary symmetric tensor $\mathbf{S}$ by

$$\mathcal{U}_\alpha, \qquad \alpha = 1, \ldots n \leq 3, \tag{2.116}$$

we can then summarize cases (i)–(iii) by decomposing an arbitrary vector into elements of distinct characteristic subspaces

$$\mathbf{v} = \sum_{\alpha=1}^{n} \mathbf{u}_\alpha, \qquad \text{where } \mathbf{u}_\alpha \text{ belongs to } \mathcal{U}_\alpha. \tag{2.117}$$

Given a set $\mathcal{F}$ of vectors, we say that a tensor $\mathbf{T}$ **leaves** $\mathcal{F}$ **invariant** if for every vector $\mathbf{v}$ in $\mathcal{F}$ the vector $\mathbf{Tv}$ also belongs to $\mathcal{F}$.

Suppose that two (not necessarily symmetric) tensors $\mathbf{S}$ and $\mathbf{T}$ *commute*:

$$\mathbf{ST} = \mathbf{TS}.$$

Assume that $\mathbf{v}$ belongs to a characteristic space $\mathcal{U}$ for $\mathbf{S}$ with ω the corresponding eigenvalue so that

$$\mathbf{Sv} = \omega\mathbf{v}.$$

Then,

$$\mathbf{S}(\mathbf{Tv}) = \mathbf{TSv} = \omega(\mathbf{Tv})$$

and $\mathbf{Tv}$ *also belongs to* $\mathcal{U}$. Thus, $\mathbf{T}$ *leaves the characteristic space* $\mathcal{U}$ *of* $\mathbf{S}$ *invariant.*

This result has a converse for $\mathbf{S}$ *symmetric* ($\mathbf{T}$ arbitrary):

(‡) *If* $\mathbf{T}$ *leaves each characteristic space* $\mathcal{U}_\alpha$ *of* $\mathbf{S}$ *invariant, then* $\mathbf{S}$ *and* $\mathbf{T}$ *commute.*

To prove (‡), assume that $\mathbf{T}$ leaves each characteristic space $\mathcal{U}_\alpha$ of $\mathbf{S}$ invariant. Thus, if $\mathbf{u}_\alpha$ belongs to $\mathcal{U}_\alpha$, so that

$$\mathbf{S}\mathbf{u}_\alpha = \omega_\alpha \mathbf{u}_\alpha,$$

then, by hypothesis, $\mathbf{Tv}_\alpha$ belongs to $\mathcal{U}_\alpha$; hence,

$$\begin{aligned}\mathbf{S}(\mathbf{Tu}_\alpha) &= \omega_\alpha(\mathbf{Tu}_\alpha)\\ &= \mathbf{T}(\omega_\alpha \mathbf{u}_\alpha)\\ &= \mathbf{T}(\mathbf{Su}_\alpha).\end{aligned}$$

Choose a vector $\mathbf{v}$ and consider its expansion $\mathbf{v} = \sum_{\alpha=1}^{n} \mathbf{u}_\alpha$ as in (2.117). Then

$$\begin{aligned}\mathbf{STv} &= \sum_{\alpha=1}^{n} \mathbf{STu}_\alpha\\ &= \sum_{\alpha=1}^{n} \mathbf{TSu}_\alpha\\ &= \mathbf{TSv}.\end{aligned}$$

Since $\mathbf{v}$ was arbitrarily chosen, $\mathbf{ST} = \mathbf{TS}$. This proves (‡).

Note that by (2.112), the matrix of $\mathbf{S}$ relative to the eigenvector basis $\{\mathbf{e}_i\}$ is diagonal:

$$[\mathbf{S}] = \begin{bmatrix} \omega_1 & 0 & 0 \\ 0 & \omega_2 & 0 \\ 0 & 0 & \omega_3 \end{bmatrix}, \tag{2.118}$$

so that by (2.107) and (2.108),

$$\operatorname{tr}\mathbf{S} = \omega_1 + \omega_2 + \omega_3 \quad \text{and} \quad \det\mathbf{S} = \omega_1\omega_2\omega_3. \tag{2.119}$$

EXERCISES

1. Let $\mathbf{f}_1$ and $\mathbf{f}_2$ be orthogonal unit vectors, so that $|\mathbf{f}_1| = |\mathbf{f}_2| = 1$ and $\mathbf{f}_1 \cdot \mathbf{f}_2 = 0$. Determine the eigenvalues and eigenvectors of the tensor $\mathbf{S} = \mathbf{1} + \mathbf{f}_1 \otimes \mathbf{f}_2 + \mathbf{f}_2 \otimes \mathbf{f}_1$.
2. Show that for case (ii) in which $\mathbf{S}$ has the form (2.113) we may replace the eigenvectors $\mathbf{e}_2$ and $\mathbf{e}_3$ in (2.112) by any two orthogonal unit vectors in the plane perpendicular to $\mathbf{e}_1$.
3. Let $\mathbf{S}$ be an invertible tensor with eigenvalue ω and corresponding eigenvector $\mathbf{e}$. Show that ω^{-1} is an eigenvalue of $\mathbf{S}^{-1}$ and determine the corresponding eigenvector.
4. Let $\mathbf{S}$ be a symmetric tensor with three distinct eigenvalues ω_1, ω_2, and ω_3. Let $\mathbf{P}_1$, $\mathbf{P}_2$, and $\mathbf{P}_3$ be defined by

$$\begin{aligned}\mathbf{P}_1 &= (\omega_1 - \omega_2)^{-1}(\omega_1 - \omega_3)^{-1}(\mathbf{S} - \omega_2\mathbf{1})(\mathbf{S} - \omega_3\mathbf{1}),\\ \mathbf{P}_2 &= (\omega_2 - \omega_1)^{-1}(\omega_2 - \omega_3)^{-1}(\mathbf{S} - \omega_1\mathbf{1})(\mathbf{S} - \omega_3\mathbf{1}),\\ \mathbf{P}_3 &= (\omega_3 - \omega_1)^{-1}(\omega_3 - \omega_2)^{-1}(\mathbf{S} - \omega_1\mathbf{1})(\mathbf{S} - \omega_2\mathbf{1}).\end{aligned}$$

Show that:

a. $\mathbf{P}_i\mathbf{u} = (\mathbf{e}_i \cdot \mathbf{u})\mathbf{e}_i$ for each $i = 1, 2, 3$ (no sum);
b. $\mathbf{P}_i\mathbf{P}_i = \mathbf{P}_i$ for each $i = 1, 2, 3$ (no sum);
c. $\mathbf{P}_1 + \mathbf{P}_2 + \mathbf{P}_3 = \mathbf{1}$;

d. given a polynomial function φ of a scalar and a corresponding polynomial tensor-valued function $\mathbf{\Phi}$ of a tensor,[20]

$$\mathbf{\Phi}(\mathbf{S}) = \varphi(\omega_i)\mathbf{P}_i.$$

e. As simple consequences of the last result,

$$\mathbf{1} = \mathbf{P}_1 + \mathbf{P}_2 + \mathbf{P}_3,$$

$$\mathbf{S} = \omega_i \mathbf{P}_i,$$

$$\mathbf{S}^{\mathrm{C}} = \omega_2\omega_3\mathbf{P}_1 + \omega_1\omega_3\mathbf{P}_2 + \omega_1\omega_2\mathbf{P}_3,$$

$$\mathbf{S}^{-1} = \omega_i^{-1}\mathbf{P}_i.$$

2.15 Square Root of a Symmetric, Positive-Definite Tensor. Polar Decomposition Theorem

A tensor $\mathbf{C}$ is **positive-definite** if

$$\mathbf{u} \cdot \mathbf{C}\mathbf{u} > 0 \tag{2.120}$$

for all vectors $\mathbf{u} \neq \mathbf{0}$.

Suppose that $\mathbf{C}$ is symmetric, positive-definite, and consider its spectral decomposition as defined by (2.112) with $\mathbf{S} = \mathbf{C}$. Then, by (2.120), for each *fixed* choice of i,

$$\mathbf{e}_i \cdot \mathbf{C}\mathbf{e}_i = \omega_i > 0 \qquad \text{(no sum on } i\text{)}. \tag{2.121}$$

The eigenvalues of a symmetric, positive-definite tensor are therefore strictly positive. The converse assertion that

(†) *a symmetric tensor with strictly positive eigenvalues is positive-definite*

is left as an exercise, as are the following properties of a symmetric, positive-definite tensor $\mathbf{C}$:

$$\det \mathbf{C} > 0 \tag{2.122}$$

and

$$\mathbf{R}\mathbf{C}\mathbf{R}^{\top} \text{ is symmetric and positive-definite for every rotation } \mathbf{R}. \tag{2.123}$$

We now use (2.121) to show that, given a symmetric, positive-definite tensor $\mathbf{C}$, there is a unique symmetric, positive-definite tensor $\mathbf{U}$ such that

$$\mathbf{U}^2 = \mathbf{C}; \tag{2.124}$$

in this case, we write

$$\mathbf{U} = \sqrt{\mathbf{C}} \tag{2.125}$$

and refer to $\sqrt{\mathbf{C}}$ as the **square root** of $\mathbf{C}$.

Since $\mathbf{C}$ is symmetric and positive-definite, the spectral theorem and (2.121) imply that $\mathbf{C}$ admits the spectral decomposition

$$\mathbf{C} = \sum_{i=1}^{3} \omega_i \mathbf{e}_i \otimes \mathbf{e}_i, \qquad \omega_i > 0. \tag{2.126}$$

[20] Cf., e.g., FRAZER, DUNCAN & COLLAR (1938, §3.8) and BOWEN & WANG (1976, §27). This result is known as Sylvester's theorem.

Let

$$\mathbf{U} \stackrel{\text{def}}{=} \sum_{i=1}^{3} \sqrt{\omega_i}\,\mathbf{e}_i \otimes \mathbf{e}_i. \tag{2.127}$$

Then, by $(1.10)_1$, (2.28), and (2.126),

$$\begin{aligned}\mathbf{U}^2 &= \left[\sum_{i=1}^{3} \sqrt{\omega_i}\,\mathbf{e}_i \otimes \mathbf{e}_i\right]\left[\sum_{j=1}^{3} \sqrt{\omega_j}\,\mathbf{e}_j \otimes \mathbf{e}_j\right] \\ &= \sum_{i,j=1}^{3} \sqrt{\omega_i}\sqrt{\omega_j}(\mathbf{e}_i \otimes \mathbf{e}_i)(\mathbf{e}_j \otimes \mathbf{e}_j) \\ &= \sum_{i=1}^{3} \omega_i \mathbf{e}_i \otimes \mathbf{e}_i \\ &= \mathbf{C}.\end{aligned}$$

Thus, $\mathbf{U}$ is *a* square root of $\mathbf{C}$.

But is $\mathbf{U}$ the only square root of $\mathbf{C}$? To show that $\mathbf{U}$ is unique,[21] suppose that there is a second symmetric, positive-definite tensor $\acute{\mathbf{U}}$ such that

$$\mathbf{U}^2 = \acute{\mathbf{U}}^2 = \mathbf{C}.$$

Let $\mathbf{e}$ be an eigenvector of $\mathbf{C}$ with $\omega > 0$ the corresponding eigenvalue. Then, letting $\lambda = \sqrt{\omega}$,

$$\mathbf{0} = (\mathbf{U}^2 - \omega\mathbf{1})\mathbf{e} = (\mathbf{U} + \lambda\mathbf{1})(\mathbf{U} - \lambda\mathbf{1})\mathbf{e}.$$

Thus, for

$$\mathbf{v} = (\mathbf{U} - \lambda\mathbf{1})\mathbf{e},$$

we find that

$$\mathbf{U}\mathbf{e} = -\lambda\mathbf{e}$$

and $\mathbf{v}$ must vanish, for otherwise $-\lambda$ would be an eigenvalue of $\mathbf{U}$, which is an impossibility, since $\mathbf{U}$ is, by definition, positive-definite.[22] Hence,

$$\mathbf{U}\mathbf{e} = \lambda\mathbf{e}.$$

Similarly,

$$\acute{\mathbf{U}}\mathbf{e} = \lambda\mathbf{e}$$

for every eigenvector $\mathbf{e}$ of $\mathbf{C}$. Since by the spectral theorem we can form a basis of eigenvectors of $\mathbf{C}$, we find that

$$\mathbf{U} = \acute{\mathbf{U}}.$$

We have therefore shown that the square root of a symmetric, positive-definite tensor $\mathbf{C}$ is well-defined; in fact, by (2.127), $\sqrt{\mathbf{C}}$ has the explicit form

$$\sqrt{\mathbf{C}} = \sum_{i=1}^{3} \sqrt{\omega_i}\,\mathbf{e}_i \otimes \mathbf{e}_i, \tag{2.128}$$

[21] This proof is due to Stephenson (1980).
[22] Cf. (2.121).

where $\{\mathbf{e}_i\}$ is a basis of eigenvectors of $\mathbf{C}$ and for each i, ω_i is the eigenvalue corresponding to $\mathbf{e}_i$. Thus, in view of (†) on page 31,

(‡) *the square root of a symmetric, positive-definite tensor is itself a symmetric, positive-definite tensor.*

Next, we show that if $\mathbf{F}$ is an invertible tensor, then

$$\mathbf{F}^{\top}\mathbf{F} \textit{ and } \mathbf{F}\mathbf{F}^{\top} \textit{are symmetric and positive-definite.} \tag{2.129}$$

To verify this result, note that by (2.17),

$$(\mathbf{F}^{\top}\mathbf{F})^{\top} = \mathbf{F}^{\top}\mathbf{F} \qquad \text{and} \qquad (\mathbf{F}\mathbf{F}^{\top})^{\top} = \mathbf{F}\mathbf{F}^{\top};$$

hence $\mathbf{F}^{\top}\mathbf{F}$ and $\mathbf{F}\mathbf{F}^{\top}$ are symmetric. Further, by (2.14),

$$\mathbf{u}\cdot\mathbf{F}^{\top}\mathbf{F}\mathbf{u} = |\mathbf{F}\mathbf{u}|^2 > 0 \qquad \text{and} \qquad \mathbf{u}\cdot\mathbf{F}\mathbf{F}^{\top}\mathbf{u} = |\mathbf{F}^{\top}\mathbf{u}|^2 > 0$$

for any vector $\mathbf{u} \neq \mathbf{0}$, and $\mathbf{F}^{\top}\mathbf{F}$ and $\mathbf{F}\mathbf{F}^{\top}$ are positive-definite.

The next result is central to the modern discussion of strain.

POLAR DECOMPOSITION THEOREM *Let* $\mathbf{F}$ *be an invertible tensor with* $\det\mathbf{F} > 0$. *Then there are symmetric, positive-definite tensors* $\mathbf{U}$ *and* $\mathbf{V}$ *and a rotation* $\mathbf{R}$ *such that*

$$\boxed{\mathbf{F} = \mathbf{R}\mathbf{U} = \mathbf{V}\mathbf{R}.} \tag{2.130}$$

Moreover, each of these decompositions is unique in the sense that if

$$\mathbf{F} = \tilde{\mathbf{R}}\tilde{\mathbf{U}} \qquad \textit{and} \qquad \mathbf{F} = \bar{\mathbf{V}}\bar{\mathbf{R}} \tag{2.131}$$

with $\tilde{\mathbf{U}}$ *and* $\bar{\mathbf{V}}$ *symmetric, positive-definite and* $\tilde{\mathbf{R}}$ *and* $\bar{\mathbf{R}}$ *rotations, then*

$$\tilde{\mathbf{U}} = \mathbf{U}, \qquad \bar{\mathbf{V}} = \mathbf{V}, \qquad \tilde{\mathbf{R}} = \bar{\mathbf{R}} = \mathbf{R}.$$

We refer to $\mathbf{F} = \mathbf{R}\mathbf{U}$ *and* $\mathbf{F} = \mathbf{V}\mathbf{R}$, *respectively, as the* **right** *and* **left polar decompositions** *of* $\mathbf{F}$. *Finally, granted these decompositions,* $\mathbf{F}$ *determines* $\mathbf{U}$ *and* $\mathbf{V}$ *through the relations*

$$\boxed{\begin{aligned}\mathbf{U} &= \sqrt{\mathbf{F}^{\top}\mathbf{F}},\\ \mathbf{V} &= \sqrt{\mathbf{F}\mathbf{F}^{\top}},\end{aligned}} \tag{2.132}$$

and

$$\mathbf{V} = \mathbf{R}\mathbf{U}\mathbf{R}^{\top}. \tag{2.133}$$

PROOF. Our first step is to show that if $\mathbf{F}$ has the right and left polar decompositions (2.130), then $\mathbf{U}$, $\mathbf{V}$, and $\mathbf{R}$ satisfy (2.132) and (2.133). Indeed, by $(2.130)_1$ with $\mathbf{U}$ symmetric and $\mathbf{R}$ a rotation,

$$\begin{aligned}\mathbf{F}^{\top}\mathbf{F} &= \mathbf{U}\mathbf{R}^{\top}\mathbf{R}\mathbf{U}\\ &= \mathbf{U}^2;\end{aligned}$$

similarly $(2.130)_2$ with $\mathbf{V}$ symmetric yields

$$\mathbf{F}\mathbf{F}^{\top} = \mathbf{V}^2.$$

Further, since by hypothesis and (2.129), $\mathbf{F}^{\top}\mathbf{F}$ and $\mathbf{F}\mathbf{F}^{\top}$ are symmetric and positive-definite, each of these tensors has a square root; hence, (2.132) is satisfied. Finally, solving $(2.130)_{2,3}$ for $\mathbf{V}$ yields (2.133).

We are now in a position to establish the existence of the right and left decompositions (2.131). Since (2.132) is necessary for existence, we begin by *defining* a symmetric, positive-definite tensor $\mathbf{U}$ by $(2.132)_1$ and a tensor $\mathbf{R}$ through $\mathbf{R} = \mathbf{F}\mathbf{U}^{-1}$. To verify that $\mathbf{F} = \mathbf{R}\mathbf{U}$ is a right polar decomposition, we have only to show that $\mathbf{R}$ is a rotation. By the first of (2.135),

$$\begin{aligned}\mathbf{R}^{\top}\mathbf{R} &= \mathbf{U}^{-1}\mathbf{F}^{\top}\mathbf{F}\mathbf{U}^{-1}\\ &= \mathbf{U}^{-1}\mathbf{U}^{2}\mathbf{U}^{-1}\\ &= \mathbf{1},\end{aligned} \tag{2.134}$$

so that $\mathbf{R}$ is orthogonal. Also, since $\mathbf{U}$ is positive-definite so that by (2.122), $\det\mathbf{U} > 0$, and since $\det\mathbf{F} > 0$, we may use (2.82) and (2.134) to show that $\det\mathbf{R} > 0$. Thus, $\mathbf{R}$ is a rotation.

We next show that $\mathbf{F} = \mathbf{V}\mathbf{R}$, with $\mathbf{V}$ symmetric and positive-definite. Since $\mathbf{U}$ is symmetric, positive-definite and $\mathbf{R}$ a rotation, (2.123) implies that $\mathbf{V} = \mathbf{R}\mathbf{U}\mathbf{R}^{\top}$ is symmetric and positive-definite; it follows that

$$\begin{aligned}\mathbf{F} &= \mathbf{R}\mathbf{U}\\ &= \mathbf{R}\mathbf{U}\mathbf{R}^{\top}\mathbf{R}\\ &= \mathbf{R}\mathbf{U}\mathbf{R}^{\top}\mathbf{R}\\ &= \mathbf{V}\mathbf{R}\end{aligned}$$

and $\mathbf{F} = \mathbf{V}\mathbf{R}$ is a left polar decomposition. Every invertible $\mathbf{F}$ with strictly positive determinant therefore has right and left polar decompositions of the form (2.131).

We next establish the uniqueness of these decompositions. To this end, assume that $\mathbf{F}$ also has the decompositions (2.131) with $\tilde{\mathbf{U}}$ and $\bar{\mathbf{V}}$ symmetric, positive-definite and $\tilde{\mathbf{R}}$ and $\bar{\mathbf{R}}$ rotations. Then, arguing as in the first paragraph of the proof, we find that $\tilde{\mathbf{U}} = \sqrt{\mathbf{F}^{\top}\mathbf{F}} = \mathbf{U}$ and $\bar{\mathbf{V}} = \sqrt{\mathbf{F}\mathbf{F}^{\top}} = \mathbf{V}$. Further, by (2.131), given $\mathbf{F}$, $\mathbf{U}$, and $\mathbf{V}$, the tensors $\tilde{\mathbf{R}}$ and $\bar{\mathbf{R}}$ satisfy

$$\tilde{\mathbf{R}} = \mathbf{F}\mathbf{U}^{-1} = \mathbf{R} \qquad \text{and} \qquad \bar{\mathbf{R}} = \mathbf{V}^{-1}\mathbf{F} = \mathbf{R}. \tag{2.135}$$

The decompositions (2.131) are therefore unique.

This completes the proof of the polar decomposition theorem.

EXERCISES

1. Show that a tensor $\mathbf{C}$ is positive-definite if and only if its symmetric part $\operatorname{sym}\mathbf{C}$ is positive-definite.
2. Show that a symmetric tensor with strictly positive eigenvalues is positive-definite.
3. Consider the polar decomposition (2.131) of an invertible tensor $\mathbf{F}$ with $\det\mathbf{F} > 0$.
   (a) Show that $\mathbf{U}$ admits a spectral decomposition of the form

$$\mathbf{U} = \sum_{i=1}^{3} \lambda_i \mathbf{r}_i \otimes \mathbf{r}_i\,, \tag{2.136}$$

with $\{\mathbf{r}_i\}$ an orthonormal basis of eigenvectors and corresponding eigenvalues $\lambda_i > 0$, and that

$$\mathbf{U}^{-1} = \sum_{i=1}^{3} \frac{1}{\lambda_i} \mathbf{r}_i \otimes \mathbf{r}_i. \tag{2.137}$$

(b) Show that the spectral decomposition of $\mathbf{V}$ has the form

$$\mathbf{V} = \sum_{i=1}^{3} \lambda_i \mathbf{l}_i \otimes \mathbf{l}_i, \tag{2.138}$$

with

$$\mathbf{l}_i = \mathbf{R}\mathbf{r}_i. \qquad i = 1, 2, 3. \tag{2.139}$$

The eigenvalues of $\mathbf{U}$ *and* $\mathbf{V}$ *are identical while the associated eigenvectors are therefore related via the rotation* $\mathbf{R}$.

(c) Show that

$$\left.\begin{aligned} \mathbf{F} &= \sum_{i=1}^{3} \lambda_i \mathbf{l}_i \otimes \mathbf{r}_i, \\ \mathbf{F}^{-1} &= \sum_{i=1}^{3} \lambda_i^{-1} \mathbf{r}_i \otimes \mathbf{l}_i, \\ \mathbf{F}^{-\top} &= \sum_{i=1}^{3} \lambda_i^{-1} \mathbf{l}_i \otimes \mathbf{r}_i. \end{aligned}\right\} \tag{2.140}$$

2.16 Principal Invariants of a Tensor. Cayley–Hamilton Equation

If ω and $\mathbf{e}$ are an eigenvalue and corresponding eigenvector for a tensor $\mathbf{S}$, then (2.110) implies that

$$(\mathbf{S} - \omega\mathbf{1})\mathbf{e} = \mathbf{0};$$

$\mathbf{S} - \omega\mathbf{1}$ is therefore not invertible, so that, by (2.81),

$$\det(\mathbf{S} - \omega\mathbf{1}) = 0.$$

Thus, in view of (2.108), the determinant of the 3×3 matrix $[\mathbf{S}] - \omega[\mathbf{1}]$ must vanish,

$$\det([\mathbf{S}] - \omega[\mathbf{1}]) = 0;$$

each eigenvalue ω of a tensor $\mathbf{S}$ must therefore be a real solution of a polynomial equation of the form $\omega^3 - a_1\omega^2 + a_2\omega - a_3 = 0$, with the coefficients being functions of $\mathbf{S}$. In fact, a tedious computation shows that this **characteristic equation** has the explicit form

$$\omega^3 - I_1(\mathbf{S})\omega^2 + I_2(\mathbf{S})\omega - I_3(\mathbf{S}) = 0, \tag{2.141}$$

where $I_1(\mathbf{S})$, $I_2(\mathbf{S})$, and $I_3(\mathbf{S})$, called the **principal invariants**[23] of $\mathbf{S}$, are given by

$$\boxed{\begin{aligned} I_1(\mathbf{S}) &= \operatorname{tr}\mathbf{S}, \\ I_2(\mathbf{S}) &= \tfrac{1}{2}[\operatorname{tr}^2(\mathbf{S}) - \operatorname{tr}(\mathbf{S}^2)], \\ I_3(\mathbf{S}) &= \det\mathbf{S}. \end{aligned}} \tag{2.142}$$

Moreover, for $\mathbf{S}$ symmetric, these relations and the spectral decomposition (2.112) imply that the invariants (2.142) are easily computed from the eigenvalues ω_i of $\mathbf{S}$ via the relations

$$\left.\begin{aligned} I_1(\mathbf{S}) &= \omega_1 + \omega_2 + \omega_3, \\ I_2(\mathbf{S}) &= \omega_1\omega_2 + \omega_2\omega_3 + \omega_3\omega_1, \\ I_3(\mathbf{S}) &= \omega_1\omega_2\omega_3. \end{aligned}\right\} \tag{2.143}$$

Let $\mathbf{S}$ be a symmetric tensor. Further, let $\mathbf{e}$ be an eigenvector of $\mathbf{S}$ with ω the corresponding eigenvalue, so that $\mathbf{Se} = \omega\mathbf{e}$. Then,

$$\begin{aligned} \mathbf{S}^2\mathbf{e} &= \mathbf{S}(\mathbf{Se}) \\ &= \mathbf{S}(\omega\mathbf{e}) \\ &= \omega\mathbf{Se} \\ &= \omega^2\mathbf{e}. \end{aligned}$$

Similarly, $\mathbf{S}^3\mathbf{e} = \omega^3\mathbf{e}$ and, therefore, by (2.141),

$$\begin{aligned} [\mathbf{S}^3 - I_1(\mathbf{S})\mathbf{S}^2 + I_2(\mathbf{S})\mathbf{S} - I_3(\mathbf{S})\mathbf{1}]\mathbf{e} &= [\omega^3 - I_1(\mathbf{S})\omega^2 + I_2(\mathbf{S})\omega - I_3(\mathbf{S})]\mathbf{e} \\ &= \mathbf{0}. \end{aligned}$$

Thus, *for every eigenvector* $\mathbf{e}$ *of* $\mathbf{S}$,

$$[\mathbf{S}^3 - I_1(\mathbf{S})\mathbf{S}^2 + I_2(\mathbf{S})\mathbf{S} - I_3(\mathbf{S})\mathbf{1}]\mathbf{e} = \mathbf{0}.$$

But, since $\mathbf{S}$ is symmetric, we may conclude from the Spectral Theorem (page 28) that there is an orthonormal basis of eigenvectors of $\mathbf{S}$. Hence,

$$\mathbf{S}^3 - I_1(\mathbf{S})\mathbf{S}^2 + I_2(\mathbf{S})\mathbf{S} - I_3(\mathbf{S})\mathbf{1} = \mathbf{0}, \tag{2.144}$$

and $\mathbf{S}$ *satisfies its characteristic equation.*[24] The relation (2.144), which is generally referred to as the **Cayley–Hamilton equation** for $\mathbf{S}$, is a tensorial analogue of the characterisic equation (2.141).

EXERCISES

1. Show that the principal invariants (2.142) of a tensor $\mathbf{S}$ satisfy

$$I_k(\mathbf{QSQ}^\top) = I_k(\mathbf{S}) \qquad \text{for all orthogonal tensors } \mathbf{Q}. \tag{2.145}$$

[23] $I_k(\mathbf{S})$ are called *invariants* because of the way they transform under the group of orthogonal tensors: $I_k(\mathbf{QSQ}^\top) = I_k(\mathbf{S})$ for each orthogonal tensor $\mathbf{Q}$.

[24] Actually, *every tensor* satisfies its characteristic equation. This result does not require that $\mathbf{S}$ be symmetric. Cf., e.g., Gantmacher (1959, p. 83) and Bowen & Wang (1976, Theorem 26.1).

2. Show that, for any tensor $\mathbf{S}$,

$$\begin{aligned} \operatorname{tr}(\mathbf{S}^2) &= I_1^2(\mathbf{S}) - 2I_2(\mathbf{S}), \\ \operatorname{tr}(\mathbf{S}^3) &= I_1^3(\mathbf{S}) - 3I_1(\mathbf{S})I_2(\mathbf{S}) + 3I_3(\mathbf{S}). \end{aligned} \tag{2.146}$$

3. Suppose that $\mathbf{U}$ and $\mathbf{C}$ are symmetric, positive-definite tensors with $\mathbf{U}^2 = \mathbf{C}$. Use the definitions (2.142) to show that the principal invariants of $\mathbf{C}$ are related to those of $\mathbf{U}$ via

$$\begin{aligned} I_1(\mathbf{C}) &= I_1^2(\mathbf{U}) - 2I_2(\mathbf{U}), \\ I_2(\mathbf{C}) &= I_2^2(\mathbf{U}) - 2I_1(\mathbf{U})I_3(\mathbf{U}), \\ I_3(\mathbf{C}) &= I_3^2(\mathbf{U}). \end{aligned}$$

4. Use the characteristic equation (2.141) to show that if the principal invariants of a symmetric tensor $\mathbf{S}$ obey $I_1(\mathbf{S})I_2(\mathbf{S}) = I_3(\mathbf{S})$, then $\mathbf{S}$ cannot be positive-definite.
5. Given an invertible tensor $\mathbf{F}$ with right and left polar decompositions $\mathbf{F} = \mathbf{RU}$ and $\mathbf{F} = \mathbf{VR}$, use the Cayley–Hamilton equation (2.144) to show that $\mathbf{R}$ can be obtained via

$$\begin{aligned} \mathbf{R} &= \frac{1}{I_3(\mathbf{U})}[\mathbf{FU}^2 - I_1(\mathbf{U})\mathbf{FU} + I_2(\mathbf{U})\mathbf{F}] \\ &= \frac{1}{I_3(\mathbf{V})}[\mathbf{V}^2\mathbf{F} - I_1(\mathbf{V})\mathbf{VF} + I_2(\mathbf{V})\mathbf{F}]. \end{aligned}$$

6. Suppose that $\mathbf{U}$, $\mathbf{V}$, $\mathbf{C}$, and $\mathbf{B}$ are symmetric, positive-definite tensors with $\mathbf{U}^2 = \mathbf{C}$ and $\mathbf{V}^2 = \mathbf{B}$. Use the Cayley–Hamilton equation (2.144) to establish the identities

$$\begin{aligned} \mathbf{U} &= \frac{1}{I_3(\mathbf{U}) - I_1(\mathbf{U})I_3(\mathbf{U})}\big[\mathbf{C}^2 - [I_1^2(\mathbf{U}) - I_2(\mathbf{U})]\mathbf{C} - I_1(\mathbf{U})I_3(\mathbf{U})\mathbf{1}\big], \\ \mathbf{V} &= \frac{1}{I_3(\mathbf{V}) - I_1(\mathbf{V})I_2(\mathbf{V})}\big[\mathbf{B}^2 - [I_1^2(\mathbf{V}) - I_2(\mathbf{V})]\mathbf{B} - I_1(\mathbf{V})I_3(\mathbf{V})\mathbf{1}\big]. \end{aligned}$$

7. Use the Cayley–Hamilton equation (2.144) to show that the third principal invariant of a tensor $\mathbf{S}$ can be expressed in the form

$$I_3(\mathbf{S}) = \tfrac{1}{6}[\operatorname{tr}^3(\mathbf{S}) - 3\operatorname{tr}(\mathbf{S})\operatorname{tr}(\mathbf{S}^2) + 2\operatorname{tr}(\mathbf{S}^3)] \tag{2.147}$$

and, thus, in terms of the traces of $\mathbf{S}$, $\mathbf{S}^2$, and $\mathbf{S}^3$.
8. Establish the characteristic equation (2.141) directly using the definition (2.79) of the determinant.
9. Let $\mathbf{S}$ be invertible. Use $(2.142)_3$ and (2.144) to show that

$$(\det\mathbf{S})\mathbf{S}^{-\top} = [\mathbf{S}^2 - (\operatorname{tr}\mathbf{S})\mathbf{S} + \tfrac{1}{2}(\operatorname{tr}^2(\mathbf{S}) - \operatorname{tr}(\mathbf{S}^2))\mathbf{1}]^{\top}$$

and thereby verify that the definition (2.90) of the cofactor $\mathbf{S}^{\mathrm{c}}$ and the representation (2.94) are equivalent for $\mathbf{S}$ invertible.

PART II

VECTOR AND TENSOR ANALYSIS

We do not fuss over smoothness assumptions:

(‡) *Functions* and the *boundaries of regions* are presumed to have continuity and differentiability properties sufficient to make meaningful the underlying analysis.[25]

We work within a Euclidean space $\mathcal{E}$ so that the phrase "region of space" connotes a region contained in $\mathcal{E}$.

[25] In particular, the regions we consider are presumed to have well-defined unit normal fields over their bounding surfaces.

3 Differentiation

3.1 Differentiation of Functions of a Scalar

Given a scalar, vector, point, or tensor function $\varphi(t)$ of a scalar variable t, we write

$$\dot{\varphi}(t) \overset{\text{def}}{=} \frac{d\varphi(t)}{dt} = \lim_{h\to 0} \frac{\varphi(t+h) - \varphi(t)}{h}.$$

Then, for $\mathbf{x}(t)$ a point function,

$$\dot{\mathbf{x}}(t) = \lim_{h\to 0} \frac{\mathbf{x}(t+h) - \mathbf{x}(t)}{h},$$

and since the difference $\mathbf{x}(t+h) - \mathbf{x}(t)$ is a vector, the derivative $\dot{\mathbf{x}}(t)$ is a *vector* function.

Throughout, components of vectors and tensors are with respect to a *fixed* orthonormal basis $\{\mathbf{e}_i\}$, so that, for $\mathbf{v}(t)$ a vector function, and $\mathbf{T}(t)$ a tensor function,

$$\dot{\mathbf{v}}(t) = \dot{v}_i(t)\mathbf{e}_i,$$
$$\dot{\mathbf{T}}(t) = \dot{T}_{ij}(t)\mathbf{e}_i \otimes \mathbf{e}_j.$$

Recalling our initial agreement (page 1) that lightface letters indicate scalars, boldface lowercase letters indicate vectors, and boldface uppercase letters indicate tensors, the standard product rule for scalar functions has the following analogs:

$$\left.\begin{aligned}
\dot{\overline{\mathbf{u}\cdot\mathbf{v}}} &= \dot{\mathbf{u}}\cdot\mathbf{v} + \mathbf{u}\cdot\dot{\mathbf{v}},\\
\dot{\overline{\varphi\mathbf{v}}} &= \dot{\varphi}\mathbf{v} + \varphi\dot{\mathbf{v}},\\
\dot{\overline{\mathbf{T}\mathbf{v}}} &= \dot{\mathbf{T}}\mathbf{v} + \mathbf{T}\dot{\mathbf{v}},\\
\dot{\overline{\mathbf{u}\times\mathbf{v}}} &= \dot{\mathbf{u}}\times\mathbf{v} + \mathbf{u}\times\dot{\mathbf{v}},\\
\dot{\overline{\mathbf{u}\otimes\mathbf{v}}} &= \dot{\mathbf{u}}\otimes\mathbf{v} + \mathbf{u}\otimes\dot{\mathbf{v}},\\
\dot{\overline{\varphi\mathbf{T}}} &= \dot{\varphi}\mathbf{T} + \varphi\dot{\mathbf{T}},\\
\dot{\overline{\mathbf{T}\mathbf{S}}} &= \dot{\mathbf{T}}\mathbf{S} + \mathbf{T}\dot{\mathbf{S}}.
\end{aligned}\right\} \tag{3.1}$$

Here, a dot over a line indicates the derivative of the quantity under the line. For example,

$$\overline{\varphi(t)\mathbf{v}(t)}^{\boldsymbol{\cdot}} = \frac{d}{dt}[\varphi(t)\mathbf{v}(t)].$$

The identities (3.1) may be established using the standard product rule for scalar functions. Consider, for example, $(3.1)_5$:

$$\begin{aligned}(\overline{\mathbf{u}\otimes\mathbf{v}}^{\boldsymbol{\cdot}})_{ij} &= \overline{u_i v_j}^{\boldsymbol{\cdot}} \\ &= \dot{u}_i v_j + u_i \dot{v}_j.\end{aligned}$$

Suppose that $\mathbf{A}(t)$ is a nonvanishing tensor function. Define

$$\varphi(t) = |\mathbf{A}(t)|.$$

Then, since $\varphi^2 = \mathbf{A}:\mathbf{A}$, it follows, using the chain-rule, that

$$2\varphi\dot{\varphi} = 2\mathbf{A}:\dot{\mathbf{A}};$$

hence,

$$\dot{\varphi} = \frac{\mathbf{A}}{|\mathbf{A}|}:\dot{\mathbf{A}}.$$

We therefore have the important identity

$$\overline{|\mathbf{A}|}^{\boldsymbol{\cdot}} = \frac{\mathbf{A}}{|\mathbf{A}|}:\dot{\mathbf{A}} \tag{3.2}$$

valid for any nonvanishing tensor function $\mathbf{A}(t)$.

Two useful identities for an invertible tensor function $\mathbf{F}(t)$ with inverse $\mathbf{F}^{-1}(t)$ at each t are[26]

$$\begin{aligned}\overline{\det\mathbf{F}}^{\boldsymbol{\cdot}} &= (\det\mathbf{F})\,\mathrm{tr}\,(\dot{\mathbf{F}}\mathbf{F}^{-1}), \\ \overline{\mathbf{F}^{-1}}^{\boldsymbol{\cdot}} &= -\mathbf{F}^{-1}\dot{\mathbf{F}}\mathbf{F}^{-1}.\end{aligned} \tag{3.3}$$

To verify $(3.3)_2$ we use the product rule $(3.1)_7$ to differentiate the identity[27]

$$\mathbf{F}^{-1}(t)\mathbf{F}(t) = \mathbf{1}. \tag{3.4}$$

Since the right side of (3.4) is constant,

$$\begin{aligned}\mathbf{0} &= \overline{\mathbf{F}^{-1}\mathbf{F}}^{\boldsymbol{\cdot}} \\ &= \overline{\mathbf{F}^{-1}}^{\boldsymbol{\cdot}}\mathbf{F} + \mathbf{F}^{-1}\dot{\mathbf{F}},\end{aligned}$$

solving this relation for $\overline{\mathbf{F}^{-1}}^{\boldsymbol{\cdot}}$ yields $(3.3)_2$.

An important identity that is used repeatedly is that, for $\mathbf{Q}(t)$ an *orthogonal-tensor function*,

$$\mathbf{Q}^{\top}(t)\dot{\mathbf{Q}}(t) \text{ and } \dot{\mathbf{Q}}(t)\mathbf{Q}^{\top}(t) \text{ are skew tensors at each } t. \tag{3.5}$$

The proof follows upon differentiating the identity[28]

$$\mathbf{Q}^{\top}\mathbf{Q} = \mathbf{Q}\mathbf{Q}^{\top} = \mathbf{1}.$$

---

[26] For a proof of $(3.3)_1$, cf., e.g., Gurtin (1981, p. 27).
[27] Cf. (2.67).
[28] Cf. (2.101).

If we differentiate $\mathbf{Q}^\top\mathbf{Q} = \mathbf{1}$, we find that

$$\mathbf{Q}^\top\dot{\mathbf{Q}} + \dot{\mathbf{Q}}^\top\mathbf{Q} = \mathbf{0},$$

and, hence, that

$$\begin{aligned}\mathbf{Q}^\top\dot{\mathbf{Q}} &= -\dot{\mathbf{Q}}^\top\mathbf{Q}\\ &= -(\mathbf{Q}^\top\dot{\mathbf{Q}})^\top.\end{aligned}$$

Thus, by (2.19) with $\mathbf{S} = \mathbf{Q}^\top\dot{\mathbf{Q}}$, $\mathbf{Q}^\top\dot{\mathbf{Q}}$ is skew. To verify that

$$\dot{\mathbf{Q}}\mathbf{Q}^\top \text{ is skew,} \tag{3.6}$$

simply differentiate $\mathbf{Q}\mathbf{Q}^\top = \mathbf{1}$.

EXERCISES

1. Show that, for $\mathbf{T}(t)$ a tensor function,

$$\dot{\overline{\mathbf{T}^\top}} = \dot{\mathbf{T}}^\top. \tag{3.7}$$

2. Verify $(3.1)_{1\text{–}4}$ and $(3.1)_{6,7}$.
3. Let $\{\mathbf{e}_i\}$ denote a positively oriented orthonormal basis. Obtain $(3.3)_1$ directly by differentiating the identity[29]

$$\det\mathbf{F} = \mathbf{F}\mathbf{e}_1 \cdot (\mathbf{F}\mathbf{e}_2 \times \mathbf{F}\mathbf{e}_3).$$

4. Provide the steps in the verification of (3.6).

3.2 Differentiation of Fields. Gradient

Let $\boldsymbol{\Phi}(\mathbf{h})$ be a scalar, vector, or tensor function of a vector $\mathbf{h}$. We say that $\boldsymbol{\Phi}(\mathbf{h})$ *approaches zero faster than* $\mathbf{h}$ or that $\boldsymbol{\Phi}(\mathbf{h})$ is of *order* $o(|\mathbf{h}|)$ as $\mathbf{h}$ *approaches* $\mathbf{0}$, and we write

$$\boldsymbol{\Phi}(\mathbf{h}) = o(|\mathbf{h}|) \quad \text{as } \mathbf{h} \to \mathbf{0},$$

or, more simply,

$$\boldsymbol{\Phi}(\mathbf{h}) = o(|\mathbf{h}|),$$

if

$$\lim_{\mathbf{h}\to\mathbf{0}} \frac{\boldsymbol{\Phi}(\mathbf{h})}{|\mathbf{h}|} = 0. \tag{3.8}$$

Then, for example,

$$|\mathbf{h}|^2 = o(|\mathbf{h}|) \quad \text{as } \mathbf{h} \to \mathbf{0}, \qquad \text{but} \qquad \sqrt{|\mathbf{h}|} \neq o(|\mathbf{h}|) \quad \text{as } \mathbf{h} \to \mathbf{0}.$$

Given a region R, a **scalar field** φ with **domain** R is a mapping that assigns to each point $\mathbf{x}$ in R a *scalar* $\varphi(\mathbf{x})$ called the **value** of φ at $\mathbf{x}$. **Vector**, **point**, and **tensor fields** are defined analogously, that is, for example, a vector field $\mathbf{v}$ has the vector value $\mathbf{v}(\mathbf{x})$ at $\mathbf{x}$. A field with domain R is sometimes referred to as a **field on** R.

We find it most convenient to define the gradient of a field in terms of the Taylor expansion of the field, assuming such an expansion exists.[30] We say that φ and $\mathbf{v}$ are

[29] Cf. (2.79).
[30] We therefore define the gradient as a Fréchet derivative; cf. Gurtin (1981). While this notion might appear overly abstract to a practical reader, it is, in fact, useful in applications, for example, in scientific computing; cf. Reddy (1997).

differentiable at a point $\mathbf{x}$ in R if there are a vector $\mathbf{g}$ and a tensor $\mathbf{G}$ such that:[31]

$$\begin{aligned} \varphi(\mathbf{x}+\mathbf{h}) - \varphi(\mathbf{x}) &= \mathbf{g}\cdot\mathbf{h} + \mathrm{o}(|\mathbf{h}|) \qquad \text{as } |\mathbf{h}| \to 0, \\ \mathbf{v}(\mathbf{x}+\mathbf{h}) - \mathbf{v}(\mathbf{x}) &= \mathbf{G}\mathbf{h} + \mathrm{o}(|\mathbf{h}|) \qquad \text{as } |\mathbf{h}| \to 0. \end{aligned} \tag{3.9}$$

By (3.8), the *linear* terms in (3.9),

$$\mathbf{g}\cdot\mathbf{h} \qquad \text{and} \qquad \mathbf{G}\mathbf{h}, \tag{3.10}$$

represent approximations of

$$\varphi(\mathbf{x}+\mathbf{h}) - \varphi(\mathbf{x}) \qquad \text{and} \qquad \mathbf{v}(\mathbf{x}+\mathbf{h}) - \mathbf{v}(\mathbf{x}),$$

approximations whose error goes to zero faster than $\mathbf{h}$ and hence faster than the linear terms (3.10). If φ and $\mathbf{v}$ are differentiable at $\mathbf{x}$ we write[32]

$$\operatorname{grad}\varphi(\mathbf{x}) \stackrel{\text{def}}{=} \mathbf{g} \qquad \text{and} \qquad \operatorname{grad}\mathbf{v}(\mathbf{x}) \stackrel{\text{def}}{=} \mathbf{G}, \tag{3.11}$$

and we refer to $\operatorname{grad}\varphi(\mathbf{x})$ and $\operatorname{grad}\mathbf{v}(\mathbf{x})$ as the **gradients** of φ and $\mathbf{v}$ at $\mathbf{x}$;[33] with this agreement, the expansions (3.9) become

$$\begin{aligned} \varphi(\mathbf{x}+\mathbf{h}) - \varphi(\mathbf{x}) &= \operatorname{grad}\varphi(\mathbf{x})\cdot\mathbf{h} + \mathrm{o}(|\mathbf{h}|) \qquad \text{as } |\mathbf{h}| \to 0, \\ \mathbf{v}(\mathbf{x}+\mathbf{h}) - \mathbf{v}(\mathbf{x}) &= [\operatorname{grad}\mathbf{v}(\mathbf{x})]\mathbf{h} + \mathrm{o}(|\mathbf{h}|) \qquad \text{as } |\mathbf{h}| \to 0 \end{aligned} \tag{3.12}$$

and are referred to as **Taylor expansions** of φ and $\mathbf{v}$ at $\mathbf{x}$.

We now show that this definition of the gradient implies the conventional definition in terms of the partial derivatives

$$\begin{aligned} \frac{\partial\varphi(\mathbf{x})}{\partial x_i} &= \lim_{h\to 0} \frac{\varphi(\mathbf{x}+h\mathbf{e}_i) - \varphi(\mathbf{x})}{h}, \\ \frac{\partial\mathbf{v}(\mathbf{x})}{\partial x_i} &= \lim_{h\to 0} \frac{\mathbf{v}(\mathbf{x}+h\mathbf{e}_i) - \mathbf{v}(\mathbf{x})}{h}. \end{aligned} \tag{3.13}$$

To accomplish this, we note first that, by (3.8), (3.9) may be written equivalently as

$$\begin{aligned} &\lim_{|\mathbf{h}|\to 0} \frac{\varphi(\mathbf{x}+\mathbf{h}) - \varphi(\mathbf{x}) - \operatorname{grad}\varphi(\mathbf{x})\cdot\mathbf{h}}{|\mathbf{h}|} = 0, \\ &\lim_{|\mathbf{h}|\to 0} \frac{\mathbf{v}(\mathbf{x}+\mathbf{h}) - \mathbf{v}(\mathbf{x}) - [\operatorname{grad}\mathbf{v}(\mathbf{x})]\mathbf{h}}{|\mathbf{h}|} = \mathbf{0}. \end{aligned} \tag{3.14}$$

Thus, taking $\mathbf{h} = h\mathbf{e}_i$, so that $|\mathbf{h}| = |h|$, and multiplying both of (3.14) by the sign of h, we arrive at

$$\begin{aligned} &\lim_{h\to 0} \frac{\varphi(\mathbf{x}+h\mathbf{e}_i) - \varphi(\mathbf{x}) - h\operatorname{grad}\varphi(\mathbf{x})\cdot\mathbf{e}_i}{h} = 0, \\ &\lim_{h\to 0} \frac{\mathbf{v}(\mathbf{x}+h\mathbf{e}_i) - \mathbf{v}(\mathbf{x}) - h[\operatorname{grad}\mathbf{v}(\mathbf{x})]\mathbf{e}_i}{h} = \mathbf{0}, \end{aligned}$$

---

[31] Recall the convention, set forth at the outset of §1.1, that the sum of a point and a vector is a point.

[32] In our discussion of kinematics, a motion takes the form of a time-dependent mapping from material points $\mathbf{X}$ to spatial points $\mathbf{x}$. There, it will be advantageous to have a notation that allows us to distinguish between the gradients with respect to $\mathbf{X}$ and $\mathbf{x}$. Specifically, we will write ∇ for the gradient with respect to material points $\mathbf{X}$ and grad for the gradient with respect to spatial points $\mathbf{x}$.

[33] We do not find it necessary to define the gradient of a tensor field.

and hence, by (3.13), that

$$\mathbf{e}_i \cdot \operatorname{grad}\varphi(\mathbf{x}) = \frac{\partial\varphi(\mathbf{x})}{\partial x_i},$$

$$[\operatorname{grad}\mathbf{v}(\mathbf{x})]\mathbf{e}_j = \frac{\partial\mathbf{v}(\mathbf{x})}{\partial x_j}.$$

The component forms of $\operatorname{grad}\varphi(\mathbf{x})$ and $\operatorname{grad}\mathbf{v}(\mathbf{x})$, namely,

$$\begin{aligned} [\operatorname{grad}\varphi(\mathbf{x})]_i &= \frac{\partial\varphi(\mathbf{x})}{\partial x_i}, \\ [\operatorname{grad}\mathbf{v}(\mathbf{x})]_{ij} &= \frac{\partial v_i(\mathbf{x})}{\partial x_j}, \end{aligned} \tag{3.15}$$

therefore represent conventional definitions of the gradient in terms of partial derivatives.

If φ and $\mathbf{v}$ are differentiable at each $\mathbf{x}$ in R, then $\operatorname{grad}\varphi$ and $\operatorname{grad}\mathbf{v}$ represent *fields*, and it is clear from the foregoing discussion that

- *the gradient of a scalar field is a vector field; the gradient of a vector field is a tensor field.*

To differentiate *composite functions* such as

$$\varphi(\mathbf{x}(t)) \qquad \text{and} \qquad \mathbf{v}(\mathbf{x}(t)),$$

— where $\mathbf{x}(t)$ is a point function with values in R, the domain of φ and $\mathbf{v}$ — we use the **chain-rule**:

$$\overline{\varphi(\mathbf{x})}^{\boldsymbol{\cdot}} = \operatorname{grad}\varphi(\mathbf{x}) \cdot \dot{\mathbf{x}}, \qquad \overline{\varphi(\mathbf{x})}^{\boldsymbol{\cdot}} = \frac{\partial\varphi(\mathbf{x})}{\partial x_i}\dot{x}_i, \tag{3.16}$$

and

$$\overline{\mathbf{v}(\mathbf{x})}^{\boldsymbol{\cdot}} = [\operatorname{grad}\mathbf{v}(\mathbf{x})]\dot{\mathbf{x}}, \qquad \overline{v_i(\mathbf{x})}^{\boldsymbol{\cdot}} = \frac{\partial v_i(\mathbf{x})}{\partial x_j}\dot{x}_j. \tag{3.17}$$

EXERCISES

1. Show that the function $\varphi(\mathbf{h}) = \mathbf{h}$ is not of order $o(|\mathbf{h}|)$ as $\mathbf{h}$ approaches $\mathbf{0}$.
2. Show that

$$\operatorname{grad}\mathbf{v} = \frac{\partial v_i}{\partial x_j}\mathbf{e}_i \otimes \mathbf{e}_j.$$

3. Given a vector $\mathbf{a}$ and a tensor $\mathbf{A}$ (each constant), let

$$\varphi(\mathbf{x}) = (\mathbf{x} - \mathbf{o}) \cdot \mathbf{a}, \qquad \mathbf{v}(\mathbf{x}) = \mathbf{A}(\mathbf{x} - \mathbf{o}),$$

and

$$\mathbf{x}(t) = \sin t\,\mathbf{e}_1 + \cos t\,\mathbf{e}_2.$$

Compute the derivatives, with respect to t, of $\varphi(\mathbf{x}(t))$ and $\mathbf{v}(\mathbf{x}(t))$.

4. Working in components, one can derive other forms of the chain-rule. For example, derive the chain-rule for a vector function of a tensor that depends on a scalar variable.

3.3 Divergence and Curl. Vector and Tensor Identities

The **divergence** and **curl** of a vector field **v** and a tensor field **T** may be defined as follows:[34]

$$
\begin{aligned}
\operatorname{div}\mathbf{v} &= \frac{\partial v_i}{\partial x_i},\\
(\operatorname{curl}\mathbf{v})_i &= \epsilon_{ijk}\frac{\partial v_k}{\partial x_j},\\
(\operatorname{div}\mathbf{T})_i &= \frac{\partial T_{ij}}{\partial x_j},\\
(\operatorname{curl}\mathbf{T})_{ij} &= \epsilon_{ipq}\frac{\partial T_{jq}}{\partial x_p};
\end{aligned}
\tag{3.18}
$$

thus, $\operatorname{div}\mathbf{v}$ is a scalar field, $\operatorname{curl}\mathbf{v}$ and $\operatorname{div}\mathbf{T}$ are vector fields, and $\operatorname{curl}\mathbf{T}$ is a tensor field. These fields may also be defined without recourse to components

$$
\begin{aligned}
\operatorname{div}\mathbf{v} &= \operatorname{tr}(\operatorname{grad}\mathbf{v}),\\
\mathbf{a}\cdot\operatorname{curl}\mathbf{v} &= (\mathbf{a}\times):\operatorname{grad}\mathbf{v} \quad \text{for every constant vector } \mathbf{a},\\
\mathbf{a}\cdot\operatorname{div}\mathbf{T} &= \operatorname{div}(\mathbf{T}^{\top}\mathbf{a}) \quad \text{for every constant vector } \mathbf{a},\\
(\operatorname{curl}\mathbf{T})\mathbf{a} &= \operatorname{curl}(\mathbf{T}^{\top}\mathbf{a}) \quad \text{for every constant vector } \mathbf{a}.
\end{aligned}
\tag{3.19}
$$

We leave it as an exercise to show that the definitions (3.18) and (3.19) are equivalent.

The next set of identities gives counterparts, for fields, of the standard product rule for scalar functions of a scalar variable.

PRODUCT IDENTITIES

$$
\begin{aligned}
\operatorname{grad}(\varphi\mathbf{v}) &= \varphi\operatorname{grad}\mathbf{v} + \mathbf{v}\otimes\operatorname{grad}\varphi,\\
\operatorname{grad}(\mathbf{u}\cdot\mathbf{v}) &= (\operatorname{grad}\mathbf{u})^{\top}\mathbf{v} + (\operatorname{grad}\mathbf{v})^{\top}\mathbf{u},\\
\operatorname{grad}(\mathbf{u}\times\mathbf{v}) &= (\mathbf{u}\times)\operatorname{grad}\mathbf{v} - (\mathbf{v}\times)\operatorname{grad}\mathbf{u},\\
\operatorname{div}(\varphi\mathbf{v}) &= \varphi\operatorname{div}\mathbf{v} + \mathbf{v}\cdot\operatorname{grad}\varphi,\\
\operatorname{div}(\mathbf{u}\times\mathbf{v}) &= \mathbf{v}\cdot\operatorname{curl}\mathbf{u} - \mathbf{u}\cdot\operatorname{curl}\mathbf{v},\\
\operatorname{div}(\mathbf{u}\otimes\mathbf{v}) &= (\operatorname{div}\mathbf{v})\mathbf{u} + (\operatorname{grad}\mathbf{u})\mathbf{v},\\
\operatorname{div}(\mathbf{T}^{\top}\mathbf{v}) &= \mathbf{T}:\operatorname{grad}\mathbf{v} + \mathbf{v}\cdot\operatorname{div}\mathbf{T},\\
\operatorname{div}(\varphi\mathbf{T}) &= \varphi\operatorname{div}\mathbf{T} + \mathbf{T}\operatorname{grad}\varphi,\\
\operatorname{curl}(\varphi\mathbf{v}) &= \varphi\operatorname{curl}\mathbf{v} + (\operatorname{grad}\varphi)\times\mathbf{v},\\
\operatorname{curl}(\mathbf{u}\times\mathbf{v}) &= \operatorname{div}(\mathbf{u}\otimes\mathbf{v} - \mathbf{v}\otimes\mathbf{u}),\\
\operatorname{curl}(\mathbf{u}\otimes\mathbf{v}) &= [(\operatorname{grad}\mathbf{u})\mathbf{v}\times]^{\top} + (\operatorname{curl}\mathbf{v})\otimes\mathbf{u},\\
\operatorname{curl}(\varphi\mathbf{T}) &= \varphi\operatorname{curl}\mathbf{T} + [(\operatorname{grad}\varphi)\times]\mathbf{T}.
\end{aligned}
\tag{3.20}
$$

---

[34] Referring to Footnote 32, we will later use Div and Curl to denote the divergence and curl with respect to material points **X** and div and curl for the divergence and curl with respect to spatial points **x**.

The identities $(3.20)_{1,7}$ may be derived as follows:

$$[\mathrm{grad}\,(\varphi\mathbf{v})]_{ij} = \frac{\partial(\varphi v_i)}{\partial x_j}$$

$$= \varphi\frac{\partial v_i}{\partial x_j} + v_i\frac{\partial \varphi}{\partial x_j},$$

$$\mathrm{div}(\mathbf{T}^{\mathsf{T}}\mathbf{v}) = \frac{\partial(T_{ji}v_j)}{\partial x_i}$$

$$= T_{ji}\frac{\partial v_j}{\partial x_i} + \frac{\partial T_{ji}}{\partial x_i}v_j.$$

The verification of the remaining identities is left as an exercise.

Next, by $(3.18)_2$

$$\mathrm{div}\,\mathrm{curl}\,\mathbf{v} = \epsilon_{ijk}\underbrace{\frac{\partial^2 v_k}{\partial x_j \partial x_i}}_{S_{kji}},$$

and, arguing as in the steps leading to (1.13), we find that, since $S_{kji} = S_{kij}$,

$$\epsilon_{ijk}S_{kji} = \epsilon_{ijk}S_{kij}$$

$$= -\epsilon_{jik}S_{kij}$$

$$= -\epsilon_{ijk}S_{kji};$$

thus,

$$\mathrm{div}\,\mathrm{curl}\,\mathbf{v} = 0, \tag{3.21}$$

which is a standard identity for the curl.

We define the **Laplace operator** Δ for scalar fields φ and vector fields $\mathbf{v}$ as follows:

$$\begin{aligned} \Delta\varphi &= \mathrm{div}\,\mathrm{grad}\,\varphi, & \Delta\varphi &= \frac{\partial^2\varphi}{\partial x_i \partial x_i}, \\ \Delta\mathbf{v} &= \mathrm{div}\,\mathrm{grad}\,\mathbf{v}, & \Delta v_i &= \frac{\partial^2 v_i}{\partial x_j \partial x_j}; \end{aligned} \tag{3.22}$$

thus, $\Delta\varphi$ is a scalar field and $\Delta\mathbf{v}$ is a vector field. Further, for a tensor field $\mathbf{T}$, $\Delta\mathbf{T}$ is the tensor field defined such that

$$(\Delta\mathbf{T})\mathbf{a} = \Delta(\mathbf{T}\mathbf{a}) \quad \text{for every constant vector } \mathbf{a} \tag{3.23}$$

or, equivalently, using components,

$$\Delta T_{ij} = \frac{\partial^2 T_{ij}}{\partial x_k \partial x_k}. \tag{3.24}$$

The operator Δ is often called the **Laplacian**.

EXERCISES

1. Show that (3.18) and (3.19) are equivalent.
2. Verify $(3.20)_{2\text{–}6,8\text{–}12}$.

3. Let $\mathbf{r} = \mathbf{x} - \mathbf{x}_0$, with $\mathbf{x}_0$ a fixed point, and let $r = |\mathbf{r}|$. Establish the following identities:

$$\operatorname{div}\mathbf{r} = 3,$$

$$\operatorname{div}\frac{\mathbf{r}}{r} = \frac{2}{r},$$

$$\operatorname{div}(\mathbf{r}\otimes\mathbf{r}) = 4\mathbf{r},$$

$$\operatorname{div}\left(\frac{\mathbf{r}\otimes\mathbf{r}}{r^2}\right) = \frac{2\mathbf{r}}{r^2},$$

$$\operatorname{div}(\mathbf{r}\times) = \mathbf{0},$$

$$\operatorname{curl}\mathbf{r} = \mathbf{0},$$

$$\operatorname{curl}\frac{\mathbf{r}}{r} = \mathbf{0},$$

$$\operatorname{curl}(\mathbf{r}\otimes\mathbf{r}) = -\mathbf{r}\times,$$

$$\operatorname{curl}\left(\frac{\mathbf{r}\otimes\mathbf{r}}{r^2}\right) = -\frac{\mathbf{r}\times}{r^2},$$

$$\operatorname{curl}(\mathbf{r}\times) = 2\mathbf{1},$$

$$\operatorname{grad} r = \frac{\mathbf{r}}{r},$$

$$\operatorname{grad}\frac{1}{r} = -\frac{\mathbf{r}}{r^2},$$

$$\operatorname{grad}\mathbf{r} = \mathbf{1},$$

$$\operatorname{grad}\frac{\mathbf{r}}{r} = -\frac{(\mathbf{r}\times)^2}{r^3}.$$

4. Defining a vector field $\mathbf{v}$ by

$$\mathbf{v}(\mathbf{x}) = kr^p\mathbf{r}, \qquad \mathbf{r} \neq \mathbf{0},$$

with $\mathbf{r}$ and r as defined in Exercise 3 and p constant, show that

$$\operatorname{div}\mathbf{v}(\mathbf{x}) = k(3+p)r^p,$$

$$\operatorname{curl}\mathbf{v}(\mathbf{x}) = \mathbf{0},$$

$$\Delta\mathbf{v}(\mathbf{x}) = kp(3+p)r^{p-2}\mathbf{r}.$$

Provide a physical interpretation of these results for the special case $p = -3$.

5. Defining scalar and vector fields φ and $\mathbf{v}$ by

$$\varphi(\mathbf{x}) = \frac{\mathbf{c}\cdot\mathbf{r}}{r^3} \quad \text{and} \quad \mathbf{v}(\mathbf{x}) = \frac{\mathbf{c}\times\mathbf{r}}{r^3}, \qquad \mathbf{r} \neq \mathbf{0},$$

with $\mathbf{r}$ and r as defined in Exercise 3 and $\mathbf{c}$ a constant vector, show that

$$\operatorname{grad}\varphi + \operatorname{curl}\mathbf{v} = \mathbf{0},$$

$$\Delta\varphi = 0,$$

$$\operatorname{div}\mathbf{v} = 0.$$

6. Establish the identities

$$
\begin{aligned}
\operatorname{div}(\operatorname{grad}\varphi \times \operatorname{grad}\vartheta) &= 0,\\
\operatorname{div}(\mathbf{v}\times) &= -\operatorname{curl}\mathbf{v},\\
\operatorname{div}(\varphi\mathbf{1}) &= \operatorname{grad}\varphi,\\
\operatorname{curl}(\operatorname{grad}\varphi) &= \mathbf{0},\\
\operatorname{curl}\operatorname{curl}\mathbf{v} &= \operatorname{grad}(\operatorname{div}\mathbf{v}) - \Delta\mathbf{v},\\
\operatorname{curl}\operatorname{grad}\mathbf{v} &= \mathbf{0},\\
\operatorname{curl}[(\operatorname{grad}\mathbf{v})^{\top}] &= \operatorname{grad}\operatorname{curl}\mathbf{v},\\
\operatorname{div}\operatorname{curl}\mathbf{T} &= \operatorname{curl}\operatorname{div}(\mathbf{T}^{\top}),\\
\operatorname{div}[(\operatorname{curl}\mathbf{T})^{\top}] &= \mathbf{0},\\
\operatorname{curl}\operatorname{curl}\mathbf{T} &= [\operatorname{curl}\operatorname{curl}(\mathbf{T}^{\top})]^{\top},\\
\operatorname{curl}(\varphi\mathbf{1}) &= (\operatorname{grad}\varphi)\times,\\
\operatorname{curl}(\mathbf{v}\times) &= (\operatorname{div}\mathbf{v})\mathbf{1} - \operatorname{grad}\mathbf{v}.
\end{aligned}
\tag{3.25}
$$

7. Verify that

$$
\begin{aligned}
\operatorname{curl}\operatorname{curl}\mathbf{T} = {} & -\Delta\mathbf{T} + \operatorname{grad}\operatorname{div}\mathbf{T} + (\operatorname{grad}\operatorname{div}\mathbf{T})^{\top} - \operatorname{grad}\operatorname{grad}(\operatorname{tr}\mathbf{T})\\
& + (\Delta(\operatorname{tr}\mathbf{T}) - \operatorname{div}\operatorname{div}\mathbf{T})\mathbf{1}.
\end{aligned}
$$

8. Provided that $\mathbf{T}$ is symmetric, show that

$$\operatorname{tr}(\operatorname{curl}\mathbf{T}) = 0$$

and that

$$\operatorname{curl}\operatorname{curl}\mathbf{T} = -\Delta\mathbf{T} + 2\operatorname{grad}\operatorname{div}\mathbf{T} - \operatorname{grad}\operatorname{grad}(\operatorname{tr}\mathbf{T}) + (\Delta(\operatorname{tr}\mathbf{T}) - \operatorname{div}\operatorname{div}\mathbf{T})\mathbf{1}.$$

3.4 Differentiation of a Scalar Function of a Tensor

For a scalar function $\Psi(\mathbf{T})$ of a tensor variable $\mathbf{T}$, the derivative $\partial\Psi(\mathbf{T})/\partial\mathbf{T}$ is the tensor function defined by

$$\left[\frac{\partial\Psi(\mathbf{T})}{\partial\mathbf{T}}\right]_{ij} = \frac{\partial\Psi(\mathbf{T})}{\partial T_{ij}}.$$

In computing this derivative, care must be taken to respect the tensor space within which the domain of Ψ lies. For example, if this space is the space of symmetric tensors, then, since $\Psi(\mathbf{T})$ is defined only for symmetric $\mathbf{T}$, $\partial\Psi(\mathbf{T})/\partial\mathbf{T}$ is a symmetric tensor, and so forth.

The chain-rule

$$
\begin{aligned}
\overline{\Psi(\mathbf{T})}^{\boldsymbol{\cdot}} &= \frac{\partial\Psi(\mathbf{T})}{\partial T_{ij}}\dot{T}_{ij}\\
&= \frac{\partial\Psi(\mathbf{T})}{\partial\mathbf{T}} : \dot{\mathbf{T}}
\end{aligned}
\tag{3.26}
$$

affords a simple and efficient method of computing the derivative $\partial\Psi(\mathbf{T})/\partial\mathbf{T}$. To demonstrate this use of the chain-rule, let Ψ be defined on the space of *symmetric tensors*, and consider the function $\Lambda(\mathbf{F})$ defined on the space of *all tensors* through the requirement that

$$\Lambda(\mathbf{F}) = \Psi(\mathbf{C}), \qquad \mathbf{C} = \mathbf{F}^{\top}\mathbf{F};$$

that is,

$$\Lambda(\mathbf{F}) = \Psi(\mathbf{F}^{\top}\mathbf{F}) \quad \text{for all tensors } \mathbf{F}.$$

Choose an arbitrary function $\mathbf{F}(t)$ with values in the domain of Λ, so that $\mathbf{C}(t)$ is also a function of t. Then

$$\overline{\Lambda(\mathbf{F})}^{\boldsymbol{\cdot}} = \overline{\Psi(\mathbf{C})}^{\boldsymbol{\cdot}},$$

and hence,

$$\begin{aligned}\frac{\partial \Lambda(\mathbf{F})}{\partial \mathbf{F}} : \dot{\mathbf{F}} &= \frac{\partial \Psi(\mathbf{C})}{\partial \mathbf{C}} : \dot{\mathbf{C}} \\ &= \frac{\partial \Psi(\mathbf{C})}{\partial \mathbf{C}} : (\dot{\mathbf{F}}^{\top}\mathbf{F} + \mathbf{F}^{\top}\dot{\mathbf{F}}).\end{aligned} \tag{3.27}$$

Since $\partial\Psi(\mathbf{C})/\partial\mathbf{C}$ is symmetric, it follows from $(2.61)_1$ that

$$\begin{aligned}\frac{\partial \Psi(\mathbf{C})}{\partial \mathbf{C}} : (\dot{\mathbf{F}}^{\top}\mathbf{F}) &= \frac{\partial \Psi(\mathbf{C})}{\partial \mathbf{C}} : (\dot{\mathbf{F}}^{\top}\mathbf{F})^{\top} \\ &= \frac{\partial \Psi(\mathbf{C})}{\partial \mathbf{C}} : (\mathbf{F}^{\top}\dot{\mathbf{F}}).\end{aligned} \tag{3.28}$$

Therefore, by (2.52),

$$\begin{aligned}\frac{\partial \Lambda(\mathbf{F})}{\partial \mathbf{F}} : \dot{\mathbf{F}} &= 2\frac{\partial \Psi(\mathbf{C})}{\partial \mathbf{C}} : (\mathbf{F}^{\top}\dot{\mathbf{F}}) \\ &= 2\left[\mathbf{F}\frac{\partial \Psi(\mathbf{C})}{\partial \mathbf{C}}\right] : \dot{\mathbf{F}},\end{aligned}$$

and hence,

$$\left[\frac{\partial \Lambda(\mathbf{F})}{\partial \mathbf{F}} - 2\mathbf{F}\frac{\partial \Psi(\mathbf{C})}{\partial \mathbf{C}}\right] : \dot{\mathbf{F}} = 0 \quad \text{for any choice of the function } \mathbf{F}(t). \tag{3.29}$$

We now present an argument used repeatedly throughout this book to show, as a consequence of (3.29), that

$$\frac{\partial \Lambda(\mathbf{F})}{\partial \mathbf{F}} = 2\mathbf{F}\frac{\partial \Psi(\mathbf{C})}{\partial \mathbf{C}}. \tag{3.30}$$

Choose, arbitrarily, (constant) tensors $\mathbf{F}_0$ and $\mathbf{A}$, and let

$$\mathbf{F}(t) = \mathbf{F}_0 + (t - t_0)\mathbf{A},$$

so that

$$\mathbf{F}(t_0) = \mathbf{F}_0 \qquad \text{and} \qquad \dot{\mathbf{F}}(t_0) = \mathbf{A}.$$

Further, let $\mathbf{C}_0 = \mathbf{F}_0^{\top}\mathbf{F}_0$ and define new functions

$$\mathbf{S}(\mathbf{F}) = \frac{\partial \Lambda(\mathbf{F})}{\partial \mathbf{F}} \qquad \text{and} \qquad \mathbf{M}(\mathbf{C}) = \frac{\partial \Psi(\mathbf{C})}{\partial \mathbf{C}};$$

(3.29) evaluated at $t = t_0$ then yields

$$[\mathbf{S}(\mathbf{F}_0) - 2\,\mathbf{F}_0\mathbf{M}(\mathbf{C}_0)] : \mathbf{A} = 0. \tag{3.31}$$

Since $\mathbf{A}$ was arbitrarily chosen, we may use (2.56) to conclude that

$$\mathbf{S}(\mathbf{F}_0) - 2\,\mathbf{F}_0\mathbf{M}(\mathbf{C}_0) = \mathbf{0};$$

(3.30) therefore holds for $\mathbf{F} = \mathbf{F}_0$ and $\mathbf{C} = \mathbf{C}_0$. Since $\mathbf{F}_0$ is also arbitrary, (3.30) holds for all tensors $\mathbf{F}$.

As an important example of this procedure, we now derive the identity

$$\frac{\partial|\mathbf{A}|}{\partial\mathbf{A}} = \frac{\mathbf{A}}{|\mathbf{A}|}, \tag{3.32}$$

which holds for tensors $\mathbf{A} \neq \mathbf{0}$. We begin the proof by considering an arbitrary nonzero function $\mathbf{A}(t)$. The chain-rule then implies that[35]

$$\dot{\overline{|\mathbf{A}|}} = \frac{\partial|\mathbf{A}|}{\partial\mathbf{A}} : \dot{\mathbf{A}}.$$

Thus, by (3.2),

$$\left(\frac{\partial|\mathbf{A}|}{\partial\mathbf{A}} - \frac{\mathbf{A}}{|\mathbf{A}|}\right) : \dot{\mathbf{A}} = 0. \tag{3.33}$$

Finally, arguing as in the steps leading to (3.31), we conclude that (3.33) can hold for all nonzero functions $\mathbf{A}(t)$ only if (3.32) is valid.

EXERCISES

1. Let $\mathbf{A}$ be invertible. Bearing in mind (3.3), mimic the argument used to establish (3.32) to arrive at the identity

$$\frac{\partial(\det\mathbf{A})}{\partial\mathbf{A}} = (\det\mathbf{A})\mathbf{A}^{-\top}. \tag{3.34}$$

2. Use (3.34) to show that, for $\mathbf{A}$ invertible,

$$\frac{\partial(\ln\det\mathbf{A})}{\partial\mathbf{A}} = \mathbf{A}^{-\top},$$

$$\frac{\partial[\ln\det(\mathbf{A}^{-1})]}{\partial\mathbf{A}} = -\mathbf{A}^{-\top}.$$

3. Let $\mathbf{A}$ and $\mathbf{B}$ be constant. Establish the identities

$$\frac{\partial[\mathrm{tr}\,(\mathbf{A}\mathbf{S}\mathbf{B}^{\top})]}{\partial\mathbf{S}} = \mathbf{A}^{\top}\mathbf{B},$$

$$\frac{\partial[\mathrm{tr}\,(\mathbf{A}\mathbf{S}^{\top}\mathbf{B}^{\top})]}{\partial\mathbf{S}} = \mathbf{B}^{\top}\mathbf{A},$$

$$\frac{\partial(\mathrm{tr}\,[\mathbf{A}\mathbf{S}\mathbf{B}^{\top}\mathbf{S}^{\top})]}{\partial\mathbf{S}} = \mathbf{A}^{\top}\mathbf{S}\mathbf{B} + \mathbf{A}\mathbf{S}\mathbf{B}^{\top}.$$

4. Let $\mathbf{S}$ be symmetric and positive-definite. Establish the identities

$$\left.\begin{aligned} \frac{\partial I_1(\mathbf{S})}{\partial\mathbf{S}} &= \mathbf{1}, \\ \frac{\partial I_2(\mathbf{S})}{\partial\mathbf{S}} &= I_1(\mathbf{S})\mathbf{1} - \mathbf{S}, \\ \frac{\partial I_3(\mathbf{S})}{\partial\mathbf{S}} &= I_3(\mathbf{S})\mathbf{S}^{-\top}. \end{aligned}\right\} \tag{3.35}$$

[35] Cf. (3.26).

4 Integral Theorems

4.1 The Divergence Theorem

The divergence theorem and Stokes' theorem are deep mathematical results central to the formulation of the basic laws of balance and imbalance for continua. Here, we state these theorems without proof and without encumbering the presentation with smoothness assumptions regarding the underlying functions and regularity assumptions regarding the region in question.[36]

DIVERGENCE THEOREM *Let R be a bounded region with boundary ∂R. Assume we are given a scalar field φ, a vector field $\mathbf{v}$, and a tensor field $\mathbf{T}$, with R the domain of each of these fields. Let $\mathbf{n}$ denote the outward unit normal field on the boundary ∂R of R. Then,*

$$\left.\begin{aligned} \int\limits_{\partial R} \varphi \mathbf{n}\, da &= \int\limits_{R} \operatorname{grad}\varphi\, dv, \\ \int\limits_{\partial R} \mathbf{v}\cdot\mathbf{n}\, da &= \int\limits_{R} \operatorname{div}\mathbf{v}\, dv, \\ \int\limits_{\partial R} \mathbf{T}\mathbf{n}\, da &= \int\limits_{R} \operatorname{div}\mathbf{T}\, dv. \end{aligned}\right\} \tag{4.1}$$

The result $(4.1)_2$ is classical. We now establish $(4.1)_3$ as a consequence of $(4.1)_2$. Let $\mathbf{a}$ be a (constant) vector and define a vector field $\mathbf{v}$ by

$$\mathbf{v} = \mathbf{T}^{\top}\mathbf{a}. \tag{4.2}$$

Then, bearing in mind $(3.19)_3$,

$$\begin{aligned} \mathbf{a}\cdot\int\limits_{\partial R} \mathbf{T}\mathbf{n}\, da &= \int\limits_{\partial R} \mathbf{a}\cdot\mathbf{T}\mathbf{n}\, da \\ &= \int\limits_{\partial R} (\mathbf{T}^{\top}\mathbf{a})\cdot\mathbf{n}\, da \end{aligned}$$

[36] Within a classical framework, precise statements of the divergence theorem and Stokes' theorem are given by KELLOGG (1953).

$$= \int_{\partial R} \mathbf{v} \cdot \mathbf{n}\, da$$

$$= \int_{R} \operatorname{div}\mathbf{v}\, dv$$

$$= \int_{R} \operatorname{div}(\mathbf{T}^{\mathsf{T}}\mathbf{a})\, dv$$

$$= \int_{R} \mathbf{a} \cdot \operatorname{div}\mathbf{T}\, dv$$

$$= \mathbf{a} \cdot \int_{R} \operatorname{div}\mathbf{T}\, dv, \tag{4.3}$$

which implies $(4.1)_3$, since the vector $\mathbf{a}$ is arbitrary. Next, $(4.1)_1$ follows from $(4.1)_3$ on setting $\mathbf{T} = \varphi\mathbf{1}$ and using the identity $\operatorname{div}(\varphi\mathbf{1}) = \operatorname{grad}\varphi$.

The identities (4.1) have the component forms:

$$\left.\begin{aligned} \int_{\partial R} \varphi n_i\, da &= \int_{R} \frac{\partial \varphi}{\partial x_i}\, dv, \\ \int_{\partial R} v_i n_i\, da &= \int_{R} \frac{\partial v_i}{\partial x_i}\, dv, \\ \int_{\partial R} T_{ij} n_j\, da &= \int_{R} \frac{\partial T_{ij}}{\partial x_j}\, dv. \end{aligned}\right\} \tag{4.4}$$

Note that these identities follow a general rule: The n_i in the surface integral results in the partial derivative $\partial/\partial x_i$ in the volume integral; thus, for a tensor $\mathbf{T}$ of any order, with components $T_{ij\cdots k}$,

$$\int_{\partial R} T_{ij\cdots k} n_r\, da = \int_{R} \frac{\partial T_{ij\cdots k}}{\partial x_r}\, dv.$$

4.2 Line Integrals. Stokes' Theorem

Let C be a **curve** described by the parametrization

$$\mathbf{x} = \hat{\mathbf{x}}(\lambda), \qquad \lambda_0 \le \lambda \le \lambda_1.$$

Then, C is the curve traced out by the point $\mathbf{x} = \hat{\mathbf{x}}(\lambda)$ as the parameter λ increases, while the vector function

$$\mathbf{t}(\lambda) \stackrel{\text{def}}{=} \frac{d\hat{\mathbf{x}}(\lambda)}{d\lambda} \tag{4.5}$$

is tangent to C and points in the direction of increasing λ (Figure 4.1). The curve C is said to be **closed** if

$$\hat{\mathbf{x}}(\lambda_0) = \hat{\mathbf{x}}(\lambda_1).$$

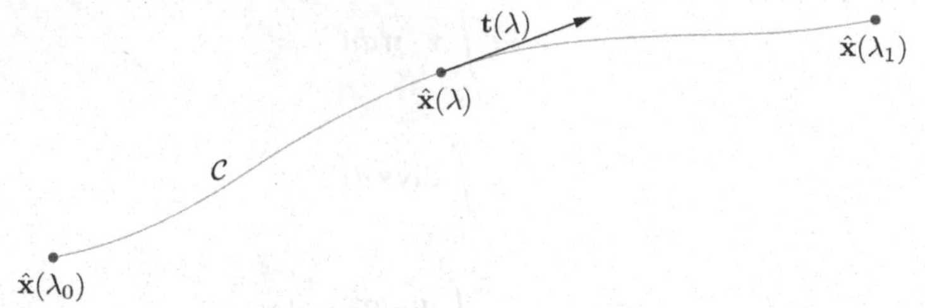

Figure 4.1. A curve $\mathcal{C}$ between two points $\hat{\mathbf{x}}(\lambda_0)$ and $\hat{\mathbf{x}}(\lambda_1)$. Also shown are an intermediate point $\hat{\mathbf{x}}(\lambda)$ and the unit tangent $\mathbf{t}(\lambda)$ at $\hat{\mathbf{x}}(\lambda)$.

For $\mathbf{v}$ a vector field with domain a region R, the **integral of v along** a curve $\mathcal{C}$ in R is the *line integral*

$$\begin{aligned}\int_{\mathcal{C}} \mathbf{v}\cdot d\mathbf{x} &\stackrel{\text{def}}{=} \int_{\lambda_0}^{\lambda_1} \mathbf{v}(\hat{\mathbf{x}}(\lambda))\cdot \frac{d\hat{\mathbf{x}}(\lambda)}{d\lambda}\, d\lambda \\ &= \int_{\lambda_0}^{\lambda_1} \mathbf{v}(\hat{\mathbf{x}}(\lambda))\cdot \mathbf{t}(\lambda)\, d\lambda. \end{aligned} \tag{4.6}$$

For φ a scalar field on R, it follows from the chain-rule that

$$\begin{aligned}\int_{\mathcal{C}} \operatorname{grad}\varphi\cdot d\mathbf{x} &= \int_{\lambda_0}^{\lambda_1} \operatorname{grad}\varphi(\hat{\mathbf{x}}(\lambda))\cdot \frac{d\hat{\mathbf{x}}(\lambda)}{d\lambda}\, d\lambda \\ &= \int_{\lambda_0}^{\lambda_1} \frac{\partial\varphi(\hat{\mathbf{x}}(\lambda))}{\partial\lambda}\, d\lambda \\ &= \varphi(\hat{\mathbf{x}}(\lambda_1)) - \varphi(\hat{\mathbf{x}}(\lambda_0)).\end{aligned}$$

Hence, for $\mathcal{C}$ a closed curve and φ a scalar field on R,

$$\int_{\mathcal{C}} \operatorname{grad}\varphi\cdot d\mathbf{x} = 0. \tag{4.7}$$

Let $\mathcal{S}$ be a surface in R, so that the boundary of $\mathcal{S}$ is a closed curve $\mathcal{C}$. A choice of orientation for $\mathcal{S}$ consists in a choice of the unit normal field $\mathbf{n}$ for $\mathcal{S}$.[37] Granted a choice of $\mathbf{n}$, $\mathcal{S}$ is referred to as a **positively oriented surface** if $\mathbf{n}$ and the tangent $\mathbf{t}$ to $\mathcal{C}$, defined in (4.5), satisfy

$$\mathbf{n}(\mathbf{x})\cdot(\mathbf{t}(\lambda)\times\mathbf{t}(\lambda_0)) > 0$$

for some choice of $\mathbf{x}$ and λ (Figure 4.2).

[37] Except in pathological cases (such as the Möbius strip) there are two possible choices, and one is the negative of the other.

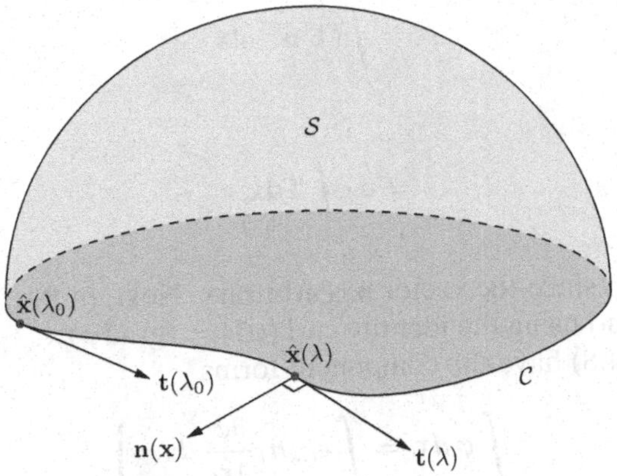

Figure 4.2. A positively oriented surface $\mathcal{S}$ with boundary being a closed curve $\mathcal{C}$. Also shown are two points $\hat{\mathbf{x}}(\lambda_0) = \hat{\mathbf{x}}(\lambda_1)$ on $\mathcal{C}$ and $\hat{\mathbf{x}}(\lambda)$, the unit tangent vectors $\mathbf{t}(\lambda_0)$ and $\mathbf{t}(\lambda)$ at those points, and the unit normal $\mathbf{n}(\mathbf{x})$ at $\hat{\mathbf{x}}(\lambda)$.

STOKES' THEOREM *Let φ, $\mathbf{v}$, and $\mathbf{T}$ be scalar, vector, and tensor fields with common domain a region R. Then given any positively oriented surface $\mathcal{S}$, with boundary $\mathcal{C}$ a closed curve, in R,*

$$\left.\begin{aligned} \int_{\mathcal{C}} \varphi\,\mathbf{dx} &= \int_{\mathcal{S}} \mathbf{n} \times \operatorname{grad}\varphi\, da, \\ \int_{\mathcal{C}} \mathbf{v}\cdot\mathbf{dx} &= \int_{\mathcal{S}} \mathbf{n}\cdot\operatorname{curl}\mathbf{v}\, da, \\ \int_{\mathcal{C}} \mathbf{T}\mathbf{dx} &= \int_{\mathcal{S}} (\operatorname{curl}\mathbf{T})^{\top}\mathbf{n}\, da. \end{aligned}\right\} \tag{4.8}$$

The result $(4.8)_2$ is classical. To establish $(4.8)_3$ as a consequence of $(4.8)_2$, let $\mathbf{a}$ be a (constant) vector and define $\mathbf{v}$ as in (4.2). Then, bearing in mind $(3.19)_4$,

$$\begin{aligned} \mathbf{a}\cdot\int_{\mathcal{S}} (\operatorname{curl}\mathbf{T})^{\top}\mathbf{n}\, da &= \int_{\mathcal{S}} \mathbf{n}\cdot[(\operatorname{curl}\mathbf{T})\mathbf{a}]\, da \\ &= \int_{\mathcal{S}} \mathbf{n}\cdot\operatorname{curl}(\mathbf{T}^{\top}\mathbf{a})\, da \\ &= \int_{\mathcal{S}} \mathbf{n}\cdot\operatorname{curl}\mathbf{v}\, da \\ &= \int_{\mathcal{C}} \mathbf{v}\cdot\mathbf{dx} \end{aligned}$$

$$= \int_C (\mathbf{T}^{\top}\mathbf{a}) \cdot \mathbf{dx}$$

$$= \mathbf{a} \cdot \int_C \mathbf{T}\mathbf{dx}, \tag{4.9}$$

which implies $(4.8)_2$, since the vector **a** is arbitrary. Next, $(4.8)_1$ follows from $(4.8)_3$ on setting $\mathbf{T} = \varphi\mathbf{1}$ and using the identity $\operatorname{curl}(\varphi\mathbf{1}) = (\operatorname{grad}\varphi)\times$.

The identities (4.8) have the component forms

$$\left.\begin{aligned} \int_C \varphi\, dx_i &= \int_S \epsilon_{ijk} n_j \frac{\partial \varphi}{\partial x_k}\, da, \\ \int_C v_i\, dx_i &= \int_S n_i \epsilon_{ijk} \frac{\partial v_k}{\partial x_j}\, da, \\ \int_C T_{ij}\, dx_j &= \int_S \epsilon_{jpq} \frac{\partial T_{iq}}{\partial x_p} n_j\, da. \end{aligned}\right\} \tag{4.10}$$

EXERCISES

1. Establish the following identities using the divergence theorem (4.1):

$$\left.\begin{aligned} &\int_{\partial R} \mathbf{n} \cdot \operatorname{curl} \mathbf{v}\, da = 0, \\ &\int_{\partial R} \mathbf{n} \times \mathbf{v}\, da = \int_R \operatorname{curl} \mathbf{v}\, dv, \\ &\int_{\partial R} \mathbf{v} \otimes \mathbf{n}\, da = \int_R \operatorname{grad} \mathbf{v}\, dv, \\ &\int_{\partial R} \mathbf{Tn} \otimes \mathbf{v}\, da = \int_R [(\operatorname{div}\mathbf{T}) \otimes \mathbf{v} + \mathbf{T}(\operatorname{grad}\mathbf{v})^{\top}]\, dv, \\ &\int_{\partial R} \mathbf{v} \cdot \mathbf{Tn}\, da = \int_R (\mathbf{v} \cdot \operatorname{div}\mathbf{T} + \mathbf{T} : \operatorname{grad}\mathbf{v})\, dv, \\ &\int_{\partial R} \mathbf{u}(\mathbf{v} \cdot \mathbf{n})\, da = \int_R (\mathbf{u} \operatorname{div}\mathbf{v} + (\operatorname{grad}\mathbf{u})\mathbf{v})\, dv. \end{aligned}\right. \tag{4.11}$$

2. Given a bounded region R with volume $\operatorname{vol}(R)$, use the divergence theorem (4.1) to show that

$$\int_{\partial R} (\mathbf{x} - \mathbf{o}) \otimes \mathbf{n}\, da = \operatorname{vol}(R)\mathbf{1}.$$

3. Verify the identity (4.7).

4. Show that Stokes' theorem $(4.8)_2$ for a vector field $\mathbf{v}$ may be written in the form

$$\int_{\mathcal{C}} \mathbf{v} \cdot \mathbf{dx} = \int_{\mathcal{S}} (\mathbf{n}\times) : \operatorname{grad} \mathbf{v} \, da.$$

5. Establish $(4.11)_1$ as a consequence of $(4.8)_1$.
6. Establish the following identities using Stokes' theorem (4.8):

$$\int_{\mathcal{C}} (\mathbf{u} \otimes \mathbf{v}) \mathbf{dx} = \int_{\mathcal{S}} [(\operatorname{grad} \mathbf{u}) \mathbf{v} \times \mathbf{n} + (\mathbf{n} \cdot \operatorname{curl} \mathbf{v}) \mathbf{u}] \, da,$$

$$\int_{\mathcal{C}} \mathbf{v} \times \mathbf{dx} = \int_{\mathcal{S}} [(\operatorname{div} \mathbf{v}) \mathbf{n} - (\operatorname{grad} \mathbf{v})^{\mathsf{T}} \mathbf{n}] \, da. \tag{4.12}$$

7. Suppose that $\mathbf{v}$ obeys $\mathbf{v} \cdot \mathbf{n} = 0$ on $\mathcal{S}$. Use Stokes' theorem (4.8) to show that

$$\int_{\mathcal{C}} (\mathbf{n} \times \mathbf{v}) \cdot \mathbf{dx} = \int_{\mathcal{S}} (\mathbf{1} - \mathbf{n} \otimes \mathbf{n}) : \operatorname{grad} \mathbf{v} \, da.$$

PART III

KINEMATICS

We generally omit smoothness assumptions in sections not dealing specifically with fields that suffer jump discontinuities, which is the only loss of continuity that we discuss.[38]

[38] Cf. §32, in which shock waves are discussed.

5 Motion of a Body

5.1 Reference Body. Material Points

In continuum mechanics the basic property of a body is that it may occupy regions of Euclidean point space $\mathcal{E}$. We may, if we wish, identify the body with the region[39] B of $\mathcal{E}$ it occupies in some fixed configuration, called a **reference configuration**, but

- *the choice of a reference configuration is arbitrary.*

For specificity, we henceforth consider a body identified with the region B it occupies in a fixed reference configuration and refer to B as the **reference body** and to a point $\mathbf{X}$ in B as a **material point** or **particle.**

5.2 Basic Quantities Associated with the Motion of a Body

We restrict attention to a given open time-interval; to avoid cumbersome statements the phrase "all t" signifies "all t in that interval." A **motion** of B is a smooth function χ that assigns to each material point $\mathbf{X}$ and time t a point

$$\boxed{\mathbf{x} = \chi(\mathbf{X}, t);} \tag{5.1}$$

$\mathbf{x}$ is referred to as the **spatial point occupied by X** at time t.

Consistent with the notational conventions set forth in §3.2, we use grad, div, and curl to denote the gradient, divergence, and curl with respect to spatial points $\mathbf{x}$.

For the moment, we restrict attention to a fixed time t. Then $\chi(\mathbf{X}, t)$ considered as a function of $\mathbf{X}$ is called the **deformation** at time t; for convenience, we write this function in the form

$$\chi_t(\mathbf{X}) = \chi(\mathbf{X}, t). \tag{5.2}$$

A basic hypothesis of continuum mechanics is that $\chi_t(\mathbf{X})$ be *one-to-one* in $\mathbf{X}$, so that no two material points may occupy the same spatial point at a given time, or, more descriptively, so that the body cannot penetrate itself. In addition, writing $\nabla\chi_t(\mathbf{X})$ for the gradient of $\chi_t(\mathbf{X})$ with respect to the material point $\mathbf{X}$, we require that

$$J(\mathbf{X}, t) \stackrel{\text{def}}{=} \det \nabla\chi_t(\mathbf{X}) > 0; \tag{5.3}$$

[39] Throughout this book the term *region* connotes a *closed region* with sufficiently regular boundary.

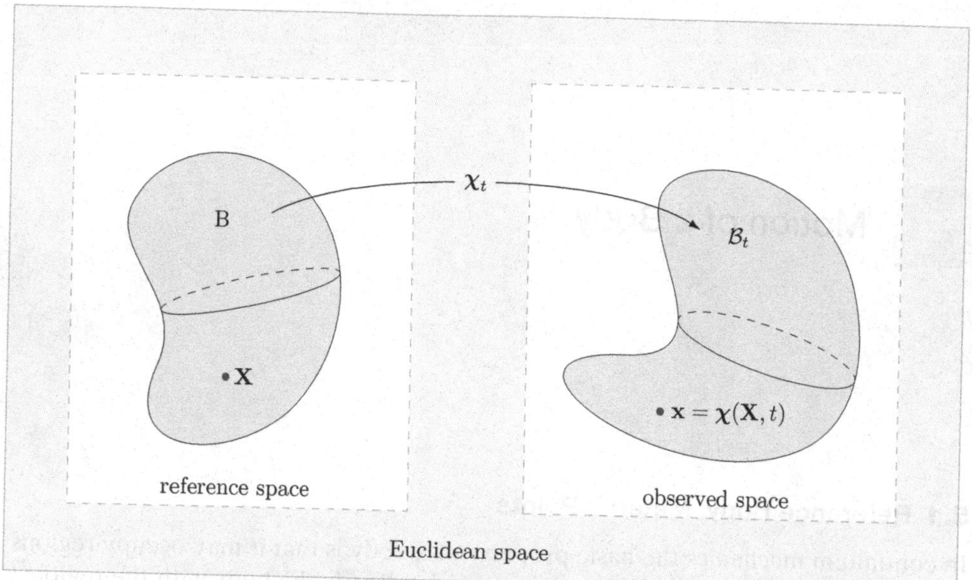

Figure 5.1. The reference body B and the deformed body $\mathcal{B}_t$.

$J(\mathbf{X}, t)$ is the *volumetric Jacobian* of the mapping χ_t at the material point $\mathbf{X}$.[40]

The region of space **occupied by** the body at time t,

$$\mathcal{B}_t = \chi_t(\mathrm{B}),$$

is referred to as the **deformed body** at time t; $\mathcal{B}_t$ is the set of points $\mathbf{x}$ such that $\mathbf{x} = \chi(\mathbf{X}, t)$ for some $\mathbf{X}$ in B. $\mathcal{B}_t$ is the region actually observed during the motion: the reference body B serves only to label material points. For this reason, while we work within the framework of a single Euclidean point space, it is useful to differentiate — at least conceptually — between the ambient space for the reference body B and the ambient space through which $\mathcal{B}_t$ evolves (Figure 5.1). In accord with this:

(i) vectors associated with the ambient space through which $\mathcal{B}_t$ evolves are referred to as **spatial vectors**;
(ii) vectors associated with the ambient space for the reference body B are referred to as **material vectors.**

Examples of spatial and material vectors are provided by the differences $\mathbf{x}_1 - \mathbf{x}_2$ and $\mathbf{X}_1 - \mathbf{X}_2$ of two spatial points $\mathbf{x}_1$ and $\mathbf{x}_2$ and two material points $\mathbf{X}_1$ and $\mathbf{X}_2$.

The time-parameterized family of sets $\mathcal{B}_t$ represents the body during an actual motion, a motion that could, in principle, be seen or felt by any one of us. On the other hand, the set B, while essential to a careful treatment of continuum mechanics, is virtual; the body need never occupy B, although it might. For many, but not all, applications it is convenient to choose the initial configuration as reference. But, because solids generally possess natural ("virgin") reference configurations, it is often convenient to choose such a configuration as reference; in fact, such a choice is essential to a rational discussion of material symmetry. In summary: allowing the choice of reference configuration to be arbitrary allows one to choose that configuration most suitable to a given application.

The spatial vectors

$$\dot{\chi}(\mathbf{X}, t) = \frac{\partial \chi(\mathbf{X}, t)}{\partial t} \quad \text{and} \quad \ddot{\chi}(\mathbf{X}, t) = \frac{\partial^2 \chi(\mathbf{X}, t)}{\partial t^2} \tag{5.4}$$

[40] We add the adjective *volumetric* because we also consider *areal* Jacobians related to the deformation of surfaces.

represent the **velocity** and **acceleration** of the material point $\mathbf{X}$ at time t.[41]

Since the mapping $\mathbf{x} = \chi(\mathbf{X}, t)$ is one-to-one in $\mathbf{X}$ for *fixed* t, it has an inverse

$$\mathbf{X} = \chi^{-1}(\mathbf{x}, t); \tag{5.5}$$

at each t, the **fixed-time inverse** $\chi^{-1}(\cdot, t)$ is a mapping of the deformed body $\mathcal{B}_t$ onto the reference body B with the following property:

$$\mathbf{x} = \chi(\mathbf{X}, t) \qquad \text{if and only if} \qquad \mathbf{X} = \chi^{-1}(\mathbf{x}, t). \tag{5.6}$$

We refer to χ^{-1} as defined by (5.5) as the **reference map**; this map associates with every time t and spatial point $\mathbf{x}$ in $\mathcal{B}_t$ a material point $\mathbf{X} = \chi^{-1}(\mathbf{x}, t)$ in B; $\mathbf{X}$ is the material point that occupies the spatial point $\mathbf{x}$ at time t.

Finally, to express material and spatial vectors in component form, we introduce an *orthonormal basis*

$$\{\mathbf{e}_i\} = \{\mathbf{e}_1, \mathbf{e}_2, \mathbf{e}_3\} \tag{5.7}$$

assumed to be positively oriented in the sense that[42]

$$(\mathbf{e}_1 \times \mathbf{e}_2) \cdot \mathbf{e}_3 = 1. \tag{5.8}$$

5.3 Convection of Sets with the Body

Let A denote a set of material points or, more simply, a **material set**.[43] Then,

$$\mathcal{A}_t = \chi_t(\mathrm{A}) \tag{5.9}$$

represents the set of spatial points **occupied by** the material points of A at time t, and we say that A **deforms to** $\mathcal{A}_t$ at time t. Consistent with this definition, we say that a time-dependent spatial set $\mathcal{A}_t$ **convects with the body** — or, more simply, is **convecting** — if there is a set A of material points such that $\mathcal{A}_t = \chi_t(\mathrm{A})$ for all t.

Let $\mathcal{P}_t$ be a spatial region that convects with the body, so that, by definition, there is a material region P such that

$$\mathcal{P}_t = \chi_t(\mathrm{P}) \tag{5.10}$$

for all t. Then,[44]

$$\partial\mathcal{P}_t = \chi_t(\partial\mathrm{P}) \tag{5.11}$$

and, therefore, if $\chi(\mathbf{X}, \tau)$ is on $\partial\mathrm{P}_\tau$ at some time τ, then $\mathbf{X}$ is on $\partial\mathrm{P}$ and, hence, $\chi(\mathbf{X}, t)$ is on $\partial\mathcal{P}_t$ for all time t: $\partial\mathcal{P}_t$ is therefore occupied by the same set of material points for all time. Thus, in terms more suggestive than precise,

(†) *material cannot cross the boundary $\partial\mathcal{P}_t$ of a spatial region convecting with the body.*

In particular, choosing $\mathrm{P} = \mathrm{B}$ in (5.11), we have the following important fact:

(‡) If $\mathbf{X}$ is on $\partial\mathrm{B}$ then $\chi(\mathbf{X}, t)$ is on $\partial\mathcal{B}_t$ for all time t. Conversely, if $\chi(\mathbf{X}, t)$ is on $\partial\mathcal{B}_t$ at some time t, then $\mathbf{X}$ is on $\partial\mathrm{B}$ (and, hence, $\chi(\mathbf{X}, \tau)$ is on $\partial\mathcal{B}_\tau$ for all time τ).

---

41 The partial time derivatives in (5.4) are with respect to t holding the material point $\mathbf{X}$ fixed. In contrast, often in the fluid dynamics literature $\partial/\partial t$ is used to denote a time derivative holding the spatial point $\mathbf{x}$ fixed. The difference between these two time derivatives is discussed at length in §9.2.

42 Cf. Footnote 7.

43 Obvious meanings then apply to the terms *material surface*, *material curve*, and so on.

44 This (nontrivial) result is a consequence of our assumption that $\chi_t(\mathbf{X})$ be one-to-one in $\mathbf{X}$ and smooth; cf. the paragraph containing (5.2).

6 The Deformation Gradient

The tensor field

$$\boxed{\mathbf{F} = \nabla\chi, \qquad F_{ij} = \frac{\partial \chi_i}{\partial X_j},} \tag{6.1}$$

is referred to as the **deformation gradient**. By (5.3),

$$J = \det\mathbf{F} > 0. \tag{6.2}$$

6.1 Approximation of a Deformation by a Homogeneous Deformation

6.1.1 Homogeneous Deformations

Consider the deformation χ_t at a fixed time t. For convenience, supress the time t and write

$$\chi(\mathbf{X}) \equiv \chi_t(\mathbf{X}).$$

We refer to χ as a **homogeneous deformation** if the instantaneous deformation gradient $\mathbf{F}(\mathbf{X}) \equiv \mathbf{F}(\mathbf{X}, t)$ is independent of $\mathbf{X}$, so that[45]

$$\underbrace{\chi(\mathbf{X}) - \chi(\mathbf{Y})}_{\substack{\text{spatial}\\\text{vector}}} = \mathbf{F}\underbrace{(\mathbf{X} - \mathbf{Y})}_{\substack{\text{material}\\\text{vector}}} \tag{6.3}$$

for all material points $\mathbf{X}$ and $\mathbf{Y}$; equivalently, in components,

$$\chi_i(\mathbf{X}) - \chi_i(\mathbf{Y}) = F_{ij}(X_j - Y_j).$$

In (6.3), we have tried to emphasize that, since, the difference $\mathbf{X} - \mathbf{Y}$ between material points is a material vector, while the difference $\chi(\mathbf{X}) - \chi(\mathbf{Y})$ between spatial points is a spatial vector,

F maps material vectors to spatial vectors.

Further, (6.3) implies that $\mathbf{X} - \mathbf{Y} = \mathbf{F}^{-1}[\chi(\mathbf{X}) - \chi(\mathbf{Y})]$; hence, $\mathbf{F}^{-1}$ maps spatial vectors to material vectors. Next, consider the inner product of (6.3) with a *spatial vector*

[45] Cf. Gurtin (1981, p. 36).

s, an operation that makes sense because $\chi(\mathbf{X}) - \chi(\mathbf{Y})$ is a spatial vector; then

$$\begin{aligned}\mathbf{s}\cdot[\chi(\mathbf{X}) - \chi(\mathbf{Y})] &= \mathbf{s}\cdot[\mathbf{F}(\mathbf{X}-\mathbf{Y})]\\ &= (\mathbf{F}^{\top}\mathbf{s})\cdot(\mathbf{X}-\mathbf{Y}),\end{aligned} \tag{6.4}$$

so that $\mathbf{F}^{\top}$ maps spatial vectors to material vectors. Finally, we leave it as an exercise to show that $\mathbf{F}^{-\top}$ maps material vectors to spatial vectors. Summarizing, we have the following **mapping properties** for the deformation gradient:

(M1) $\mathbf{F}$ and $\mathbf{F}^{-\top}$ map material vectors to spatial vectors;
(M2) $\mathbf{F}^{-1}$ and $\mathbf{F}^{\top}$ map spatial vectors to material vectors.

6.1.2 General Deformations

Consider now an arbitrary deformation χ_t. If we take the Taylor expansion of the deformation χ_t about a material point **X**,[46] we find that

$$\underline{\chi_t(\mathbf{Y}) - \chi_t(\mathbf{X}) = \mathbf{F}(\mathbf{X},t)(\mathbf{Y}-\mathbf{X})} + o(|\mathbf{Y}-\mathbf{X}|) \qquad \text{as} \qquad |\mathbf{Y}-\mathbf{X}| \to 0. \tag{6.5}$$

The term $\mathbf{F}(\mathbf{X},t)(\mathbf{Y}-\mathbf{X})$ therefore represents an approximation of $\chi_t(\mathbf{Y}) - \chi_t(\mathbf{X})$, an approximation whose error goes to zero faster than the material vector $(\mathbf{Y}-\mathbf{X})$ and hence faster than the term $\mathbf{F}(\mathbf{X},t)(\mathbf{Y}-\mathbf{X})$. Further, in the expansion (6.5), the material point **X** is fixed, so that $\mathbf{F}(\mathbf{X},t)$ is constant; the underlined portion of (6.5) therefore represents a homogeneous deformation. Thus,

- in a neighborhood of a material point **X** and to within an error of $o(|\mathbf{Y}-\mathbf{X}|)$, a deformation behaves like a homogeneous deformation.

In terms more suggestive than precise, (6.5) with the term $o(|\mathbf{Y}-\mathbf{X}|)$ "as small as we wish":

(i) shows the sense in which $\mathbf{F}(\mathbf{X},t)$ may be considered as a mapping of an infinitesimal neighborhood of **X** in the reference body to an infinitesimal neighborhood of $\mathbf{x} = \chi_t(\mathbf{X})$ in the deformed body; and
(ii) gives an asymptotic meaning to the formal relation

$$d\mathbf{x} = \mathbf{F}(\mathbf{X},t)d\mathbf{X}. \tag{6.6}$$

A consequence of (6.5) is that the mapping properties (M1) and (M2) on page 65 for a homogeneous deformation hold *pointwise* for the deformation gradient in an arbitrary deformation; that is, for example, given any **X**, the linear transformation $\mathbf{F}(\mathbf{X},t)$ associates with each material vector **m** a spatial vector

$$\mathbf{s} = \mathbf{F}(\mathbf{X},t)\mathbf{m}. \tag{6.7}$$

When discussing a relation of the form (6.7), it is convenient to refer to **m** as an input for $\mathbf{F}(\mathbf{X},t)$ and to **s** as the corresponding output. The physical tensor fields considered in this book always come equipped with pointwise mapping properties dictated by the physics, just as (M1) and (M2) are each consequences of (6.5). For that reason, it is not physically meaningful to input a spatial vector to either **F** or $\mathbf{F}^{-\top}$ or a material vector to either $\mathbf{F}^{-1}$ or $\mathbf{F}^{\top}$.

Sample problem Given a tensor field **G** that maps spatial vectors to spatial vectors, determine the pointwise mapping property of

$$\mathbf{F}^{\top}\mathbf{G}\mathbf{F}.$$

Solution Since **F** is a mapping of material vectors, we may only input material vectors to the composite $\mathbf{F}^{\top}\mathbf{G}\mathbf{F}$. Then, given a material vector **m**, the vector **Fm** is spatial, and so it makes sense to "input" this

[46] Cf. $(3.12)_2$.

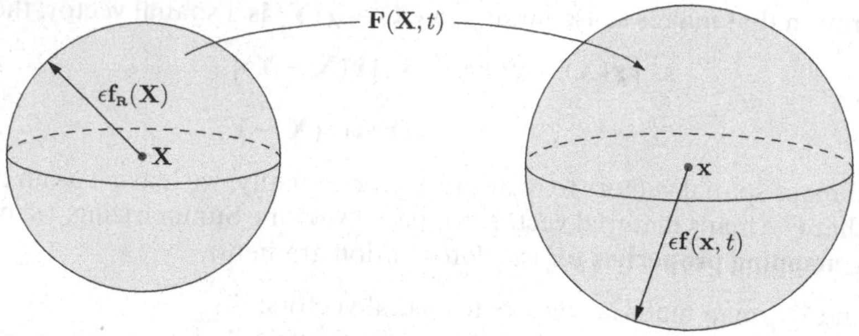

Figure 6.1. Infinitesimal neighborhoods associated with the undeformed and deformed fibers $\mathbf{f}_R(\mathbf{X})$ and $\mathbf{f}(\mathbf{x}, t) = \mathbf{F}(\mathbf{X}, t)\mathbf{f}_R(\mathbf{X})$, $\mathbf{x} = \chi_t(\mathbf{X})$.

vector to **G**, the resulting "output" **GFm** is then spatial, and if we input **GFm** to $\mathbf{F}^\top$, the output $\mathbf{F}^\top\mathbf{GFm}$ is, by (M2), a material vector. Thus, $\mathbf{F}^\top\mathbf{GF}$ maps material vectors to material vectors.

EXERCISE

1. Let **H** be a mapping of material vectors to spatial vectors. Determine the mapping properties of $\mathbf{HF}^{-1}$ and $\mathbf{F}^\top\mathbf{HF}^\top$.

6.2 Convection of Geometric Quantities

6.2.1 Infinitesimal Fibers

Let $\mathbf{f}_R$ denote a **temporally constant** *material vector field* and consider the *spatial vector field* **f** defined by

$$\mathbf{f}(\mathbf{x}, t) = \mathbf{F}(\mathbf{X}, t)\mathbf{f}_R(\mathbf{X}), \qquad \mathbf{x} = \chi_t(\mathbf{X}), \tag{6.8}$$

for all **X** and t. Consider the expressions

(i) $\underline{\chi(\mathbf{Y}, t) - \chi(\mathbf{X}, t) = \mathbf{F}(\mathbf{X}, t)(\mathbf{Y} - \mathbf{X})} + o(|\mathbf{Y} - \mathbf{X}|)$,

(ii) $d\mathbf{x} = \mathbf{F}(\mathbf{X}, t)d\mathbf{X}$,

discussed in the paragraphs containing (6.5) and (6.6). These expressions show that the mapping (6.8), at a material point **X** and time t, may be viewed as a linear transformation of an infinitesimal neighborhood of **X** in the reference body B to an infinitesimal neighborhood of **x** in the deformed body $\mathcal{B}_t$. In this regard, note that, given any $\epsilon > 0$, (6.8) is equivalent to the mapping

$$\epsilon\mathbf{f}(\mathbf{x}, t) = \mathbf{F}(\mathbf{X}, t)(\epsilon\mathbf{f}_R(\mathbf{X})); \tag{6.9}$$

thus, for $\epsilon > 0$ as small as we wish, (6.8) may be considered as describing the local deformation (6.9) when the neighborhood of **X** under consideration is magnified by a factor of ϵ^{-1} (Figure 6.1).

Based on the foregoing interpretation of (6.8), we refer to $\mathbf{f}_R(\mathbf{X})$ as an **infinitesimal undeformed fiber** and to $\mathbf{f}(\mathbf{x}, t) = \mathbf{F}(\mathbf{X}, t)\mathbf{f}_R(\mathbf{X})$, $\mathbf{x} = \chi_t(\mathbf{X})$ as the corresponding (infinitesimal) **deformed fiber**.[47] Since the undeformed fiber is material and independent of time, the deformed fiber $\mathbf{f}(\mathbf{x}, t)$ may be viewed as **embedded** in — and, hence, moving with — the deforming body $\mathcal{B}_t$, and we say that $\mathbf{f}(\mathbf{x}, t)$ convects with the body.

[47] The notion of a fiber is central to the analysis of deformation in Chapter I of Podio–Guidugli (2000).

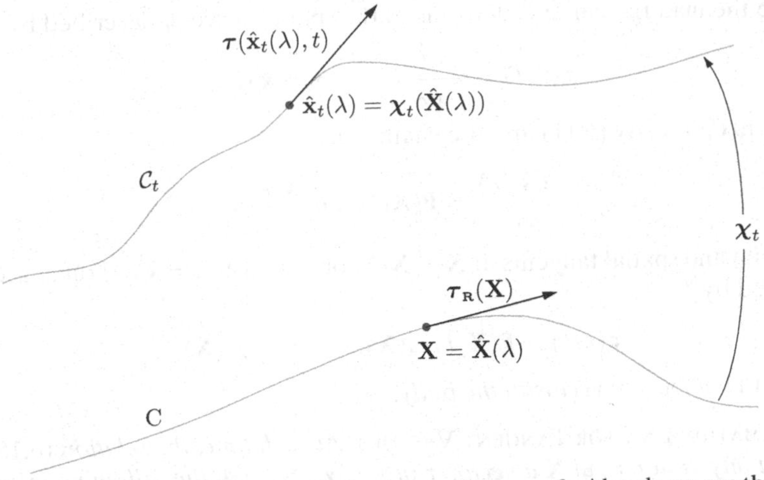

Figure 6.2. A material curve C and its image the spatial curve $\mathcal{C}_t$. Also shown are the tangents $\tau_R(\mathbf{X})$ and $\tau(\hat{\mathbf{x}}_t(\lambda), t)$ at a material point $\mathbf{X} = \hat{\mathbf{X}}(\lambda)$ and the spatial image $\hat{\mathbf{x}}_t(\lambda) = \chi_t(\hat{\mathbf{X}}(\lambda))$ of that point.

More generally, given an *arbitrary spatial vector field* $\mathbf{f}(\mathbf{x}, t)$, the phrases **f convects with the body** and **f is convecting** are meant to imply that there is a time-independent material vector field $\mathbf{f}_R(\mathbf{X})$ such that (6.8) is satisfied.[48]

6.2.2 Curves

Let C be a **material curve** described by the parametrization $\hat{\mathbf{X}}(\lambda)$, so that C is the curve traced out by the material point $\mathbf{X} = \hat{\mathbf{X}}(\lambda)$ as the parameter λ increases over its domain $\lambda_0 \le \lambda \le \lambda_1$. We restrict attention to curves that do not intersect themselves. Then, using the terminology of §5.3, the material curve C **deforms** to the **spatial curve** (Figure 6.2)

$$\mathcal{C}_t = \chi_t(\mathrm{C}). \tag{6.10}$$

As is natural, we assume that $\mathcal{C}_t$ is described by the time-dependent parametrization

$$\hat{\mathbf{x}}_t(\lambda) = \chi_t(\hat{\mathbf{X}}(\lambda)), \tag{6.11}$$

so that $\mathcal{C}_t$ is the curve traced out by the spatial point $\mathbf{x} = \hat{\mathbf{x}}_t(\lambda)$ as the parameter λ increases over its domain. In this case, $\mathcal{C}_t$ may be viewed as a curve **embedded** in — and hence moving with — the deforming body $\mathcal{B}_t$.

6.2.3 Tangent Vectors

Next, given a material point $\mathbf{X}$ on C, the vector defined by

$$\tau_R(\mathbf{X}) = \frac{d\hat{\mathbf{X}}(\lambda)}{d\lambda}, \qquad \mathbf{X} = \hat{\mathbf{X}}(\lambda), \tag{6.12}$$

is **tangent** to C at $\mathbf{X}$ (Figure 6.2).

[48] This definition should be viewed as *provisional*; in §13.1 we use the more specific phrase *convects as a tangent* to describe vector fields that convect via a relation of the form (6.8).

Since the material curve C deforms to the spatial curve $\mathcal{C}_t$ described by (6.11),

$$\boldsymbol{\tau}(\mathbf{x}, t) = \frac{\partial \hat{\mathbf{x}}_t(\lambda)}{\partial \lambda}, \qquad \mathbf{x} = \hat{\mathbf{x}}_t(\lambda), \tag{6.13}$$

is *tangent to* $\mathcal{C}_t$ *at* $\mathbf{x}$. By (6.11) and the chain-rule,

$$\frac{d\hat{\mathbf{x}}_t(\lambda)}{d\lambda} = \mathbf{F}(\hat{\mathbf{X}}(\lambda), t)\frac{d\hat{\mathbf{X}}(\lambda)}{d\lambda}; \tag{6.14}$$

the material and spatial tangents at $\mathbf{X} = \hat{\mathbf{X}}(\lambda)$ on C and at $\mathbf{x} = \hat{\mathbf{x}}_t(\alpha)$ on $\mathcal{C}_t$ are therefore related by[49]

$$\boldsymbol{\tau}(\mathbf{x}, t) = \mathbf{F}(\mathbf{X}, t)\boldsymbol{\tau}_{\mathrm{R}}(\mathbf{X}), \qquad \mathbf{x} = \boldsymbol{\chi}_t(\mathbf{X}), \tag{6.15}$$

and tangent vectors *convect with the body*.

TRANSFORMATION LAW FOR TANGENT VECTORS *At each time, the relation* (6.15) *associates with any vector* $\boldsymbol{\tau}_R$ *at* $\mathbf{X}$ *a vector* $\boldsymbol{\tau}$ *at* $\mathbf{x} = \boldsymbol{\chi}_t(\mathbf{X})$ *with the following property: if* $\boldsymbol{\tau}_R$ *is tangent to a material curve at* $\mathbf{X}$*, then* $\boldsymbol{\tau}$ *is tangent to the corresponding deformed curve through* $\mathbf{x}$.

6.2.4 Bases

Consider a fixed (not necessarily orthonormal) material basis (field)

$$\{\mathbf{m}_i(\mathbf{X})\} = \{\mathbf{m}_1(\mathbf{X}), \mathbf{m}_2(\mathbf{X}), \mathbf{m}_3(\mathbf{X})\}.$$

Since $\mathbf{F}$ is invertible, the vectors $\mathbf{s}_i(\mathbf{x}, t) = \mathbf{F}(\mathbf{X}, t)\mathbf{m}_i(\mathbf{X})$, $i = 1, 2, 3$, form a spatial basis

$$\{\mathbf{s}_i(\mathbf{x}, t)\} = \{\mathbf{F}(\mathbf{X}, t)\mathbf{m}_i(\mathbf{X})\} \tag{6.16}$$

at $\mathbf{x} = \boldsymbol{\chi}(\mathbf{X}, t)$. In view of the discussion in the paragraph containing (6.8), we may consider $\{\mathbf{s}_i\}$ as a basis that **convects with the body** or, equivalently, as a basis **embedded** in the deforming body $\mathcal{B}_t$.[50] In fact, at any time, the spatial field $\mathbf{s}_i$ (i fixed) represents a field of tangent vectors for the spatial coordinate curves that deform from material coordinate curves defined by the tangent field $\mathbf{m}_i$.

[49] A transformation of the form (6.15) of a material vector $\boldsymbol{\tau}_{\mathrm{R}}$ at $\mathbf{X}$ to a spatial vector $\boldsymbol{\tau}$ at $\mathbf{x}$ is often referred to as a *covariant transformation*.

[50] A detailed discussion of convecting bases is given in the paragraph containing (13.8).

7 Stretch, Strain, and Rotation

7.1 Stretch and Rotation Tensors. Strain

Consider the (pointwise) polar decomposition

$$\boxed{\mathbf{F} = \mathbf{R}\mathbf{U} = \mathbf{V}\mathbf{R}} \tag{7.1}$$

of the deformation gradient **F** into a **rotation R** and positive-definite symmetric tensors **U** and **V**;[51] using terminology motivated in §7.3, we refer to **U** as the **right stretch tensor** and to **V** as the **left stretch tensor**. The tensors **U** and **V**, which have the explicit representations

$$\mathbf{U} = \sqrt{\mathbf{F}^{\top}\mathbf{F}} \quad \text{and} \quad \mathbf{V} = \sqrt{\mathbf{F}\mathbf{F}^{\top}}, \tag{7.2}$$

are useful in theoretical discussions but are often problematic to apply because of the square root. For that reason, we introduce the **right and left Cauchy–Green (deformation) tensors C** and **B** defined by[52]

$$\boxed{\begin{aligned} \mathbf{C} &= \mathbf{U}^2 = \mathbf{F}^{\top}\mathbf{F}, \qquad & C_{ij} &= F_{ki}F_{kj} = \frac{\partial \chi_k}{\partial X_i}\frac{\partial \chi_k}{\partial X_j}, \\ \mathbf{B} &= \mathbf{V}^2 = \mathbf{F}\mathbf{F}^{\top}, \qquad & B_{ij} &= F_{ik}F_{jk} = \frac{\partial \chi_i}{\partial X_k}\frac{\partial \chi_j}{\partial X_k}. \end{aligned}} \tag{7.3}$$

Then, by (7.1),

$$\mathbf{V} = \mathbf{R}\mathbf{U}\mathbf{R}^{\top} \quad \text{and} \quad \mathbf{B} = \mathbf{R}\mathbf{C}\mathbf{R}^{\top}. \tag{7.4}$$

For future reference, we list the properties of the stretch and Cauchy–Green tensors:

$$\mathbf{U},\ \mathbf{V},\ \mathbf{C}, \ \text{and} \ \mathbf{B} \ \text{are symmetric and positive-definite.} \tag{7.5}$$

---

51 Cf. page 33.

52 Cf. Truesdell & Toupin (1960, §§31–33a) for a thorough discussion of strain measures.

A tensor useful in applications is the **Green–St. Venant strain tensor**

$$\mathbf{E} = \tfrac{1}{2}(\mathbf{F}^{\top}\mathbf{F} - \mathbf{1}), \tag{7.6}$$

$$= \tfrac{1}{2}(\mathbf{C} - \mathbf{1}), \tag{7.7}$$

$$= \tfrac{1}{2}(\mathbf{U}^2 - \mathbf{1}). \tag{7.8}$$

Note that $\mathbf{E}$ vanishes when $\mathbf{F}$ is a rotation, for then $\mathbf{F}^{\top}\mathbf{F} = \mathbf{1}$. This property of $\mathbf{E}$ is often adopted as one necessary for a tensor to qualify as a meaningful measure of strain.

Next, by (M1) and (M2) on page 65 and (7.3) and (7.6),

(M3) $\mathbf{U}$, $\mathbf{C}$, *and* $\mathbf{E}$ *map material vectors to material vectors;*
(M4) $\mathbf{V}$ *and* $\mathbf{B}$ *map spatial vectors to spatial vectors;*
(M5) $\mathbf{R}$ *maps material vectors to spatial vectors.*

To verify (M3), consider $\mathbf{F}^{\top}\mathbf{F}$: (M1) implies that $\mathbf{F}$ maps a material vector $\mathbf{f}_{\mathrm{R}}$ to a spatial vector $\mathbf{F}\mathbf{f}_{\mathrm{R}}$ and, by (M2), this spatial vector is mapped back to a material vector. Thus $\mathbf{C} = \mathbf{F}^{\top}\mathbf{F}$ maps material vectors to material vectors and since $\mathbf{U}\mathbf{U} = \mathbf{C}$, with $\mathbf{U}$ symmetric, $\mathbf{U}$ must also map material vectors to material vectors. A strictly analogous proof applies to (M4). Finally, by (7.1), $\mathbf{R} = \mathbf{F}\mathbf{U}^{-1}$, and, since $\mathbf{U}^{-1}$, like $\mathbf{U}$, carries material vectors to material vectors, while $\mathbf{F}$ carries material vectors to spatial vectors, (M5) must hold.

EXERCISE

1. Using the definitions (2.142) of the principal invariants of a tensor, show that

$$I_1(\mathbf{B}) = I_1(\mathbf{C}) = 2I_1(\mathbf{E}) + 3,$$

$$I_1(\mathbf{B}) = I_1(\mathbf{C}) = 4I_2(\mathbf{E}) + 4I_1(\mathbf{E}) + 3,$$

$$I_1(\mathbf{B}) = I_1(\mathbf{C}) = 8I_3(\mathbf{E}) + 4I_2(\mathbf{E}) + 2I_1(\mathbf{E}) + 1.$$

7.2 Fibers. Properties of the Tensors U and C

7.2.1 Infinitesimal Fibers

Consider now infinitesimal undeformed fibers $\mathbf{f}_{\mathrm{R}}$ and $\bar{\mathbf{f}}_{\mathrm{R}}$ and corresponding deformed fibers

$$\mathbf{f} = \mathbf{F}\mathbf{f}_{\mathrm{R}} \qquad \text{and} \qquad \bar{\mathbf{f}} = \mathbf{F}\bar{\mathbf{f}}_{\mathrm{R}}. \tag{7.9}$$

Then, since $\mathbf{R}\mathbf{R}^{\top} = \mathbf{1}$, $\mathbf{U} = \mathbf{U}^{\top}$, and $\mathbf{C} = \mathbf{U}^2$, it follows that

$$\mathbf{f}\cdot\bar{\mathbf{f}} = (\mathbf{R}\mathbf{U}\mathbf{f}_{\mathrm{R}})\cdot(\mathbf{R}\mathbf{U}\bar{\mathbf{f}}_{\mathrm{R}}),$$

$$= \mathbf{U}\mathbf{f}_{\mathrm{R}}\cdot\mathbf{U}\bar{\mathbf{f}}_{\mathrm{R}}, \tag{7.10}$$

$$= \mathbf{f}_{\mathrm{R}}\cdot\mathbf{U}^2\bar{\mathbf{f}}_{\mathrm{R}},$$

$$= \mathbf{f}_{\mathrm{R}}\cdot\mathbf{C}\bar{\mathbf{f}}_{\mathrm{R}}. \tag{7.11}$$

A consequence of (7.10) is that

$$|\mathbf{f}| = |\mathbf{U}\mathbf{f}_{\mathrm{R}}|; \tag{7.12}$$

the right stretch tensor $\mathbf{U}$ therefore characterizes the deformed length of infinitesimal fibers. We now determine the angle

$$\theta = \angle(\mathbf{f}_R, \bar{\mathbf{f}}_R)$$

between infinitesimal deformed fibers $\mathbf{f}$ and $\bar{\mathbf{f}}$.[53] By (7.10) and (7.12),

$$\frac{\mathbf{f}\cdot\bar{\mathbf{f}}}{|\mathbf{f}|\,|\bar{\mathbf{f}}|} = \frac{\mathbf{U}\mathbf{f}_R\cdot\mathbf{U}\bar{\mathbf{f}}_R}{|\mathbf{U}\mathbf{f}_R|\,|\mathbf{U}\bar{\mathbf{f}}_R|}.$$

Further, (1.2) represents a one-to-one mapping between θ and the right side of (1.2); thus,

$$\angle(\mathbf{f}, \bar{\mathbf{f}}) = \angle(\mathbf{U}\mathbf{f}_R, \mathbf{U}\bar{\mathbf{f}}_R), \tag{7.13}$$

and, hence, $\mathbf{U}$ characterizes the angle between infinitesimal deformed fibers.

7.2.2 Finite Fibers

Fix the time t and for convenience suppress it in what follows. The results for infinitesimal fibers have asymptotic conterparts. To derive these, we introduce material and spatial line segments[54]

$$\Delta\mathbf{X} = \mathbf{Y} - \mathbf{X} \qquad \text{and} \qquad \Delta\mathbf{x} = \chi(\mathbf{Y}) - \chi(\mathbf{X}),$$

with $|\Delta\mathbf{X}| > 0$, and rewrite (6.5) in the form

$$\Delta\mathbf{x} = \mathbf{F}(\mathbf{X})\Delta\mathbf{X} + \mathrm{o}(|\Delta\mathbf{X}|) \qquad \text{as} \qquad |\Delta\mathbf{X}| \to 0. \tag{7.14}$$

It is helpful to think of $\Delta\mathbf{X}$ as an **undeformed fiber** of (finite) **length** L and **direction** $\mathbf{e}$ at $\mathbf{X}$, so that

$$\Delta\mathbf{X} = L\mathbf{e}, \qquad |\mathbf{e}| = 1. \tag{7.15}$$

Then, by (7.14), the corresonding **deformed fiber** is given by

$$\Delta\mathbf{x} = L\mathbf{F}(\mathbf{X})\mathbf{e} + \mathrm{o}(L) \qquad \text{as} \qquad L \to 0,$$

and, dividing by L and using the fact that, by definition, $L^{-1}\mathrm{o}(L) \to 0$ as $L \to 0$, we find that

$$\lim_{L\to 0}\frac{\Delta\mathbf{x}}{L} = \mathbf{F}(\mathbf{X})\mathbf{e}. \tag{7.16}$$

We refer to this limit as a **stretch vector** because it represents the limiting value of the deformed fiber measured per unit length of the undeformed fiber.[55] What is most important, the scalar λ defined by

$$\lambda = \lim_{L\to 0}\frac{|\Delta\mathbf{x}|}{L} = |\mathbf{F}(\mathbf{X})\mathbf{e}| \tag{7.17}$$

represents the **stretch** at $\mathbf{x} = \chi(\mathbf{X})$ relative to the direction $\mathbf{e}$ at $\mathbf{X}$, as it measures the limiting length of the deformed fiber measured per unit length of an undeformed fiber in the direction $\mathbf{e}$. Bearing this in mind, we refer to fibers at $\mathbf{X}$ in the direction

---

[53] Cf. (1.1) and (1.2).

[54] The symbol Δ, which here denotes a difference, should not be confused with the symbol Δ, introduced on page 47, which denotes the Laplacian.

[55] Since $\Delta\mathbf{x} = \chi(\mathbf{X} + L\mathbf{e}) - \chi(\mathbf{X})$, (7.16) is the directional derivative of χ at $\mathbf{X}$ in the direction $\mathbf{e}$.

e as *stretched* or *unstretched* according to $\lambda \neq 1$ or $\lambda = 1$. Thus, appealing to (7.11) and (7.12) with $\mathbf{f}_R = \mathbf{e}$,

$$\begin{aligned} \lambda &= |\mathbf{U}(\mathbf{X})\mathbf{e}|, \\ \lambda^2 &= \mathbf{e} \cdot \mathbf{C}(\mathbf{X})\mathbf{e}. \end{aligned} \tag{7.18}$$

We therefore have the following result:

- *The stretch* λ *at* **X** *relative to any given material direction* **e** *is determined by the right stretch tensor* $\mathbf{U}(\mathbf{X})$ *through the relation* $\lambda = |\mathbf{U}(\mathbf{X})\mathbf{e}|$.

Consider now fibers $(\Delta\mathbf{X})_1$ and $(\Delta\mathbf{X})_2$ of the same length L but possibly different directions $\mathbf{e}_1$ and $\mathbf{e}_2$,

$$(\Delta\mathbf{X})_\alpha = L\mathbf{e}_\alpha, \qquad \alpha = 1, 2, \tag{7.19}$$

so that the corresponding stretch vectors are given by

$$\lim_{L\to 0} \frac{(\Delta\mathbf{x})_1}{L} = \mathbf{F}(\mathbf{X})\mathbf{e}_1 \quad \text{and} \quad \lim_{L\to 0} \frac{(\Delta\mathbf{x})_2}{L} = \mathbf{F}(\mathbf{X})\mathbf{e}_2. \tag{7.20}$$

Then, appealing to (7.10) and (7.11) with $\mathbf{f}_R = \mathbf{e}_1$ and with $\mathbf{f}_R = \mathbf{e}_2$,

$$\begin{aligned} \lim_{L\to 0} \left(\frac{(\Delta\mathbf{x})_1}{L} \cdot \frac{(\Delta\mathbf{x})_2}{L} \right) &= \lim_{L\to 0} \left(\frac{(\Delta\mathbf{x})_1}{L} \right) \cdot \lim_{L\to 0} \left(\frac{(\Delta\mathbf{x})_2}{L} \right) \\ &= \mathbf{U}(\mathbf{X})\mathbf{e}_1 \cdot \mathbf{U}(\mathbf{X})\mathbf{e}_2 && (7.21) \\ &= \mathbf{e}_1 \cdot \mathbf{C}(\mathbf{X})\mathbf{e}_2. && (7.22) \end{aligned}$$

An important consequence of (7.22) is that

- *the right Cauchy–Green tensor* $\mathbf{C}(\mathbf{X})$ *characterizes inner products of stretch vectors at* **x**; *in fact, in terms of the orthonormal basis* $\{\mathbf{e}_i\}$, *the component* $C_{ij}(\mathbf{X})$ *is the inner product of the stretch vectors at* **x** *relative to the directions* $\mathbf{e}_i$ *and* $\mathbf{e}_j$ *at* **X**. (We do not rule out the special case $\mathbf{e}_i = \mathbf{e}_j$.)

Consider again the fibers $(\Delta\mathbf{X})_1$ and $(\Delta\mathbf{X})_2$ defined in (7.19) and let θ_L denote the angle between the corresponding deformed fibers $(\Delta\mathbf{x})_1$ and $(\Delta\mathbf{x})_2$, so that, by (1.2),[56]

$$\begin{aligned} \theta_L &= \angle((\Delta\mathbf{x})_1, (\Delta\mathbf{x})_2) \\ &= \cos^{-1}\left(\frac{(\Delta\mathbf{x})_1 \cdot (\Delta\mathbf{x})_2}{|(\Delta\mathbf{x})_1||(\Delta\mathbf{x})_2|} \right). \end{aligned}$$

Then, by $(7.18)_1$, (7.19), and (7.21),

$$\begin{aligned} \lim_{L\to 0} \theta_L &= \lim_{L\to 0} \cos^{-1}\left(\frac{(\Delta\mathbf{x})_1 \cdot (\Delta\mathbf{x})_2}{L^2} \frac{L}{|(\Delta\mathbf{x})_1|} \frac{L}{|(\Delta\mathbf{x})_2|} \right) \\ &= \cos^{-1}\left(\frac{\mathbf{U}(\mathbf{X})\mathbf{e}_1 \cdot \mathbf{U}(\mathbf{X})\mathbf{e}_2}{|\mathbf{U}(\mathbf{X})\mathbf{e}_1||\mathbf{U}(\mathbf{X})\mathbf{e}_2|} \right) \\ &= \angle(\mathbf{U}(\mathbf{X})\mathbf{e}_1, \mathbf{U}(\mathbf{X})\mathbf{e}_2). \end{aligned} \tag{7.23}$$

[56] Here, consistent with (1.2), the domain of $\cos^{-1}$ is $[0, \pi]$.

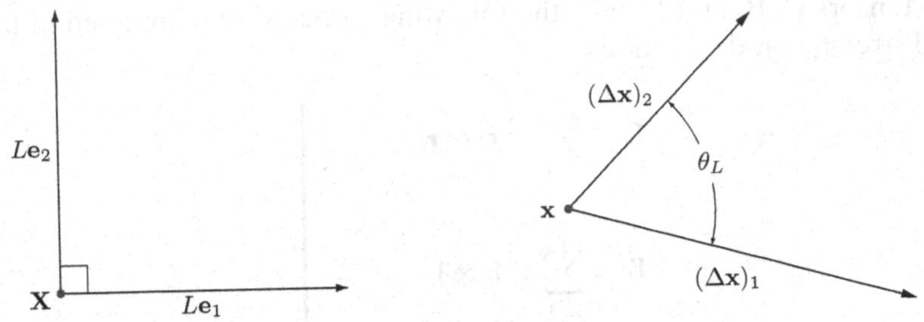

Figure 7.1. A pair of orthogonal undeformed fibers $L\mathbf{e}_1$ and $L\mathbf{e}_2$ and the angle between the corresponding deformed fibers $(\Delta\mathbf{x})_1$ and $(\Delta\mathbf{x})_2$.

We therefore have the following counterpart of (7.13):

- *Let* $(\Delta\mathbf{x})_1$ *and* $(\Delta\mathbf{x})_2$ *denote deformed fibers corresponding to fibers at* $\mathbf{X}$ *of finite length* L *in the directions* $\mathbf{e}_1$ *and* $\mathbf{e}_2$*. Then, as* $L \to 0$*, the angle between*

$$\frac{(\Delta\mathbf{x})_1}{L} \quad \text{and} \quad \frac{(\Delta\mathbf{x})_2}{L}$$

tends to the angle between $\mathbf{U}(\mathbf{X})\mathbf{e}_1$ *and* $\mathbf{U}(\mathbf{X})\mathbf{e}_2$ (Figure 7.1).

7.3 Principal Stretches and Principal Directions

Being symmetric and positive-definite, **U** and **V** admit spectral representations of the form

$$\mathbf{U} = \sum_{i=1}^{3} \lambda_i \mathbf{r}_i \otimes \mathbf{r}_i, \qquad \mathbf{V} = \sum_{i=1}^{3} \lambda_i \mathbf{l}_i \otimes \mathbf{l}_i, \tag{7.24}$$

where

(i) $\lambda_1 > 0, \lambda_2 > 0$, and $\lambda_3 > 0$, the **principal stretches**, are the eigenvalues of **U** and, by (7.4), also of **V**;

(ii) $\mathbf{r}_1, \mathbf{r}_2$, and $\mathbf{r}_3$, the **right principal directions**, are the eigenvectors of **U**

$$\mathbf{U}\mathbf{r}_i = \lambda_i \mathbf{r}_i \qquad \text{(no sum on } i\text{)}; \tag{7.25}$$

(iii) $\mathbf{l}_1, \mathbf{l}_2$, and $\mathbf{l}_3$, the **left principal directions**, are the eigenvectors of **V**[57]

$$\mathbf{V}\mathbf{l}_i = \lambda_i \mathbf{l}_i \qquad \text{(no sum on } i\text{)}. \tag{7.26}$$

Since **U** and **V** are related by $\mathbf{V} = \mathbf{R}\mathbf{U}\mathbf{R}^{\top}$, (7.24) yields

$$\sum_{i=1}^{3} \lambda_i \mathbf{R}\mathbf{r}_i \otimes \mathbf{R}\mathbf{r}_i = \sum_{i=1}^{3} \lambda_i \mathbf{l}_i \otimes \mathbf{l}_i$$

and therefore implies that the principal directions are related via

$$\mathbf{l}_i = \mathbf{R}\mathbf{r}_i, \qquad i = 1, 2, 3. \tag{7.27}$$

[57] Cf. (2.136) and (2.138).

The tensors **C**, **B**, and **E** have the following forms when expressed in terms of principal stretches and directions:

$$\left.\begin{aligned} \mathbf{C} &= \sum_{i=1}^{3} \lambda_i^2 \mathbf{r}_i \otimes \mathbf{r}_i, \\ \mathbf{B} &= \sum_{i=1}^{3} \lambda_i^2 \mathbf{l}_i \otimes \mathbf{l}_i, \\ \mathbf{E} &= \sum_{i=1}^{3} \tfrac{1}{2}(\lambda_i^2 - 1)\mathbf{r}_i \otimes \mathbf{r}_i. \end{aligned}\right\} \tag{7.28}$$

Further, since $\mathbf{F} = \mathbf{R}\mathbf{U}$,

$$\mathbf{F} = \sum_{i=1}^{3} \lambda_i \mathbf{l}_i \otimes \mathbf{r}_i. \tag{7.29}$$

Other strain measures found in the literature are the logarithmic strain tensors of HENCKY:

$$\begin{aligned} \ln \mathbf{U} &= \sum_{i=1}^{3} (\ln \lambda_i)\mathbf{r}_i \otimes \mathbf{r}_i, \\ \ln \mathbf{V} &= \sum_{i=1}^{3} (\ln \lambda_i)\mathbf{l}_i \otimes \mathbf{l}_i. \end{aligned} \tag{7.30}$$

EXERCISES

1. Express $\mathbf{F}^{-1}$ and $\mathbf{F}^{\top}$ in terms of principal stretches.

2. Show that

$$\mathbf{R}(\ln \mathbf{U})\mathbf{R}^{\top} = \ln \mathbf{V}. \tag{7.31}$$

8 Deformation of Volume and Area

Fix the time t and suppress it throughout this section.

8.1 Deformation of Normals

Choose a material point $\mathbf{X}$ and let $\mathbf{x} = \boldsymbol{\chi}(\mathbf{X})$. Further, consider a material surface S with $\mathbf{X}$ on S and write $\mathcal{S} = \boldsymbol{\chi}(\mathrm{S})$ for the deformed surface, so that $\mathbf{x}$ lies on $\mathcal{S}$. Let $\mathbf{n}_{\mathrm{R}}$, assumed nonzero, be normal to S at $\mathbf{X}$, so that

$$\mathbf{n}_{\mathrm{R}} \cdot \mathbf{t}_{\mathrm{R}} = 0 \text{ for every vector } \mathbf{t}_{\mathrm{R}} \text{ tangent to S at } \mathbf{X}. \tag{8.1}$$

We seek a vector $\mathbf{n}$ normal to $\mathcal{S}$ at $\mathbf{x}$; that is, we seek a nonzero vector $\mathbf{n}$ such that

$$\mathbf{n} \cdot \mathbf{t} = 0 \text{ for every vector } \mathbf{t} \text{ tangent to } \mathcal{S} \text{ at } \mathbf{x}. \tag{8.2}$$

To find such a vector $\mathbf{n}$ (which can be unique only up to a nonzero multiplicative scalar), choose an arbitrary vector $\mathbf{t}$ tangent to $\mathcal{S}$ at $\mathbf{x}$. There is then a curve $\mathcal{C}$ that lies on $\mathcal{S}$, passes through $\mathbf{x}$, and has $\mathbf{t}$ as its tangent at $\mathbf{x}$. Let $\hat{\mathbf{x}}(\lambda)$ be the parametrization of $\mathcal{C}$. Consider the reference map $\mathbf{X} = \boldsymbol{\chi}^{-1}(\mathbf{x})$ defined in (5.5) (bearing in mind that the time is fixed). In view of (5.6),

$$\hat{\mathbf{X}}(\lambda) \stackrel{\text{def}}{=} \boldsymbol{\chi}^{-1}(\hat{\mathbf{x}}(\lambda))$$

is a parametrization of a curve C on S passing through $\mathbf{X}$, and this curve must be consistent with (6.11). Hence $\mathbf{t}$ and the tangent $\mathbf{t}_{\mathrm{R}}$ to C at $\mathbf{X}$ as defined by (6.12) must be consistent with (6.15), so that

$$\mathbf{t}_{\mathrm{R}} = \mathbf{F}^{-1}(\mathbf{X})\mathbf{t}. \tag{8.3}$$

Further, since C lies on S, its tangent $\mathbf{t}_{\mathrm{R}}$ at $\mathbf{X}$ must be tangent to S at $\mathbf{X}$, so that $\mathbf{n}_{\mathrm{R}} \cdot \mathbf{t}_{\mathrm{R}} = 0$ and, by (8.3),

$$\begin{aligned} 0 &= \mathbf{n}_{\mathrm{R}} \cdot (\mathbf{F}^{-1}(\mathbf{X})\mathbf{t}) \\ &= (\mathbf{F}^{-\top}(\mathbf{X})\mathbf{n}_{\mathrm{R}}) \cdot \mathbf{t}; \end{aligned}$$

thus, since $\mathbf{t}$ represents an arbitrary vector tangent to $\mathcal{S}$ at $\mathbf{x}$,

$$\mathbf{n} = \mathbf{F}^{-\top}(\mathbf{X})\mathbf{n}_{\mathrm{R}} \tag{8.4}$$

is normal to S *at* $\mathbf{x}$.[58]

[58] A transformation of the form (8.4) of a material vector $\mathbf{n}_{\mathrm{R}}$ at $\mathbf{X}$ to a spatial vector $\mathbf{n}$ at $\mathbf{x}$ is often referred to as a *contravariant transformation*. In the discussion leading to (#) on page 77, we explain the sense in which (8.4) preserves orientation.

TRANSFORMATION LAW FOR NORMAL VECTORS *At each fixed time the relation* (8.4) *associates with any vector* $\mathbf{n}_R$ *at* $\mathbf{X}$ *a vector* $\mathbf{n}$ *at* $\mathbf{x}$ *with the following property: If* $\mathbf{n}_R$ *is normal to a material surface at* $\mathbf{X}$*, then* $\mathbf{n}$ *is normal to the corresponding deformed surface through* $\mathbf{x}$.

8.2 Deformation of Volume

Consider the (generally nonorthonormal) basis $\{\mathbf{F}(\mathbf{X})\mathbf{e}_i\}$ defined in the paragraph containing (6.16); as noted there, the basis $\{\mathbf{F}(\mathbf{X})\mathbf{e}_i\}$ may be viewed as one that convects with the body. By (2.79) and (6.2), the volume spanned by the basis vectors is given by

$$\begin{aligned}(\mathbf{F}\mathbf{e}_1 \times \mathbf{F}\mathbf{e}_2)\cdot\mathbf{F}\mathbf{e}_3 &= \det\mathbf{F}\\ &= J > 0,\end{aligned} \tag{8.5}$$

where for convenience we have written $\mathbf{F}$ for $\mathbf{F}(\mathbf{X})$.

To better understand the geometrical meaning of the field J, write

$$\boldsymbol{\delta}_i(\ell) = \boldsymbol{\chi}(\mathbf{X}+\ell\mathbf{e}_i) - \boldsymbol{\chi}(\mathbf{X}) \tag{8.6}$$

for the deformed fiber at $\mathbf{x} = \boldsymbol{\chi}(\mathbf{X})$ that has deformed from the fiber of length ℓ in the direction $\mathbf{e}_i$ at $\mathbf{X}$. Then,

$$\Delta v_{\mathrm{R}}(\ell) \stackrel{\text{def}}{=} \ell^3(\mathbf{e}_1\times\mathbf{e}_2)\cdot\mathbf{e}_3 = \ell^3 \qquad \text{and} \qquad \Delta v(\ell) \stackrel{\text{def}}{=} (\boldsymbol{\delta}_1(\ell)\times\boldsymbol{\delta}_2(\ell))\cdot\boldsymbol{\delta}_3(\ell) \tag{8.7}$$

represent the respective volumes spanned by the undeformed and deformed fibers (Figure 8.1). In view of the Taylor expansion (6.5),

$$\boldsymbol{\delta}_i(\ell) = \ell\mathbf{F}\mathbf{e}_i + \mathrm{o}(\ell), \tag{8.8}$$

where, by (3.8), $\mathrm{o}(\ell)$ represents a term that *goes to zero faster than* ℓ:

$$\ell^{-1}\mathrm{o}(\ell) \to 0 \qquad \text{as} \qquad \ell \to 0.$$

Thus, by (8.5) and $(8.7)_2$, as $\ell \to 0$,

$$\begin{aligned}\Delta v(\ell) &= \ell^3([\mathbf{F}\mathbf{e}_1 + \mathrm{o}(1)]\times[\mathbf{F}\mathbf{e}_2+\mathrm{o}(1)])\cdot[\mathbf{F}\mathbf{e}_3+\mathrm{o}(1)]\\ &= \ell^3[(\mathbf{F}\mathbf{e}_1\times\mathbf{F}\mathbf{e}_2)\cdot\mathbf{F}\mathbf{e}_3] + \mathrm{o}(\ell^3)\\ &= J\ell^3 + \mathrm{o}(\ell^3),\end{aligned}$$

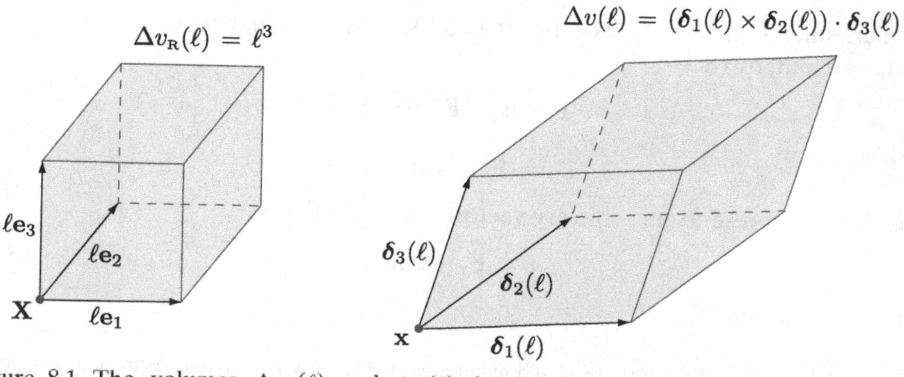

Figure 8.1. The volumes $\Delta v_{\mathrm{R}}(\ell)$ and $\Delta v(\ell)$ formed by the triples $\{\ell\mathbf{e}_1, \ell\mathbf{e}_2, \ell\mathbf{e}_3\}$ and $\{\boldsymbol{\delta}_1(\ell), \boldsymbol{\delta}_2(\ell), \boldsymbol{\delta}_3(\ell)\}$.

where $o(\ell^3)$ represents a term that goes to zero faster than ℓ^3. We therefore have the important estimate

$$\Delta v = J\,\Delta v_R + o(\Delta v_R) \qquad \text{as} \qquad \Delta v_R \to 0. \tag{8.9}$$

Thus

- *given an undeformed fiber cube of volume Δv_R, the corresponding deformed fiber parallelepiped has volume $J\,\Delta v_R$ to within an error of $o(\Delta v_R)$.*

8.3 Deformation of Area

Consider next a plane Π_R at a material point $\mathbf{X}$, let $\mathbf{n}_R$, with

$$|\mathbf{n}_R| = 1, \tag{8.10}$$

denote a unit normal to Π_R, and, without loss in generality, assume that the basis $\{\mathbf{e}_i\}$ is chosen such that

$$\mathbf{e}_3 = \mathbf{n}_R.$$

Then, trivially, the volume spanned by the vectors $\mathbf{F}\mathbf{e}_1$, $\mathbf{F}\mathbf{e}_2$, and $\mathbf{F}\mathbf{n}_R$ is

$$(\mathbf{F}\mathbf{e}_1 \times \mathbf{F}\mathbf{e}_2) \cdot \mathbf{F}\mathbf{n}_R = J. \tag{8.11}$$

The vector $\mathbf{F}\mathbf{e}_1 \times \mathbf{F}\mathbf{e}_2$ is normal to the deformed plane at $\mathbf{x}$ and, by (2.89) and (8.4),

$$\begin{aligned}\mathbf{F}\mathbf{e}_1 \times \mathbf{F}\mathbf{e}_2 &= J\mathbf{F}^{-\top}\underbrace{(\mathbf{e}_1 \times \mathbf{e}_2)}_{\mathbf{e}_3=\mathbf{n}_R}\\ &= J\,\underbrace{\mathbf{F}^{-\top}\mathbf{n}_R}_{\mathbf{n}}\\ &= J\mathbf{n}.\end{aligned} \tag{8.12}$$

Then, since $\mathbf{e}_1$ and $\mathbf{e}_2$ are tangent to Π_R at $\mathbf{X}$, the vectors $\mathbf{F}\mathbf{e}_1$ and $\mathbf{F}\mathbf{e}_2$ are tangent to the plane Π with normal $\mathbf{n}$ at $\mathbf{x}$. Further, $\{\mathbf{F}\mathbf{e}_1, \mathbf{F}\mathbf{e}_2, \mathbf{n}\}$ is a basis at $\mathbf{x}$, because

$$(\mathbf{F}\mathbf{e}_1 \times \mathbf{F}\mathbf{e}_2) \cdot \mathbf{n} = J\,|\mathbf{n}|^2 > 0.$$

Thus, since $(\mathbf{e}_1 \times \mathbf{e}_2) \cdot \mathbf{e}_3 > 0$, this basis and $\{\mathbf{F}\mathbf{e}_1, \mathbf{F}\mathbf{e}_2, \mathbf{n}\}$ have the same orientation. This is the sense in which

(#) the transformation $\mathbf{n} = \mathbf{F}^{-\top}(\mathbf{X})\mathbf{n}_R$ (as a transformation of normals) preserves orientation.

Next, by (2.91), the *area* spanned by the vectors $\mathbf{F}(\mathbf{X})\mathbf{e}_1$ and $\mathbf{F}(\mathbf{X})\mathbf{e}_2$ at $\mathbf{x}$ is

$$\jmath(\mathbf{X}) \stackrel{\text{def}}{=} |\mathbf{F}(\mathbf{X})\mathbf{e}_1 \times \mathbf{F}(\mathbf{X})\mathbf{e}_2|, \tag{8.13}$$

the *areal Jacobian*. But, by (8.12),

$$\begin{aligned}|\mathbf{F}\mathbf{e}_1 \times \mathbf{F}\mathbf{e}_2| &= J\,|\mathbf{n}|\\ &= J\,|\mathbf{F}^{-\top}\mathbf{n}_R|,\end{aligned}$$

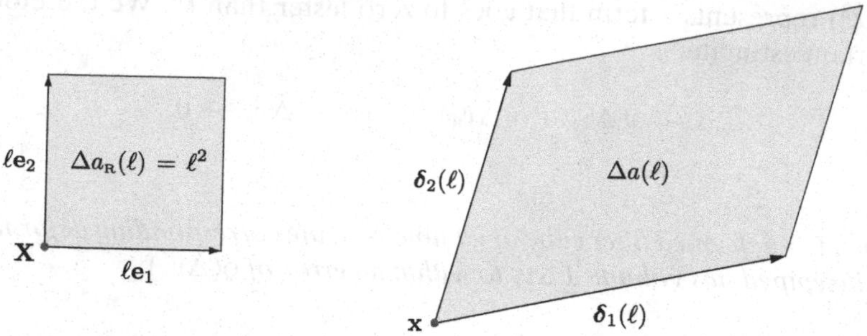

Figure 8.2. The areas $\Delta a_{\rm R}(\ell)$ and $\Delta a(\ell)$ spanned, respectively, by undeformed fibers $\ell\mathbf{e}_1$ and $\ell\mathbf{e}_2$ and deformed fibers $\boldsymbol{\delta}_1(\ell)$ and $\boldsymbol{\delta}_2(\ell)$.

and we arrive at an important result,

$$\jmath = J|\mathbf{F}^{-\top}\mathbf{n}_{\rm R}|, \tag{8.14}$$

relating the areal and volumetric Jacobians $\jmath$ and J. Note that

$$\jmath \text{ depends on the choice of normal vector } \mathbf{n}_{\rm R}. \tag{8.15}$$

Further, since $\mathbf{n} = \mathbf{F}^{-\top}\mathbf{n}_{\rm R}$, (8.14) implies that

$$\begin{aligned}\jmath\frac{\mathbf{n}}{|\mathbf{n}|} &= J|\mathbf{F}^{-\top}\mathbf{n}_{\rm R}|\frac{\mathbf{n}}{|\mathbf{n}|}\\ &= J\mathbf{F}^{-\top}\mathbf{n}_{\rm R}.\end{aligned} \tag{8.16}$$

Thus we have the important relation[59]

$$\jmath\frac{\mathbf{n}}{|\mathbf{n}|} = \mathbf{F}^{\rm C}\mathbf{n}_{\rm R}, \tag{8.17}$$

in which

$$\mathbf{F}^{\rm C} = (\det\mathbf{F})\mathbf{F}^{-\top} \tag{8.18}$$

is the **cofactor** of $\mathbf{F}$.[60]

Finally, mimicking the steps that led from (8.7) to (8.9), we note that

$$\Delta a_{\rm R}(\ell) \overset{\rm def}{=} \ell^2|\mathbf{e}_1\times\mathbf{e}_2| = \ell^2 \qquad \text{and} \qquad \Delta a(\ell) \overset{\rm def}{=} |\boldsymbol{\delta}_1(\ell)\times\boldsymbol{\delta}_2(\ell)|, \tag{8.19}$$

respectively, represent areas spanned by undeformed fibers at $\mathbf{X}$ of length ℓ in the directions $\mathbf{e}_1$ and $\mathbf{e}_2$ and the corresponding deformed fibers at $\mathbf{x}$ (Figure 8.2). Then, by (8.13), as $\ell \to 0$,

$$\begin{aligned}\Delta a(\ell) &= \ell^2\big|[\mathbf{F}\mathbf{e}_1 + {\rm o}(1)]\times[\mathbf{F}\mathbf{e}_2 + {\rm o}(1)]\big|\\ &= \ell^2|\mathbf{F}\mathbf{e}_1\times\mathbf{F}\mathbf{e}_2| + {\rm o}(\ell^2)\\ &= \jmath\ell^2 + {\rm o}(\ell^2),\end{aligned}$$

and we have the estimate

$$\Delta a = \jmath\Delta a_{\rm R} + {\rm o}(\Delta a_{\rm R}) \qquad \text{as} \qquad \Delta a_{\rm R} \to 0. \tag{8.20}$$

[59] Bear in mind (8.10).
[60] Cf. (2.90).

Thus,

- *given an undeformed fiber square of area Δa_R, the corresponding deformed fiber parallelogram has area $J\,\Delta a_R$ to within an error of $o(\Delta a_R)$.*

Further, (8.17) and (8.20) yield an important estimate for the corresponding undeformed and deformed *vector* areas

$$\Delta a \frac{\mathbf{n}}{|\mathbf{n}|} = \Delta a_{\mathrm{R}} \mathbf{F}^{\mathrm{c}} \mathbf{n}_{\mathrm{R}} + o(\Delta a_{\mathrm{R}}). \tag{8.21}$$

9 Material and Spatial Descriptions of Fields

Consider a motion χ. Since the mapping $\mathbf{x} = \chi(\mathbf{X}, t)$ is invertible in $\mathbf{X}$ for fixed t, it has an inverse $\mathbf{X} = \chi^{-1}(\mathbf{x}, t)$, the *reference map* defined in (5.5) and (5.6). χ^{-1} associates with each time t and spatial point $\mathbf{x}$ in $\mathcal{B}_t$, a material point $\mathbf{X} = \chi^{-1}(\mathbf{x}, t)$ in B.

Using the reference map, we can describe the velocity $\dot{\chi}(\mathbf{X}, t)$ as a function $\mathbf{v}(\mathbf{x}, t)$ of the spatial point $\mathbf{x}$ and t:

$$\mathbf{v}(\mathbf{x}, t) = \dot{\chi}(\chi^{-1}(\mathbf{x}, t), t) \qquad \text{or equivalently} \qquad \dot{\chi}(\mathbf{X}, t) = \mathbf{v}(\chi(\mathbf{X}, t), t). \tag{9.1}$$

The field $\mathbf{v}$ represents the **spatial description of the velocity**; $\mathbf{v}(\mathbf{x}, t)$ is the velocity of the material point that at time t occupies the spatial point $\mathbf{x}$.

More generally, let φ denote a scalar, vector, or tensor field defined on the body for all time. We generally consider φ to be a function $\varphi(\mathbf{X}, t)$ of the material point $\mathbf{X}$ and the time t; this is called the **material description** of φ. But, as with the velocity, we may also consider φ to be a function $\phi(\mathbf{x}, t)$ of the spatial point $\mathbf{x}$ and t; this is called the **spatial description** and is related to the material description through

$$\phi(\mathbf{x}, t) = \varphi(\chi^{-1}(\mathbf{x}, t), t).$$

Similarly, a field $\phi(\mathbf{x}, t)$ described spatially may be considered as a function $\varphi(\mathbf{X}, t)$ of the material point $\mathbf{X}$ and t; this is called the material description and is given by

$$\varphi(\mathbf{X}, t) = \phi(\chi(\mathbf{X}, t), t).$$

- *When there is no danger of confusion we use the same symbol for both the material and spatial descriptions.*

9.1 Gradient, Divergence, and Curl

We write

$$\nabla, \quad \text{Div}, \quad \text{and} \quad \text{Curl},$$

respectively, for the **material gradient**, the **material divergence**, and the **material curl**; that is, the gradient, divergence, and curl with respect to the material point $\mathbf{X}$ in the reference body; for example, for $\mathbf{h}$ a vector field

$$(\nabla\mathbf{h})_{ij} = \frac{\partial h_i}{\partial X_j}, \qquad \text{Div}\,\mathbf{h} = \frac{\partial h_i}{\partial X_i}, \qquad \text{and} \qquad (\text{Curl}\,\mathbf{h})_i = \epsilon_{ijk}\frac{\partial h_k}{\partial X_j}.$$

Analogously,

$$\text{grad}, \quad \text{div}, \quad \text{and} \quad \text{curl},$$

the **spatial gradient, spatial divergence**, and **spatial curl**, are the gradient, divergence, and curl with respect to the spatial point $\mathbf{x} = \boldsymbol{\chi}(\mathbf{X}, t)$ in the *deformed body*, so that, for $\mathbf{g}$ a vector field,

$$(\text{grad}\,\mathbf{g})_{ij} = \frac{\partial g_i}{\partial x_j}, \qquad \text{div}\,\mathbf{g} = \frac{\partial g_i}{\partial x_i}, \qquad \text{and} \qquad (\text{curl}\,\mathbf{g})_i = \epsilon_{ijk}\frac{\partial g_k}{\partial x_j}.$$

By the chain-rule, for φ a scalar field and $\mathbf{g}$ a vector field,

$$\frac{\partial \varphi}{\partial X_i} = \frac{\partial \varphi}{\partial x_j}\frac{\partial \chi_j}{\partial X_i} = F_{ji}\frac{\partial \varphi}{\partial x_j} \qquad \text{and} \qquad \frac{\partial g_i}{\partial X_j} = \frac{\partial g_i}{\partial x_k}\frac{\partial \chi_k}{\partial X_j} = \frac{\partial g_i}{\partial x_k}F_{kj}$$

or, equivalently,

$$\nabla \varphi = \mathbf{F}^{\top}\text{grad}\,\varphi \qquad \text{and} \qquad \nabla \mathbf{g} = (\text{grad}\,\mathbf{g})\mathbf{F}. \tag{9.2}$$

EXERCISE

1. Show that

$$\text{Div}\,\mathbf{g} = \mathbf{F}^{\top} : \text{grad}\,\mathbf{g} \qquad \text{and} \qquad \text{div}\,\mathbf{g} = \mathbf{F}^{-\top} : \nabla \mathbf{g}.$$

9.2 Material and Spatial Time Derivatives

Given any field φ, we write

$$\dot{\varphi}(\mathbf{X}, t) = \frac{\partial \varphi(\mathbf{X}, t)}{\partial t} \quad \text{(holding } \mathbf{X} \text{ fixed)} \tag{9.3}$$

for its **material time-derivative** and

$$\varphi'(\mathbf{x}, t) = \frac{\partial \varphi(\mathbf{x}, t)}{\partial t} \quad \text{(holding } \mathbf{x} \text{ fixed)} \tag{9.4}$$

for its **spatial time-derivative.**

The existence of two time derivatives begs the question as to their relationship. To answer this question, let φ be a spatial scalar field. Since φ is described spatially, to compute its material time-derivative we must first convert its description to material, take its time derivative, and then convert the result back to spatial:[61]

$$\dot{\varphi}(\mathbf{x}, t) = \left[\frac{\partial}{\partial t}\bigg|_{\mathbf{X}} \varphi(\boldsymbol{\chi}(\mathbf{X}, t), t)\right]_{\mathbf{X}=\boldsymbol{\chi}^{-1}(\mathbf{x},t)}. \tag{9.5}$$

Thus, using the chain-rule,

$$\dot{\varphi}(\mathbf{x}, t) = [\text{grad}\,\varphi(\mathbf{x}, t) \cdot \dot{\boldsymbol{\chi}}(\mathbf{X}, t)]_{\mathbf{X}=\boldsymbol{\chi}^{-1}(\mathbf{x},t)} + \varphi'(\mathbf{x}, t),$$

and, by $(9.1)_1$, $\dot{\boldsymbol{\chi}}(\mathbf{X}, t)$ evaluated at $\mathbf{X} = \boldsymbol{\chi}^{-1}(\mathbf{x}, t)$ is $\mathbf{v}(\mathbf{x}, t)$, the spatial description of the velocity; hence,

$$\dot{\varphi}(\mathbf{x}, t) = \varphi'(\mathbf{x}, t) + \text{grad}\,\varphi(\mathbf{x}, t) \cdot \mathbf{v}(\mathbf{x}, t),$$

61 When dealing with composite functions of the form $\varphi(\boldsymbol{\chi}(\mathbf{X}, t), t)$, the notation

$$\frac{\partial}{\partial t}\bigg|_{\mathbf{X}}$$

is used to denote the partial derivative with respect to t holding $\mathbf{X}$ fixed.

Similarly, for $\mathbf{g}$ a spatial vector field,

$$\begin{aligned}\dot{\mathbf{g}}(\mathbf{x}, t) &= \left[\frac{\partial}{\partial t}\bigg|_{\mathbf{X}} \mathbf{g}(\chi(\mathbf{X}, t), t)\right]_{\mathbf{X}=\chi^{-1}(\mathbf{x},t)} \\ &= [\operatorname{grad}\mathbf{g}(\mathbf{x}, t)]\dot{\chi}(\mathbf{X}, t)\big|_{\mathbf{X}=\chi^{-1}(\mathbf{x},t)} + \mathbf{g}'(\mathbf{x}, t) \\ &= [\operatorname{grad}\mathbf{g}(\mathbf{x}, t)]\mathbf{v}(\mathbf{x}, t) + \mathbf{g}'(\mathbf{x}, t).\end{aligned}$$

Note that, by (9.1),

$$\begin{aligned}\dot{\mathbf{v}}(\mathbf{x}, t) &= \left[\frac{\partial}{\partial t}\bigg|_{\mathbf{X}} \mathbf{v}(\chi(\mathbf{X}, t), t)\right]_{\mathbf{X}=\chi^{-1}(\mathbf{x},t)} \\ &= \frac{\partial^2 \chi(\mathbf{X}, t)}{\partial t^2}\bigg|_{\mathbf{X}=\chi^{-1}(\mathbf{x},t)} \\ &= \ddot{\chi}(\mathbf{X}, t)\big|_{\mathbf{X}=\chi^{-1}(\mathbf{x},t)};\end{aligned} \tag{9.6}$$

thus, $\dot{\mathbf{v}}(\mathbf{x}, t)$ *represents the spatial description of the acceleration.* We therefore have the

TIME-DERIVATIVE IDENTITIES *The material and spatial time derivatives are related through*

$$\begin{aligned}\dot{\varphi} &= \varphi' + \mathbf{v}\cdot\operatorname{grad}\varphi, \\ \dot{\mathbf{g}} &= \mathbf{g}' + (\operatorname{grad}\mathbf{g})\mathbf{v}.\end{aligned} \tag{9.7}$$

In particular, the acceleration is given by

$$\dot{\mathbf{v}} = \mathbf{v}' + (\operatorname{grad}\mathbf{v})\mathbf{v}. \tag{9.8}$$

Sample problem Find the material time-derivative of the spatial position vector

$$\mathbf{r}(\mathbf{x}) = \mathbf{x} - \mathbf{o}.$$

Solution We could simply use $(9.7)_2$: since $\mathbf{r}' = \mathbf{0}$ and $\operatorname{grad}\mathbf{r} = \mathbf{1}$,

$$\dot{\mathbf{r}} = \mathbf{v}. \tag{9.9}$$

A more instructive procedure would be to note that the material description of $\mathbf{r}(\mathbf{x})$ is the field $\chi(\mathbf{X}, t) - \mathbf{o}$ whose material time-derivative $\dot{\chi}(\mathbf{X}, t)$ has spatial description $\mathbf{v}(\mathbf{x}, t)$.

9.3 Velocity Gradient

The spatial tensor field

$$\mathbf{L} = \operatorname{grad}\mathbf{v} \tag{9.10}$$

is called the **velocity gradient**. By $(9.2)_2$,

$$\begin{aligned}\dot{\mathbf{F}}(\mathbf{X}, t) &= \frac{\partial}{\partial t}\nabla\chi(\mathbf{X}, t) \\ &= \nabla\dot{\chi}(\mathbf{X}, t) \\ &= \operatorname{grad}\mathbf{v}(\mathbf{x}, t)\big|_{\mathbf{x}=\chi(\mathbf{X},t)}\mathbf{F}(\mathbf{X}, t),\end{aligned}$$

so that, omitting arguments,

$$\dot{\mathbf{F}} = \mathbf{L}\mathbf{F}; \tag{9.11}$$

(9.11) represents a (tensorial) evolution equation for the deformation gradient **F**, granted a knowledge of **L**. Since **F** is invertible, (9.11) is easily solved for **L**; the result, which is basic to much of what follows, is

$$\boxed{\mathbf{L} = \dot{\mathbf{F}}\mathbf{F}^{-1}.} \tag{9.12}$$

Thus, by (M1) and (M2) on page 65,

$$\mathbf{L} \textit{ maps spatial vectors to spatial vectors.} \tag{9.13}$$

Next, as a consequence of (9.10),

$$\begin{aligned} \operatorname{tr}\mathbf{L} &= \operatorname{tr}(\operatorname{grad}\mathbf{v}) \\ &= \operatorname{div}\mathbf{v}; \end{aligned} \tag{9.14}$$

on the other hand, by $(3.3)_1$,

$$\dot{J} = J\operatorname{tr}\mathbf{L}; \tag{9.15}$$

therefore,

$$\dot{J} = J\operatorname{div}\mathbf{v}, \tag{9.16}$$

which is an important identity expressing transport of volume.

Next, by (9.11),

$$\dot{\mathbf{F}}^{\top} = \mathbf{F}^{\top}\mathbf{L}^{\top}, \tag{9.17}$$

which represents an evolution equation for the transpose $\mathbf{F}^{\top}$ of the deformation gradient. Further, since $\mathbf{F}\mathbf{F}^{-1} = \mathbf{1}$, it follows that $\dot{\mathbf{F}}\mathbf{F}^{-1} = -\mathbf{F}\overline{\mathbf{F}^{-1}}^{\centerdot}$, and, hence, that

$$\overline{\mathbf{F}^{-1}}^{\centerdot} = -\mathbf{F}^{-1}\dot{\mathbf{F}}\mathbf{F}^{-1}. \tag{9.18}$$

Thus, since $\dot{\mathbf{F}}\mathbf{F}^{-1} = \mathbf{L}$,

$$\overline{\mathbf{F}^{-1}}^{\centerdot} = -\mathbf{F}^{-1}\mathbf{L}, \tag{9.19}$$

which represents an evolution equation for the inverse deformation gradient $\mathbf{F}^{-1}$. If we take the transpose of (9.19), we arrive at an evolution equation for $\mathbf{F}^{-\top}$:

$$\overline{\mathbf{F}^{-\top}}^{\centerdot} = -\mathbf{L}^{\top}\mathbf{F}^{-\top}. \tag{9.20}$$

Thus, summarizing, we have the following **evolution equations associated with the deformation gradient**:

$$\begin{aligned} \dot{\mathbf{F}} &= \mathbf{L}\mathbf{F}, \\ \dot{\mathbf{F}}^{\top} &= \mathbf{F}^{\top}\mathbf{L}^{\top}, \\ \overline{\mathbf{F}^{-1}}^{\centerdot} &= -\mathbf{F}^{-1}\mathbf{L}, \\ \overline{\mathbf{F}^{-\top}}^{\centerdot} &= -\mathbf{L}^{\top}\mathbf{F}^{-\top}. \end{aligned} \tag{9.21}$$

Sample problem Show that

$$\operatorname{grad}\dot{\mathbf{v}} = \ddot{\mathbf{F}}\mathbf{F}^{-1}. \tag{9.22}$$

Solution By $(9.2)_2$,

$$\begin{aligned} \operatorname{grad}\dot{\mathbf{v}}(\mathbf{x},t)\big|_{\mathbf{x}=\boldsymbol{\chi}(\mathbf{X},t)} &= \nabla\ddot{\boldsymbol{\chi}}(\mathbf{X},t)\mathbf{F}^{-1}(\mathbf{X},t) \\ &= \ddot{\mathbf{F}}(\mathbf{X},t)\mathbf{F}^{-1}(\mathbf{X},t). \end{aligned}$$

EXERCISE

1. Show that

$$\operatorname{grad} \overset{(n)}{\mathbf{v}} = \nabla \overset{(n+1)}{\boldsymbol{\chi}} \mathbf{F}^{-1},$$

where $\overset{(n)}{\mathbf{v}}$ denotes the n-th material time-derivative of $\mathbf{v}$.

9.4 Commutator Identities

Consider the identity $(9.7)_1$ relating the material and spatial time derivatives of a spatial scalar field φ. Bearing in mind that the spatial time-derivative and the spatial gradient commute, making use of the identity $(3.20)_3$, and noting that the second gradient $\operatorname{grad}^2\varphi = \operatorname{grad}\operatorname{grad}\varphi$ of a scalar field φ is symmetric, computing the gradient on both sides of $(9.7)_1$ yields

$$\begin{aligned}\operatorname{grad}\dot{\varphi} &= \operatorname{grad}(\varphi') + \operatorname{grad}(\mathbf{v}\cdot\operatorname{grad}\varphi)\\ &= (\operatorname{grad}\varphi)' + (\operatorname{grad}^2\varphi)^{\top}\mathbf{v} + (\operatorname{grad}\mathbf{v})^{\top}\operatorname{grad}\varphi\\ &= (\operatorname{grad}\varphi)' + (\operatorname{grad}^2\varphi)\mathbf{v} + \mathbf{L}^{\top}\operatorname{grad}\varphi;\end{aligned}$$

further, choosing $\mathbf{g} = \operatorname{grad}\varphi$ in $(9.7)_2$ to yield

$$\overline{\operatorname{grad}\varphi}^{\boldsymbol{\cdot}} = (\operatorname{grad}\varphi)' + (\operatorname{grad}^2\varphi)\mathbf{v},$$

we find that

$$\operatorname{grad}\dot{\varphi} = \overline{\operatorname{grad}\varphi}^{\boldsymbol{\cdot}} + \mathbf{L}^{\top}\operatorname{grad}\varphi, \qquad \frac{\partial\dot{\varphi}}{\partial x_i} = \overline{\frac{\partial\varphi}{\partial x_i}}^{\boldsymbol{\cdot}} + L_{ji}\frac{\partial\varphi}{\partial x_j}. \tag{9.23}$$

Next, let $\varphi = \mathbf{g}\cdot\mathbf{c}$, with $\mathbf{g}$ a spatial vector field and $\mathbf{c} \neq \mathbf{0}$ an arbitrary constant vector. Then, by $(3.1)_4$, $(3.20)_3$, and (3.7),

$$\dot{\varphi} = \dot{\mathbf{g}}\cdot\mathbf{c}, \qquad \operatorname{grad}\dot{\varphi} = (\operatorname{grad}\dot{\mathbf{g}})^{\top}\mathbf{c}, \qquad \operatorname{grad}\varphi = (\operatorname{grad}\mathbf{g})^{\top}\mathbf{c},$$

and

$$\begin{aligned}\overline{\operatorname{grad}\varphi}^{\boldsymbol{\cdot}} &= \overline{(\operatorname{grad}\mathbf{g})^{\top}\mathbf{c}}^{\boldsymbol{\cdot}}\\ &= \left(\overline{(\operatorname{grad}\mathbf{g})^{\top}}^{\boldsymbol{\cdot}}\right)\mathbf{c}\\ &= \left(\overline{\operatorname{grad}\mathbf{g}}^{\boldsymbol{\cdot}}\right)^{\top}\mathbf{c}.\end{aligned}$$

It therefore follows from (9.23) that

$$\left[\operatorname{grad}\dot{\mathbf{g}} - \overline{\operatorname{grad}\mathbf{g}}^{\boldsymbol{\cdot}} - (\operatorname{grad}\mathbf{g})\mathbf{L}\right]^{\top}\mathbf{c} = \mathbf{0}$$

for all $\mathbf{c} \neq \mathbf{0}$. Thus,

$$\operatorname{grad}\dot{\mathbf{g}} = \overline{\operatorname{grad}\mathbf{g}}^{\boldsymbol{\cdot}} + (\operatorname{grad}\mathbf{g})\mathbf{L}, \qquad \frac{\partial\dot{g}_j}{\partial x_i} = \overline{\frac{\partial g_j}{\partial x_i}}^{\boldsymbol{\cdot}} + \frac{\partial g_j}{\partial x_k}L_{ki}. \tag{9.24}$$

The relations (9.23) and (9.24) show that, unlike the spatial time derivative and the spatial gradient, the material time-derivative and the spatial gradient do not generally commute. Moreover, (9.23) and (9.24) determine explicitly the discrepancies that ensue on interchanging the material time derivative and the spatial gradient. For that reason, (9.23) and (9.24) are referred to as *commutator identities*.

EXERCISES

1. Show that
$$\operatorname{div}\dot{\mathbf{g}} = \overline{\operatorname{div}\mathbf{g}}^{\,\cdot} + \mathbf{L}^{\top} : \operatorname{grad}\mathbf{g}. \tag{9.25}$$

2. Show that
$$\operatorname{grad}\dot{\mathbf{v}} = \dot{\mathbf{L}} + \mathbf{L}^2. \tag{9.26}$$

9.5 Particle Paths

Given a material point $\mathbf{X}$, the function $\mathbf{p}(t)$ defined by
$$\mathbf{p}(t) = \chi(\mathbf{X}, t) \tag{9.27}$$
is called the **particle path** of $\mathbf{X}$; $\mathbf{p}(t)$ describes the *spatial trajectory* of the particle $\mathbf{X}$ during the motion. Differentiating (9.27) with respect to time, bearing in mind that the material point $\mathbf{X}$ is fixed, we find, using (9.1), that $\mathbf{p}(t)$ is a solution of the differential equation
$$\dot{\mathbf{p}}(t) = \mathbf{v}(\mathbf{p}(t), t). \tag{9.28}$$
Suppose that $\chi(\mathbf{X}, \tau)$ is given at some time τ. Granted sufficient smoothness, given $\mathbf{v}$, the differential equations (9.28) for the particle paths may then be integrated to give the motion near $t = \tau$.

9.6 Stretching of Deformed Fibers

The scalar
$$\delta(t) = |\chi(\mathbf{X}_1, t) - \chi(\mathbf{X}_2, t)| \tag{9.29}$$
represents the distance between the particle paths of the material points $\mathbf{X}_1$ and $\mathbf{X}_2$. What is more important, for $\mathbf{X}_1$ close to $\mathbf{X}_2$,

(‡) $\delta(t)$ *represents the length at time* t *of the deformed fiber corresponding to the undeformed fiber*
$$\Delta\mathbf{X} = \mathbf{X}_2 - \mathbf{X}_1$$
and $\dot{\delta}(t)$ *represents the rate at which the deformed fiber is being stretched.*[62]

Differentiating $\frac{1}{2}\delta^2(t)$ yields
$$\delta(t)\dot{\delta}(t) = [\chi(\mathbf{X}_1, t) - \chi(\mathbf{X}_2, t)] \cdot [\dot{\chi}(\mathbf{X}_1, t) - \dot{\chi}(\mathbf{X}_2, t)],$$
so that, for $\mathbf{x}_1$ and $\mathbf{x}_2$ the spatial points occupied by $\mathbf{X}_1$ and $\mathbf{X}_2$ at time t, we have the useful identity
$$\delta(t)\dot{\delta}(t) = (\mathbf{x}_1 - \mathbf{x}_2) \cdot [\mathbf{v}(\mathbf{x}_1, t) - \mathbf{v}(\mathbf{x}_2, t)]. \tag{9.30}$$

[62] Cf. §7.2.

10 Special Motions

10.1 Rigid Motions

A motion χ is **rigid** if, at each time t,

$$\frac{\partial}{\partial t}|\chi(\mathbf{X},t)-\chi(\mathbf{Y},t)|=0$$

for all material points $\mathbf{X}$ and $\mathbf{Y}$, or equivalently, by (9.30),

$$(\mathbf{x}-\mathbf{y})\cdot[\mathbf{v}(\mathbf{x},t)-\mathbf{v}(\mathbf{y},t)]=0$$

for all $\mathbf{x}$ and $\mathbf{y}$ in $\mathcal{B}_t$. The spatial gradient of this relation yields

$$\mathbf{v}(\mathbf{x},t)=\mathbf{v}(\mathbf{y},t)-[\operatorname{grad}\mathbf{v}(\mathbf{x},t)]^{\top}(\mathbf{x}-\mathbf{y}),$$

and a second differentiation, this time with respect to $\mathbf{y}$, results in

$$\operatorname{grad}\mathbf{v}(\mathbf{y},t)=-[\operatorname{grad}\mathbf{v}(\mathbf{x},t)]^{\top}.$$

Setting $\mathbf{x}=\mathbf{y}$, we conclude that $\operatorname{grad}\mathbf{v}$ is skew. The last relation therefore implies that $\operatorname{grad}\mathbf{v}(\mathbf{y},t)=\operatorname{grad}\mathbf{v}(\mathbf{x},t)$ for all $\mathbf{x}$ and $\mathbf{y}$; hence, $\operatorname{grad}\mathbf{v}(\mathbf{x},t)$ is independent of $\mathbf{x}$. Let $\mathbf{W}(t)$, which we call the **spin**, denote this spatially constant field, so that

$$\mathbf{v}(\mathbf{x},t)=\mathbf{v}(\mathbf{y},t)+\mathbf{W}(t)(\mathbf{x}-\mathbf{y}).$$

We may therefore conclude from (9.12) that the velocity gradient of a rigid motion is the spin:

$$\mathbf{L}=\mathbf{W}. \tag{10.1}$$

Next, let $\mathbf{w}(t)$ denote the axial vector associated with $\mathbf{W}(t)$. Then

$$\mathbf{W}=\mathbf{w}\times$$

and

$$\mathbf{v}(\mathbf{x},t)=\mathbf{v}(\mathbf{y},t)+\mathbf{w}(t)\times(\mathbf{x}-\mathbf{y}); \tag{10.2}$$

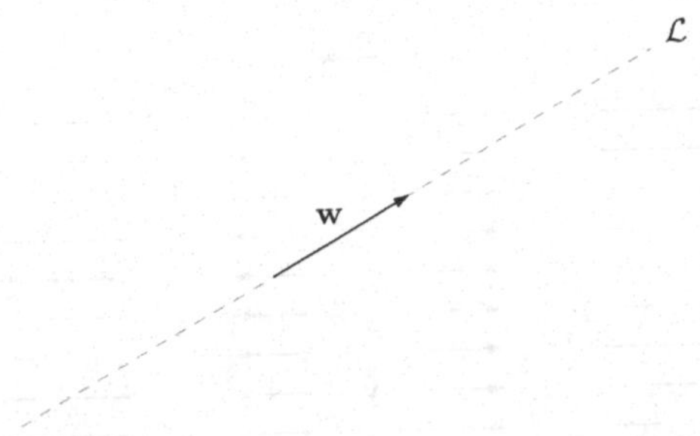

Figure 10.1. A vector $\mathbf{w}$ and its spin axis $\mathcal{L}$.

$\mathbf{w}(t)$ represents the *angular velocity* of the motion. Using $(3.20)_{6,10}$ to compute the curl of $\mathbf{v}$ as defined by (10.2), we find that

$$\begin{aligned}\operatorname{curl}\mathbf{v} &= \operatorname{div}(\mathbf{w}\otimes(\mathbf{x}-\mathbf{y})-(\mathbf{x}-\mathbf{y})\otimes\mathbf{w})\\ &= \mathbf{w}\operatorname{div}(\mathbf{x}-\mathbf{y})+(\operatorname{grad}\mathbf{w})(\mathbf{x}-\mathbf{y})-(\mathbf{x}-\mathbf{y})\operatorname{div}\mathbf{w}-[\operatorname{grad}(\mathbf{x}-\mathbf{y})]\mathbf{w}\\ &= 3\mathbf{w}-\mathbf{w}\\ &= 2\mathbf{w},\end{aligned} \tag{10.3}$$

which gives a physical interpretation of curl $\mathbf{v}$, at least for a rigid motion.

Assume that $\mathbf{w}\neq\mathbf{0}$. Then, for any vector $\mathbf{a}$, $\mathbf{w}\times\mathbf{a}=\mathbf{0}$ if and only if $\mathbf{Wa}=\mathbf{0}$. For that reason we use the term **spin axis** to denote the subspace $\mathcal{L}$ of vectors $\mathbf{a}$ such that

$$\mathbf{Wa}=\mathbf{0}. \tag{10.4}$$

More generally, we use the term **rigid velocity field** for a spatial field of the form $\mathbf{v}(\mathbf{x},t)=\mathbf{v}(\mathbf{y},t)+\boldsymbol{\lambda}(t)\times(\mathbf{x}-\mathbf{y})$, or equivalently,

$$\mathbf{v}(\mathbf{x},t)=\boldsymbol{\alpha}(t)+\boldsymbol{\lambda}(t)\times(\mathbf{x}-\mathbf{o}). \tag{10.5}$$

10.2 Motions Whose Velocity Gradient is Symmetric and Spatially Constant

In this case, writing $\mathbf{D}(=\mathbf{L})$ for the symmetric, spatially constant velocity gradient and suppressing the argument t, the velocity field has the form

$$\mathbf{v}(\mathbf{x})=\mathbf{D}(\mathbf{x}-\mathbf{y})$$

for some choice of $\mathbf{y}$. Since $\mathbf{D}$ is symmetric, it possesses a spectral representation of the form[63]

$$\mathbf{D}=\sum_{i=1}^{3}\alpha_i\mathbf{m}_i\otimes\mathbf{m}_i,$$

with the three "axes" being mutually orthogonal. It therefore suffices to limit our discussion to the velocity field

$$\mathbf{v}(\mathbf{x})=\alpha(\mathbf{e}\otimes\mathbf{e})(\mathbf{x}-\mathbf{y}). \tag{10.6}$$

Consider the fixed orthonormal basis $\{\mathbf{e}_i\}$ introduced in the paragraph containing (5.7). Suppose that $\mathbf{e}=\mathbf{e}_i$. Then, $\mathbf{v}$ as defined by (10.6) has only a single nontrivial

[63] Cf. (2.112).

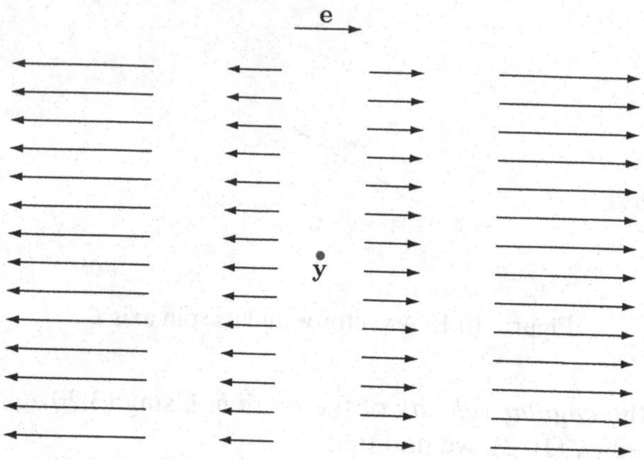

Figure 10.2. Schematic of the velocity field **v** defined in (10.6).

component $v_i = \mathbf{v} \cdot \mathbf{e} = \mathbf{v} \cdot \mathbf{e}_i$ with (Figure 10.2)

$$v_i(\mathbf{x}) = \alpha(x_i - y_i).$$

Up to an additive constant, every velocity field with gradient symmetric and constant is therefore the sum of three velocity fields of the form (10.6) with "axes" mutually orthogonal.

EXERCISE

1. Show that a motion whose velocity field is rigid is itself rigid.

11 Stretching and Spin in an Arbitrary Motion

11.1 Stretching and Spin as Tensor Fields

Consider now a general velocity field $\mathbf{v}$. Since $\mathbf{L} = \operatorname{grad}\mathbf{v}$, if we take the Taylor expansion the velocity $\mathbf{v}$ about an arbitrarily prescribed spatial point $\mathbf{y}$,[64] we find, suppressing the argument t, that

$$\mathbf{v}(\mathbf{x}) - \mathbf{v}(\mathbf{y}) = \mathbf{L}(\mathbf{y})(\mathbf{x} - \mathbf{y}) + \mathrm{o}(|\mathbf{x} - \mathbf{y}|) \tag{11.1}$$

as $\mathbf{x} \to \mathbf{y}$, where, as before, $\mathrm{o}(|\mathbf{x} - \mathbf{y}|)$ represents a term that goes to zero faster than $|\mathbf{x} - \mathbf{y}|$ and hence faster than $\mathbf{L}(\mathbf{y})(\mathbf{x} - \mathbf{y})$. Let $\mathbf{D}$ and $\mathbf{W}$, respectively, represent the symmetric and skew parts of $\mathbf{L}$:

$$\begin{aligned} \mathbf{D} &= \tfrac{1}{2}(\mathbf{L} + \mathbf{L}^{\top}) = \tfrac{1}{2}(\operatorname{grad}\mathbf{v} + (\operatorname{grad}\mathbf{v})^{\top}), \\ \mathbf{W} &= \tfrac{1}{2}(\mathbf{L} - \mathbf{L}^{\top}) = \tfrac{1}{2}(\operatorname{grad}\mathbf{v} - (\operatorname{grad}\mathbf{v})^{\top}). \end{aligned} \tag{11.2}$$

Then

$$\mathbf{L} = \mathbf{D} + \mathbf{W} \tag{11.3}$$

and (11.1) becomes

$$\mathbf{v}(\mathbf{x}) - \mathbf{v}(\mathbf{y}) = \mathbf{W}(\mathbf{y})(\mathbf{x} - \mathbf{y}) + \mathbf{D}(\mathbf{y})(\mathbf{x} - \mathbf{y}) + \mathrm{o}(|\mathbf{x} - \mathbf{y}|).$$

In a neighborhood of a given point $\mathbf{y}$ *and to within an error of* $\mathrm{o}(|\mathbf{x} - \mathbf{y}|)$ *a general velocity field is therefore the sum of a rigid velocity field*

$$\mathbf{v}(\mathbf{y}) + \mathbf{W}(\mathbf{y})(\mathbf{x} - \mathbf{y})$$

and a velocity field of the form

$$\mathbf{D}(\mathbf{y})(\mathbf{x} - \mathbf{y}).$$

Thus, bearing in mind the discussions in §§10.1–10.2, we refer to the tensor fields $\mathbf{W}$ and $\mathbf{D}$, respectively, as the **spin** and the **stretching**, and we use the term **spin axis** at $(\mathbf{y}, t)$ for the subspace $\mathcal{L}$ of vectors $\mathbf{a}$ such that

$$\mathbf{W}(\mathbf{y}, t)\mathbf{a} = \mathbf{0}. \tag{11.4}$$

($\mathcal{L}$ has dimension one when $\mathbf{W}(\mathbf{y}, t) \neq \mathbf{0}$.) An immediate consequence of (9.13) and (11.2) is that

$$\mathbf{D} \textit{ and } \mathbf{W} \textit{ map spatial vectors to spatial vectors.} \tag{11.5}$$

[64] Cf. $(3.12)_2$.

Our next step is to relate **L** to the right and left stretch tensors **U** and **V** and the rotation **R** appearing in the polar decompositions $\mathbf{F} = \mathbf{RU} = \mathbf{VR}$.[65] Focusing first on the right polar decomposition, we substitute $\mathbf{F} = \mathbf{RU}$ into (9.12) to obtain

$$\begin{aligned}\mathbf{L} &= \dot{\mathbf{F}}\mathbf{F}^{-1}\\ &= (\dot{\mathbf{R}}\mathbf{U} + \mathbf{R}\dot{\mathbf{U}})\mathbf{U}^{-1}\mathbf{R}^{\top}\\ &= \dot{\mathbf{R}}\mathbf{R}^{\top} + \mathbf{R}\dot{\mathbf{U}}\mathbf{U}^{-1}\mathbf{R}^{\top}.\end{aligned} \tag{11.6}$$

Since $\mathbf{R}\mathbf{R}^{\top} = \mathbf{1}$, $\dot{\overline{\mathbf{R}\mathbf{R}^{\top}}} = \mathbf{0}$, and it follows from $(3.1)_7$ that

$$\begin{aligned}\dot{\mathbf{R}}\mathbf{R}^{\top} &= -\mathbf{R}\dot{\mathbf{R}}^{\top}\\ &= -(\dot{\mathbf{R}}\mathbf{R}^{\top})^{\top}.\end{aligned} \tag{11.7}$$

Thus, $\dot{\mathbf{R}}\mathbf{R}^{\top}$ is skew and the symmetric and skew parts of (11.6) yield relations for the stretching **D** and the spin **W**:

$$\begin{aligned}\mathbf{D} &= \mathbf{R}[\operatorname{sym}(\dot{\mathbf{U}}\mathbf{U}^{-1})]\mathbf{R}^{\top},\\ \mathbf{W} &= \dot{\mathbf{R}}\mathbf{R}^{\top} + \mathbf{R}[\operatorname{skw}(\dot{\mathbf{U}}\mathbf{U}^{-1})]\mathbf{R}^{\top}.\end{aligned} \tag{11.8}$$

Interestingly, the spin **W** is therefore the sum of a rotational spin $\mathbf{W}_{\mathrm{rot}}$ induced by the rotation **R** and a stretch-spin $\mathbf{W}_{\mathrm{str}}$ induced by the stretch **U**:

$$\begin{aligned}&\mathbf{W} = \mathbf{W}_{\mathrm{rot}} + \mathbf{W}_{\mathrm{str}},\\ &\mathbf{W}_{\mathrm{rot}} = \dot{\mathbf{R}}\mathbf{R}^{\top}, \qquad \mathbf{W}_{\mathrm{str}} = \mathbf{R}[\operatorname{skw}(\dot{\mathbf{U}}\mathbf{U}^{-1})]\mathbf{R}^{\top}.\end{aligned} \tag{11.9}$$

An important identity

$$\mathbf{F}^{\top}\mathbf{D}\mathbf{F} = \dot{\mathbf{E}} \tag{11.10}$$

relating the stretching **D** and the material time-derivative of the Green–St. Venant strain **E** arises on noting that, by (7.6) and (9.12),

$$\begin{aligned}2\mathbf{F}^{\top}\mathbf{D}\mathbf{F} &= \mathbf{F}^{\top}(\mathbf{L} + \mathbf{L}^{\top})\mathbf{F}\\ &= \mathbf{F}^{\top}(\dot{\mathbf{F}}\mathbf{F}^{-1} + \mathbf{F}^{-\top}\dot{\mathbf{F}}^{\top})\mathbf{F}\\ &= \mathbf{F}^{\top}\dot{\mathbf{F}} + \dot{\mathbf{F}}^{\top}\mathbf{F}\\ &= 2\dot{\mathbf{E}}.\end{aligned}$$

11.2 Properties of D

Let **e** be a spatial unit vector and let $\delta_\ell(\tau)$ denote the distance at time τ between the particle paths that pass through the spatial points **x** and $\mathbf{x} + \ell\mathbf{e}$ at time t; hence $\delta_\ell(\tau)$ gives the length at time τ of the deformed fiber that at time t emanates from **x** and has length ℓ and direction **e** (Figure 11.1).[66] Then, by (9.30),

$$\frac{\dot{\delta}_\ell(t)}{\delta_\ell(t)} = \frac{\mathbf{e}\cdot[\mathbf{v}(\mathbf{x} + \ell\mathbf{e}, t) - \mathbf{v}(\mathbf{x}, t)]}{\ell}.$$

[65] Cf. (7.1).
[66] Cf. §7.2.

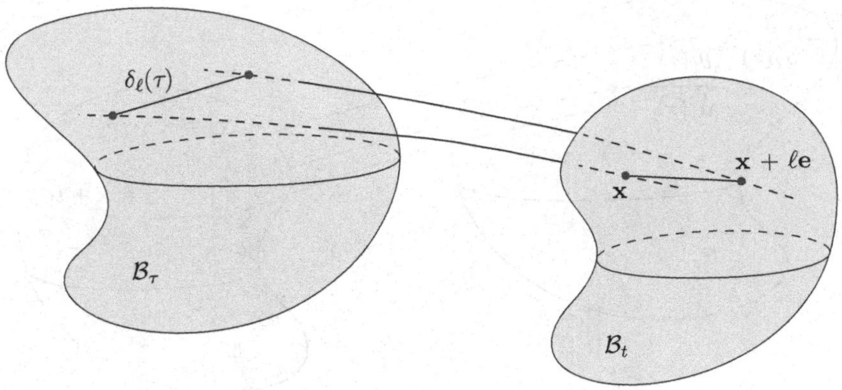

Figure 11.1. Schematic showing the length $\delta_\ell(\tau)$ at time τ of a deformed fiber that at time t emanates from **x** and has length ℓ and direction **e**.

But

$$\lim_{\ell\to 0}\frac{\mathbf{v}(\mathbf{x}+\ell\mathbf{e},t)-\mathbf{v}(\mathbf{x},t)}{\ell} = \operatorname{grad}\mathbf{v}(\mathbf{x},t)\mathbf{e} = \mathbf{L}(\mathbf{x},t)\mathbf{e} \tag{11.11}$$

and, by (11.3) and the fact that **W** is skew, $\mathbf{e}\cdot\mathbf{Le}=\mathbf{e}\cdot\mathbf{De}$. Thus,

$$\lim_{\ell\to 0}\frac{\dot{\delta}_\ell(t)}{\delta_\ell(t)} = \mathbf{e}\cdot\mathbf{D}(\mathbf{x},t)\mathbf{e} \tag{11.12}$$

and, in view of (‡) on page 85,

- **D** *characterizes the rate at which deforming fibers are stretched.*

Next, we turn to a discussion of the rate at which angles between deformed fibers are changing. Let **X**, $\mathbf{Y}_1$, and $\mathbf{Y}_2$ denote the material points that occupy the spatial points

$$\mathbf{x}, \qquad \mathbf{y}_1=\mathbf{x}+\ell\mathbf{e}_1, \qquad \text{and} \qquad \mathbf{y}_2=\mathbf{x}+\ell\mathbf{e}_2$$

at time t (Figure 11.2).

Further, let $\theta_\ell(\tau)$ denote the angle between the deformed fibers

$$\mathbf{u}_\alpha(\tau)=\boldsymbol{\chi}(\mathbf{Y}_\alpha,\tau)-\boldsymbol{\chi}(\mathbf{X},\tau), \quad \alpha=1,2,$$

so that $\theta_\ell(\tau)$ is the angle subtended by the spatial points

$$\boldsymbol{\chi}(\mathbf{Y}_1,\tau), \qquad \boldsymbol{\chi}(\mathbf{X},\tau), \qquad \text{and} \qquad \boldsymbol{\chi}(\mathbf{Y}_2,\tau).$$

Then,

$$\mathbf{u}_\alpha(t)=\mathbf{y}_\alpha-\mathbf{x}=\ell\mathbf{e}_\alpha, \qquad \dot{\mathbf{u}}_\alpha(t)=\mathbf{v}(\mathbf{y}_\alpha,t)-\mathbf{v}(\mathbf{x},t),$$

and, hence,

$$\ell^{-1}\overline{(\mathbf{u}_1\cdot\mathbf{u}_2)}^{\boldsymbol{\cdot}}(t)=\mathbf{e}_1\cdot[\mathbf{v}(\mathbf{y}_2,t)-\mathbf{v}(\mathbf{x},t)]+\mathbf{e}_2\cdot[\mathbf{v}(\mathbf{y}_1,t)-\mathbf{v}(\mathbf{x},t)].$$

Further, by (1.2),

$$\cos\theta_\ell=\frac{\mathbf{u}_1\cdot\mathbf{u}_2}{|\mathbf{u}_1||\mathbf{u}_2|},$$

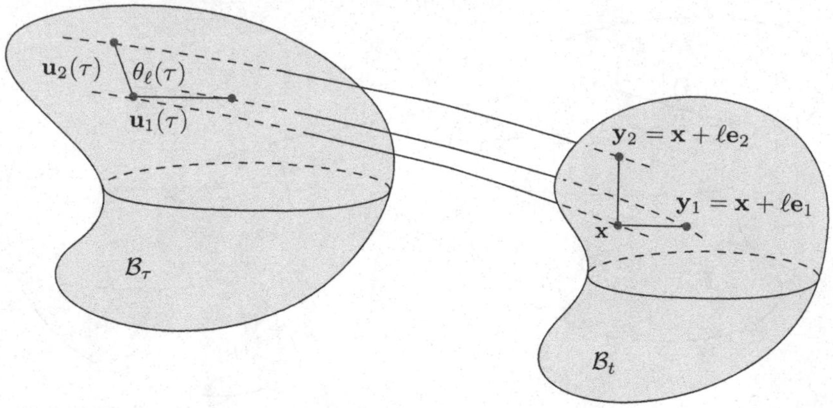

Figure 11.2. Schematic showing the angle $\theta_\ell(\tau)$ at time τ between two deformed fibers that at time t emanate from $\mathbf{x}$ and have lengths ℓ and directions $\mathbf{e}_1$ and $\mathbf{e}_2$.

and, because $\mathbf{u}_1$ and $\mathbf{u}_2$ are orthogonal at time t,

$$\overline{\cos\theta_\ell}^{\,\centerdot}(t) = \frac{\overline{(\mathbf{u}_1 \cdot \mathbf{u}_2)}^{\,\centerdot}(t)}{|\mathbf{u}_1(t)||\mathbf{u}_2(t)|} - \frac{(\mathbf{u}_1 \cdot \mathbf{u}_2)(t)\overline{|\mathbf{u}_1(t)||\mathbf{u}_2(t)|}^{\,\centerdot}}{|\mathbf{u}_1(t)|^2|\mathbf{u}_2(t)|^2}$$

$$= \frac{\overline{(\mathbf{u}_1 \cdot \mathbf{u}_2)}^{\,\centerdot}(t)}{|\mathbf{u}_1(t)||\mathbf{u}_2(t)|}.$$

On the other hand, since $\sin\theta_\ell(t) = 1$,

$$\overline{(\cos\theta_\ell)}^{\,\centerdot}(t) = -\dot{\theta}_\ell(t),$$

and we may conclude that

$$-\ell\dot{\theta}_\ell(t) = \mathbf{e}_1 \cdot [\mathbf{v}(\mathbf{x} + \ell\mathbf{e}_2, t) - \mathbf{v}(\mathbf{x}, t)] + \mathbf{e}_2 \cdot [\mathbf{v}(\mathbf{x} + \ell\mathbf{e}_1, t) - \mathbf{v}(\mathbf{x}, t)].$$

Finally, if we divide by ℓ and let $\ell \to 0$, we conclude, with the aid of $(11.2)_1$ and (11.11), that

$$\lim_{\ell\to 0}\dot{\theta}_\ell(t) = -\mathbf{e}_1 \cdot \mathbf{L}(\mathbf{x}, t)\mathbf{e}_2 - \mathbf{e}_2 \cdot \mathbf{L}(\mathbf{x}, t)\mathbf{e}_1$$

$$= -\mathbf{e}_1 \cdot (\mathbf{L}(\mathbf{x}, t) + \mathbf{L}(\mathbf{x}, t)^{\top})\mathbf{e}_2.$$

We are therefore led to a relation,

$$\lim_{\ell\to 0}\dot{\theta}_\ell(t) = -2\mathbf{e}_1 \cdot \mathbf{D}(\mathbf{x}, t)\mathbf{e}_2,$$

that makes precise a second basic property of the stretching $\mathbf{D}$:

- $\mathbf{D}$ *characterizes rates at which angles between deforming fibers change.*

11.3 Stretching and Spin Using the Current Configuration as Reference

It is often convenient to use — as reference for the motion of the body — the configuration at a fixed time t. The material point $\mathbf{X}$ that occupies the spatial point $\mathbf{x} = \chi(\mathbf{X}, t)$ may be expressed as a function of $(\mathbf{x}, t)$ through the reference map χ^{-1} (Figure 11.3):

$$\mathbf{X} = \chi^{-1}(\mathbf{x}, t).$$

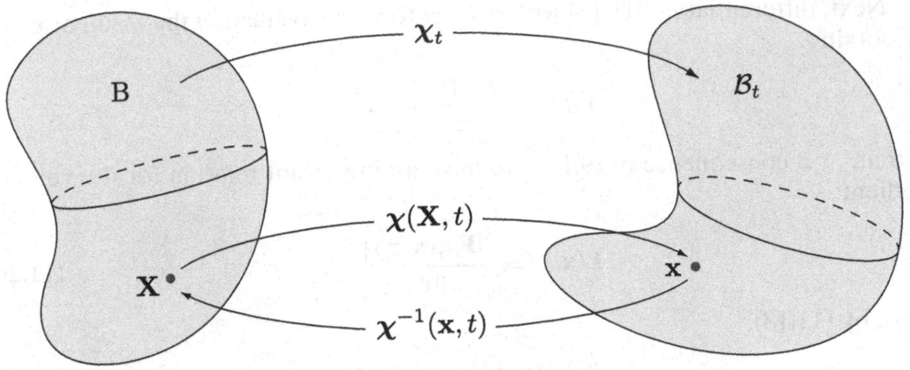

Figure 11.3. Schematic showing that a material point **X** can be expressed as a function of $(\mathbf{x}, t)$ through the reference map χ^{-1}.

The motion $\chi(\mathbf{X}, \tau)$ — expressed in terms of the spatial point **x** and the time τ — is therefore given by

$$\begin{aligned} \boldsymbol{\xi} &= \chi_{(t)}(\mathbf{x}, \tau) \\ &= \chi(\mathbf{X}, \tau)|_{\mathbf{X}=\chi^{-1}(\mathbf{x},t)} \\ &= \chi(\chi^{-1}(\mathbf{x}, t), \tau); \end{aligned} \tag{11.13}$$

$\boldsymbol{\xi}$ is the spatial point occupied at time τ by the material point that occupies the spatial point **x** at time t. The gradient of $\chi_{(t)}(\mathbf{x}, \tau)$ with respect to **x**,

$$\mathbf{F}_{(t)}(\mathbf{x}, \tau) = \nabla\chi_{(t)}(\mathbf{x}, \tau), \tag{11.14}$$

is referred to as the *relative deformation gradient*. In view of (11.13),

$$\chi(\mathbf{X}, \tau) = \chi_{(t)}(\chi(\mathbf{X}, t), \tau); \tag{11.15}$$

thus, by (11.15) and the chain-rule, suppressing nontemporal arguments,

$$\mathbf{F}(\tau) = \mathbf{F}_{(t)}(\tau)\mathbf{F}(t) \tag{11.16}$$

and, hence,[67]

$$\mathbf{F}_{(t)}(t) = \mathbf{1}. \tag{11.17}$$

Using a notation and terminology strictly analogous to that used for **F**, we may use the polar decomposition

$$\mathbf{F}_{(t)} = \mathbf{R}_{(t)}\mathbf{U}_{(t)} = \mathbf{V}_{(t)}\mathbf{R}_{(t)} \tag{11.18}$$

to define the relative rotation $\mathbf{R}_{(t)}$, the relative stretch tensors, $\mathbf{U}_{(t)}$ and $\mathbf{V}_{(t)}$, and the relative Cauchy–Green tensors $\mathbf{C}_{(t)}$ and $\mathbf{B}_{(t)}$. Then, since the polar decomposition of the identity tensor **1** is the product **(1)(1)** of the identity tensor with itself, (11.17) and the uniqueness of the polar decomposition (11.18) imply that

$$\mathbf{R}_{(t)}(t) = \mathbf{U}_{(t)}(t) = \mathbf{V}_{(t)}(t) = \mathbf{C}_{(t)}(t) = \mathbf{B}_{(t)}(t) = \mathbf{1}. \tag{11.19}$$

---

67 More precisely, since the configuration at time t is being used as reference, $\mathbf{F}_{(t)}(\tau)$ maps material vectors to spatial vectors; thus, $\mathbf{F}_{(t)}(t)$ maps each vector **c**, considered as a material vector, to **c**, considered as a spatial vector.

Next, differentiating (11.16) with respect to τ and evaluating the result at $\tau = t$, we obtain

$$\dot{\mathbf{F}}(t) = \left.\frac{\partial \mathbf{F}_{(t)}(\tau)}{\partial \tau}\right|_{\tau=t} \mathbf{F}(t),$$

so that, as a consequence of (9.11), we have an important relation for the velocity gradient:

$$\mathbf{L}(\mathbf{x}, t) = \left.\frac{\partial \mathbf{F}_{(t)}(\mathbf{x}, \tau)}{\partial \tau}\right|_{\tau=t}. \tag{11.20}$$

Thus, by (11.18),

$$\mathbf{L}(\mathbf{x}, t) = \underbrace{\left.\frac{\partial \mathbf{U}_{(t)}(\mathbf{x}, \tau)}{\partial \tau}\right|_{\tau=t}}_{\text{symmetric}} + \underbrace{\left.\frac{\partial \mathbf{R}_{(t)}(\mathbf{x}, \tau)}{\partial \tau}\right|_{\tau=t}}_{\text{skew}} \tag{11.21}$$

and, appealing to the uniqueness of the decomposition $\mathbf{L} = \mathbf{D} + \mathbf{W}$ into symmetric and skew parts,

$$\mathbf{D}(\mathbf{x}, t) = \left.\frac{\partial \mathbf{U}_{(t)}(\mathbf{x}, \tau)}{\partial \tau}\right|_{\tau=t} \quad \text{and} \quad \mathbf{W}(\mathbf{x}, t) = \left.\frac{\partial \mathbf{R}_{(t)}(\mathbf{x}, \tau)}{\partial \tau}\right|_{\tau=t}. \tag{11.22}$$

EXERCISES

1. Express $\mathbf{D}$ and $\mathbf{W}$ in terms of the tensor fields comprising the left polar decomposition $\mathbf{F} = \mathbf{VR}$.
2. Using the spectral decomposition $(7.24)_1$ for $\mathbf{U}$, prove that

$$\operatorname{skw}(\dot{\mathbf{U}}\mathbf{U}^{-1}) = \operatorname{skw}\left(\sum_{i=1}^{3} \dot{\mathbf{r}}_i \otimes \mathbf{r}_i + \sum_{i,j=1}^{3} \frac{\lambda_i}{\lambda_j}(\dot{\mathbf{r}}_i \cdot \mathbf{r}_j)\mathbf{r}_i \otimes \mathbf{r}_j\right),$$

   and, hence, that the stretch-spin $\mathbf{W}_{\rm str}$ defined in (11.9) is caused by the spin of the principal directions $\mathbf{r}_i$.
3. The tensors

$$\mathbf{A}_n(\mathbf{x}, t) = \left.\frac{\partial^n \mathbf{C}_{(t)}(\mathbf{x}, \tau)}{\partial \tau^n}\right|_{\tau=t}, \qquad n = 1, 2, \ldots, \tag{11.23}$$

   are called the **Rivlin–Ericksen tensors**. Show that
   (a) $\mathbf{A}_1 = 2\mathbf{D}$.
   (b) $\overset{(n)}{\mathbf{C}} = \mathbf{F}^{\top}\mathbf{A}_n\mathbf{F}$.
   (c) $\mathbf{A}_{n+1} = \dot{\mathbf{A}}_n + \mathbf{A}_n\mathbf{L} + \mathbf{L}^{\top}\mathbf{A}_n$.
   (d) $\mathbf{A}_n = \mathbf{A}_n^{\top}$.

12 Material and Spatial Tensor Fields. Pullback and Pushforward Operations

12.1 Material and Spatial Tensor Fields

We define material and spatial tensors in terms of mapping properties. To be more specific, given a tensor field $\mathbf{G}$, we say that:

- $\mathbf{G}$ is a **spatial tensor field** if, pointwise, $\mathbf{G}$ *maps spatial vectors to spatial vectors*;
- $\mathbf{G}$ is a **material tensor field** if, pointwise, $\mathbf{G}$ *maps material vectors to material vectors*;
- $\mathbf{G}$ is a **mixed tensor field** if, pointwise, $\mathbf{G}$ *maps either spatial vectors to material vectors or material vectors to spatial vectors.*

Then, by (9.13) and (M1)–(M5) on pages 65 and 70,

(a) $\mathbf{V}$, $\mathbf{B}$, $\mathbf{L}$, $\mathbf{D}$, and $\mathbf{W}$ are *spatial tensor fields*;
(b) $\mathbf{U}$, $\mathbf{C}$, and $\mathbf{E}$ are *material tensor fields*;
(c) $\mathbf{F}$ and $\mathbf{R}$ are *mixed tensor fields.*

12.2 Pullback and Pushforward Operations

These operations are based on the mapping properties of the deformation gradient and its inverse, transpose, and inverse-transpose as discussed on page 65; to facilitate the ensuing discussion, we recall these properties here:

(M1) $\mathbf{F}$ *and* $\mathbf{F}^{-\top}$ *map material vectors to spatial vectors*;
(M2) $\mathbf{F}^{-1}$ *and* $\mathbf{F}^{\top}$ *map spatial vectors to material vectors.*

The deformation gradient can be used to transform a spatial tensor field $\mathbf{G}$ to a material tensor field. Indeed, consider the tensor field

$$\mathbb{P}[\mathbf{G}] \stackrel{\text{def}}{=} \mathbf{F}^{\top}\mathbf{G}\mathbf{F}. \tag{12.1}$$

Bearing in mind that

- $\mathbf{F}$ maps material vectors to spatial vectors;
- $\mathbf{G}$ maps spatial vectors to spatial vectors; and
- $\mathbf{F}^{\top}$ maps spatial vectors to material vectors.

The combination $\mathbf{F}^{\top}\mathbf{G}\mathbf{F}$ maps material vectors to material vectors; thus $\mathbb{P}$ represents a material tensor field. For that reason $\mathbb{P}$ is referred to as a **pullback** of $\mathbf{G}$.[68] Another

[68] More precisely, $\mathbb{P}$ is a tensorial pullback. Pullbacks of vectors are discussed in exercises.

pullback of $\mathbf{G}$ is given by

$$\underline{\mathbb{P}}[\mathbf{G}] \stackrel{\text{def}}{=} \mathbf{F}^{-1}\mathbf{G}\mathbf{F}^{-\top}. \tag{12.2}$$

The pullbacks $\mathbb{P}$ and $\underline{\mathbb{P}}$ preserve the operations "sym" and "skw":

$$\operatorname{sym}\mathbb{P}[\mathbf{G}] = \mathbb{P}[\operatorname{sym}\mathbf{G}] \qquad \text{and} \qquad \operatorname{skw}\mathbb{P}[\mathbf{G}] = \mathbb{P}[\operatorname{skw}\mathbf{G}] \tag{12.3}$$

and analogously for $\underline{\mathbb{P}}$. There are two other pullbacks of $\mathbf{G}$ using $\mathbf{F}$, namely

$$\mathbf{F}^{-1}\mathbf{G}\mathbf{F} \qquad \text{and} \qquad \mathbf{F}^{\top}\mathbf{G}\mathbf{F}^{-\top}; \tag{12.4}$$

these pullbacks are trace preserving.

The pullbacks $\mathbb{P}$ and $\underline{\mathbb{P}}$ are linear transformations of tensors to tensors and, hence, may be viewed as fourth-order tensors; for example, since

$$(\mathbb{P}[\mathbf{G}])_{ij} = F_{pi}G_{pr}F_{rj},$$

$\mathbb{P}$ expressed in components has the form

$$(\mathbb{P})_{ijpr} = F_{pi}F_{rj}.$$

The deformation gradient can be used to transform a material tensor field $\mathbf{M}$ to a spatial tensor field: Two such operations are defined as follows:

$$\mathbb{P}^{-1}[\mathbf{M}] = \mathbf{F}^{-\top}\mathbf{M}\mathbf{F}^{-1} \qquad \text{and} \qquad \underline{\mathbb{P}}^{-1}[\mathbf{M}] \stackrel{\text{def}}{=} \mathbf{F}\mathbf{M}\mathbf{F}^{\top}. \tag{12.5}$$

As the notation suggests, $\mathbb{P}^{-1}$ is the inverse of $\mathbb{P}$, while $\underline{\mathbb{P}}^{-1}$ is the inverse of $\underline{\mathbb{P}}$; indeed,

$$\begin{aligned}\mathbb{P}^{-1}\big[\mathbb{P}[\mathbf{G}]\big] &= \mathbb{P}^{-1}\big[\mathbf{F}^{\top}\mathbf{G}\mathbf{F}\big]\\ &= \mathbf{F}^{-\top}\mathbf{F}^{\top}\mathbf{G}\mathbf{F}\mathbf{F}^{-1}\\ &= \mathbf{G},\end{aligned}$$

and similarly for $\underline{\mathbb{P}}^{-1}$. The inverses $\mathbb{P}^{-1}$ and $\underline{\mathbb{P}}^{-1}$, which are referred to as **pushforward operations**, also preserve the operations "sym" and "skw."

An example of the pullback in action is the kinematical identity

$$\mathbb{P}[\mathbf{D}] = \dot{\mathbf{E}}, \tag{12.6}$$

which follows from (11.10) and (12.1), asserting that the pullback of the stretching tensor is the material time-derivative of the Green–St. Venant strain tensor.

EXERCISES

1. Establish the component form of $\underline{\mathbb{P}}$.
2. Verify that $\underline{\mathbb{P}}^{-1}$ is the inverse of $\underline{\mathbb{P}}$.
3. Show that

$$\operatorname{sym}\big(\mathbb{P}[\mathbf{G}]\big) = \mathbf{F}^{\top}(\operatorname{sym}\mathbf{G})\mathbf{F}.$$

4. Verify (12.3) and its analog for $\underline{\mathbb{P}}$.
5. Prove that the operations defined in (12.4) preserve the trace.
6. The pullback operations for vector fields are defined as follows: Given a spatial vector field $\mathbf{g}$,[69]

$$\mathbf{P}[\mathbf{g}] \stackrel{\text{def}}{=} \mathbf{F}^{\top}\mathbf{g} \qquad \text{and} \qquad \underline{\mathbf{P}}[\mathbf{g}] \stackrel{\text{def}}{=} \mathbf{F}^{-1}\mathbf{g}. \tag{12.7}$$

[69] Trivially, $\mathbf{P} = \mathbf{F}^{\top}$ and $\underline{\mathbf{P}} = \mathbf{F}^{-1}$; the added notation and terminology allows for comparisons between vector and tensor pullbacks.

By (M2) on page 95, the linear transformations $\mathbf{P}$ and $\underline{\mathbf{P}}$ map spatial vectors to material vectors. The inverses of these transformations, which correspond to pushforward operations, are defined on a material vector field $\mathbf{m}$ by

$$\mathbf{P}^{-1}[\mathbf{m}] \stackrel{\text{def}}{=} \mathbf{F}^{-\top}\mathbf{m} \qquad \text{and} \qquad \underline{\mathbf{P}}^{-1}[\mathbf{m}] \stackrel{\text{def}}{=} \mathbf{F}\mathbf{m}. \tag{12.8}$$

Show that the operations defined by (12.8) are inverses of the operations defined in (12.7). Show, further, that given arbitrary spatial vector fields $\mathbf{g}$ and $\mathbf{h}$,

$$\mathbb{P}[\mathbf{g} \otimes \mathbf{h}] = \mathbf{P}[\mathbf{g}] \otimes \mathbf{P}[\mathbf{h}] \qquad \text{and} \qquad \underline{\mathbb{P}}[\mathbf{g} \otimes \mathbf{h}] = \underline{\mathbf{P}}[\mathbf{g}] \otimes \underline{\mathbf{P}}[\mathbf{h}]. \tag{12.9}$$

13 Modes of Evolution for Vector and Tensor Fields

In this section we discuss various ways in which vector and tensor fields may evolve with the body.

13.1 Vector and Tensor Fields That Convect With the Body

Our discussion of fibers and tangent vectors in §6.2 led us to lay down the following definition: A spatial vector field $\mathbf{f}$ convects with the body if there is a time-independent material vector field $\mathbf{f}_R$ such that[70]

$$\mathbf{f}(\mathbf{x}, t) = \mathbf{F}(\mathbf{X}, t)\mathbf{f}_R(\mathbf{X}). \tag{13.1}$$

On the other hand, in light of our discussion of normal vectors in §8.1, we might also apply the phrase "convecting with the body" to a vector field $\mathbf{f}$ that satisfies[71]

$$\mathbf{f}(\mathbf{x}, t) = \mathbf{F}^{-\top}(\mathbf{X}, t)\mathbf{f}_R(\mathbf{X}). \tag{13.2}$$

To differentiate between these two types of "vector convection," we refer to the former as "*tangential* convection" and to the latter as "*normal* convection."

13.1.1 Vector Fields That Convect as Tangents

Guided by (13.1), we say that a spatial vector field $\mathbf{f}$ **convects as a tangent** if

$$\overline{\mathbf{F}^{-1}\mathbf{f}}^{\boldsymbol{\cdot}} = \mathbf{0}. \tag{13.3}$$

The defining relation (13.3) leads to a simple evolution equation for a vector field $\mathbf{f}$ that convects as a tangent, an evolution equation that we now derive. Since, by $(9.21)_3$,

$$\begin{aligned}\overline{\mathbf{F}^{-1}\mathbf{f}}^{\boldsymbol{\cdot}} &= \overline{\mathbf{F}^{-1}}^{\boldsymbol{\cdot}}\mathbf{f} + \mathbf{F}^{-1}\dot{\mathbf{f}} \\ &= -\mathbf{F}^{-1}\mathbf{L}\mathbf{f} + \mathbf{F}^{-1}\dot{\mathbf{f}} \\ &= \mathbf{F}^{-1}(-\mathbf{L}\mathbf{f} + \dot{\mathbf{f}}),\end{aligned}$$

it follows that a spatial vector field $\mathbf{f}$ convects as a tangent if and only if it evolves according to

$$\dot{\mathbf{f}} = \mathbf{L}\mathbf{f}. \tag{13.4}$$

[70] Cf. (6.8) and (6.15).
[71] Cf. (8.4).

13.1.2 Vector Fields That Convect as Normals

Based on (13.2), we say that a spatial vector field $\mathbf{f}$ **convects as a normal** if

$$\overline{\mathbf{F}^{\top}\mathbf{f}}^{\boldsymbol{\cdot}} = \mathbf{0}. \tag{13.5}$$

Then, arguing as before, by (9.21)₃,

$$\begin{aligned}\overline{\mathbf{F}^{\top}\mathbf{f}}^{\boldsymbol{\cdot}} &= \dot{\overline{\mathbf{F}^{\top}}}\mathbf{f} + \mathbf{F}^{\top}\dot{\mathbf{f}} \\ &= \mathbf{F}^{\top}\mathbf{L}^{\top}\mathbf{f} + \mathbf{F}^{\top}\dot{\mathbf{f}} \\ &= \mathbf{F}^{\top}(\mathbf{L}^{\top}\mathbf{f} + \dot{\mathbf{f}})\end{aligned}$$

and we arrive at the conclusion that *a spatial vector field* $\mathbf{f}$ *convects as a normal if and only if it evolves according to*

$$\dot{\mathbf{f}} = -\mathbf{L}^{\top}\mathbf{f}. \tag{13.6}$$

Let $\mathbf{f}$ and $\mathbf{g}$ be spatial vector fields. Then,

$$\overline{\mathbf{f}\cdot\mathbf{g}}^{\boldsymbol{\cdot}} = 0 \tag{13.7}$$

if $\mathbf{f}$ convects as a tangent while $\mathbf{g}$ convects as a normal. This result, which is useful, is a consequence of (13.4) and (13.6) and is verified as follows:

$$\begin{aligned}\overline{\mathbf{f}\cdot\mathbf{g}}^{\boldsymbol{\cdot}} &= \dot{\mathbf{f}}\cdot\mathbf{g} + \mathbf{f}\cdot\dot{\mathbf{g}} \\ &= \mathbf{L}\mathbf{f}\cdot\mathbf{g} - \underbrace{\mathbf{f}\cdot\mathbf{L}^{\top}\mathbf{g}}_{\mathbf{L}\mathbf{f}\cdot\mathbf{g}} \\ &= 0.\end{aligned}$$

13.1.3 Tangentially Convecting Basis and Its Dual Basis. Covariant and Contravariant Components of Spatial Fields

Let $\{\mathbf{f}_i\}$ be a triad of spatial vector fields that convect tangentially. Then, there is an *associated triad* $\{\mathbf{m}_i\}$ of material vector fields such that

$$\mathbf{f}_i = \mathbf{F}\mathbf{m}_i \qquad \text{and} \qquad \dot{\mathbf{m}}_i = \mathbf{0}. \tag{13.8}$$

Assume that, at some time, $\{\mathbf{f}_i\}$ is a basis. Then, since $\mathbf{F}$ is invertible, $\{\mathbf{m}_i\}$ is a basis, and since $\mathbf{f}_i = \mathbf{F}\mathbf{m}_i$ for all time, $\{\mathbf{f}_i\}$ is a basis for all time. Thus, *if a triad* $\{\mathbf{f}_i\}$ *of tangentially convecting vector fields is a basis at some time, then* $\{\mathbf{f}_i\}$ *is a basis for all time*, and we refer to $\{\mathbf{f}_i\}$ as a **tangentially convecting basis** and to $\{\mathbf{m}_i\}$ as its associated **material basis**. By (13.4), the vector fields of this basis evolve according to

$$\dot{\mathbf{f}}_i = \mathbf{L}\mathbf{f}_i. \tag{13.9}$$

In view of the discussion on page 66, we view a tangentially convecting basis as a basis **embedded** in — and, hence, moving with — the deforming body $\mathcal{B}_t$.

Given a tangentially convecting basis $\{\mathbf{f}_i\}$, let $\{\mathbf{f}^i\}$ denote the corresponding **dual basis**, so that $\{\mathbf{f}^i\}$ is the triad of spatial vector fields defined uniquely by the relations

$$\mathbf{f}_i\cdot\mathbf{f}^j = \delta_i{}^j, \tag{13.10}$$

with $\delta_i{}^j$, the Kronecker delta, defined by

$$\delta_i{}^j = \begin{cases}1, & \text{if } i = j,\\ 0, & \text{if } i \neq j.\end{cases} \tag{13.11}$$

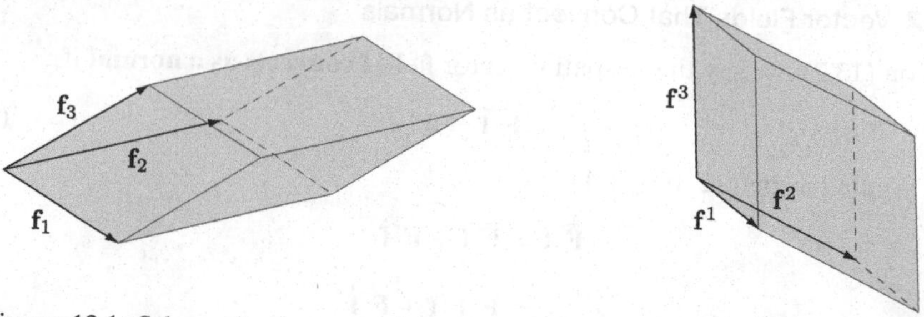

Figure 13.1. Schematic of a tangentially convecting basis $\{\mathbf{f}_i\}$ and its dual basis $\{\mathbf{f}^i\}$.

Thus, $\mathbf{f}^1$ is orthogonal to the plane spanned by the vectors $\mathbf{f}_2$ and $\mathbf{f}_3$, and so forth (Figure 13.1).

To verify that $\{\mathbf{f}^i\}$ is a basis, we have only to show that the vectors $\mathbf{f}^i$ are linearly independent, or, equivalently, that an arbitrary linear combination $c_i\mathbf{f}^i$ can vanish only when each of the scalar coefficients vanishes. Thus, assume that $c_i\mathbf{f}^i = \mathbf{0}$ and take the inner product of this relation with an arbitrary basis vector $\mathbf{f}_j$; in view of (13.11), this gives $c_j = 0$, which is what we set out to prove. Thus, $\{\mathbf{f}^i\}$ is a basis.

We now establish the differential equation that characterizes the evolution of the dual basis. By (13.10),

$$\overline{\mathbf{f}_i \cdot \mathbf{f}^j}^{\boldsymbol{\cdot}} = 0.$$

On the other hand, making use of (13.9),

$$\begin{aligned}\overline{\mathbf{f}_i \cdot \mathbf{f}^j}^{\boldsymbol{\cdot}} &= \mathbf{f}_i \cdot \dot{\mathbf{f}}^j + \dot{\mathbf{f}}_i \cdot \mathbf{f}^j \\ &= \mathbf{f}_i \cdot \dot{\mathbf{f}}^j + \mathbf{L}\mathbf{f}_i \cdot \mathbf{f}^j \\ &= \mathbf{f}_i \cdot \dot{\mathbf{f}}^j + \mathbf{f}_i \cdot \mathbf{L}^{\top}\mathbf{f}^j \\ &= \mathbf{f}_i \cdot (\dot{\mathbf{f}}^j + \mathbf{L}^{\top}\mathbf{f}^j).\end{aligned} \tag{13.12}$$

Thus,

$$\mathbf{f}_i \cdot (\dot{\mathbf{f}}^j + \mathbf{L}^{\top}\mathbf{f}^j) = 0$$

and, since $\{\mathbf{f}_i\}$ is a basis, we are led to the following evolution equation for the vector fields of the dual basis:

$$\dot{\mathbf{f}}^i = -\mathbf{L}^{\top}\mathbf{f}^i, \tag{13.13}$$

and, by (13.6), the vector fields $\mathbf{f}^i$ $(i = 1, 2, 3)$ convect normally. Thus,

(†) *if $\{\mathbf{f}^i\}$ is the basis dual to a tangentially convecting basis, then the vector fields $\mathbf{f}^i$ convect normally.*

Further, in view of (†) and (13.5), and the argument given in the paragraph containing (13.8), there is an **associated material basis** $\{\mathbf{m}^i\}$ such that

$$\mathbf{f}^i = \mathbf{F}^{-\top}\mathbf{m}^i \qquad \text{and} \qquad \dot{\mathbf{m}}^i = \mathbf{0}. \tag{13.14}$$

A consequence of (13.8) and (13.14) is that $\{\mathbf{m}^i\}$ is the basis dual to $\{\mathbf{m}_i\}$:

$$\mathbf{m}_i \cdot \mathbf{m}^j = \delta_i{}^j; \tag{13.15}$$

the verification of this assertion is left as an exercise.

We continue to assume that $\{\mathbf{f}_i\}$ is a tangentially convecting basis with $\{\mathbf{f}^i\}$ the corresponding dual basis. Then, because there are two bases, there are two sets of components for a vector **g** and two sets of components for a tensor **G**, and these are given by[72]

$$\begin{aligned} g_i &= \mathbf{f}_i \cdot \mathbf{g}, & g^i &= \mathbf{f}^i \cdot \mathbf{g}, \\ G_{ij} &= \mathbf{f}_i \cdot \mathbf{G}\mathbf{f}_j, & G^{ij} &= \mathbf{f}^i \cdot \mathbf{G}\mathbf{f}^j; \end{aligned} \tag{13.16}$$

g_i and G_{ij} are referred to as **covariant components**, while g^i and G^{ij} are called **contravariant components**; these components generate **g** and **G** through

$$\begin{aligned} \mathbf{g} &= g_i \mathbf{f}^i = g^i \mathbf{f}_i, \\ \mathbf{G} &= G_{ij} \mathbf{f}^i \otimes \mathbf{f}^j = G^{ij} \mathbf{f}_i \otimes \mathbf{f}_j. \end{aligned} \tag{13.17}$$

To verify the expansion $\mathbf{g} = g_i \mathbf{f}^i$ note that, since $\{\mathbf{f}_i\}$ is a basis there are unique scalars c_i such that $\mathbf{g} = c_i \mathbf{f}^i$ and, by (13.10), the inner product of this relation with $\mathbf{f}_j$ yields $c_j = g_j$, which establishes the desired expansion $\mathbf{g} = g_i \mathbf{f}^i$.

Our next step is to show that the tensors $\mathbf{f}^i \otimes \mathbf{f}^j$ form a basis for the nine-dimensional space of all tensors. As before, we have only to show that these tensors are linearly independent. Let A_{ij} denote arbitrary scalars, assume that

$$A_{ij} \mathbf{f}^i \otimes \mathbf{f}^j = 0, \tag{13.18}$$

and take the inner product of (13.18) with $\mathbf{f}_k \otimes \mathbf{f}_l$; this yields, by virtue of (2.66), $A_{kl} = 0$; the triad $\mathbf{f}^i \otimes \mathbf{f}^j$ is thus linearly independent and hence a basis. Next, to establish the expansion $\mathbf{G} = G_{ij} \mathbf{f}^i \otimes \mathbf{f}^j$, we use the fact that, since $\mathbf{f}^i \otimes \mathbf{f}^j$ is linearly independent, there are scalars K_{ij} such that $\mathbf{G} = K_{ij} \mathbf{f}^i \otimes \mathbf{f}^j$ and taking the inner product of this relation with $\mathbf{f}_k \otimes \mathbf{f}_l$ yields, again by (2.66), $K_{kl} = G_{kl}$. The remaining steps in the verification of (13.17) are left as an exercise.

Consider the material tensor fields

$$\mathbf{M} = \mathbb{P}[\mathbf{G}] = \mathbf{F}^{\top}\mathbf{G}\mathbf{F} \qquad \text{and} \qquad \mathbf{N} = \underline{\mathbb{P}}[\mathbf{G}] = \mathbf{F}^{-1}\mathbf{G}\mathbf{F}^{-\top} \tag{13.19}$$

obtained by pulling **G** back to the reference space using the pullbacks $\mathbb{P}$ and $\underline{\mathbb{P}}$.[73] Then

$$M_{ij} = \mathbf{m}_i \cdot \mathbf{M}\mathbf{m}_j \qquad \text{and} \qquad N^{ij} = \mathbf{m}^i \cdot \mathbf{N}\mathbf{m}^j$$

represent the covariant and contravariant components of **M** and **N** with respect to the material basis $\{\mathbf{m}_i\}$, and (13.8) and (13.14) imply that

$$\begin{aligned} M_{ij} &= \mathbf{m}_i \cdot \mathbf{F}^{\top}\mathbf{G}\mathbf{F}\mathbf{m}_j \\ &= (\mathbf{F}^{-1}\mathbf{f}_i) \cdot (\mathbf{F}^{\top}\mathbf{G}\mathbf{F})(\mathbf{F}^{-1}\mathbf{f}_j) \\ &= \mathbf{f}_i \cdot \mathbf{G}\mathbf{f}_j \end{aligned}$$

and

$$\begin{aligned} N^{ij} &= \mathbf{m}^i \cdot \mathbf{F}^{-1}\mathbf{G}\mathbf{F}^{-\top}\mathbf{m}^j \\ &= (\mathbf{F}^{\top}\mathbf{f}^i) \cdot (\mathbf{F}^{-1}\mathbf{G}\mathbf{F}^{-\top})(\mathbf{F}^{\top}\mathbf{f}^j) \\ &= \mathbf{f}^i \cdot \mathbf{G}\mathbf{f}^j. \end{aligned}$$

---

72 Heretofore we have, for the most part, restricted attention to orthonormal bases and any such basis coincides with its dual basis. Here, because we do not have this coincidence, it is customary to label components, as shown, using both subscripts and superscripts.

73 Cf. (12.1) and (12.2).

Hence, $M_{ij} = G_{ij}$ and $N^{ij} = G^{ij}$, or, equivalently, by (13.19),

$$(\mathbb{P}[\mathbf{G}])_{ij} = G_{ij} \qquad \text{and} \qquad (\underline{\mathbb{P}}[\mathbf{G}])^{ij} = G^{ij}; \tag{13.20}$$

thus, the covariant components of the tensor $\mathbb{P}[\mathbf{G}]$ are equal to the covariant components of $\mathbf{G}$, and the contravariant components of the tensor $\underline{\mathbb{P}}[\mathbf{G}]$ are equal to the contravariant components of $\mathbf{G}$. Hence, roughly speaking, $\mathbb{P}$ preserves covariant components, while $\underline{\mathbb{P}}$ preserves contravariant components. For that reason, we refer to $\mathbb{P}$ as the **covariant pullback** and to $\underline{\mathbb{P}}$ as the **contravariant pullback**.[74]

13.1.4 Covariant and Contravariant Convection of Tensor Fields

We continue to assume that $\{\mathbf{f}_i\}$ is a tangentially convecting basis with $\{\mathbf{f}^i\}$ the corresponding dual basis; hence, there is a material basis $\{\mathbf{m}_i\}$ (with dual basis $\{\mathbf{m}^i\}$) such that[75]

$$\begin{aligned} \mathbf{f}_i &= \mathbf{F}\mathbf{m}_i, \qquad & \dot{\mathbf{m}}_i &= \mathbf{0}, \\ \mathbf{f}^i &= \mathbf{F}^{-\top}\mathbf{m}^i, \qquad & \dot{\mathbf{m}}^i &= \mathbf{0}. \end{aligned} \tag{13.21}$$

Then, by $(13.16)_3$, for $\mathbf{G}$ an arbitrary spatial tensor field,[76]

$$\begin{aligned} \overline{G_{ij}}^{\boldsymbol{\cdot}} &= \overline{\mathbf{f}_i \cdot \mathbf{G}\mathbf{f}_j}^{\boldsymbol{\cdot}} \\ &= \overline{(\mathbf{F}\mathbf{m}_i) \cdot (\mathbf{G}\mathbf{F}\mathbf{m}_j)}^{\boldsymbol{\cdot}} \\ &= \mathbf{m}_i \cdot \overline{\mathbf{F}^{\top}\mathbf{G}\mathbf{F}}^{\boldsymbol{\cdot}}\mathbf{m}_j. \end{aligned}$$

Similarly, by (13.9),

$$\begin{aligned} \overline{G_{ij}}^{\boldsymbol{\cdot}} &= \overline{\mathbf{f}_i \cdot \mathbf{G}\mathbf{f}_j}^{\boldsymbol{\cdot}} \\ &= \mathbf{f}_i \cdot \dot{\mathbf{G}}\mathbf{f}_j + \dot{\mathbf{f}}_i \cdot \mathbf{G}\mathbf{f}_j + \mathbf{f}_i \cdot \mathbf{G}\dot{\mathbf{f}}_j \\ &= \mathbf{f}_i \cdot \dot{\mathbf{G}}\mathbf{f}_j + (\mathbf{L}\mathbf{f}_i) \cdot \mathbf{G}\mathbf{f}_j + \mathbf{f}_i \cdot \mathbf{G}(\mathbf{L}\mathbf{f}_j) \\ &= \mathbf{f}_i \cdot (\dot{\mathbf{G}} + \mathbf{G}\mathbf{L} + \mathbf{L}^{\top}\mathbf{G})\mathbf{f}_j. \end{aligned}$$

Thus,

$$\overline{G_{ij}}^{\boldsymbol{\cdot}} = \overline{\mathbf{f}_i \cdot \mathbf{G}\mathbf{f}_j}^{\boldsymbol{\cdot}} \tag{13.22}$$

$$= \mathbf{m}_i \cdot \overline{\mathbf{F}^{\top}\mathbf{G}\mathbf{F}}^{\boldsymbol{\cdot}}\mathbf{m}_j \tag{13.23}$$

$$= \mathbf{f}_i \cdot (\dot{\mathbf{G}} + \mathbf{G}\mathbf{L} + \mathbf{L}^{\top}\mathbf{G})\mathbf{f}_j \tag{13.24}$$

and, since the triads $\{\mathbf{f}_i\}$ and $\{\mathbf{m}_i\}$ are each a basis, we may conclude from $(13.19)_1$ and the identities (13.23) and (13.24) that the assertion of "equivalence" in the following definition is valid.

COVARIANT CONVECTION OF A TENSOR FIELD *We say that a spatial tensor field* **G convects covariantly** *if any one of the following three equivalent conditions is satisfied*:

---

[74] These are the unique pullbacks with the properties (13.20); the two other pullbacks defined by (12.4) preserve mixed components.

[75] Cf. (13.8) and (13.14).

[76] Bear in mind that, because $\{\mathbf{f}_i\}$ is time-dependent, $\overline{G_{ij}}^{\boldsymbol{\cdot}} = \overline{\mathbf{f}_i \cdot \mathbf{G}\mathbf{f}_j}^{\boldsymbol{\cdot}}$ and $\dot{G}_{ij} = \mathbf{f}_i \cdot \dot{\mathbf{G}}\mathbf{f}_j$ are not equal: the former represents the time-derivative of the components of $\mathbf{G}$, the latter the components of $\dot{\mathbf{G}}$.

(i) **G** *evolves according to the differential equation*

$$\dot{\mathbf{G}} + \mathbf{G}\mathbf{L} + \mathbf{L}^{\top}\mathbf{G} = \mathbf{0}. \tag{13.25}$$

(ii) *The covariant pullback of* **G** *is materially time-independent,*

$$\begin{aligned}\overline{\mathbb{P}[\mathbf{G}]}^{\,\boldsymbol{\cdot}} &= \overline{\mathbf{F}^{\top}\mathbf{G}\mathbf{F}}^{\,\boldsymbol{\cdot}} \\ &= \mathbf{0}. \end{aligned}\tag{13.26}$$

(iii) *The covariant components of* **G** *(with respect to the basis* $\{\mathbf{f}_i\}$*) are materially time-independent,*[77]

$$\dot{\overline{G_{ij}}} = 0. \tag{13.27}$$

Our next step is to derive contravariant counterparts of (13.22)–(13.24). By $(13.16)_4$, for **G** an arbitrary spatial tensor field,

$$\begin{aligned}\dot{\overline{G^{ij}}} &= \overline{\mathbf{f}^i \cdot \mathbf{G}\mathbf{f}^j}^{\,\boldsymbol{\cdot}} \\ &= \overline{(\mathbf{F}^{-\top}\mathbf{m}^i)\cdot(\mathbf{G}\mathbf{F}^{-\top}\mathbf{m}^j)}^{\,\boldsymbol{\cdot}} \\ &= \mathbf{m}^i \cdot \overline{\mathbf{F}^{-1}\mathbf{G}\mathbf{F}^{-\top}}^{\,\boldsymbol{\cdot}}\,\mathbf{m}^j. \end{aligned}$$

Similarly, by (13.13),

$$\begin{aligned}\dot{\overline{G^{ij}}} &= \overline{\mathbf{f}^i \cdot \mathbf{G}\mathbf{f}^j}^{\,\boldsymbol{\cdot}} \\ &= \mathbf{f}^i \cdot \dot{\mathbf{G}}\mathbf{f}^j + \dot{\mathbf{f}}^i \cdot \mathbf{G}\mathbf{f}^j + \mathbf{f}^i \cdot \mathbf{G}\dot{\mathbf{f}}^j \\ &= \mathbf{f}^i \cdot \dot{\mathbf{G}}\mathbf{f}^j - (\mathbf{L}^{\top}\mathbf{f}^i)\cdot\mathbf{G}\mathbf{f}^j - \mathbf{f}^i \cdot \mathbf{G}(\mathbf{L}^{\top}\mathbf{f}^j) \\ &= \mathbf{f}^i \cdot (\dot{\mathbf{G}} - \mathbf{L}\mathbf{G} - \mathbf{G}\mathbf{L}^{\top})\mathbf{f}^j. \end{aligned}$$

Thus,

$$\dot{\overline{G^{ij}}} = \overline{\mathbf{f}^i \cdot \mathbf{G}\mathbf{f}^j}^{\,\boldsymbol{\cdot}} \tag{13.28}$$

$$= \mathbf{m}^i \cdot \overline{\mathbf{F}^{-1}\mathbf{G}\mathbf{F}^{-\top}}^{\,\boldsymbol{\cdot}}\,\mathbf{m}^j \tag{13.29}$$

$$= \mathbf{f}^i \cdot (\dot{\mathbf{G}} - \mathbf{L}\mathbf{G} - \mathbf{G}\mathbf{L}^{\top})\mathbf{f}^j, \tag{13.30}$$

identities that, with $(13.19)_2$, represent the basis for the next definition.

Contravariant Convection of a Tensor Field *We say that a spatial tensor field* **G convects contravariantly** *if any one of the following three equivalent conditions is satisfied:*

(i) **G** *evolves according to the differential equation*

$$\dot{\mathbf{G}} - \mathbf{L}\mathbf{G} - \mathbf{G}\mathbf{L}^{\top} = \mathbf{0}. \tag{13.31}$$

(ii) *The contravariant pullback of* **G** *is materially time-independent,*

$$\begin{aligned}\overline{\underline{\mathbb{P}}[\mathbf{G}]}^{\,\boldsymbol{\cdot}} &= \overline{\mathbf{F}^{-1}\mathbf{G}\mathbf{F}^{-\top}}^{\,\boldsymbol{\cdot}} \\ &= \mathbf{0}. \end{aligned}\tag{13.32}$$

[77] That is, the components of **G** are materially time-independent when computed relative to a basis embedded in the deforming body.

(iii) *The contravariant components of* $\mathbf{G}$ *(with respect to the basis* $\{\mathbf{f}^i\}$*) are materially time-independent,*

$$\overline{G^{ij}}^{\boldsymbol{\cdot}} = 0. \tag{13.33}$$

EXERCISES

1. Let $\mathbf{f}_1$ and $\mathbf{f}_2$ be spatial vector fields that convect both as tangents. Show that

$$\overline{\mathbf{f}_1 \cdot \mathbf{f}_2}^{\boldsymbol{\cdot}} = 2\mathbf{f}_1 \cdot \mathbf{D}\mathbf{f}_2. \tag{13.34}$$

Show further that if θ is the angle between $\mathbf{f}_1$ and $\mathbf{f}_2$ as defined by (1.2) and if, at some time, $\mathbf{f}_1$ and $\mathbf{f}_2$ are orthogonal, then, at that time,

$$\dot{\theta} = -2\frac{\mathbf{f}_1}{|\mathbf{f}_1|} \cdot \mathbf{D}\frac{\mathbf{f}_2}{|\mathbf{f}_2|}.$$

What happens if, instead of convecting as tangents, $\mathbf{f}_1$ and $\mathbf{f}_2$ convect as normals?

2. Let $\mathbf{f}$ be a spatial vector field that convects as a tangent. Show that

$$\overline{\ln|\mathbf{f}|}^{\boldsymbol{\cdot}} = \frac{\mathbf{f}}{|\mathbf{f}|} \cdot \mathbf{D}\frac{\mathbf{f}}{|\mathbf{f}|}. \tag{13.35}$$

Hint: (13.34) is helpful.

3. Establish (13.15).
4. Derive the component representations

$$\mathbf{g} = g^i\mathbf{f}_i \qquad \text{and} \qquad \mathbf{G} = G^{ij}\mathbf{f}_i \otimes \mathbf{f}_j.$$

5. Show that
   (a) $\mathbf{g}$ convects as a tangent if and only if its contravariant components g^i are materially time-independent,

$$\overline{g^i}^{\boldsymbol{\cdot}} = 0;$$

   (b) $\mathbf{g}$ convects as a normal if and only if its covariant components g_i are materially time-independent,

$$\overline{g_i}^{\boldsymbol{\cdot}} = 0.$$

6. The pullback operations for vector fields are defined in (12.7). Show that

$$(\mathbf{P}[\mathbf{g}])_i = g_i \qquad \text{and} \qquad (\underline{\mathbf{P}}[\mathbf{g}])^i = g^i, \tag{13.36}$$

and, hence, that the covariant components of the vector $\mathbf{P}[\mathbf{g}]$ are equal to the covariant components of $\mathbf{g}$ and the contravariant components of the vector $\underline{\mathbf{P}}[\mathbf{g}]$ are equal to the contravariant components of $\mathbf{g}$. Thus, $\mathbf{P}$ preserves covariant components, while $\underline{\mathbf{P}}$ preserves contravariant components. For that reason, when discussing vector fields, we refer to $\mathbf{P}$ as the **covariant pullback** and to $\underline{\mathbf{P}}$ as the **contravariant pullback**.

7. Given a spatial vector field $\mathbf{g}$, show that the three conditions

$$\text{(i)}\ \dot{\mathbf{g}} + \mathbf{L}^{\scriptscriptstyle\top}\mathbf{g} = \mathbf{0}, \qquad \text{(ii)}\ \overline{\mathbf{P}[\mathbf{g}]}^{\boldsymbol{\cdot}} = \mathbf{0}, \qquad \text{(iii)}\ \overline{g_i}^{\boldsymbol{\cdot}} = 0 \tag{13.37}$$

are equivalent, as are the three conditions

$$\text{(i)}\ \dot{\mathbf{g}} - \mathbf{L}\mathbf{g} = \mathbf{0}, \qquad \text{(ii)}\ \overline{\underline{\mathbf{P}}[\mathbf{g}]}^{\boldsymbol{\cdot}} = \mathbf{0}, \qquad \text{(iii)}\ \overline{g^i}^{\boldsymbol{\cdot}} = 0. \tag{13.38}$$

When any one (and hence all) of the conditions (13.37) are satisfied, we say that **g convects covariantly**; when any one (and hence all) of the conditions (13.38) are satisfied, we say that **g convects contravariantly**.[78] A consequence of the conditions (i) of the foregoing definitions:

- *A spatial vector field* **g** *convects covariantly if and only if* **g** *convects as a normal, contravariantly if and only if* **g** *convects as a tangent.*

13.2 Corotational Vector and Tensor Fields

Vector fields may also evolve by spinning with the body; such vector fields are called *corotational*. Precisely, a spatial vector field **k** is termed **corotational** if **k** evolves according to the differential equation

$$\dot{\mathbf{k}} = \mathbf{W}\mathbf{k}. \tag{13.39}$$

Let **k** and **l** be corotational vector fields. Then, since the spin **W** is skew,

$$\begin{aligned}\overline{\mathbf{k}\cdot\mathbf{l}}^{\centerdot} &= \dot{\mathbf{k}}\cdot\mathbf{l} + \mathbf{k}\cdot\dot{\mathbf{l}} \\ &= (\mathbf{W}\mathbf{k})\cdot\mathbf{l} + \mathbf{k}\cdot(\mathbf{W}\mathbf{l}) \\ &= -\mathbf{k}\cdot(\mathbf{W}\mathbf{l}) + \mathbf{k}\cdot(\mathbf{W}\mathbf{l});\end{aligned}$$

hence,

$$\overline{\mathbf{k}\cdot\mathbf{l}}^{\centerdot} = 0. \tag{13.40}$$

Thus, if a triad $\{\mathbf{k}_i\}$ of corotational vector fields is an orthonormal basis at some time, then $\{\mathbf{k}_i\}$ is an orthonormal basis for all time. Such a basis is called an **orthonormal corotational basis.**[79]

Consider the components

$$G_{ij} = \mathbf{k}_i \cdot \mathbf{G}\mathbf{k}_j$$

of a spatial tensor field **G** with respect to an orthonormal corotational basis $\{\mathbf{k}_i\}$. Then, by (13.39),

$$\begin{aligned}\dot{\overline{G_{ij}}} &= \overline{\mathbf{k}_i\cdot\mathbf{G}\mathbf{k}_j}^{\centerdot} \\ &= \mathbf{k}_i\cdot\dot{\mathbf{G}}\mathbf{k}_j + \dot{\mathbf{k}}_i\cdot\mathbf{G}\mathbf{k}_j + \mathbf{k}_i\cdot\mathbf{G}\dot{\mathbf{k}}_j \\ &= \mathbf{k}_i\cdot\dot{\mathbf{G}}\mathbf{k}_j + (\mathbf{W}\mathbf{k}_i)\cdot\mathbf{G}\mathbf{k}_j + \mathbf{k}_i\cdot(\mathbf{G}\mathbf{W})\mathbf{k}_j\end{aligned}$$

and, since $\mathbf{W}^{\top} = -\mathbf{W}$, we have the identities

$$\dot{\overline{G_{ij}}} = \overline{\mathbf{k}_i\cdot\mathbf{G}\mathbf{k}_j}^{\centerdot} \tag{13.41}$$

$$= \mathbf{k}_i\cdot(\dot{\mathbf{G}} + \mathbf{G}\mathbf{W} - \mathbf{W}\mathbf{G})\mathbf{k}_j, \tag{13.42}$$

which we take as the basis for the next definition.

Corotational Convection of a Tensor Field *We say that a spatial tensor field* **G convects corotationally** *if any one of the following two equivalent conditions is satisfied:*

(i) **G** *evolves according to the differential equation*

$$\dot{\mathbf{G}} + \mathbf{G}\mathbf{W} - \mathbf{W}\mathbf{G} = \mathbf{0}. \tag{13.43}$$

[78] Cf. the analogous defining conditions (13.25)–(13.27) and (13.31)–(13.33) for tensor fields.

[79] Here, as opposed to §13.1, the discussion is best restricted to bases that are orthonormal.

(ii) *The components of* **G** *relative to an orthonormal corotational basis are materially time-independent,*

$$\overline{G_{ij}}^{\boldsymbol{\cdot}} = 0. \tag{13.44}$$

EXERCISES

1. Provide a geometrical interpretation of (13.40).
2. Show that if **k** is corotational, then so also is $\mathbf{k} \otimes \mathbf{k}$.
3. The term corotational may be misleading, as it might also be thought to refer to a spatial vector field **f** such that

$$\mathbf{f} = \mathbf{R}\mathbf{f}_{\mathrm{R}}, \qquad \dot{\mathbf{f}}_{\mathrm{R}} = \mathbf{0}, \tag{13.45}$$

for some a material vector field $\mathbf{f}_{\mathrm{R}}$. Show that this generally cannot describe a corotational vector field by showing that **f** evolves according to the differential equation

$$\dot{\mathbf{f}} = (\dot{\mathbf{R}}\mathbf{R}^{\top})\mathbf{f}. \tag{13.46}$$

Since, by $(11.8)_2$, $\dot{\mathbf{R}}\mathbf{R}^{\top}$ is not generally equal to **W**, (13.45) does not generally define a corotational vector field. (For this reason, we say that a corotational vector spins with the body rather than rotates with the body.)
4. Show that if **k** and **l** are corotational vector fields, then

$$\overline{\mathbf{k} \cdot \mathbf{G}\mathbf{l}}^{\boldsymbol{\cdot}} = \mathbf{k} \cdot (\dot{\mathbf{G}} + \mathbf{G}\mathbf{W} - \mathbf{W}\mathbf{G})\mathbf{l},$$

and, hence, **G** is corotational if and only if

$$\overline{\mathbf{k} \cdot \mathbf{G}\mathbf{l}}^{\boldsymbol{\cdot}} = \mathbf{0}$$

for all such vector fields **k** and **l**.

14 Motions with Constant Velocity Gradient

The general theory of such motions is based on a consideration of the function $e^{t\mathbf{A}}$, with $\mathbf{A}$ a tensor, as a solution of a tensorial ordinary differential equation, as presented in §114.

14.1 Motions

We seek motions of the body in which the velocity gradient

$$\mathbf{L} = \operatorname{grad} \mathbf{v}$$

satisfies

$$\mathbf{L} = \text{constant}.$$

Granted this, the velocity has the form

$$\mathbf{v}(\mathbf{x}, t) = \mathbf{v}_0(t) + \mathbf{L}(\mathbf{x} - \mathbf{x}_0),$$

with $\mathbf{v}_0(t)$ an arbitrary function of time.

Next, by (9.12), the deformation gradient is a solution of the ordinary differential equation

$$\dot{\mathbf{F}}(t) = \mathbf{L}\mathbf{F}(t), \tag{14.1}$$

which we consider with the initial condition

$$\mathbf{F}(t_0) = \mathbf{F}_0, \qquad \det \mathbf{F}_0 > 0. \tag{14.2}$$

The differential equation (14.1), subject to (14.2), has the unique solution[80]

$$\mathbf{F}(t) = e^{(t-t_0)\mathbf{L}}\mathbf{F}_0, \qquad -\infty < t < \infty. \tag{14.3}$$

To verify this, we note that, by (114.1) and (114.2), $\mathbf{F}(t)$ defined by (14.3) satisfies

$$\begin{aligned}\dot{\mathbf{F}}(t) &= \dot{\mathbf{Z}}(t - t_0)\mathbf{F}_0\\ &= \mathbf{L}\mathbf{Z}(t - t_0)\mathbf{F}_0\\ &= \mathbf{L}\mathbf{F}(t)\end{aligned}$$

[80] See Appendix 114 for an overview of the exponential of a tensor.

and

$$\begin{aligned}\mathbf{F}(t_0) &= \mathbf{Z}(0)\mathbf{F}_0\\ &= \mathbf{F}_0.\end{aligned}$$

Hence, $\mathbf{F}(t)$ satisfies (14.1) and (14.2). Moreover, by (114.3), $\mathbf{F}(t)$ satisfies[81]

$$\overline{\det \mathbf{F}(t)} = (\det \mathbf{F}_0)e^{(\mathrm{tr}\,\mathbf{L})t}. \tag{14.4}$$

Since the deformation gradient $\mathbf{F}(t)$ is independent of the material point, the equation $\mathbf{F} = \nabla\boldsymbol{\chi}$ may be solved for the motion $\boldsymbol{\chi}$: By (6.1), assuming that the material point $\mathbf{X}_0$ goes through the spatial point $\mathbf{x}_0$ at time t_0, this motion has the explicit form

$$\boldsymbol{\chi}(\mathbf{X}, t) = \mathbf{x}_0 + e^{(t-t_0)\mathbf{L}}\mathbf{F}_0(\mathbf{X} - \mathbf{X}_0), \qquad -\infty < t < \infty. \tag{14.5}$$

This solution is well defined for a reference body B occupying all of space, and hence represents a motion for any reference body, no matter the shape. Some properties of this motion, useful in what follows, are associated with the choice $\mathbf{X} = \mathbf{X}_0$ and $t = t_0$:

$$\left.\begin{aligned}\mathbf{F} &= \mathbf{F}_0,\\ \dot{\mathbf{F}} &= \mathbf{L}\mathbf{F}_0,\\ \mathbf{C} &= \mathbf{F}_0^{\top}\mathbf{F}_0,\\ \dot{\mathbf{C}} &= 2\mathbf{F}_0^{\top}\mathbf{D}\mathbf{F}_0.\end{aligned}\right\} \tag{14.6}$$

[81] Cf. (9.15).

15 Material and Spatial Integration

15.1 Line Integrals

Let $\boldsymbol{\phi}(\mathbf{x},t)$ and $\boldsymbol{\varphi}(\mathbf{X},t)$ denote the spatial and material descriptions of a vector field. Further, let C denote a material curve and $\mathcal{C}_t$ the corresponding spatial curve at each time t. Then, using the notation spelled out in the paragraphs containing (6.10) and (6.14) and making explicit use of (6.14), we have the following derivation of the transformation law for line integrals:[82]

$$\begin{aligned}
\int_{\mathcal{C}_t} \boldsymbol{\phi}(\mathbf{x},t)\cdot d\mathbf{x} &= \int_{\lambda_0}^{\lambda_1} \boldsymbol{\phi}(\hat{\mathbf{x}}_t(\lambda),t)\cdot \frac{\partial \hat{\mathbf{x}}_t(\lambda)}{\partial \lambda}\,d\lambda \\
&= \int_{\lambda_0}^{\lambda_1} \boldsymbol{\varphi}(\hat{\mathbf{X}}(\lambda))\cdot \mathbf{F}(\hat{\mathbf{X}}(\lambda),t)\frac{d\hat{\mathbf{X}}(\lambda)}{d\lambda}\,d\lambda \\
&= \int_{\mathrm{C}} \boldsymbol{\varphi}(\mathbf{X},t)\cdot \mathbf{F}(\mathbf{X},t)d\mathbf{X} \\
&= \int_{\mathrm{C}} \mathbf{F}^{\top}(\mathbf{X},t)\boldsymbol{\varphi}(\mathbf{X},t)\cdot d\mathbf{X}.
\end{aligned} \tag{15.1}$$

15.2 Volume and Surface Integrals

Let P be a *material region* and let

$$\mathcal{P}_t = \boldsymbol{\chi}(\mathrm{P},t), \tag{15.2}$$

so that, by definition, $\mathcal{P}_t$ *convects with the body*;[83] then

(i) we write $\mathbf{n}_{\mathrm{R}}$ for the outward unit normal to $\partial\mathrm{P}$, and
(ii) $\mathbf{n}$ for the outward unit normal to $\partial\mathcal{P}_t$,

82 Cf. §4.2.
83 Cf. §5.3.

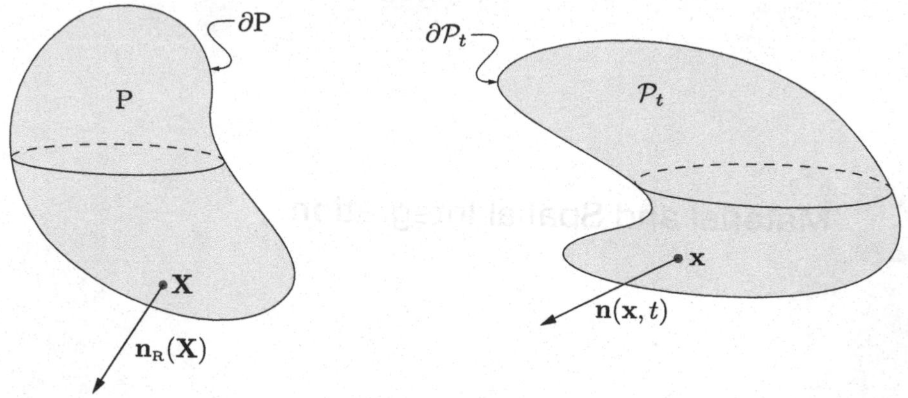

Figure 15.1. The outward unit normals $\mathbf{n}_R(\mathbf{X})$ and $\mathbf{n}(\mathbf{x}, t)$ at points $\mathbf{X}$ and $\mathbf{x} = \boldsymbol{\chi}(\mathbf{X}, t)$ on the boundaries ∂P and $\partial\mathcal{P}_t$ of a material region P and the corresponding spatial region $\mathcal{P}_t$.

so that, by (8.17)

$$\jmath\,\mathbf{n} = \mathbf{F}^{\mathrm{C}}\mathbf{n}_{\mathrm{R}}, \tag{15.3}$$

where $\jmath$ is the areal surface Jacobian related to mapping from ∂P to $\partial\mathcal{P}_t$ and

$$\mathbf{F}^{\mathrm{C}} = J\,\mathbf{F}^{-\top} \tag{15.4}$$

is the **cofactor** of **F**.[84]

Notation: *Throughout this book* P *denotes a material region and* $\mathcal{P}_t$ *denotes a spatial region convecting with the body.*

We write dv_{R} and dv for the volume elements in the reference and deformed bodies, and da_{R} and da for the analogous area elements. The estimates (8.9), (8.20), and (8.21) together with conventional definitions of integrals as limits of suitably subdivided domains lead to basic relations for transforming volume elements and scalar and vector area elements. These transformation laws are most easily remembered in terms of the *formal* relations[85]

$$\begin{aligned} dv &= J\,dv_{\mathrm{R}}, \qquad da = \jmath\,da_{\mathrm{R}}, \\ \mathbf{n}\,da &= \mathbf{F}^{\mathrm{C}}\mathbf{n}_{\mathrm{R}}\,da_{\mathrm{R}} = J\,\mathbf{F}^{-\top}\mathbf{n}_{\mathrm{R}}\,da_{\mathrm{R}}. \end{aligned} \tag{15.5}$$

15.2.1 Volume Integrals

Let φ and φ_{R} be the spatial and material descriptions of a field. Given any **material region** P, the transformation law for the integral of φ over $\mathcal{P}_t$ to an integral of φ_{R} over P is then

$$\int\limits_{\mathcal{P}_t} \varphi(\mathbf{x}, t)\,dv(\mathbf{x}) = \int\limits_{\mathrm{P}} \varphi_{\mathrm{R}}(\mathbf{X}, t)\,J(\mathbf{X}, t)\,dv_{\mathrm{R}}(\mathbf{X}). \tag{15.6}$$

Here and in what follows we write $dv_{\mathrm{R}}(\mathbf{X})$ when we wish to emphasize that the variable of integration is the material point $\mathbf{X}$; $dv(\mathbf{x})$ when we wish to emphasize that the variable of integration is the spatial point $\mathbf{x}$. More succintly, suppressing

[84] Cf. (8.14) and (2.91).

[85] These relations are meant only to be suggestive; they are not meant to replace actual proofs of the results (15.6) and (15.8).

arguments and dropping the subscript S on φ, we write (15.6) in the form

$$\int_{\mathcal{P}_t} \varphi\, dv = \int_{\mathrm{P}} \varphi J\, dv_{\mathrm{R}}. \tag{15.7}$$

15.2.2 Surface Integrals

In this case, the transformation laws are more complicated.

TRANSFORMATION LAWS FOR SURFACE INTEGRALS[86] *Given a material region* P, *a vector field* **u**, *and a tensor field* **G**,

$$\left.\begin{aligned} \int_{\partial\mathcal{P}_t} \mathbf{u}\cdot\mathbf{n}\, da &= \int_{\partial\mathrm{P}} \mathbf{u}\cdot\mathbf{F}^{\mathrm{C}}\mathbf{n}_{\mathrm{R}}\, da_{\mathrm{R}}, \\ \int_{\partial\mathcal{P}_t} \mathbf{G}\mathbf{n}\, da &= \int_{\partial\mathrm{P}} \mathbf{G}\mathbf{F}^{\mathrm{C}}\mathbf{n}_{\mathrm{R}}\, da_{\mathrm{R}}, \\ \int_{\partial\mathcal{P}_t} \mathbf{u}\cdot\mathbf{G}\mathbf{n}\, da &= \int_{\partial\mathrm{P}} \mathbf{u}\cdot\mathbf{G}\mathbf{F}^{\mathrm{C}}\mathbf{n}_{\mathrm{R}}\, da_{\mathrm{R}}, \end{aligned}\right\} \tag{15.8}$$

where $\mathbf{n}_R$ *and* $\mathbf{n}$, *respectively, are the outward unit normals to* $\partial\mathrm{P}$ *and* $\partial\mathcal{P}_t$, *and* $\mathbf{F}^{\mathrm{C}} = J\,\mathbf{F}^{-\top}$ *is the cofactor of* **F**.

Thus, when transforming volume and surface integrals from the deformed body to the reference body

(i) one replaces the volume element dv by $J\, dv_{\mathrm{R}}$;
(ii) one replaces the vectorial area element $\mathbf{n}\,da$ by $\mathbf{F}^{\mathrm{C}}\mathbf{n}_{\mathrm{R}}\, da_{\mathrm{R}}$.

EXERCISE

1. Show that

$$\int_{\partial\mathrm{P}} \mathbf{F}^{\mathrm{C}}\mathbf{n}_{\mathrm{R}}\, da_{\mathrm{R}} = \mathbf{0}.$$

15.3 Localization of Integrals

Often in continuum mechanics basic laws that are written in integral form for arbitrary subregions P of the reference body B — or arbitrary regions $\mathcal{P}_t$ convecting with the body — lead to relations of the form

$$\int_{\mathrm{P}} (\cdots)\, dv_{\mathrm{R}} = 0 \qquad \text{and} \qquad \int_{\mathrm{P}} (\cdots)\, dv_{\mathrm{R}} \ge 0$$

or

$$\int_{\mathcal{P}_t} (\cdots)\, dv = 0 \qquad \text{and} \qquad \int_{\mathcal{P}_t} (\cdots)\, dv \ge 0.$$

86 Cf., for example, TRUESDELL & TOUPIN (1960, eq. (20.8)).

The following results, which we use repeatedly, often without comment, allow us to localize such laws: given a spatial field φ,[87]

$$\int_{\mathrm{P}} \varphi \, dv_{\mathrm{R}} = 0 \text{ for all material regions } \mathrm{P} \subset \mathrm{B} \quad \Longrightarrow \quad \varphi = 0,$$

$$\int_{\mathrm{P}} \varphi \, dv_{\mathrm{R}} \geq 0 \text{ for all material regions } \mathrm{P} \subset \mathrm{B} \quad \Longrightarrow \quad \varphi \geq 0, \tag{15.9}$$

and

$$\int_{\mathcal{P}_t} \varphi \, dv = 0 \text{ for all convecting regions } \mathcal{P}_t \text{ and all } t \quad \Longrightarrow \quad \varphi = 0,$$

$$\int_{\mathcal{P}_t} \varphi \, dv \geq 0 \text{ for all convecting regions } \mathcal{P}_t \text{ and all } t \quad \Longrightarrow \quad \varphi \geq 0. \tag{15.10}$$

Another result and one that we find useful asserts that

(†) given any time t_0 and any spatial region $\mathcal{P}$ in $\mathcal{B}_{t_0}$, there is a convecting region $\mathcal{P}_t$ such that

$$\mathcal{P}_t|_{t=t_0} = \mathcal{P}. \tag{15.11}$$

15.3.1 Verification of (15.9), (15.10), and (†)

To verify (15.9), assume first that $(15.9)_2$ is not true; that is, assume that the left side of the implication is satisfied, but that there is a point $\mathbf{X}$ in B with $\varphi(\mathbf{X}) < 0$. Then, granted continuity, there is a neighborhood P of $\mathbf{X}$ in B such that $\varphi < 0$ everywhere in P. This yields $\int_{\mathrm{P}} \varphi \, dv_{\mathrm{R}} < 0$, which is a contradiction. Hence, $(15.9)_2$ is true. Further, replacing φ by $-\varphi$ in $(15.9)_2$, we see that the implication $(15.9)_2$ holds with ≥ 0 replaced by ≤ 0. Finally, if the left side of the implication $(15.9)_1$ is satisfied, then the integral $\int_{\mathrm{P}} \varphi \, dv_{\mathrm{R}}$ must be both ≥ 0 and ≤ 0; thus, $(15.9)_2$ implies that φ is both ≥ 0 and ≤ 0, and, hence, equal to zero. Thus, $(15.9)_1$ is valid.

In view of (15.9) and the strict positivity of the determinant J, (15.10) follows from (15.7).

Finally, to verify (†), choose an arbitrary time t_0 and an arbitrary spatial region $\mathcal{P}$ in $\mathcal{B}_{t_0}$. To find the convecting region $\mathcal{P}_t$ consistent with (15.11), we let P denote the material region that deforms to $\mathcal{P}_t$ at time t_0[88] and then take $\mathcal{P}_t = \boldsymbol{\chi}_t(\mathrm{P})$ for all t. Thus, (†) is valid.

[87] More precisely, if the left side of, say, $(15.9)_1$ holds for all time, then the right side holds on B for all time; and if the left side of, say, $(15.10)_1$ holds for all t, then the right side holds on $\mathcal{B}_t$ for all t.

[88] I.e. $\mathrm{P} = \boldsymbol{\chi}_{t_0}^{-1}(\mathcal{P})$; cf. (5.5).

16 Reynolds' Transport Relation. Isochoric Motions

To see the conventions and integral transformations in §15.2 in action, note that, by (9.16) and (15.7), granted sufficient smoothness,

$$\begin{aligned}\overline{\int_{\mathcal{P}_t} \varphi\, dv} &= \overline{\int_{\mathrm{P}} \varphi J\, dv_{\mathrm{R}}}^{\boldsymbol{\cdot}} \\ &= \int_{\mathrm{P}} (\dot{\varphi} J + \varphi \dot{J}\,)\, dv_{\mathrm{R}} \\ &= \int_{\mathrm{P}} (\dot{\varphi} + \varphi\,\mathrm{div}\mathbf{v}) J\, dv_{\mathrm{R}} \\ &= \int_{\mathcal{P}_t} (\dot{\varphi} + \varphi\,\mathrm{div}\mathbf{v})\, dv. \end{aligned} \tag{16.1}$$

We therefore have **Reynolds' transport relation**

$$\overline{\int_{\mathcal{P}_t} \varphi\, dv}^{\boldsymbol{\cdot}} = \int_{\mathcal{P}_t} (\dot{\varphi} + \varphi\mathrm{div}\mathbf{v})\, dv, \tag{16.2}$$

an integral identity valid whenever the spatial region $\mathcal{P}_t$ convects with the body; this identity is central to much of continuum mechanics.

A motion is **isochoric** if

$$\overline{\mathrm{vol}(\mathcal{P}_t)}^{\boldsymbol{\cdot}} = 0$$

for all spatial regions $\mathcal{P}_t$ convecting with the body, so that the deformed volume of any part is independent of time. Since

$$\mathrm{vol}(\mathcal{P}_t) = \int_{\mathcal{P}_t} dv,$$

the choice $\varphi \equiv 1$ in Reynolds' transport relation (16.2) yields

$$\begin{aligned}\overline{\mathrm{vol}(\mathcal{P}_t)}^{\cdot} &= \overline{\int_{\mathcal{P}_t} dv}^{\cdot} \\ &= \int_{\mathrm{P}} \dot{J}\, dv_{\mathrm{R}} \\ &= \int_{\mathcal{P}_t} \mathrm{div}\mathbf{v}\, dv. \end{aligned} \tag{16.3}$$

Thus, since the part P (and hence the spatial region $\mathcal{P}_t$) may be arbitrarily chosen, we may conclude that

(‡) the following three statements are equivalent:

(i) The motion is isochoric.
(ii) $\dot{J} = 0$.
(iii) $\mathrm{div}\mathbf{v} = 0$.

Further, a direct consequence of (16.2) and (iii) is that

$$\overline{\int_{\mathcal{P}_t} \varphi\, dv}^{\cdot} = \int_{\mathcal{P}_t} \dot{\varphi}\, dv \quad \text{in an isochoric motion.}$$

EXERCISES

1. Show that, since $\mathbf{L} = \dot{\mathbf{F}}\mathbf{F}^{-1}$ and $\mathbf{D} = \mathrm{sym}\,\mathbf{L}$,[89]

$$\mathbf{F}^{-\top}\dot{\mathbf{C}}\mathbf{F}^{-1} = 2\mathbf{D}$$

and that, therefore, a motion is isochoric if and only if

$$\mathrm{tr}\,(\mathbf{F}^{-\top}\dot{\mathbf{C}}\mathbf{F}^{-1}) = 0,$$

or, since[90]

$$\mathbf{D} = \mathbf{R}[\mathrm{sym}\,(\dot{\mathbf{U}}\mathbf{U}^{-1})]\mathbf{R}^{\top},$$

if and only if

$$\mathrm{tr}\,(\dot{\mathbf{U}}\mathbf{U}^{-1}) = 0.$$

2. Show that

$$\mathbf{A}\!:\!\dot{\mathbf{C}} = 2(\mathbf{F}\mathbf{A}\mathbf{F}^{\top})\!:\!\mathbf{D}.$$

[89] Cf. (11.10).
[90] Cf. $(11.8)_1$.

17 More Kinematics

The kinematics discussed below, while general and applicable to both solids and fluids, is particularly important in the study of fluids.

17.1 Vorticity

The **vorticity** $\boldsymbol{\omega}$ is defined by

$$\boldsymbol{\omega} = \operatorname{curl} \mathbf{v}. \tag{17.1}$$

Let $\mathbf{w}$ denote the axial vector of the spin $\mathbf{W}$ so that, by (2.36), $\mathbf{W} = \mathbf{w}\times$. Then, choosing an arbitrary vector $\mathbf{a}$ and bearing in mind that the tensor $\mathbf{a}\times$ is skew, it follows from $(2.61)_2$, (2.64), and $(3.19)_2$ that

$$\begin{aligned} \mathbf{a} \cdot \boldsymbol{\omega} &= (\mathbf{a}\times) : \mathbf{L} \\ &= (\mathbf{a}\times) : \mathbf{W} \\ &= 2\mathbf{a} \cdot \mathbf{w}. \end{aligned}$$

Thus, $\boldsymbol{\omega} = 2\mathbf{w}$ and, since $\mathbf{W} = \mathbf{w}\times$,[91] we find that the spin and vorticity in an arbitrary motion are related by

$$\mathbf{W} = \tfrac{1}{2}\boldsymbol{\omega}\times. \tag{17.2}$$

This shows that the vorticity serves as a vectorial counterpart to the spin. Recall from (11.4) that the **spin axis** denotes the subspace $\mathcal{L}$ of vectors $\mathbf{a}$ such that

$$\mathbf{W}\mathbf{a} = \mathbf{0}. \tag{17.3}$$

EXERCISE

1. Use (17.2) to show that

$$|\boldsymbol{\omega}|^2 = 2|\mathbf{W}|^2.$$

17.2 Transport Relations for Spin and Vorticity

Kinematical relations central to a discussion of fluids consist of[92]

(i) the relation

$$\dot{\mathbf{v}} = \mathbf{v}' + \mathbf{L}\mathbf{v} \tag{17.4}$$

[91] Cf. the result (10.3) obtained in our discussion of rigid motions.

[92] Cf. (9.8), (11.2), and (11.3).

between the acceleration $\dot{\mathbf{v}}$, the spatial time derivative $\mathbf{v}'$, and the spatial gradient $\mathbf{L} = \operatorname{grad}\mathbf{v}$ of the velocity (§9.2);

(ii) the decomposition

$$\mathbf{L} = \operatorname{grad}\mathbf{v} = \mathbf{D} + \mathbf{W}, \qquad \mathbf{D} = \operatorname{sym}\mathbf{L}, \qquad \mathbf{W} = \operatorname{skw}\mathbf{L}, \tag{17.5}$$

of the velocity gradient $\mathbf{L}$ into a symmetric stretching tensor $\mathbf{D}$ and a skew spin tensor $\mathbf{W}$;

(iii) the decomposition

$$\mathbf{D} = \mathbf{D}_0 + \tfrac{1}{3}(\operatorname{tr}\mathbf{D})\mathbf{1}, \qquad \operatorname{tr}\mathbf{D}_0 = 0, \tag{17.6}$$

of the stretching $\mathbf{D}$ into a deviatoric part $\mathbf{D}_0$ and a spherical part $\frac{1}{3}(\operatorname{tr}\mathbf{D})\mathbf{1}$.

We now establish several important relations for the transport of spin and vorticity. Since

$$\mathbf{L}^{\top}\mathbf{v} = (\operatorname{grad}\mathbf{v})^{\top}\mathbf{v} = \tfrac{1}{2}\operatorname{grad}(|\mathbf{v}|^2),$$

it follows that

$$2\mathbf{W}\mathbf{v} = \mathbf{L}\mathbf{v} - \tfrac{1}{2}\operatorname{grad}(|\mathbf{v}|^2)$$

and (9.8) yields the first of

$$\dot{\mathbf{v}} = \mathbf{v}' + \tfrac{1}{2}\operatorname{grad}(|\mathbf{v}|^2) + 2\mathbf{W}\mathbf{v}, \qquad \dot{\mathbf{v}} = \mathbf{v}' + \tfrac{1}{2}\operatorname{grad}(|\mathbf{v}|^2) + \boldsymbol{\omega}\times\mathbf{v}, \tag{17.7}$$

while the second follows from the first using (17.1) and (17.2).

Consider the vorticity and note that, by $(3.25)_6$ and $(17.7)_2$,

$$\operatorname{curl}\dot{\mathbf{v}} = \boldsymbol{\omega}' + \operatorname{curl}(\boldsymbol{\omega}\times\mathbf{v}).$$

Thus, since, by $(3.20)_{6,10}$,

$$\begin{aligned}\operatorname{curl}(\boldsymbol{\omega}\times\mathbf{v}) &= (\operatorname{grad}\boldsymbol{\omega})\mathbf{v} + (\operatorname{div}\mathbf{v})\boldsymbol{\omega} - (\operatorname{div}\boldsymbol{\omega})\mathbf{v} - (\operatorname{grad}\mathbf{v})\boldsymbol{\omega}\\ &= (\operatorname{grad}\boldsymbol{\omega})\mathbf{v} + (\operatorname{div}\mathbf{v})\boldsymbol{\omega} - \mathbf{L}\boldsymbol{\omega},\end{aligned} \tag{17.8}$$

and since $\dot{\boldsymbol{\omega}} = \boldsymbol{\omega}' + (\operatorname{grad}\boldsymbol{\omega})\mathbf{v}$, we have an important transport relation for the vorticity:

$$\dot{\boldsymbol{\omega}} - \mathbf{L}\boldsymbol{\omega} + (\operatorname{div}\mathbf{v})\,\boldsymbol{\omega} = \operatorname{curl}\dot{\mathbf{v}}. \tag{17.9}$$

Next, since $\mathbf{W}$ is the skew part of $\mathbf{L}$, (9.12) yields

$$\begin{aligned}2\mathbf{F}^{\top}\mathbf{W}\mathbf{F} &= \mathbf{F}^{\top}(\dot{\mathbf{F}}\mathbf{F}^{-1} - \mathbf{F}^{-\top}\dot{\mathbf{F}}^{\top})\mathbf{F},\\ &= \mathbf{F}^{\top}\dot{\mathbf{F}} - \dot{\mathbf{F}}^{\top}\mathbf{F},\end{aligned}$$

and hence, using (9.22),

$$\begin{aligned}2\overline{\mathbf{F}^{\top}\mathbf{W}\mathbf{F}}^{\boldsymbol{\cdot}} &= \mathbf{F}^{\top}\ddot{\mathbf{F}} - \ddot{\mathbf{F}}^{\top}\mathbf{F}\\ &= \mathbf{F}^{\top}(\ddot{\mathbf{F}}\mathbf{F}^{-1} - \mathbf{F}^{-\top}\ddot{\mathbf{F}}^{\top})\mathbf{F}\\ &= \mathbf{F}^{\top}(\operatorname{grad}\dot{\mathbf{v}} - (\operatorname{grad}\dot{\mathbf{v}})^{\top})\mathbf{F}.\end{aligned}$$

Thus, if we define

$$\mathbf{J} = \tfrac{1}{2}(\operatorname{grad}\dot{\mathbf{v}} - (\operatorname{grad}\dot{\mathbf{v}})^{\top}), \tag{17.10}$$

steps analogous to that leading to (17.2) yield

$$\mathbf{J} = (\operatorname{curl}\dot{\mathbf{v}})\times, \tag{17.11}$$

which implies a material transport relation for the spin

$$\overline{\mathbf{F}^{\top}\mathbf{W}\mathbf{F}}^{\boldsymbol{\cdot}} = \mathbf{F}^{\top}\mathbf{J}\mathbf{F}. \tag{17.12}$$

We use the adjective "material" because tensors of the form $\mathbf{F}^{\top}(\cdots)\mathbf{F}$ map material vectors to material vectors.[93]

Continuing, by (9.11), $\dot{\mathbf{F}} = \mathbf{L}\mathbf{F}$, so that

$$\begin{aligned} 2\overline{\mathbf{F}^{\top}\mathbf{W}\mathbf{F}}^{\boldsymbol{\cdot}} &= \mathbf{F}^{\top}\dot{\mathbf{W}}\mathbf{F} + \dot{\mathbf{F}}^{\top}\mathbf{W}\mathbf{F} + \mathbf{F}^{\top}\mathbf{W}\dot{\mathbf{F}} \\ &= \mathbf{F}^{\top}\dot{\mathbf{W}}\mathbf{F} + \mathbf{F}^{\top}\mathbf{L}^{\top}\mathbf{W}\mathbf{F} + \mathbf{F}^{\top}\mathbf{W}\mathbf{L}\mathbf{F} \\ &= \mathbf{F}^{\top}(\dot{\mathbf{W}} + \mathbf{L}^{\top}\mathbf{W} + \mathbf{W}\mathbf{L})\mathbf{F}, \end{aligned} \tag{17.13}$$

and, by (17.12), we have a spatial transport relation for the spin

$$\dot{\mathbf{W}} + \mathbf{W}\mathbf{L} + \mathbf{L}^{\top}\mathbf{W} = \mathbf{J}. \tag{17.14}$$

Further, since $\mathbf{L} = \mathbf{D} + \mathbf{W}$,

$$\begin{aligned} \mathbf{W}\mathbf{L} + \mathbf{L}^{\top}\mathbf{W} &= \mathbf{W}(\mathbf{D} + \mathbf{W}) + (\mathbf{D} - \mathbf{W})\mathbf{W} \\ &= \mathbf{W}\mathbf{D} + \mathbf{D}\mathbf{W}, \end{aligned}$$

and, by (17.14), we may write (17.14) in the form

$$\dot{\mathbf{W}} + \mathbf{W}\mathbf{D} + \mathbf{D}\mathbf{W} = \mathbf{J}. \tag{17.15}$$

17.3 Irrotational Motions

A motion is **irrotational** if

$$\mathbf{W} = \mathbf{0},$$

or, equivalently, by (17.2), if

$$\operatorname{curl}\mathbf{v} = \mathbf{0}.$$

A spatial vector field $\mathbf{g}$ is the **gradient of a potential** if there is a spatial scalar field φ such that

$$\mathbf{g} = \operatorname{grad}\varphi. \tag{17.16}$$

For a large class of fluids, in particular, for inviscid fluids subject to conservative conventional body forces, the acceleration is the gradient of a potential. When this is so, then (17.10) and (17.16) imply that

$$\mathbf{J} = \mathbf{0}, \tag{17.17}$$

and, hence, by virtue of (17.14), that

$$\overline{\mathbf{F}^{\top}\mathbf{W}\mathbf{F}}^{\boldsymbol{\cdot}} = \mathbf{0}. \tag{17.18}$$

---

93 In fact, the tensor $\mathbf{F}^{\top}\mathbf{W}\mathbf{F}$ is the covariant pullback of the spin from the deformed body to the reference body. Cf. (13.19) and the terminology introduced in the paragraph containing (13.20).

Granted this — expressing all fields materially — $(\mathbf{F}^{\top}\mathbf{W}\mathbf{F})(\mathbf{X},t)$ is independent of t. Thus, if the motion is irrotational at some time τ, then $\mathbf{W}(\mathbf{X},\tau)=\mathbf{0}$ for all $\mathbf{X}$ in B, so that $(\mathbf{F}^{\top}\mathbf{W}\mathbf{F})(\mathbf{X},\tau)=\mathbf{0}$ for all $\mathbf{X}$ in B. Thus, by (17.18), $(\mathbf{F}^{\top}\mathbf{W}\mathbf{F})(\mathbf{X},t)=\mathbf{0}$ for all $\mathbf{X}$ in B and all t, and, since $\mathbf{F}$ is invertible, $\mathbf{W}(\mathbf{X},t)=\mathbf{0}$ for all $\mathbf{X}$ in B and all t and the motion is irrotational for all time:

$$\operatorname{curl}\mathbf{v}\equiv\mathbf{0}. \tag{17.19}$$

We therefore have the following important result of LAGRANGE and CAUCHY:

(‡) *A motion with acceleration the gradient of a potential is irrotational if it is irrotational at some time.*

Consider now a plane motion. The stretching and spin, $\mathbf{D}$ and $\mathbf{W}$, then have matrices of the form

$$[\mathbf{D}]=\begin{bmatrix} a & b & 0\\ b & c & 0\\ 0 & 0 & 0\end{bmatrix},\qquad [\mathbf{W}]=\begin{bmatrix} 0 & g & 0\\ -g & 0 & 0\\ 0 & 0 & 0\end{bmatrix},$$

so that

$$[\mathbf{W}\mathbf{D}+\mathbf{D}\mathbf{W}]=\begin{bmatrix} 0 & g(a+c) & 0\\ -g(a+c) & 0 & 0\\ 0 & 0 & 0\end{bmatrix}=(a+c)[\mathbf{W}].$$

Thus, since

$$\begin{aligned}\operatorname{div}\mathbf{v}&=\operatorname{tr}\mathbf{L}\\ &=\operatorname{tr}\mathbf{D}\\ &=a+c,\end{aligned}$$

it follows that

$$\mathbf{W}\mathbf{D}+\mathbf{D}\mathbf{W}=(\operatorname{div}\mathbf{v})\mathbf{W}.$$

Thus, by (17.14) and (17.10), for a plane, isochoric motion with acceleration the gradient of a potential,

$$\dot{\mathbf{W}}\equiv\mathbf{0}.$$

17.4 Circulation

Consider a motion of the body. Let C be a **material curve** described by the parametrization

$$\hat{\mathbf{X}}(\lambda),\qquad \lambda_0\le\lambda\le\lambda_1,$$

and let

$$\mathcal{C}_t=\chi_t(\mathrm{C}) \tag{17.20}$$

denote the corresponding spatial curve described by the time-dependent parametrization

$$\hat{\mathbf{x}}_t(\lambda)=\chi_t(\hat{\mathbf{X}}(\lambda)),\qquad \lambda_0\le\lambda\le\lambda_1, \tag{17.21}$$

described in the paragraph containing (6.11). Assume that C and, hence, $\mathcal{C}_t$ are closed, so that

$$\hat{\mathbf{x}}_t(\lambda_1)=\hat{\mathbf{x}}_t(\lambda_0). \tag{17.22}$$

The line integral[94]

$$\int_{\mathcal{C}_t} \mathbf{v}(\mathbf{x}, t) \cdot \mathbf{dx} \tag{17.23}$$

then gives the **circulation** around $\mathcal{C}_t$ at time t. An important result, which we now derive, involves the time derivative

$$\overline{\int_{\mathcal{C}_t} \mathbf{v}(\mathbf{x}, t) \cdot \mathbf{dx}}^{\cdot} \tag{17.24}$$

of the circulation. By definition,

$$\int_{\mathcal{C}_t} \mathbf{v}(\mathbf{x}, t) \cdot \mathbf{dx} = \int_{\lambda_0}^{\lambda_1} \mathbf{v}(\hat{\mathbf{x}}_t(\lambda), t) \cdot \frac{\partial \hat{\mathbf{x}}_t(\lambda)}{\partial \lambda} d\lambda, \tag{17.25}$$

and our evaluation of (17.24) involves differentiating the right side of (17.25) under the integral. We begin with some identities. By (17.21),

$$\begin{aligned} \frac{\partial}{\partial t}\hat{\mathbf{x}}_t(\lambda) &= \frac{\partial}{\partial t}\chi_t(\hat{\mathbf{X}}(\lambda), t) \\ &= \dot{\chi}(\hat{\mathbf{X}}(\lambda), t) \\ &= \mathbf{v}(\hat{\mathbf{x}}_t(\lambda), t); \end{aligned} \tag{17.26}$$

further,

$$\begin{aligned} \frac{\partial^2}{\partial t \partial \lambda}\hat{\mathbf{x}}_t(\lambda) &= \frac{\partial^2}{\partial \lambda \partial t}\hat{\mathbf{x}}_t(\lambda) \\ &= \frac{\partial}{\partial \lambda}\mathbf{v}(\hat{\mathbf{x}}_t(\lambda), t) \end{aligned} \tag{17.27}$$

and

$$\begin{aligned} \frac{\partial}{\partial t}\mathbf{v}(\hat{\mathbf{x}}_t(\lambda), t) &= \ddot{\chi}(\hat{\mathbf{X}}(\lambda), t) \\ &= \dot{\mathbf{v}}(\hat{\mathbf{x}}_t(\lambda), t). \end{aligned} \tag{17.28}$$

The identities (17.26)–(17.28) can be used to transform (17.24) as follows:

$$\begin{aligned} \overline{\int_{\mathcal{C}_t} \mathbf{v}(\mathbf{x}, t) \cdot \mathbf{dx}}^{\cdot} &= \frac{d}{dt}\int_{\lambda_0}^{\lambda_1} \mathbf{v}(\hat{\mathbf{x}}_t(\lambda), t) \cdot \frac{\partial \hat{\mathbf{x}}_t(\lambda)}{\partial \lambda} d\lambda \\ &= \int_{\lambda_0}^{\lambda_1} \frac{\partial}{\partial t}\Big(\mathbf{v}(\hat{\mathbf{x}}_t(\lambda), t)\Big) \cdot \frac{\partial \hat{\mathbf{x}}_t(\lambda)}{\partial \lambda} d\lambda + \int_{\lambda_0}^{\lambda_1} \mathbf{v}(\hat{\mathbf{x}}_t(\lambda), t) \cdot \frac{\partial}{\partial t}\left(\frac{\partial \hat{\mathbf{x}}_t(\lambda)}{\partial \lambda}\right) d\lambda \end{aligned}$$

---

[94] Cf. §4.2.

$$= \int_{\lambda_0}^{\lambda_1} \dot{\mathbf{v}}(\hat{\mathbf{x}}_t(\lambda), t) \cdot \frac{\partial \hat{\mathbf{x}}_t(\lambda)}{\partial \lambda} d\lambda + \int_{\lambda_0}^{\lambda_1} \mathbf{v}(\hat{\mathbf{x}}_t(\lambda), t) \cdot \frac{\partial}{\partial \lambda} \mathbf{v}(\hat{\mathbf{x}}_t(\lambda), t) d\lambda$$

$$= \int_{\mathcal{C}_t} \dot{\mathbf{v}}(\mathbf{x}, t) \cdot \mathbf{dx} + \underbrace{\tfrac{1}{2}\left[|\mathbf{v}(\hat{\mathbf{x}}_t(\lambda_1), t)|^2 - |\mathbf{v}(\hat{\mathbf{x}}_t(\lambda_0), t)|^2\right]}_{=0 \;\text{ since }\; \hat{\mathbf{x}}_t(\lambda_0) = \hat{\mathbf{x}}_t(\lambda_1)}.$$

Thus, we have the **circulation-transport relation**: For the spatial image $\mathcal{C}_t$ of a closed material curve C,

$$\overline{\int_{\mathcal{C}_t} \mathbf{v}(\mathbf{x}, t) \cdot \mathbf{dx}}^{\boldsymbol{\cdot}} = \int_{\mathcal{C}_t} \dot{\mathbf{v}}(\mathbf{x}, t) \cdot \mathbf{dx}. \tag{17.29}$$

We say that the motion *preserves circulation* if

$$\overline{\int_{\mathcal{C}_t} \mathbf{v}(\mathbf{x}, t) \cdot \mathbf{dx}}^{\boldsymbol{\cdot}} = 0$$

for every *closed material curve* and all t.

Assume that the acceleration is the gradient of a potential, then by (4.7), for $\mathcal{C}_t$ a closed curve,

$$\int_{\mathcal{C}_t} \dot{\mathbf{v}}(\mathbf{x}, t) \cdot \mathbf{dx} = \int_{\mathcal{C}_t} \nabla\varphi(\mathbf{x}, t) \cdot \mathbf{dx} = 0.$$

Appealing to the circulation transport relation (17.29), we therefore have a famous result of KELVIN:

(‡) *If the acceleration is the gradient of a potential, then the motion preserves circulation.*

17.5 Vortex Lines

Let C be a material curve with $\mathcal{C}_t$ the corresponding spatial curve at each time t. Further, let $\boldsymbol{\tau}_{\mathrm{R}}(\mathbf{X})$, defined for every $\mathbf{X}$ on C, and $\boldsymbol{\tau}(\mathbf{x}, t)$, defined for every $\mathbf{x}$ on $\mathcal{C}_t$, denote respective tangents to C and $\mathcal{C}_t$, so that, by (6.15),

$$\boldsymbol{\tau}(\mathbf{x}, t) = \mathbf{F}(\mathbf{X}, t)\boldsymbol{\tau}_{\mathrm{R}}(\mathbf{X}), \qquad \mathbf{x} = \boldsymbol{\chi}_t(\mathbf{X}). \tag{17.30}$$

We say that $\mathcal{C}_t$ is a **vortex line** at time $t = s$ if the tangent to $\mathcal{C}_s$ at each spatial point $\mathbf{x}$ on $\mathcal{C}_s$ lies on the spin axis of the motion at $(\mathbf{x}, s)$. By (17.3), the spin axis at $(\mathbf{x}, s)$ is the subspace $\mathcal{L}$ of all vectors $\mathbf{a}$ such that $\mathbf{W}(\mathbf{x}, s)\mathbf{a} = \mathbf{0}$. Thus, $\mathcal{C}_s$ is a vortex line if and only if

$$\mathbf{W}(\mathbf{x}, s)\boldsymbol{\tau}(\mathbf{x}, s) = \mathbf{0},$$

or, equivalently,

$$\mathbf{W}(\mathbf{x}, s)\mathbf{F}(\mathbf{X}, s)\boldsymbol{\tau}_{\mathrm{R}}(\mathbf{X}) = \mathbf{0}, \qquad \mathbf{x} = \boldsymbol{\chi}_s(\mathbf{X}),$$

so that, trivially,

$$\mathbf{F}^{\top}(\mathbf{X}, s)\mathbf{W}(\mathbf{x}, s)\mathbf{F}(\mathbf{X}, s)\boldsymbol{\tau}_{\mathrm{R}}(\mathbf{X}) = \mathbf{0}, \qquad \mathbf{x} = \boldsymbol{\chi}_s(\mathbf{X}), \tag{17.31}$$

for all $\mathbf{x}$ on $\mathcal{C}_s$. Assume that the acceleration is the gradient of a potential, so that, by (17.18),

$$\overline{\mathbf{F}^{\top}\mathbf{W}\mathbf{F}}^{\cdot} = \mathbf{0}$$

and (17.31) must hold for all t:

$$\mathbf{F}^{\top}(\mathbf{X},t)\mathbf{W}(\mathbf{x},t)\mathbf{F}(\mathbf{X},t)\boldsymbol{\tau}_{\mathrm{R}}(\mathbf{X}) = \mathbf{0}, \qquad \mathbf{x} = \chi_t(\mathbf{X}),$$

for all $\mathbf{x}$ on $\mathcal{C}_t$. Equivalently, by (17.30),

$$\mathbf{W}(\mathbf{x},t)\boldsymbol{\tau}(\mathbf{x},t) = \mathbf{0}$$

and $\mathcal{C}_t$ is a vortex line at all times t. We therefore have the important result:

(‡) *Assume that the acceleration is the gradient of a potential. Vortex lines then convect with the body; that is, for any material curve* C*, if* $\mathcal{C}_s$ *is a vortex line at some time s, then* $\mathcal{C}_t$ *is a vortex line for all time t.*

17.6 Steady Motions

We say that a motion is **steady** if

(i) the deformed body is independent of time,

$$\mathcal{B}_t = \mathcal{B}_0 \;\text{ for all time } t; \tag{17.32}$$

(ii) the velocity $\mathbf{v}$ is independent of time,

$$\mathbf{v}' = \mathbf{0},$$

and, hence, a field $\mathbf{v}(\mathbf{x})$ with domain $\mathcal{B}_0$.

Given a steady motion, consider the differential equation

$$\frac{d\mathbf{s}(t)}{dt} = \mathbf{v}(\mathbf{s}(t)). \tag{17.33}$$

Solutions $\mathbf{s}$ of this differential equation are referred to as **streamlines** of the motion.

Two important properties of a steady motion are that all particles that pass through a given spatial point $\mathbf{x}$ do with velocity $\mathbf{v}(\mathbf{x})$; and particle paths and streamlines satisfy the same differential equation (17.33), so that *every streamline is a particle path and every particle path is a streamline.*

Given a steady motion, we say that φ is a **steady field** if

$$\varphi' = 0, \tag{17.34}$$

so that φ is independent of time; by (9.7) and (17.34), φ is steady if and only if

$$\dot{\varphi} = \mathbf{v}\cdot\operatorname{grad}\varphi. \tag{17.35}$$

We say that a steady field φ is **constant on streamlines** if, given any streamline $\mathbf{s}$,

$$\frac{d}{dt}\varphi(\mathbf{s}(t)) = 0 \tag{17.36}$$

for all t. But, for φ steady and $\mathbf{s}$ an arbitrary streamline, (17.33) yields

$$\begin{aligned}\frac{d}{dt}\varphi(\mathbf{s}(t)) &= \operatorname{grad}\varphi(\mathbf{s}(t))\dot{\mathbf{s}}(t)\\ &= \operatorname{grad}\varphi(\mathbf{s}(t))\mathbf{v}(\mathbf{s}(t))\\ &= \dot{\varphi}(\mathbf{s}(t)).\end{aligned} \tag{17.37}$$

Thus, since every point of $\mathcal{B}_0$ has a streamline passing through it,[95]

- *a steady field φ is constant on streamlines if and only if $\dot{\varphi} = 0$.*

Consider now a (not necessarily steady) motion χ and choose a material point $\mathbf{X}$ with $\chi(\mathbf{X}, \tau)$ in $\mathcal{B}_\tau$ at some time τ. Then, by (‡) on page 63, $\chi(\mathbf{X}, t)$ lies on $\partial\mathcal{B}_t$ for all time t. If, in particular, the motion is steady, so that $\mathcal{B}_t \equiv \mathcal{B}_0$, then $\chi(\mathbf{X}, t)$, as a function of t, describes a curve on $\partial\mathcal{B}_0$; hence $\dot{\chi}(\mathbf{X}, t)$ is tangent to $\partial\mathcal{B}_0$. Thus,

- *in a steady motion the velocity field is tangent to the boundary; specifically, $\mathbf{v}(\mathbf{x})$ is tangent to $\partial\mathcal{B}_0$ at each point $\mathbf{x}$ on $\partial\mathcal{B}_0$.*

17.7 A Class of Natural Reference Configurations for Fluids

Standard problems involving fluids are initial-value problems that generally begin with a prescription of

(i) the spatial region B_{t_0} occupied by the fluid at a given time t_0; and
(ii) the fluid's specific volume $\upsilon_0(\mathbf{X})$ at each $\mathbf{X}$ in B_{t_0}.

Bearing this in mind, and bearing in mind the need for a reference configuration with respect to which material points are labeled, we stipulate that

(‡) *the reference configuration of the fluid is the configuration at a prescribed time t_0.*

Granted this, motions χ of the fluid are always reckoned with respect to the configuration of the fluid at time t_0; specifically,

(M1) χ associates with each material point $\mathbf{X}$ in B_{t_0} and each time t a spatial point

$$\mathbf{x} = \chi(\mathbf{X}, t); \tag{17.38}$$

(M2) χ satisfies the time-t_0 condition

$$\chi(\mathbf{X}, t_0) = \mathbf{X} \tag{17.39}$$

for all $\mathbf{X}$ in B_{t_0}.

Motions consistent with (M1) and (M2) are referred to as **motions relative to time t_0**.

17.8 The Motion Problem

17.8.1 Kinematical Boundary Conditions

Consider a motion χ and choose a material point $\mathbf{X}$ with $\chi(\mathbf{X}, \tau)$ in $\mathcal{B}_\tau$ at some time τ. Then, by (‡) on page 63, $\chi(\mathbf{X}, t)$ lies on $\partial\mathcal{B}_t$ for all time t. In particular, if $\mathcal{B}_t \equiv \mathcal{B}_0$ — which would be the case if the motion were steady — then $\chi(\mathbf{X}, t)$, as a function of t, describes a curve on $\partial\mathcal{B}_0$; hence $\dot{\chi}(\mathbf{X}, t)$ is tangent to $\partial\mathcal{B}_0$. Thus, letting $\mathbf{n}(\mathbf{x}, t)$ denote the outward (say) normal to the boundary, the velocity must obey the **boundary condition**

$$\mathbf{v}(\mathbf{x}, t) \cdot \mathbf{n}(\mathbf{x}, t) = 0 \tag{17.40}$$

for all $\mathbf{x}$ on $\partial\mathcal{B}_0$ and all t.

More generally, the boundary condition (17.40) would be applicable at all times t at which a portion $(\partial\mathcal{B}_t)_{\mathrm{abut}}$ of $\partial\mathcal{B}_t$ abuts a *fixed* solid surface. In this case we would

[95] Granted $\mathbf{v}$ is smooth, given any spatial point $\mathbf{x}$ and any time t_0, the differential equation (17.36) always has a (unique) solution curve $\mathbf{s}$ passing through $\mathbf{x}$ at time t_0.

have the boundary condition

$$\mathbf{v}(\mathbf{x}, t) \cdot \mathbf{n}(\mathbf{x}, t) = 0 \qquad \text{for all } \mathbf{x} \text{ on } (\partial\mathcal{B}_t)_{\text{abut}}. \tag{17.41}$$

17.8.2 The Motion Problem in a Fixed Container

When discussing fluids, it is often most convenient to work with the velocity field $\mathbf{v}$ rather than with the motion $\boldsymbol{\chi}$. If the flow region is known *a priori*—for example, for problems involving flows in which the fluid fills a fixed container occupying a region $\mathcal{B} \equiv \mathcal{B}_t$, then identifying material points by their positions at some "fixed time", say t_0, the motion at all other times is the solution of the **motion problem**

$$\left.\begin{aligned} \dot{\boldsymbol{\chi}}(\mathbf{X}, t) &= \mathbf{v}(\boldsymbol{\chi}(\mathbf{X}, t), t), \\ \boldsymbol{\chi}(\mathbf{X}, t_0) &= \mathbf{X}, \end{aligned}\right\} \tag{17.42}$$

for all $\mathbf{X}$ in $\mathcal{B}$ and all t. To be guaranteed a solution for all time, the velocity $\mathbf{v}$ on the boundary cannot be arbitrary, as it must be consistent with (17.42). Granted this, if $\mathbf{v}$ is smooth and bounded on $\mathcal{B}$ for all t, then the motion problem has a smooth solution $\boldsymbol{\chi}$ with $\boldsymbol{\chi}_t(\mathbf{X})$ one-to-one in $\mathbf{X}$ for all t.[96]

The motion problem is difficult when a portion of the boundary is a free surface, as the problem then is not purely kinematical: the conventional free-surface condition has $\mathbf{Tn} = p_{\text{en}}\mathbf{n}$, with p_{en} the environmental pressure, and the complete system of field equations would be needed to track the evolution of the free surface.

17.8.3 The Motion Problem in All of Space. Solution with Constant Velocity Gradient

Consider now a motion $\boldsymbol{\chi}$ relative to time t_0, so that (17.38) and (17.39) hold. Then,

$$\mathbf{F}_0 = \mathbf{1}, \qquad \dot{\mathbf{F}} = \mathbf{L}, \qquad \mathbf{C} = \mathbf{1}, \qquad \dot{\mathbf{C}} = 2\mathbf{D},$$

and the results of §14 specialize to yield the following solution of the motion problem

$$\boldsymbol{\chi}(\mathbf{X}, t) = \mathbf{x}_0 + e^{\mathbf{L}(t-t_0)}(\mathbf{X} - \mathbf{x}_0). \tag{17.43}$$

[96] A proof of this assertion, which may be based on a corollary of a theorem of PEANO (cf. Theorem 3.1 of HARTMAN (1964)), is beyond the scope of this book.

PART IV

BASIC MECHANICAL PRINCIPLES

We initially formulate balance laws globally for regions that convect with the body. Using the requirement that the underlying regions be arbitrary, we then derive local balance laws in the form of partial-differential equations (and jump conditions when the velocity and deformation gradient suffer jump discontinuities). Thereafter, we derive equivalent global balance laws for material regions and for control volumes. We therefore utilize three types of regions:

(i) *Material regions* P.
(ii) *Spatial regions* $\mathcal{P}_t$ *that convect with the body.* In this case there is a material region P such that

$$\mathcal{P}_t = \chi_t(\mathrm{P})$$

for all t.
(iii) *Control volumes* R that are *fixed spatial* regions that lie in the deformed body $\mathcal{B}_t$ *for all* t *in some time interval.* In contrast to a region $\mathcal{P}_t$ that convects with the body (for which material cannot cross $\partial\mathcal{P}_t$),[97]
   (†) *material generally flows into and out of a control volume* R *across its boundary.*

Throughout this book the symbols P, $\mathcal{P}_t$, and R, respectively, always have the meanings specified in (i), (ii), and (iii).

We consistently use the notation

$$\overline{\int_{\mathrm{P}} \varphi(\mathbf{X},t)\,dv_{\mathrm{R}}(\mathbf{X})}^{\boldsymbol{\cdot}} = \frac{d}{dt}\underbrace{\int_{\mathrm{P}} \varphi(\mathbf{X},t)\,dv_{\mathrm{R}}(\mathbf{X})}_{\Phi(t)}, \tag{17.44}$$

$$\overline{\int_{\mathcal{P}_t} \varphi(\mathbf{x},t)\,dv(\mathbf{x})}^{\boldsymbol{\cdot}} = \frac{d}{dt}\underbrace{\int_{\mathcal{P}_t} \varphi(\mathbf{x},t)\,dv(\mathbf{x})}_{\Theta(t)}, \tag{17.45}$$

[97] Cf. (†) on page 63.

and

$$\overline{\int_R \varphi(\mathbf{x},t)\,dv(\mathbf{x})}^{\,\boldsymbol{\cdot}} = \frac{d}{dt}\underbrace{\int_R \varphi(\mathbf{x},t)\,dv(\mathbf{x})}_{\Lambda(t)}. \tag{17.46}$$

Warning: The notation for time derivatives of material and spatial integrals used by many authors is often ambiguous, confusing, and in conflict with what a student is taught in calculus. In contrast, the notation defined in (17.44), (17.45), and (17.46) suffers none of these deficiencies. In (17.44), (17.45), and (17.46), the respective variables **X**, **x**, and **x** are "integrated out," leaving functions $\Phi(t)$, $\Theta(t)$, and $\Lambda(t)$ that **depend only on** t; the derivatives $d\Phi(t)/dt$, $d\Theta(t)/dt$, and $d\Lambda(t)/dt$ are therefore well-defined.

For convenience, we usually omit arguments when writing integrals of the form (17.44)–(17.46) and write instead

$$\overline{\int_{\mathrm{P}} \varphi\,dv_{\mathrm{R}}}^{\,\boldsymbol{\cdot}}, \qquad \overline{\int_{\mathcal{P}_t} \varphi\,dv}^{\,\boldsymbol{\cdot}}, \qquad \text{and} \qquad \overline{\int_R \varphi\,dv}^{\,\boldsymbol{\cdot}}. \tag{17.47}$$

This notation, while somewhat less transparent than (17.44)–(17.46), is consistent: In (17.44) the volume element dv_{R} is material and so the integration is with respect to **X** holding t fixed; in (17.45) and (17.46) the volume element dv is spatial and so the integration is with respect to **x** holding t fixed.

18 Balance of Mass

We assume throughout this section that the motion is smooth. Balance of mass in the presence of a shock wave is discussed in §33.1.

18.1 Global Form of Balance of Mass

Let $\mathcal{P}_t$ be a spatial region that convects with the body, so that $\mathcal{P}_t = \chi_t(\mathrm{P})$ for some material region P. We write $\rho_{\mathrm{R}}(\mathbf{X}) > 0$ for the mass density at the material point $\mathbf{X}$ in the reference body B, so that

$$\int_{\mathrm{P}} \rho_{\mathrm{R}}(\mathbf{X})\, dv_{\mathrm{R}}(\mathbf{X})$$

represents the mass of the material region P. We refer to $\rho_{\mathrm{R}}(\mathbf{X})$ as the **reference density**. Analogously, given a motion, we write $\rho(\mathbf{x}, t) > 0$ for the (mass) **density** at the spatial point $\mathbf{x}$ in the deformed body $\mathcal{B}_t$, so that

$$\int_{\mathcal{P}_t} \rho(\mathbf{x}, t)\, dv(\mathbf{x})$$

represents the mass of the spatial region $\mathcal{P}_t$ occupied by P at time t. **Balance of mass** is then the requirement that, given any motion,

$$\int_{\mathrm{P}} \rho_{\mathrm{R}}(\mathbf{X})\, dv_{\mathrm{R}}(\mathbf{X}) = \int_{\mathcal{P}_t} \rho(\mathbf{x}, t)\, dv(\mathbf{x}) \tag{18.1}$$

for every material region P. The left side of (18.1) is independent of t; differentiating (18.1) with respect to t, thus, yields an expression for balance of mass appropriate to a convecting spatial region $\mathcal{P}_t$:

$$\boxed{\overline{\int_{\mathcal{P}_t} \rho\, dv}^{\,\boldsymbol{\cdot}} = 0.} \tag{18.2}$$

This form of mass balance is independent of the choice of reference configuration and is often used in discussions of fluids. More importantly, because (18.2) does not involve derivatives of the integrand, *it is valid even in the presence of a shock wave.*

Note that (18.1) applied to the body yields

$$\int_{\mathrm{B}} \rho_{\mathrm{R}}(\mathbf{X})\, dv_{\mathrm{R}}(\mathbf{X}) = \int_{\mathcal{B}_t} \rho(\mathbf{x},t)\, dv(\mathbf{x})$$

$$\stackrel{\text{def}}{=} M, \tag{18.3}$$

with M the (fixed) **mass of the body**.

18.2 Local Forms of Balance of Mass

In view of Reynolds' transport relation (16.2), balance of mass (18.2) implies that

$$\int_{\mathcal{P}_t} (\dot{\rho} + \rho\, \mathrm{div}\mathbf{v})\, dv = 0 \tag{18.4}$$

for every convecting region $\mathcal{P}_t$, and, in view (15.10)$_1$, yields the local **mass balance**

$$\boxed{\dot{\rho} + \rho\, \mathrm{div}\mathbf{v} = 0, \qquad \dot{\rho} + \rho\frac{\partial v_i}{\partial x_i} = 0.} \tag{18.5}$$

By (9.7)$_1$, this balance may be written equivalently as

$$\rho' + \mathrm{div}(\rho\mathbf{v}) = 0. \tag{18.6}$$

The field υ defined by

$$\upsilon = \frac{1}{\rho} \tag{18.7}$$

is referred to as the **specific volume**; by (18.5) and (18.6),

$$\dot{\upsilon} = \upsilon\, \mathrm{div}\mathbf{v}. \tag{18.8}$$

We now provide a derivation of the local mass balance (18.5) that utilizes the reference configuration. By (15.7) applied to the right side of (18.1),

$$\int_{\mathcal{P}_t} \rho\, dv = \int_{\mathrm{P}} \rho J\, dv_{\mathrm{R}},$$

so that, by (18.1),

$$\int_{\mathrm{P}} (\rho_{\mathrm{R}} - \rho J)\, dv_{\mathrm{R}} = 0.$$

Thus, since P is arbitrary,

$$\boxed{\rho = \frac{\rho_{\mathrm{R}}}{J},} \tag{18.9}$$

which is an alternative relation expressing balance of mass. Note that, unlike mass balance in the form (18.5), this relation, which is algebraic, is based on a choice of reference configuration. An immediate consequence of this relation and (15.7) is the transformation law

$$\int_{\mathcal{P}_t} \varphi\rho\, dv = \int_{\mathrm{P}} \varphi\rho_{\mathrm{R}}\, dv_{\mathrm{R}}, \tag{18.10}$$

which with arguments has the form

$$\int_{\mathcal{P}_t} \varphi(\mathbf{x},t)\rho(\mathbf{x},t)\,dv(\mathbf{x}) = \int_{\mathrm{P}} \varphi(\boldsymbol{\chi}(\mathbf{X},t),t)\rho_{\mathrm{R}}(\mathbf{X})\,dv_{\mathrm{R}}(\mathbf{X}). \tag{18.11}$$

Like the "volume elements" dv and dv_{R}, the quantities $dm = \rho\,dv$ and $dm_{\mathrm{R}} = \rho_{\mathrm{R}}\,dv_{\mathrm{R}}$ represent well-defined mathematical objects called *measures*. The relation (18.11) asserts that integrating a spatial field φ over the spatial region $\mathcal{P}_t$ with respect to the spatial mass measure dm is equivalent to integrating the material description of φ over the corresponding material region P with respect to the referential mass measure dm_{R}. In this regard note that, formally,

$$dm = dm_{\mathrm{R}},$$

which is a relation comparable to the corresponding relation $dv = J\,dv_{\mathrm{R}}$ for volume.[98]

EXERCISE

1. Derive the local mass balance (18.5) by taking the material time-derivative of (18.9) and using (9.16).

18.3 Simple Consequences of Mass Balance

Let φ be a spatial scalar field. Then, by $(9.7)_1$ and (18.6),

$$\begin{aligned}(\rho\varphi)' &= \rho\varphi' + \rho'\varphi\\ &= \rho\varphi' - \varphi\mathrm{div}(\rho\mathbf{v})\\ &= \rho\varphi' - \mathrm{div}(\rho\varphi\mathbf{v}) + \rho\mathbf{v}\cdot\mathrm{grad}\varphi\\ &= \rho(\varphi' + \mathbf{v}\cdot\mathrm{grad}\varphi) - \mathrm{div}(\rho\varphi\mathbf{v})\\ &= \rho\dot{\varphi} - \mathrm{div}(\rho\varphi\mathbf{v})\end{aligned}$$

and, hence,

$$\rho\dot{\varphi} = (\rho\varphi)' + \mathrm{div}(\rho\varphi\mathbf{v}). \tag{18.12}$$

Similarly, for a spatial vector field $\mathbf{g}$, by $(9.7)_2$ and (18.6),

$$\begin{aligned}(\rho\mathbf{g})' &= \rho\mathbf{g}' + \rho'\mathbf{g}\\ &= \rho\mathbf{g}' - \mathbf{g}\,\mathrm{div}(\rho\mathbf{v})\\ &= \rho\mathbf{g}' - \mathrm{div}(\rho\mathbf{g}\otimes\mathbf{v}) + \rho(\mathrm{grad}\,\mathbf{g})\mathbf{v}\\ &= \rho\dot{\mathbf{g}} - \mathrm{div}(\rho\mathbf{g}\otimes\mathbf{v}),\end{aligned}$$

and we have

$$\rho\dot{\mathbf{g}} = (\rho\mathbf{g})' + \mathrm{div}(\rho\mathbf{g}\otimes\mathbf{v}). \tag{18.13}$$

---

[98] Cf. $(15.5)_1$.

Also, by Reynolds' transport relation (16.2) and the mass balance (18.5), given any spatial field φ and any spatial region $\mathcal{P}_t$ convecting with the body,

$$\overline{\int_{\mathcal{P}_t} \rho\varphi\, dv} = \int_{\mathcal{P}_t} (\dot{\overline{\rho\varphi}} + \rho\varphi \operatorname{div}\mathbf{v})\, dv$$

$$= \int_{\mathcal{P}_t} (\rho\dot{\varphi} + (\dot{\rho} + \rho\operatorname{div}\mathbf{v})\varphi)\, dv$$

$$= \int_{\mathcal{P}_t} \rho\dot{\varphi}\, dv. \tag{18.14}$$

In other words,

(‡) *to differentiate the integral*

$$\int_{\mathcal{P}_t} \rho\varphi\, dv$$

with respect to time we simply differentiate under the integral sign using the material time-derivative while treating the "mass measure" $\rho\, dv$ as constant, thus giving

$$\int_{\mathcal{P}_t} \rho\dot{\varphi}\, dv.$$

This result is valid when the field φ is replaced by a spatial vector field or a spatial tensor field.

EXERCISE

1. Derive (18.14) by differentiating (18.11) with respect to time.

19 Forces and Moments. Balance Laws for Linear and Angular Momentum

The current entrenched, facile conception of force in terms of "pushes" and "pulls" has fostered a view of force as a "real quantity" rather than a mathematical concept. In the words of PIERCE (1934, p. 262): [Force is] "the great conception which, developed in the early part of the seventeenth century from the rude idea of a cause, and constantly improved upon since, has shown us how to explain all the changes of motion which bodies experience, and how to think about physical phenomena; which has given birth to modern science; and which ... has played a principal part in directing the course of modern thought ... It is, therefore, worth some pains to comprehend it."

Those who believe the notion of force is obvious should read the scientific literature of the period following Newton. TRUESDELL (1966) notes that "D'Alembert spoke of Newtonian forces as 'obscure and metaphysical beings, capable of nothing but spreading darkness over a science clear by itself,'" while JAMMER (1957, pp. 209, 215) paraphrases a remark of Maupertis, "we speak of forces only to conceal our ignorance," and one of Carnot, "an obscure metaphysical notion, that of force."

Within the framework of continuum mechanics, the basic balance laws for linear and angular momentum assert that, given any spatial region $\mathcal{P}_t$ convecting with the body,

(i) the net force on $\mathcal{P}_t$ is balanced by temporal changes in the linear momentum of $\mathcal{P}_t$;
(ii) the net moment on $\mathcal{P}_t$ is balanced by temporal changes in the angular momentum of $\mathcal{P}_t$.

We assume throughout this section that the motion is smooth. Momentum balance in the presence of a shock wave is discussed in §31.4.

19.1 Inertial Frames. Linear and Angular Momentum

According to THORNE (1994, p. 80): "Einstein expressed his principle of relativity not in terms of arbitrary reference frames, but in terms of rather special ones: frames ... that move freely under their own inertia, neither pushed nor pulled by any forces, and that therefore continue always onward in the same state of uniform motion that they began. Such frames Einstein called *inertial* because their motion is governed solely by their own inertia."

In classical particle mechanics an inertial frame is one within which motion is governed by Newton's law

$$\mathbf{f} = m\mathbf{a}. \tag{19.1}$$

Astronomers often choose an inertial frame attached to certain stars ("fixed stars") that appear at rest relative to one another, while engineers when discussing earthly applications often use an inertial frame attached to the earth. What is important is that Newtonian mechanics, however,[99] "merely assumes that there are inertial frames and does not enter into the question of how they should be interpreted in nature."

[99] TRUESDELL (1991, p. 68).

We assume throughout this book that

- we are working in an **inertial frame**.

Given a spatial point $\mathbf{x}$, it is convenient to write $\mathbf{r}$ for the position vector

$$\mathbf{r}(\mathbf{x}) = \mathbf{x} - \mathbf{o}. \tag{19.2}$$

Given a spatial region $\mathcal{P}_t$ that convects with the body, the integrals

$$\mathbf{l}(\mathcal{P}_t) = \int_{\mathcal{P}_t} \rho\mathbf{v}\,dv \qquad \text{and} \qquad \mathbf{a}(\mathcal{P}_t) = \int_{\mathcal{P}_t} \mathbf{r}\times(\rho\mathbf{v})\,dv, \tag{19.3}$$

respectively, represent the **linear and angular momentum** of $\mathcal{P}_t$. By $(1.23)_3$ and (9.9),

$$\begin{aligned}\mathbf{r}\times\dot{\mathbf{v}} &= \overline{\mathbf{r}\times\mathbf{v}}^{\boldsymbol{\cdot}} - \mathbf{v}\times\mathbf{v}\\ &= \overline{\mathbf{r}\times\mathbf{v}}^{\boldsymbol{\cdot}},\end{aligned} \tag{19.4}$$

Thus, by (19.3) and (‡) on page 130,

$$\begin{aligned}\overline{\mathbf{l}(\mathcal{P}_t)}^{\boldsymbol{\cdot}} &= \overline{\int_{\mathcal{P}_t} \rho\mathbf{v}\,dv}^{\boldsymbol{\cdot}}\\ &= \int_{\mathcal{P}_t} \rho\dot{\mathbf{v}}\,dv\end{aligned} \tag{19.5}$$

and

$$\begin{aligned}\overline{\mathbf{a}(\mathcal{P}_t)}^{\boldsymbol{\cdot}} &= \overline{\int_{\mathcal{P}_t} \mathbf{r}\times(\rho\mathbf{v})\,dv}^{\boldsymbol{\cdot}}\\ &= \int_{\mathcal{P}_t} \mathbf{r}\times(\rho\dot{\mathbf{v}})\,dv.\end{aligned} \tag{19.6}$$

19.2 Surface Tractions. Body Forces

Motions are accompanied by forces. Classically, forces in continuum mechanics are described spatially by

(i) contact forces between adjacent spatial regions; that is, spatial regions that intersect along their boundaries;
(ii) contact forces exerted on the boundary of the body by its environment;
(iii) body forces exerted on the interior points of a body by the environment.

Contact forces and body forces may be measured per unit area and volume in the reference body or per unit volume and area in the deformed body. Both descriptions are important, but the latter seems more natural when introducing basic principles.

One of the most important and far-reaching axioms of continuum mechanics is **Cauchy's hypothesis** concerning the form of the contact forces. Cauchy introduced a **surface-traction field** $\mathbf{t}(\mathbf{n}, \mathbf{x}, t)$ — defined for each unit vector $\mathbf{n}$, each $\mathbf{x}$ in $\mathcal{B}_t$, and

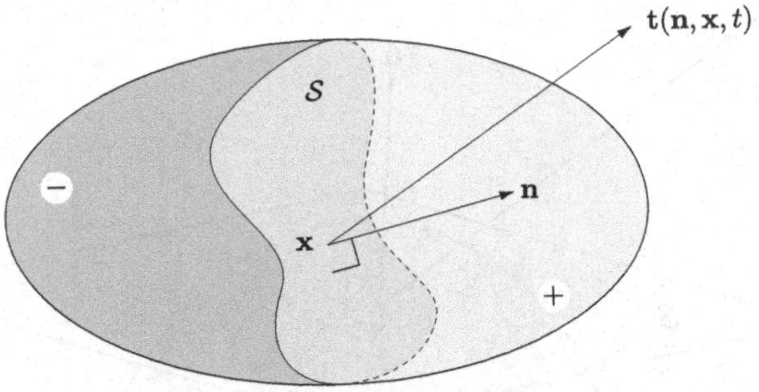

Figure 19.1. The normal $\mathbf{n}(\mathbf{x}, t)$ and surface traction $\mathbf{t}(\mathbf{n}, \mathbf{x}, t)$ at a point $\mathbf{x}$ on a surface $\mathcal{S}$ dividing a region in $\mathcal{B}_t$ into two subregions.

each t (Figure 19.1) — assumed to have the following property:

- Given any oriented spatial surface $\mathcal{S}$ in $\mathcal{B}_t$, $\mathbf{t}(\mathbf{n}, \mathbf{x}, t)$ represents the force, per unit area, exerted across $\mathcal{S}$ *upon* the material on the negative side of $\mathcal{S}$ by the material on the positive side.[100]

Cauchy's hypothesis has a strong consequence: If $\mathcal{C}$ is an oriented surface tangent to $\mathcal{S}$ at $\mathbf{x}$ and having the same positive unit normal there, then the force per unit area at $\mathbf{x}$ is the same on $\mathcal{C}$ as on $\mathcal{S}$ (Figure 19.2).

To determine the contact force between *adjacent* spatial regions $\mathcal{P}$ and $\mathcal{D}$, one simply integrates the traction over the surface

$$\mathcal{S} = \mathcal{P} \cap \mathcal{D}$$

of contact; thus, on introducing the shorthand

$$\int_{\mathcal{S}} \mathbf{t}(\mathbf{n})\, da = \int_{\mathcal{S}} \mathbf{t}(\mathbf{n}(\mathbf{x}, t), \mathbf{x}, t)\, da(\mathbf{x}),$$

granted the orienting unit normal to $\mathcal{S}$ coincides with that of $\mathcal{P}$, the integral

$$\int_{\mathcal{S}} \mathbf{t}(\mathbf{n})\, da \tag{19.7}$$

gives the force exerted on $\mathcal{S}$ by $\mathcal{D}$ (Figure 19.1). Similarly,

$$\int_{\partial\mathcal{P}} \mathbf{t}(\mathbf{n})\, da \tag{19.8}$$

represents the net contact force exerted on the spatial region $\mathcal{P}$ at time t.[101]

For points on the boundary of $\mathcal{B}_t$, $\mathbf{t}(\mathbf{n}, \mathbf{x}, t)$ — with $\mathbf{n}$ the outward unit normal to $\partial\mathcal{B}_t$ at $\mathbf{x}$ — gives the surface force, per unit area, exerted on the body at $\mathbf{x}$ by contact with the environment.

The environment can also exert forces on interior point of $\mathcal{B}_t$, with a classical example of such a force being that due to gravity. Such forces are determined by a vector field $\mathbf{b}_0(\mathbf{x}, t)$ giving the force, per unit volume, exerted by the environment on $\mathbf{x}$. For any spatial region $\mathcal{P}$, the integral

$$\int_{\mathcal{P}} \mathbf{b}_0\, dv$$

[100] By the positive side of $\mathcal{S}$ we mean the portion of $\mathcal{B}_t$ into which $\mathbf{n}$ points; similarly, the negative side of $\mathcal{S}$ is the portion of $\mathcal{B}_t$ out of which $\mathbf{n}$ points.

[101] When time-differentiation is not involved, the result (†) on page 112 allows us to work with an arbitrary subregion $\mathcal{P}$ of the deformed body at a fixed time, say t_0, and know that there is a corresponding convecting spatial region $\mathcal{P}_t$ such that $\mathcal{P}_{t_0} = \mathcal{P}$ (Figure 19.3).

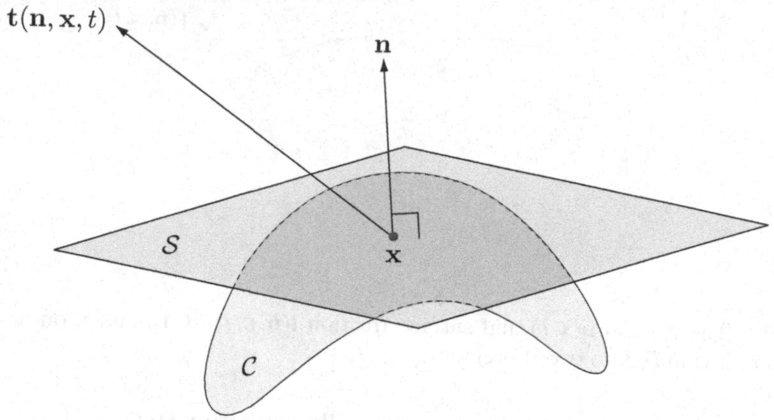

Figure 19.2. Schematic depicting Cauchy's hypothesis concerning the form of the constant forces.

gives that part of the environmental force not due to contact. We thus refer to $\mathbf{b}_0$ as the **conventional body force.**[102]

In the same vein, given any spatial region $\mathcal{P}$,

$$\int_{\partial\mathcal{P}} \mathbf{r} \times \mathbf{t}(\mathbf{n})\, da \qquad \text{and} \qquad \int_{\mathcal{P}} \mathbf{r} \times \mathbf{b}_0\, dv \tag{19.9}$$

represent net moments exerted on that region by contact and body forces.

For $\mathcal{P}_t$ a spatial region convecting with the body, the **net force** $\mathbf{f}(\mathcal{P}_t)$ and the **net moment** $\mathbf{m}(\mathcal{P}_t)$ exerted on $\mathcal{P}_t$ are therefore given by

$$\begin{aligned} \mathbf{f}(\mathcal{P}_t) &= \int_{\partial\mathcal{P}_t} \mathbf{t}(\mathbf{n})\, da + \int_{\mathcal{P}_t} \mathbf{b}_0\, dv, \\ \mathbf{m}(\mathcal{P}_t) &= \int_{\partial\mathcal{P}_t} \mathbf{r} \times \mathbf{t}(\mathbf{n})\, da + \int_{\mathcal{P}_t} \mathbf{r} \times \mathbf{b}_0\, dv. \end{aligned} \tag{19.10}$$

19.3 Balance Laws for Linear and Angular Momentum

The basic mechanical balance laws, which are assumed to hold at each time for all spatial regions $\mathcal{P}_t$ convecting with the body, are the balance laws for linear and angular momentum

$$\mathbf{f}(\mathcal{P}_t) = \overline{\mathbf{l}(\mathcal{P}_t)}^{\boldsymbol{\cdot}} \qquad \text{and} \qquad \mathbf{m}(\mathcal{P}_t) = \overline{\mathbf{a}(\mathcal{P}_t)}^{\boldsymbol{\cdot}}, \tag{19.11}$$

which, by (19.3) and (19.10), have the less succinct form

$$\boxed{\int_{\partial\mathcal{P}_t} \mathbf{t}(\mathbf{n})\, da + \int_{\mathcal{P}_t} \mathbf{b}_0\, dv = \overline{\int_{\mathcal{P}_t} \rho\mathbf{v}\, dv}^{\boldsymbol{\cdot}}} \tag{19.12}$$

[102] We use the adjective "conventional" to differentiate this body force from a body force that accounts also for inertia (i.e., also for the d'Alembert force).

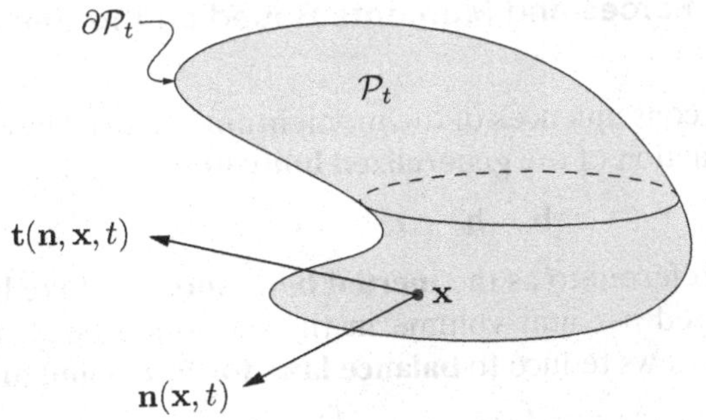

Figure 19.3. Schematic showing the traction $\mathbf{t}(\mathbf{n}, \mathbf{x}, t)$ acting at a spatial point $\mathbf{x}$ on the boundary $\partial\mathcal{P}_t$ of $\mathcal{P}_t$.

expressing **balance of linear momentum** and

$$\boxed{\int_{\partial\mathcal{P}_t} \mathbf{r} \times \mathbf{t}(\mathbf{n})\, da + \int_{\mathcal{P}_t} \mathbf{r} \times \mathbf{b}_0\, dv = \overline{\int_{\mathcal{P}_t} \mathbf{r} \times (\rho\mathbf{v})\, dv}^{\,\cdot}} \tag{19.13}$$

expressing **balance of angular momentum**. Further, by (19.5) and (19.6), these laws have alternative forms

$$\begin{gathered}
\int_{\partial\mathcal{P}_t} \mathbf{t}(\mathbf{n})\, da + \int_{\mathcal{P}_t} \mathbf{b}_0\, dv = \int_{\mathcal{P}_t} \rho\dot{\mathbf{v}}\, dv, \\
\int_{\partial\mathcal{P}_t} \mathbf{r} \times \mathbf{t}(\mathbf{n})\, da + \int_{\mathcal{P}_t} \mathbf{r} \times \mathbf{b}_0\, dv = \int_{\mathcal{P}_t} \mathbf{r} \times (\rho\dot{\mathbf{v}})\, dv,
\end{gathered} \tag{19.14}$$

useful in localization arguments.

Recall the expression (18.3) for the mass M of the body. The **center of mass** $\mathbf{r}_c(t)$ at time t, relative to the origin $\mathbf{o}$, is defined by

$$\mathbf{r}_c(t) = \frac{1}{M} \int_{\mathcal{B}_t} \rho\mathbf{r}\, dv.$$

By (18.14),

$$\dot{\mathbf{r}}_c(t) = \frac{1}{M} \int_{\mathcal{B}_t} \rho\mathbf{v}\, dv,$$

and $\dot{\mathbf{r}}_c$ represents the average velocity of the body. Further,

$$\mathbf{l}(\mathcal{B}_t) = M\dot{\mathbf{r}}_c(t);$$

therefore *the linear momentum of a body* B *is the same as that of a particle of mass* M *attached to the center of mass of* B. Finally, by (19.11),

$$\mathbf{f}(\mathcal{B}_t) = M\ddot{\mathbf{r}}_c;$$

thus, consistent with the classical statement (19.1),

- *the total force on a bounded body is equal to its mass times the acceleration of its center of mass.*

19.4 Balance of Forces and Moments Based on the Generalized Body Force

The task of finding consequences of the momentum balance laws is simplified somewhat by the introduction of the **generalized body force b** defined by

$$\mathbf{b} = \mathbf{b}_0 + \boldsymbol{\imath}, \qquad \boldsymbol{\imath} = -\rho\dot{\mathbf{v}}, \tag{19.15}$$

in which $\boldsymbol{\imath}$ may be referred to as the **inertial body force.**[103] Like $\mathbf{b}_0$, the body forces $\mathbf{b}$ and $\boldsymbol{\imath}$ are measured per unit volume in the deformed body. Using (19.15), the momentum balance laws reduce to **balance laws for forces and moments**

$$\begin{aligned} \int\limits_{\partial\mathcal{P}} \mathbf{t}(\mathbf{n})\,da + \int\limits_{\mathcal{P}} \mathbf{b}\,dv &= \mathbf{0}, \\ \int\limits_{\partial\mathcal{P}} \mathbf{r}\times\mathbf{t}(\mathbf{n})\,da + \int\limits_{\mathcal{P}} \mathbf{r}\times\mathbf{b}\,dv &= \mathbf{0}. \end{aligned} \tag{19.16}$$

These balances are to hold for every spatial region $\mathcal{P}$ and all time.[104]

For convenience, we let

$$\begin{aligned} \mathfrak{f}(\mathcal{P}) &= \int\limits_{\partial\mathcal{P}} \mathbf{t}(\mathbf{n})\,da + \int\limits_{\mathcal{P}} \mathbf{b}\,dv, \\ \mathfrak{m}(\mathcal{P}) &= \int\limits_{\partial\mathcal{P}} \mathbf{r}\times\mathbf{t}(\mathbf{n})\,da + \int\limits_{\mathcal{P}} \mathbf{r}\times\mathbf{b}\,dv; \end{aligned} \tag{19.17}$$

then, trivially, the force and moment balances take the simple forms

$$\mathfrak{f}(\mathcal{P}) = \mathbf{0} \qquad \text{and} \qquad \mathfrak{m}(\mathcal{P}) = \mathbf{0}. \tag{19.18}$$

Given any rigid velocity field[105]

$$\mathbf{w}(\mathbf{x},t) = \boldsymbol{\alpha}(t) + \boldsymbol{\lambda}(t)\times\mathbf{r} \tag{19.19}$$

and any spatial region $\mathcal{P}$,

$$\mathcal{W}_{\text{rig}}(\mathcal{P},\mathbf{w}) \overset{\text{def}}{=} \int\limits_{\partial\mathcal{P}} \mathbf{t}(\mathbf{n})\cdot\mathbf{w}\,da + \int\limits_{\mathcal{P}} \mathbf{b}\cdot\mathbf{w}\,dv \tag{19.20}$$

represents *power expended on $\mathcal{P}$ over the rigid velocity field* $\mathbf{w}$.[106] This notion has an interesting and important consequence that we now deduce. By (19.19),

$$\begin{aligned} \mathbf{b}\cdot\mathbf{w} &= \mathbf{b}\cdot(\boldsymbol{\alpha} + \boldsymbol{\lambda}\times\mathbf{r}) \\ &= \boldsymbol{\alpha}\cdot\mathbf{b} + \boldsymbol{\lambda}\cdot(\mathbf{r}\times\mathbf{b}), \end{aligned}$$

and a similar identity applies to $\mathbf{t}(\mathbf{n})\cdot\mathbf{w}$; hence, (19.17) and (19.20) imply that

$$\mathcal{W}_{\text{rig}}(\mathcal{P},\mathbf{w}) = \boldsymbol{\alpha}\cdot\mathfrak{f}(\mathcal{P}) + \boldsymbol{\lambda}\cdot\mathfrak{m}(\mathcal{P}). \tag{19.21}$$

A direct consequence of this identity is the equivalency of the following three assertions:

(i) $\mathcal{W}_{\text{rig}}(\mathcal{P},\mathbf{w}) = 0$ for all rigid velocity fields $\mathbf{w}$;
(ii) $\mathcal{W}_{\text{rig}}(\mathcal{P},\mathbf{w}) = 0$ for all vectors $\boldsymbol{\alpha}$ and $\boldsymbol{\lambda}$;
(iii) $\mathfrak{f}(\mathcal{P}) = \mathfrak{m}(\mathcal{P}) = \mathbf{0}$.

---

[103] Generally called the d'Alembert body force.
[104] Cf. Footnote 101.
[105] Cf. (10.5).
[106] The general notion of expended power is discussed in detail in §19.7.

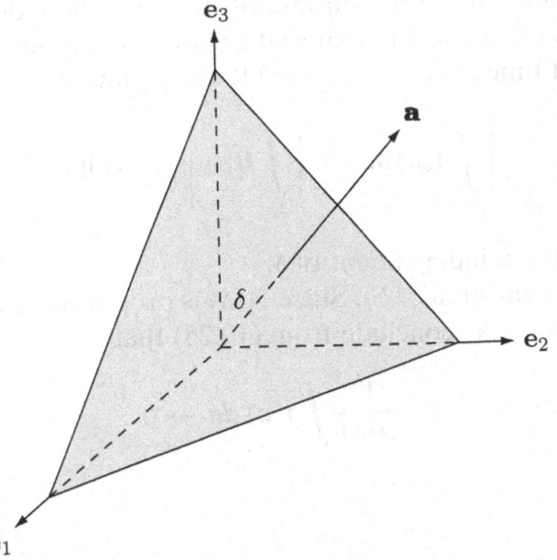

Figure 19.4. The tetrahedron $\mathcal{T}_\delta$ used in the proof of Cauchy's theorem.

Thus,

(‡) *the power expended over every rigid velocity field* **w** *vanishes if and only if the force and moment balances are satisfied.*

19.5 Cauchy's Theorem for the Existence of Stress

A deep result central to all of continuum mechanics is

CAUCHY'S THEOREM[107] *A consequence of balance of forces* (19.16)$_1$ *is that there exists a spatial tensor field* **T**, *called the* **Cauchy stress**, *such that*

$$\boxed{\mathbf{t}(\mathbf{n}) = \mathbf{T}\mathbf{n}.} \tag{19.22}$$

PROOF.[108] The proof proceeds in a series of steps.

Assertion 1. Given any **x** in $\mathcal{B}_t$, any orthonormal basis $\{\mathbf{e}_i\}$, and any unit vector **a** with

$$\mathbf{a}\cdot\mathbf{e}_i > 0, \quad i = 1, 2, 3, \tag{19.23}$$

it follows, on suppressing the argument t, that

$$\mathbf{t}(\mathbf{a},\mathbf{x}) = -\sum_{i=1}^{3}(\mathbf{a}\cdot\mathbf{e}_i)\mathbf{t}(-\mathbf{e}_i,\mathbf{x}). \tag{19.24}$$

PROOF. Let **x** belong to the interior of $\mathcal{B}_t$. Choose $\delta > 0$ and consider the (spatial) tetrahedron $\mathcal{T}_\delta$ with the following properties: The faces of $\mathcal{T}_\delta$ are $\mathcal{S}_\delta$, $\mathcal{S}_{1\delta}$, $\mathcal{S}_{2\delta}$, and $\mathcal{S}_{3\delta}$, where **a** and $-\mathbf{e}_i$ are the outward unit normals on $\mathcal{S}_\delta$ and $\mathcal{S}_{i\delta}$, respectively; the vertex opposite to $\mathcal{S}_\delta$ is **x**; the distance from **x** to $\mathcal{S}_\delta$ is δ (Figure 19.4). Then, $\mathcal{T}_\delta$ is contained in the interior of $\mathcal{B}_t$ for all sufficiently small δ, say $\delta \le \delta_0$.

[107] Cf., e.g., GURTIN (1981, §14).

[108] For completeness, we present a detailed proof; we suggest that readers more interested in applications bypass the proof, which, unfortunately, is technical.

Next, if we assume that $\mathbf{b}$ is continuous, then $\mathbf{b}$ is bounded on $\mathcal{T}_\delta$. If we apply the force balance $(19.16)_1$ to the material region P occupying the region $\mathcal{T}_\delta$ in the deformed region at time t, we are then led to the estimate

$$\left| \int_{\partial \mathcal{T}_\delta} \mathbf{t}(\mathbf{n})\, da \right| = \left| \int_{\mathcal{T}_\delta} \mathbf{b}\, dv \right| \leq \kappa \,\mathrm{vol}(\mathcal{T}_\delta) \tag{19.25}$$

for all $\delta \leq \delta_0$, where κ is independent of δ.

Let $A(\delta)$ denote the area of $\mathcal{S}_\delta$. Since $A(\delta)$ is proportional to δ^2, while $\mathrm{vol}(\mathcal{T}_\delta)$ is proportional to δ^3, we may conclude from (19.25) that

$$\frac{1}{A(\delta)} \int_{\partial \mathcal{T}_\delta} \mathbf{t}(\mathbf{n})\, da \to 0$$

as $\delta \to 0$. But

$$\int_{\partial \mathcal{T}_\delta} \mathbf{t}(\mathbf{n})\, da = \int_{\mathcal{S}_\delta} \mathbf{t}(\mathbf{a})\, da + \sum_{i=1}^{3} \int_{\mathcal{S}_{i\delta}} \mathbf{t}(-\mathbf{e})\, da$$

and, assuming that $\mathbf{t}(\mathbf{n}, \mathbf{x})$ is continuous in $\mathbf{x}$ for each $\mathbf{n}$, since the area of $\mathcal{S}_{i\delta}$ is $A(\delta)(\mathbf{a} \cdot \mathbf{e}_i)$,

$$\frac{1}{A(\delta)} \int_{\partial \mathcal{S}_\delta} \mathbf{t}(\mathbf{a})\, da \to \mathbf{t}(\mathbf{a}, \mathbf{x})$$

and

$$\frac{1}{A(\delta)} \int_{\partial \mathcal{S}_{i\delta}} \mathbf{t}(-\mathbf{e}_i)\, da \to (\mathbf{a} \cdot \mathbf{e}_i)\mathbf{t}(-\mathbf{e}_i, \mathbf{x}).$$

Combining the relations above we conclude that (19.24) is satisfied.

Assertion 2. (Newton's law of action and reaction).

$$\mathbf{t}(\mathbf{n}) = -\mathbf{t}(-\mathbf{n}). \tag{19.26}$$

PROOF. In (19.24), let $\mathbf{a} \to \mathbf{e}_i$; then, $\mathbf{t}(\mathbf{e}_i, \mathbf{x}) = -\mathbf{t}(-\mathbf{e}_i, \mathbf{x})$ and, since the basis $\{\mathbf{e}_i\}$ is arbitrary, (19.26) must hold.

Assertion 3. Given any $\mathbf{x}$ in $\mathcal{B}_t$ and any orthonormal basis $\{\mathbf{e}_i\}$,

$$\mathbf{t}(\mathbf{a}, \mathbf{x}) = \sum_{i=1}^{3} (\mathbf{a} \cdot \mathbf{e}_i)\mathbf{t}(\mathbf{e}_i, \mathbf{x}) \tag{19.27}$$

for all unit vectors $\mathbf{a}$.

PROOF. Choose an orthonormal basis $\{\mathbf{e}_i\}$ and a unit vector $\mathbf{a}$ that does not lie in a coordinate plane (that is, in a plane spanned by two elements $\mathbf{e}_i$ of the basis). Then, there is no i such that $\mathbf{a} \cdot \mathbf{e}_i = 0$ and we can define a new orthonormal basis $\{\mathbf{e}_i^\star\}$ such that[109]

$$\mathbf{e}_i^\star = [\mathrm{sgn}\,(\mathbf{a} \cdot \mathbf{e}_i)]\mathbf{e}_i.$$

[109] For any scalar β, $\mathrm{sgn}\beta$ denotes the sign of β.

It follows that $\mathbf{a}\cdot\mathbf{e}_i^\star > 0$ for all i and Assertion 1, together with (19.26) applied to the basis $\{\mathbf{e}_i^\star\}$, yields

$$\begin{aligned}\mathbf{t}(\mathbf{a},\mathbf{x}) &= -\sum_{i=1}^{3}(\mathbf{a}\cdot\mathbf{e}_i^\star)\mathbf{t}(-\mathbf{e}_i^\star,\mathbf{x})\\ &= \sum_{i=1}^{3}(\mathbf{a}\cdot\mathbf{e}_i^\star)\mathbf{t}(\mathbf{e}_i^\star,\mathbf{x})\\ &= \sum_{i=1}^{3}(\mathbf{a}\cdot\mathbf{e}_i)\mathbf{t}(\mathbf{e}_i,\mathbf{x}).\end{aligned}$$

Thus, (19.27) holds as long as $\mathbf{a}$ does not lie in a coordinate plane and, assuming continuity of $\mathbf{t}(\mathbf{n},\mathbf{x})$ in $\mathbf{n}$, (19.27) must hold for all unit vectors $\mathbf{a}$, and Assertion 3 is proved.

Next, define the Cauchy stress $\mathbf{T}$ by

$$\mathbf{T}(\mathbf{x},t) = \sum_{i=1}^{3}\mathbf{t}(\mathbf{e}_i,\mathbf{x},t)\otimes\mathbf{e}_i.$$

Given any unit vector $\mathbf{n}$, we may then use (19.27) to conclude that

$$\begin{aligned}\mathbf{T}\mathbf{n} &= \sum_{i=1}^{3}\mathbf{t}(\mathbf{e}_i)(\mathbf{n}\cdot\mathbf{e}_i)\\ &= \mathbf{t}(\mathbf{n}),\end{aligned}$$

which completes the proof of Cauchy's Theorem.

It is important to note that, since $\mathbf{n}(\mathbf{x},t)$ is a spatial vector, as is the traction $\mathbf{T}(\mathbf{x},t)\mathbf{n}(\mathbf{x},t)$,

$$\mathbf{T}\ \textit{maps spatial vectors to spatial vectors.} \tag{19.28}$$

19.6 Local Forms of the Force and Moment Balances

In view of Cauchy's theorem (19.22), the force and moment balances (19.16) — for $\mathcal{P}$ an arbitrary spatial region — become

$$\begin{aligned}\int_{\partial\mathcal{P}}\mathbf{T}\mathbf{n}\,da + \int_{\mathcal{P}}\mathbf{b}\,dv &= \mathbf{0},\\ \int_{\partial\mathcal{P}}\mathbf{r}\times\mathbf{T}\mathbf{n}\,da + \int_{\mathcal{P}}\mathbf{r}\times\mathbf{b}\,dv &= \mathbf{0}.\end{aligned} \tag{19.29}$$

Assume that the force and moment balances are satisfied. We can then use the divergence theorem to rewrite balance of forces as

$$\int_{\mathcal{P}}(\mathrm{div}\mathbf{T}+\mathbf{b})\,dv = \mathbf{0}$$

and, since this must hold for all spatial regions $\mathcal{P}$, we are led to the local **force balance**

$$\mathrm{div}\mathbf{T}+\mathbf{b} = \mathbf{0}, \qquad \frac{\partial T_{ij}}{\partial x_j} + b_i = 0. \tag{19.30}$$

Next, we may use (19.15) to rewrite (19.30) as a local **balance of linear momentum**:

$$\rho\dot{\mathbf{v}} = \operatorname{div}\mathbf{T} + \mathbf{b}_0, \qquad \rho\dot{v}_i = \frac{\partial T_{ij}}{\partial x_j} + b_{0i}. \tag{19.31}$$

In view of (18.13),

$$\rho\dot{\mathbf{v}} = (\rho\mathbf{v})' + \operatorname{div}(\rho\mathbf{v}\otimes\mathbf{v}),$$

and it follows that balance of linear momentum may be written equivalently as

$$(\rho\mathbf{v})' = \operatorname{div}(\mathbf{T} - \rho\mathbf{v}\otimes\mathbf{v}) + \mathbf{b}_0. \tag{19.32}$$

To derive the local moment balance, let $\mathbf{\Lambda}$ denote an arbitrary (constant) skew tensor and let $\boldsymbol{\lambda}$ denote its axial vector, so that $\mathbf{\Lambda} = \boldsymbol{\lambda}\times$ (i.e., $\Lambda_{ij} = \epsilon_{ipj}\lambda_p$) and consider the (arbitrary) rigid velocity field defined by[110]

$$\begin{aligned}\mathbf{w}(\mathbf{x}) &= \boldsymbol{\lambda}\times\mathbf{r}\\ &= \mathbf{\Lambda}\mathbf{r},\end{aligned}$$

so that

$$\operatorname{grad}(\mathbf{\Lambda}\mathbf{r}) = \mathbf{\Lambda} \tag{19.33}$$

for all rigid velocity fields $\mathbf{w}$ and, hence, we may conclude from (‡) on page 137 and Cauchy's theorem (19.22) that, for any spatial region $\mathcal{P}$,[111]

$$\int\limits_{\partial\mathcal{P}} \mathbf{T}\mathbf{n}\cdot\mathbf{w}\,da + \int\limits_{\mathcal{P}} \mathbf{b}\cdot\mathbf{w}\,dv = 0.$$

Since $\mathbf{w}$ is rigid, by (19.33) and the divergence theorem in the form $(4.11)_5$,

$$\int\limits_{\partial\mathcal{P}} \mathbf{T}\mathbf{n}\cdot\mathbf{w}\,da = \int\limits_{\mathcal{P}} (\operatorname{div}\mathbf{T}\cdot\mathbf{w} + \mathbf{T}\colon\mathbf{\Lambda})\,dv.$$

The last two equations and the local balance $\operatorname{div}\mathbf{T} + \mathbf{b} = \mathbf{0}$ imply that

$$\int\limits_{\mathcal{P}} \mathbf{T}\colon\mathbf{\Lambda}\,dv = 0;$$

thus, since $\mathcal{P}$ is arbitrary,

$$\mathbf{T}\colon\mathbf{\Lambda} = 0.$$

But since $\mathbf{\Lambda}$ is an arbitrarily chosen skew tensor, the Cauchy stress $\mathbf{T}$ must therefore be symmetric,

$$\mathbf{T} = \mathbf{T}^{\top}, \qquad T_{ij} = T_{ji}, \tag{19.34}$$

which is the local **moment balance**.

EXERCISES

1. Show that the local balances (19.30) and (19.34) imply the global balances (19.29).

---

[110] Cf. (10.5).
[111] Cf. (19.20).

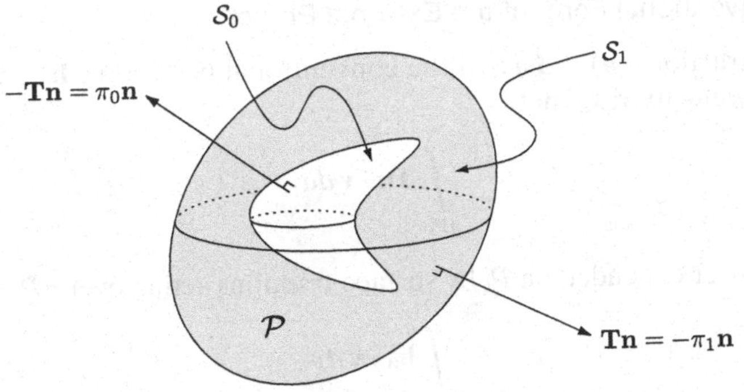

Figure 19.5. A region $\mathcal{P}$ with boundary $\partial\mathcal{P}$ consisting of two closed surfaces $\mathcal{S}_0$ and $\mathcal{S}_1$.

2. Define the mean stress $\bar{\mathbf{T}}$ via

$$\mathrm{vol}(\mathcal{P})\bar{\mathbf{T}} = \int\limits_{\mathcal{P}} \mathbf{T}\,dv.$$

   (a) (Signorini's theorem) Show that $\bar{\mathbf{T}}$ is determined by the traction $\mathbf{Tn}$ on $\partial\mathcal{P}$ and the generalized body force $\mathbf{b}$ via

$$\mathrm{vol}(\mathcal{P})\bar{\mathbf{T}} = \int\limits_{\partial\mathcal{P}} \mathbf{Tn}\otimes\mathbf{r}\,da + \int\limits_{\mathcal{P}} \mathbf{b}\otimes\mathbf{r}\,dv.$$

   (b) Assume that $\mathbf{b} = \mathbf{0}$ and that $\partial\mathcal{P}$ consists of two closed surfaces $\mathcal{S}_0$ and $\mathcal{S}_1$ with $\mathcal{S}_1$ enclosing $\mathcal{S}_0$ (Figure 19.5). Assume further that $\mathcal{S}_0$ and $\mathcal{S}_1$ are acted on by uniform pressures π_0 and π_1, so that

$$\mathbf{Tn} = \begin{cases} -\pi_0\mathbf{n} & \text{on } \mathcal{S}_0, \\ -\pi_1\mathbf{n} & \text{on } \mathcal{S}_1. \end{cases}$$

   Show that $\bar{\mathbf{T}}$ is spherical with pressure

$$\bar{p} = -\frac{\pi_1 V_1 - \pi_0 V_0}{V_1 - V_0},$$

   where V_0 and V_1 are the volumes enclosed by $\mathcal{S}_0$ and $\mathcal{S}_1$.

3. Provide a physical interpretation of the tensor $\mathbf{T} - \rho\mathbf{v}\otimes\mathbf{v}$ entering (19.32).
4. Show that (19.34) can be derived alternatively by taking the inner product of $(19.29)_2$ with an arbitrary constant vector $\mathbf{c} \neq \mathbf{0}$.

19.7 Kinetic Energy. Conventional and Generalized External Power Expenditures

In classical mechanics, the **power expended** by a force f acting on a particle in motion with velocity v has the form $f \cdot v$, in which case we say that f is **power conjugate** to v. As we now show, this classical notion has an immediate counterpart in continuum mechanics.

Throughout this subsection $\mathcal{P}_t$ denotes a spatial region convecting with the body.

19.7.1 Conventional Form of the External Power

The surface traction $\mathbf{t}(\mathbf{n}) = \mathbf{T}\mathbf{n}$ and the conventional body force $\mathbf{b}_0$ are power conjugate to the velocity $\mathbf{v}$, so that

$$\int_{\partial\mathcal{P}_t} \mathbf{T}\mathbf{n} \cdot \mathbf{v}\, da$$

represents power expended on $\mathcal{P}_t$ by surface tractions acting over $\partial\mathcal{P}_t$, while

$$\int_{\mathcal{P}_t} \mathbf{b}_0 \cdot \mathbf{v}\, dv$$

represents power expended on points interior to $\mathcal{P}_t$ by the environment. We refer to

$$\mathcal{W}_0(\mathcal{P}_t) \stackrel{\text{def}}{=} \int_{\partial\mathcal{P}_t} \mathbf{T}\mathbf{n} \cdot \mathbf{v}\, da + \int_{\mathcal{P}_t} \mathbf{b}_0 \cdot \mathbf{v}\, dv \tag{19.35}$$

as the **conventional external power**[112] because it represents the net power expended on $\mathcal{P}_t$ by "agencies" external to $\mathcal{P}_t$.

19.7.2 Kinetic Energy and Inertial Power

The spatial field

$$\tfrac{1}{2}\rho|\mathbf{v}|^2$$

is the *kinetic energy* measured per unit volume in the deformed body; hence,

$$\mathcal{K}(\mathcal{P}_t) \stackrel{\text{def}}{=} \int_{\mathcal{P}_t} \tfrac{1}{2}\rho|\mathbf{v}|^2\, dv \tag{19.36}$$

represents the **kinetic energy** of $\mathcal{P}_t$. The following calculation establishes a basic relation between the kinetic-energy rate and the power expended by the inertial body force $\boldsymbol{\imath}$. Since, by (‡) on page 130 and $(19.15)_2$,

$$\overline{\int_{\mathcal{P}_t} \tfrac{1}{2}\rho|\mathbf{v}|^2\, dv}^{\boldsymbol{\cdot}} = \int_{\mathcal{P}_t} \tfrac{1}{2}\rho\overline{|\mathbf{v}|^2}^{\boldsymbol{\cdot}}\, dv$$

$$= \int_{\mathcal{P}_t} \underbrace{\rho\dot{\mathbf{v}}}_{-\boldsymbol{\imath}} \cdot \mathbf{v}\, dv,$$

it follows that

$$\int_{\mathcal{P}_t} \boldsymbol{\imath} \cdot \mathbf{v}\, dv = -\overline{\mathcal{K}(\mathcal{P}_t)}^{\boldsymbol{\cdot}}. \tag{19.37}$$

The term $\int_{\mathcal{P}_t} \boldsymbol{\imath} \cdot \mathbf{v}\, dv$ represents the **inertial power** expended on $\mathcal{P}_t$ and (19.37) asserts that *this inertial power expenditure is balanced by the negative kinetic-energy rate of* $\mathcal{P}_t$.[113]

[112] Cf. Footnote 102.

[113] Cf. Podio-Guidugli (1997).

Next, we give a series of useful identities: by $(19.15)_1$ and (19.37),

$$\int_{\mathcal{P}_t} \mathbf{b}\cdot\mathbf{v}\,dv = \int_{\mathcal{P}_t} \mathbf{b}_0\cdot\mathbf{v}\,dv - \overline{\mathcal{K}(\mathcal{P}_t)}^{\boldsymbol{\cdot}}; \tag{19.38}$$

also, by $(4.11)_5$ and (19.30),

$$\begin{aligned}\int_{\partial\mathcal{P}_t} \mathbf{Tn}\cdot\mathbf{v}\,da &= \int_{\mathcal{P}_t} (\mathbf{v}\cdot\operatorname{div}\mathbf{T} + \mathbf{T}\colon\operatorname{grad}\mathbf{v})\,dv \\ &= \int_{\mathcal{P}_t} (\mathbf{T}\colon\operatorname{grad}\mathbf{v} - \mathbf{b}\cdot\mathbf{v})\,dv;\end{aligned} \tag{19.39}$$

by (9.12), $(11.2)_2$, and the symmetry of $\mathbf{T}$,

$$\begin{aligned}\mathbf{T}\colon\operatorname{grad}\mathbf{v} &= \mathbf{T}\colon\mathbf{L} \\ &= \mathbf{T}\colon\mathbf{D}.\end{aligned} \tag{19.40}$$

The far right side of (19.40) establishes the stretching $\mathbf{D}$ as the appropriate power-conjugate variable for the Cauchy stress $\mathbf{T}$, and we say that $\mathbf{T}$ is **power conjugate** to $\mathbf{D}$.[114] The field

$$\mathbf{T}\colon\mathbf{D} \tag{19.41}$$

represents the power expended *within* $\mathcal{P}_t$, measured per unit volume, and is referred to as the **stress power**; the integral

$$\int_{\mathcal{P}_t} \mathbf{T}\colon\mathbf{D}\,dv \tag{19.42}$$

represents the *net power expended within* $\mathcal{P}_t$ and is referred to as the **internal power**. Combining (19.38)–(19.40), we arrive at the **conventional power balance**

$$\underbrace{\int_{\partial\mathcal{P}_t} \mathbf{Tn}\cdot\mathbf{v}\,da + \int_{\mathcal{P}_t} \mathbf{b}_0\cdot\mathbf{v}\,dv}_{\text{conventional external power}} = \underbrace{\int_{\mathcal{P}_t} \mathbf{T}\colon\mathbf{D}\,dv}_{\text{internal power}} + \underbrace{\overline{\int_{\mathcal{P}_t} \tfrac{1}{2}\rho|\mathbf{v}|^2\,dv}^{\boldsymbol{\cdot}}}_{\text{kinetic-energy rate}}, \tag{19.43}$$

a relation which asserts that

- *for a spatial region $\mathcal{P}_t$ convecting with the body, the conventional external power expended on $\mathcal{P}_t$ is balanced by the sum of the internal power expended within $\mathcal{P}_t$ and the temporal change in kinetic energy of $\mathcal{P}_t$.*

This balance therefore relates the conventional external power to temporal changes in the kinetic energy.

19.7.3 Generalized Power Balance

If we replace the conventional body force $\mathbf{b}_0$ in (19.35) by the generalized body force $\mathbf{b}$, which is defined in (19.15) and which includes the inertial body force $\boldsymbol{\imath}$ — a body

[114] Power-conjugate pairings such as (19.41) are common in continuum mechanics. The notion of power-conjugate fields is useful in discussing applications involving force systems other than the classical Newtonian system associated with the motion of material points.

force we view as external in nature — we arrive at a quantity,

$$\mathcal{W}(\mathcal{P}_t) = \int_{\partial \mathcal{P}_t} \mathbf{Tn} \cdot \mathbf{v}\, da + \int_{\mathcal{P}_t} \mathbf{b} \cdot \mathbf{v}\, dv, \tag{19.44}$$

which we refer to as the **generalized external power**. In view of (19.39) and (19.40), we have the **generalized power balance**

$$\underbrace{\int_{\partial \mathcal{P}_t} \mathbf{Tn} \cdot \mathbf{v}\, da + \int_{\mathcal{P}_t} \mathbf{b} \cdot \mathbf{v}\, dv}_{\mathcal{W}(\mathcal{P}_t)} = \underbrace{\int_{\mathcal{P}_t} \mathbf{T} : \mathbf{D}\, dv}_{\mathcal{I}(\mathcal{P}_t)}, \tag{19.45}$$

in which

$$\mathcal{I}(\mathcal{P}_t) = \int_{\mathcal{P}_t} \mathbf{T} : \mathbf{D}\, dv \tag{19.46}$$

is the **internal power**. The generalized power balance asserts that

- *the power expended on $\mathcal{P}_t$ by material and agencies external to $\mathcal{P}_t$ and by inertia is balanced by the power expended within $\mathcal{P}_t$.*

Finally, we note for future use that (19.35), (19.38), and (19.45) yield the following relations between the conventional and generalized power expenditures, the internal power, and the kinetic-energy rate:

$$\begin{aligned} \mathcal{W}(\mathcal{P}_t) &= \mathcal{W}_0(\mathcal{P}_t) - \overline{\mathcal{K}(\mathcal{P}_t)}^{\boldsymbol{\cdot}}, \\ \mathcal{W}_0(\mathcal{P}_t) &= \mathcal{I}(\mathcal{P}_t) + \overline{\mathcal{K}(\mathcal{P}_t)}^{\boldsymbol{\cdot}}. \end{aligned} \tag{19.47}$$

19.7.4 The Assumption of Negligible Inertial Forces

Following Noll (1995) we view the relation

$$\boldsymbol{\imath} = -\rho \dot{\mathbf{v}} \tag{19.48}$$

as a *constitutive law for inertia*. As such, we allow ourselves the luxury of considering, as constitutive, the relation[115]

$$\boldsymbol{\imath} \equiv \mathbf{0}, \tag{19.49}$$

introduced to account for — and to connote — situations in which **inertial forces are negligible.**[116] In view of (19.15), (19.37), and (19.47), consequences of (19.49)

[115] (19.49) is meant to replace (19.48); as such it places no restrictions on ρ and $\dot{\mathbf{v}}$.

[116] Noll (1995) writes: "When dealing with deformable bodies, inertia plays very often a secondary role. In some situations it is even appropriate to neglect inertia altogether. For example, when analyzing the forces and deformations that occur when one squeezes toothpaste out of a tube, inertial forces are generally negligible." In practice the constitutive assumption (19.49) is often used, for example, to characterize the slow flow of plastic and highly viscous materials.

are that

$$\left.\begin{aligned}
\mathbf{b} &\equiv \mathbf{b}_0, \\
\tfrac{1}{2}\rho\overline{|\mathbf{v}|^2}^{\boldsymbol{\cdot}} &\equiv 0, \\
\overline{\mathcal{K}(\mathcal{P}_t)}^{\boldsymbol{\cdot}} &\equiv 0, \\
\mathcal{W}(\mathcal{P}_t) &\equiv \mathcal{W}_0(\mathcal{P}_t), \\
\mathcal{W}_0(\mathcal{P}_t) &\equiv \mathcal{I}(\mathcal{P}_t),
\end{aligned}\right\} \tag{19.50}$$

where $\mathcal{P}_t$ is an arbitrary spatial region convecting with the body.

EXERCISE

1. Assuming that (19.49) holds, verify $(19.50)_2$.

20 Frames of Reference

According to TRUESDELL & NOLL (1965, §41): "The position of an event can be specified only if a frame of reference, or observer, is given. Physically, a frame of reference is a set of objects whose mutual distances change comparatively little in time, like the walls of a laboratory [or] the fixed stars... Only if such a frame is given for all times does it make sense to compare the positions of a particle at different times, and only then can we speak about velocities, accelerations, etc. of a particle..."

20.1 Changes of Frame

As noted in section (5.2), $\mathcal{B}_t$ is the region actually *observed during the motion*: The reference body B serves only to label material points. For that reason, to discuss a notion of invariance under observer changes, it is useful to differentiate conceptually between the ambient space for B and the space through which $\mathcal{B}_t$ evolves (Figure 20.1). In accord with this:

(i) the ambient space through which $\mathcal{B}_t$ evolves is termed the **observed space**;
(ii) the ambient space for the reference body B is termed the **reference space**.

Granted this dichotomy, spatial vectors belong to the observed space, while material vectors belong to the reference space.

Suppose that a frame of reference $\mathcal{F}$ for the observed space is prescribed, an assumption tacit in the discussion thus far. Then, roughly speaking, a change of frame is, at each time, a rotation and translation of the observed space. Precisely, a **change of frame** $\mathcal{F} \to \mathcal{F}^*$ is, at each fixed time t, defined by a rotation $\mathbf{Q}(t)$ and a spatial point $\mathbf{y}(t)$ and transforms spatial points $\mathbf{x}$ to spatial points[117]

$$\mathbf{x}^* = \mathbf{y}(t) + \mathbf{Q}(t)(\mathbf{x} - \mathbf{o}). \tag{20.1}$$

(No loss in generality is incurred by assuming that the spatial origin $\mathbf{o}$ is fixed.)

We refer to $\mathbf{Q}$ as the **frame-rotation.** A consequence of (3.5) is that the tensor $\dot{\mathbf{Q}}\mathbf{Q}^{\top}$ is skew. We refer to the *skew tensor*

$$\boldsymbol{\Omega} = \dot{\mathbf{Q}}\mathbf{Q}^{\top} \tag{20.2}$$

as the **frame-spin**; $\boldsymbol{\Omega}$ represents the rate at which the new frame $\mathcal{F}^*$ is spinning.

[117] Changes of frame are often referred to as *changes of observer.*

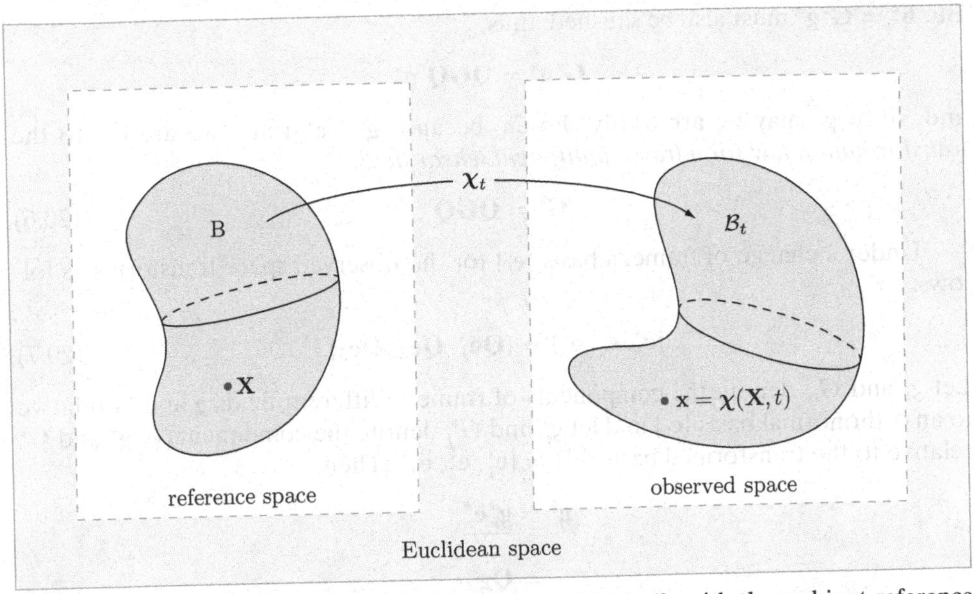

Figure 20.1. The reference body B and the deformed body $\mathcal{B}_t$, with the ambient reference and observed spaces and the underlying Euclidean space depicted.

It is important to emphasize that while a change of frame affects the observed space through which the deformed body evolves, it does not affect the reference space; thus

- *material points and material vectors are invariant under changes in frame.*

Given a field Φ, we write Φ^* for Φ as measured in the new frame $\mathcal{F}^*$. In the material description, the argument $(\mathbf{X}, t)$ of Φ is invariant, so that both $\Phi(\mathbf{X}, t)$ and $\Phi^*(\mathbf{X}, t)$ have $(\mathbf{X}, t)$ as their argument. On the other hand, in the spatial description, $\Phi(\mathbf{x}, t)$ transforms to $\Phi^*(\mathbf{x}^*, t)$.

Scalar fields such as the density ρ and, hence, the specific volume $\upsilon = 1/\rho$ are **invariant**:

$$\rho^* = \rho, \qquad \upsilon^* = \upsilon.$$

20.2 Frame-Indifferent Fields

Assume that a change of frame is prescribed. We then say that a vector field $\mathbf{g}$ is **frame-indifferent** if it simply rotates with the frame-rotation:

$$\mathbf{g}^* = \mathbf{Q}\mathbf{g}. \tag{20.3}$$

On the other hand, we say that a tensor field $\mathbf{G}$ is **frame-indifferent** if, given *frame-indifferent vector fields* $\mathbf{g}$ and $\mathbf{h}$, with $\mathbf{g}$ arbitrary, and any change in frame,

$$\mathbf{h} = \mathbf{G}\mathbf{g} \quad \text{implies that} \quad \mathbf{h}^* = \mathbf{G}^*\mathbf{g}^*. \tag{20.4}$$

To determine the transformation law implied by this definition, note that, if $\mathbf{h} = \mathbf{G}\mathbf{g}$,

$$\begin{aligned} \mathbf{h}^* &= \mathbf{Q}\mathbf{h} \\ &= \mathbf{Q}\mathbf{G}\mathbf{g} \\ &= \mathbf{Q}\mathbf{G}\mathbf{Q}^{\top}\mathbf{g}^*. \end{aligned} \tag{20.5}$$

But $\mathbf{h}^* = \mathbf{G}^*\mathbf{g}^*$ must also be satisfied; thus,

$$\mathbf{G}^*\mathbf{g}^* = \mathbf{Q}\mathbf{G}\mathbf{Q}^\top\mathbf{g}^*$$

and, since $\mathbf{g}^*$ may be arbitrarily chosen, because $\mathbf{g}$ is abitrary, we are led to the *transformation law for a frame-indifferent tensor field*:[118]

$$\mathbf{G}^* = \mathbf{Q}\mathbf{G}\mathbf{Q}^\top. \tag{20.6}$$

Under a change of frame, a basis $\{\mathbf{e}_i\}$ for the observed space transforms as follows:

$$\{\mathbf{e}_1^*, \mathbf{e}_2^*, \mathbf{e}_3^*\} = \{\mathbf{Q}\mathbf{e}_1, \mathbf{Q}\mathbf{e}_2, \mathbf{Q}\mathbf{e}_3\}. \tag{20.7}$$

Let g_i and G_{ij} denote the components of frame-indifferent fields $\mathbf{g}$ and $\mathbf{G}$ relative to an orthonormal basis $\{\mathbf{e}_i\}$ and let g_i^* and G_{ij}^* denote the components of $\mathbf{g}^*$ and $\mathbf{G}^*$ relative to the transformed basis $\{\mathbf{e}_i^*\} = \{\mathbf{e}_1^*, \mathbf{e}_2^*, \mathbf{e}_3^*\}$. Then,

$$\begin{aligned}
\mathbf{g}^* &= \underline{g_i^*\mathbf{e}_i^*} \\
&= \mathbf{Q}\mathbf{g} \\
&= \mathbf{Q}(g_i\mathbf{e}_i) \\
&= g_i\mathbf{Q}\mathbf{e}_i \\
&= \underline{g_i\mathbf{e}_i^*},
\end{aligned} \tag{20.8}$$

and equality of the underlined terms implies that

$$g_i^* = g_i.$$

Similarly

$$\begin{aligned}
\mathbf{G}^* &= \underline{G_{ij}^*\mathbf{e}_i^* \otimes \mathbf{e}_j^*} \\
&= \mathbf{Q}\mathbf{G}\mathbf{Q}^\top \\
&= \mathbf{Q}(G_{ij}\mathbf{e}_i \otimes \mathbf{e}_j)\mathbf{Q}^\top \\
&= G_{ij}\mathbf{Q}\mathbf{e}_i \otimes \mathbf{Q}\mathbf{e}_j \\
&= \underline{G_{ij}\mathbf{e}_i^* \otimes \mathbf{e}_j^*},
\end{aligned} \tag{20.9}$$

so that

$$G_{ij}^* = G_{ij}.$$

The components of a frame-indifferent field relative to a basis that rotates with the frame are therefore independent of the frame.

20.3 Transformation Rules for Kinematic Fields

Next, since $\mathbf{x} = \chi(\mathbf{X}, t)$ is a spatial point, but $\mathbf{X}$ is a material point (and hence invariant), the motion χ transforms according to

$$\chi^*(\mathbf{X}, t) = \mathbf{y}(t) + \mathbf{Q}(t)(\chi(\mathbf{X}, t) - \mathbf{o}). \tag{20.10}$$

[118] Frame-indifferent fields are also referred to as objective fields.

Taking the gradient of this relation with respect to $\mathbf{X}$ yields a transformation law

$$\mathbf{F}^*(\mathbf{X},t) = \mathbf{Q}(t)\mathbf{F}(\mathbf{X},t) \tag{20.11}$$

for the deformation gradient. Further, by (20.10) and (20.11),

$$\begin{aligned} \overline{\chi^*}^{\boldsymbol{\cdot}}(\mathbf{X},t) &= \dot{\mathbf{y}}(t) + \mathbf{Q}(t)\dot{\chi}(\mathbf{X},t) + \dot{\mathbf{Q}}(t)(\chi(\mathbf{X},t) - \mathbf{o}), \\ \dot{\overline{\mathbf{F}^*}}(\mathbf{X},t) &= \mathbf{Q}(t)\dot{\mathbf{F}}(\mathbf{X},t) + \dot{\mathbf{Q}}(t)\mathbf{F}(\mathbf{X},t). \end{aligned} \tag{20.12}$$

Care must be taken when working with the spatial description $\varphi(\mathbf{x},t)$ of a field, since the argument $\mathbf{x}$ is no longer invariant. Consider the spatial description $\mathbf{v}(\mathbf{x},t)$ of the velocity field. Then $\mathbf{v}(\mathbf{x},t)$ transforms to $\mathbf{v}^*(\mathbf{x}^*,t)$ with $\mathbf{x}^* = \chi^*(\mathbf{X},t)$. Thus, since $\mathbf{x}^*$ and $\mathbf{x}$ correspond to the same material point $\mathbf{X}$, and since $\mathbf{v}(\mathbf{x},t) = \dot{\chi}(\mathbf{X},t)$, while $\mathbf{v}^*(\mathbf{x}^*,t) = \dot{\overline{\chi^*}}(\mathbf{X},t)$, $(20.12)_1$ yields

$$\mathbf{v}^*(\mathbf{x}^*,t) = \mathbf{Q}(t)\mathbf{v}(\mathbf{x},t) + \dot{\mathbf{y}}(t) + \dot{\mathbf{Q}}(t)(\mathbf{x} - \mathbf{o}). \tag{20.13}$$

The velocity gradient $\mathbf{L}(\mathbf{x},t) = \operatorname{grad}\mathbf{v}(\mathbf{x},t)$ transforms to

$$\mathbf{L}^*(\mathbf{x}^*,t) = \operatorname{grad}^*\mathbf{v}^*(\mathbf{x}^*,t),$$

with grad^* the gradient with respect to $\mathbf{x}^*$. Using the chain-rule to differentiate (20.13) with respect to $\mathbf{x}$, while bearing in mind the relation (20.1) between $\mathbf{x}^*$ and $\mathbf{x}$ and the definition (20.2) of the frame spin $\boldsymbol{\Omega}$, yields

$$\begin{aligned} (\operatorname{grad}^*\mathbf{v}^*)\mathbf{Q} &= \mathbf{Q}(\operatorname{grad}\mathbf{v}) + \dot{\mathbf{Q}} \\ &= (\mathbf{Q}(\operatorname{grad}\mathbf{v})\mathbf{Q}^{\top} + \boldsymbol{\Omega})\mathbf{Q}; \end{aligned}$$

the velocity gradient therefore transforms according to

$$\mathbf{L}^*(\mathbf{x}^*,t) = \mathbf{Q}(t)\mathbf{L}(\mathbf{x},t)\mathbf{Q}^{\top}(t) + \boldsymbol{\Omega}(t). \tag{20.14}$$

Since the frame-spin $\boldsymbol{\Omega}$ is skew, the transformed stretching and spin,

$$\mathbf{D}^* = \operatorname{sym}\mathbf{L}^* \quad \text{and} \quad \mathbf{W}^* = \operatorname{skw}\mathbf{L}^*$$

are given by

$$\begin{aligned} \mathbf{D}^*(\mathbf{x}^*,t) &= \mathbf{Q}(t)\mathbf{D}(\mathbf{x},t)\mathbf{Q}^{\top}(t), \\ \mathbf{W}^*(\mathbf{x}^*,t) &= \mathbf{Q}(t)\mathbf{W}(\mathbf{x},t)\mathbf{Q}^{\top}(t) + \boldsymbol{\Omega}(t). \end{aligned} \tag{20.15}$$

Of the three fields $\mathbf{L}$, $\mathbf{D}$, and $\mathbf{W}$, it is therefore only the stretching $\mathbf{D}$ that is frame-indifferent.

Consider, next, the stretch fields $\mathbf{U}$ and $\mathbf{C}$ and the strain field $\mathbf{E}$. By (7.3) and (7.6), $\mathbf{E} = \frac{1}{2}(\mathbf{C} - \mathbf{1})$ and $\mathbf{C} = \mathbf{U}^2 = \mathbf{F}^{\top}\mathbf{F}$, and, by (20.11),

$$\begin{aligned} \mathbf{C}^* &= (\mathbf{QF})^{\top}\mathbf{QF} \\ &= \mathbf{F}^{\top}\mathbf{F} \\ &= \mathbf{C}. \end{aligned}$$

The fields $\mathbf{C}$, $\mathbf{U}$, *and* $\mathbf{E}$ *are thus invariant.* These results are special cases of a more general result. Let $\mathbf{A}$ be a tensor field that maps material vectors to material vectors. Let $\mathbf{a}$ and $\mathbf{b}$ be material vector fields. Then, the scalar field $\mathbf{a}\cdot\mathbf{A}\mathbf{b}$ must be invariant under a change of frame: $\mathbf{a}^*\cdot\mathbf{A}^*\mathbf{b}^* = \mathbf{a}\cdot\mathbf{A}\mathbf{b}$. But, $\mathbf{a}$ and $\mathbf{b}$ must be invariant, as they are material vector fields, so that $\mathbf{a}\cdot\mathbf{A}^*\mathbf{b} = \mathbf{a}\cdot\mathbf{A}\mathbf{b}$ for all (material) vector fields $\mathbf{a}$

and **b**. Thus, $\mathbf{A}^* = \mathbf{A}$ and we have the following result:

- *Tensor fields that map material vectors to material vectors are invariant under changes in frame.*

Further, since

$$\dot{\mathbf{A}}(\mathbf{X}, t) = \lim_{\tau \to 0} \frac{\mathbf{A}(\mathbf{X}, t + \tau) - \mathbf{A}(\mathbf{X}, t)}{\tau},$$

$\dot{\mathbf{A}}$ must also be invariant.

Next, since $\mathbf{U}$ is invariant, $\mathbf{F}^* = \mathbf{QF}$, and $\mathbf{R} = \mathbf{FU}^{-1}$, it follows that $\mathbf{R}^* = \mathbf{QR}$. In summary, we have the following **transformation laws**:

$$\begin{aligned}
\mathbf{F}^* &= \mathbf{QF}, \\
\mathbf{R}^* &= \mathbf{QR}, \\
\mathbf{U}^* &= \mathbf{U}, \\
\mathbf{C}^* &= \mathbf{C}, \\
\mathbf{E}^* &= \mathbf{E}, \\
\mathbf{V}^* &= \mathbf{QVQ}^{\top}, \\
\mathbf{B}^* &= \mathbf{QBQ}^{\top}, \\
\mathbf{L}^* &= \mathbf{QLQ}^{\top} + \mathbf{\Omega}, \\
\mathbf{D}^* &= \mathbf{QDQ}^{\top}, \\
\mathbf{W}^* &= \mathbf{QWQ}^{\top} + \mathbf{\Omega}.
\end{aligned} \tag{20.16}$$

In (20.16) it is tacit that, when described spatially, the transformed fields should be evaluated at $(\mathbf{x}^*, t)$, while the original fields should be evaluated at $(\mathbf{x}, t)$. Summarizing,

- the tensors **V**, **B**, and **D** are *frame-indifferent*;
- the tensors **U**, **C**, and **E** are *invariant*;
- the tensors **F**, **R**, **L**, and **W** are neither frame-indifferent nor invariant.

Consider the relative deformation gradient $\mathbf{F}_{(t)}(\mathbf{x}, \tau)$ defined in §11.3. By (11.16), suppressing nontemporal arguments,

$$\mathbf{F}_{(t)}(\tau) = \mathbf{F}(\tau)\mathbf{F}^{-1}(t), \tag{20.17}$$

and one might be tempted to assume, as a consequence of $(20.16)_1$, that $\mathbf{F}^*_{(t)}(\tau) = \mathbf{Q}(\tau)\mathbf{F}_{(t)}(\tau)\mathbf{Q}^{\top}(t)$. But that assumption would be incorrect: Because (20.17) was derived using the configuration at time t as *reference*, the transformation law for $\mathbf{F}_{(t)}(\tau)$ should be the same as that for $\mathbf{F}(\tau)$, namely $\mathbf{F}^*_{(t)}(\tau) = \mathbf{Q}(\tau)\mathbf{F}_{(t)}(\tau)$.

EXERCISES

1. Choose an arbitrary fixed nonzero vector **e** in the observed space. Under a change of frame **e** transforms to a vector

$$\mathbf{e}^*(t) = \mathbf{Q}(t)\mathbf{e}, \tag{20.18}$$

which rotates. Show that

$$\dot{\overline{\mathbf{e}^*}} = \mathbf{\Omega}\mathbf{e}^*. \tag{20.19}$$

2. Show that, under a change of frame with frame rotation $\mathbf{Q}$, the spatial gradient $\mathbf{g} = \operatorname{grad}\varphi$ of a spatial scalar field φ is frame-indifferent:
$$\mathbf{g}^* = \mathbf{Q}\mathbf{g}. \tag{20.20}$$
3. Show that the spatial gradient $\operatorname{grad}\boldsymbol{\omega}$ of the vorticity $\boldsymbol{\omega} = \operatorname{curl}\mathbf{v}$ is frame-indifferent.
4. Show that the tensors $\mathbf{L} - \boldsymbol{\Omega}^{\top}$ and $\mathbf{W} - \boldsymbol{\Omega}^{\top}$ are frame-indifferent.
5. Use induction to show that each Rivlin–Ericksen tensor $\mathbf{A}_n$, as defined via (11.23), is frame-indifferent.

20.3.1 Material Time-Derivatives of Frame-Indifferent Tensor Fields are Not Frame-Indifferent

Let $\mathbf{G}$ be a tensor field that — like the stretching $\mathbf{D}$ or the left Cauchy–Green tensor $\mathbf{B}$ — is frame-indifferent. Then, in any change of frame,
$$\mathbf{G}^* = \mathbf{Q}\mathbf{G}\mathbf{Q}^{\top}, \tag{20.21}$$
so that, by (20.2),
$$\begin{aligned}\overline{\mathbf{G}^*}^{\boldsymbol{\cdot}} &= \mathbf{Q}\dot{\mathbf{G}}\mathbf{Q}^{\top} + \dot{\mathbf{Q}}\mathbf{G}\mathbf{Q}^{\top} + \mathbf{Q}\mathbf{G}\dot{\mathbf{Q}}^{\top}\\ &= \mathbf{Q}\dot{\mathbf{G}}\mathbf{Q}^{\top} + \boldsymbol{\Omega}\mathbf{Q}\mathbf{G}\mathbf{Q}^{\top} - \mathbf{Q}\mathbf{G}\mathbf{Q}^{\top}\boldsymbol{\Omega}\\ &= \mathbf{Q}\dot{\mathbf{G}}\mathbf{Q}^{\top} + \boldsymbol{\Omega}\mathbf{G}^* - \mathbf{G}^*\boldsymbol{\Omega}.\end{aligned} \tag{20.22}$$
Thus, in general,
$$\dot{\mathbf{G}}\ \textit{is not frame-indifferent.}$$

20.3.2 The Corotational, Covariant, and Contravariant Rates of a Tensor Field

We continue to assume that $\mathbf{G}$ is frame-indifferent. The failure of $\dot{\mathbf{G}}$ to be frame-indifferent is due to the presence of the frame-spin $\boldsymbol{\Omega}$ in (20.22). Guided by this, we note that, by $(20.16)_{10}$,
$$\boldsymbol{\Omega} = \mathbf{W}^* - \mathbf{Q}\mathbf{W}\mathbf{Q}^{\top}, \tag{20.23}$$
and using this relation to eliminate $\boldsymbol{\Omega}$ from (20.22) yields
$$\overline{\mathbf{G}^*}^{\boldsymbol{\cdot}} - (\mathbf{W}^* - \mathbf{Q}\mathbf{W}\mathbf{Q}^{\top})\mathbf{G}^* + \mathbf{G}^*(\mathbf{W}^* - \mathbf{Q}\mathbf{W}\mathbf{Q}^{\top}) = \mathbf{Q}\dot{\mathbf{G}}\mathbf{Q}^{\top}.$$
Thus, by (20.21),
$$\begin{aligned}\overline{\mathbf{G}^*}^{\boldsymbol{\cdot}} + \mathbf{G}^*\mathbf{W}^* - \mathbf{W}^*\mathbf{G}^* &= \mathbf{Q}\dot{\mathbf{G}}\mathbf{Q}^{\top} + (\mathbf{Q}\mathbf{G}\mathbf{Q}^{\top})(\mathbf{Q}\mathbf{W}\mathbf{Q}^{\top}) - (\mathbf{Q}\mathbf{W}\mathbf{Q}^{\top})(\mathbf{Q}\mathbf{G}\mathbf{Q}^{\top})\\ &= \mathbf{Q}(\dot{\mathbf{G}} + \mathbf{G}\mathbf{W} - \mathbf{W}\mathbf{G})\mathbf{Q}^{\top},\end{aligned}$$
showing that $\dot{\mathbf{G}} + \mathbf{G}\mathbf{W} - \mathbf{W}\mathbf{G}$ is frame-indifferent. We are therefore led to the frame-indifferent rate
$$\overset{\circ}{\mathbf{G}} \stackrel{\text{def}}{=} \dot{\mathbf{G}} + \mathbf{G}\mathbf{W} - \mathbf{W}\mathbf{G}. \tag{20.24}$$

Other frame-indifferent rates may be derived from (20.24): Since $\mathbf{D}$ and $\mathbf{G}$ are frame-indifferent, so also are $\mathbf{G}\mathbf{D}$ and $\mathbf{D}\mathbf{G}$; adding either

(a) $\mathbf{G}\mathbf{D} + \mathbf{D}\mathbf{G}$, or
(b) $-(\mathbf{G}\mathbf{D} + \mathbf{D}\mathbf{G})$

to the right side of (20.24) therefore yields a new frame-indifferent rate. Since $\mathbf{L} = \mathbf{D} + \mathbf{W}$ with $\mathbf{D} = \mathbf{D}^{\top}$ and $\mathbf{W} = -\mathbf{W}^{\top}$, adding $\mathbf{GD} + \mathbf{DG}$ yields the rate

$$\overset{\vartriangle}{\mathbf{G}} \overset{\text{def}}{=} \dot{\mathbf{G}} + \mathbf{GL} + \mathbf{L}^{\top}\mathbf{G}, \tag{20.25}$$

while adding $-(\mathbf{GD} + \mathbf{DG})$ yields the rate[119]

$$\overset{\diamond}{\mathbf{G}} \overset{\text{def}}{=} \dot{\mathbf{G}} - \mathbf{LG} - \mathbf{GL}^{\top}. \tag{20.26}$$

Next, the definitions of covariant and contravariant convection as expressed in (13.25) and (13.31) and the definition of a corotational tensor field in (13.43) imply that

$$\left.\begin{aligned} \overset{\circ}{\mathbf{G}} = \mathbf{0} \qquad & \text{if and only if } \mathbf{G} \text{ is corotational},\\ \overset{\vartriangle}{\mathbf{G}} = \mathbf{0} \qquad & \text{if and only if } \mathbf{G} \text{ convects covariantly},\\ \overset{\diamond}{\mathbf{G}} = \mathbf{0} \qquad & \text{if and only if } \mathbf{G} \text{ convects contravariantly}; \end{aligned}\right\} \tag{20.27}$$

guided by (20.27), we refer to[120]

$\overset{\circ}{\mathbf{G}}$ as the **corotational rate** of $\mathbf{G}$,

$\overset{\vartriangle}{\mathbf{G}}$ as the **covariant rate** of $\mathbf{G}$,

$\overset{\diamond}{\mathbf{G}}$ as the **contravariant rate** of $\mathbf{G}$.

The corotational, covariant, and contravariant tensorial rates were deduced assuming that $\mathbf{G}$ is frame-indifferent, but the rates themselves are well defined for any tensor $\mathbf{G}$ (although for $\mathbf{G}$ not frame-indifferent the resulting rate is generally not frame-indifferent).

20.3.3 Other Relations for the Corotational Rate

Let $\mathbf{G}$ be a (not necessarily frame-indifferent) spatial tensor field. Then, by (13.41) and (13.42), for $\{\mathbf{k}_i\}$ an orthonormal corotational basis,

$$\begin{aligned} \overline{G_{ij}}^{\,\boldsymbol{\cdot}} &= \overline{\mathbf{k}_i \cdot \mathbf{G}\mathbf{k}_j}^{\,\boldsymbol{\cdot}} \\ &= \mathbf{k}_i \cdot \overset{\circ}{\mathbf{G}}\mathbf{k}_j; \end{aligned}$$

hence,

$$(\overset{\circ}{\mathbf{G}})_{ij} = \overline{G_{ij}}^{\,\boldsymbol{\cdot}}. \tag{20.28}$$

Relative to an orthonormal basis spinning with the deforming body, the components of the corotational rate of $\mathbf{G}$ *are therefore equal to the material time-derivatives of the components of* $\mathbf{G}$. Hence, by (2.11),

$$\overset{\circ}{\mathbf{G}} = \overline{G_{ij}}^{\,\boldsymbol{\cdot}}\, \mathbf{k}_i \otimes \mathbf{k}_j, \tag{20.29}$$

[119] There are two other simple possibilities: Adding $\mathbf{GD} - \mathbf{DG}$ yields the rate $\dot{\mathbf{G}} + \mathbf{GL} - \mathbf{LG}$; adding $\mathbf{DG} - \mathbf{GD}$ yields $\dot{\mathbf{G}} - \mathbf{GL}^{\top} + \mathbf{L}^{\top}\mathbf{G}$; neither of these rates appears in the literature. For the particular choice $\mathbf{G} = \mathbf{D}$, these rates coincide and yield an expression identical to that determined by (20.24).

[120] $\overset{\circ}{\mathbf{G}}$ is also referred to as the Jaumann–Zaremba rate; $\overset{\vartriangle}{\mathbf{G}}$ as the Cotter–Rivlin rate and the lower-convected rate; $\overset{\diamond}{\mathbf{G}}$ as the Oldroyd rate and the upper-convected rate.

an expression that with (20.28) lends further credence to our use of the term *corotational rate*.

An alternative expression for the corotational rate of $\mathbf{G}$ may be derived using notions and results of §11.3. We begin with the material description $\mathbf{G}(\mathbf{X}, t)$ of $\mathbf{G}$ and define

$$\mathbf{G}_{(t)}(\mathbf{x}, \tau) = \mathbf{G}(\chi^{-1}(\mathbf{x}, t), \tau), \qquad \mathbf{X} = \chi^{-1}(\mathbf{x}, t), \tag{20.30}$$

so that

$$\left.\begin{aligned} \mathbf{G}_{(t)}(\mathbf{x}, t) &= \mathbf{G}(\mathbf{X}, t), \\ \frac{\partial \mathbf{G}_{(t)}(\mathbf{x}, \tau)}{\partial \tau}\bigg|_{\tau=t} &= \dot{\mathbf{G}}(\mathbf{X}, t). \end{aligned}\right\} \tag{20.31}$$

Then, suppressing nontemporal arguments, $(11.22)_2$ and (20.24) imply that

$$\begin{aligned} \overset{\circ}{\mathbf{G}}(t) &= \dot{\mathbf{G}}(t) + \mathbf{G}(t)\frac{\partial \mathbf{R}_{(t)}(\tau)}{\partial \tau}\bigg|_{\tau=t} - \frac{\partial \mathbf{R}_{(t)}(\tau)}{\partial \tau}\bigg|_{\tau=t}\mathbf{G}(t) \\ &= \left[\frac{\partial \mathbf{G}_{(t)}(\tau)}{\partial \tau} + \mathbf{G}_{(t)}(\tau)\frac{\partial \mathbf{R}_{(t)}(\tau)}{\partial \tau} - \frac{\partial \mathbf{R}_{(t)}(\tau)}{\partial \tau}\mathbf{G}_{(t)}(\tau)\right]_{\tau=t} \end{aligned}$$

and, by $(11.19)_1$, we have an alternative relation for $\overset{\circ}{\mathbf{G}}$:

$$\overset{\circ}{\mathbf{G}}(t) = \left[\frac{\partial}{\partial \tau}\Big(\mathbf{R}^{\top}_{(t)}(\tau)\mathbf{G}_{(t)}(\tau)\mathbf{R}_{(t)}(\tau)\Big)\right]_{\tau=t}. \tag{20.32}$$

20.3.4 Other Relations for the Covariant Rate

Let $\mathbf{G}$ be a (not necessarily frame-indifferent) spatial tensor field and let $\{\mathbf{f}_i\}$ be a tangentially convecting basis with corresponding dual basis $\{\mathbf{f}^i\}$.[121] Then, by (13.8) and the identities (13.22)–(13.24),

$$\overline{G_{ij}}^{\,\boldsymbol{\cdot}} = \overline{\mathbf{f}_i \cdot \mathbf{G}\mathbf{f}_j}^{\,\boldsymbol{\cdot}} \tag{20.33}$$

$$= \mathbf{m}_i \cdot \overline{\mathbf{F}^{\top}\mathbf{G}\mathbf{F}}^{\,\boldsymbol{\cdot}}\mathbf{m}_j$$

$$= (\mathbf{F}^{-1}\mathbf{f}_i) \cdot \overline{\mathbf{F}^{\top}\mathbf{G}\mathbf{F}}^{\,\boldsymbol{\cdot}}(\mathbf{F}^{-1}\mathbf{f}_j)$$

$$= \mathbf{f}_i \cdot \big[\mathbf{F}^{-\top}\overline{\mathbf{F}^{\top}\mathbf{G}\mathbf{F}}^{\,\boldsymbol{\cdot}}\mathbf{F}^{-1}\big]\mathbf{f}_j \tag{20.34}$$

$$= \mathbf{f}_i \cdot (\dot{\mathbf{G}} + \mathbf{G}\mathbf{L} + \mathbf{L}^{\top}\mathbf{G})\mathbf{f}_j \tag{20.35}$$

$$= \mathbf{f}_i \cdot \overset{\triangle}{\mathbf{G}}\mathbf{f}_j. \tag{20.36}$$

The relations (20.33) and (20.36) imply that

$$\underbrace{\mathbf{f}_i \cdot \overset{\triangle}{\mathbf{G}}\mathbf{f}_j}_{(\overset{\triangle}{\mathbf{G}})_{ij}} = \underbrace{\overline{\mathbf{f}_i \cdot \mathbf{G}\mathbf{f}_j}^{\,\boldsymbol{\cdot}}}_{\overline{G_{ij}}^{\,\boldsymbol{\cdot}}} \tag{20.37}$$

and, hence, that, *relative to a basis embedded in and deforming with the body, the components of the covariant rate of* $\mathbf{G}$ *are equal to the material time-derivatives of the*

[121] Cf. the material following (13.8).

components of $\mathbf{G}$.[122] A consequence of $(13.17)_4$ and (20.37) is the expansion

$$\overset{\triangle}{\mathbf{G}} = \overline{G_{ij}}^{\boldsymbol{\cdot}}\,\mathbf{f}^i \otimes \mathbf{f}^j. \tag{20.38}$$

Next, (20.34) and (20.35) yield an alternative expression for $\overset{\triangle}{\mathbf{G}}$:

$$\overset{\triangle}{\mathbf{G}} = \mathbf{F}^{-\top}\,\overline{\mathbf{F}^{\top}\mathbf{G}\mathbf{F}}^{\boldsymbol{\cdot}}\,\mathbf{F}^{-1}. \tag{20.39}$$

This result — expressed in terms of the covariant pullback and pushforward operations $\mathbb{P}$ and $\mathbb{P}^{-1}$ introduced in (12.1) and (12.5) — has the form

$$\overset{\triangle}{\mathbf{G}} = \mathbb{P}^{-1}[\overline{\mathbb{P}[\mathbf{G}]}^{\boldsymbol{\cdot}}]. \tag{20.40}$$

Thus, to compute the covariant rate of $\mathbf{G}$, one pulls $\mathbf{G}$ back from the observed space to the reference space, takes the material time-derivative, and then pushes the differentiated term forward to the observed space.

The result (20.39), applied using the configuration at a fixed time t as reference, as was done in the derivation of (20.32), yields yet another relation for the covariant rate. Indeed, by (11.14), $(20.31)_1$, and (20.39),

$$\mathbf{F}_{(t)}^{-\top}(\tau)\frac{\partial}{\partial\tau}\Big(\mathbf{F}_{(t)}^{\top}(\tau)\mathbf{G}_{(t)}(\tau)\mathbf{F}_{(t)}(\tau)\Big)\mathbf{F}_{(t)}^{-1}(\tau)$$

represents the covariant rate of $\mathbf{G}_{(t)}(\tau)$ with respect to τ (holding t fixed), so that

$$\overset{\triangle}{\mathbf{G}}(t) = \left[\mathbf{F}_{(t)}^{-\top}(\tau)\frac{\partial}{\partial\tau}\Big(\mathbf{F}_{(t)}^{\top}(\tau)\mathbf{G}_{(t)}(\tau)\mathbf{F}_{(t)}(\tau)\Big)\mathbf{F}_{(t)}^{-1}(\tau)\right]_{\tau=t} \tag{20.41}$$

and, by (11.17),

$$\overset{\triangle}{\mathbf{G}}(t) = \left[\frac{\partial}{\partial\tau}\Big(\mathbf{F}_{(t)}^{\top}(\tau)\mathbf{G}_{(t)}(\tau)\mathbf{F}_{(t)}(\tau)\Big)\right]_{\tau=t}. \tag{20.42}$$

Summarizing, the covariant rate of a spatial tensor field $\mathbf{G}$ may be expressed in the following forms:

$$\left.\begin{aligned} \overset{\triangle}{\mathbf{G}} &= \overline{G_{ij}}^{\boldsymbol{\cdot}}\,\mathbf{f}^i \otimes \mathbf{f}^j, \\ \overset{\triangle}{\mathbf{G}} &= \mathbf{F}^{-\top}\,\overline{\mathbf{F}^{\top}\mathbf{G}\mathbf{F}}^{\boldsymbol{\cdot}}\,\mathbf{F}^{-1}, \\ \overset{\triangle}{\mathbf{G}}(t) &= \left[\frac{\partial}{\partial\tau}\Big(\mathbf{F}_{(t)}^{\top}(\tau)\mathbf{G}_{(t)}(\tau)\mathbf{F}_{(t)}(\tau)\Big)\right]_{\tau=t}. \end{aligned}\right\} \tag{20.43}$$

20.3.5 Other Relations for the Contravariant Rate

The following counterparts of (20.43) hold for the *contravariant rate* of $\mathbf{G}$:

$$\left.\begin{aligned} \overset{\diamond}{\mathbf{G}} &= \overline{G^{ij}}^{\boldsymbol{\cdot}}\,\mathbf{f}_i \otimes \mathbf{f}_j, \\ \overset{\diamond}{\mathbf{G}} &= \mathbf{F}\,\overline{\mathbf{F}^{-1}\mathbf{G}\mathbf{F}^{-\top}}^{\boldsymbol{\cdot}}\,\mathbf{F}^{\top}, \\ \overset{\diamond}{\mathbf{G}} &= \left[\frac{\partial}{\partial\tau}\Big(\mathbf{F}_{(t)}^{-1}(\tau)\mathbf{G}_{(t)}(\tau)\mathbf{F}_{(t)}^{-\top}(\tau)\Big)\right]_{\tau=t}; \end{aligned}\right\} \tag{20.44}$$

the verification of these relations is left as an exercise.

---

[122] Cf. Footnote 77.

20.3.6 General Tensorial Rate

The results (20.24), (20.25), and (20.26) are of the general form

$$\mathbf{G}^{\flat} = \dot{\mathbf{G}} + \boldsymbol{\Phi}(\mathbf{G}, \mathbf{L}) \tag{20.45}$$

and, granted that $\mathbf{G}$ is frame-indifferent, it seems reasonable to ask: What is the most general frame-indifferent field of this form? If we take an arbitrary change of frame such that, at some time $\mathbf{Q} = \mathbf{1}$ and $\dot{\mathbf{Q}} = \boldsymbol{\Omega} = -\boldsymbol{\Omega}^{\top}$, then, for $\mathbf{G}^{\flat}$ frame-indifferent, we may use $(20.16)_6$ to conclude that, at that time,

$$\begin{aligned}\mathbf{G}^{\flat} &= (\mathbf{G}^{\flat})^{*}\\ &= \dot{\mathbf{G}} + \boldsymbol{\Omega}\mathbf{G} - \mathbf{G}\boldsymbol{\Omega} + \boldsymbol{\Phi}(\mathbf{G}, \mathbf{L} + \boldsymbol{\Omega}).\end{aligned}$$

Since the skew tensor $\boldsymbol{\Omega}$ is arbitrary, we may therefore take $\boldsymbol{\Omega} = -\mathbf{W}$ to obtain

$$\mathbf{G}^{\flat} = \dot{\mathbf{G}} + \mathbf{G}\mathbf{W} - \mathbf{W}\mathbf{G} + \boldsymbol{\Phi}(\mathbf{G}, \mathbf{D}),$$

or equivalently, by (20.24),

$$\mathbf{G}^{\flat} = \overset{\circ}{\mathbf{G}} + \boldsymbol{\Phi}(\mathbf{G}, \mathbf{D}). \tag{20.46}$$

Further, since $\overset{\circ}{\mathbf{G}}$ is frame-indifferent, if we choose an arbitrary change of frame with rotation $\mathbf{Q}$, we find that, in the new frame,

$$\mathbf{Q}\mathbf{G}^{\flat}\mathbf{Q}^{\top} - \mathbf{Q}\overset{\circ}{\mathbf{G}}\mathbf{Q}^{\top} = \boldsymbol{\Phi}(\mathbf{Q}\mathbf{G}\mathbf{Q}^{\top}, \mathbf{Q}\mathbf{D}\mathbf{Q}^{\top}),$$

so that, by (20.46),

$$\mathbf{Q}\boldsymbol{\Phi}(\mathbf{G}, \mathbf{D})\mathbf{Q} = \boldsymbol{\Phi}(\mathbf{Q}\mathbf{D}\mathbf{Q}^{\top}, \mathbf{Q}\mathbf{D}\mathbf{Q}^{\top})$$

and $\boldsymbol{\Phi}$ must be an isotropic function.[123] Conversely, if $\boldsymbol{\Phi}$ is isotropic, and if $\mathbf{G}$ is frame-indifferent, then $\mathbf{G}^{\flat}$ is frame-indifferent, an assertion whose proof we leave as an exercise. Thus, for $\mathbf{G}$ frame-indifferent, the rate (20.45) is frame-indifferent if and only if it has the specific form (20.46) with $\boldsymbol{\Phi}$ an isotropic function.[124] Therefore,

- *the corotational rate is generic up to an arbitrary isotropic tensor function of* $\mathbf{G}$ *and* $\mathbf{D}$.

EXERCISES

1. Show that

$$\overset{\circ}{\mathbf{1}} = \mathbf{0} \qquad \text{and} \qquad \overset{\vartriangle}{\mathbf{1}} = -\overset{\diamond}{\mathbf{1}} = 2\mathbf{D}.$$

2. Show that

$$\overset{\vartriangle}{\mathbf{D}} = \overset{\circ}{\mathbf{D}} + 2\mathbf{D}^2 \qquad \text{and} \qquad \overset{\diamond}{\mathbf{D}} = \overset{\circ}{\mathbf{D}} - 2\mathbf{D}^2.$$

3. Show that

$$\overset{\circ}{\mathbf{G}} = \tfrac{1}{2}(\overset{\vartriangle}{\mathbf{G}} + \overset{\diamond}{\mathbf{G}}).$$

4. Establish (20.44).
5. Give an alternative derivation of (20.43) via a direct computation of the term

$$\mathbf{F}^{-\top}\,\overline{\mathbf{F}^{\top}\mathbf{G}\mathbf{F}}^{\boldsymbol{\cdot}}\,\mathbf{F}^{-1}.$$

---

123 Cf. §113.2.
124 Cf. Noll (1955, §7).

6. Express $(20.44)_2$ in terms of the contravariant pullback (12.2) and its inverse $(12.5)_2$.

7. Verify that (20.46), with $\boldsymbol{\Phi}$ isotropic, is frame-indifferent.

8. Bearing in mind (17.14) and (20.25), show that

$$\overset{\triangle}{\mathbf{W}} = \mathbf{J}.$$

9. Let $\mathbf{g}$ be a frame-indifferent spatial vector field. Show that $\dot{\mathbf{g}}$ is not frame-indifferent. Deduce the following frame-indifferent rates:

$$\left. \begin{aligned} \overset{\circ}{\mathbf{g}} &= \dot{\mathbf{g}} - \mathbf{Wg}, \\[1mm] \overset{\triangle}{\mathbf{g}} &= \dot{\mathbf{g}} + \mathbf{L}^{\mathsf{T}}\mathbf{g}, \\[1mm] \overset{\diamond}{\mathbf{g}} &= \dot{\mathbf{g}} - \mathbf{Lg}. \end{aligned} \right\} \tag{20.47}$$

Further, give an argument in support of referring to

$$\left. \begin{aligned} &\overset{\circ}{\mathbf{g}} \text{ as the } \textbf{corotational rate} \text{ of } \mathbf{g}, \\[1mm] &\overset{\triangle}{\mathbf{g}} \text{ as the } \textbf{covariant rate} \text{ of } \mathbf{g}, \\[1mm] &\overset{\diamond}{\mathbf{g}} \text{ as the } \textbf{contravariant rate} \text{ of } \mathbf{g}. \end{aligned} \right\} \tag{20.48}$$

10. Show that

$$\overset{\circ}{\mathbf{g}} = \tfrac{1}{2}(\overset{\triangle}{\mathbf{g}} + \overset{\diamond}{\mathbf{g}}).$$

11. Use the definitions (20.24)–(20.26) and (20.47) of the corotational, covariant, and contravariant rates of tensor and vector fields to show that

$$\overline{\overset{\circ}{\mathbf{g} \otimes \mathbf{h}}} = \overset{\circ}{\mathbf{g}} \otimes \mathbf{h} + \mathbf{g} \otimes \overset{\circ}{\mathbf{h}},$$

$$\overline{\overset{\triangle}{\mathbf{g} \otimes \mathbf{h}}} = \overset{\triangle}{\mathbf{g}} \otimes \mathbf{h} + \mathbf{g} \otimes \overset{\triangle}{\mathbf{h}},$$

$$\overline{\overset{\diamond}{\mathbf{g} \otimes \mathbf{h}}} = \overset{\diamond}{\mathbf{g}} \otimes \mathbf{h} + \mathbf{g} \otimes \overset{\diamond}{\mathbf{h}}.$$

12. Assuming that $\mathbf{g}$ is a frame-indifferent spatial vector field, show that any frame-indifferent rate

$$\mathbf{g}^{\flat} = \dot{\mathbf{g}} + \varphi(\mathbf{g}, \mathbf{L}) \tag{20.49}$$

necessarily has the specific form

$$\mathbf{g}^{\flat} = \overset{\circ}{\mathbf{g}} + \varphi(\mathbf{g}, \mathbf{D}), \tag{20.50}$$

with φ isotropic.

21 Frame-Indifference Principle

A basic principle underlying most of physics is that

- *physical laws be independent of the frame of reference.*

We refer to this principle as **frame-indifference**. Within the rubric of continuum mechanics this principle is based on the notion of a change in frame — introduced in §20.1 — as a mapping of *spatial points* **x** to *spatial points*

$$\mathbf{x}^* = \mathbf{y}(t) + \mathbf{Q}(t)(\mathbf{x} - \mathbf{o}), \tag{21.1}$$

in which $\mathbf{Q}(t)$ is a rotation and $\mathbf{y}(t)$ a spatial point at each *fixed time* t.[125]

There is some disagreement as to whether only rotations or all orthogonal tensors should be employed in the statement of the frame-indifference principle. For example, TRUESDELL & NOLL (1965, §19) state the principle with $\mathbf{Q}(t)$ an *orthogonal tensor*, while CHADWICK (1976, p.130) and GURTIN (1981, pp. 139–145) require only that $\mathbf{Q}(t)$ be a rotation. This issue has been settled by MURDOCH (2003): Using a rigorous argument, Murdoch concludes that the statement of the principle should involve only rotations. In addition, Murdoch notes that inclusion of the orthogonal tensor $\mathbf{Q} = -\mathbf{1}$ in the frame-indifference principle would preclude one from characterizing optically-active sugar solutions that rotate plane-polarized light in opposing senses. Further, Murdoch (private communication (2007)) credits Rivlin with visualizing a material made of identical parallel helical molecules that have a substructure which makes the ends different; if these all "point" locally in the same direction then inversion would yield a material with different structure and hence, presumably, different response.

21.1 Transformation Rules for Stress and Body Force

We find it most convenient to work with balance of forces using the generalized body force $\mathbf{b} = \mathbf{b}_0 - \rho\dot{\mathbf{v}}$.[126]

Consistent with this principle, we assume that balance of forces is frame-indifferent: Given any change of frame and any spatial region $\mathcal{P}_t$ convecting with the body,

$$\int_{\partial\mathcal{P}_t^*} \mathbf{T}^*\mathbf{n}^*\,da + \int_{\mathcal{P}_t^*} \mathbf{b}^*\,dv = \mathbf{0}, \tag{21.2}$$

where $\mathcal{P}_t^* = \chi^*(\mathrm{P}, t)$ is the deformed material region $\mathcal{P}_t$ as viewed with respect to the new frame.

[125] Cf. TRUESDELL & NOLL (1965, §19a) for an interesting and illuminating history of the frame-indifference principle.

[126] Cf. (19.15).

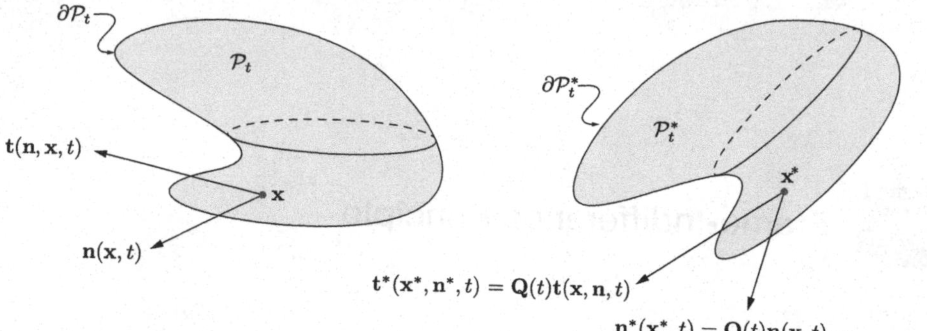

Figure 21.1. Transformations of a material region $\mathcal{P}_t$ along with the traction $\mathbf{t}(\mathbf{n},\mathbf{x},t)$ and the outward unit normal $\mathbf{n}(\mathbf{x},t)$ at a point $\mathbf{x}$ on $\partial\mathcal{P}_t$ under a change of frame with rotation $\mathbf{Q}$.

Consider the Cauchy stress $\mathbf{T}$ at a particular point and time. As noted in (19.28), the mapping

$$\mathbf{t} = \mathbf{T}\mathbf{n} \tag{21.3}$$

carries *spatial* vectors $\mathbf{n}$ (unit normals) to *spatial* vectors $\mathbf{t}$ (tractions). Consistent with Figure 21.1, we assume that $\mathbf{t}$ and $\mathbf{n}$ transform according to

$$\mathbf{t}^* = \mathbf{Q}\mathbf{t} \qquad \text{and} \qquad \mathbf{n}^* = \mathbf{Q}\mathbf{n} \tag{21.4}$$

and are, therefore, frame-indifferent. If $\mathbf{T}^*$ denotes the Cauchy stress in the new frame, then $\mathbf{t}^* = \mathbf{T}^*\mathbf{n}^*$; thus, $\mathbf{Q}\mathbf{t} = \mathbf{T}^*\mathbf{Q}\mathbf{n}$ and, hence,

$$\mathbf{t} = (\mathbf{Q}^{\top}\mathbf{T}^*\mathbf{Q})\mathbf{n}. \tag{21.5}$$

Combining (21.3) and (21.5), we obtain

$$(\mathbf{T} - \mathbf{Q}^{\top}\mathbf{T}^*\mathbf{Q})\mathbf{n} = \mathbf{0},$$

and, since $\mathbf{n}$ is an arbitrary unit vector, we find that

$$\mathbf{T}^* = \mathbf{Q}\mathbf{T}\mathbf{Q}^{\top}; \tag{21.6}$$

the Cauchy stress $\mathbf{T}$ *is therefore frame-indifferent.*

Next, bearing in mind that a change of frame is a rigid mapping and that $\mathbf{Q}$ depends only on t, a consequence of (21.4) and (21.6) is that

$$\begin{aligned}\int\limits_{\partial\mathcal{P}_t^*} \mathbf{T}^*\mathbf{n}^*\,da &= \int\limits_{\partial\mathcal{P}_t} (\mathbf{Q}\mathbf{T}\mathbf{Q}^{\top})\mathbf{Q}\mathbf{n}\,da \\ &= \mathbf{Q}\int\limits_{\partial\mathcal{P}_t} \mathbf{T}\mathbf{n}\,da.\end{aligned}$$

Thus, appealing to the balances (19.29) and (21.2), we find that

$$\int\limits_{\mathcal{P}_t^*} \mathbf{b}^*\,dv = \mathbf{Q}\int\limits_{\mathcal{P}_t} \mathbf{b}\,dv,$$

and changing the variable of integration on the left side from $\mathbf{x}^*$ to $\mathbf{x}$, so that the region of integration $\mathcal{P}_t^*$ changes to $\mathcal{P}_t$, we obtain

$$\int\limits_{\mathcal{P}_t} (\mathbf{b}^* - \mathbf{Q}\mathbf{b})\,dv = \mathbf{0}.$$

Since the convecting spatial region $\mathcal{P}_t$ is arbitrary,

$$\mathbf{b}^* = \mathbf{Q}\mathbf{b} \tag{21.7}$$

and we conclude that *the generalized body force* $\mathbf{b}$ *is frame-indifferent.*

EXERCISE

1. Assume that $\mathbf{T}$ and $\mathbf{b}$ are frame-indifferent, so that (21.6) and (21.7) are satisfied.
   (a) Show that (21.2) is satisfied, so that balance of forces is frame-indifferent.
   (b) Show that balance of moments is frame-indifferent in the sense that

$$\int\limits_{\partial\mathcal{P}_t^*} \mathbf{r}^* \times \mathbf{T}^*\mathbf{n}^*\, da + \int\limits_{\mathcal{P}_t^*} \mathbf{r}^* \times \mathbf{b}^*\, dv = \mathbf{0}, \tag{21.8}$$

   with $\mathbf{r}^* = \mathbf{x}^* - \mathbf{o}$.

21.2 Inertial Body Force in a Frame That Is Not Inertial

One often sees the assertion that "inertial forces are not frame-indifferent," when, in fact, what is meant is that inertial forces do not have the form (mass)(acceleration) in all frames. Were inertial forces not frame-indifferent, then one would not be able to determine the form taken by an inertial force relative to a particular frame that is not inertial. Indeed, according to TRUESDELL (1991, p. 70), "When for purposes of interpretation in a particular application we need to employ some frame that is not inertial, as for example in problems referred to a rotating earth, we formulate the laws of mechanics first in an inertial frame and then transform them to the other frame of interest. Such is the traditional approach, which derives from CLAIRAUT and EULER."

Consistent with (21.7), we assume that the conventional body force $\mathbf{b}_0$ is frame-indifferent and this, in turn, implies that *the inertial force* $\boldsymbol{\iota}$ *is frame-indifferent*:

$$\boldsymbol{\iota}^* = \mathbf{Q}\boldsymbol{\iota}. \tag{21.9}$$

We now use the transformation law (21.9) to compute the inertial body force in a noninertial frame F^*. By (20.13),

$$\mathbf{Q}\mathbf{v} = \mathbf{v}^* - \dot{\mathbf{y}} - \dot{\mathbf{Q}}(\mathbf{x} - \mathbf{o}). \tag{21.10}$$

On the other hand, by (21.1), $\mathbf{x} - \mathbf{o} = \mathbf{Q}^{\top}(\mathbf{x}^* - \mathbf{y})$, and, since $\dot{\mathbf{Q}}\mathbf{Q}^{\top}$ is the frame-spin $\boldsymbol{\Omega}$,[127]

$$\mathbf{Q}\mathbf{v} = \mathbf{v}^* - \dot{\mathbf{y}} - \boldsymbol{\Omega}(\mathbf{x}^* - \mathbf{y}). \tag{21.11}$$

Thus,

$$\mathbf{v} = \mathbf{Q}^{\top}\mathbf{v}^* - \mathbf{Q}^{\top}\dot{\mathbf{y}} - \mathbf{Q}^{\top}\boldsymbol{\Omega}(\mathbf{x}^* - \mathbf{y}). \tag{21.12}$$

Next, since $\mathbf{x}^* = \boldsymbol{\chi}^*(\mathbf{X}, t)$, the material time-derivative of $\mathbf{x}^* - \mathbf{y}$ is $\mathbf{v}^* - \dot{\mathbf{y}}$; taking the material time-derivative of (21.12) and then premultiplying by $\mathbf{Q}$ therefore gives

$$\mathbf{Q}\dot{\mathbf{v}} = \dot{\mathbf{v}}^* + \mathbf{Q}\dot{\mathbf{Q}}^{\top}(\mathbf{v}^* - \dot{\mathbf{y}}) - \ddot{\mathbf{y}} - \mathbf{Q}\dot{\mathbf{Q}}^{\top}\boldsymbol{\Omega}(\mathbf{x}^* - \mathbf{y}) - \dot{\boldsymbol{\Omega}}(\mathbf{x}^* - \mathbf{y}) - \boldsymbol{\Omega}(\mathbf{v}^* - \dot{\mathbf{y}}), \tag{21.13}$$

and, since $\boldsymbol{\Omega} = -\boldsymbol{\Omega}^{\top} = -\mathbf{Q}\dot{\mathbf{Q}}^{\top}$,

$$\mathbf{Q}\dot{\mathbf{v}} = \dot{\mathbf{v}}^* - \ddot{\mathbf{y}} + \boldsymbol{\Omega}^2(\mathbf{x}^* - \mathbf{y}) - \dot{\boldsymbol{\Omega}}(\mathbf{x}^* - \mathbf{y}) - 2\boldsymbol{\Omega}(\mathbf{v}^* - \dot{\mathbf{y}}). \tag{21.14}$$

Finally, by (19.15) and (21.9), $\boldsymbol{\iota}^* = -\rho\mathbf{Q}\dot{\mathbf{v}}$; therefore,

$$\boldsymbol{\iota}^* = -\rho[\dot{\mathbf{v}}^* - \ddot{\mathbf{y}} + (\boldsymbol{\Omega}^2 - \dot{\boldsymbol{\Omega}})(\mathbf{x}^* - \mathbf{y}) - 2\boldsymbol{\Omega}(\mathbf{v}^* - \dot{\mathbf{y}})].$$

---

[127] Cf. (20.2).

Equivalently, writing $\boldsymbol{\Omega} = \boldsymbol{\zeta}\times$, with $\boldsymbol{\zeta}$ the angular velocity of the frame,

$$\boldsymbol{\imath}^* = -\rho[\dot{\mathbf{v}}^* - \ddot{\mathbf{y}} + \boldsymbol{\zeta} \times [\boldsymbol{\zeta} \times (\mathbf{x}^* - \mathbf{y})] - \dot{\boldsymbol{\zeta}} \times (\mathbf{x}^* - \mathbf{y}) - 2\boldsymbol{\zeta} \times (\mathbf{v}^* - \dot{\mathbf{y}})], \tag{21.15}$$

with $\ddot{\mathbf{y}}$ the relative translational acceleration, $\boldsymbol{\zeta} \times [\boldsymbol{\zeta} \times (\mathbf{x}^* - \mathbf{y})]$ the centripetal acceleration, $-\dot{\boldsymbol{\zeta}} \times (\mathbf{x}^* - \mathbf{y})$ the Euler acceleration, and $-2\boldsymbol{\zeta} \times (\mathbf{v}^* - \dot{\mathbf{y}})$ the Coriolis acceleration.[128]

EXERCISES

1. Show that a frame $\mathcal{F}^*$ in which the inertial body force $\boldsymbol{\imath}^*$ has the form given in (21.15) is inertial if and only if the relative translational acceleration $\ddot{\mathbf{y}}$ and the angular velocity $\boldsymbol{\zeta}$ vanish and determine the restricted form of the transformation law (21.9) arising from these restrictions.
2. Given φ defined by

$$\varphi(\mathbf{x}^*) = \tfrac{1}{2}(|\boldsymbol{\zeta}|^2|\mathbf{x}^* - \mathbf{y}|^2 - (\boldsymbol{\zeta} \cdot (\mathbf{x}^* - \mathbf{y}))^2),$$

show that

$$\operatorname{grad}^*\varphi(\mathbf{x}^*) = \boldsymbol{\zeta} \times [\boldsymbol{\zeta} \times (\mathbf{x}^* - \mathbf{y})]$$

and, thus, that the centripetal acceleration can be expressed as the gradient of a potential.

[128] TRUESDELL & TOUPIN (1960, §143); bear in mind that $\boldsymbol{\Omega} = \boldsymbol{\zeta}\times$ is the frame-spin.

22 Alternative Formulations of the Force and Moment Balances

In this section — working with the generalized body force $\mathbf{b}$ — we revisit the derivation of the force and moment balances. Specifically,

- *we do not assume a priori that the force and moment balance laws are satisfied, but instead show that they are derivable starting from other hypotheses.*

Consistent with this, *we do not assume that Cauchy's relation* $\mathbf{t}(\mathbf{n}) = \mathbf{T}\mathbf{n}$ *is satisfied.*

22.1 Force and Moment Balances as a Consequence of Frame-Indifference of the Expended Power

In this section we show that frame-indifference of the expended power has, as interesting and unexpected consequences, the balance laws for forces and moments.[129] Throughout the time t is fixed and $\mathcal{P}$ is an arbitrary spatial region.[130]

Consider a force system defined by a surface traction $\mathbf{t}(\mathbf{n})$ and a generalized body force $\mathbf{b}$ that enter the theory through the generalized external power[131]

$$\mathcal{W}(\mathcal{P}, \mathbf{v}) = \int_{\partial\mathcal{P}} \mathbf{t}(\mathbf{n}) \cdot \mathbf{v}\, da + \int_{\mathcal{P}} \mathbf{b} \cdot \mathbf{v}\, dv, \tag{22.1}$$

written to make explicit its dependence on the velocity field $\mathbf{v}$. Consider a frame-change $\mathcal{F} \to \mathcal{F}^*$ and assume that both the traction $\mathbf{t}(\mathbf{n})$ and the body force $\mathbf{b}$ are frame-indifferent,

$$\mathbf{t}^*(\mathbf{n}^*) = \mathbf{Q}\mathbf{t}(\mathbf{n}), \qquad \mathbf{b}^* = \mathbf{Q}\mathbf{b}, \tag{22.2}$$

and that the expended power in the new frame has the form

$$\mathcal{W}^*(\mathcal{P}^*, \mathbf{v}^*) = \int_{\partial\mathcal{P}^*} \mathbf{t}^*(\mathbf{n}^*) \cdot \mathbf{v}^*\, da + \int_{\mathcal{P}^*} \mathbf{b}^* \cdot \mathbf{v}^*\, dv. \tag{22.3}$$

129 In §26 we show that frame-indifference of the expended power is a simple consequence of frame-indifference of the first law of thermodynamics.

130 Cf. Footnote 101.

131 Cf. (19.44), which account for inertia.

Suppressing arguments, we may use (20.13), (21.1), and $(21.4)_2$ to conclude that[132]

$$\mathbf{t}^*(\mathbf{n}^*)\cdot\mathbf{v}^* = [\mathbf{Q}\mathbf{t}(\mathbf{n})]\cdot[\mathbf{Q}\mathbf{v}+\dot{\mathbf{y}}+\dot{\mathbf{Q}}(\mathbf{x}-\mathbf{o})] \tag{22.4}$$

$$= \mathbf{t}(\mathbf{n})\cdot\mathbf{Q}^{\top}(\mathbf{Q}\mathbf{v}+\dot{\mathbf{y}}+\dot{\mathbf{Q}}\mathbf{r}) \tag{22.5}$$

$$= \mathbf{t}(\mathbf{n})\cdot(\mathbf{v}+\mathbf{Q}^{\top}\dot{\mathbf{y}}+\mathbf{Q}^{\top}\dot{\mathbf{Q}}\mathbf{r}), \tag{22.6}$$

with $\mathbf{r}(\mathbf{x}) = \mathbf{x}-\mathbf{o}$. Let

$$\mathbf{w}(\mathbf{x},t) = \mathbf{Q}^{\top}\dot{\mathbf{y}}(t)+\mathbf{Q}^{\top}(t)\dot{\mathbf{Q}}(t)\mathbf{r},$$

let $\boldsymbol{\alpha} = \mathbf{Q}^{\top}\dot{\mathbf{y}}$, and let $\boldsymbol{\lambda}$ denote the axial vector corresponding to the skew tensor $\mathbf{Q}^{\top}\dot{\mathbf{Q}}$;[133] then, since

$$\mathbf{w}(\mathbf{x},t) = \boldsymbol{\alpha}(t)+\boldsymbol{\lambda}(t)\times\mathbf{r}, \tag{22.7}$$

$\mathbf{w}$ is a rigid velocity field;[134] (22.6) therefore takes the form

$$\mathbf{t}^*(\mathbf{n}^*)\cdot\mathbf{v}^* = \mathbf{t}(\mathbf{n})\cdot(\mathbf{v}+\mathbf{w}).$$

Similarly,

$$\mathbf{b}^*\cdot\mathbf{v}^* = \mathbf{Q}\mathbf{b}\cdot(\mathbf{Q}\mathbf{v}+\dot{\mathbf{y}}+\dot{\mathbf{Q}}\mathbf{r})$$

$$= \mathbf{b}\cdot(\mathbf{v}+\mathbf{w}).$$

Thus, changing the variable of integration in (22.3) from $\mathbf{x}^*$ to $\mathbf{x}$, we obtain

$$\mathcal{W}^*(\mathcal{P}^*,\mathbf{v}^*) = \int\limits_{\partial\mathcal{P}}\mathbf{t}(\mathbf{n})\cdot(\mathbf{v}+\mathbf{w})\,da+\int\limits_{\mathcal{P}}\mathbf{b}\cdot(\mathbf{v}+\mathbf{w})\,dv;$$

hence, by (22.1),

$$\mathcal{W}^*(\mathcal{P}^*,\mathbf{v}^*) = \mathcal{W}(\mathcal{P},\mathbf{v})+\mathcal{W}(\mathcal{P},\mathbf{w}), \tag{22.8}$$

where

$$\mathcal{W}(\mathcal{P},\mathbf{w}) = \int\limits_{\partial\mathcal{P}}\mathbf{t}(\mathbf{n})\cdot\mathbf{w}\,da+\int\limits_{\mathcal{P}}\mathbf{b}\cdot\mathbf{w}\,dv$$

$$= \mathcal{W}_{\rm rig}(\mathcal{P},\mathbf{w}), \tag{22.9}$$

with $\mathcal{W}_{\rm rig}(\mathcal{P},\mathbf{w})$ given by (19.20). By (22.8) and (22.9), $\mathcal{W}_{\rm rig}(\mathcal{P},\mathbf{w}) = 0$ if and only if

$$\mathcal{W}^*(\mathcal{P}^*,\mathbf{v}^*) = \mathcal{W}(\mathcal{P},\mathbf{v}) \quad \text{for every change of frame;} \tag{22.10}$$

that is, if and only if the expended power is frame-indifferent. A consequence of this observation and (‡) on page 137 is the following result due to NOLL (1963):

- *The generalized external power is frame-indifferent if and only if the force and moment balances* (19.16) *are satisfied.*

---

132 Note that $\mathbf{Q}^{\top}\dot{\mathbf{Q}} = \mathbf{Q}^{\top}\boldsymbol{\Omega}\mathbf{Q}$, where $\boldsymbol{\Omega}$ as defined in (20.2) is the frame-spin.
133 Cf. (3.5).
134 Cf. (10.5).

22.2 Principle of Virtual Power

Our starting point is the generalized power balance (19.45) in a more general form that makes no use of Cauchy's relation (19.22):

$$\underbrace{\int_{\partial\mathcal{P}} \mathbf{t}(\mathbf{n})\cdot\mathbf{v}\,da + \int_{\mathcal{P}} \mathbf{b}\cdot\mathbf{v}\,dv}_{\mathcal{W}(\mathcal{P})} = \underbrace{\int_{\mathcal{P}} \mathbf{T}\colon\mathrm{grad}\,\mathbf{v}\,dv}_{\mathcal{I}(\mathcal{P})}. \tag{22.11}$$

As before, $\mathcal{W}(\mathcal{P})$ and $\mathcal{I}(\mathcal{P})$ represent the external and internal power expenditures. The essential feature of (22.11) is that the Cauchy stress, here neither necessarily symmetric nor frame-indifferent, enters the theory through its internal expenditure of power.

Consider the power balance (22.11), but with the velocity considered as *virtual* and not related to the actual velocity field $\mathbf{v}$. We make this explicit by replacing $\mathbf{v}$ in (22.11) by its virtual counterpart $\tilde{\mathbf{v}}$. Further, we assume that the time t is fixed and, hence, treat the virtual fields as arbitrary functions $\tilde{\mathbf{v}}(\mathbf{x})$ and the spatial region $\mathcal{P}$ as an arbitrary subregion of the deformed body $\mathcal{B} = \mathcal{B}_t$. Using

$$\begin{aligned} \mathcal{W}(\mathcal{P},\tilde{\mathbf{v}}) &= \int_{\partial\mathcal{P}} \mathbf{t}(\mathbf{n})\cdot\tilde{\mathbf{v}}\,da + \int_{\mathcal{P}} \mathbf{b}\cdot\tilde{\mathbf{v}}\,dv, \\ \mathcal{I}(\mathcal{P},\tilde{\mathbf{v}}) &= \int_{\mathcal{P}} \mathbf{T}\colon\mathrm{grad}\,\tilde{\mathbf{v}}\,dv, \end{aligned} \tag{22.12}$$

to denote the virtual expenditures of external and internal power, we therefore rewrite (22.11) in the form of a **virtual power balance:**

$$\boxed{\underbrace{\int_{\partial\mathcal{P}} \mathbf{t}(\mathbf{n})\cdot\tilde{\mathbf{v}}\,da + \int_{\mathcal{P}} \mathbf{b}\cdot\tilde{\mathbf{v}}\,dv}_{\mathcal{W}(\mathcal{P},\tilde{\mathbf{v}})} = \underbrace{\int_{\mathcal{P}} \mathbf{T}\colon\mathrm{grad}\,\tilde{\mathbf{v}}\,dv}_{\mathcal{I}(\mathcal{P},\tilde{\mathbf{v}})}.} \tag{22.13}$$

The virtual power balance (22.13) and, hence, the principle of virtual power always uses the generalized external power (22.1).

We now show that the virtual power balance (22.13) encapsulates the local force balance $\mathrm{div}\mathbf{T} + \mathbf{b} = \mathbf{0}$ as well as Cauchy's relation $\mathbf{t}(\mathbf{n}) = \mathbf{T}\mathbf{n}$. Assume that (22.13) is satisfied for all choices of $\mathcal{P}$ and $\tilde{\mathbf{v}}$. Then, using the divergence theorem in the form $(4.11)_5$,

$$\int_{\mathcal{P}} \mathbf{T}\colon\mathrm{grad}\,\tilde{\mathbf{v}}\,dv = \int_{\partial\mathcal{P}} \mathbf{T}\mathbf{n}\cdot\tilde{\mathbf{v}}\,da - \int_{\mathcal{P}} \mathrm{div}\mathbf{T}\cdot\tilde{\mathbf{v}}\,dv; \tag{22.14}$$

thus, by (22.13),

$$\int_{\mathcal{P}} (\mathrm{div}\mathbf{T} + \mathbf{b})\cdot\tilde{\mathbf{v}}\,dv + \int_{\partial\mathcal{P}} (\mathbf{t}(\mathbf{n}) - \mathbf{T}\mathbf{n})\cdot\tilde{\mathbf{v}}\,da = 0 \tag{22.15}$$

for every choice of the field $\tilde{\mathbf{v}}$. At this point we appeal to the fundamental lemma of the calculus of variations.[135] This lemma with $\mathcal{S} = \partial\mathcal{P}$ and $\boldsymbol{\phi} = \tilde{\mathbf{v}}$ applied to (22.15) yields the conclusion that $\mathbf{t}(\mathbf{n}) = \mathbf{T}\mathbf{n}$ on $\partial\mathcal{P}$ and $\mathrm{div}\mathbf{T} + \mathbf{b} = \mathbf{0}$ within $\mathcal{P}$. Further, since the choice of subregion $\mathcal{P}$ is arbitrary, these relations must hold throughout the deformed body.

[135] Cf. page 167.

We next show that the requirement that the virtual internal power $\mathcal{I}(\mathcal{P}, \tilde{\mathbf{v}})$ be invariant under changes of frame implies that $\mathbf{T}$ is both symmetric and frame-indifferent. Bearing in mind that the time is fixed and suppressed, we consider an arbitrary change in frame and write $\boldsymbol{\phi}$ for the corresponding function (21.1) mapping spatial points $\mathbf{x}$ to spatial points $\mathbf{x}^*$; viz.

$$\begin{aligned}\mathbf{x}^* &= \boldsymbol{\phi}(\mathbf{x})\\ &\overset{\text{def}}{=} \mathbf{y} + \mathbf{Q}(\mathbf{x} - \mathbf{o}),\end{aligned} \tag{22.16}$$

with $\mathbf{Q}$ the frame rotation. In discussing the manner in which the internal power transforms it is most convenient to not suppress spatial arguments; for that reason, we rewrite the internal power as follows:

$$\mathcal{I}(\mathcal{P}, \tilde{\mathbf{v}}) = \int_{\mathcal{P}} \mathbf{T}(\mathbf{x}) \colon \operatorname{grad} \tilde{\mathbf{v}}(\mathbf{x})\, dv(\mathbf{x}).$$

By (20.14), the virtual velocity gradient

$$\tilde{\mathbf{L}}(\mathbf{x}) = \operatorname{grad} \tilde{\mathbf{v}}(\mathbf{x})$$

transforms as follows under the change of frame:

$$\tilde{\mathbf{L}}^*(\mathbf{x}^*) = \mathbf{Q}\tilde{\mathbf{L}}(\mathbf{x})\mathbf{Q}^{\top} + \boldsymbol{\Omega}, \tag{22.17}$$

with $\boldsymbol{\Omega}$ the frame-spin. Thus, for $\mathcal{P}^*$ and $\mathcal{I}^*(\mathcal{P}^*, \tilde{\mathbf{v}}^*)$ the region and internal power in the new frame, frame-indifference requires that

$$\underbrace{\int_{\mathcal{P}} \mathbf{T}(\mathbf{x}) \colon \tilde{\mathbf{L}}(\mathbf{x})\, dv(\mathbf{x})}_{\mathcal{I}(\mathcal{P}, \tilde{\mathbf{v}})} = \underbrace{\int_{\mathcal{P}^*} \mathbf{T}^*(\mathbf{x}^*) \colon \tilde{\mathbf{L}}^*(\mathbf{x}^*)\, dv(\mathbf{x}^*)}_{\mathcal{I}^*(\mathcal{P}^*, \tilde{\mathbf{v}}^*)} \tag{22.18}$$

with $\mathbf{T}^*$ the stress $\mathbf{T}$ in the new frame. Using (22.16) and (22.17) we can express the right side of (22.18) in the form

$$\mathcal{I}^*(\mathcal{P}^*, \tilde{\mathbf{v}}^*) = \int_{\mathcal{P}^*} \mathbf{T}^*(\mathbf{x}^*) \colon [\mathbf{Q}\tilde{\mathbf{L}}(\boldsymbol{\phi}^{-1}(\mathbf{x}^*))\mathbf{Q}^{\top} + \boldsymbol{\Omega}]\, dv(\mathbf{x}^*). \tag{22.19}$$

By (22.16), since the Jacobian of the transformation $\boldsymbol{\phi}$ satisfies $\det(\operatorname{grad}\boldsymbol{\phi}) = \det \mathbf{Q} = 1$, if we change the variable of integration in (22.19) from $\mathbf{x}^*$ to $\mathbf{x}$, we find that

$$\mathcal{I}^*(\mathcal{P}^*, \tilde{\mathbf{v}}^*) = \int_{\mathcal{P}} \mathbf{T}^*(\boldsymbol{\phi}(\mathbf{x})) \colon (\mathbf{Q}\tilde{\mathbf{L}}(\mathbf{x})\mathbf{Q}^{\top} + \boldsymbol{\Omega})\, dv(\mathbf{x});$$

hence, (22.18) implies that

$$\int_{\mathcal{P}} \mathbf{T}(\mathbf{x}) \colon \tilde{\mathbf{L}}(\mathbf{x})\, dv(\mathbf{x}) = \int_{\mathcal{P}} \mathbf{T}^*(\boldsymbol{\phi}(\mathbf{x})) \colon (\mathbf{Q}\tilde{\mathbf{L}}(\mathbf{x})\mathbf{Q}^{\top} + \boldsymbol{\Omega})\, dv(\mathbf{x})$$

or, equivalently, since the spatial region $\mathcal{P}$ is arbitrary, that

$$\begin{aligned}\mathbf{T} \colon \tilde{\mathbf{L}} &= \mathbf{T}^* \colon (\mathbf{Q}\tilde{\mathbf{L}}\mathbf{Q}^{\top} + \boldsymbol{\Omega})\\ &= (\mathbf{Q}^{\top}\mathbf{T}^*\mathbf{Q}) \colon \tilde{\mathbf{L}} + \mathbf{T}^* \colon \boldsymbol{\Omega},\end{aligned} \tag{22.20}$$

where we have once again suppressed arguments. Without loss of generality, we may consider a change of frame in which the frame rotation $\mathbf{Q}$ is constant. The frame-spin $\mathbf{\Omega}$ then vanishes and (22.20) implies that

$$(\mathbf{T} - \mathbf{Q}^{\top}\mathbf{T}^{*}\mathbf{Q}):\tilde{\mathbf{L}} = 0$$

for all choices of $\tilde{\mathbf{L}}$; hence,

$$\mathbf{T}^{*} = \mathbf{Q}\mathbf{T}\mathbf{Q}^{\top} \tag{22.21}$$

for all rotations $\mathbf{Q}$; *the Cauchy stress* $\mathbf{T}$ *is therefore frame-indifferent.* Although we previously reached this conclusion in §21.1, doing so rested on a somewhat stronger hypothesis — namely that the traction $\mathbf{t}$ be frame-indifferent.

Next, making use of (22.21) in (22.20), we find that

$$\mathbf{T}^{*}:\mathbf{\Omega} = 0;$$

on the other hand, by (20.2) we may assume that $\mathbf{\Omega}$ is an arbitrary skew tensor; thus, $\mathbf{T}^{*} = \mathbf{T}^{*\top}$, so that, by (22.21),

$$\mathbf{T} = \mathbf{T}^{\top}; \tag{22.22}$$

the Cauchy stress $\mathbf{T}$ *is therefore symmetric.* Previously, we reached this conclusion on the basis of balance of moments. Requiring the internal power to be frame-indifferent therefore obviates the need to impose moment balance. As a direct consequence of (22.22), the internal power must furthermore have the form

$$\mathcal{I}(\mathcal{P}) = \int\limits_{\mathcal{P}} \mathbf{T}:\mathbf{D}\,dv. \tag{22.23}$$

CONSEQUENCES OF THE PRINCIPLE OF VIRTUAL POWER *Assume that for any subregion* $\mathcal{P}$ *of the deformed body and any choice of the virtual velocity* $\tilde{\mathbf{v}}$ *the virtual power balance* (22.13) *is satisfied. Then, at all points of the deformed body:*

(i) *The traction* $\mathbf{t}(\mathbf{n})$ *and the stress* $\mathbf{T}$ *are related through Cauchy's relation*

$$\mathbf{t}(\mathbf{n}) = \mathbf{T}\mathbf{n} \tag{22.24}$$

for every unit vector $\mathbf{n}$.

(ii) $\mathbf{T}$ *and* $\mathbf{b}$ *satisfy the local force and moment balances*

$$\operatorname{div}\mathbf{T} + \mathbf{b} = \mathbf{0} \quad \text{and} \quad \mathbf{T} = \mathbf{T}^{\top}. \tag{22.25}$$

(iii) $\mathbf{T}$ *is frame-indifferent, viz.*

$$\mathbf{T}^{*} = \mathbf{Q}\mathbf{T}\mathbf{Q}^{\top} \tag{22.26}$$

under any change of frame with frame rotation $\mathbf{Q}$.

The foregoing results demonstrate the all-encompassing nature of the principle of virtual power. The value of this principle becomes particularly clear in nonclassical settings involving additional kinematical degrees of freedom. In particular, we use the principle of virtual power repeatedly in our treatment of plasticity.

EXERCISE

1. Show that if (i) and (ii) are satisfied, then the virtual power balance (22.13) holds for all choices of $\mathcal{P}$ and $\tilde{\mathbf{v}}$.

22.2.1 Application to Boundary-Value Problems

A minor modification of the virtual-power principle is of great importance in the solution of boundary-value problems as it encapsulates both the balance of forces and the standard boundary condition for tractions. We therefore consider the deformed body $\mathcal{B} = \mathcal{B}_t$ (at some fixed time t) augmented by the traction condition

$$\mathbf{Tn} = \mathbf{t}_{\mathcal{S}} \quad \text{on } \mathcal{S}, \tag{22.27}$$

in which $\mathbf{t}_{\mathcal{S}}$ is a prescribed function on a subsurface $\mathcal{S}$ of $\partial\mathcal{B}$.[136] In this application we restrict attention to virtual fields, termed *admissible*, that satisfy

$$\tilde{\mathbf{v}} = \mathbf{0} \quad \text{on } \partial\mathcal{B} \setminus \mathcal{S}. \tag{22.28}$$

Granted (22.27) and (22.28), the virtual power balance (22.13) has the form

$$\int_{\mathcal{S}} \mathbf{t}_{\mathcal{S}} \cdot \tilde{\mathbf{v}}\, da + \int_{\mathcal{B}} \mathbf{b} \cdot \tilde{\mathbf{v}}\, dv = \int_{\mathcal{B}} \mathbf{T} : \operatorname{grad} \tilde{\mathbf{v}}\, dv. \tag{22.29}$$

Assume that the virtual power balance (22.29) holds for all admissible $\tilde{\mathbf{v}}$. Then, by (22.28),

$$\begin{aligned}\int_{\mathcal{B}} \mathbf{T} : \operatorname{grad} \tilde{\mathbf{v}}\, dv &= \int_{\partial\mathcal{B}} \mathbf{Tn} \cdot \tilde{\mathbf{v}}\, da - \int_{\mathcal{B}} \operatorname{div}\mathbf{T} \cdot \tilde{\mathbf{v}}\, dv \\ &= \int_{\mathcal{S}} \mathbf{Tn} \cdot \tilde{\mathbf{v}}\, da - \int_{\mathcal{B}} \operatorname{div}\mathbf{T} \cdot \tilde{\mathbf{v}}\, dv,\end{aligned}$$

and from (22.29) we obtain

$$\int_{\mathcal{B}} (\operatorname{div}\mathbf{T} + \mathbf{b}) \cdot \tilde{\mathbf{v}}\, dv + \int_{\mathcal{S}} (\mathbf{t}_{\mathcal{S}} - \mathbf{Tn}) \cdot \tilde{\mathbf{v}}\, da = 0. \tag{22.30}$$

Since (22.30) must hold for all admissible $\tilde{\mathbf{v}}$, we may conclude from the fundamental lemma of the calculus of variations (page 167) that the balance $\operatorname{div}\mathbf{T} + \mathbf{b} = \mathbf{0}$ is satisfied in $\mathcal{B}$ and the traction condition $\mathbf{Tn} = \mathbf{t}_{\mathcal{S}}$ holds on $\mathcal{S}$.

Conversely, if this balance and traction condition are satisfied, then, again applying the divergence theorem, we obtain

$$\begin{aligned}\int_{\mathcal{B}} \mathbf{T} : \operatorname{grad} \tilde{\mathbf{v}}\, dv &= \int_{\partial\mathcal{B}} \mathbf{Tn} \cdot \tilde{\mathbf{v}}\, da - \int_{\mathcal{B}} \operatorname{div}\mathbf{T} \cdot \tilde{\mathbf{v}}\, dv \\ &= \int_{\mathcal{S}} \mathbf{t}_{\mathcal{S}} \cdot \tilde{\mathbf{v}}\, da + \int_{\mathcal{B}} \mathbf{b} \cdot \tilde{\mathbf{v}}\, dv,\end{aligned}$$

which is the virtual power balance in the form (22.29). We have therefore established the following result:

Weak Form of the Force Balance and Traction Condition *The virtual-power balance*

$$\int_{\mathcal{S}} \mathbf{t}_{\mathcal{S}} \cdot \tilde{\mathbf{v}}\, da + \int_{\mathcal{B}} \mathbf{b} \cdot \tilde{\mathbf{v}}\, dv = \int_{\mathcal{B}} \mathbf{T} : \operatorname{grad} \tilde{\mathbf{v}}\, dv \tag{22.31}$$

is satisfied for all virtual fields $\tilde{\mathbf{v}}$ on $\mathcal{B}$ that vanish on $\partial\mathcal{B} \setminus \mathcal{S}$ if and only if

$$\operatorname{div}\mathbf{T} + \mathbf{b} = \mathbf{0}$$

[136] The traction $\mathbf{t}_{\mathcal{S}}$ represents the contact force, per unit area of $\mathcal{S}$, exerted on the body by its environment.

in $\mathcal{B}$ and

$$\mathbf{Tn} = \mathbf{t}_S$$

on $\mathcal{S}$.

REMARKS

(i) To apply (22.31) one need not assume that $\mathbf{T}$ is differentiable.
(ii) The virtual power balance (22.31), being global rather than local, is directly amenable to numerical schemes such as the finite-element method.
(iii) Extensions of this principle are useful in more general situations in which the form of the underlying balance(s) is not known.

22.2.2 Fundamental Lemma of the Calculus of Variations

This lemma may be stated as follows: *Let* $\mathbf{f}$ *be a continuous vector field on $\mathcal{P}$, let* $\mathbf{h}$ *be a continuous vector field on a smooth subsurface $\mathcal{S}$ of $\partial\mathcal{P}$, and assume that*

$$\int_{\mathcal{P}} \mathbf{f}\cdot\boldsymbol{\phi}\, dv + \int_{\mathcal{S}} \mathbf{h}\cdot\boldsymbol{\phi}\, da = 0 \tag{22.32}$$

for all continuous vector fields $\boldsymbol{\phi}$ on $\mathcal{P}$ that vanish on $\partial\mathcal{B}\setminus\mathcal{S}$. Then,

$$\mathbf{f} \equiv \mathbf{0} \ \text{ in } \mathcal{P}, \qquad \mathbf{h} \equiv \mathbf{0} \ \text{ on } \mathcal{S}. \tag{22.33}$$

To verify this lemma, consider first the special case in which $\boldsymbol{\phi}$ vanishes also on $\mathcal{S}$, so that $\int_{\mathcal{P}} \mathbf{f}\cdot\boldsymbol{\phi}\, dv = 0$. Assume that $\mathbf{f}$ does not vanish identically. Then, granted continuity, there is an open ball $\mathcal{N} \subset \mathcal{P}$ such that $\mathbf{f}$ never vanishes on $\mathcal{N}$. Choose $\boldsymbol{\phi}$ in the form

$$\boldsymbol{\phi}(\mathbf{x}) = g(\mathbf{x})\mathbf{f}(\mathbf{x}),$$

where g is strictly positive inside of $\mathcal{N}$ and vanishes outside of $\mathcal{N}$.[137] For this choice of $\boldsymbol{\phi}$, $\int_{\mathcal{P}} \mathbf{f}\cdot\boldsymbol{\phi}\, dv = \int_{\mathcal{N}} |\mathbf{f}|^2 g(\mathbf{x})\, dv > 0$, which is a contradiction; hence, $\mathbf{f} \equiv \mathbf{0}$. We are therefore left with showing that if $\int_{\mathcal{S}} \mathbf{h}\cdot\boldsymbol{\phi}\, da = 0$ for all $\boldsymbol{\phi}$ that vanish on $\partial\mathcal{B}\setminus\mathcal{S}$, then $\mathbf{h} \equiv \mathbf{0}$; the proof follows as above: We simply work on the subsurface $\mathcal{S}$ rather than in the region $\mathcal{P}$.

---

137 E.g., in terms of spherical coordinates with $r = 0$ at the center of $\mathcal{N}$ and $r = r_0$ the radius of $\mathcal{N}$, take $g(r) = (r_0 - r)^2$ for $r \le r_0$ and $g(r) = 0$ otherwise.

23 Mechanical Laws for a Spatial Control Volume

Recall that a control volume is a fixed region R that lies in the deformed body $\mathcal{B}_t$ for all t in some time interval.[138] Because our notation (17.45) for the time derivative of an integral over a control volume is nonstandard, we repeat it here:

$$\overline{\int_R \varphi(\mathbf{x}, t)\, dv(\mathbf{x})}^{\boldsymbol{\cdot}} = \frac{d}{dt}\int_R \varphi(\mathbf{x}, t)\, dv(\mathbf{x}). \tag{23.1}$$

Since R is independent of time,

$$\overline{\int_R \varphi\, dv}^{\boldsymbol{\cdot}} = \int_R \varphi'\, dv, \tag{23.2}$$

which is an identity basic to the ensuing analysis.

The following identities are used repeatedly here and in subsequent sections: For φ a spatial scalar field and $\mathbf{g}$ a spatial vector field,

$$\begin{aligned} \int_R \rho\dot{\varphi}\, dv &= \overline{\int_R \rho\varphi\, dv}^{\boldsymbol{\cdot}} + \int_{\partial R} (\rho\varphi)\mathbf{v}\cdot\mathbf{n}\, dv, \\ \int_R \rho\dot{\mathbf{g}}\, dv &= \overline{\int_R \rho\mathbf{g}\, dv}^{\boldsymbol{\cdot}} + \int_{\partial R} (\rho\mathbf{g})\mathbf{v}\cdot\mathbf{n}\, dv, \end{aligned} \tag{23.3}$$

where $\mathbf{n}$ is the outward unit normal to the boundary ∂R of R. These identities are direct consequences of (18.12), (18.13), and (23.2).

[138] Cf. (iii) on page 125.

23.1 Mass Balance for a Control Volume

In view of the local mass balance (18.6), the transport identity (23.2), and the divergence theorem,

$$\overline{\int_R \rho\,dv}^{\boldsymbol{\cdot}} = \int_R \rho'\,dv$$

$$= -\int_R \operatorname{div}(\rho\mathbf{v})\,dv$$

$$= -\int_{\partial R} \rho\mathbf{v}\cdot\mathbf{n}\,da;$$

we, thus, have **balance of mass** for a control volume R:

$$\overline{\int_R \rho\,dv}^{\boldsymbol{\cdot}} = -\int_{\partial R} \rho\mathbf{v}\cdot\mathbf{n}\,da. \tag{23.4}$$

The term

$$-\int_{\partial R} \rho\mathbf{v}\cdot\mathbf{n}\,da$$

(and others like it) has an important physical interpretation. Since $-\mathbf{n}$ is the inward unit normal to ∂R, $-\mathbf{v}\cdot\mathbf{n}$ represents the rate at which material is entering R across ∂R; hence, $-\rho\,\mathbf{v}\cdot\mathbf{n}$ represents the **inflow-rate** of mass across ∂R. The balance (23.4) therefore asserts that the mass of R increases at a rate equal to the inflow-rate of mass across ∂R.

23.2 Momentum Balances for a Control Volume

Consider the force and moment balances (19.29) for $\mathcal{P}$ a spatial region. Since a spatial control volume R is trivially a spatial region, we may rewrite (19.29) as follows:

$$\begin{aligned}\int_{\partial R} \mathbf{Tn}\,da + \int_R \mathbf{b}\,dv &= \mathbf{0},\\ \int_{\partial R} \mathbf{r}\times\mathbf{Tn}\,da + \int_R \mathbf{r}\times\mathbf{b}\,dv &= \mathbf{0}.\end{aligned} \tag{23.5}$$

By (19.15), the generalized body force is related to the inertial force and the conventional body force $\mathbf{b}_0$ via the expression

$$\mathbf{b} = \mathbf{b}_0 - \rho\dot{\mathbf{v}}. \tag{23.6}$$

Thus, using $(23.3)_2$ with $\mathbf{g} = \mathbf{v}$, we can convert the force balance $(23.5)_1$ to a momentum balance

$$\int_{\partial R} \mathbf{Tn}\,da + \int_R \mathbf{b}_0\,dv = \overline{\int_R \rho\mathbf{v}\,dv}^{\boldsymbol{\cdot}} + \int_{\partial R} (\rho\mathbf{v})\mathbf{v}\cdot\mathbf{n}\,da. \tag{23.7}$$

To establish the corresponding balance for angular momentum we apply $(23.3)_2$ with $\mathbf{g} = \mathbf{r} \times \mathbf{v}$ in the last step of the following calculation: Since $\dot{\mathbf{r}} = \mathbf{v}$,

$$\int_R \mathbf{r} \times \rho\dot{\mathbf{v}}\, dv = \int_R \rho\,\dot{\overline{\mathbf{r} \times \mathbf{v}}}\, dv$$

$$= \overline{\int_R \mathbf{r} \times \rho\mathbf{v} dv}^{\,\boldsymbol{\cdot}} + \int_{\partial R} (\mathbf{r} \times \rho\mathbf{v})\mathbf{v} \cdot \mathbf{n}\, da.$$

Therefore, using (23.6), we can convert $(23.5)_2$ to an angular momentum balance

$$\int_{\partial R} \mathbf{r} \times \mathbf{T}\mathbf{n}\, da + \int_R \mathbf{r} \times \mathbf{b}_0\, dv = \overline{\int_R \mathbf{r} \times \rho\mathbf{v}\, dv}^{\,\boldsymbol{\cdot}} + \int_{\partial R} (\mathbf{r} \times \rho\mathbf{v})\mathbf{v} \cdot \mathbf{n}\, da. \tag{23.8}$$

Rearranging terms in (23.7) and (23.8), we have the **linear** and **angular momentum balances** for a control volume R:

$$\begin{aligned} \overline{\int_R \rho\mathbf{v}\, dv}^{\,\boldsymbol{\cdot}} &= -\int_{\partial R} (\rho\mathbf{v})\mathbf{v} \cdot \mathbf{n}\, da + \int_{\partial R} \mathbf{T}\mathbf{n}\, da + \int_R \mathbf{b}_0\, dv, \\ \overline{\int_R \mathbf{r} \times \rho\mathbf{v}\, dv}^{\,\boldsymbol{\cdot}} &= -\int_{\partial R} (\mathbf{r} \times \rho\mathbf{v})\mathbf{v} \cdot \mathbf{n}\, da + \int_{\partial R} \mathbf{r} \times \mathbf{T}\mathbf{n}\, da + \int_R \mathbf{r} \times \mathbf{b}_0\, dv. \end{aligned} \tag{23.9}$$

In (23.9), the terms

$$-\int_{\partial R} (\rho\mathbf{v})\mathbf{v} \cdot \mathbf{n}\, da \qquad \text{and} \qquad -\int_{\partial R} (\mathbf{r} \times \rho\mathbf{v})\mathbf{v} \cdot \mathbf{n}\, da$$

represent respective inflow-rates of linear and angular momentum into R across ∂R; $(23.9)_1$ therefore asserts that the rate at which the linear momentum of R is increasing is equal to the inflow-rate of linear momentum across ∂R plus the net force exerted on R; analogously, $(23.9)_2$ asserts that the rate at which the angular momentum of R is increasing is equal to the inflow-rate of angular momentum across ∂R plus the net moment exerted on R.

Sample problem for a control volume Consider the flow of a fluid through a curved pipe (Figure 23.1). Suppose that the flow is steady in the sense that

$$\rho' = 0 \qquad \text{and} \qquad \mathbf{v}' = \mathbf{0},$$

and let R be the control volume bounded by the pipe walls and the cross-sections marking the ends of the pipe. Assume that the stress is a pressure ($\mathbf{T} = -p\mathbf{1}$), that the conventional body force vanishes, and that the density, velocity, and pressure at the entrance and exit have constant values

$$\rho_1,\ v_1\mathbf{e}_1,\ p_1 \qquad \text{and} \qquad \rho_2,\ v_2\mathbf{e}_2,\ p_2,$$

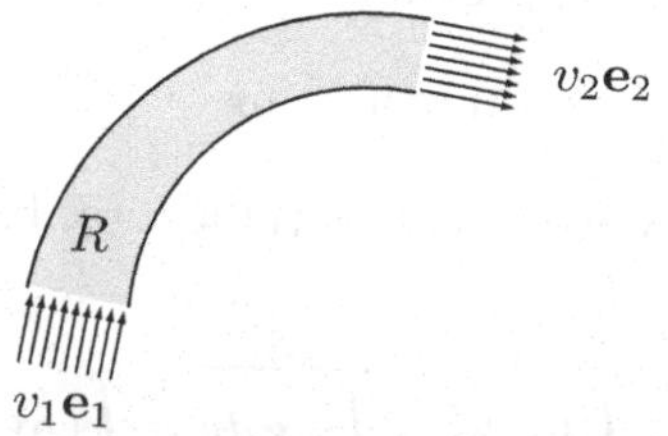

Figure 23.1. Fluid flow through a curved pipe R.

respectively, with $\mathbf{e}_1$ a unit vector perpendicular to the entrance cross-section and $\mathbf{e}_2$ a unit vector perpendicular to the exit cross-section. Using the statements of mass and linear momentum balance for R, determine the mass flow $\mathcal{M}$ through R and the net force $\mathbf{f}$ exerted by the fluid on the pipe walls.

Solution Since the flow is steady,

$$\overline{\int\limits_R \rho\mathbf{v}\,dv}^{\,\boldsymbol{\cdot}} = 0,$$

and, since $\mathbf{v}\cdot\mathbf{n}$ must vanish at the pipe walls,

$$\int\limits_{\partial R} (\rho\mathbf{v})\mathbf{v}\cdot\mathbf{e}\,da = (\rho_2 v_2^2 A_2)\mathbf{e}_2 - (\rho_1 v_1^2 A_1)\mathbf{e}_1,$$

where A_1 and A_2 denote the respective areas of the entrance and exit cross-sections. Next, since $\mathbf{f}$ is the net force exerted by the fluid on the pipe walls, the total force on the control volume R is

$$-\mathbf{f} + p_1 A_1 \mathbf{e}_1 - p_2 A_2 \mathbf{e}_2;$$

hence, the linear momentum balance (23.7) yields

$$\mathbf{f} = (p_1 + \rho_1 v_1^2)A_1\mathbf{e}_1 - (p_2 + \rho_2 v_2^2)A_2\mathbf{e}_2.$$

A similar analysis based on balance of mass (23.4) yields an expression

$$\rho_1 v_1 A_1 = \rho_2 v_2 A_2 = \mathcal{M} \tag{23.10}$$

for the mass flow $\mathcal{M}$ through the pipe. Thus,

$$\mathbf{f} = (p_1 A_1 + \mathcal{M} v_1)\mathbf{e}_1 - (p_2 A_2 + \mathcal{M} v_2)\mathbf{e}_2,$$

giving an expression for the force on the pipe walls in terms of conditions at the entrance and exit. Although the assumptions underlying this example are restrictive, they are, in fact, a good approximation for a large class of applications.

Consider the generalized power balance (19.45) written in its equivalent form for an arbitrary control volume R:

$$\int\limits_{\partial R} \mathbf{Tn}\cdot\mathbf{v}\,da + \int\limits_R \mathbf{b}\cdot\mathbf{v}\,dv = \int\limits_R \mathbf{T}\!:\!\mathbf{D}\,dv. \tag{23.11}$$

Since

$$\rho\mathbf{v}\cdot\dot{\mathbf{v}} = \tfrac{1}{2}\rho\overline{|\mathbf{v}^2|}^{\,\boldsymbol{\cdot}},$$

$(23.3)_1$ with $\varphi = \frac{1}{2}\overline{|\mathbf{v}^2|}^{\,\boldsymbol{\cdot}}$ and (23.6) may be used to convert (23.11) to a **conventional power balance** for a spatial control volume:[139]

$$\underbrace{\int\limits_{\partial R} \mathbf{Tn}\cdot\mathbf{v}\,da + \int\limits_R \mathbf{b}_0\cdot\mathbf{v}\,dv}_{\text{conventional external power}} = \underbrace{\int\limits_R \mathbf{T}\!:\!\mathbf{D}\,dv}_{\text{internal power}} + \underbrace{\overline{\int\limits_R \tfrac{1}{2}\rho|\mathbf{v}|^2\,dv}^{\,\boldsymbol{\cdot}}}_{\text{kinetic-energy rate}} + \underbrace{\int\limits_{\partial R} \tfrac{1}{2}\rho|\mathbf{v}|^2\mathbf{v}\cdot\mathbf{n}\,da}_{\text{kinetic-energy outflow-rate}}. \tag{23.12}$$

This power balance therefore asserts that

(‡) *the conventional external power expended on R is balanced by the sum of three terms: the internal power expended within R, the rate at which the kinetic energy of R is increasing, and the outflow-rate of kinetic energy across ∂R.*

EXERCISES

1. Establish the mass flow relation (23.10).
2. Establish the power balance (23.12) for a control volume.

[139] Cf. 19.43.

3. Show that the linear and angular momentum balances for a control volume R can be expressed as[140]

$$\overline{\int_R \rho\mathbf{v}\,dv}^{\,\boldsymbol{\cdot}} = \int_{\partial R} (\mathbf{T} - \rho\mathbf{v}\otimes\mathbf{v})\mathbf{n}\,da + \int_R \mathbf{b}_0\,dv,$$

$$\overline{\int_R \mathbf{r}\times\rho\mathbf{v}\,dv}^{\,\boldsymbol{\cdot}} = \int_{\partial R} \mathbf{r}\times(\mathbf{T} - \rho\mathbf{v}\otimes\mathbf{v})\mathbf{n}\,da + \int_R \mathbf{r}\times\mathbf{b}_0\,dv.$$

[140] Cf. (19.32).

24 Referential Forms for the Mechanical Laws

When working with solids, the use of a purely spatial description can be problematic. On the other hand, solids typically possess stress-free reference configurations with respect to which one may measure strain and develop constitutive equations. For that reason, we now reformulate the basic laws in a referential setting.

24.1 Piola Stress. Force and Moment Balances

If we define

$$\boxed{\mathbf{T}_R = J\mathbf{T}\mathbf{F}^{-\top},} \tag{24.1}$$

then, by $(15.8)_2$, for P a subregion of the reference body B and $\mathcal{P}_t = \chi_t(\mathrm{P})$ the corresponding convecting spatial region,

$$\int_{\partial\mathcal{P}_t} \mathbf{T}\mathbf{n}\, da = \int_{\partial \mathrm{P}} \mathbf{T}_R\mathbf{n}_R\, da_R. \tag{24.2}$$

Thus, $\mathbf{T}_R$ represents the stress measured per unit area in the reference body; $\mathbf{T}_R$ is referred to as the **Piola stress**. The tensor $\mathbf{T}_R$ carries the material vector $\mathbf{n}_R$ to the traction $\mathbf{T}_R\mathbf{n}_R$, which is a spatial vector;[141] thus, like the deformation gradient $\mathbf{F}$,

$$\mathbf{T}_R \textit{ maps material vectors to spatial vectors.} \tag{24.3}$$

Next, we define

$$\mathbf{b}_{0R} = J\mathbf{b}_0, \tag{24.4}$$

so that, by (15.7),

$$\int_{\mathcal{P}_t} \mathbf{b}_0\, dv = \int_{\mathrm{P}} \mathbf{b}_{0R}\, dv_R; \tag{24.5}$$

$\mathbf{b}_{0R}$ represents the conventional body force measured per unit volume in the reference body.

Since $\ddot{\chi}$ represents the referential description of $\dot{\mathbf{v}}$, we conclude using (18.10) that

$$\overline{\int_{\mathcal{P}_t} \rho\mathbf{v}\, dv}^{\,\cdot} = \int_{\mathcal{P}_t} \rho\dot{\mathbf{v}}\, dv = \int_{\mathrm{P}} \rho_R\ddot{\chi}\, dv_R = \overline{\int_{\mathrm{P}} \rho_R\, \dot{\chi}\, dv_R}^{\,\cdot}. \tag{24.6}$$

[141] By $(24.1)_1$ the "output" of $\mathbf{T}_R$, like that of $\mathbf{T}$, consists of spatial vectors.

Bearing in mind that $\mathbf{b} = \mathbf{b}_0 - \rho\dot{\mathbf{v}}$, we define the referential counterpart of the generalized body force by

$$\mathbf{b}_{\mathrm{R}} = \mathbf{b}_{0\mathrm{R}} - \rho_{\mathrm{R}}\ddot{\boldsymbol{\chi}}, \tag{24.7}$$

in which case (24.5) and the central equation in (24.6) yield

$$\int_{\mathcal{P}_t} \mathbf{b}\, dv = \int_{\mathrm{P}} \mathbf{b}_{\mathrm{R}}\, dv_{\mathrm{R}} \tag{24.8}$$

(and $\mathbf{b} = J\mathbf{b}_{\mathrm{R}}$). Using (24.2), (24.5), and (24.6), the spatial linear momentum balance (19.12) yields **balance of linear momentum**, expressed referentially:

$$\int_{\partial\mathrm{P}} \mathbf{T}_{\mathrm{R}}\mathbf{n}_{\mathrm{R}}\, da_{\mathrm{R}} + \int_{\mathrm{P}} \mathbf{b}_{0\mathrm{R}}\, dv_{\mathrm{R}} = \overline{\int_{\mathrm{P}} \rho_{\mathrm{R}}\, \dot{\boldsymbol{\chi}}\, dv_{\mathrm{R}}}^{\boldsymbol{\cdot}}. \tag{24.9}$$

Next, since P is a subregion of the reference body B and Div represents the referential divergence operator, the divergence theorem yields that

$$\int_{\partial\mathrm{P}} \mathbf{T}_{\mathrm{R}}\mathbf{n}_{\mathrm{R}}\, da_{\mathrm{R}} = \int_{\mathrm{P}} \mathrm{Div}\,\mathbf{T}_{\mathrm{R}}\, dv_{\mathrm{R}}$$

and (24.9) implies that

$$\int_{\mathrm{P}} (\mathrm{Div}\,\mathbf{T}_{\mathrm{R}} + \mathbf{b}_{0\mathrm{R}} - \rho_{\mathrm{R}}\ddot{\boldsymbol{\chi}})\, dv_{\mathrm{R}} = \mathbf{0}.$$

Thus, since the material region P is arbitrary, we arrive at the referential relation expressing the local form of **balance of linear momentum**:

$$\rho_{\mathrm{R}}\ddot{\boldsymbol{\chi}} = \mathrm{Div}\,\mathbf{T}_{\mathrm{R}} + \mathbf{b}_{0\mathrm{R}}, \qquad \rho_{\mathrm{R}}\ddot{\chi}_i = \frac{\partial T_{\mathrm{R}ij}}{\partial X_j} + b_{0\mathrm{R}i}. \tag{24.10}$$

Further, since the Cauchy stress $\mathbf{T}$ is symmetric, the Piola stress $\mathbf{T}_{\mathrm{R}}$ as defined in (24.1) must satisfy

$$\mathbf{T}_{\mathrm{R}}\mathbf{F}^{\top} = \mathbf{F}\mathbf{T}_{\mathrm{R}}^{\top}, \qquad T_{\mathrm{R}ik}F_{jk} = F_{ik}T_{\mathrm{R}kj}; \tag{24.11}$$

(24.11) represents the local **angular moment balance**, expressed referentially.

EXERCISES

1. Show that

$$\int_{\partial\mathrm{P}} \mathbf{r}\times\mathbf{T}_{\mathrm{R}}\mathbf{n}_{\mathrm{R}}\, da_{\mathrm{R}} + \int_{\mathrm{P}} \mathbf{r}\times\mathbf{b}_{0\mathrm{R}}\, dv = \overline{\int_{\mathrm{P}} \mathbf{r}\times(\rho_{\mathrm{R}}\,\dot{\boldsymbol{\chi}})\, dv_{\mathrm{R}}}^{\boldsymbol{\cdot}} \tag{24.12}$$

with[142]

$$\mathbf{r}(\mathbf{X}, t) = \boldsymbol{\chi}(\mathbf{X}, t) - \mathbf{o};$$

(24.12) represents the referential counterpart of the angular momentum balance (19.13).

[142] Cf. (19.2).

2. Show that the Piola stress $\mathbf{T}_R$ and the associated body force $\mathbf{b}_R$ transform according to

$$\mathbf{T}_R^* = \mathbf{Q}\mathbf{T}_R \qquad \text{and} \qquad \mathbf{b}_R^* = \mathbf{Q}\mathbf{b}_R. \tag{24.13}$$

3. Show, as a consequence of $(24.1)_1$ and (24.11), that the Cauchy stress $\mathbf{T}$ is symmetric.

24.2 Expended Power

Using (15.7) and $(15.8)_3$ we are led to the following identity, which transforms the spatial relation (19.35) to one that is referential:[143]

$$\begin{aligned}\mathcal{W}_0(\mathcal{P}_t) &= \int\limits_{\partial\mathcal{P}_t} \mathbf{T}\mathbf{n}\cdot\mathbf{v}\,da + \int\limits_{\mathcal{P}_t} \mathbf{b}_0\cdot\mathbf{v}\,dv \\ &= \int\limits_{\partial \mathrm{P}} \mathbf{T}_R\mathbf{n}_R\cdot\dot{\boldsymbol{\chi}}\,da_R + \int\limits_{\mathrm{P}} \mathbf{b}_{0R}\cdot\dot{\boldsymbol{\chi}}\,dv_R & (24.14)\\ &= \mathcal{W}_0(\mathrm{P}). & (24.15)\end{aligned}$$

Similarly, we use (18.10) to transform the kinetic energy:

$$\begin{aligned}\mathcal{K}(\mathcal{P}_t) &= \int\limits_{\mathcal{P}_t} \tfrac{1}{2}\rho|\mathbf{v}|^2\,dv \\ &= \int\limits_{\mathrm{P}} \tfrac{1}{2}\rho_R|\dot{\boldsymbol{\chi}}|^2\,dv_R \\ &= \mathcal{K}(\mathrm{P}). & (24.16)\end{aligned}$$

Transformations of the form (24.15) and (24.16) of set functions like $\mathcal{W}_0$ and $\mathcal{K}$ are called **spatial to referential transformations**; such transformations play a basic role in converting fundamental laws expressed spatially to equivalent laws expressed referentially.

Next, since $\mathbf{T}$ is symmetric and by (9.12) and (24.1),

$$\begin{aligned}\mathbf{T}:\mathbf{D} &= \mathbf{T}:\mathbf{L} \\ &= \mathbf{T}:(\dot{\mathbf{F}}\mathbf{F}^{-1}) \\ &= (\mathbf{T}\mathbf{F}^{-\top}):\dot{\mathbf{F}} \\ &= J^{-1}\mathbf{T}_R:\dot{\mathbf{F}};\end{aligned}$$

hence,

$$\mathbf{T}:\mathbf{D} = J^{-1}\mathbf{T}_R:\dot{\mathbf{F}}. \tag{24.17}$$

[143] Here, with a minor abuse of notation, we use the same symbol $\mathcal{W}_0$ for $\mathcal{W}_0(\mathcal{P}_t)$, which is a function of subregions of B_t, and for $\mathcal{W}_0(\mathrm{P})$, which is a function of subregions of B. Since we always include the argument there should be no danger of confusion.

Thus, (15.7) yields

$$\int_{\mathcal{P}_t} \mathbf{T}\!:\!\mathbf{D}\, dv = \int_{\mathrm{P}} \mathbf{T}_{\mathrm{R}}\!:\!\dot{\mathbf{F}}\, dv_{\mathrm{R}}$$

and the conventional power balance (19.43), when expressed referentially, has the form

$$\underbrace{\int_{\partial \mathrm{P}} \mathbf{T}_{\mathrm{R}}\mathbf{n}_{\mathrm{R}} \cdot \dot{\boldsymbol{\chi}}\, da_{\mathrm{R}} + \int_{\mathrm{P}} \mathbf{b}_{0\mathrm{R}} \cdot \dot{\boldsymbol{\chi}}\, dv_{\mathrm{R}}}_{\mathcal{W}_0(\mathrm{P})} = \int_{\mathrm{P}} \mathbf{T}_{\mathrm{R}}\!:\!\dot{\mathbf{F}}\, dv_{\mathrm{R}} + \underbrace{\overline{\int_{\mathrm{P}} \tfrac{1}{2}\rho_{\mathrm{R}} |\dot{\boldsymbol{\chi}}|^2\, dv_{\mathrm{R}}}^{\,\cdot}}_{\dot{\mathcal{K}}(\mathrm{P})}. \tag{24.18}$$

The field $\mathbf{T}_{\mathrm{R}}\!:\!\dot{\mathbf{F}}$ represents the stress power measured per unit volume in the reference body.

EXERCISE

1. Establish the following referential form of the generalized power balance (19.45):

$$\underbrace{\int_{\partial \mathrm{P}} \mathbf{T}_{\mathrm{R}}\mathbf{n}_{\mathrm{R}} \cdot \dot{\boldsymbol{\chi}}\, da_{\mathrm{R}} + \int_{\mathrm{P}} \mathbf{b}_{\mathrm{R}} \cdot \dot{\boldsymbol{\chi}}\, dv_{\mathrm{R}}}_{\mathcal{W}(\mathrm{P})} = \underbrace{\int_{\mathrm{P}} \mathbf{T}_{\mathrm{R}}\!:\!\dot{\mathbf{F}}\, dv_{\mathrm{R}}}_{\mathcal{I}(\mathrm{P})}. \tag{24.19}$$

25 Further Discussion of Stress

25.1 Power-Conjugate Pairings. Second Piola Stress

As noted in the paragraph containing (19.41), the stress power $\mathbf{T}:\mathbf{D}$ is an example of a power-conjugate pairing. The relation (24.17) exhibits the correspondence between this pairing and another pairing $\mathbf{T}_{\mathrm{R}}:\dot{\mathbf{F}}$, which measures the stress power per unit volume in the reference body.

Whereas $\mathbf{D}$ is purely a measure of the rate at which material elements stretch, $\dot{\mathbf{F}}$ carries information concerning the rates at which material elements stretch and rotate. One might therefore ask whether there are alternative referential measures of stress power in which a measure of stress is paired with a pure strain-rate. A multitude of such pairings exist,[144] but for our purposes only one of these is important. Specifically, we seek a pairing involving a measure $\mathbf{T}_{\mathrm{RR}}$ of stress that is conjugate to the time-rate $\dot{\mathbf{C}}$ of the right Cauchy–Green tensor $\mathbf{C} = \mathbf{F}^{\top}\mathbf{F}$. To determine the form of $\mathbf{T}_{\mathrm{RR}}$, note that, by (2.52), (2.53), (9.12), and (24.1),

$$\begin{aligned}
\mathbf{T}:\mathbf{D} &= \tfrac{1}{2}\mathbf{T}:(\mathbf{L}+\mathbf{L}^{\top}) \\
&= \tfrac{1}{2}\mathbf{T}:(\dot{\mathbf{F}}\mathbf{F}^{-1}+\mathbf{F}^{-\top}\dot{\mathbf{F}}^{\top}) \\
&= \tfrac{1}{2}\mathbf{T}:\mathbf{F}^{-\top}(\mathbf{F}^{\top}\dot{\mathbf{F}}+\dot{\mathbf{F}}^{\top}\mathbf{F})\mathbf{F}^{-1} \\
&= \tfrac{1}{2}(\mathbf{F}^{-1}\mathbf{T}\mathbf{F}^{-\top}):(\mathbf{F}^{\top}\dot{\mathbf{F}}+\dot{\mathbf{F}}^{\top}\mathbf{F}) \\
&= \tfrac{1}{2}(\mathbf{F}^{-1}\mathbf{T}\mathbf{F}^{-\top}):\dot{\mathbf{C}} \\
&= \tfrac{1}{2}J^{-1}(\mathbf{F}^{-1}\mathbf{T}_{\mathrm{R}}):\dot{\mathbf{C}}.
\end{aligned} \tag{25.1}$$

Thus, since we seek a measure of stress power per unit reference volume, we define

$$\boxed{\mathbf{T}_{\mathrm{RR}} = \mathbf{F}^{-1}\mathbf{T}_{\mathrm{R}}} \tag{25.2}$$

and note that, by (25.1),

$$\mathbf{T}:\mathbf{D} = \tfrac{1}{2}J^{-1}\mathbf{T}_{\mathrm{RR}}:\dot{\mathbf{C}}. \tag{25.3}$$

[144] Cf. Hill (1968).

As an important consequence of (25.2) we see that

$$\begin{aligned}\mathbf{T}_{RR}^{\top} &= (J\mathbf{F}^{-1}\mathbf{T}\mathbf{F}^{-\top})^{\top}\\ &= J\mathbf{F}^{-1}\mathbf{T}^{\top}\mathbf{F}^{-\top}\\ &= J\mathbf{F}^{-1}\mathbf{T}\mathbf{F}^{-\top}\\ &= \mathbf{T}_{RR}.\end{aligned} \tag{25.4}$$

Consistent with its pairing with $\dot{\mathbf{C}}$, the stress $\mathbf{T}_{RR}$ is thus symmetric.

By (24.3) and (19.28), the Cauchy stress $\mathbf{T}$ maps spatial vectors to spatial vectors, while the Piola stress $\mathbf{T}_R$ maps material vectors to spatial vectors. On the other hand, the symmetric tensor $\mathbf{T}_{RR}$ maps material vectors to material vectors.[145] The tensor field $\mathbf{T}_{RR}$ is usually referred to as the **second Piola stress.**

Finally, we note that, by (24.17) and (25.3),

$$\mathbf{T}_R : \dot{\mathbf{F}} = \tfrac{1}{2}\mathbf{T}_{RR} : \dot{\mathbf{C}}. \tag{25.5}$$

The pairings established above and in (24.17) are summarized as follows:

Power-conjugate Pairings The power expenditures of the Cauchy stress, the Piola stress, and the second Piola stress are related as follows:

$$\left.\begin{aligned}\mathbf{T}:\mathbf{D} &= J^{-1}\mathbf{T}_R : \dot{\mathbf{F}},\\ \mathbf{T}:\mathbf{D} &= \tfrac{1}{2}J^{-1}\mathbf{T}_{RR} : \dot{\mathbf{C}},\\ \mathbf{T}_R : \dot{\mathbf{F}} &= \tfrac{1}{2}\mathbf{T}_{RR} : \dot{\mathbf{C}}.\end{aligned}\right\} \tag{25.6}$$

Finally, using $(25.6)_3$, we can write the conventional power balance (24.18) in the alternative, and often useful, form:

$$\underbrace{\int_{\partial P} \mathbf{T}_R\mathbf{n}_R \cdot \dot{\chi}\, da_R + \int_P \mathbf{b}_{0R} \cdot \dot{\chi}\, dv_R}_{\mathcal{W}_0(P)} = \int_P \tfrac{1}{2}\mathbf{T}_{RR} : \dot{\mathbf{C}}\, dv_R + \underbrace{\overline{\int_P \tfrac{1}{2}\rho_R |\dot{\chi}|^2\, dv_R}}_{\dot{\overline{\mathcal{K}(P)}}}. \tag{25.7}$$

25.2 Transformation Laws for the Piola Stresses

From (21.6), we recall that the Cauchy stress $\mathbf{T}$ is frame-indifferent. Thus, by (20.16) and (24.1)

$$\begin{aligned}\mathbf{T}_R^* &= (J\mathbf{T}\mathbf{F}^{-\top})^*\\ &= J^*\mathbf{T}^*(\mathbf{F}^*)^{-\top}\\ &= J\mathbf{Q}\mathbf{T}\mathbf{Q}^{\top}(\mathbf{Q}\mathbf{F})^{-\top}\\ &= J\mathbf{Q}\mathbf{T}\mathbf{Q}^{\top}\mathbf{Q}\mathbf{F}^{-\top}\\ &= \mathbf{Q}(J\mathbf{T}\mathbf{F}^{-\top})\\ &= \mathbf{Q}\mathbf{T}_R,\end{aligned} \tag{25.8}$$

[145] The double subscript in $\mathbf{T}_{RR}$ is meant to underline this mapping property. Cf. (M1) and (M2) on page 65; recall, from (12.2), that the term $\mathbf{F}^{-1}\mathbf{T}_R = \mathbf{F}^{-1}\mathbf{T}\mathbf{F}^{-\top} = \underline{\mathbb{P}}[\mathbf{T}]$ represents the contravariant pullback of $\mathbf{T}$ from the deformed body to the reference body.

from which we conclude that, like its power-conjugate $\dot{\mathbf{F}}$, the Piola stress $\mathbf{T}_R$ is neither frame-indifferent nor invariant. Similarly, by (25.2),

$$\begin{aligned}\mathbf{T}^*_{RR} &= (\mathbf{F}^{-1}\mathbf{T}_R)^* \\ &= (\mathbf{F}^*)^{-1}\mathbf{T}^*_R \\ &= (\mathbf{QF})^{-1}\mathbf{QT}_R \\ &= \mathbf{F}^{-1}\mathbf{Q}^\top\mathbf{QT}_R \\ &= \mathbf{T}_{RR},\end{aligned} \tag{25.9}$$

from which we conclude that, like its power-conjugate $\dot{\mathbf{C}}$, the second Piola stress $\mathbf{T}_{RR}$ is invariant. Summarizing:

- the Cauchy stress tensor $\mathbf{T}$ is *frame-indifferent*;
- the second Piola stress tensors $\mathbf{T}_{RR}$ is *invariant*;
- the Piola stress tensor $\mathbf{T}_R$ is neither frame-indifferent nor invariant.

We emphasize that each of the three expressions $\mathbf{T}:\mathbf{D}$, $\mathbf{T}_R:\dot{\mathbf{F}}$, and $\frac{1}{2}\mathbf{T}_{RR}:\dot{\mathbf{C}}$ for the stress power involves a pairing of quantities with shared transformation properties. Furthermore, as a scalar, each of the resulting pairings is invariant.

EXERCISES

1. Use (24.1) and (25.2) to show that

$$\mathbf{T} = J^{-1}\mathbf{F}\mathbf{T}_{RR}\mathbf{F}^\top. \tag{25.10}$$

2. Show that

$$\mathbf{T}:\mathbf{D} = J^{-1}\mathbf{T}_{RR}:\dot{\mathbf{E}},$$

and, thus, that the second Piola stress can also be viewed as conjugate to the time-rate $\dot{\mathbf{E}}$ of the Green–St. Venant strain tensor.

3. Show that the **Biot stress** defined via

$$\tfrac{1}{2}(\mathbf{R}^\top\mathbf{T}_R + \mathbf{T}_R^\top\mathbf{R})$$

is symmetric. Show also that the Biot stress and the rate $\dot{\mathbf{U}}$ of the right stretch tensor provide a power-conjugate pairing:

$$\tfrac{1}{2}J^{-1}(\mathbf{R}^\top\mathbf{T}_R + \mathbf{T}_R^\top\mathbf{R}):\dot{\mathbf{U}} = \mathbf{T}:\mathbf{D}.$$

4. Take the material time-derivative of the relation (25.10) and use $(20.44)_2$ to show that

$$J\overset{\diamond}{\mathbf{T}} + \dot{J}\mathbf{T} = \mathbf{F}\dot{\mathbf{T}}_{RR}\mathbf{F}^\top. \tag{25.11}$$

PART V

BASIC THERMODYNAMICAL PRINCIPLES

In introducing basic versions of the first two laws of thermodynamics appropriate to continua, we emphasize that

- like force, we view energy, entropy, heat flow, and entropy flow as primitive objects;[146]
- *a priori* notions of "equilibrium" and "state" are *not* employed.

We find it most convenient to use a spatial formulation — that is, a formulation in terms of quantities measured per unit volume and area in the observed space.

[146] Cf. the discussion of TRUESDELL (1966, pp. 99–100).

26 The First Law: Balance of Energy

We assume throughout this section that the motion is smooth. Balance of energy in the presence of a shock wave is discussed in §33.2.

As before, $\mathcal{P}_t$ represents a spatial region convecting with the body, so that $\mathcal{P}_t = \chi(\mathrm{P}, t)$ for some material region P.

The first law of thermodynamics represents a detailed balance describing the interplay between the internal energy of $\mathcal{P}_t$, the kinetic energy of $\mathcal{P}_t$, the rate at which power is expended on $\mathcal{P}_t$, and the heat transferred to $\mathcal{P}_t$. Specifically, introducing:

(i) the *net internal energy* $\mathcal{E}(\mathcal{P}_t)$ of $\mathcal{P}_t$,[147] and
(ii) the *heat flow* $\mathcal{Q}(\mathcal{P}_t)$, which is the rate at which energy — in the form of heat — is transferred to $\mathcal{P}_t$,

the energy balance has the form

$$\overline{\mathcal{E}(\mathcal{P}_t) + \mathcal{K}(\mathcal{P}_t)}^{\boldsymbol{\cdot}} = \mathcal{W}_0(\mathcal{P}_t) + \mathcal{Q}(\mathcal{P}_t), \tag{26.1}$$

where

(iii) $\mathcal{K}(\mathcal{P}_t)$, the *kinetic energy*, is given by[148]

$$\mathcal{K}(\mathcal{P}_t) \stackrel{\text{def}}{=} \int_{\mathcal{P}_t} \tfrac{1}{2}\rho|\mathbf{v}|^2\,dv; \tag{26.2}$$

(iv) $\mathcal{W}_0(\mathcal{P}_t)$, the *conventional external power*, is given by[149]

$$\begin{aligned}\mathcal{W}_0(\mathcal{P}_t) &= \int_{\partial\mathcal{P}_t} \mathbf{Tn}\cdot\mathbf{v}\,da + \int_{\mathcal{P}_t} \mathbf{b}_0\cdot\mathbf{v}\,dv \\ &= \int_{\mathcal{P}_t} \mathbf{T}:\mathbf{D}\,dv + \overline{\mathcal{K}(\mathcal{P}_t)}^{\boldsymbol{\cdot}}.\end{aligned} \tag{26.3}$$

147 We use the same letter, $\mathcal{E}$, to denote the net internal energy as that used to denote three-dimensional Euclidean point space. However, because we always explicitly indicate the material region over which the net internal energy is evaluated there should be no reason for confusion.

148 Cf. (19.36).

149 Cf. (19.35) and (19.43).

26.1 Global and Local Forms of Energy Balance

We assume there is a scalar field ε, the **specific internal-energy**, such that

$$\mathcal{E}(\mathcal{P}_t) = \int\limits_{\mathcal{P}_t} \rho\varepsilon \, dv. \tag{26.4}$$

The term "specific" indicates that ε is measured per unit mass. The hypothesis (26.4) renders the set function $\mathcal{E}(\mathcal{P})$ additive (over the collection of spatial regions) in the sense that, given any pair of disjoint spatial regions $\mathcal{P}_1$ and $\mathcal{P}_2$, the net internal-energy of the region $\mathcal{P}_1 \cup \mathcal{P}_2$ is the net internal energy of $\mathcal{P}_1$ plus that of $\mathcal{P}_2$:

$$\mathcal{E}(\mathcal{P}_1 \cup \mathcal{P}_2) = \mathcal{E}(\mathcal{P}_1) + \mathcal{E}(\mathcal{P}_2).$$

The net internal-energy is therefore an "extensive parameter" in the sense used by thermodynamicists.[150]

Additivity, by itself, does not imply that $\mathcal{E}(\mathcal{P})$ has the form (26.4); an additional assumption, such as $\mathcal{E}(\mathcal{P}) \to 0$ as $\mathrm{vol}(\mathcal{P}) \to 0$, asserting that the internal energy be distributed continuously over the volume of the body, is needed. Questions of this form lie in the realm of measure theory.[151]

We presume that heat flow is described by a vector **heat flux q** and a scalar **heat supply** q; these fields determine $\mathcal{Q}(\mathcal{P}_t)$ as follows:

$$\mathcal{Q}(\mathcal{P}_t) = -\int\limits_{\partial\mathcal{P}_t} \mathbf{q}\cdot\mathbf{n}\, da + \int\limits_{\mathcal{P}_t} q\, dv. \tag{26.5}$$

The term

$$-\int\limits_{\partial\mathcal{P}_t} \mathbf{q}\cdot\mathbf{n}\, da$$

gives the rate at which heat is transferred to $\mathcal{P}_t$ across $\partial\mathcal{P}_t$; because $\mathbf{n}$ is the outward unit normal to $\partial\mathcal{P}_t$, the minus sign renders this term nonnegative when the flux $\mathbf{q}$ points into $\mathcal{P}_t$. Since $\mathbf{n}$ is spatial, $\mathbf{q}$ is a spatial vector field. The term

$$\int\limits_{\mathcal{P}_t} q\, dv$$

represents the rate at which heat is transferred to $\mathcal{P}_t$ by agencies *external* to the deforming body $\mathcal{B}_t$, for example by radiation.

Substituting the explicit forms for $\mathcal{E}(\mathcal{P}_t)$, $\mathcal{K}(\mathcal{P}_t)$, $\mathcal{Q}(\mathcal{P}_t)$, and $\mathcal{W}_0(\mathcal{P}_t)$ into (26.1) yields the basic expression for **balance of energy**:

$$\boxed{\overline{\int\limits_{\mathcal{P}_t} \rho(\varepsilon + \tfrac{1}{2}|\mathbf{v}|^2)\, dv}^{\,\boldsymbol{\cdot}} = -\int\limits_{\partial\mathcal{P}_t} \mathbf{q}\cdot\mathbf{n}\, da + \int\limits_{\mathcal{P}_t} q\, dv + \int\limits_{\partial\mathcal{P}_t} \mathbf{T}\mathbf{n}\cdot\mathbf{v}\, da + \int\limits_{\mathcal{P}_t} \mathbf{b}_0\cdot\mathbf{v}\, dv.} \tag{26.6}$$

To localize this balance we note first that, in view of the identity (18.14),

$$\overline{\int\limits_{\mathcal{P}_t} \rho\varepsilon\, dv}^{\,\boldsymbol{\cdot}} = \int\limits_{\mathcal{P}_t} \rho\dot{\varepsilon}\, dv.$$

150 Cf., e.g., CALLEN (1960, p. 9).
151 Cf., e.g., HALMOS (1950).

Thus, if we appeal to the power balance in (26.3) and apply the divergence theorem to the term involving the heat flux, we find that

$$\int_{\mathcal{P}_t} (\rho\dot{\varepsilon} - \mathbf{T}:\mathbf{D} + \operatorname{div}\mathbf{q} - q)\, dv = 0; \tag{26.7}$$

since (26.7) must hold for all convecting regions $\mathcal{P}_t$, we have the local **energy balance**

$$\rho\dot{\varepsilon} = \mathbf{T}:\mathbf{D} - \operatorname{div}\mathbf{q} + q, \qquad \rho\dot{\varepsilon} = T_{ij}D_{ij} - \frac{\partial q_i}{\partial x_i} + q. \tag{26.8}$$

In view of (18.12),

$$\rho\dot{\varepsilon} = (\rho\varepsilon)' + \operatorname{div}(\rho\varepsilon\mathbf{v})$$

and balance of energy may be written equivalently as

$$(\rho\varepsilon)' = \mathbf{T}:\mathbf{D} - \operatorname{div}(\mathbf{q} + \rho\varepsilon\mathbf{v}) + q. \tag{26.9}$$

Finally, we may combine (19.38) and (26.6) to obtain the form the energy balance takes when the conventional body force is replaced by the generalized body force **b**:

$$\overline{\int_{\mathcal{P}_t} \rho\varepsilon\, dv}^{\centerdot} = -\int_{\partial\mathcal{P}_t} \mathbf{q}\cdot\mathbf{n}\, da + \int_{\mathcal{P}_t} q\, dv + \int_{\partial\mathcal{P}_t} \mathbf{Tn}\cdot\mathbf{v}\, da + \int_{\mathcal{P}_t} \mathbf{b}\cdot\mathbf{v}\, dv. \tag{26.10}$$

EXERCISE

1. Provide a physical interpretation for the quantity $\mathbf{q} + \rho\varepsilon\mathbf{v}$ appearing in the energy balance (26.9).

26.2 Terminology for “Extensive” Quantities

When discussing “thermodynamic” set functions of the form (26.4) describing a physical quantity (for example, internal energy), we consistently use the adjective *net* to denote the set function (for example, $\mathcal{E}$), *no adjective* to denote the density per unit volume in the reference body, and the adjective *specific* to denote the density per unit mass (e.g., ε).

27 The Second Law: Nonnegative Production of Entropy

Power expenditures represent a macroscopic transfer of energy as they are reckoned using the velocity of material points. We view heat as representing an additional transfer of energy due to the fluctuations of atoms and/or molecules, and entropy as a measure of the disorder in the system induced by these fluctuations: the higher the degree of disorder, the higher the entropy.[152] As with energy, regions that convect with the body are allowed to possess entropy, and entropy is allowed to flow from region to region and into a region from the external world. But unlike energy, *regions convecting with the body are allowed to produce entropy*. Specifically, introducing

(i) the *net internal-entropy* $\mathcal{S}(\mathcal{P}_t)$ of $\mathcal{P}_t$, and
(ii) the *entropy flow* $\mathcal{J}(\mathcal{P}_t)$, which is the rate at which entropy is transferred to $\mathcal{P}_t$,

we let $\mathcal{H}(\mathcal{P}_t)$ denote the **net entropy production** in $\mathcal{P}_t$; that is, $\mathcal{H}(\mathcal{P}_t)$ is the rate at which the net entropy of $\mathcal{P}_t$ is increasing minus the rate at which entropy is transferred to $\mathcal{P}_t$:

$$\mathcal{H}(\mathcal{P}_t) \stackrel{\text{def}}{=} \overline{\mathcal{S}(\mathcal{P}_t)}^{\cdot} - \mathcal{J}(\mathcal{P}_t). \tag{27.1}$$

The basic premise that systems tend to increase their degree of disorder manifests itself in the requirement that

(‡) *the net entropy production in each convecting spatial region $\mathcal{P}_t$ be nonnegative*:

$$\mathcal{H}(\mathcal{P}_t) \geq 0. \tag{27.2}$$

A simple rearrangement of terms in (27.1) yields the *entropy balance*

$$\overline{\mathcal{S}(\mathcal{P}_t)}^{\cdot} = \mathcal{J}(\mathcal{P}_t) + \mathcal{H}(\mathcal{P}_t) \tag{27.3}$$

asserting that the rate at which the net entropy of $\mathcal{P}_t$ is changing is balanced by the entropy flow into $\mathcal{P}_t$ plus the rate at which entropy is produced in $\mathcal{P}_t$.

Next, on combining (27.2) and (27.3), we are led to the **entropy imbalance**[153]

$$\begin{aligned}\mathcal{H}(\mathcal{P}_t) &= \overline{\mathcal{S}(\mathcal{P}_t)}^{\cdot} - \mathcal{J}(\mathcal{P}_t)\\ &\geq 0,\end{aligned} \tag{27.4}$$

asserting that the rate of increase of the internal entropy of a convecting region $\mathcal{P}_t$ be at least as great as the rate at which entropy flows into $\mathcal{P}_t$.

[152] Cf. Penrose (1989, pp. 304–322) for a penetrating discussion of entropy.

[153] Although (27.4) can be viewed as a balance determining $\mathcal{H}(\mathcal{P}_t)$, the inequality $\overline{\mathcal{S}(\mathcal{P}_t)}^{\cdot} - \mathcal{J}(\mathcal{P}_t) \geq 0$ is of far greater importance to the theory.

27.1 Global Form of the Entropy Imbalance

As with energy, we assume there is a scalar field η, the **specific entropy**, such that

$$\mathcal{S}(\mathcal{P}_t) = \int\limits_{\mathcal{P}_t} \rho\eta \, dv. \tag{27.5}$$

Further, we suppose that, like the heat flow, the entropy flow is characterized by an **entropy flux** $\jmath$ and an **entropy supply** $\jmath$ as follows

$$\mathcal{J}(\mathcal{P}_t) = -\int\limits_{\partial\mathcal{P}_t} \boldsymbol{\jmath} \cdot \mathbf{n} \, da + \int\limits_{\mathcal{P}_t} \jmath \, dv. \tag{27.6}$$

Trivially, using (27.5) and (27.6) in the entropy imbalance (27.4) we obtain

$$\overline{\int\limits_{\mathcal{P}_t} \rho\eta \, dv}^{\displaystyle\cdot} \geq -\int\limits_{\partial\mathcal{P}_t} \boldsymbol{\jmath} \cdot \mathbf{n} \, da + \int\limits_{\mathcal{P}_t} \jmath \, dv. \tag{27.7}$$

27.2 Temperature and the Entropy Imbalance

A fundamental hypothesis of the theory relates entropy flow to heat flow and asserts that there is a scalar field

$$\vartheta > 0,$$

the (absolute) **temperature**, such that

$$\boldsymbol{\jmath} = \frac{\mathbf{q}}{\vartheta} \qquad \text{and} \qquad \jmath = \frac{q}{\vartheta}. \tag{27.8}$$

Thus,

- *entropy and heat flow in the same direction, and neither can vanish without the other.*

We can use (27.8) to write the entropy imbalance (27.7) in a form

$$\boxed{\overline{\int\limits_{\mathcal{P}_t} \rho\eta \, dv}^{\displaystyle\cdot} \geq -\int\limits_{\partial\mathcal{P}_t} \frac{\mathbf{q}}{\vartheta} \cdot \mathbf{n} \, da + \int\limits_{\mathcal{P}_t} \frac{q}{\vartheta} \, dv} \tag{27.9}$$

often referred to as the **Clausius–Duhem inequality**.[154]

Next, arguing as we did in the steps leading to (26.7), we find, as a consequence of (27.4), (27.5), (27.6), and (27.8), that

$$\mathcal{H}(\mathcal{P}_t) = \int\limits_{\mathcal{P}_t} \left(\rho\dot{\eta} + \operatorname{div}\left(\frac{\mathbf{q}}{\vartheta}\right) - \frac{q}{\vartheta} \right) dv. \tag{27.10}$$

Tacit in the derivation of (27.10) is

(‡) *the assumption that the underlying fields be smooth*;

[154] Cf. Truesdell & Toupin (1960, p. 644). More general forms of the entropy imbalance were proposed by Müller (1967), who drops (27.8), and Gurtin & Williams (1966, 1967), who allow for two temperatures in (27.8).

granted this, the net entropy production $\mathcal{H}$ has a *density* Γ, measured per unit volume in the deformed body, such that

$$\mathcal{H}(\mathcal{P}_t) = \int_{\mathcal{P}_t} \Gamma\, dv, \qquad \Gamma = \rho\dot{\eta} + \operatorname{div}\left(\frac{\mathbf{q}}{\vartheta}\right) - \frac{q}{\vartheta}. \tag{27.11}$$

Further, since (27.11) must hold for all convecting regions $\mathcal{P}_t$, we may conclude from (27.2) that

$$\Gamma \geq 0. \tag{27.12}$$

Rearranging terms in $(27.11)_2$ and bearing in mind (27.12), we arrive at the local **entropy imbalance** in either of two forms:

$$\boxed{\rho\dot{\eta} \geq -\operatorname{div}\left(\frac{\mathbf{q}}{\vartheta}\right) + \frac{q}{\vartheta}} \tag{27.13}$$

and

$$\Gamma = \rho\dot{\eta} + \operatorname{div}\left(\frac{\mathbf{q}}{\vartheta}\right) - \frac{q}{\vartheta} \geq 0. \tag{27.14}$$

In view of (18.12),

$$\rho\dot{\eta} = (\rho\eta)' + \operatorname{div}(\rho\eta\mathbf{v}),$$

and the entropy imbalance may be written equivalently as

$$(\rho\eta)' \geq -\operatorname{div}\left(\frac{\mathbf{q} + \rho\eta\vartheta\mathbf{v}}{\vartheta}\right) + q. \tag{27.15}$$

Remark. *The entropy production Γ is a derived quantity whose derivation is based on the tacit assumption that the underlying fields be smooth.* But, for a shock wave, the velocity, entropy, and heat flux suffer jump discontinuities and the localization process that led to (27.10) is invalid. In fact, when a shock wave $\mathcal{S}(t)$ is present, the net entropy production $\mathcal{H}(\mathcal{P}_t)$ generally includes an *explicit* contribution from the portion of $\mathcal{S}(t)$ in $\mathcal{P}_t$.[155]

27.3 Free-Energy Imbalance. Dissipation

In view of the local energy balance (26.8),

$$\begin{aligned} -\operatorname{div}\left(\frac{\mathbf{q}}{\vartheta}\right) + \frac{q}{\vartheta} &= \frac{1}{\vartheta}\Big(-\operatorname{div}\mathbf{q} + q\Big) + \frac{1}{\vartheta^2}\mathbf{q}\cdot\operatorname{grad}\vartheta \\ &= \frac{1}{\vartheta}\left(\rho\dot{\varepsilon} - \mathbf{T}\colon\mathbf{D} + \frac{1}{\vartheta}\mathbf{q}\cdot\operatorname{grad}\vartheta\right) \end{aligned}$$

and this with (27.14) implies that

$$\rho(\dot{\varepsilon} - \vartheta\dot{\eta}) - \mathbf{T}\colon\mathbf{D} + \frac{1}{\vartheta}\mathbf{q}\cdot\operatorname{grad}\vartheta = -\vartheta\Gamma \leq 0. \tag{27.16}$$

Thus, on defining the **specific free-energy** through

$$\boxed{\psi = \varepsilon - \vartheta\eta,} \tag{27.17}$$

[155] Cf. The remark containing (33.25).

(27.16) yields the local **free-energy imbalance**

$$\boxed{\rho(\dot{\psi} + \eta\dot{\vartheta}) - \mathbf{T}\!:\!\mathbf{D} + \frac{1}{\vartheta}\mathbf{q}\cdot \operatorname{grad}\vartheta = -\vartheta\Gamma \le 0,} \tag{27.18}$$

a result basic to much of what follows. The terms of (27.18) have the dimensions of (energy)/time and for that reason we view $\vartheta\Gamma$ as representing *dissipation* per unit volume, a view strengthened by the remark below.

Next, by (18.14),

$$\overline{\int_{\mathcal{P}_t} \rho\psi\, dv}^{\,\boldsymbol{\cdot}} = \int_{\mathcal{P}_t} \rho\dot{\psi}\, dv. \tag{27.19}$$

Thus, if we integrate (27.18) over $\mathcal{P}_t$, we conclude, with the aid of (26.3), that

$$\underbrace{\int_{\mathcal{P}_t} \vartheta\Gamma\, dv}_{\substack{\text{dissipation}\\ \ge 0}} = \underbrace{\int_{\partial\mathcal{P}_t} \mathbf{T}\mathbf{n}\cdot\mathbf{v}\, da + \int_{\mathcal{P}_t} \mathbf{b}_0\cdot\mathbf{v}\, dv}_{\substack{\text{conventional external}\\ \text{power expenditure}}} - \underbrace{\overline{\int_{\mathcal{P}_t} \rho(\psi + \tfrac{1}{2}|\mathbf{v}|^2)\, dv}^{\,\boldsymbol{\cdot}}}_{\substack{\text{rate of free and}\\ \text{kinetic energies}}} - \underbrace{\int_{\mathcal{P}_t} \left(\rho\eta\dot{\vartheta} + \frac{1}{\vartheta}\mathbf{q}\cdot\operatorname{grad}\vartheta\right) dv}_{\substack{\text{thermal production}\\ \text{of energy}}}. \tag{27.20}$$

Remark. For situations in which thermal influences are negligible the last integral vanishes and (27.20) formally reduces to a *free-energy imbalance* requiring that *the rate of change of the net free- and kinetic-energy be balanced by the conventional expended power plus the dissipation.* As we show in §29, the resulting free-energy imbalance plays a major role in our discussion of mechanical theories.

EXERCISES

1. Localize (27.7) to obtain

$$\rho\dot{\eta} \ge -\operatorname{div}\jmath + j$$

and show that

$$(\rho\eta)' \ge -\operatorname{div}(\jmath + \rho\eta\mathbf{v}) + j.$$

2. Provide a physical interpretation for the quantity $(\mathbf{q} + \rho\vartheta\eta\mathbf{v})/\vartheta$ appearing in the entropy imbalance (27.15).

28 General Results

28.1 Invariant Nature of the First Two Laws

As the next result shows, the laws of thermodynamics afford a degree of flexibility in the specification of the fields ε, η, and $\mathbf{q}$.

INVARIANCE PROPERTIES *The energy balance* (26.6) *and the entropy imbalance* (27.9) *are invariant under transformations of the form*

$$\left.\begin{aligned} \varepsilon &\to \varepsilon + \varepsilon_0, & \dot{\varepsilon}_0 &= 0, \\ \eta &\to \eta + \eta_0, & \dot{\eta}_0 &= 0, \\ \mathbf{q} &\to \mathbf{q} + \boldsymbol{\omega} \times \operatorname{grad} \vartheta, & \operatorname{grad} \boldsymbol{\omega} &= \mathbf{0}. \end{aligned}\right\} \tag{28.1}$$

The internal energy and the entropy of each material point may, at will, thus be additively scaled by terms dependent only on the material point. In a sense, a choice of scalings amounts to a choice of material reference scales for the energy and entropy.

To show that (26.10) and (27.9) are invariant under (28.1) it suffices to show that, given any region $\mathcal{P}_t$ convecting with the body,

$$\left.\begin{aligned} \overline{\int_{\mathcal{P}_t} \rho\varepsilon\, dv}^{\bullet} &= \overline{\int_{\mathcal{P}_t} \rho(\varepsilon + \varepsilon_0)\, dv}^{\bullet}, \\ \overline{\int_{\mathcal{P}_t} \rho\eta\, dv}^{\bullet} &= \overline{\int_{\mathcal{P}_t} \rho(\eta + \eta_0)\, dv}^{\bullet}, \\ \int_{\partial\mathcal{P}_t} \mathbf{q}\cdot\mathbf{n}\, da &= \int_{\partial\mathcal{P}_t} (\mathbf{q} + \boldsymbol{\omega} \times \operatorname{grad}\vartheta)\cdot\mathbf{n}\, da, \\ \int_{\partial\mathcal{P}_t} \frac{\mathbf{q}\cdot\mathbf{n}}{\vartheta}\, da &= \int_{\partial\mathcal{P}_t} \frac{\mathbf{q} + \boldsymbol{\omega} \times \operatorname{grad}\vartheta}{\vartheta}\cdot\mathbf{n}\, da. \end{aligned}\right\} \tag{28.2}$$

Clearly,

$$\begin{aligned} \overline{\int_{\mathcal{P}_t} \rho\varepsilon_0\, dv}^{\bullet} &= \int_{\mathcal{P}_t} \rho\dot{\varepsilon}_0\, dv \\ &= 0, \end{aligned}$$

and similarly with ε_0 replaced by η_0; hence, $(28.2)_{1,2}$ are satisfied. Further, since

$$\boldsymbol{\omega} \times \operatorname{grad} \vartheta = -\operatorname{curl}(\vartheta \boldsymbol{\omega}),$$

$$\frac{\boldsymbol{\omega} \times \operatorname{grad} \vartheta}{\vartheta} = -\operatorname{curl}[(\ln \vartheta)\boldsymbol{\omega}],$$

it follows that

$$\int_{\partial \mathcal{P}_t} (\boldsymbol{\omega} \times \operatorname{grad} \vartheta) \cdot \mathbf{n}\, da = -\int_{\mathcal{P}_t} \operatorname{div}[\operatorname{curl}(\vartheta \boldsymbol{\omega})]\, dv$$
$$= 0,$$

and similarly with ϑ replaced by $\ln \vartheta$; thus, $(28.2)_{3,4}$ are satisfied.

28.2 Decay Inequalities for the Body Under Passive Boundary Conditions

We now show that granted simple boundary conditions of a passive nature there are integrals of the form

$$\int_{\mathcal{B}_t} (\cdots)\, dv,$$

often called Lyapunov functions, *that decrease with time*. These results are a direct consequence of the underlying thermodynamical framework — they do not rely in any way on constitutive equations.

We assume throughout this section that there are no conventional body forces and that the heat supply vanishes:

$$\mathbf{b}_0 \equiv \mathbf{0}, \qquad q \equiv 0. \tag{28.3}$$

Then, by (28.3), the first two laws in the forms $(26.6)_1$ and (27.9) become

$$\overline{\int_{\mathcal{P}_t} \rho(\varepsilon + \tfrac{1}{2}|\mathbf{v}|^2)\, dv}^{\displaystyle\cdot} = -\int_{\partial \mathcal{P}_t} \mathbf{q} \cdot \mathbf{n}\, da + \int_{\partial \mathcal{P}_t} \mathbf{T}\mathbf{n} \cdot \mathbf{v}\, da,$$
$$\overline{\int_{\mathcal{P}_t} \rho \eta\, dv}^{\displaystyle\cdot} \geq -\int_{\partial \mathcal{P}_t} \frac{\mathbf{q}}{\vartheta} \cdot \mathbf{n}\, da. \tag{28.4}$$

28.2.1 Isolated Body

Assume that the **body is isolated** in the sense that, at each time,

$$\mathbf{T}\mathbf{n} = \mathbf{0} \text{ on a portion of } \partial \mathcal{B}_t \text{ and } \mathbf{v} = \mathbf{0} \text{ on the remainder of } \partial \mathcal{B}_t,$$
$$\mathbf{q} \cdot \mathbf{n} = 0 \text{ on } \partial \mathcal{B}_t. \tag{28.5}$$

Then, by (28.4),

$$\begin{aligned} \overline{\int\limits_{\mathcal{B}_t} \rho(\varepsilon + \tfrac{1}{2}|\mathbf{v}|^2)\,dv}^{\bullet} &= 0, \\ \overline{\int\limits_{\mathcal{B}_t} \rho\eta\,dv}^{\bullet} &\geq 0, \end{aligned} \tag{28.6}$$

and *the net energy is constant, while the net entropy is nondecreasing.*

28.2.2 Boundary Essentially at Constant Pressure and Temperature

These boundary conditions may be stated precisely as follows: There are a constant pressure p_0 and a constant temperature $\vartheta_0 > 0$ such that

$$\begin{aligned} &\mathbf{Tn} = -p_0\mathbf{n} \text{ on a portion of } \partial\mathcal{B}_t \text{ and } \mathbf{v} = \mathbf{0} \text{ on the remainder of } \partial\mathcal{B}_t, \\ &\vartheta = \vartheta_0 \text{ on a portion of } \partial\mathcal{B}_t \text{ and } \mathbf{q}\cdot\mathbf{n} = 0 \text{ on the remainder of } \partial\mathcal{B}_t. \end{aligned} \tag{28.7}$$

To determine the consequences of (28.7), note that $(28.4)_2$ and the thermal conditions $(28.7)_2$ imply that

$$\begin{aligned} -\int\limits_{\partial\mathcal{B}_t} \mathbf{q}\cdot\mathbf{n}\,da &= -\vartheta_0 \int\limits_{\partial\mathcal{B}_t} \frac{\mathbf{q}}{\vartheta}\cdot\mathbf{n}\,da \\ &= \vartheta_0 \overline{\int\limits_{\mathcal{B}_t} \rho\eta\,dv}^{\bullet} - \underbrace{\vartheta_0 \int\limits_{\mathcal{B}_t} \Gamma\,dv}_{\geq 0}, \end{aligned} \tag{28.8}$$

while the mechanical conditions $(28.7)_1$, the divergence theorem, and the identities (18.7), (18.8), and (18.14) yield

$$\begin{aligned} \int\limits_{\partial\mathcal{B}_t} \mathbf{Tn}\cdot\mathbf{v}\,da &= -p_0 \int\limits_{\partial\mathcal{B}_t} \mathbf{n}\cdot\mathbf{v}\,da \\ &= -p_0 \int\limits_{\mathcal{B}_t} \operatorname{div}\mathbf{v}\,dv \\ &= -p_0 \int\limits_{\mathcal{B}_t} \rho\dot{\upsilon}\,dv \\ &= -\overline{\int\limits_{\mathcal{B}_t} p_0\upsilon\rho\,dv}^{\bullet} \\ &= -\overline{\int\limits_{\mathcal{B}_t} p_0\,dv}^{\bullet}. \end{aligned} \tag{28.9}$$

The identities (28.8) and (28.9) reduce the first law $(28.4)_1$ to the decay inequality

$$\overline{\int\limits_{\mathcal{B}_t} \rho(\varepsilon - \vartheta_0\eta + p_0\upsilon + \tfrac{1}{2}|\mathbf{v}|^2)\,dv}^{\boldsymbol{\cdot}} = -\vartheta_0 \int\limits_{\mathcal{B}_t} \Gamma\,dv \leq 0. \tag{28.10}$$

The decay inequalities (28.6) and (28.10) involve no constitutive assumptions and are hence valid for all materials whose behavior is consistent with the first two laws as described here. Such decay inequalities are important as they furnish a formal justification, within a dynamical setting, of standard variational principles used to characterize equilibrium. Equally important, such results yield *a priori* estimates for solutions of initial/boundary-value problems.

29 A Free-Energy Imbalance for Mechanical Theories

Much of continuum mechanics involves theories that are purely mechanical. Such theories neglect all thermal influences: Fields such as temperature, entropy, and heat flux are not mentioned, the basic balances being those for mass and momentum. Even so, most mechanical theories of continua are consistent with a "thermodynamic-like" inequality involving an "energy-like" field and a nonnegative quantity representing the dissipation of energy. We refer to the underlying energy as *free energy*, a choice motivated by the remark following (27.20). In fact, that remark formally justifies our taking the "thermodynamic-like" inequality to be an *imbalance* for free energy.

Bearing this in mind, the purely mechanical theories we discuss are based on a free-energy imbalance that embodies the intuitive notion that, for dissipative processes,

- *not all conventional power expended on a convecting spatial region $\mathcal{P}_t$ can be converted into changes in the net free- and kinetic-energy of $\mathcal{P}_t$, because a portion of that power must go into dissipation.*

We view such a balance as a mechanical manifestation of the first two laws of thermodynamics.

29.1 Free-Energy Imbalance. Dissipation

Introducing the net **free energy** $\mathcal{F}(\mathcal{P}_t)$ of $\mathcal{P}_t$, we let $\mathcal{D}(\mathcal{P}_t)$ denote the net **dissipation** in $\mathcal{P}_t$; that is, the conventional power expenditure $\mathcal{W}_0(\mathcal{P}_t)$ minus the rate at which the free- and kinetic-energy of $\mathcal{P}_t$ are increasing:

$$\mathcal{D}(\mathcal{P}_t) \overset{\text{def}}{=} \mathcal{W}_0(\mathcal{P}_t) - \overline{\mathcal{F}(\mathcal{P}_t) + \mathcal{K}(\mathcal{P}_t)}^{\cdot}. \tag{29.1}$$

A central hypothesis of the mechanical theory is the requirement that

(‡) *the net dissipation in each convecting spatial region $\mathcal{P}_t$ be nonnegative:*

$$\mathcal{D}(\mathcal{P}_t) \geq 0. \tag{29.2}$$

Rearranging the terms in (29.1), we have the *free-energy imbalance*

$$\overline{\mathcal{F}(\mathcal{P}_t) + \mathcal{K}(\mathcal{P}_t)}^{\cdot} \leq \mathcal{W}(\mathcal{P}_t). \tag{29.3}$$

We posit the existence of a scalar field ψ, the **specific free-energy**, such that

$$\mathcal{F}(\mathcal{P}_t) = \int\limits_{\mathcal{P}_t} \rho\psi \, dv \tag{29.4}$$

for every convecting spatial region $\mathcal{P}_t$. By (26.2), (26.3), and (29.4) applied to (29.1),

$$\mathcal{D}(\mathcal{P}_t) = \int\limits_{\mathcal{P}_t} (\mathbf{T}:\mathbf{D} - \rho\dot{\psi})\, dv.$$

Granted sufficient smoothness,[156] the net dissipation $\mathcal{D}$ therefore has a density δ, measured per unit volume in the deformed body, such that

$$\mathcal{D}(\mathcal{P}_t) = \int\limits_{\mathcal{P}_t} \delta\, dv, \qquad \delta = \mathbf{T}:\mathbf{D} - \rho\dot{\psi}, \tag{29.5}$$

and, hence, since $\mathcal{D}(\mathcal{P}_t) \geq 0$ and $\mathcal{P}_t$ is arbitrary, such that

$$\delta \geq 0. \tag{29.6}$$

Thus, (29.5) yields the **local free-energy imbalance**

$$\boxed{\rho\dot{\psi} - \mathbf{T}:\mathbf{D} = -\delta \leq 0, \qquad \rho\dot{\psi} - T_{ij}D_{ij} = -\delta \leq 0.} \tag{29.7}$$

On the other hand, (26.2), (26.3), and (29.7) yield the global **free-energy imbalance**:

$$\boxed{\int\limits_{\mathcal{P}_t} \delta\, dv = \int\limits_{\partial\mathcal{P}_t} \mathbf{Tn}\cdot\mathbf{v}\, da + \int\limits_{\mathcal{P}_t} \mathbf{b}_0\cdot\mathbf{v}\, dv - \overline{\int\limits_{\mathcal{P}_t} \rho(\psi + \tfrac{1}{2}|\mathbf{v}|^2)\, dv}^{\,\centerdot} \geq 0.} \tag{29.8}$$

The general thermodynamical invariance properties (28.1) have an immediate counterpart within the purely mechanical framework under consideration: *The free-energy imbalance* (29.8) *is invariant under transformations of the form*

$$\psi \to \psi + \psi_0, \qquad \dot{\psi}_0 = 0. \tag{29.9}$$

The specific free-energy of each material point may, at will, thus be additively scaled by a term dependent only on the material point.

29.2 Digression: Role of the Free-Energy Imbalance within the General Thermodynamic Framework

Consider, for the moment, the general thermodynamic theory discussed in §26 and §27. Consider an **isothermal process**; that is a process consistent with the restriction

$$\vartheta = \vartheta_0 \equiv \text{constant}. \tag{29.10}$$

In this case, the free-energy imbalance (27.20) of the general theory becomes

$$\int\limits_{\mathcal{P}_t} \vartheta_0\Gamma\, dv = \int\limits_{\partial\mathcal{P}_t} \mathbf{Tn}\cdot\mathbf{v}\, da + \int\limits_{\mathcal{P}_t} \mathbf{b}_0\cdot\mathbf{v}\, dv - \overline{\int\limits_{\mathcal{P}_t} \rho(\varepsilon - \vartheta_0\eta + \tfrac{1}{2}|\mathbf{v}|^2)\, dv}^{\,\centerdot} \geq 0, \tag{29.11}$$

which is the free-energy imbalance (29.8) with

$$\psi = \varepsilon - \vartheta_0\eta, \qquad \delta = \vartheta_0\Gamma. \tag{29.12}$$

Thus, since the free-energy imbalance (27.20) of the general theory arises upon combining the balance for energy and the imbalance for entropy,

[156] Cf. (‡) on page 187 and the paragraph in petite type on page 188.

(‡) *in an isothermal process with temperature* ϑ_0, *the energy balance* (26.10) *and the entropy imbalance* (27.9) *of the general theory together reduce to the free-energy imbalance* (29.8) *of the mechanical theory with dissipation* δ *expressed in terms of the entropy production* Γ *by*

$$\delta = \vartheta_0 \Gamma. \tag{29.13}$$

29.3 Decay Inequalities

Assume now that the conventional body force vanishes:

$$\mathbf{b}_0 = \mathbf{0}.$$

Then, by (29.8), if the body is isolated in the sense of $(28.5)_1$,

$$\overline{\int_{\mathcal{B}_t} \rho\,(\psi + \tfrac{1}{2}\,|\mathbf{v}|^2)\,dv}^{\bullet} = -\int_{\mathcal{B}_t} \delta\,dv \le 0 \tag{29.14}$$

and the total energy cannot increase.

Consider a boundary essentially at constant pressure p_0 in the sense of the boundary condition $(28.7)_1$. Then (28.9) remains valid and the free-energy imbalance (29.8) yields the decay inequality

$$\overline{\int_{\mathcal{B}_t} \rho(\psi + p_0 \upsilon + \tfrac{1}{2}|\mathbf{v}|^2)\,dv}^{\bullet} = -\int_{\mathcal{B}_t} \delta\,dv \le 0. \tag{29.15}$$

30 The First Two Laws for a Spatial Control Volume

Let R be a spatial control volume. Integrating the local energy balance and entropy imbalance[157]

$$\rho\dot{\varepsilon} = \mathbf{T}:\mathbf{D} - \mathrm{div}\,\mathbf{q} + q \qquad \text{and} \qquad \rho\dot{\eta} \geq -\mathrm{div}\left(\frac{\mathbf{q}}{\vartheta}\right) + \frac{q}{\vartheta} \tag{30.1}$$

over R using the power balance (26.3) (and the divergence theorem for $\mathrm{div}\,\mathbf{q}$ and $\mathrm{div}(\mathbf{q}/\vartheta)$) we find that

$$\begin{aligned}\int\limits_R \rho\,\overline{(\varepsilon + \tfrac{1}{2}|\mathbf{v}|^2)}^{\cdot}\,dv &= \int\limits_{\partial R} \mathbf{Tn}\cdot\mathbf{v}\,da + \int\limits_R \mathbf{b}_0\cdot\mathbf{v}\,dv - \int\limits_{\partial R} \mathbf{q}\cdot\mathbf{n}\,da + \int\limits_R q\,dv,\\ \int\limits_R \rho\dot{\eta}\,dv &\geq \int\limits_{\partial R} \frac{\mathbf{q}}{\vartheta}\cdot\mathbf{n}\,da + \int\limits_R \frac{q}{\vartheta}\,dv.\end{aligned} \tag{30.2}$$

Thus, $(23.3)_1$ with $\varphi = \varepsilon + \frac{1}{2}|\mathbf{v}|^2$ applied to $(30.2)_1$ yields **balance of energy**

$$\begin{aligned}&\overline{\int\limits_R \rho\,(\varepsilon + \tfrac{1}{2}|\mathbf{v}|^2)\,dv}^{\cdot} + \int\limits_{\partial R} \rho(\varepsilon + \tfrac{1}{2}|\mathbf{v}|^2)\mathbf{v}\cdot\mathbf{n}\,da\\ &\qquad = \int\limits_{\partial R} \mathbf{Tn}\cdot\mathbf{v}\,da + \int\limits_R \mathbf{b}\cdot\mathbf{v}\,dv - \int\limits_{\partial R} \mathbf{q}\cdot\mathbf{n}\,da + \int\limits_R q\,dv\end{aligned} \tag{30.3}$$

for a control volume R, while $(23.3)_1$ with $\varphi = \eta$ applied to $(30.2)_2$ yields the **entropy imbalance**

$$\overline{\int\limits_R \rho\eta\,dv}^{\cdot} + \int\limits_{\partial R} \rho\eta\mathbf{v}\cdot\mathbf{n}\,da \geq -\int\limits_{\partial R} \frac{\mathbf{q}}{\vartheta}\cdot\mathbf{n}\,da + \int\limits_R \frac{q}{\vartheta}\,dv \tag{30.4}$$

for R. In (30.3) and (30.4) the terms

$$\int\limits_{\partial R} \rho(\varepsilon + \tfrac{1}{2}|\mathbf{v}|^2)\mathbf{v}\cdot\mathbf{n}\,da \qquad \text{and} \qquad \int\limits_{\partial R} \rho\eta\mathbf{v}\cdot\mathbf{n}\,da$$

[157] Cf. (26.8) and (27.13).

represent respective outflow-rates of internal energy and entropy across the boundary ∂R of the control volume R.

EXERCISE

1. Derive a counterpart — for a spatial control volume R — of the free-energy imbalance (29.8).

31 The First Two Laws Expressed Referentially

In this section we determine referential counterparts of the thermodynamic laws and associated relations discussed in §§26–29.

Let P denote an arbitrary subregion of the reference body B and let

$$\mathcal{P}_t = \chi_t(\mathrm{P})$$

denote the corresponding convecting subregion of the deforming $\mathcal{B}_t$. Important to the present discussion are

(i) the spatial to referential transformation

$$\begin{aligned}\mathcal{W}_0(\mathcal{P}_t) &= \int\limits_{\partial\mathcal{P}_t} \mathbf{Tn}\cdot\mathbf{v}\,da + \int\limits_{\mathcal{P}_t} \mathbf{b}_0\cdot\mathbf{v}\,dv \\ &= \int\limits_{\partial\mathrm{P}} \mathbf{T}_{\mathrm{R}}\mathbf{n}_{\mathrm{R}}\cdot\dot{\chi}\,da_{\mathrm{R}} + \int\limits_{\mathrm{P}} \mathbf{b}_{0\mathrm{R}}\cdot\dot{\chi}\,dv_{\mathrm{R}} \\ &= \mathcal{W}_0(\mathrm{P})\end{aligned} \tag{31.1}$$

of the conventional external power $\mathcal{W}_0$;[158]

(ii) the spatial to referential transformation

$$\mathcal{K}(\mathcal{P}_t) = \int\limits_{\mathcal{P}_t} \tfrac{1}{2}\rho|\mathbf{v}|^2\,dv = \int\limits_{\mathrm{P}} \tfrac{1}{2}\rho_{\mathrm{R}}|\dot{\chi}|^2\,dv_{\mathrm{R}} = \mathcal{K}(\mathrm{P}) \tag{31.2}$$

of the kinetic energy $\mathcal{K}$;[159]

(iii) the conventional power balance,[160]

$$\underbrace{\int\limits_{\partial\mathrm{P}} \mathbf{T}_{\mathrm{R}}\mathbf{n}_{\mathrm{R}}\cdot\dot{\chi}\,da_{\mathrm{R}} + \int\limits_{\mathrm{P}} \mathbf{b}_{0\mathrm{R}}\cdot\dot{\chi}\,dv_{\mathrm{R}}}_{\mathcal{W}_0(\mathrm{P})} = \int\limits_{\mathrm{P}} \mathbf{T}_{\mathrm{R}}:\dot{\mathbf{F}}\,dv_{\mathrm{R}} + \underbrace{\overline{\int\limits_{\mathrm{P}} \tfrac{1}{2}\rho_{\mathrm{R}}|\dot{\chi}|^2\,dv_{\mathrm{R}}}^{\boldsymbol{\cdot}}}_{\dot{\overline{\mathcal{K}(\mathrm{P})}}}; \tag{31.3}$$

158 Cf. (24.15), Footnote 143.
159 Cf. (24.16).
160 Cf. (25.7).

(iv) the transformation laws[161]

$$\int_{\mathcal{P}_t} \varphi\, dv = \int_{\mathrm{P}} \varphi J\, dv_{\mathrm{R}}, \qquad \int_{\mathcal{P}_t} \rho\Phi\, dv = \int_{\mathrm{P}} \rho_{\mathrm{R}}\Phi\, dv_{\mathrm{R}},$$
$$\int_{\partial\mathcal{P}_t} \mathbf{h}\cdot\mathbf{n}\, da = \int_{\partial\mathrm{P}} (J\mathbf{F}^{-1}\mathbf{h})\cdot\mathbf{n}_{\mathrm{R}}\, da_{\mathrm{R}}. \tag{31.4}$$

31.1 Global Forms of the First Two Laws

Our first step is to determine the spatial to referential transformations of the net internal energy and entropy $\mathcal{E}(\mathcal{P}_t)$ and $\mathcal{S}(\mathcal{P}_t)$, the heat and entropy flows $\mathcal{Q}(\mathcal{P}_t)$ and $\mathcal{J}(\mathcal{P}_t)$, and the net entropy production $\mathcal{H}(\mathcal{P}_t)$.[162]

We define referential energy and entropy densities ε_{R} and η_{R}, heat flux and heat supply $\mathbf{q}_{\mathrm{R}}$ and q_{R}, and entropy production (density) by Γ_{R} by

$$\varepsilon_{\mathrm{R}} = \rho_{\mathrm{R}}\varepsilon, \qquad \eta_{\mathrm{R}} = \rho_{\mathrm{R}}\eta,$$
$$\mathbf{q}_{\mathrm{R}} = J\mathbf{F}^{-1}\mathbf{q}, \qquad q_{\mathrm{R}} = Jq, \qquad \Gamma_{\mathrm{R}} = J\Gamma \geq 0, \tag{31.5}$$

so that, by (31.4),[163]

$$\left.\begin{aligned}
\mathcal{E}(\mathcal{P}_t) &= \int_{\mathcal{P}_t} \rho\varepsilon\, dv = \int_{\mathrm{P}} \varepsilon_{\mathrm{R}}\, dv_{\mathrm{R}} = \mathcal{E}(\mathrm{P}),\\
\mathcal{S}(\mathcal{P}_t) &= \int_{\mathcal{P}_t} \rho\eta\, dv = \int_{\mathrm{P}} \eta_{\mathrm{R}}\, dv_{\mathrm{R}} = \mathcal{S}(\mathrm{P}),\\
\mathcal{Q}(\mathcal{P}_t) &= -\int_{\partial\mathcal{P}_t} \mathbf{q}\cdot\mathbf{n}\, da + \int_{\mathcal{P}_t} q\, dv = -\int_{\partial\mathrm{P}} \mathbf{q}_{\mathrm{R}}\cdot\mathbf{n}_{\mathrm{R}}\, da_{\mathrm{R}} + \int_{\mathrm{P}} q_{\mathrm{R}}\, dv_{\mathrm{R}} = \mathcal{Q}(\mathrm{P}),\\
\mathcal{J}(\mathcal{P}_t) &= -\int_{\partial\mathcal{P}_t} \frac{\mathbf{q}}{\vartheta}\cdot\mathbf{n}\, da + \int_{\mathcal{P}_t} \frac{q}{\vartheta}\, dv = -\int_{\partial\mathrm{P}} \frac{\mathbf{q}_{\mathrm{R}}}{\vartheta}\cdot\mathbf{n}_{\mathrm{R}}\, da_{\mathrm{R}} + \int_{\mathrm{P}} \frac{q_{\mathrm{R}}}{\vartheta}\, dv_{\mathrm{R}} = \mathcal{J}(\mathrm{P}),\\
\mathcal{H}(\mathcal{P}_t) &= \int_{\mathcal{P}_t} \Gamma\, dv = \int_{\mathrm{P}} \Gamma_{\mathrm{R}}\, dv_{\mathrm{R}} = \mathcal{H}(\mathrm{P}).
\end{aligned}\right\} \tag{31.6}$$

The internal energy and entropy densities ε_{R} and η_{R} are therefore measured per volume in the reference body B, as are the heat supply q_{R} and the entropy production Γ_{R}, while the heat flux $\mathbf{q}_{\mathrm{R}}$ is measured per unit area, also in B. Because the referential heat flux arises via the inner product $\mathbf{q}_{\mathrm{R}}\cdot\mathbf{n}_{\mathrm{R}}$, $\mathbf{q}_{\mathrm{R}}$ is a material vector field. Note that the temperature ϑ, not being a density, is invariant under the spatial to material transformations $(31.6)_{4,5}$.

Using the material to spatial transformations (31.1), (31.2), and (31.6) in conjunction with the basic forms (26.1) and (27.4) of the first two laws, we obtain the

[161] Cf. (15.7), (18.10), and $(15.8)_1$.
[162] Cf. §26, §27.
[163] Cf. §26, §27.

global form

$$\overline{\int_{\mathrm{P}} (\varepsilon_{\mathrm{R}} + \tfrac{1}{2}\rho_{\mathrm{R}}|\dot{\chi}|^2)\,dv_{\mathrm{R}}}^{\boldsymbol{\cdot}} = -\int_{\partial\mathrm{P}} \mathbf{q}_{\mathrm{R}} \cdot \mathbf{n}_{\mathrm{R}}\,da_{\mathrm{R}} + \int_{\mathrm{P}} q_{\mathrm{R}}\,dv_{\mathrm{R}} + \int_{\partial\mathrm{P}} \mathbf{T}_{\mathrm{R}}\mathbf{n}_{\mathrm{R}} \cdot \dot{\chi}\,da_{\mathrm{R}} + \int_{\mathrm{P}} \mathbf{b}_{0\mathrm{R}} \cdot \dot{\chi}\,dv_{\mathrm{R}} \tag{31.7}$$

of **balance of energy** and the global form

$$\int_{\mathrm{P}} \Gamma_{\mathrm{R}}\,dv_{\mathrm{R}} = \overline{\int_{\mathrm{P}} \eta_{\mathrm{R}}\,dv_{\mathrm{R}}}^{\boldsymbol{\cdot}} + \int_{\partial\mathrm{P}} \frac{\mathbf{q}_{\mathrm{R}}}{\vartheta} \cdot \mathbf{n}_{\mathrm{R}}\,da_{\mathrm{R}} - \int_{\mathrm{P}} \frac{q_{\mathrm{R}}}{\vartheta}\,dv_{\mathrm{R}} \geq 0 \tag{31.8}$$

of the **entropy imbalance.**

EXERCISE

1. Show that the energy balance (31.7) and the entropy balance (31.8) are invariant under transformations of the form

$$\left.\begin{aligned} \varepsilon_{\mathrm{R}} &\to \varepsilon_{\mathrm{R}} + \varepsilon_{0\mathrm{R}}, & \dot{\varepsilon}_{0\mathrm{R}} &= 0,\\ \eta_{\mathrm{R}} &\to \eta_{\mathrm{R}} + \eta_{0\mathrm{R}}, & \dot{\eta}_{0\mathrm{R}} &= 0,\\ \mathbf{q}_{\mathrm{R}} &\to \mathbf{q}_{\mathrm{R}} + \boldsymbol{\omega} \times \nabla\vartheta, & \nabla\boldsymbol{\omega} &= \mathbf{0}. \end{aligned}\right\} \tag{31.9}$$

31.2 Local Forms of the First Two Laws

Since the material region P is independent of time, as is ρ_{R},

$$\overline{\int_{\mathrm{P}} \varepsilon_{\mathrm{R}}\,dv_{\mathrm{R}}}^{\boldsymbol{\cdot}} = \int_{\mathrm{P}} \dot{\varepsilon}_{\mathrm{R}}\,dv_{\mathrm{R}}, \qquad \overline{\int_{\mathrm{P}} \eta_{\mathrm{R}}\,dv_{\mathrm{R}}}^{\boldsymbol{\cdot}} = \int_{\mathrm{P}} \dot{\eta}_{\mathrm{R}}\,dv_{\mathrm{R}},$$

$$\overline{\int_{\mathrm{P}} \tfrac{1}{2}\rho_{\mathrm{R}}|\dot{\chi}|^2\,dv_{\mathrm{R}}}^{\boldsymbol{\cdot}} = \int_{\mathrm{P}} \rho_{\mathrm{R}}\ddot{\chi} \cdot \dot{\chi}\,dv_{\mathrm{R}}.$$

Thus, using the conventional power balance (31.3) and the divergence theorem, we find that the energy balance (31.7) and the entropy imbalance (31.8) reduce to

$$\int_{\mathrm{P}} (\dot{\varepsilon}_{\mathrm{R}} + \mathrm{Div}\,\mathbf{q}_{\mathrm{R}} - q_{\mathrm{R}} - \mathbf{T}_{\mathrm{R}} : \dot{\mathbf{F}})\,dv_{\mathrm{R}} = 0,$$

$$\int_{\mathrm{P}} \Gamma_{\mathrm{R}}\,dv_{\mathrm{R}} = \int_{\mathrm{P}} \left(\dot{\eta}_{\mathrm{R}} + \mathrm{Div}\,\frac{\mathbf{q}_{\mathrm{R}}}{\vartheta} - \frac{q_{\mathrm{R}}}{\vartheta}\right) dv_{\mathrm{R}} \geq 0.$$

Since these relations are to be satisfied for all subregions P of B, we are led to the local forms of the **energy balance** and **entropy imbalance**

$$\begin{aligned}
\dot{\varepsilon}_{\mathrm{R}} &= \mathbf{T}_{\mathrm{R}}:\dot{\mathbf{F}} - \mathrm{Div}\,\mathbf{q}_{\mathrm{R}} + q_{\mathrm{R}}, & \dot{\varepsilon}_{\mathrm{R}} &= T_{\mathrm{R}ij}\dot{F}_{ij} - \frac{\partial q_{\mathrm{R}i}}{\partial X_i} + q_{\mathrm{R}}, \\
\Gamma_{\mathrm{R}} &= \dot{\eta}_{\mathrm{R}} + \mathrm{Div}\left(\frac{\mathbf{q}_{\mathrm{R}}}{\vartheta}\right) - \frac{q_{\mathrm{R}}}{\vartheta} \geq 0, & \Gamma_{\mathrm{R}} &= \dot{\eta}_{\mathrm{R}} + \frac{\partial}{\partial X_i}\left(\frac{q_{\mathrm{R}i}}{\vartheta}\right) - \frac{q_{\mathrm{R}}}{\vartheta} \geq 0.
\end{aligned} \tag{31.10}$$

Next, the *free energy* (per unit referential volume) is defined by

$$\psi_{\mathrm{R}} = \rho_{\mathrm{R}}\psi, \tag{31.11}$$

so that, by (81.8),

$$\psi_{\mathrm{R}} = \varepsilon_{\mathrm{R}} - \vartheta\eta_{\mathrm{R}} \tag{31.12}$$

and, by (31.10),

$$\begin{aligned}
\dot{\psi}_{\mathrm{R}} + \eta_{\mathrm{R}}\dot{\vartheta} &= \dot{\varepsilon}_{\mathrm{R}} - \vartheta\dot{\eta}_{\mathrm{R}}, \\
&= \mathbf{T}_{\mathrm{R}}:\dot{\mathbf{F}} - \mathrm{Div}\,\mathbf{q}_{\mathrm{R}} + q_{\mathrm{R}} - \vartheta\left(-\mathrm{Div}\left(\frac{\mathbf{q}_{\mathrm{R}}}{\vartheta}\right) + \frac{q_{\mathrm{R}}}{\vartheta} + \Gamma_{\mathrm{R}}\right), \\
&= \mathbf{T}_{\mathrm{R}}:\dot{\mathbf{F}} - \frac{1}{\vartheta}\mathbf{q}_{\mathrm{R}}\cdot\nabla\vartheta + \vartheta\Gamma_{\mathrm{R}},
\end{aligned}$$

and we arrive at the local **free-energy imbalance**

$$\dot{\psi}_{\mathrm{R}} + \eta_{\mathrm{R}}\dot{\vartheta} - \mathbf{T}_{\mathrm{R}}:\dot{\mathbf{F}} + \frac{1}{\vartheta}\mathbf{q}_{\mathrm{R}}\cdot\nabla\vartheta = -\vartheta\Gamma_{\mathrm{R}} \leq 0. \tag{31.13}$$

31.3 Decay Inequalities for the Body Under Passive Boundary Conditions

We assume that the conventional body force and the heat supply vanish:

$$\mathbf{b}_{0\mathrm{R}} = \mathbf{0}, \qquad q_{\mathrm{R}} = 0;$$

then, (31.7) and (31.8) become

$$\begin{aligned}
\overline{\int_{\mathrm{P}} (\varepsilon_{\mathrm{R}} + \tfrac{1}{2}\rho_{\mathrm{R}}|\dot{\boldsymbol{\chi}}|^2)\,dv_{\mathrm{R}}}^{\boldsymbol{\cdot}} &= -\int_{\partial\mathrm{P}} \mathbf{q}_{\mathrm{R}}\cdot\mathbf{n}_{\mathrm{R}}\,da_{\mathrm{R}} + \int_{\partial\mathrm{P}} \mathbf{T}_{\mathrm{R}}\mathbf{n}_{\mathrm{R}}\cdot\dot{\boldsymbol{\chi}}\,da_{\mathrm{R}}, \\
\overline{\int_{\mathrm{P}} \eta_{\mathrm{R}}\,dv_{\mathrm{R}}}^{\boldsymbol{\cdot}} &= -\int_{\partial\mathrm{P}} \frac{\mathbf{q}_{\mathrm{R}}}{\vartheta}\cdot\mathbf{n}_{\mathrm{R}}\,da_{\mathrm{R}} + \underbrace{\int_{\mathrm{P}} \Gamma_{\mathrm{R}}\,dv_{\mathrm{R}}}_{\geq 0}.
\end{aligned} \tag{31.14}$$

Assume, in addition, that the boundary is essentially referentially dead-loaded and at constant temperature in the sense that there are a constant stress tensor $\mathbf{S}_0$

and a constant temperature $\vartheta_0 > 0$ such that

$$
\begin{aligned}
&\mathbf{T}_R\mathbf{n}_R = \mathbf{S}_0\mathbf{n}_R \text{ on a portion of } \partial B \text{ and } \mathbf{v} = \mathbf{0} \text{ on the remainder of } \partial B,\\
&\vartheta = \vartheta_0 \text{ on a portion of } \partial B \text{ and } \mathbf{q}_R \cdot \mathbf{n}_R = 0 \text{ on the remainder of } \partial B.
\end{aligned}
\tag{31.15}
$$

Consequences of the mechanical boundary conditions $(31.15)_1$ and the divergence theorem are then that

$$
\begin{aligned}
\int_{\partial B} \mathbf{T}_R\mathbf{n}_R \cdot \dot{\boldsymbol{\chi}}\, da_R &= \int_{\partial B} \mathbf{S}_0\mathbf{n}_R \cdot \dot{\boldsymbol{\chi}}\, da_R\\
&= \overline{\int_{\partial B} \mathbf{S}_0\mathbf{n}_R \cdot (\boldsymbol{\chi} - \mathbf{o})\, da_R}^{\,\cdot}\\
&= \overline{\int_{B} \mathbf{S}_0 : \operatorname{grad} \boldsymbol{\chi}\, dv_R}^{\,\cdot} \qquad (31.16)\\
&= \overline{\int_{B} \mathbf{S}_0 : \mathbf{F}\, dv_R}^{\,\cdot}, \qquad (31.17)
\end{aligned}
$$

while the thermal boundary conditions $(31.15)_2$ and the second law $(31.14)_2$ imply that

$$
\begin{aligned}
-\int_{\partial B} \mathbf{q}_R \cdot \mathbf{n}_R\, da &= -\vartheta_0 \int_{\partial B} \frac{\mathbf{q}_R}{\vartheta} \cdot \mathbf{n}_R\, da_R\\
&= \vartheta_0 \overline{\int_B \eta_R\, dv}^{\,\cdot} - \vartheta_0 \int_P \Gamma_R\, dv_R.
\end{aligned}
\tag{31.18}
$$

By (31.17), (31.18), and the energy balance in the form $(31.14)_1$, we arrive at the decay relation

$$
\overline{\int_B (\varepsilon_R - \vartheta_0\eta_R + \mathbf{S}_0 : \mathbf{F} + \tfrac{1}{2}\rho_R|\dot{\boldsymbol{\chi}}|^2)\, dv_R}^{\,\cdot} = -\vartheta_0 \int_B \Gamma_R\, dv_R \le 0,
\tag{31.19}
$$

or equivalently, in terms of the free energy $\psi_R = \varepsilon_R - \vartheta_0\eta_R$,

$$
\overline{\int_B (\psi_R + (\vartheta - \vartheta_0)\eta_R + \mathbf{S}_0 : \mathbf{F} + \tfrac{1}{2}\rho_R|\dot{\boldsymbol{\chi}}|^2)\, dv_R}^{\,\cdot} = -\vartheta_0 \int_B \Gamma_R\, dv_R \le 0.
\tag{31.20}
$$

The stress tensor $\mathbf{S}_0$ represents a Piola stress; by (24.1), the corresponding Cauchy stress is $\mathbf{T}_0 = J^{-1}\mathbf{S}_0\mathbf{F}^{\top}$, and

$$
\begin{aligned}
P &\overset{\text{def}}{=} -\tfrac{1}{3}\operatorname{tr}(\mathbf{S}_0\mathbf{F}^{\top})\\
&= -\tfrac{1}{3}\mathbf{S}_0 : \mathbf{F}
\end{aligned}
\tag{31.21}
$$

represents the **effective pressure** (within B). Note that, in general, $P = P(\mathbf{X}, t)$.

31.4 Mechanical Theory: Free-Energy Imbalance

Consider now the purely mechanical theory discussed in §29. The free energy and dissipation, measured per unit referential volume, are given by

$$\psi_R = \rho_R \psi, \qquad \delta_R = J\delta \geq 0, \tag{31.22}$$

so that, by (29.4), (29.5), and $(31.4)_{1,2}$,

$$\begin{aligned} \mathcal{F}(\mathcal{P}_t) &= \int_{\mathcal{P}_t} \rho\psi \, dv = \int_{\mathrm{P}} \psi_R \, dv_R = \mathcal{F}(\mathrm{P}), \\ \mathcal{D}(\mathcal{P}_t) &= \int_{\mathcal{P}_t} \delta \, dv = \int_{\mathrm{P}} \delta_R \, dv_R = \mathcal{D}(\mathrm{P}). \end{aligned} \tag{31.23}$$

Using the general free-energy balance (29.1) and the power balance (31.3), we then find that

$$\overline{\int_{\mathrm{P}} \psi_R \, dv_R}^{\,\cdot} = \int_{\mathrm{P}} \mathbf{T}_R : \dot{\mathbf{F}} \, dv_R - \int_{\mathrm{P}} \delta_R \, dv_R.$$

Thus, since $\overline{\int_{\mathrm{P}} \psi_R \, dv_R}^{\,\cdot} = \int_{\mathrm{P}} \dot{\psi}_R \, dv_R$,

$$\int_{\mathrm{P}} (\dot{\psi}_R - \mathbf{T}_R : \dot{\mathbf{F}}) \, dv_R = -\int_{\mathrm{P}} \delta_R \, dv_R \leq 0$$

for all subregions P of B, which yields the local **free-energy imbalance**:

$$\boxed{\dot{\psi}_R - \mathbf{T}_R : \dot{\mathbf{F}} = -\delta_R \leq 0, \qquad \dot{\psi}_R - T_{Rij} \dot{F}_{ij} = -\delta_R \leq 0.} \tag{31.24}$$

Next, if we use the general free-energy balance (29.1) in conjunction with (31.1), (31.2), and (31.23), we obtain the global free-energy imbalance:

$$\boxed{\int_{\mathrm{P}} \delta_R \, dv_R = \int_{\partial \mathrm{P}} \mathbf{T}_R \mathbf{n}_R \cdot \dot{\boldsymbol{\chi}} \, da_R + \int_{\mathrm{P}} \mathbf{b}_{0R} \cdot \dot{\boldsymbol{\chi}} \, dv_R - \overline{\int_{\mathrm{P}} (\psi_R + \tfrac{1}{2}\rho_R |\dot{\boldsymbol{\chi}}|^2) \, dv_R}^{\,\cdot} \geq 0.} \tag{31.25}$$

EXERCISES

1. Show that, under a change of frame,

$$\begin{aligned} \mathrm{grad}^* \vartheta^* &= \mathbf{Q} \, \mathrm{grad} \, \vartheta, \\ \mathbf{q}^* &= \mathbf{Q}\mathbf{q}. \end{aligned} \tag{31.26}$$

2. Establish the referential form of the invariance properties (28.2).

3. Show that $\nabla\vartheta$ and $\mathbf{q}_R$ are invariant under a change of frame.
4. Within the purely mechanical framework of §31.4, assume that $\mathbf{b}_{0R} = \mathbf{0}$ and that the boundary is essentially dead-loaded in the sense of $(31.15)_1$ with $\mathbf{S}_0$ a constant stress tensor. Show that

$$\overline{\int_{B} (\psi_R + 3P + \tfrac{1}{2}\rho_R|\dot{\boldsymbol{\chi}}|^2)\, dv_R}^{\,\boldsymbol{\cdot}} = -\int_{B} \delta_R \, dv_R \leq 0 \tag{31.27}$$

with P the boundary pressure (31.21).

PART VI

MECHANICAL AND THERMODYNAMICAL LAWS AT A SHOCK WAVE

A tacit assumption of the discussion to this point has been that the basic mechanical and thermodynamical fields be smooth. But there are important examples such as shock waves and phase transitions in which, although the motion is continuous, the velocity and deformation gradient as well as the stress, temperature, and entropy suffer jump discontinuities across a surface moving through the material.

32 Shock Wave Kinematics

32.1 Notation. Terminology

Throughout this section, χ is a motion of the body. Let $S(t)$ denote an oriented surface evolving smoothly through the reference body B. Assume that $S(t)$ **separates** B into complementary subregions $B^+(t)$ and $B^-(t)$ in the sense that (Figure 32.1):

(i) B (considered as a closed region) is the union of closed regions $B^+(t)$ and $B^-(t)$;
(ii) $S(t)$ is the intersection of $B^+(t)$ and $B^-(t)$.

We write $\mathbf{m}_R(\mathbf{X}, t)$ for the unit normal field on $S(t)$ directed outward from $B^-(t)$, so that $\mathbf{m}_R(\mathbf{X}, t)$ points into $B^+(t)$. Further, we write $V_R(\mathbf{X}, t)$ for the (scalar) normal velocity of $S(t)$ in the direction $\mathbf{m}_R(\mathbf{X}, t)$ and

$$\mathbf{P}_R(\mathbf{X}, t) \stackrel{\text{def}}{=} \mathbf{1} - \mathbf{m}_R(\mathbf{X}, t) \otimes \mathbf{m}_R(\mathbf{X}, t)$$

for the projection onto the plane tangent to $S(t)$ at $\mathbf{X}$.[164] Suppressing arguments, we say that S is a **shock wave** if[165]

(S1) the motion χ is continuous across S;
(S2) the velocity $\dot{\chi}$ and the deformation gradient $\mathbf{F}$ suffer **jump discontinuities** across S; that is, $\dot{\chi}$ and $\mathbf{F}$ are continuous up to S from either side but not across S.

An immediate consequence of (S2) is that $J = \det \mathbf{F}$ may suffer a jump discontinuity across S.

For Φ a material field that suffers a **jump discontinuity** across S, we write Φ^+ for the limit of Φ as S is approached from B^+ and Φ^- for the limit of Φ as S is approached from B^-; that is, for $\mathbf{X}$ on $S(t)$ and $h > 0$,

$$\begin{aligned} \Phi^+(\mathbf{X}, t) &= \lim_{h \to 0} \Phi(\mathbf{X} + h\mathbf{m}_R(\mathbf{X}, t), t), \\ \Phi^-(\mathbf{X}, t) &= \lim_{h \to 0} \Phi(\mathbf{X} - h\mathbf{m}_R(\mathbf{X}, t), t). \end{aligned} \tag{32.1}$$

Further, $[\![\Phi]\!]$ denotes the **jump** of Φ across S:

$$[\![\Phi]\!] = \Phi^+ - \Phi^-. \tag{32.2}$$

164 For any vector $\mathbf{a}$, $\mathbf{P}_R(\mathbf{X}, t)\mathbf{a}$ represents the component of $\mathbf{a}$ relative to the tangent plane to $S(t)$ at $\mathbf{X}$. Cf. $(2.6)_2$.

165 Such a surface is also referred to as a singular surface of order one and, in the materials literature, as a coherent phase interface.

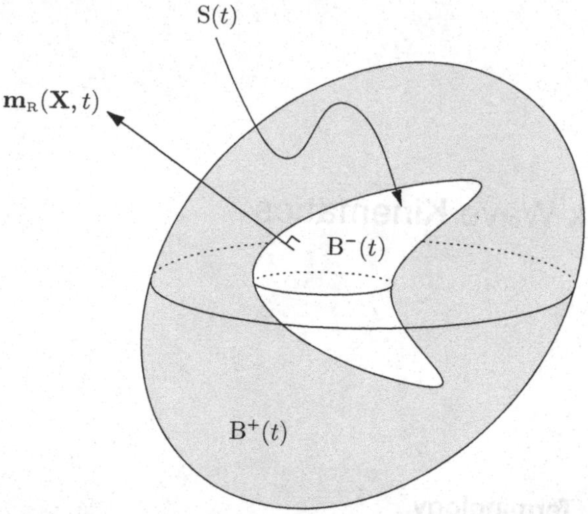

Figure 32.1. A smoothly evolving, oriented surface S(t) that at time t divides B into regions B$^-(t)$. The orientation $\mathbf{m}_R(\mathbf{X}, t)$ of S(t) is chosen to point from B$^-(t)$ into B$^+(t)$.

32.2 Hadamard's Compatibility Conditions

Basic to a discussion of shock waves are **Hadamard's compatibility conditions:**

$$\boxed{[\![\dot{\chi}]\!] = -V_R[\![\mathbf{F}]\!]\mathbf{m}_R \qquad \text{and} \qquad [\![\mathbf{F}]\!]\mathbf{P}_R = \mathbf{0}.} \tag{32.3}$$

As a consequence of $(32.3)_2$, $[\![\mathbf{F}]\!]\mathbf{t} = \mathbf{0}$ for any vector field $\mathbf{t}$ tangent to S, so that, for any vector field $\mathbf{f}$

$$[\![\mathbf{F}]\!]\mathbf{f} = (\mathbf{m}_R \cdot \mathbf{f})[\![\mathbf{F}]\!]\mathbf{m}_R. \tag{32.4}$$

To verify (32.3), choose a point $\mathbf{X}_0$ on S(t_0) and let $\bar{\mathbf{t}}_0$ be an arbitrary vector tangent to S(t_0) at $\mathbf{X}_0$:

$$\bar{\mathbf{t}}_0 \cdot \mathbf{m}_R(\mathbf{X}_0, t_0) = 0.$$

For each t, let $C_R(t)$ denote a curve on S(t) traced out by the material point

$$\mathbf{X} = \hat{\mathbf{X}}(\lambda, t) \tag{32.5}$$

as the scalar parameter λ varies.[166] Assume that the curve $C_R(t)$ passes through the point $\mathbf{X}_0$ at time t_0 and that $\bar{\mathbf{t}}_0$ is tangent to $C_R(t_0)$ at $\mathbf{X}_0$. (We can always find such a curve.) Then

$$\hat{\mathbf{X}}(\lambda_0, t_0) = \mathbf{X}_0,$$

and if we let

$$\left.\frac{\partial \hat{\mathbf{X}}}{\partial \lambda}\right|_0 = \left.\frac{\partial \hat{\mathbf{X}}(\lambda, t)}{\partial \lambda}\right|_{(\lambda,t)=(\lambda_0,t_0)},$$

then the vector

$$\mathbf{t}_0 \overset{\text{def}}{=} \left.\frac{\partial \hat{\mathbf{X}}}{\partial \lambda}\right|_0 \tag{32.6}$$

is tangent to $C_R(t_0)$ at $\mathbf{X}_0$ and hence parallel to $\bar{\mathbf{t}}_0$.

166 Cf. the paragraph containing (6.10).

At each t, $\mathrm{S}(t)$ deforms to a surface

$$\mathcal{S}(t) = \chi_t(\mathrm{S}(t)) \tag{32.7}$$

in the deformed body $\mathcal{B}_t$; $\mathcal{S}(t)$ represents the **spatial description of the shock wave.** Further, $\mathrm{C_R}(t)$ deforms to a curve $\mathcal{C}(t) = \chi_t(\mathrm{C_R}(t))$ on $\mathcal{S}(t)$, with $\mathcal{C}(t)$ traced out by

$$\begin{aligned}\mathbf{x} &= \hat{\mathbf{x}}(\lambda, t)\\ &= \chi_t(\hat{\mathbf{X}}(\lambda, t)).\end{aligned} \tag{32.8}$$

In view of (S1) and (S2), the deformation χ is continuously differentiable up to S from either side. If we restrict attention to the plus side we find — using (32.6), (32.8), and the chain-rule — that

$$\left.\frac{\partial \hat{\mathbf{x}}}{\partial \lambda}\right|_0 = \mathbf{F}^{+}(\mathbf{X}_0, t_0)\mathbf{t}_0. \tag{32.9}$$

A second relation, strictly analogous to (32.9), obtains if we restrict attention to the minus side of S; if we subtract this second relation from the first we find that

$$[\![\mathbf{F}]\!](\mathbf{X}_0, t_0)\mathbf{t}_0 = \mathbf{0} \qquad \text{and hence that} \qquad [\![\mathbf{F}]\!](\mathbf{X}_0, t_0)\bar{\mathbf{t}}_0 = \mathbf{0}.$$

Thus, since the pair $(\mathbf{X}_0, t_0)$ was arbitrarily chosen, as was the vector $\bar{\mathbf{t}}_0$ tangent to $\mathrm{S}(t_0)$ at $\mathbf{X}_0$, the second of (32.3) is satisfied.

Next, by (8.4) the unit vector field $\mathbf{m}$ defined on $\mathcal{S}$ by

$$\mathbf{m} = \frac{(\mathbf{F}^{-\top})^{\pm}\mathbf{m}_{\mathrm{R}}}{|(\mathbf{F}^{-\top})^{\pm}\mathbf{m}_{\mathrm{R}}|} \tag{32.10}$$

is normal to $\mathcal{S}$. Further, the vectors

$$\left.\frac{\partial \hat{\mathbf{X}}}{\partial \lambda}\right|_0 \qquad \text{and} \qquad \left.\frac{\partial \hat{\mathbf{x}}}{\partial \lambda}\right|_0,$$

respectively, represent velocities for the surfaces $\mathrm{S}(t_0)$ and $\mathcal{S}(t_0)$ at the points $\mathbf{X}_0$ and $\mathbf{x}_0 = \chi_{t_0}(\mathbf{X}_0)$; hence,

$$\mathbf{m}_{\mathrm{R}}(\mathbf{X}_0, t_0) \cdot \left.\frac{\partial \hat{\mathbf{X}}}{\partial \lambda}\right|_0 = V_{\mathrm{R}}(\mathbf{X}_0, t_0), \tag{32.11}$$

while

$$V(\mathbf{x}_0, t_0) \stackrel{\text{def}}{=} \mathbf{m}(\mathbf{x}_0, t_0) \cdot \left.\frac{\partial \hat{\mathbf{x}}}{\partial \lambda}\right|_0 \tag{32.12}$$

represents the scalar normal velocity of $\mathcal{S}(t_0)$ at x_0.[167]

On the other hand, by the chain-rule

$$\left.\frac{\partial \hat{\mathbf{x}}}{\partial \lambda}\right|_{=(\lambda_0, t_0)} = \mathbf{F}^{\pm}(\mathbf{X}_0, t_0)\left.\frac{\partial \hat{\mathbf{X}}}{\partial \lambda}\right|_0 + \dot{\chi}^{\pm}(\mathbf{X}_0, t_0)$$

and, hence, (32.4) and (32.11) yield

$$\left.\frac{\partial \hat{\mathbf{x}}}{\partial \lambda}\right|_0 = V_{\mathrm{R}}(\mathbf{X}_0, t_0)\mathbf{F}^{\pm}(\mathbf{X}_0, t_0)\mathbf{m}_{\mathrm{R}}(\mathbf{X}_0, t_0) + \dot{\chi}^{\pm}(\mathbf{X}_0, t_0). \tag{32.13}$$

---

167 The tangential component of the velocity $\partial\hat{\mathbf{X}}/\partial t$ depends on the parametrization $\hat{\mathbf{X}}$, but the normal component is independent of $\hat{\mathbf{X}}$ and hence intrinsic to the evolution of S; a similar assertion applies to $\partial\hat{\mathbf{x}}/\partial t$; cf. GURTIN & STRUTHERS (1990, Lemma 2A).

Thus, since $\mathbf{X}_0$ and t_0 were chosen arbitrarily,

$$V_R [\![\mathbf{F}]\!] \mathbf{m}_R + [\![\dot{\chi}]\!] = \mathbf{0},$$

which is $(32.3)_1$. This completes the verification of the Hadamard relations.

32.3 Relation Between the Scalar Normal Velocities V_R and V

If we take the inner product of (32.13) with $\mathbf{m}(\mathbf{x}_0, t_0)$ and use (32.12) we obtain

$$V = V_R(\mathbf{m} \cdot \mathbf{F}^{\pm}\mathbf{m}_R) + \mathbf{m} \cdot \dot{\chi}^{\pm},$$

where for convenience we have suppressed arguments. On the other hand, by (8.14) and (32.10),

$$\begin{aligned}\mathbf{m} \cdot \mathbf{F}^{\pm}\mathbf{m}_R &= \frac{(\mathbf{F}^{-\top})^{\pm}\mathbf{m}_R \cdot \mathbf{F}^{\pm}\mathbf{m}_R}{|(\mathbf{F}^{-\top})^{\pm}\mathbf{m}_R|} \\ &= \frac{\mathbf{m}_R \cdot (\mathbf{F}^{-1})^{\pm}\mathbf{F}^{\pm}\mathbf{m}_R}{|(\mathbf{F}^{-\top})^{\pm}\mathbf{m}_R|} \\ &= \frac{1}{|(\mathbf{F}^{-\top})^{\pm}\mathbf{m}_R|} \\ &= \frac{J^{\pm}}{\jmath},\end{aligned} \tag{32.14}$$

where $J^{\pm}/\jmath$ represents the ratio of the volumetric Jacobians $J^{\pm}$ of the material on the two sides of the shock wave to the areal Jacobian $\jmath$ of the mapping between the undeformed and deformed surfaces S and $\mathcal{S}$. Thus writing $\mathbf{v}^{\pm} = \dot{\chi}^{\pm}$ (for the spatial descriptions of the velocities $\dot{\chi}^{\pm}$), (32.14) becomes

$$V - \mathbf{m} \cdot \mathbf{v}^{\pm} = \frac{J^{\pm}}{\jmath} V_R. \tag{32.15}$$

The normal velocity V_R at which the shock wave moves through the reference body (and hence through the material) — scaled by the factor $J^{\pm}/\jmath$ — is therefore equal to the "observed velocity" V of the wave measured relative to the normal velocities $\mathbf{m} \cdot \mathbf{v}^{\pm}$ of those material points instantaneously situated at the two sides of the wave.

32.4 Transport Relations in the Presence of a Shock Wave

Basic to what follows is a well-known transport relation. To state this relation, let R(t) denote a bounded region evolving smoothly through the reference body B and let $V_{\partial R}(\mathbf{X}, t)$ denote the scalar normal outward velocity of the time-dependent boundary ∂R(t). Then, given a smooth scalar function $\varphi(\mathbf{X}, t)$ defined for all $\mathbf{X}$ in R(t) and all t, we have the transport relation:

$$\overline{\int_{R(t)} \varphi(\mathbf{X}, t)\, dv_R(\mathbf{X})}^{\boldsymbol{\cdot}} = \int_{R(t)} \dot{\varphi}(\mathbf{X}, t)\, dv_R(\mathbf{X}) + \int_{\partial R(t)} \varphi(\mathbf{X}, t) V_{\partial R}(\mathbf{X}, t)\, dv_R(\mathbf{X}). \tag{32.16}$$

Consider a motion χ containing a shock wave S. The derivation of balance laws requires the calculation of time-derivatives of the form

$$I(t) = \overline{\int_{P} \Phi(\mathbf{X}, t)\, dv_R(\mathbf{X})}^{\boldsymbol{\cdot}}, \tag{32.17}$$

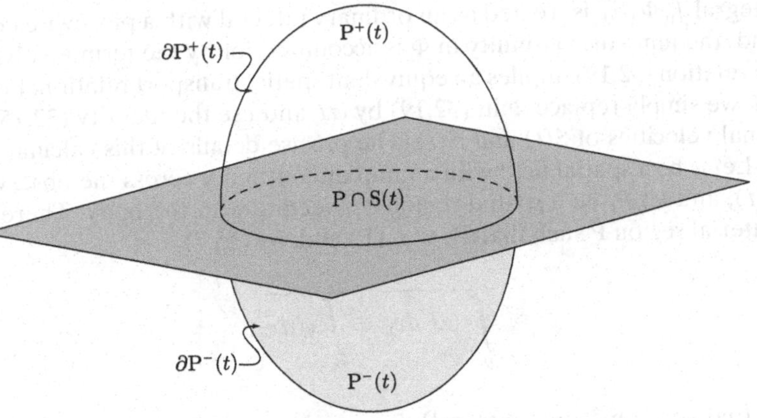

Figure 32.2. Fixed material region p divided by $S(t)$ into time-dependent regions $P^{\pm}(t)$.

where Φ is a material field with a jump discontinuity across the shock wave and P is an arbitrary fixed subregion of the reference body B. We assume that (on some time interval) P contains a (nontrivial) portion

$$P \cap S(t) \tag{32.18}$$

of the shock wave in its interior; $P \cap S(t)$ then separates P into time-dependent complementary subregions $P^+(t)$ and $P^-(t)$ such that (Figure 32.2):

(i) the boundary $\partial P^+(t)$ of $P^+(t)$ is the union of the shock surface $S(t)$ and a subsurface of ∂P, and the outward velocity of $\partial P^+(t)$ is equal to $-V_R(t)$ on the shock surface, but otherwise vanishes;
(ii) the boundary $\partial P^-(t)$ of $P^-(t)$ has properties strictly analogous to those of $\partial P^+(t)$, except that the outward velocity of $\partial P^-(t)$ is equal to $+V_R(t)$ on the shock surface.

Were Φ smooth, we could simply commute the time derivative and the integral, but we cannot do this because of the jump discontinuity. However, because P is the union of $P^+(t)$ and $P^-(t)$,

$$I(t) = \overline{\int_{P^+(t)} \Phi(\mathbf{X}, t)\, dv_R(\mathbf{X})}^{\boldsymbol{\cdot}} + \overline{\int_{P^-(t)} \Phi(\mathbf{X}, t)\, dv_R(\mathbf{X})}^{\boldsymbol{\cdot}},$$

with $\Phi(\mathbf{X}, t)$ smooth on $P^+(t)$ and on $P^-(t)$. Thus, bearing in mind (i) and (ii) and using (32.16), we find that

$$\overline{\int_{P^+(t)} \Phi(\mathbf{X}, t)\, dv_R(\mathbf{X})}^{\boldsymbol{\cdot}} = \int_{P^+(t)} \dot{\Phi}(\mathbf{X}, t)\, dv_R(\mathbf{X}) - \int_{P\cap S(t)} \Phi(\mathbf{X}, t) V_R(\mathbf{X}, t)\, da_R(\mathbf{X}),$$

$$\overline{\int_{P^-(t)} \Phi(\mathbf{X}, t)\, dv_R(\mathbf{X})}^{\boldsymbol{\cdot}} = \int_{P^-(t)} \dot{\Phi}(\mathbf{X}, t)\, dv_R(\mathbf{X}) + \int_{P\cap S(t)} \Phi(\mathbf{X}, t) V_R(\mathbf{X}, t)\, da_R(\mathbf{X}).$$

Thus, suppressing arguments, we have the **referential transport relation**

$$\overline{\int_{P} \Phi\, dv_R}^{\boldsymbol{\cdot}} = \int_{P} \dot{\Phi}\, dv_R - \int_{P\cap S} [\![\Phi]\!] V_R\, da_R. \tag{32.19}$$

(The integral $\int_{\mathrm{P}} \dot{\Phi}\, dv_{\mathrm{R}}$ is treated as an ordinary integral with a piecewise continuous integrand; the jump discontinuity in Φ is accounted for by the term involving $[\![\Phi]\!]$.)

The relation (32.19) implies an equivalent spatial transport relation: For φ a spatial field, we simply replace Φ in (32.19) by φJ and use the identity (32.15) relating the normal velocities of $\mathrm{S}(t)$ and $\mathcal{S}(t)$. The precise details of this calculation are as follows. Let φ be a spatial field with a jump discontinuity across the observed shock wave $\mathcal{S}(t)$, and let $\mathcal{P}_t$ be a spatial region convecting with the body. There is then a fixed material region P such that $\mathcal{P}_t = \boldsymbol{\chi}_t(\mathrm{P})$ and, by (15.7),

$$\overline{\int_{\mathrm{P}} \varphi J\, dv_{\mathrm{R}}}^{\cdot} = \overline{\int_{\mathcal{P}_t} \varphi\, dv.}^{\cdot} \tag{32.20}$$

Assume that (on some time interval) $\mathcal{P}_t$ contains a portion of $\mathcal{S}(t)$ in its interior. Then, since $\mathrm{P} = \boldsymbol{\chi}_t^{-1}(\mathcal{P}_t)$ and $\mathrm{S}(t) = \boldsymbol{\chi}_t^{-1}(\mathcal{S}(t))$, it follows that

$$\mathrm{P} \cap \mathrm{S}(t) = \boldsymbol{\chi}_t^{-1}(\mathcal{P}_t \cap \mathcal{S}(t))$$

and, hence, that P contains a portion of $\mathrm{S}(t)$ in its interior. Then, by (9.16),[168]

$$\begin{aligned} \int_{\mathrm{P}} \dot{\overline{\varphi J}}\, dv_{\mathrm{R}} &= \int_{\mathrm{P}} (\dot{\varphi} J + \varphi \dot{J})\, dv_{\mathrm{R}} \\ &= \int_{\mathrm{P}} (\dot{\varphi} + \varphi\, \mathrm{div}\mathbf{v}) J\, dv_{\mathrm{R}} \\ &= \int_{\mathcal{P}_t} (\dot{\varphi} + \varphi\, \mathrm{div}\mathbf{v})\, dv \end{aligned} \tag{32.21}$$

and, by $(15.5)_2$ and (32.15),

$$\begin{aligned} \int_{\mathrm{P}\cap\mathrm{S}(t)} [\![\varphi J]\!] V_{\mathrm{R}}\, da_{\mathrm{R}} &= \int_{\mathcal{P}_t\cap\mathcal{S}(t)} [\![\varphi J]\!] V_{\mathrm{R}}\, J^{-1}\, da \\ &= \int_{\mathcal{P}_t\cap\mathcal{S}(t)} \left[\!\!\left[\varphi \frac{J}{J} V_{\mathrm{R}} \right]\!\!\right] da \\ &= \int_{\mathcal{P}_t\cap\mathcal{S}(t)} [\![\varphi(V - \mathbf{m}\cdot\mathbf{v})]\!]\, da. \end{aligned} \tag{32.22}$$

Substituting (32.20)–(32.22) into (32.19) with $\Phi = \varphi J$, we are led to the **spatial transport relation**

$$\overline{\int_{\mathcal{P}_t} \varphi\, dv}^{\cdot} = \int_{\mathcal{P}_t} (\dot{\varphi} + \varphi\, \mathrm{div}\mathbf{v})\, dv - \int_{\mathcal{P}_t\cap\mathcal{S}(t)} [\![\varphi(V - \mathbf{m}\cdot\mathbf{v})]\!]\, da; \tag{32.23}$$

(32.23) generalizes Reynolds' transport relation (16.2) to account for the presence of a shock wave.

---

[168] Cf. the parenthetical remark following (32.19).

32.5 The Divergence Theorem in the Presence of a Shock Wave

As a discussion of the divergence theorem does not involve time-differentiation, we may confine our attention to a fixed time.

Let $\mathbf{G}$ be a material tensor field with a jump discontinuity across a shock wave S, and let P — with outward unit normal $\mathbf{n}$ on ∂P — be a subregion of the reference body B. Then, S separates P into subregions P^+ and P^- such that

(i) the boundary $\partial\mathrm{P}^+$ of P^+ is the union of the shock surface S and a subsurface of ∂P, and the outward unit normal $\mathbf{n}_\mathrm{R}^+$ of $\partial\mathrm{P}^+$ satisfies

$$\mathbf{n}_\mathrm{R}^+ = \begin{cases} -\mathbf{m}_\mathrm{R} & \text{on S}, \\ \mathbf{n}_\mathrm{R} & \text{otherwise}; \end{cases} \tag{32.24}$$

(ii) the boundary $\partial\mathrm{P}^-$ of P^- has properties strictly analogous to those of $\partial\mathrm{P}^+$, except that the outward unit normal of $\partial\mathrm{P}^-$ is equal to $+\mathbf{m}_\mathrm{R}$ on the shock surface.

Then,

$$\begin{aligned}\int\limits_{\partial\mathrm{P}} \mathbf{G}\mathbf{n}_\mathrm{R}\,da_\mathrm{R} &= \int\limits_{\partial\mathrm{P}^+} \mathbf{G}\mathbf{n}_\mathrm{R}^+\,da_\mathrm{R} + \int\limits_{\partial\mathrm{P}^-} \mathbf{G}\mathbf{n}_\mathrm{R}^-\,da_\mathrm{R} - \int\limits_{\mathrm{P}\cap\mathrm{S}} \mathbf{G}^+\mathbf{n}_\mathrm{R}^+\,da_\mathrm{R} - \int\limits_{\mathrm{P}\cap\mathrm{S}} \mathbf{G}^-\mathbf{n}_\mathrm{R}^-\,da_\mathrm{R} \\ &= \int\limits_{\partial\mathrm{P}^+} \mathbf{G}\mathbf{n}_\mathrm{R}^+\,da_\mathrm{R} + \int\limits_{\partial\mathrm{P}^-} \mathbf{G}\mathbf{n}_\mathrm{R}^-\,da_\mathrm{R} - \int\limits_{\mathrm{P}\cap\mathrm{S}} \mathbf{G}^+(-\mathbf{m}_\mathrm{R})\,da_\mathrm{R} - \int\limits_{\mathrm{P}\cap\mathrm{S}} \mathbf{G}^-\mathbf{m}_\mathrm{R}\,da_\mathrm{R} \\ &= \int\limits_{\mathrm{P}^+} \mathrm{Div}\,\mathbf{G}\,dv_\mathrm{R} + \int\limits_{\mathrm{P}^-} \mathrm{Div}\,\mathbf{G}\,dv_\mathrm{R} + \int\limits_{\mathrm{P}\cap\mathrm{S}} [\![\mathbf{G}]\!]\mathbf{m}_\mathrm{R}\,da_\mathrm{R};\end{aligned}$$

therefore, we obtain an identity,

$$\int\limits_{\partial\mathrm{P}} \mathbf{G}\mathbf{n}_\mathrm{R}\,da_\mathrm{R} = \int\limits_{\mathrm{P}} \mathrm{Div}\,\mathbf{G}\,dv_\mathrm{R} + \int\limits_{\mathrm{P}\cap\mathrm{S}} [\![\mathbf{G}]\!]\mathbf{m}_\mathrm{R}\,da_\mathrm{R}, \tag{32.25}$$

representing the divergence theorem for a material region containing a shock wave.

An (obvious) strictly analogous argument yields the counterpart of (32.25) for a spatial region $\mathcal{P}$:

$$\int\limits_{\partial\mathcal{P}} \mathbf{G}\mathbf{n}\,da = \int\limits_{\mathcal{P}} \mathrm{div}\mathbf{G}\,dv + \int\limits_{\mathcal{P}\cap\mathcal{S}} [\![\mathbf{G}]\!]\mathbf{m}\,da. \tag{32.26}$$

Counterparts of the results (32.25) and (32.26) for a vector field $\mathbf{g}$ are

$$\begin{aligned}\int\limits_{\partial\mathrm{P}} \mathbf{g}\cdot\mathbf{n}_\mathrm{R}\,da_\mathrm{R} &= \int\limits_{\mathrm{P}} \mathrm{Div}\,\mathbf{g}\,dv_\mathrm{R} + \int\limits_{\mathrm{P}\cap\mathrm{S}} [\![\mathbf{g}]\!]\cdot\mathbf{m}_\mathrm{R}\,da_\mathrm{R}, \\ \int\limits_{\partial\mathcal{P}} \mathbf{g}\cdot\mathbf{n}\,da &= \int\limits_{\mathcal{P}} \mathrm{div}\,\mathbf{g}\,dv + \int\limits_{\mathcal{P}\cap\mathcal{S}} [\![\mathbf{g}]\!]\cdot\mathbf{m}\,da.\end{aligned} \tag{32.27}$$

EXERCISE

1. Establish (32.27) as well as corresponding results for a scalar field φ.

33 Basic Laws at a Shock Wave: Jump Conditions

In most discussions of shock waves, the stress, temperature, and entropy suffer jump discontinuities across the shock surface $\mathcal{S}(t)$. Such discussions generally, but not always, neglect heat flow. Here, for completeness, we allow the heat flux to suffer a jump discontinuity across $\mathcal{S}(t)$.

In the absence of shock waves, the basic laws reduce to partial-differential relations that hold locally. In the presence of a shock wave, these local relations are satisfied away from the shock surface and — what is most cogent — are supplemented by jump conditions that hold at the shock surface.

33.1 Balance of Mass and Momentum

Since the density ρ and velocity $\mathbf{v}$ suffer jump discontinuities across a shock wave,[169] to determine the forms taken by the mechanical laws at a shock wave, we need to consider these laws in forms that do not involve pointwise derivatives of ρ and $\mathbf{v}$; for that reason, we return to the general integrated relations

$$\overline{\int_{\mathcal{P}_t} \rho\, dv}^{\,\boldsymbol{\cdot}} = 0 \tag{33.1}$$

for balance of mass[170] and, using Cauchy's relation $\mathbf{t}(\mathbf{n}) = \mathbf{T}\mathbf{n}$,[171]

$$\begin{aligned} \overline{\int_{\mathcal{P}_t} \rho\mathbf{v}\, dv}^{\,\boldsymbol{\cdot}} &= \int_{\partial\mathcal{P}_t} \mathbf{T}\mathbf{n}\, da + \int_{\mathcal{P}_t} \mathbf{b}_0\, dv, \\ \overline{\int_{\mathcal{P}_t} \mathbf{r}\times(\rho\mathbf{v})\, dv}^{\,\boldsymbol{\cdot}} &= \int_{\partial\mathcal{P}_t} \mathbf{r}\times\mathbf{T}\mathbf{n}\, da + \int_{\mathcal{P}_t} \mathbf{r}\times\mathbf{b}_0\, dv \end{aligned} \tag{33.2}$$

for the balances of linear and angular momentum.[172] Away from the shock surface, these balances have the local forms[173]

$$\dot{\rho} + \rho\,\mathrm{div}\mathbf{v} = 0, \qquad \rho\dot{\mathbf{v}} = \mathrm{div}\mathbf{T} + \mathbf{b}_0, \qquad \text{and} \qquad \mathbf{T} = \mathbf{T}^{\top}; \tag{33.3}$$

---

[169] By (18.9), $\rho J = \rho_{\mathrm{R}}$ with ρ_{R} independent of time; thus, since $J = \det\mathbf{F}$ suffers a jump discontinuity, ρ should also.

[170] Cf. (18.2).

[171] Cf. (19.22).

[172] Cf. (19.12) and (19.13). The balances (19.16) for forces and moments are no longer valid because $\dot{\mathbf{v}}$ and, hence, the generalized body force $\mathbf{b}$ are not defined at the wave; cf. (19.15).

[173] Cf. (18.5), (19.31), and (19.34).

we now turn to a derivation of corresponding jump conditions that hold at the shock surface.

Since the balances (33.1)–(33.3) are spatial, we work with the spatial description of the shock wave as represented by the surface $\mathcal{S}(t)$ in $\mathcal{B}_t$. Basic identities that we use in localizing (33.1) and (33.2) to $\mathcal{S}(t)$ are the relation (32.23) for a scalar field φ and its (obvious) counterpart for a vector field **h**:

$$\overline{\int_{\mathcal{P}_t} \varphi\, dv}^{\boldsymbol{\cdot}} = \int_{\mathcal{P}_t} (\dot{\varphi} + \varphi\,\mathrm{div}\mathbf{v})\, dv - \int_{\mathcal{P}_t \cap \mathcal{S}(t)} [\![\varphi(V - \mathbf{m}\cdot\mathbf{v})]\!]\, da,$$

$$\overline{\int_{\mathcal{P}_t} \mathbf{h}\, dv}^{\boldsymbol{\cdot}} = \int_{\mathcal{P}_t} (\dot{\mathbf{h}} + \mathbf{h}\,\mathrm{div}\mathbf{v})\, dv - \int_{\mathcal{P}_t \cap \mathcal{S}(t)} [\![\mathbf{h}(V - \mathbf{m}\cdot\mathbf{v})]\!]\, da. \tag{33.4}$$

To localize the mass balance (33.1), we use $(33.4)_1$ with $\varphi = \rho$ and $(33.3)_1$, which holds away from the shock surface; the result is

$$\int_{\mathcal{P}_t \cap \mathcal{S}(t)} [\![\rho(V - \mathbf{m}\cdot\mathbf{v})]\!]\, da = 0.$$

Since this relation must be satisfied for every convecting region $\mathcal{P}_t$, we arrive at a jump condition expressing *mass balance at a shock wave*:

$$\boxed{[\![\rho(V - \mathbf{m}\cdot\mathbf{v})]\!] = 0.} \tag{33.5}$$

The limits $\rho^{+}(V - \mathbf{m}\cdot\mathbf{v}^{+})$ and $\rho^{-}(V - \mathbf{m}\cdot\mathbf{v}^{-})$ represent the mass flow-rates at the two sides of the shock surface, measured relative to the material; (33.5) is therefore the requirement that the relative mass flow be continuous across a shock wave. The quantity

$$m \stackrel{\text{def}}{=} \rho^{\pm}(V - \mathbf{m}\cdot\mathbf{v}^{\pm}) \tag{33.6}$$

therefore represents the **relative mass flow-rate** across the shock wave.

Consider next the balance $(33.2)_1$ of linear momentum. By $(33.4)_2$ with $\mathbf{h} = \rho\mathbf{v}$,

$$\overline{\int_{\mathcal{P}_t} \rho\mathbf{v}\, dv}^{\boldsymbol{\cdot}} = \int_{\mathcal{P}_t} \underbrace{(\rho\dot{\mathbf{v}} + \dot{\rho}\mathbf{v} + \rho\mathbf{v}\,\mathrm{div}\mathbf{v})}_{=\rho\dot{\mathbf{v}} \text{ by } (33.3)_1}\, dv - \int_{\mathcal{P}_t \cap \mathcal{S}(t)} [\![(V - \mathbf{m}\cdot\mathbf{v})\rho\mathbf{v}]\!]\, da. \tag{33.7}$$

Further, (32.26) implies that

$$\int_{\partial\mathcal{P}_t} \mathbf{T}\mathbf{n}\, da = \int_{\mathcal{P}_t} \mathrm{div}\mathbf{T}\, dv + \int_{\mathcal{P}_t \cap \mathcal{S}(t)} [\![\mathbf{T}]\!]\mathbf{m}\, da. \tag{33.8}$$

Substituting (33.7) and (33.8) into (33.2), we find that

$$\int_{\mathcal{P}_t} \underbrace{(\mathrm{div}\mathbf{T} + \mathbf{b}_0 - \rho\dot{\mathbf{v}})}_{=\mathbf{0} \text{ by } (33.3)_2}\, dv = -\int_{\mathcal{P}_t \cap \mathcal{S}(t)} ([\![\mathbf{T}]\!]\mathbf{m} + [\![(V - \mathbf{m}\cdot\mathbf{v})\rho\mathbf{v}]\!])\, da; \tag{33.9}$$

since (33.9) must hold for all convecting regions $\mathcal{P}_t$, we arrive at a jump condition expressing *balance of linear momentum at a shock wave*:

$$[\![(V - \mathbf{m}\cdot\mathbf{v})\rho\mathbf{v}]\!] = -[\![\mathbf{T}]\!]\mathbf{m}. \tag{33.10}$$

Granted (33.10), the corresponding localization of the global angular momentum relation $(33.2)_2$ yields "$\mathbf{0} = \mathbf{0}$"; thus,

(‡) *angular momentum is balanced across a shock wave provided linear momentum is so balanced.*

The proof of this result is left as an exercise.

MASS AND MOMENTUM BALANCE AT A SHOCK WAVE *The jump conditions expressing mass and momentum balance at a shock wave have the respective forms*

$$\begin{aligned} [\![\rho(V - \mathbf{m}\cdot\mathbf{v})]\!] &= 0, \\ [\![(V - \mathbf{m}\cdot\mathbf{v})\rho\mathbf{v}]\!] &= -[\![\mathbf{T}]\!]\mathbf{m}. \end{aligned} \tag{33.11}$$

In view of (33.6), *the momentum balance* $(33.11)_2$ *has the simple form*

$$\boxed{m[\![\mathbf{v}]\!] = -[\![\mathbf{T}]\!]\mathbf{m}.} \tag{33.12}$$

33.2 Balance of Energy and the Entropy Imbalance

First of all, away from the shock surface, we must have[174]

$$\rho\dot{\varepsilon} = \mathbf{T}:\mathbf{D} - \operatorname{div}\mathbf{q} + q \qquad \text{and} \qquad \rho\dot{\eta} \geq -\operatorname{div}\left(\frac{\mathbf{q}}{\vartheta}\right) + \frac{q}{\vartheta}. \tag{33.13}$$

To derive the corresponding jump conditions that hold at the shock wave, consider first the balance of energy in the form (26.6); viz.

$$\overline{\int_{\mathcal{P}_t} \rho(\varepsilon + \tfrac{1}{2}|\mathbf{v}|^2)\,dv}^{\boldsymbol{\cdot}} = -\int_{\partial\mathcal{P}_t} \mathbf{q}\cdot\mathbf{n}\,da + \int_{\mathcal{P}_t} q\,dv + \int_{\partial\mathcal{P}_t} \mathbf{Tn}\cdot\mathbf{v}\,da + \int_{\mathcal{P}_t} \mathbf{b}_0\cdot\mathbf{v}\,dv. \tag{33.14}$$

By (33.4) with $\varphi = \rho k$, $k = \varepsilon + \frac{1}{2}|\mathbf{v}|^2$,

$$\overline{\int_{\mathcal{P}_t} \rho k\,dv}^{\boldsymbol{\cdot}} = \int_{\mathcal{P}_t} \underbrace{(\rho\dot{k} + \dot{\rho}k + \rho k\operatorname{div}\mathbf{v})}_{=\rho\dot{k}\text{ by }(33.3)_1}\,dv - \int_{\mathcal{P}_t\cap\mathcal{S}(t)} [\![\rho k]\!]\,da \tag{33.15}$$

and, since $\rho\dot{k} = \rho(\dot{\varepsilon} + \mathbf{v}\cdot\dot{\mathbf{v}})$, (33.15) becomes

$$\overline{\int_{\mathcal{P}_t} \rho(\varepsilon + \tfrac{1}{2}|\mathbf{v}|^2)\,dv}^{\boldsymbol{\cdot}} = \int_{\mathcal{P}_t} \rho(\dot{\varepsilon} + \dot{\mathbf{v}}\cdot\mathbf{v})\,dv - \int_{\mathcal{P}_t\cap\mathcal{S}(t)} [\![\rho(\varepsilon + \tfrac{1}{2}|\mathbf{v}|^2)(V - \mathbf{m}\cdot\mathbf{v})]\!]\,da. \tag{33.16}$$

Further, since the stress $\mathbf{T}$ is symmetric,

$$\int_{\partial\mathcal{P}_t} \mathbf{Tn}\cdot\mathbf{v}\,da = \int_{\partial\mathcal{P}_t} \mathbf{Tv}\cdot\mathbf{n}\,da$$

[174] Cf. (26.8) and (27.13).

and $(32.27)_2$ with $\mathbf{g} = \mathbf{Tv}$ and $(33.3)_2$ yield

$$\begin{aligned}\int_{\partial\mathcal{P}_t} \mathbf{Tv}\cdot\mathbf{n}\,da &= \int_{\mathcal{P}_t} \mathrm{div}(\mathbf{Tv})\,dv + \int_{\mathcal{P}_t\cap\mathcal{S}(t)} [\![\mathbf{Tv}]\!]\cdot\mathbf{m}\,da \\ &= \int_{\mathcal{P}_t} (\mathbf{v}\cdot\mathrm{div}\mathbf{T} + \mathbf{T}\colon\mathbf{D})\,da + \int_{\mathcal{P}_t\cap\mathcal{S}(t)} [\![\mathbf{Tv}]\!]\cdot\mathbf{m}\,da \\ &= \int_{\mathcal{P}_t} (\mathbf{v}\cdot(\rho\dot{\mathbf{v}} - \mathbf{b}_0) + \mathbf{T}\colon\mathbf{D})\,da + \int_{\mathcal{P}_t\cap\mathcal{S}(t)} [\![\mathbf{Tv}]\!]\cdot\mathbf{m}\,da\end{aligned}$$

Therefore (33.14) can be written equivalently as

$$\begin{aligned}&\overline{\int_{\mathcal{P}_t} \rho(\varepsilon + \tfrac{1}{2}|\mathbf{v}|^2)\,dv}^{\boldsymbol{\cdot}} - \int_{\partial\mathcal{P}_t} \mathbf{Tn}\cdot\mathbf{v}\,da - \int_{\mathcal{P}_t} \mathbf{b}_0\cdot\mathbf{v}\,dv \\ &\quad = \int_{\mathcal{P}_t} \rho\dot{\varepsilon}\,dv - \int_{\mathcal{P}_t\cap\mathcal{S}(t)} [\![\rho(\varepsilon + \tfrac{1}{2}|\mathbf{v}|^2)(V - \mathbf{m}\cdot\mathbf{v})]\!]\,da - \int_{\mathcal{P}_t\cap\mathcal{S}(t)} [\![\mathbf{Tv}]\!]\cdot\mathbf{m}\,da. \qquad (33.17)\end{aligned}$$

Further, $(32.27)_2$ with $\mathbf{g} = \mathbf{q}$ yields

$$\int_{\mathcal{P}_t} q\,dv - \int_{\partial\mathcal{P}_t} \mathbf{q}\cdot\mathbf{n}\,da = -\int_{\mathcal{P}_t} (q - \mathrm{div}\mathbf{q})\,dv - \int_{\mathcal{P}_t\cap\mathcal{S}(t)} [\![\mathbf{q}]\!]\cdot\mathbf{m}\,da \qquad (33.18)$$

and adding (33.17) and (33.18) we find using $(33.13)_1$ that

$$\int_{\mathcal{P}_t\cap\mathcal{S}(t)} [\![\rho(\varepsilon + \tfrac{1}{2}|\mathbf{v}|^2)(V - \mathbf{m}\cdot\mathbf{v})]\!]\,da = -\int_{\mathcal{P}_t\cap\mathcal{S}(t)} [\![\mathbf{Tv}]\!]\cdot\mathbf{m}\,da - \int_{\mathcal{P}_t\cap\mathcal{S}(t)} [\![\mathbf{q}]\!]\cdot\mathbf{m}\,da.$$

Since the convecting spatial region $\mathcal{P}_t$ is arbitrary, we therefore have the jump condition expressing *balance of energy* at a shock wave:

$$[\![\rho(\varepsilon + \tfrac{1}{2}|\mathbf{v}|^2)(V - \mathbf{m}\cdot\mathbf{v})]\!] = -[\![\mathbf{Tv}]\!]\cdot\mathbf{m} - [\![\mathbf{q}]\!]\cdot\mathbf{m}. \qquad (33.19)$$

The jump condition expressing *entropy imbalance* at a shock wave is given by

$$[\![\rho\eta(V - \mathbf{m}\cdot\mathbf{v})]\!] \geq -\left[\!\!\left[\frac{\mathbf{q}}{\vartheta}\right]\!\!\right]\cdot\mathbf{m} \qquad (33.20)$$

and its derivation is left as an exercise.

ENERGY BALANCE AND ENTROPY IMBALANCE AT A SHOCK WAVE *The jump conditions expressing energy balance and entropy imbalance at a shock wave have the respective forms*

$$\begin{aligned}&[\![\rho(\varepsilon + \tfrac{1}{2}|\mathbf{v}|^2)(V - \mathbf{m}\cdot\mathbf{v})]\!] = -[\![\mathbf{Tv}]\!]\cdot\mathbf{m} - [\![\mathbf{q}]\!]\cdot\mathbf{m}, \\ &[\![\rho\eta(V - \mathbf{m}\cdot\mathbf{v})]\!] \geq -\left[\!\!\left[\frac{\mathbf{q}}{\vartheta}\right]\!\!\right]\cdot\mathbf{m}.\end{aligned} \qquad (33.21)$$

Further, using (33.6) *these jump conditions take the simple form*

$$\boxed{\begin{aligned} m[\![\varepsilon + \tfrac{1}{2}|\mathbf{v}|^2]\!] &= -[\![\mathbf{T}\mathbf{v}]\!]\cdot\mathbf{m} - [\![\mathbf{q}]\!]\cdot\mathbf{m},\\ m[\![\eta]\!] &\geq -\left[\!\!\left[\frac{\mathbf{q}}{\vartheta}\right]\!\!\right]\cdot\mathbf{m}. \end{aligned}} \tag{33.22}$$

Discussions of shock waves often restrict attention to **nonconductors**, which are bodies for which the heat flux vanishes:

$$\mathbf{q} \equiv \mathbf{0} \tag{33.23}$$

In this case, the jump conditions (33.21) reduce to

$$\begin{aligned} m[\![\varepsilon + \tfrac{1}{2}|\mathbf{v}|^2]\!] &= -[\![\mathbf{T}\mathbf{v}]\!]\cdot\mathbf{m},\\ m[\![\eta]\!] &\geq 0. \end{aligned} \tag{33.24}$$

Remark. The quantity $m[\![\eta]\!] \geq 0$ represents the entropy production per unit area at the shock wave: For $\mathcal{P}_t$ a convecting region that contains a portion of the shock surface $\mathcal{S}(t)$ in its interior, the net entropy-production $\mathcal{H}(\mathcal{P}_t)$ is given by[175]

$$\mathcal{H}(\mathcal{P}_t) = \int\limits_{\mathcal{P}_t} \Gamma\, dv + \int\limits_{\mathcal{P}_t\cap\mathcal{S}(t)} m[\![\eta]\!]\, da. \tag{33.25}$$

EXERCISES

1. Show that, for $\mathcal{P}_t$ a region convecting with the body, if $\mathcal{P}_t$ contains a portion of a shock wave $\mathcal{S}(t)$ in its interior, then
$$\overline{\int\limits_{\mathcal{P}_t} \mathbf{r}\times(\rho\mathbf{v})\, dv}^{\,\boldsymbol{\cdot}} = \int\limits_{\mathcal{P}_t} \mathbf{r}\times(\rho\dot{\mathbf{v}})\, dv - \int\limits_{\mathcal{P}_t\cap\mathcal{S}(t)} [\![(V - \mathbf{m}\cdot\mathbf{v})\mathbf{r}\times(\rho\mathbf{v})]\!]\, da.$$
Use this result to establish (‡) on page 135.
2. Derive the jump condition (33.20) expressing entropy imbalance at a shock wave.

[175] Cf. The remark on page 188.

PART VII

INTERLUDE: BASIC HYPOTHESES FOR DEVELOPING PHYSICALLY MEANINGFUL CONSTITUTIVE THEORIES

34 General Considerations

The balance laws for mass, momentum, and energy and the imbalance law for entropy represent fundamental principles of continuum thermomechanics and as such are presumed to hold for *all bodies*, whether they be solid, liquid, or gas. In contrast, the constitution of a class of bodies composed of a particular material is specified by *constitutive equations*. Such equations limit the class of "processes" that bodies comprised of a given material may undergo.

In the words of TRUESDELL & NOLL (1965, §1). "The general physical laws in themselves do not suffice to determine the deformation or motion of a body subject to given loading. Before a determinate problem can be formulated, it is usually necessary to specify the *material* of which the body is made. In the program of continuum mechanics, such specification is stated by *constitutive equations*, which relate the stress tensor... to the motion. For example, the classical theory of elasticity rests upon the assumption that the stress tensor at a point depends linearly on the changes in length and mutual angle suffered by elements at that point..., while the classical theory of viscosity is based on the assumption that the stress tensor depends linearly on the instantaneous rates of change of length and mutual angle. These statements... are *definitions of ideal materials*. The former expresses in words the constitutive equation that defines a linearly and infinitesimally elastic material; the latter a *linearly viscous fluid*. Each... represents in ideal form an *aspect*, and a different one, of the mechanical behavior of nearly all natural materials, and... each does predict with considerable... accuracy the observed response of many different natural materials in certain restricted situations."

35 Constitutive Response Functions

Typically, a constitutive equation[176] gives the pointwise values of a physical field Φ, say, in terms of the values of another such field (or list of such fields) Λ:[177]

$$\Phi(\mathbf{X}, t) = \hat{\Phi}(\Lambda(\mathbf{X}, t), \mathbf{X}). \tag{35.1}$$

The function $\hat{\Phi}$ is referred to as a **constitutive response function**. We distinguish between constitutive response functions and the fields they deliver using a superposed symbol such as a "hat," "tilde," or "bar." The explicit dependence on $\mathbf{X}$ allows for a dependence on the material point in question; when this dependence is absent from all constitutive equations describing a given body, the body is termed **homogeneous.**[178] We generally write constitutive equations of the form (35.1) in the succinct form

$$\Phi = \hat{\Phi}(\Lambda). \tag{35.2}$$

This notation is not meant to imply homogeneity. In the equation (35.2), Λ denotes the independent constitutive variable and Φ denotes the dependent constitutive variable.

[176] We use the phrases "constitutive equation" and "constitutive relation" interchangeably.

[177] As we shall see, for classical viscous fluids, which generally do not possess natural reference configurations, (35.1) is replaced by a relation of the form $\Phi(\mathbf{x}, t) = \hat{\Phi}(\Lambda(\mathbf{x}, t))$.

[178] More precisely, the body is homogeneous, as is the choice of reference configuration. For a homogeneous body, the density ρ_R is constant.

36 Frame-Indifference and Compatibility with Thermodynamics

As a general principle,

(‡) *we require that constitutive equations be invariant under changes of frame and compatible with thermodynamics.*

The first hypothesis is referred to as **frame-indifference.**[179] Specifically, if Φ and Λ are related through a constitutive response function $\hat{\Phi}$, then so also are Φ^* and Λ^* for all changes in frame:

$$\text{if} \quad \Phi = \hat{\Phi}(\Lambda) \quad \text{then} \quad \Phi^* = \hat{\Phi}(\Lambda^*). \tag{36.1}$$

To satisfy the second hypothesis, we utilize a procedure invented by COLEMAN & NOLL (1964), a procedure under which only constitutive equations that are consistent with the free-energy inequality in all processes are deemed physically viable.

These ideas, while broadly applicable in continuum thermomechanics, are most easily conveyed in restricted contexts. We begin, in the next chapter, by developing a theory for the conduction of heat in a rigid medium. The assumed rigidity renders moot issues of invariance and allows us to focus on the use of thermodynamics to restrict constitutive equations. In the subsequent chapter, we consider the purely mechanical theory of viscous fluids. This topic provides a simple and beautiful context for the application of frame-indifference.

179 Often referred to as "material frame-indifference" when used to restrict constitutive relations. The frame-indifference principle is discussed in §21.

PART VIII

RIGID HEAT CONDUCTORS

We consider a rigid medium occupying a (fixed) region B, the body. Material regions P are then subregions of B. We allow for the conduction of heat — in agreement with the first and second laws of thermodynamics — within B. Our notation differs only slightly from that introduced in §26 and §27: $\mathbf{x}$ denotes a point in B, grad and div denote the gradient and divergence with respect to $\mathbf{x}$, ε is the *internal energy* and η the *internal entropy*, each measured per unit volume in B;[180] ϑ is the (absolute) *temperature*; $\mathbf{q}$ is the *heat flux* measured per unit area in B; q and Γ, with

$$\Gamma \geq 0,$$

are the *heat supply* and *entropy production*, measured per unit volume in B.[181]

180 Everywhere else in this book, ε, η, and $\psi = \varepsilon - \vartheta\eta$ denote the specific internal energy, entropy, and free energy. Here, because the density of a rigid medium is fixed, it is advantageous to absorb the density and work with internal energy, entropy, and free energy measured per unit volume.

181 Cf. the discussions in §26.1 and §27.1.

37 Basic Laws

The first two laws, the **balance of energy** and the **imbalance of entropy**, require that, for each material region P,[182]

$$\overline{\int_{\mathrm{P}} \varepsilon\, dv} = -\int_{\partial \mathrm{P}} \mathbf{q}\cdot\mathbf{n}\, da + \int_{\mathrm{P}} q\, dv,$$

$$\overline{\int_{\mathrm{P}} \eta\, dv}^{\boldsymbol{\cdot}} \geq -\int_{\partial \mathrm{P}} \frac{\mathbf{q}}{\vartheta}\cdot\mathbf{n}\, da + \int_{\mathrm{P}} \frac{q}{\vartheta}\, dv. \tag{37.1}$$

The divergence theorem and the arbitrary nature of P together yield the local forms of these laws:[183]

$$\begin{aligned} \dot{\varepsilon} &= -\mathrm{div}\mathbf{q} + q, \\ \dot{\eta} &\geq -\mathrm{div}\left(\frac{\mathbf{q}}{\vartheta}\right) + \frac{q}{\vartheta}. \end{aligned} \tag{37.2}$$

By (37.2),

$$\begin{aligned} \dot{\varepsilon} - \vartheta\dot{\eta} &\leq -\mathrm{div}\mathbf{q} + \vartheta\,\mathrm{div}\left(\frac{\mathbf{q}}{\vartheta}\right) \\ &= -\frac{1}{\vartheta}\mathbf{q}\cdot\mathrm{grad}\,\vartheta; \end{aligned}$$

thus, introducing the *free energy*

$$\psi = \varepsilon - \vartheta\eta, \tag{37.3}$$

we are led to the **free-energy imbalance**

$$\dot{\psi} + \eta\dot{\vartheta} + \frac{1}{\vartheta}\mathbf{q}\cdot\mathrm{grad}\,\vartheta = -\vartheta\Gamma \leq 0, \tag{37.4}$$

where Γ is the entropy production and $\vartheta\Gamma$ represents the *dissipation*, measured per unit volume.[184] In fact,

(‡) *granted balance of energy* $(37.2)_1$ *and the expression* (37.3) *for the free energy, the entropy imbalance* $(37.2)_2$ *is equivalent to the free-energy imbalance* (37.4).

---

182 Cf. (26.10) and (27.9). We have assumed from the outset that the underlying fields are smooth, an assumption that ensures the existence of a smooth entropy production Γ. Were we discussing, e.g., solidification, this assumption would be invalid.

183 Here, e.g., $\dot{\varepsilon} = \partial\varepsilon/\partial t$.

184 Cf. (27.18).

38 General Constitutive Equations

The balance of energy and imbalance of entropy represent fundamental principles of continuum thermodynamics and as such are presumed to hold for *all bodies.* In contrast, the constitution of a particular class of bodies composed of a particular material is specified by *constitutive assumptions.* Such assumptions limit the class of "processes" that bodies comprised of a given material may undergo.

Guided by the free-energy imbalance (37.4),[185] we consider constitutive equations giving the free energy, the entropy, and the heat flux when the temperature and its gradient are known:

$$\left.\begin{aligned}\psi &= \hat{\psi}(\vartheta, \operatorname{grad}\vartheta),\\ \eta &= \hat{\eta}(\vartheta, \operatorname{grad}\vartheta),\\ \mathbf{q} &= \hat{\mathbf{q}}(\vartheta, \operatorname{grad}\vartheta).\end{aligned}\right\} \tag{38.1}$$

These relations yield an auxiliary constitutive relation

$$\begin{aligned}\varepsilon &= \hat{\varepsilon}(\vartheta, \operatorname{grad}\vartheta)\\ &= \hat{\psi}(\vartheta, \operatorname{grad}\vartheta) + \vartheta\,\hat{\eta}(\vartheta, \operatorname{grad}\vartheta)\end{aligned} \tag{38.2}$$

for the internal energy.

The constitutive equations (38.1) might be motivated as follows: Most theories of heat conduction are based on constitutive equations for the internal energy and heat flux of the form

$$\varepsilon = \hat{\varepsilon}(\vartheta), \qquad \mathbf{q} = -\mathbf{K}\operatorname{grad}\vartheta, \tag{38.3}$$

with $\mathbf{K}$ the symmetric, positive-definite conductivity tensor; classical thermodynamical treatments are based on equations of state that — within a theory such as ours in which the medium is rigid — have the form

$$\psi = \hat{\psi}(\vartheta), \qquad \eta = \hat{\eta}(\vartheta) = -\frac{d\hat{\psi}(\vartheta)}{d\vartheta}. \tag{38.4}$$

While one cannot deny the broad applicability of theories based on (38.3) and (38.4), these relations do beg the question: Is it possible to have a thermodynamically consistent theory in which the free energy and entropy depend on the temperature gradient? Indeed, in a general theory of the type desired, one might expect constitutive interactions between the fields ϑ and $\operatorname{grad}\vartheta$. Such a starting point is motivated

[185] The free-energy imbalance is equivalent to the requirement that the rate of entropy production within the body be nonnegative and, as is clear from §39, represents the sole restriction placed by the basic laws on the internal constitution of the body. This imbalance might therefore be viewed as indicating which of the basic fields should have constitutive descriptions.

by the belief that the underlying physics should determine which of the constitutive interactions are inappropriate, at least in a general theory. This, in essence, is *Truesdell's principle of equipresence*: "A quantity present as an independent variable in one constitutive equation should be so present in all, unless...its presence contradicts some law of physics or rule of invariance." Cf. TRUESDELL and NOLL (1965, §96), who assert that: "This principle forbids us to eliminate any of the 'causes' present from interacting with any other as regards a particular 'effect.' It reflects on the scale of gross phenomena the fact that all observed effects result from a common structure such as the motions of molecules." Physics, not caprice, should rule out interactions.

39 Thermodynamics and Constitutive Restrictions: The Coleman–Noll Procedure

Consider an arbitrary **constitutive process**; that is, a temperature field ϑ together with fields ψ, η, and $\mathbf{q}$ determined by ϑ through the constitutive equations (38.1). The local relation $(37.2)_1$ expressing balance of energy then determines the heat supply q needed to support the process. We assume that

- *q is arbitrarily assignable.*

This hypothesis is important: because of it, balance of energy in no way restricts the class of constitutive equations under consideration. On the other hand, unless these constitutive equations are suitably restricted, not all constitutive processes will be compatible with the second law as embodied in the entropy balance $(37.2)_2$ or equivalently in the free-energy imbalance (37.4).[186] Thus, we take as a basic hypothesis the requirement that *all constitutive processes be consistent with the free-energy imbalance* (37.4).

As noted by GURTIN,[187] The Coleman–Noll procedure[188] "...is based on the premise that the second law be satisfied in all conceivable processes, irrespective of the difficulties involved in producing such processes in the laboratory. The rational application of this procedure requires [possibly virtual] external forces and supplies that may be assigned arbitrarily to ensure satisfaction of the underlying balances in all processes. This may seem artificial, but it is no more artificial than theories based on virtual power, a paradigm that requires arbitrary variations, [variations] not guaranteed to be consistent with the resulting evolution equations, granted a constitutive description. The Coleman–Noll procedure makes explicit the external fields needed to support the 'virtual processes' used, and in so doing ensures that these external fields, whether virtual or not, enter the theory in a thermodynamically consistent manner."

The free-energy imbalance severely restricts the constitutive equations. To see this, consider an arbitrary constitutive process and let

$$\mathbf{g} = \operatorname{grad}\vartheta \tag{39.1}$$

denote the **temperature gradient**. Then, by (38.1),

$$\dot{\psi} = \frac{\partial\hat{\psi}(\vartheta,\mathbf{g})}{\partial\vartheta}\dot{\vartheta} + \frac{\partial\hat{\psi}(\vartheta,\mathbf{g})}{\partial\mathbf{g}}\cdot\dot{\mathbf{g}};$$

[186] Cf. (‡) on page 229.

[187] (2000, p. 53).

[188] COLEMAN & NOLL (1963) introduced this procedure for the study of elastic materials with heat conduction and viscosity. Subsequently, the proceedure has been used to restrict constitutive equations for more general theories of solids and fluids.

hence, the free-energy imbalance (37.4) is equivalent to the requirement that

$$\left(\frac{\partial\hat{\psi}(\vartheta,\mathbf{g})}{\partial\vartheta}+\hat{\eta}(\vartheta,\mathbf{g})\right)\dot{\vartheta}+\frac{\partial\hat{\psi}(\vartheta,\mathbf{g})}{\partial\mathbf{g}}\cdot\dot{\mathbf{g}}+\frac{1}{\vartheta}\hat{\mathbf{q}}(\vartheta,\mathbf{g})\cdot\mathbf{g}\le 0 \tag{39.2}$$

for all temperature fields. Given any point $\mathbf{x}_0$ in B and any time t_0, it is possible to find a temperature field such that

$$\vartheta,\ \mathbf{g},\ \dot{\vartheta},\ \text{and}\ \dot{\mathbf{g}} \tag{39.3}$$

have arbitrarily prescribed values at $(\mathbf{x}_0, t_0)$.[189] Granted this, the coefficients of $\dot{\vartheta}$ and $\dot{\mathbf{g}}$ must vanish, for otherwise these rates may be chosen to violate the inequality (39.2). We therefore have the following **thermodynamic restrictions**:

(i) *The free energy and the entropy are independent of the temperature gradient.*
(ii) *The free energy determines the entropy through the* **entropy relation**

$$\boxed{\hat{\eta}(\vartheta)=-\frac{d\hat{\psi}(\vartheta)}{d\vartheta}.} \tag{39.4}$$

(iii) *The heat flux satisfies the* **heat-conduction inequality**

$$\boxed{\hat{\mathbf{q}}(\vartheta,\mathbf{g})\cdot\mathbf{g}\le 0} \tag{39.5}$$

for all $(\vartheta, \mathbf{g})$.

We refer to (i) and (ii) as **state restrictions**. The restriction (iii), which arises on applying the state restrictions to the constitutively augmented version (39.2) of the free-energy imbalance (37.4), is referred to as a **reduced dissipation inequality**.

Proof of the thermodynamic restrictions (i)–(iii) requires finding a temperature field consistent with the assertion in the sentence containing (39.3). With this in mind, arbitrarily choose: scalar constants $\vartheta_0 > 0$ and α; vector constants $\mathbf{g}_0$ and $\mathbf{h}$. We now construct a (strictly positive) temperature field ϑ such that

$$\begin{aligned}\vartheta(\mathbf{x}_0,t_0)&=\vartheta_0, & \dot{\vartheta}(\mathbf{x}_0,t_0)&=\alpha,\\ \mathbf{g}(\mathbf{x}_0,t_0)&=\mathbf{g}_0, & \dot{\mathbf{g}}(\mathbf{x}_0,t_0)&=\mathbf{h}.\end{aligned} \tag{39.6}$$

Note first that

$$\vartheta(\mathbf{x},t)=\vartheta_0+\mathbf{g}_0\cdot(\mathbf{x}-\mathbf{x}_0)+\alpha(t-t_0)+(t-t_0)\mathbf{h}\cdot(\mathbf{x}-\mathbf{x}_0)$$

satisfies (39.6) but is inadmissible because it does not ensure that the resulting temperature field be strictly positive. To remedy this, choose, arbitrarily, a scalar constant p such that

$$0<p<\vartheta_0.$$

Then, we can always find a smooth strictly positive scalar function $\Theta_0(\mathbf{x})$ and a scalar function $\Theta_1(\mathbf{x})$ such that

$$\begin{aligned}\Theta_0(\mathbf{x}_0)&=\vartheta_0, & \operatorname{grad}\Theta_0(\mathbf{x}_0)&=\mathbf{g}_0,\\ \operatorname{grad}\Theta_1(\mathbf{x}_0)&=\mathbf{h}, & |\Theta_1(\mathbf{x})|&<\sqrt{p},\end{aligned}$$

for all $\mathbf{x}$ in B. Similarly, we can find a positive scalar function $\Phi_0(t)$ and a scalar function $\Phi_1(t)$ such that

$$\begin{aligned}\Phi_0(t_0)&=0, & \dot{\Phi}_0(t_0)&=\alpha,\\ \Phi_1(t_0)&=0, & \dot{\Phi}_1(t_0)&=1, & |\Phi_1(t)|&<\sqrt{p}\end{aligned}$$

for all t. The field

$$\vartheta(\mathbf{x},t)\stackrel{\text{def}}{=}\Theta_0(\mathbf{x})+\Phi_0(t)+\Theta_1(\mathbf{x})\Phi_1(t)$$

is then strictly positive for all $\mathbf{x}$ in B and all t and satisfies (39.6).

189 This assertion, which is intuitively plausible, is proved at the end of this subsection.

40 Consequences of the State Restrictions

A consequence of the state restriction (i), the constitutive relation (38.2) for the internal energy becomes

$$\begin{aligned}\varepsilon &= \hat{\varepsilon}(\vartheta)\\ &= \hat{\psi}(\vartheta) + \vartheta\,\hat{\eta}(\vartheta)\\ &= \hat{\psi}(\vartheta) - \vartheta\,\frac{d\hat{\psi}(\vartheta)}{d\vartheta}.\end{aligned} \tag{40.1}$$

We now list two important relations typically called *Gibbs relations* and typically *assumed to hold a priori* in conventional discussions of thermodynamics.[190] The first Gibbs relation is a direct consequence of the entropy relation; the second follows from the first and the identity $\psi = \varepsilon - \theta\eta$. Stated precisely, the Gibbs relations assert that in any constitutive process

$$\dot{\psi} = -\eta\dot{\vartheta} \quad \text{and} \quad \dot{\varepsilon} = \vartheta\dot{\eta}. \tag{40.2}$$

Note that the second Gibbs relation allows us to rewrite the energy balance $(37.2)_1$ as an **entropy balance**

$$\vartheta\dot{\eta} = -\mathrm{div}\mathbf{q} + q. \tag{40.3}$$

The temperature-dependent constitutive modulus defined by

$$c(\vartheta) = \frac{d\hat{\varepsilon}(\vartheta)}{d\vartheta} \tag{40.4}$$

is called the **specific heat**. By (39.4) and (40.1),

$$c(\vartheta) = \vartheta\,\frac{d\hat{\eta}(\vartheta)}{d\vartheta} \tag{40.5}$$

$$= -\vartheta\,\frac{d^2\hat{\psi}(\vartheta)}{d\vartheta^2}. \tag{40.6}$$

Thus, *if the specific heat satisfies $c(\vartheta) > 0$ for all ϑ, then — as functions of ϑ — the entropy $\hat{\eta}(\vartheta)$ is strictly increasing and the free energy $\hat{\psi}(\vartheta)$ is strictly concave.*

190 Cf., e.g., DE GROOT & MAZUR (1962). Here, the Gibbs relations (40.2) arise as a consequence of the underlying thermodynamic structure.

Note that, by (37.4) and $(40.2)_1$, the entropy production Γ in any constitutive process has the particular form

$$\Gamma = -\frac{1}{\vartheta^2}\mathbf{q}\cdot\operatorname{grad}\vartheta. \tag{40.7}$$

We refer to the heat-conduction inequality as **strict** if

$$\hat{\mathbf{q}}(\vartheta,\mathbf{g})\cdot\mathbf{g} < 0, \qquad \text{if } \mathbf{g}\neq\mathbf{0}, \tag{40.8}$$

or equivalently, if Γ as defined in (40.7) is strictly positive at any point with nonvanishing temperature gradient.

41 Consequences of the Heat-Conduction Inequality

Let

$$\varphi(\vartheta, \mathbf{g}) = \hat{\mathbf{q}}(\vartheta, \mathbf{g}) \cdot \mathbf{g};$$

then, since $\varphi(\vartheta, \mathbf{g}) \leq 0$ and $\varphi(\vartheta, \mathbf{0}) = \mathbf{0}$,

$$\varphi(\vartheta, \mathbf{g}), \text{ as a function of } \mathbf{g}, \text{ must have a maximum at } \mathbf{g} = \mathbf{0}, \tag{41.1}$$

and, hence,

$$\left.\frac{\partial \varphi(\vartheta, \mathbf{g})}{\partial \mathbf{g}}\right|_{\mathbf{g}=\mathbf{0}} = \mathbf{0}. \tag{41.2}$$

Thus, since

$$\frac{\partial \varphi(\vartheta, \mathbf{g})}{\partial \mathbf{g}} = \hat{\mathbf{q}}(\vartheta, \mathbf{g}) + \left(\frac{\partial \hat{\mathbf{q}}(\vartheta, \mathbf{g})}{\partial \mathbf{g}}\right)^{\top} \mathbf{g}, \tag{41.3}$$

that is,

$$\frac{\partial \varphi}{\partial g_i} = \hat{q}_i + \frac{\partial \hat{q}_j}{\partial g_i} g_j, \tag{41.4}$$

it follows that

$$\hat{\mathbf{q}}(\vartheta, \mathbf{0}) = \mathbf{0}, \tag{41.5}$$

and we have an important result: *The heat flux vanishes when the temperature gradient vanishes, independent of the value of the temperature*

$$\mathbf{q} = \mathbf{0} \qquad \text{whenever} \qquad \operatorname{grad} \vartheta = \mathbf{0}. \tag{41.6}$$

42 Fourier's Law

Classical linear theories of heat conduction are based on **Fourier's law**

$$\mathbf{q} = -\mathbf{K}\mathbf{g},$$

with $\mathbf{K}$, the **conductivity tensor**, symmetric, positive-definite, and constant. We now establish the sense in which the constitutive equation $\mathbf{q} = \hat{\mathbf{q}}(\vartheta, \mathbf{g})$ is approximated by Fourier's law in situations close to a state of uniform temperature ϑ_0.

With a view toward expanding $\hat{\mathbf{q}}(\vartheta, \mathbf{g})$ in a Taylor series about the state $(\vartheta_0, \mathbf{0})$, note that, since, by (41.5), $\hat{\mathbf{q}}(\vartheta, \mathbf{0}) = \mathbf{0}$,

$$\frac{\partial \hat{\mathbf{q}}(\vartheta, \mathbf{0})}{\partial \vartheta} = \mathbf{0}. \tag{42.1}$$

Further, (41.4) yields

$$\frac{\partial^2 \varphi}{\partial g_i \partial g_k} = \frac{\partial \hat{q}_i}{\partial g_k} + \frac{\partial \hat{q}_j}{\partial g_i}\delta_{jk} + \frac{\partial^2 \hat{q}_j}{\partial g_i \partial g_k} g_j, \tag{42.2}$$

and, hence,

$$\left.\frac{\partial^2 \varphi}{\partial \mathbf{g}^2}\right|_{\mathbf{g}=\mathbf{0}} = -(\mathbf{K} + \mathbf{K}^{\top}), \tag{42.3}$$

where $\mathbf{K}$ is the constant tensor defined by

$$\mathbf{K} = -\left.\frac{\partial \hat{\mathbf{q}}}{\partial \mathbf{g}}\right|_{\mathbf{g}=\mathbf{0}}. \tag{42.4}$$

Further, by (41.1) and (42.3), $\mathbf{K}$ is positive-semidefinite. Finally, expanding $\mathbf{q} = \hat{\mathbf{q}}(\vartheta, \mathbf{g})$ in a Taylor series about $(\vartheta_0, \mathbf{0})$, we conclude, with the aid of (42.1) and (42.3), that

$$\hat{\mathbf{q}}(\vartheta, \mathbf{g}) = -\mathbf{K}\mathbf{g} + o(|\mathbf{g}|) \qquad \text{as } \mathbf{g} \to \mathbf{0}.$$

Thus, *to within terms of* $o(|\mathbf{g}|)$ *as* $\mathbf{g} \to \mathbf{0}$*, the heat flux is given approximately by the linear constitutive equation*

$$\mathbf{q} = -\mathbf{K}\mathbf{g},$$

with $\mathbf{K}$ *positive-definite, but not necessarily symmetric. To within this approximation, the heat flux is therefore independent of temperature.*

There is a large literature[191] concerning the symmetry of the conductivity tensor **K** in Fourier's law and, in particular, whether there is a sense in which this symmetry follows from basic principles. For most, but not all, problems of interest and for many, but not all, classes of crystals this question has little relevance. To begin with, the heat flux enters the local energy balance only through the term

$$-\mathrm{div}\,\mathbf{q} = K_{ij}\frac{\partial^2\vartheta}{\partial X_i \partial X_j}$$

and hence only through the symmetric part of **K**.[192] Furthermore, in the case of crystals, symmetry renders **K** symmetric in the rhombic system, in four classes of the tetragonal system, in the cubic system, and in seven classes of the hexagonal system.[193]

Further, the result $(28.1)_3$ for the heat flux — which holds also in the present theory — asserts that the first two laws (37.1) are invariant under transformations of the form

$$\mathbf{q} \to \mathbf{q} + \boldsymbol{\omega} \times \mathrm{grad}\,\vartheta.$$

This result has an interesting consequence regarding Fourier's law

$$\mathbf{q} = -\mathbf{K}\,\mathrm{grad}\,\vartheta, \tag{42.5}$$

a constitutive equation in which **K**, a constant tensor, represents the thermal conductivity of the material. Let $\mathbf{K}_+$ and $\mathbf{K}_-$ denote the symmetric and skew parts of **K**, and let $\boldsymbol{\omega}$ denote the axial vector corresponding to $\mathbf{K}_-$. Then,

$$\mathbf{q} = -\mathbf{K}_+\,\mathrm{grad}\,\vartheta - \boldsymbol{\omega}\times\mathrm{grad}\,\vartheta;$$

thus, dropping the skew part of **K** would alter neither balance of energy nor the entropy balance, a fact that renders the basic theory within the body independent of $\mathbf{K}_-$. However, boundary conditions involving the heat flux might still require consideration of $\mathbf{K}_-$.

EXERCISES

1. Suppress time as an argument, assume that, at some point $\mathbf{x}_0$, the temperature gradient obeys $\mathrm{grad}\,\vartheta(\mathbf{x}_0) \neq \mathbf{0}$, and define

$$\mathbf{e} = \frac{\mathrm{grad}\,\vartheta(\mathbf{x}_0)}{|\mathrm{grad}\,\vartheta(\mathbf{x}_0)|}. \tag{42.6}$$

   (a) Show that, for all sufficiently small $h > 0$,

$$\vartheta(\mathbf{x}_0 + h\mathbf{e}) > \vartheta(\mathbf{x}_0) \tag{42.7}$$

   and, thus, that the point $\mathbf{x}_0 + h\mathbf{e}$ is hotter than $\mathbf{x}_0$.

   (b) Using the heat-conduction inequality (39.5) in the strict form and the definition (42.6) of **e**, show that

$$\mathbf{q}(\mathbf{x}_0)\cdot\mathbf{e} \leq 0. \tag{42.8}$$

   (c) Argue on the basis of (42.7) and (42.8) that heat flows from hot to cold.

2. Establish a counterpart, for a rigid heat conductor, of the decay relation (28.10), granted the boundary condition $(28.7)_2$.
3. (Project) Determine the consequences of thermocompatibility for a rigid heat conductor (no deformation or stress) defined by the constitutive equations

$$\psi = \hat{\psi}(\vartheta, \mathrm{grad}\,\vartheta, \mathrm{grad}\,\mathrm{grad}\,\vartheta),$$

$$\eta = \hat{\eta}(\vartheta, \mathrm{grad}\,\vartheta, \mathrm{grad}\,\mathrm{grad}\,\vartheta),$$

$$\mathbf{q} = \hat{\mathbf{q}}(\vartheta, \mathrm{grad}\,\vartheta, \mathrm{grad}\,\mathrm{grad}\,\vartheta).$$

---

[191] Cf. the expositions of Meixner & Reik (1959) and de Groot & Mazur (1962) and the critical comments of Truesdell (1969).

[192] Cf. the second paragraph in petite type on page 238.

[193] Cf. the Appendix by Wang in Truesdell (1969).

PART IX

THE MECHANICAL THEORY OF COMPRESSIBLE AND INCOMPRESSIBLE FLUIDS

This section discusses the theories of elastic fluids and compressible and incompressible viscous fluids, neglecting thermal effects. These theories are the cornerstone of classical fluid mechanics and provide a background for a discussion of more modern theories.

43 Brief Review

43.1 Basic Kinematical Relations

Kinematical relations central to a discussion of fluids consist of[194]

(i) the relation

$$\dot{\mathbf{v}} = \mathbf{v}' + (\operatorname{grad}\mathbf{v})\mathbf{v} \tag{43.1}$$

between the acceleration $\dot{\mathbf{v}}$ and the spatial time derivative $\mathbf{v}'$ of the velocity (§9.2);

(ii) the decomposition

$$\begin{gathered}\mathbf{L} = \operatorname{grad}\mathbf{v} = \mathbf{D} + \mathbf{W},\\ \mathbf{D} = \operatorname{sym}\mathbf{L}, \qquad \mathbf{W} = \operatorname{skw}\mathbf{L}\end{gathered} \tag{43.2}$$

of the velocity gradient $\mathbf{L} = \operatorname{grad}\mathbf{v}$ into a symmetric stretching tensor $\mathbf{D}$ and a skew spin tensor $\mathbf{W}$;

(iii) the decomposition

$$\mathbf{D} = \mathbf{D}_0 + \tfrac{1}{3}(\operatorname{tr}\mathbf{D})\mathbf{1}, \qquad \operatorname{tr}\mathbf{D}_0 = 0, \tag{43.3}$$

of the stretching $\mathbf{D}$ into a deviatoric part $\mathbf{D}_0$ and a spherical part $\frac{1}{3}(\operatorname{tr}\mathbf{D})\mathbf{1}$.

(iv) the relation

$$\mathbf{W} = \tfrac{1}{2}\boldsymbol{\omega}\times \tag{43.4}$$

between the spin and the vorticity $\boldsymbol{\omega} = \operatorname{curl}\mathbf{v}$.

43.2 Basic Laws

Fluids do not possess natural reference configurations, and for that reason it is most convenient to work with the basic laws — balance of mass, balance of momentum, and the free-energy imbalance — in *spatial form*. We now review these laws and summarize relevant frame-change relations.[195]

194 Cf. (2.48), (9.8), (11.2), (11.3), and (17.2).

195 Cf. (18.5), (18.7), (19.34), (19.31), (20.16), and (29.7).

In spatial form, the mechanical laws **balance of mass**, **balance of linear momentum**, and **balance of moments** are[196]

$$\left.\begin{aligned}\dot{\rho} + \rho\,\mathrm{div}\mathbf{v} &= 0,\\ \rho\dot{\mathbf{v}} &= \mathrm{div}\mathbf{T} + \mathbf{b}_0,\\ \mathbf{T} &= \mathbf{T}^{\top}.\end{aligned}\right\}\tag{43.5}$$

As the theory under consideration is mechanical,[197] we take as our basic thermodynamical law the **free-energy imbalance**

$$\rho\dot{\psi} - \mathbf{T}{:}\mathbf{D} = -\delta \le 0. \tag{43.6}$$

43.3 Transformation Rules and Objective Rates

The tensor fields most relevant to our discussion — namely **L**, **D**, **W**, and **T** — transform as follows under a *change of frame*:

$$\begin{aligned}\mathbf{L}^* &= \mathbf{Q}\mathbf{L}\mathbf{Q}^{\top} + \dot{\mathbf{Q}}\mathbf{Q}^{\top},\\ \mathbf{D}^* &= \mathbf{Q}\mathbf{D}\mathbf{Q}^{\top},\\ \mathbf{W}^* &= \mathbf{Q}\mathbf{W}\mathbf{Q}^{\top} + \dot{\mathbf{Q}}\mathbf{Q}^{\top},\\ \mathbf{T}^* &= \mathbf{Q}\mathbf{T}\mathbf{Q}^{\top}.\end{aligned}\tag{43.7}$$

The other relevant fields, ψ, p, and ρ, being scalars, are invariant under a change of frame.

Given a spatial vector field **g**, we recall from (20.47) that its corotational, covariant, and contravariant rates are respectively defined by

$$\left.\begin{aligned}\overset{\circ}{\mathbf{g}} &= \dot{\mathbf{g}} - \mathbf{W}\mathbf{g},\\ \overset{\vartriangle}{\mathbf{g}} &= \dot{\mathbf{g}} + \mathbf{L}^{\top}\mathbf{g},\\ \overset{\diamond}{\mathbf{g}} &= \dot{\mathbf{g}} - \mathbf{L}\mathbf{g}.\end{aligned}\right\}\tag{43.8}$$

Referring to (13.4), we note that **g** convects like a tangent if and only if $\dot{\mathbf{g}} = \mathbf{L}\mathbf{g}$. Similarly, referring to (13.6), we note that **g** convects like a normal if and only if $\dot{\mathbf{g}} = -\mathbf{L}^{\top}\mathbf{g}$. Thus, by $(43.8)_{2,3}$

$$\begin{aligned}&\overset{\vartriangle}{\mathbf{g}} = \mathbf{0} \qquad \text{if and only if } \mathbf{g} \text{ convects like a normal,}\\ &\overset{\diamond}{\mathbf{g}} = \mathbf{0} \qquad \text{if and only if } \mathbf{g} \text{ convects like a tangent.}\end{aligned}\tag{43.9}$$

---

[196] Cf. (18.5), (19.34), and (19.31).

[197] The role of a free-energy imbalance in a mechanical theory is discussed in §29. The local imbalance (43.6) is derived in the steps leading to (29.7).

EXERCISE

1. Show that the corotational, covariant, and contravariant rates of the vorticity $\omega = \text{curl}\,\mathbf{v}$ are given by

$$\left.\begin{aligned} \overset{\circ}{\omega} &= \dot{\omega}, \\ \overset{\vartriangle}{\omega} &= \dot{\omega} + \mathbf{D}\omega, \\ \overset{\diamond}{\omega} &= \dot{\omega} - \mathbf{D}\omega, \end{aligned}\right\} \qquad (43.10)$$

and provide a geometrical interpretation of $(43.10)_1$.

44 Elastic Fluids

A simple but important model for the isothermal behavior of a gas stems from the observation that a gas exhibits elasticity in its response to compression. On the basis of this observation, the model starts from a constitutive assumption asserting that the stress be a pressure that depends upon density.[198] In discussing the constitutive assumptions underlying that model, we find it most convenient to start with the Cauchy stress **T** rather than with a pressure.

44.1 Constitutive Theory

An **elastic fluid** is a homogeneous body governed by constitutive equations giving the specific free-energy and stress when the density is known:

$$\begin{aligned} \psi &= \hat{\psi}(\rho), \\ \mathbf{T} &= \hat{\mathbf{T}}(\rho). \end{aligned} \tag{44.1}$$

The specific free-energy ψ is measured per unit mass, while the density ρ and the Cauchy stress **T** are measured per unit volume and per unit area in the deformed configuration; thus,[199]

(‡) *the constitutive equations* (44.1) *of an elastic fluid are independent of the choice of reference configuration.*

Constitutive theories for fluids typically share property (‡). As an illustration of this property: if you pour water out of a pitcher into a glass, it shows no tendency to return to the pitcher.

44.2 Consequences of Frame-Indifference

Frame-indifference severely restricts the equation $(44.1)_2$ for the stress, which is the sole field in the constitutive relations (44.1) not invariant under a change of frame.

[198] The simplest model of a fluid is the *perfect fluid*, which is both incompressible and inviscid. Despite its rudimentary nature, this model provides a remarkable amount of insight regarding fluid flow. Within our framework, a perfect fluid is most naturally viewed as the inviscid limit of an incompressible, viscous fluid. We therefore postpone treatment of perfect fluids until §46.

[199] This stands in contrast to what occurs for an elastic *solid*; cf. (48.1)

In fact, the stress, being frame-indifferent, transforms according to $(43.7)_3$:

$$\mathbf{T}^* = \mathbf{Q}\mathbf{T}\mathbf{Q}^\top.$$

Thus, since ρ is invariant, the stress in the new frame must satisfy

$$\mathbf{T}^* = \hat{\mathbf{T}}(\rho),$$

so that, to be frame-indifferent, the constitutive equation for the stress must satisfy

$$\hat{\mathbf{T}}(\rho) = \mathbf{Q}\hat{\mathbf{T}}(\rho)\mathbf{Q}^\top$$

for all rotations $\mathbf{Q}$ and any value of ρ. Thus, for each fixed ρ, $\hat{\mathbf{T}}(\rho)$ must be a spherical tensor,[200] and hence the stress reduces to a pressure

$$\boxed{\mathbf{T} = -p\mathbf{1}, \quad p = \hat{p}(\rho).} \tag{44.2}$$

Trivially, by (44.2),

$$\mathbf{T} = \mathbf{T}^\top \quad \text{and} \quad \mathbf{T}:\mathbf{D} = -p\,\mathrm{tr}\,\mathbf{D}. \tag{44.3}$$

By (44.2), the linear momentum balance $(43.5)_2$ simplifies to

$$\rho\dot{\mathbf{v}} = -\mathrm{grad}\,p + \mathbf{b}_0. \tag{44.4}$$

Further, by $(44.3)_1$, the moment balance $(43.5)_3$ is satisfied and, bearing in mind the mass balance $(43.5)_1$ and that ρ cannot be negative, the free-energy imbalance (43.6) becomes

$$\rho\dot{\psi} - \frac{p\dot{\rho}}{\rho} = -\delta \le 0. \tag{44.5}$$

44.3 Consequences of Thermodynamics

With a view toward applying the Coleman–Noll procedure, consider a **constitutive process** for the fluid, which here consists of

(C1) a motion χ relative to some time t_0;
(C2) a field $\rho(\mathbf{x}, t)$ for the mass density defined for all $\mathbf{x}$ in $\mathcal{B}_t$ and all time t and consistent with balance of mass $(43.5)_1$;
(C3) fields $\psi(\mathbf{x}, t)$ and $p(\mathbf{x}, t)$ for the specific free-energy and pressure defined by the constitutive relations $(44.1)_1$ and (44.2).

Note that in the definition of a constitutive process we do not require that the density at t_0 be prescribed; such an initial condition, while necessary for the prescription of a well-posed initial-value problem for the fluid, is not intrinsic to the fluid itself.

Given an arbitrary **constitutive process**, the local force balance (44.4) gives the conventional body force

$$\mathbf{b}_0 = \mathrm{grad}\,p + \rho\dot{\mathbf{v}} \tag{44.6}$$

needed to support the process. *We presume that the body force is arbitrarily assignable*; thus, by (44.3), the momentum balance law in no way restricts the class of processes the material may undergo. Further, balance of mass is automatically satisfied by virtue of (C2). On the other hand, unless the constitutive equations are suitably restricted, not all constitutive processes will be compatible with the laws of thermodynamics as embodied in the free-energy imbalance in the reduced form (44.5). We require that *all constitutive processes be consistent with the free-energy imbalance*, here in the form (44.5).

[200] Cf., e.g., Gurtin (1981) — the Corollary on p. 13.

Thus, choose an arbitrary constitutive process. Then, by $(44.1)_1$,

$$\dot{\psi} = \frac{d\hat{\psi}(\rho)}{d\rho}\dot{\rho};$$

therefore, appealing to (44.2), we see that satisfaction of (44.5) is equivalent to the requirement that all constitutive processes be consistent with the inequality

$$\left(-\rho^2 \frac{d\hat{\psi}(\rho)}{d\rho} + \hat{p}(\rho) \right)\dot{\rho} \le 0. \tag{44.7}$$

Given any spatial point and any time, it is possible to find a constitutive process such that

$$\rho \text{ and } \dot{\rho} \text{ have arbitrarily prescribed values at that point and time.} \tag{44.8}$$

Postponing, for the moment, the verification of (44.8), we note that, granted (44.8), the coefficient of $\dot{\rho}$ must vanish, for otherwise $\dot{\rho}$ may be chosen to violate the inequality (44.7). We therefore have the following **thermodynamic restrictions**:

(i) *The specific free-energy determines the pressure through the* **pressure relation**

$$\boxed{\hat{p}(\rho) = \rho^2 \frac{d\hat{\psi}(\rho)}{d\rho}.} \tag{44.9}$$

(ii) *The dissipation* (44.5) *vanishes in smooth motions*[201]

$$\delta = 0. \tag{44.10}$$

Granted (i), the result (ii) is a direct consequence of (44.5). The verification of (i) is not so simple, as it requires establishing (44.8). In this regard, the result (17.43) represents a motion χ on all space for all time, a motion that has *constant* velocity gradient $\mathbf{L}$ and that uses the configuration at time t_0 as reference. This solution is valid for any choice of the tensor $\mathbf{L}$. In view of $(43.5)_1$, the density ρ associated with this motion satisfies

$$\dot{\rho} + \theta\rho = 0, \qquad \theta = \operatorname{tr}\mathbf{L},$$

and — restricting attention to spatially constant ρ — must have the specific form

$$\rho(t) = \rho_0 e^{-\theta(t-t_0)}.$$

Since the constant tensor $\mathbf{L}$ may be arbitrarily chosen, we may consider θ to be an arbitrary scalar constant; thus, since the scalar ρ_0 may also be arbitrarily chosen, we have a motion with the property that, at time t_0,

$$\rho(t_0) = \rho_0, \qquad \dot{\rho}(t_0) = -\rho_0\theta;$$

hence $\rho(t)$ and $\dot{\rho}(t)$ may be arbitrarily specified at $t = t_0$. This completes the verification of the result (i).

Note that, as a consequence of the pressure relation (44.9), the constitutive equation for the specific free-energy may be determined by quadrature.

44.4 Evolution Equations

Consider the flow of an elastic fluid in an inertial frame. The basic equations consist of the balances $(43.5)_1$ and $(43.5)_2$ for mass and linear momentum supplemented by

201 Shock waves in elastic fluids generally dissipate energy, a phenomenon that does not contradict (ii), since such waves render the resulting flow nonsmooth. Shock waves are discussed in §32.

the constitutive relation $(44.2)_2$ for the pressure

$$\begin{aligned}\dot{\rho} + \rho\,\mathrm{div}\,\mathbf{v} &= 0,\\ \rho\dot{\mathbf{v}} = -\mathrm{grad}\,p + \mathbf{b}_0, \qquad p &= \hat{p}(\rho).\end{aligned} \tag{44.11}$$

Using (43.1), these equations have the alternative form

$$\begin{aligned}\rho' + \mathbf{v}\cdot\mathrm{grad}\,\rho + \rho\,\mathrm{div}\,\mathbf{v} &= 0,\\ \rho\mathbf{v}' + \rho(\mathrm{grad}\,\mathbf{v})\mathbf{v} = -\mathrm{grad}\,p + \mathbf{b}_0, \qquad p &= \hat{p}(\rho).\end{aligned} \tag{44.12}$$

The quantity

$$\alpha(\rho) = \sqrt{\frac{d\hat{p}(\rho)}{d\rho}} \tag{44.13}$$

is called the **wave speed**.[202] If

$$\alpha(\rho) > 0 \tag{44.14}$$

for all $\rho > 0$, then (44.11) (or (44.12)) constitutes a nonlinear hyperbolic system for ρ and $\mathbf{v}$ and is generally difficult to solve, as solutions may not be smooth.[203]

To derive some basic properties of solutions, note that, by (44.9),

$$\begin{aligned}\frac{\mathrm{grad}\,p}{\rho} &= \mathrm{grad}\left(\frac{p}{\rho}\right) - p\,\mathrm{grad}\,\frac{1}{\rho}\\ &= \mathrm{grad}\left(\frac{p}{\rho}\right) + \frac{d\hat{\psi}(\rho)}{d\rho}\,\mathrm{grad}\,\rho\\ &= \mathrm{grad}\left(\psi + \frac{p}{\rho}\right).\end{aligned} \tag{44.15}$$

Thus, assuming that the body force is **conservative** with potential β in the sense that

$$\mathbf{b}_0 = \rho\,\mathrm{grad}\,\beta \tag{44.16}$$

with β a spatial scalar field, we may rewrite $(44.11)_2$ in the form

$$\dot{\mathbf{v}} = -\mathrm{grad}\left(\psi + \frac{p}{\rho} - \beta\right). \tag{44.17}$$

The field $\psi + p/\rho$ is referred to as the **specific enthalpy**. A consequence of (44.17) is that

$$\mathrm{curl}\,\dot{\mathbf{v}} = \mathbf{0}$$

and we have the important result: In the motion of an elastic fluid under a conservative conventional body force, the acceleration is the gradient of a potential. Thus, granted this, we may conclude from the results of Lagrange and Cauchy on page 118 and Kelvin on page 120, each labeled by a (‡), that if the acceleration is the gradient of a potential, then: (a) *the motion preserves circulation*, and (b) *a flow that is once irrotational is always irrotational.*

Consider next a **steady flow** as defined by the requirement that both the motion and the density be steady; thus in a steady flow

$$\mathbf{v}' = \mathbf{0}, \qquad \rho' = 0,$$

and $\mathcal{B}_t$ is independent of t, so that the flow region, $\mathcal{B}_0$ say, is fixed.[204]

---

[202] Cf., e.g, GURTIN (1981, pp. 131–133).

[203] Cf. Footnote 201.

[204] Cf. §17.6.

For a steady flow, (17.7) yields

$$\dot{\mathbf{v}} = \tfrac{1}{2}\,\mathrm{grad}\,(|\mathbf{v}|^2) + \boldsymbol{\omega} \times \mathbf{v},$$

and hence, by (44.17),

$$\tfrac{1}{2}\,\mathrm{grad}\,(|\mathbf{v}|^2) + \boldsymbol{\omega} \times \mathbf{v} = -\mathrm{grad}\left(\psi + \frac{p}{\rho} - \beta\right). \tag{44.18}$$

Thus, for

$$\varphi \stackrel{\mathrm{def}}{=} \psi + \frac{p}{\rho} + \tfrac{1}{2}|\mathbf{v}|^2 - \beta,$$

it follows that

$$\mathrm{grad}\,\varphi = \mathbf{0} \qquad \text{when} \qquad \boldsymbol{\omega} = \mathbf{0}. \tag{44.19}$$

Next, bear in mind that the field ρ is steady, so that, since $\psi = \hat{\psi}(\rho)$ and $p = \hat{p}(\rho)$, p and ψ are also steady. Hence, φ is steady. Thus, if we take the inner product of (44.18) with $\mathbf{v}$ and use (17.35), we arrive at the conclusion

$$\dot{\varphi} = 0. \tag{44.20}$$

Therefore, φ is constant on streamlines. Further, as a consequence of (44.19) and (44.20), if the flow is irrotational, then φ is identically constant. These results are generally referred to as *Bernoulli's theorem.*

The results of this section may be summarized as follows:

PROPERTIES OF ELASTIC FLUIDS *Consider the flow of an elastic fluid under a conservative conventional body force with potential* β*.*

(i) *If the flow is once irrotational, then it is always irrotational.*
(i) *The flow preserves circulation.*
(ii) *If the flow is steady, then the field*

$$\psi + \frac{p}{\rho} + \tfrac{1}{2}|\mathbf{v}|^2 - \beta \tag{44.21}$$

is constant on streamlines. If, in addition, the flow is irrotational, then (44.21) *is constant everywhere and for all time.*

EXERCISES

1. Show that, for an elastic fluid, the specific vorticity $\boldsymbol{\zeta} = \boldsymbol{\omega}/\rho$ is transported according to $\dot{\boldsymbol{\zeta}} = \mathbf{L}\boldsymbol{\zeta} = \mathbf{D}\boldsymbol{\zeta}$ or, equivalently,

$$\overset{\diamond}{\boldsymbol{\zeta}} = \mathbf{0}$$

and thus, by $(43.9)_2$, convects like a tangent.

2. Consider an elastic fluid subject to a conservative conventional body force. Assuming that the flow is irrotational at some time, show that the covariant rate of $\mathbf{W}$ obeys

$$\overset{\vartriangle}{\mathbf{W}} = \dot{\mathbf{W}} + \mathbf{W}\mathbf{L} + \mathbf{L}^{\top}\mathbf{W} = \mathbf{0}.$$

Further, using (20.37), show the material time-derivative of the covariant components of $\mathbf{W}$ relative to any basis $\{\mathbf{f}_i\}$ convecting with the body vanish:

$$\dot{W}_{ij} = \overline{\mathbf{f}_i \cdot \mathbf{W}\mathbf{f}_j}^{\boldsymbol{\cdot}} \equiv 0 \tag{44.22}$$

and thus conclude that the components of $\mathbf{W}$ relative to a basis embedded in the material do not change with time.

3. Consider a flow of an elastic fluid that is close to a given rest state with density $\bar{\rho}$; that is, introducing characteristic length and time scales L and T, assume that

$$\frac{|\rho - \bar{\rho}|}{\bar{\rho}} = \mathrm{O}(h), \quad \frac{L|\operatorname{grad}\rho|}{\bar{\rho}} = \mathrm{O}(h), \quad \frac{T|\mathbf{v}|}{L} = \mathrm{O}(h), \quad \text{and} \quad T|\operatorname{grad}\mathbf{v}| = \mathrm{O}(h),$$

with $h \ll 1$. Assume further that (44.14) is satisfied and define the wave speed $c^2 = \alpha(\bar{\rho})$. Show that when terms of $\mathrm{o}(h)$ are neglected, the system (44.12) reduces to the classical wave equation

$$\rho'' = c^2 \Delta \rho,$$

with Δ the spatial Laplacian.

45 Compressible, Viscous Fluids

The elastic fluid just discussed is inviscid. In contrast to such an idealized model, fluids generally exhibit internal friction that retards the relative motion of fluid particles. A local measure of this relative motion is the velocity gradient $\mathbf{L} = \mathrm{grad}\,\mathbf{v}$, and we now consider constitutive relations in which the $\mathbf{L}$ joins the density ρ as an independent constitutive variable.

45.1 General Constitutive Equations

By a **compressible viscous fluid**[205] we mean a homogeneous body governed by constitutive equations of the form

$$\begin{aligned}\psi &= \hat{\psi}(\rho, \mathbf{L}),\\ \mathbf{T} &= \hat{\mathbf{T}}(\rho, \mathbf{L}).\end{aligned} \tag{45.1}$$

The response functions $\hat{\psi}$ and $\hat{\mathbf{T}}$ are defined on all pairs $(\rho, \mathbf{L})$ with $\rho > 0$ and $\mathbf{L}$ a tensor, and we require that $\hat{\mathbf{T}} = \hat{\mathbf{T}}^{\scriptscriptstyle\top}$ to ensure that $\mathbf{T}$ be symmetric in all constitutive processes.

Note that, as for an elastic fluid, the constitutive relations (45.1) are independent of the choice of reference configuration.

While it might seem unnatural to include the variable $\mathbf{L}$ (or even the stretching $\mathbf{D}$) as an independent variable in the constitutive relation *for the specific free-energy*, we believe it more reasonable to allow for such a dependence at the outset rather than to rule it out by fiat. In fact, an energetic dependence on $\mathbf{D}$ was shown by Dunn & Fosdick (1974) to arise in certain types of Rivlin–Ericksen fluids and is intrinsic to the Navier–Stokes-α model for turbulence developed by Foais, Holm, & Titi (2001).

The constitutive equations (45.1) are therefore consistent with the *principle of equipresence* as discussed by Truesdell & Toupin (1960, §293) and Truesdell & Noll (1965, §96).[206] This principle asserts that "a quantity present as an independent variable in one constitutive equation should be so present in all, unless ... its presence contradicts some law of physics or rule of invariance." According to Truesdell and Noll, "This principle forbids us to eliminate any of the 'causes' present from interacting with any other as regards a particular 'effect.' It reflects on the scale of gross phenomena the fact that all observed effects result from a common

[205] The term "compressible simple fluid without memory" might be more appropriate. The term "compressible fluid" as used here connotes only standard inviscid and viscous fluids.

[206] Cf. the paragraph in petite type on page 230.

structure such as the motions of molecules." Physics, not caprice, should rule out interactions.

45.2 Consequences of Frame-Indifference

The fields ψ and ρ, being scalars, are invariant under a change of frame, while the tensor fields $\mathbf{T}$ and $\mathbf{L}$ transform according to $(43.7)_{1,4}$; thus, by (36.1) and since $\psi^* = \psi$ and $\rho^* = \rho$, the constitutive relations (45.1) must transform according to

$$\begin{aligned} \psi &= \hat{\psi}(\rho, \mathbf{QLQ}^\top + \dot{\mathbf{Q}}\mathbf{Q}^\top), \\ \underbrace{\mathbf{QTQ}^\top}_{\mathbf{T}^*} &= \hat{\mathbf{T}}(\rho, \underbrace{\mathbf{QLQ}^\top + \dot{\mathbf{Q}}\mathbf{Q}^\top}_{\mathbf{L}^*}). \end{aligned} \tag{45.2}$$

Frame-indifference than requires that the response functions $\hat{\psi}$ and $\hat{\mathbf{T}}$ satisfy

$$\begin{aligned} \hat{\psi}(\rho, \mathbf{L}) &= \hat{\psi}(\rho, \mathbf{QLQ}^\top + \dot{\mathbf{Q}}\mathbf{Q}^\top), \\ \mathbf{Q}\hat{\mathbf{T}}(\rho, \mathbf{L})\mathbf{Q}^\top &= \hat{\mathbf{T}}(\rho, \mathbf{QLQ}^\top + \dot{\mathbf{Q}}\mathbf{Q}^\top), \end{aligned} \tag{45.3}$$

at each time under any change of frame.

Choose an *arbitrary* skew tensor $\boldsymbol{\Omega}_0$ and let $\mathbf{Q}(t)$ be the unique solution of the initial-value problem

$$\begin{aligned} \dot{\mathbf{Q}}(t) &= \boldsymbol{\Omega}_0\mathbf{Q}(t), \\ \mathbf{Q}(0) &= \mathbf{1}. \end{aligned}$$

Then, as discussed in §114, $\mathbf{Q}(t)$ is a rotation at each t, and since (45.3) must hold for all such $\mathbf{Q}(t)$, if we apply this choice of $\mathbf{Q}(t)$ at $t = 0$, we find that

$$\begin{aligned} \hat{\psi}(\rho, \mathbf{L}) &= \hat{\psi}(\rho, \mathbf{L} + \boldsymbol{\Omega}_0), \\ \hat{\mathbf{T}}(\rho, \mathbf{L}) &= \hat{\mathbf{T}}(\rho, \mathbf{L} + \boldsymbol{\Omega}_0), \end{aligned}$$

or equivalently, since $\mathbf{L} = \mathbf{D} + \mathbf{W}$, with $\mathbf{D}$ the stretching and $\mathbf{W}$ the spin,

$$\begin{aligned} \hat{\psi}(\rho, \mathbf{L}) &= \hat{\psi}(\rho, \mathbf{D} + \mathbf{W} + \boldsymbol{\Omega}_0), \\ \hat{\mathbf{T}}(\rho, \mathbf{L}) &= \hat{\mathbf{T}}(\rho, \mathbf{D} + \mathbf{W} + \boldsymbol{\Omega}_0). \end{aligned} \tag{45.4}$$

These relations must hold for all skew tensors $\boldsymbol{\Omega}_0$ and all $(\rho, \mathbf{L})$ in the domain of the response functions. Fix $(\rho, \mathbf{L})$ (and hence $\mathbf{D}$ and $\mathbf{W}$); the choice

$$\boldsymbol{\Omega}_0 = -\mathbf{W}$$

in (45.4) then yields

$$\begin{aligned} \hat{\psi}(\rho, \mathbf{L}) &= \hat{\psi}(\rho, \mathbf{D}), \\ \hat{\mathbf{T}}(\rho, \mathbf{L}) &= \hat{\mathbf{T}}(\rho, \mathbf{D}). \end{aligned} \tag{45.5}$$

The constitutive relations for a compressible viscous fluid thus cannot include the spin as an independent constitutive variable.

Finally, taking $\mathbf{Q}$ in (45.3) to be constant yields the additional restriction

$$\begin{aligned} \hat{\psi}(\rho, \mathbf{D}) &= \hat{\psi}(\rho, \mathbf{QDQ}^\top), \\ \mathbf{Q}\hat{\mathbf{T}}(\rho, \mathbf{D})\mathbf{Q}^\top &= \hat{\mathbf{T}}(\rho, \mathbf{QDQ}^\top). \end{aligned} \tag{45.6}$$

Summarizing, we have the following restrictions imposed by material frame-indifference:

(i) *the dependence of the specific free-energy and stress on the velocity gradient must be through the stretching* $\mathbf{D}$

$$\psi = \hat{\psi}(\rho, \mathbf{D}),$$
$$\mathbf{T} = \hat{\mathbf{T}}(\rho, \mathbf{D}); \tag{45.7}$$

(ii) *the response functions* $\hat{\psi}$ *and* $\hat{\mathbf{T}}$ *must be isotropic* (*that is* (45.6) *must hold for all rotations* $\mathbf{Q}$).

The function $\hat{\mathbf{T}}(\rho, \mathbf{0})$ represents the stress in the fluid in the absence of flow, while

$$\hat{\mathbf{T}}_{\text{vis}}(\rho, \mathbf{D}) = \hat{\mathbf{T}}(\rho, \mathbf{D}) - \hat{\mathbf{T}}(\rho, \mathbf{0}),$$

which might be termed the **viscous stress**, represents that part of the stress due to flow. By (45.6),

$$\mathbf{Q}\hat{\mathbf{T}}(\rho, \mathbf{0})\mathbf{Q}^{\top} = \hat{\mathbf{T}}(\rho, \mathbf{0}),$$

so that, for each ρ, $\hat{\mathbf{T}}(\rho, \mathbf{0})$ is an isotropic tensor. Thus $\hat{\mathbf{T}}(\rho, \mathbf{0})$ must have the specific form[207]

$$\hat{\mathbf{T}}(\rho, \mathbf{0}) = -\hat{p}_{\text{eq}}(\rho)\mathbf{1}$$

with $\hat{p}_{\text{eq}}(\rho)$ a scalar function. We refer to $\hat{p}_{\text{eq}}(\rho)$ as the **equilibrium pressure**. Summarizing, we have the decomposition

$$\hat{\mathbf{T}}(\rho, \mathbf{D}) = -\hat{p}_{\text{eq}}(\rho)\mathbf{1} + \hat{\mathbf{T}}_{\text{vis}}(\rho, \mathbf{D}), \tag{45.8}$$

with

$$\hat{\mathbf{T}}_{\text{vis}}(\rho, \mathbf{D}) \ \textit{an isotropic function}. \tag{45.9}$$

The total pressure p in the fluid is defined by

$$p = -\tfrac{1}{3}\text{tr}\,\mathbf{T}$$

and is given by

$$p = \hat{p}_{\text{eq}}(\rho) - \tfrac{1}{3}\text{tr}\,\hat{\mathbf{T}}_{\text{vis}}(\rho, \mathbf{D}). \tag{45.10}$$

The total pressure p in a compressible viscous fluid therefore includes both an equilibrium contribution $\hat{p}_{\text{eq}}(\rho)$ and a dynamical contribution $-\tfrac{1}{3}\text{tr}\,\hat{\mathbf{T}}_{\text{vis}}(\rho, \mathbf{D})$ generated by internal friction and, therefore, of a dissipative nature.

Not all workers agree with our use of frame-indifference as expressed in the requirement that

(FI) constitutive relations be invariant under *all* changes in frame.[208]

---

[207] Cf. Footnote 200.

[208] Cf. MÜLLER (1972) and EDELEN & MCLENNAN (1973), who noted that equations for the stress and heat flux, in a moderately rarefied gas, derived as *approximations* to the kinetic theory are — when expressed relative to a general frame — *frame-dependent* starting with the third approximation (i.e., the BURNETT (1932) approximation). In deciding whether or not we should enlarge the structure upon which our book is based, we agree with TRUESDELL (1976) that kinetic theory is not a continuum theory and that conclusions based on kinetic theory, and especially those involving major approximations, should not necessitate a complete reworking of the axiomatic foundation of continuum mechanics.

Specifically, MURDOCH (2006) argued that — granted the underlying frame is inertial — frame-indifference should require that constitutive relations be invariant, *not under all* changes in frame (21.1), but only *under Galilean changes in frame* as defined by the transformation law

$$\mathbf{x}^* = \mathbf{x}_0 + (t - t_0)\mathbf{v}_0 + \mathbf{Q}_0(\mathbf{x} - \mathbf{o})$$

with $\mathbf{v}_0$ a constant translational velocity and $\mathbf{Q}_0$ a *constant* rotation. Under *Galilean invariance* the constitutive relation $(45.1)_1$, written for convenience in the form

$$\mathbf{T} = \hat{\mathbf{T}}(\rho, \mathbf{L}) = \hat{\mathbf{T}}(\rho, \mathbf{D}, \mathbf{W}),$$

need only be isotropic in the sense that

$$\mathbf{Q}_0\hat{\mathbf{T}}(\rho, \mathbf{D}, \mathbf{W})\mathbf{Q}_0^{\top} = \hat{\mathbf{T}}(\rho, \mathbf{Q}_0\mathbf{D}\mathbf{Q}_0^{\top}, \mathbf{Q}_0\mathbf{W}\mathbf{Q}_0^{\top})$$

for all rotations $\mathbf{Q}_0$. Thus in contrast to the result $(45.7)_1$, which asserts that $\mathbf{T}$ *cannot* depend on the spin $\mathbf{W}$,

- *Galilean invariance allows for spin-dependent constitutive relations for the stress.*

While physically relevant theories that incorporate spin dependence often arise from approximation and averaging schemes,[209] no continuum-based theory of which we are aware allows for spin-dependence in the constitutive relation for the stress in a fluid.[210] And a similar assertion applies to solids.

Were we to replace the hypothesis (FI) by the requirement of Galilean invariance together with the stipulation that constitutive relations be independent of spin, then the theories we discuss would be unchanged; but such a replacement would becloud the far reaching implications of (FI), and it would grant to Galilean invariance an undeserved status.

We believe that the hypothesis (FI) is at the level of our tacit hypothesis that there exist an inertial frame; it leads, using only mathematics, to theories that allow us to "face, explain, and in varying amount control, our daily environment."[211] When comparing continuum mechanics to other physical theories it is essential to realize that *all inertial effects* in continuum mechanics emanate from the *single* inertial body-force term $-\rho\dot{\mathbf{v}}$; for that reason constitutive equations, which are *independent* of inertial effects, seem more naturely based on the requirement (FI) rather than Galilean invariance.

Our use of (FI) is not limited to obtaining restrictions on constitutive relations. We use it in §22: (i) to establish an important relation between a fundamental balance law for power and the basic balance laws for forces and moments, and (ii) to show that this power balance underlies a discussion of the principle of virtual power, a principle that serves as a precursor to a variational statement of the local force balance.

45.3 Consequences of Thermodynamics

Consider a **constitutive process** for the fluid, which here consists of a motion χ relative to some time t_0, a density ρ consistent with balance of mass $(43.5)_1$, and fields ψ and $\mathbf{T}$ for the specific free-energy and stress defined by the constitutive relations (43.6). (Given the motion χ, $\mathbf{F} = \nabla\chi$ and we may compute the velocity gradient by $\mathbf{L} = \dot{\mathbf{F}}\mathbf{F}^{-1}$; the field $\mathbf{D}$ is then the symmetric part of $\mathbf{L}$.) Arguing as we did for an elastic fluid in the paragraph containing (44.6), we conclude that, unless the constitutive equations are suitably restricted, not all constitutive processes will be compatible with the free-energy imbalance in the form (43.6); we therefore require that all constitutive processes be consistent with (43.6). By the chain-rule and $(43.6)_1$, given any constitutive process,

$$\dot{\psi} = \frac{\partial\hat{\psi}(\rho, \mathbf{D})}{\partial\rho}\dot{\rho} + \frac{\partial\hat{\psi}(\rho, \mathbf{D})}{\partial\mathbf{D}} : \dot{\mathbf{D}},$$

209 Aside from expressions for the stress and heat flux that arise from approximations applied to the kinetic theory of gases, second-order closure models for stresses and heat fluxes arising in Reynolds-averaged turbulence models also display spin dependence; cf., e.g., LUMLEY (1970, 1983), SPEZIALE (1998), WEISS & HUTTER (2003), and GATSKI & WALLIN (2004).

210 Of course, constitutive relations involving spin via a frame-indifferent rate such as the corotational rate (20.24) are allowed.

211 The passage in quotes is taken from TRUESDELL & NOLL (1965, p. 3).

and, hence, this process satisfies the free-energy imbalance (43.6) if and only if

$$\rho\left(\frac{\partial\hat{\psi}(\rho,\mathbf{D})}{\partial\rho}\dot{\rho}+\frac{\partial\hat{\psi}(\rho,\mathbf{D})}{\partial\mathbf{D}}:\dot{\mathbf{D}}\right)-\hat{\mathbf{T}}(\rho,\mathbf{D}):\mathbf{D}\leq 0. \tag{45.11}$$

Next, by (45.8),

$$\begin{aligned}\hat{\mathbf{T}}(\rho,\mathbf{D}):\mathbf{D}&=-\hat{p}_{\mathrm{eq}}(\rho)\mathbf{1}:\mathbf{D}+\hat{\mathbf{T}}_{\mathrm{vis}}(\rho,\mathbf{D}):\mathbf{D},\\ &=-\hat{p}_{\mathrm{eq}}(\rho)\,\mathrm{tr}\,\mathbf{D}+\hat{\mathbf{T}}_{\mathrm{vis}}(\rho,\mathbf{D}):\mathbf{D},\end{aligned} \tag{45.12}$$

so that, $(43.5)_1$, (45.11) becomes

$$\rho\frac{\partial\hat{\psi}(\rho,\mathbf{D})}{\partial\mathbf{D}}:\dot{\mathbf{D}}+\left(-\rho^2\frac{\partial\hat{\psi}(\rho,\mathbf{D})}{\partial\rho}+\hat{p}_{\mathrm{eq}}(\rho)\right)\mathrm{tr}\,\mathbf{D}-\hat{\mathbf{T}}_{\mathrm{vis}}(\rho,\mathbf{D}):\mathbf{D}\leq 0. \tag{45.13}$$

This inequality is to hold for all motions and all density fields consistent with the mass balance $(43.5)_1$.

Given any point of the body and any time, it is possible to find a motion such that, at that point and time

$$\rho,\ \mathbf{D},\ \text{and}\ \dot{\mathbf{D}}\ \text{have arbitrarily prescribed values.} \tag{45.14}$$

Granted assertion (45.14) — whose proof we postpone — it is clear that, since $\dot{\mathbf{D}}$ appears linearly in (45.13), we can choose its value to violate the inequality unless its coefficient vanishes. Thus $\partial\hat{\psi}/\partial\mathbf{D}\equiv\mathbf{0}$ and the specific free-energy is independent of $\mathbf{D}$; the inequality (45.13) therefore takes the form

$$\left(-\rho^2\frac{d\hat{\psi}(\rho)}{d\rho}+\hat{p}_{\mathrm{eq}}(\rho)\right)\mathrm{tr}\,\mathbf{D}-\hat{\mathbf{T}}_{\mathrm{vis}}(\rho,\mathbf{D}):\mathbf{D}\leq 0. \tag{45.15}$$

Fix ρ. Since (45.15) must hold for all symmetric tensors $\mathbf{D}$, we may then, without loss in generality, replace $\mathbf{D}$ by $a\mathbf{D}$ with $a>0$. Dividing by a, we are thus left with the inequality

$$\left(-\rho^2\frac{d\hat{\psi}(\rho)}{d\rho}+\hat{p}_{\mathrm{eq}}(\rho)\right)\mathrm{tr}\,\mathbf{D}-\hat{\mathbf{T}}_{\mathrm{vis}}(\rho,a\mathbf{D}):\mathbf{D}\leq 0.$$

Since $\hat{\mathbf{T}}_{\mathrm{vis}}(\rho,\mathbf{0})=\mathbf{0}$, if we let $a\to 0$, we find that

$$\left(-\rho^2\frac{d\hat{\psi}(\rho)}{d\rho}+\hat{p}_{\mathrm{eq}}(\rho)\right)\mathrm{tr}\,\mathbf{D}\leq 0,$$

and since this must hold for all $\mathbf{D}$, we find that $\hat{p}_{\mathrm{eq}}(\rho)=\rho^2 d\hat{\psi}(\rho)/d\rho$. Moreover, it is clear from $(29.3)_2$ that the left side of (45.15) is the negative of the dissipation δ. We therefore have the following **thermodynamic restrictions**:

(i) *The specific free-energy depends only on the mass density and determines the equilibrium pressure through the* **pressure relation**

$$\boxed{\hat{p}_{\mathrm{eq}}(\rho)=\rho^2\frac{d\hat{\psi}(\rho)}{d\rho}.} \tag{45.16}$$

(ii) *The viscous stress satisfies the* **reduced dissipation inequality**

$$\boxed{\hat{\mathbf{T}}_{\mathrm{vis}}(\rho,\mathbf{D}):\mathbf{D}\geq 0} \tag{45.17}$$

for all $(\rho,\mathbf{D})$. *In fact,* $\hat{\mathbf{T}}_{\mathrm{vis}}(\rho,\mathbf{D}):\mathbf{D}$ *represents the dissipation* δ *in any motion.*

Thus, as is true for an elastic fluid, given the constitutive relation for the pressure, the specific free-energy may be determined by quadrature.

We now turn to the verification of (45.14). In this regard, the paragraph containing (17.42) asserts the existence of a motion $\boldsymbol{\chi}$ on all space for all time, a motion that uses the configuration at time t_0 (t_0 arbitrary) as reference and that has spatially constant velocity gradient $\mathbf{L}(t)$. This solution is valid for any choice of the tensor function $\mathbf{L}(t)$; hence we may specify $\mathbf{D}$ and $\dot{\mathbf{D}}$ as arbitrary symmetric tensors at, say t_0, and, since the only constraint on the density ρ is that it satisfy the mass balance $(43.5)_1$, we may, without loss in generality, specify its initial value $\rho(t_0)$ arbitrarily. Since the initial time t_0 is arbitrary, the verification of (45.14) is complete.

45.4 Compressible, Linearly Viscous Fluids

Many real fluids under a large class of operating conditions are **linearly viscous** in the sense that the viscous stress $\hat{\mathbf{T}}_{\rm vis}$ is linear in $\mathbf{D}$. Since $\hat{\mathbf{T}}_{\rm vis}$ is, by (45.9), an isotropic function, the Representation Theorem for Isotropic Linear Tensor Functions in Appendix 113.3 yields[212]

$$\hat{\mathbf{T}}_{\rm vis}(\rho,\mathbf{D}) = 2\mu(\rho)\mathbf{D} + \lambda(\rho)(\mathrm{tr}\,\mathbf{D})\mathbf{1}. \tag{45.18}$$

In view of (43.3), the deviatoric part $\mathbf{D}_0$ of the stretching $\mathbf{D} = \mathbf{D}_0 + \frac{1}{3}(\mathrm{tr}\,\mathbf{D})\mathbf{1}$ obeys

$$\mathrm{tr}\,\mathbf{D}_0 = 0 \qquad \text{and} \qquad \mathbf{D}\!:\!\mathbf{D}_0 = |\mathbf{D}_0|^2.$$

Thus, by (45.17), (45.18), and (45.20),

$$\begin{aligned}\hat{\mathbf{T}}_{\rm vis}(\rho,\mathbf{D})\!:\!\mathbf{D} &= (2\mu(\rho)\mathbf{D} + \lambda(\rho)(\mathrm{tr}\,\mathbf{D})\mathbf{1})\!:\!(\mathbf{D}_0 + \tfrac{1}{3}(\mathrm{tr}\,\mathbf{D})\mathbf{1})\\ &= 2\mu(\rho)|\mathbf{D}_0|^2 + (\lambda(\rho) + \tfrac{2}{3}\mu(\rho))(\mathrm{tr}\,\mathbf{D})^2\\ &= 2\mu(\rho)|\mathbf{D}_0|^2 + \kappa(\rho)(\mathrm{tr}\,\mathbf{D})^2\\ &\ge 0,\end{aligned} \tag{45.19}$$

where

$$\kappa = \lambda + \tfrac{2}{3}\mu. \tag{45.20}$$

The inequality (45.19) must hold for all symmetric $\mathbf{D}$. Choosing $\mathbf{D} = \mathbf{1}$ (so that $\mathrm{tr}\,\mathbf{D} = 3$ and $\mathbf{D}_0 = \mathbf{0}$) yields $2\mu + 3\lambda \ge 0$; choosing $\mathbf{D} = \mathbf{e}\otimes\mathbf{f} + \mathbf{f}\otimes\mathbf{e}$ with $\mathbf{e}$ and $\mathbf{f}$ orthonormal (so that $\mathrm{tr}\,\mathbf{D} = 0$ and $|\mathbf{D}_0|^2 = 2$) yields $\mu \ge 0$. Thus,

$$\mu \ge 0 \qquad \text{and} \qquad \kappa \ge 0. \tag{45.21}$$

Next, since

$$\mathrm{tr}\,\hat{\mathbf{T}}_{\rm vis}(\rho,\mathbf{D}) = 3\kappa(\rho)\,\mathrm{tr}\,\mathbf{D},$$

the constitutive equation (45.10) for the fluid pressure becomes

$$p = \hat{p}_{\rm eq}(\rho) - \kappa(\rho)\mathrm{tr}\,\mathbf{D}. \tag{45.22}$$

The constitutive equation for the stress in a compressible, linearly viscous fluid therefore has the form

$$\boxed{\mathbf{T} = -(\hat{p}_{\rm eq}(\rho) - \kappa(\rho)\mathrm{tr}\,\mathbf{D})\mathbf{1} + 2\mu(\rho)\mathbf{D}_0.} \tag{45.23}$$

212 Cf. (113.9).

The coefficient μ in (45.23), called the **shear viscosity**,[213] characterizes the resistance of the fluid to shear. The coefficient κ in (45.23), called the **dilatational viscosity**,[214] characterizes the nonequilibrium response of the fluid to volume changes. For gases that are rarefied enough so that intermolecular forces can be neglected it is common to invoke the STOKES (1845) relation $\kappa = 0$, that is,

$$\lambda = -\tfrac{2}{3}\mu. \tag{45.24}$$

TISZA (1942) shows that (45.24) holds for monatomic gases but that, for polyatomic gases and liquids, κ generally exceeds μ (i.e., $\lambda > \frac{1}{3}\mu$).[215]

45.5 Compressible Navier–Stokes Equations

In view of (45.23),

$$\operatorname{div}\mathbf{T} = -\operatorname{grad} p + 2\operatorname{div}(\mu(\rho)\mathbf{D}_0), \qquad p = \hat{p}_{\text{eq}}(\rho) - \kappa(\rho)\operatorname{tr}\mathbf{D},$$

and the balances $(43.5)_1$ and $(43.5)_2$ for mass and linear momentum yield the system

$$\begin{aligned} &\dot{\rho} + \rho\operatorname{div}\mathbf{v} = 0, \\ &\rho\dot{\mathbf{v}} = -\operatorname{grad} p + 2\operatorname{div}(\mu(\rho)\mathbf{D}_0) + \mathbf{b}_0, \qquad p = \hat{p}_{\text{eq}}(\rho) - \kappa(\rho)\operatorname{tr}\mathbf{D}. \end{aligned} \tag{45.25}$$

These equations are generally referred to as the **compressible Navier–Stokes equations.**

For certain gases and liquids, the shear viscosity μ shows strong dependence on the density.[216] However, it happens often that the differences in density are small enough to neglect dependence of the viscosities on the density. Then, since

$$\begin{aligned} 2\operatorname{div}\mathbf{D} &= \operatorname{div}(\operatorname{grad}\mathbf{v} + (\operatorname{grad}\mathbf{v})^{\top}) \\ &= \Delta\mathbf{v} + \operatorname{grad}\operatorname{div}\mathbf{v}, \end{aligned} \tag{45.26}$$

the flow equation $(45.25)_2$ reduces to

$$\rho\dot{\mathbf{v}} = -\operatorname{grad} p + \mu\,\Delta\mathbf{v} + (\lambda + \mu)\operatorname{grad}\operatorname{div}\mathbf{v} + \mathbf{b}_0, \qquad p = \hat{p}_{\text{eq}}(\rho), \tag{45.27}$$

where Δ denotes the spatial Laplacian. Even when the viscosities are constant, the combined effects of compressibility and viscosity render the analysis of the compressible Navier–Stokes equations extremely challenging.

45.6 Vorticity Transport Equation

Assume now that the shear and bulk viscosities are constant and that the body force is conservative with potential β. As with an elastic fluid,[217]

$$\frac{\operatorname{grad} p}{\rho} = \operatorname{grad}\left(\psi + \frac{p}{\rho}\right).$$

---

[213] μ is also referred to as the *dynamic viscosity*.

[214] κ is also referred to as the *bulk viscosity*.

[215] See also the review of KARIM & ROSENHEAD (1952) and the incisive appraisal of the status of the Stokes relation by TRUESDELL (1954).

[216] For gases, see CARR, KOBAYASHI & BURROWS (1954), whose experiments demonstrate the strong pressure-dependence of the shear viscosity of hydrocarbons; in view of the pressure relation (45.16), dependence of μ on the equilibrium pressure is tantamount to dependence on the density. For liquids, see VAN DER GULIK (1997), whose experiments demonstrate the strong density dependence of the shear viscosity of liquified carbon dioxide.

[217] Cf. (44.15).

Thus, (45.27) can be written as

$$\dot{\mathbf{v}} = -\mathrm{grad}\left(\psi + \frac{p}{\rho} - \beta - \frac{\lambda+\mu}{\rho}\mathrm{div}\mathbf{v}\right) + \frac{\mu\Delta\mathbf{v}}{\rho}.$$

Computing the curl of this relation and making use of $(3.25)_4$ now results in a nonzero source of vorticity,

$$\begin{aligned}\mathrm{curl}\,\dot{\mathbf{v}} &= \mu\,\mathrm{curl}\left(\frac{\Delta\mathbf{v}}{\rho}\right)\\ &= \mu\left(\frac{\Delta\boldsymbol{\omega}}{\rho} - \frac{1}{\rho^2}\mathrm{grad}\,\rho\times\Delta\mathbf{v}\right),\end{aligned}$$

and, by (17.9) and $(43.10)_3$, leads to the following equation for the transport of vorticity:

$$\overset{\diamond}{\boldsymbol{\omega}} + (\mathrm{div}\mathbf{v})\boldsymbol{\omega} = \mu\left(\frac{\Delta\boldsymbol{\omega}}{\rho} - \frac{1}{\rho^2}\mathrm{grad}\,\rho\times\Delta\mathbf{v}\right). \tag{45.28}$$

EXERCISE (PROJECT)

1. Consider a *compressible, heat conducting, viscous fluid* defined by the constitutive relations

$$\begin{aligned}\psi &= \hat{\psi}(\rho,\vartheta),\\ \eta &= \hat{\eta}(\rho,\vartheta),\\ \mathbf{T} &= \hat{\mathbf{T}}(\rho,\vartheta,\mathbf{L}),\\ \mathbf{q} &= \hat{\mathbf{q}}(\rho,\vartheta,\mathbf{g}).\end{aligned} \tag{45.29}$$

(a) Show as a consequence of frame-indifference that the constitutive relation for the stress must take the form

$$\mathbf{T} = \hat{\mathbf{T}}(\rho,\vartheta,\mathbf{D})$$

and that this relation and the relation for $\mathbf{q}$ are isotropic:

$$\mathbf{Q}\hat{\mathbf{T}}(\rho,\vartheta,\mathbf{D})\mathbf{Q}^{\top} = \hat{\mathbf{T}}(\rho,\vartheta,\mathbf{Q}\mathbf{D}\mathbf{Q}^{\top}), \qquad \mathbf{Q}\hat{\mathbf{q}}(\rho,\vartheta,\mathbf{g}) = \hat{\mathbf{q}}(\rho,\vartheta,\mathbf{Q}\mathbf{g}),$$

for all rotations $\mathbf{Q}$. Show that the constitutive relation for the stress has the decomposition

$$\hat{\mathbf{T}}(\rho,\vartheta,\mathbf{D}) = -\hat{p}_{\mathrm{eq}}(\rho,\vartheta)\mathbf{1} + \hat{\mathbf{T}}_{\mathrm{vis}}(\rho,\vartheta,\mathbf{D}), \qquad \text{with} \qquad \hat{\mathbf{T}}_{\mathrm{vis}}(\rho,\vartheta,\mathbf{0}) = \mathbf{0},$$

and discuss the physical meaning of the individual terms.

(b) Use the Coleman–Noll procedure in conjunction with the free-energy imbalance (27.18) to show that

$$\begin{aligned}\hat{p}_{\mathrm{eq}}(\rho,\vartheta) &= \rho^2\frac{\partial\hat{\psi}(\rho,\vartheta)}{\partial\rho}, \qquad &\hat{\eta}(\rho,\vartheta) &= -\frac{\partial\hat{\psi}(\rho,\vartheta)}{\partial\vartheta},\\ \hat{\mathbf{T}}_{\mathrm{vis}}(\rho,\vartheta,\mathbf{D})\colon\mathbf{D} &\ge 0, \qquad &\hat{\mathbf{q}}(\rho,\vartheta,\mathbf{g})\cdot\mathbf{g} &\le 0.\end{aligned} \tag{45.30}$$

Show that the total pressure defined by $p = -\frac{1}{3}\mathrm{tr}\,\mathbf{T}$ is given by

$$p = \hat{p}_{\mathrm{eq}}(\rho,\vartheta) - \tfrac{1}{3}\mathrm{tr}\,\hat{\mathbf{T}}_{\mathrm{vis}}(\rho,\vartheta,\mathbf{D})$$

and discuss the meaning of this relation.

(c) Define the equilibrium pressure p_{eq} by

$$p_{eq} = \hat{p}_{eq}(\rho, \vartheta)$$

and establish the Gibbs relations

$$\dot{\psi} = \frac{p_{eq}}{\rho^2}\dot{\rho} - \eta\dot{\vartheta}, \qquad \dot{\varepsilon} = \frac{p_{eq}}{\rho^2}\dot{\rho} + \vartheta\dot{\eta},$$

the Maxwell relation

$$\frac{\partial \hat{p}_{eq}(\rho, \vartheta)}{\partial \vartheta} = -\rho^2 \frac{\partial \hat{\eta}(\rho, \vartheta)}{\partial \rho},$$

and the identity

$$\hat{p}_{eq}(\rho, \vartheta) = \vartheta \frac{\partial \hat{p}_{eq}(\rho, \vartheta)}{\partial \vartheta} + \rho^2 \frac{\partial \hat{\varepsilon}(\rho, \vartheta)}{\partial \rho}.$$

Show that the energy balance (26.8) may be written as an entropy balance

$$\rho\vartheta\dot{\eta} = -\mathrm{div}\mathbf{q} + q + \mathbf{T}_{vis} : \mathbf{D}.$$

(d) The constitutive modulus defined by

$$c = \hat{c}(\rho, \vartheta) = \frac{\partial \hat{\varepsilon}(\rho, \vartheta)}{\partial \vartheta}$$

is called the specific heat at constant volume. Show that

$$\hat{c}(\rho, \vartheta) = \vartheta \frac{\partial \hat{\eta}(\rho, \vartheta)}{\partial \vartheta} = -\vartheta \frac{\partial^2 \hat{\psi}(\rho, \vartheta)}{\partial \vartheta^2},$$

$$\frac{\partial \hat{c}(\rho, \vartheta)}{\partial \rho} = -\frac{\vartheta}{\rho^2} \frac{\partial^2 \hat{p}_{eq}(\rho, \vartheta)}{\partial \vartheta^2},$$

and that the energy balance equation may be written in the form

$$\rho c\dot{\vartheta} = \mathbf{T}_{vis} : \mathbf{D} - \vartheta \frac{\partial \hat{p}_{eq}(\rho, \vartheta)}{\partial \vartheta} \mathrm{tr}\,\mathbf{D} - \mathrm{div}\mathbf{q} + q.$$

(e) Assume that the fluid is linearly viscous in the sense that the viscous stress $\hat{\mathbf{T}}_{vis}(\rho, \vartheta, \mathbf{D})$ is linear in $\mathbf{D}$. In this case show that

$$\hat{\mathbf{T}}_{vis}(\rho, \vartheta, \mathbf{D}) = 2\mu(\rho, \vartheta)\mathbf{D} + \lambda(\rho, \vartheta)(\mathrm{tr}\,\mathbf{D})\mathbf{1},$$

or equivalently

$$\hat{\mathbf{T}}_{vis}(\rho, \vartheta, \mathbf{D}) = 2\mu(\rho, \vartheta)\mathbf{D}_0 + \kappa(\rho, \vartheta)(\mathrm{tr}\,\mathbf{D})\mathbf{1},$$

with $\mu \geq 0$, $\kappa = \lambda + \frac{2}{3}\mu \geq 0$.

(f) Assume that the heat flux is linear in the temperature gradient

$$\hat{\mathbf{q}}(\rho, \vartheta, \mathbf{g}) = -k(\rho, \vartheta)\mathbf{g},$$

where $k(\rho, \vartheta) \geq 0$ is the thermal conductivity of the material. Show that the balances of mass, linear momentum, and energy yield the partial-differential equations

$$\dot{\rho} + \rho\mathrm{div}\mathbf{v} = 0,$$

$$\rho\dot{\mathbf{v}} = -\mathrm{grad}\, p_{eq} + \mathrm{div}(\mu\,\mathrm{grad}\,\mathbf{v}) + \mathrm{div}(\mu(\mathrm{grad}\,\mathbf{v})^{\top}) + \mathrm{grad}\,(\lambda\,\mathrm{div}\mathbf{v}) + \mathbf{b}_0,$$

$$\rho c\dot{\vartheta} = 2\mu|\mathbf{D}_0|^2 + \kappa(\mathrm{tr}\,\mathbf{D})^2 - \vartheta\frac{\partial p_{eq}}{\partial \vartheta}\mathrm{tr}\,\mathbf{D} + \mathrm{div}(k\mathrm{grad}\,\vartheta) + q.$$

46 Incompressible Fluids

Incompressibility is a constraint on the class of motions a body may undergo and its incorporation into the theory requires a modified framework, chiefly because the fluid pressure $p = -\frac{1}{3}\mathrm{tr}\,\mathbf{T}$ expends no power.

46.1 Free-Energy Imbalance for an Incompressible Body

An **incompressible body** is one for which only isochoric motions are possible, a requirement that manifests itself in the **constraint**[218]

$$\mathrm{div}\,\mathbf{v} = 0. \tag{46.1}$$

Balance of mass (18.5) therefore yields

$$\dot{\rho} = 0 \tag{46.2}$$

in all motions and the mass density ρ — when expressed referentially — is independent of time.[219] Here, we restrict attention to situations for which the density at some (and hence every) time is independent of the material point in question, so that

$$\rho \equiv \text{constant}. \tag{46.3}$$

This last assumption excludes from consideration applications such as oceanic and mantle convection where spatial variations of the density are important.

The essential change in the theory induced by the constraint (46.1) lies in the form of the stress power $\mathbf{T}:\mathbf{D}$ and, hence, in the form of the internal power[220]

$$\mathcal{I}(\mathrm{P}) = \int\limits_{\mathcal{P}_t} \mathbf{T}:\mathbf{D}\,dv.$$

Indeed, using (2.48) to decompose $\mathbf{T}$ into a deviatoric (i.e., traceless) part and a pressure p through[221]

$$\mathbf{T} = -p\mathbf{1} + \mathbf{T}_0, \tag{46.4}$$

[218] Cf. the results concerning isochoric motions in §16.

[219] Alternatively, the mass density is constant on particle paths.

[220] Cf. (19.42).

[221] For an incompressible, viscous fluid, the deviatoric part of the stress corresponds to what is often called the extra stress.

we find that

$$\mathbf{T}:\mathbf{D} = (\mathbf{T}_0 - p\mathbf{1}):\mathbf{D}$$

$$= \mathbf{T}_0:\mathbf{D} - p\,\mathrm{tr}\,\mathbf{D}.$$

But for an incompressible body, $\mathrm{tr}\,\mathbf{D} = \mathrm{div}\,\mathbf{v} = 0$, so that

$$\mathbf{T}:\mathbf{D} = \mathbf{T}_0:\mathbf{D}, \tag{46.5}$$

and the pressure does not affect the stress power. The internal power therefore reduces to

$$\mathcal{I}(\mathrm{P}) = \int\limits_{\mathcal{P}_t} \mathbf{T}_0:\mathbf{D}\,dv$$

and arguing as in the derivation of (29.7) we are led to the free-energy imbalance

$$\rho\dot{\psi} - \mathbf{T}_0:\mathbf{D} = -\delta \le 0. \tag{46.6}$$

46.2 Incompressible, Viscous Fluids

Constitutive relations describe the internal structure of materials. For an incompressible material the pressure does not expend power internally and consequently does not enter the energy imbalance. The pressure is therefore irrelevant to the internal thermodynamic structure of the theory and for that reason we consider it as **indeterminate** — that is *not specified constitutively*.[222]

An **incompressible fluid** is a homogeneous, incompressible body defined by constitutive equations of the form

$$\left.\begin{aligned}\psi &= \hat{\psi}(\mathbf{L}),\\ \mathbf{T}_0 &= \hat{\mathbf{T}}_0(\mathbf{L}),\end{aligned}\right\} \tag{46.7}$$

with response functions $\hat{\psi}$ and $\hat{\mathbf{T}}_0 = \hat{\mathbf{T}}_0^{\mathsf{T}}$ defined on the space of all *traceless* (i.e., deviatoric) tensors.[223]

The derivation of restrictions placed by material frame-indifference and thermodynamics will only be sketched, as it uses arguments that differ little from those used for compressible fluids. Frame-indifference requires that the constitutive equations (46.7) take the form

$$\left.\begin{aligned}\psi &= \hat{\psi}(\mathbf{D}),\\ \mathbf{T}_0 &= \hat{\mathbf{T}}_0(\mathbf{D}),\end{aligned}\right\} \tag{46.8}$$

with both response functions being *isotropic*.

Further, compatibility with thermodynamics as embodied in the free-energy imbalance (46.6) requires that

$$\rho\frac{\partial\hat{\psi}(\mathbf{D})}{\partial\mathbf{D}}:\dot{\mathbf{D}} - \hat{\mathbf{T}}_0(\mathbf{D}):\mathbf{D} \le 0$$

in all motions consistent with the constraint. This yields the following **thermodynamic restrictions**:

(i) *The specific free-energy is constant.*

[222] This is consistent with classical particle mechanics, where a force is indeterminate if it performs no work.

[223] Cf. §2.7.

(ii) *The viscous stress satisfies the* **reduced dissipation inequality**

$$\hat{\mathbf{T}}_0(\mathbf{D})\colon\mathbf{D} \geq 0 \tag{46.9}$$

for all traceless symmetric tensors **D**; *in fact, the dissipation in any motion is given by* $\delta = \hat{\mathbf{T}}_0(\mathbf{D})\colon\mathbf{D}$.

EXERCISES

1. For an incompressible fluid, use (2.146) to show that

$$\begin{aligned} I_2(\mathbf{D}) &= -\tfrac{1}{2}I_1(\mathbf{D}^2), \\ I_3(\mathbf{D}) &= \tfrac{1}{3}I_1(\mathbf{D}^3), \end{aligned} \tag{46.10}$$

and, hence, conclude from the Cayley–Hamilton equation (2.144) that

$$(\mathbf{D}^3)_0 = \tfrac{1}{2}I_1(\mathbf{D}^2)\mathbf{D}. \tag{46.11}$$

2. Verify the result (46.8). (To do this the transformation law for the deviatoric part of the stress under a change in frame is needed.)
3. Establish the thermodynamic restrictions (i) and (ii).

46.3 Incompressible, Linearly Viscous Fluids

If $\hat{\mathbf{T}}_0$ is linear in **D**, the Representation Theorem for Isotropic Linear Tensor Functions in Appendix 113.3 yields[224]

$$\hat{\mathbf{T}}_0(\mathbf{D}) = 2\mu\mathbf{D},$$

with μ, a constant, the **shear viscosity** which, by (46.9), satisfies

$$\mu \geq 0.$$

The constitutive equation for the stress in a linearly viscous incompressible fluid therefore has the form

$$\boxed{\mathbf{T} = -p\mathbf{1} + 2\mu\mathbf{D}.} \tag{46.12}$$

Thus, for such a fluid,

$$\begin{aligned} \delta &= \mathbf{T}\colon\mathbf{D} \\ &= -p\mathbf{1}\colon\mathbf{D} + 2\mu|\mathbf{D}|^2 \\ &= 2\mu|\mathbf{D}|^2, \end{aligned}$$

whereby the internal power is given explicitly by

$$\mathcal{I}(\mathrm{P}) = 2\mu\int_{\mathcal{P}_t}|\mathbf{D}|^2\,dv;$$

further, bearing in mind that the specific free energy ψ is constant, the power balance (19.45) for a convecting spatial region $\mathcal{P}_t$ specializes to

$$\int_{\partial\mathcal{P}_t}\mathbf{Tn}\cdot\mathbf{v}\,da + \int_{\mathcal{P}_t}\mathbf{b}\cdot\mathbf{v}\,dv = 2\mu\int_{\mathcal{P}_t}|\mathbf{D}|^2\,dv, \tag{46.13}$$

[224] Cf. (113.10).

which asserts that the rate of change the net power expended on $\mathcal{P}_t$ must equal the rate at which energy is dissipated within $\mathcal{P}_t$. Similarly, for an inertial frame, the inertial form (19.43) of the power balance for a convecting spatial region $\mathcal{P}_t$ specializes to the **kinetic energy balance**

$$\overline{\int\limits_{\mathcal{P}_t} \tfrac{1}{2}\rho|\mathbf{v}|^2\,dv}^{\,\boldsymbol{\cdot}} = \int\limits_{\partial\mathcal{P}_t} \mathbf{Tn}\cdot\mathbf{v}\,da + \int\limits_{\mathcal{P}_t} \mathbf{b}_0\cdot\mathbf{v}\,dv - 2\mu\int\limits_{\mathcal{P}_t} |\mathbf{D}|^2\,dv, \tag{46.14}$$

showing that, for a linearly viscous, incompressible fluid in an inertial frame, the rate of change of the kinetic energy within a convecting spatial region $\mathcal{P}_t$ must equal the net power (excluding inertial contributions) expended on $\mathcal{P}_t$ by external agencies minus the rate at viscous dissipation within $\mathcal{P}_t$.

In particular, for a finite body $\mathcal{B}_t$, if $\mathbf{Tn}\cdot\mathbf{v}$ vanishes on $\partial\mathcal{B}_t$ and $\mathbf{b}_0 \equiv \mathbf{0}$, (46.14) implies that

$$\overline{\int\limits_{\mathcal{P}_t} \tfrac{1}{2}\rho|\mathbf{v}|^2\,dv}^{\,\boldsymbol{\cdot}} \le 0,$$

so that the kinetic energy of the body may not increase with time. This result establishes a type of stability inherent in viscous fluids. A proof of a stronger result, showing that the kinetic energy actually decreases exponentially with time, is provided by GURTIN (1981, §24).

EXERCISE

1. Consider an incompressible, linearly viscous fluid in a fixed, bounded region R of space and assume that

$$\mathbf{v} = \mathbf{0} \qquad \text{on} \qquad \partial R$$

for all time.

a. Show that the rate at which energy is dissipated within R, that is,

$$2\mu\int_R |\mathbf{D}|^2\,dv$$

can be written in the alternative forms

$$2\mu\int_R |\mathbf{W}|^2\,dv \qquad \text{and} \qquad \mu\int_R |\boldsymbol{\omega}|^2\,dv,$$

indicating that spin (i.e., vorticity) is the only source of energy dissipation.

b. The surface force exerted on ∂R by the fluid is given by the simple expression

$$\begin{aligned} -\mathbf{Tn} &= p\mathbf{n} + 2\mu\mathbf{Wn} \\ &= p\mathbf{n} - \mu\mathbf{n}\times\boldsymbol{\omega}, \end{aligned}$$

where $\mathbf{n}$ is the outward unit normal on ∂R. Establish this relation for a plane portion of ∂R.

46.4 Incompressible Navier–Stokes Equations

Consider the flow of an incompressible, linearly viscous fluid in an inertial frame under purely inertial body forces. Then, since $\operatorname{div}\mathbf{v} = 0$, (45.26) becomes

$$2\operatorname{div}\mathbf{D} = \Delta\mathbf{v},$$

and the momentum balance yields

$$\boxed{\rho\dot{\mathbf{v}} = -\operatorname{grad} p + \mu\,\Delta\mathbf{v} + \mathbf{b}_0,} \tag{46.15}$$

which with the constraint equation

$$\operatorname{div}\mathbf{v} = 0$$

comprise the basic equations for the theory of incompressible, linearly viscous fluids, the **incompressible Navier–Stokes equations.**

Suppose, now, that the body force is conservative with potential β. Then, bearing in mind (46.3),

$$\frac{\operatorname{grad} p}{\rho} - \mathbf{b}_0 = \operatorname{grad}\left(\frac{p}{\rho} - \beta\right)$$

and the flow equation (46.15) becomes

$$\dot{\mathbf{v}} = -\operatorname{grad}\left(\frac{p}{\rho} - \beta\right) + \nu\Delta\mathbf{v}, \tag{46.16}$$

where

$$\nu = \frac{\mu}{\rho} \geq 0$$

is a constant called the **kinematic viscosity**. Since the kinematic viscosity ν has dimensions of $(\text{length})^2/(\text{time})$, (46.16) suggests an interpretation of ν as an effective diffusivity for the velocity $\mathbf{v}$.

EXERCISE

1. Show that the flow equation (46.16) for an incompressible, linearly viscous fluid may be written alternatively as

$$\mathbf{v}' - \mathbf{v} \times \boldsymbol{\omega} = -\operatorname{grad}\left(\frac{p}{\rho} + \tfrac{1}{2}|\mathbf{v}|^2 - \beta\right) + \nu\Delta\mathbf{v}.$$

46.5 Circulation. Vorticity-Transport Equation

Next, substituting (46.16) into the circulation transport relation (17.29), we find that, since $\mathcal{C}_t$ is the image of a closed material curve C,

$$\overline{\int_{\mathcal{C}_t} \mathbf{v}(\mathbf{x}, t)\cdot d\mathbf{x}}^{\,\boldsymbol{\cdot}} = \int_{\mathcal{C}_t} \dot{\mathbf{v}}(\mathbf{x}, t)\cdot d\mathbf{x}$$

$$= \nu \int_{\mathcal{C}_t} \Delta\mathbf{v}(\mathbf{x}, t)\cdot d\mathbf{x}. \tag{46.17}$$

Furthermore, (17.9) and (46.16) lead to the **vorticity-transport equation**

$$\dot{\boldsymbol{\omega}} - \mathbf{L}\boldsymbol{\omega} = \nu\Delta\boldsymbol{\omega}. \tag{46.18}$$

Importantly, (46.18) shows that the vorticity evolves independently of the pressure. Next, by (43.4),

$$\begin{aligned}\mathbf{L}\boldsymbol{\omega} &= \mathbf{D}\boldsymbol{\omega} + \mathbf{W}\boldsymbol{\omega}\\ &= \mathbf{D}\boldsymbol{\omega} + \tfrac{1}{2}\boldsymbol{\omega}\times\boldsymbol{\omega}\\ &= \mathbf{D}\boldsymbol{\omega}.\end{aligned}$$

Thus, using the identity $(43.10)_3$ for the contravariant rate of the vorticity, we may rewrite (46.18) somewhat more succinctly as

$$\overset{\diamond}{\boldsymbol{\omega}} = \nu\Delta\boldsymbol{\omega}. \tag{46.19}$$

Recall from $(43.9)_2$ that a spatial vector field convects like a tangent if and only if its contravariant rate vanishes. The vorticity-transport equation (46.19) therefore shows that, for a linearly viscous fluid, the vorticity does not generally convect like a tangent. Indeed, viscosity causes vorticity to diffuse with respect to the body.

EXERCISES

1. Show that, for an incompressible fluid,

$$\operatorname{curl}\boldsymbol{\omega} = -\Delta\mathbf{v}$$

and, thus, that the circulation transport relation (46.17) for an incompressible, linearly viscous fluid has the alternative form

$$\overline{\int_{\mathcal{C}_t}\mathbf{v}(\mathbf{x},t)\cdot d\mathbf{x}}^{\,\boldsymbol{\cdot}} = -\nu\int_{\mathcal{C}_t}\operatorname{curl}\boldsymbol{\omega}(\mathbf{x},t)\cdot d\mathbf{x}.$$

2. Show that the vorticity-transport equation (46.18) can be written alternatively as

$$\boldsymbol{\omega}' = \mathbf{D}\boldsymbol{\omega} + \nu\Delta\boldsymbol{\omega}$$

and provide a geometrical interpretation of the term $\mathbf{D}\boldsymbol{\omega}$.

3. Show that the vorticity-transport equation (46.18) can be written alternatively as

$$\boldsymbol{\omega}' + \operatorname{curl}(\boldsymbol{\omega}\times\mathbf{v}) = \nu\Delta\boldsymbol{\omega}.$$

4. Defining the enstrophy $e = \frac{1}{2}|\boldsymbol{\omega}|^2$, use the vorticity-transport equation (46.18) to arrive at the **enstrophy-transport equation**

$$\dot{e} = -\boldsymbol{\omega}\cdot\mathbf{D}\boldsymbol{\omega} - \nu\,|\operatorname{grad}\boldsymbol{\omega}|^2 + \nu\Delta e. \tag{46.20}$$

5. Show that, for a two-dimensional flow, the vorticity-transport equation (46.18) reduces to

$$\dot{\boldsymbol{\omega}} = \nu\Delta\boldsymbol{\omega}.$$

6. As a consequence of the definition (17.1) of the vorticity and the constraint $\operatorname{div}\mathbf{v} = 0$ there exists a vector potential $\boldsymbol{\xi}$ such that $\mathbf{v} = \operatorname{curl}\boldsymbol{\xi}$ and

$$\Delta\boldsymbol{\xi} = -\boldsymbol{\omega}. \tag{46.21}$$

Provided that $\boldsymbol{\omega}$ is known on a simply connected domain R, (46.21) yields

$$\boldsymbol{\xi}(\mathbf{x},t) = \int_R K(\mathbf{x}-\mathbf{y})\boldsymbol{\omega}(\mathbf{y},t)\,dv_{\mathbf{y}},$$

with

$$K(\mathbf{r}) = \frac{1}{4\pi|\mathbf{r}|}$$

the fundamental solution of the Laplace equation (in three space dimensions). Show that

$$\begin{aligned}\mathbf{v}(\mathbf{x},t) &= \operatorname{curl}\int_R K(\mathbf{x}-\mathbf{y})\boldsymbol{\omega}(\mathbf{y},t)\,dv_{\mathbf{y}} \\ &= \frac{1}{4\pi}\int_R \frac{\boldsymbol{\omega}(\mathbf{y},t)\times(\mathbf{x}-\mathbf{y})}{|\mathbf{x}-\mathbf{y}|^3}\,dv_{\mathbf{y}}\end{aligned} \tag{46.22}$$

and, thus, conclude that, when augmented by (46.22) and the conditions

$$\boldsymbol{\omega} = \operatorname{curl}\mathbf{v} \qquad \text{and} \qquad \operatorname{div}\boldsymbol{\omega} = 0,$$

the second of which is an immediate consequence of the first, the vorticity-transport equation (46.18) and the pressure Poisson equation (46.24) provide an equivalent alternative to the Navier–Stokes equations. This alternative provides the basis for the numerically powerful vortex method.

46.6 Pressure Poisson Equation

Since $\operatorname{div}\mathbf{v} = 0$, it follows from (9.25) that

$$\begin{aligned}\operatorname{div}\dot{\mathbf{v}} &= \overline{\operatorname{div}\mathbf{v}}^{\boldsymbol{\cdot}} + (\operatorname{grad}\mathbf{v})\colon(\operatorname{grad}\mathbf{v})^{\top} \\ &= \operatorname{div}[\operatorname{div}(\mathbf{v}\otimes\mathbf{v})];\end{aligned}$$

further,

$$\begin{aligned}\operatorname{div}\Delta\mathbf{v} &= \Delta\operatorname{div}\mathbf{v} \\ &= 0.\end{aligned}$$

Thus, since $\rho \equiv$ constant, multiplying the momentum balance (46.16) by ρ and computing the divergence of each term yields

$$\Delta(p - \rho\beta) = -\operatorname{div}[\operatorname{div}(\rho\mathbf{v}\otimes\mathbf{v})]. \tag{46.23}$$

In particular, when external body forces are absent, so that $\beta = 0$, (46.23) reduces to the **pressure Poisson equation**

$$\Delta p = -\operatorname{div}[\operatorname{div}(\rho\mathbf{v}\otimes\mathbf{v})]. \tag{46.24}$$

EXERCISE

1. Show that the pressure Poisson equation (46.24) can be written alternatively as

$$\Delta p = \rho(|\mathbf{W}|^2 - |\mathbf{D}|^2). \tag{46.25}$$

46.7 Transport Equations for the Velocity Gradient, Stretching, and Spin in a Linearly Viscous, Incompressible Fluid

On computing the gradient on both sides of the evolution equation (46.16) and using the identity (9.26), it follows immediately that

$$\dot{\mathbf{L}} + \mathbf{L}^2 = -\operatorname{grad}\operatorname{grad}\left(\frac{p}{\rho} - \beta\right) + \nu\Delta\mathbf{L}, \tag{46.26}$$

which provides a transport equation for the velocity gradient in a linearly viscous, incompressible fluid. Bearing in mind that $\operatorname{tr}\mathbf{L} = \operatorname{div}\mathbf{v} = 0$ and taking the trace on both sides of (46.26), we find that

$$\operatorname{tr}(\mathbf{L}^2) = -\Delta\left(\frac{p}{\rho} - \beta\right), \tag{46.27}$$

which, since

$$\begin{aligned}\operatorname{div}[\operatorname{div}(\mathbf{v}\otimes\mathbf{v})] &= \operatorname{div}(\mathbf{L}\mathbf{v} + (\operatorname{div}\mathbf{v})\mathbf{v}) \\ &= \mathbf{L}:\mathbf{L}^{\top} + \mathbf{v}\cdot(\operatorname{grad}\operatorname{div}\mathbf{v}) \\ &= \operatorname{tr}(\mathbf{L}^2),\end{aligned}$$

is equivalent to the generalization (46.23) of the pressure Poisson equation (46.24) to account for the action of conservative external body forces.

Next, by (46.27), the transport equation (46.26) for the velocity gradient simplifies somewhat to

$$\dot{\mathbf{L}} + (\mathbf{L}^2)_0 = -\left[\operatorname{grad}\operatorname{grad}\left(\frac{p}{\rho} - \beta\right)\right]_0 + \nu\Delta\mathbf{L}. \tag{46.28}$$

By (43.2),

$$\operatorname{skw}(\dot{\mathbf{L}}) = \dot{\mathbf{W}}, \qquad \operatorname{skw}(\mathbf{L}^2) = \mathbf{D}\mathbf{W} + \mathbf{W}\mathbf{D}, \qquad \operatorname{skw}(\Delta\mathbf{L}) = \Delta\mathbf{W};$$

thus, recalling, from (17.14) and (20.25), that

$$\overset{\scriptscriptstyle\triangle}{\mathbf{W}} = \dot{\mathbf{W}} + \mathbf{D}\mathbf{W} + \mathbf{W}\mathbf{D},$$

bearing in mind that $\operatorname{grad}\operatorname{grad}\varphi$ is symmetric for any scalar field φ, and taking the skew part of (46.28), we arrive at the **spin-transport equation**

$$\overset{\scriptscriptstyle\triangle}{\mathbf{W}} = \nu\Delta\mathbf{W}. \tag{46.29}$$

We leave it as an exercise to show that (46.29) is equivalent to the vorticity-transport equation (46.18). Next, by (43.2),

$$\operatorname{sym}(\dot{\mathbf{L}}) = \dot{\mathbf{D}}, \qquad \operatorname{sym}(\mathbf{L}^2) = \mathbf{D}^2 + \mathbf{W}^2, \qquad \operatorname{sym}(\Delta\mathbf{L}) = \Delta\mathbf{D},$$

and taking the symmetric part of (46.28), we arrive at **stretching-transport equation**

$$\dot{\mathbf{D}} + (\mathbf{D}^2 + \mathbf{W}^2)_0 = -\left[\operatorname{grad}\operatorname{grad}\left(\frac{p}{\rho} - \beta\right)\right]_0 + \nu\Delta\mathbf{D}. \tag{46.30}$$

EXERCISES

1. Derive the enstrophy-transport equation (46.20) directly from the spin-transport equation (46.29).
2. Use the relation (43.4) between the spin and the vorticity and the relations $(20.47)_3$ and (20.25) defining the contravariant rate of a vector field and the covariant rate of a tensor field to show that the vorticity-transport equation (46.18) and the spin-transport equation (46.29) are equivalent.
3. Using (46.18), (46.25), and (46.30) show that

$$\overline{\mathbf{D}\boldsymbol{\omega}}^{\boldsymbol{\cdot}} = -\left[\operatorname{grad}\operatorname{grad}\left(\frac{p}{\rho} - \beta\right)\right]\boldsymbol{\omega} + \nu\Delta(\mathbf{D}\boldsymbol{\omega}) - 2\nu(\operatorname{grad}\mathbf{D})\boldsymbol{\omega},$$

where, using components, $[(\operatorname{grad}\mathbf{D})\boldsymbol{\omega}]_i = D_{ij,k}\omega_{j,k}$.

4. Specialize the identities (2.146) to reflect the constraint $\operatorname{tr}\mathbf{D} = 0$ and use (46.10), (46.11), and (46.30) to show that for a linearly viscous, incompressible fluid, the second and third principal invariants of the stretching tensor obey transport equations of the form

$$\overline{I_2(\mathbf{D})}^{\boldsymbol{\cdot}} = 3I_3(\mathbf{D}) + \mathbf{D}:\mathbf{W}^2 + \mathbf{D}:\operatorname{grad}\operatorname{grad}\left(\frac{p}{\rho} - \beta\right) + \nu\mathbf{1}:\mathbf{N} + \nu\Delta I_2(\mathbf{D}),$$

$$\overline{I_3(\mathbf{D})}^{\boldsymbol{\cdot}} = -\tfrac{2}{3}I_2^2(\mathbf{D}) + \tfrac{4}{3}I_2(\mathbf{D})I_2(\mathbf{W}) - \mathbf{D}^2:\mathbf{W}^2 \\ - \mathbf{D}^2:\left[\operatorname{grad}\operatorname{grad}\left(\frac{p}{\rho} - \beta\right)\right]_0 - 2\nu\mathbf{D}:\mathbf{N} + \nu\Delta I_3(\mathbf{D}),$$

where, using components, $N_{ij} = D_{ip,q}D_{jp,q}$.

46.8 Impetus-Gauge Formulation of the Navier–Stokes Equations

Let $\mathbf{m}$ be defined by

$$\mathbf{m} = \mathbf{v} - \operatorname{grad}\varphi, \tag{46.31}$$

where φ is an arbitrary scalar field called the **gauge**. Maddocks & Pego (1995) refer to $\mathbf{m}$ as the **impetus**. Bearing in mind that $\operatorname{div}\mathbf{v} = 0$ and $\operatorname{curl}\operatorname{grad}\varphi = \mathbf{0}$, it then follows that

$$\mathbf{m}' = \mathbf{v}' - \operatorname{grad}(\varphi'),$$

$$\operatorname{div}\mathbf{m} = -\Delta\varphi,$$

$$\operatorname{curl}\mathbf{m} = \operatorname{curl}\mathbf{v},$$

$$\Delta\mathbf{m} = \Delta\mathbf{v} - \operatorname{grad}\Delta\varphi.$$

Thus, assuming that the body force is conservative with potential β and using $(17.7)_2$ to write the flow equation (46.16) as

$$\mathbf{v}' + (\operatorname{curl}\mathbf{v}) \times \mathbf{v} = -\operatorname{grad}\left(\frac{p}{\rho} + \tfrac{1}{2}|\mathbf{v}|^2 - \beta\right) + \nu\Delta\mathbf{v},$$

we find that the impetus must obey

$$\mathbf{m}' + (\operatorname{curl}\mathbf{m}) \times \mathbf{v} = -\operatorname{grad}\left(\varphi' - \nu\Delta\varphi + \frac{p}{\rho} + \tfrac{1}{2}|\mathbf{v}|^2 - \beta\right) + \nu\Delta\mathbf{m}. \tag{46.32}$$

Together with

$$\Delta\varphi = -\operatorname{div}\mathbf{m} \qquad \text{and} \qquad \mathbf{v} = \mathbf{m} + \operatorname{grad}\varphi,$$

(46.32) constitutes the **impetus-gauge formulation** of the Navier–Stokes equations.

Unlike the pressure, the gauge has no intrinsic physical meaning. An evolution equation and boundary conditions for φ may therefore be imposed for the sake of expedience. This allows for the development of novel numerical schemes. In particular, various choices for φ yield special versions of the gauge formulation. Among these, the simplest choice is the zero gauge

$$\varphi' = \nu\Delta\varphi - \frac{p}{\rho} - \tfrac{1}{2}|\mathbf{v}|^2 + \beta, \tag{46.33}$$

for which (46.32) becomes

$$\mathbf{m}' + (\operatorname{curl}\mathbf{m}) \times \mathbf{v} = \nu\Delta\mathbf{m}. \tag{46.34}$$

A more interesting choice of gauge arises on noting, by the definition $(43.8)_2$ of the covariant rate of a vector field, that

$$\begin{aligned}\mathbf{m}' + (\operatorname{curl}\mathbf{m}) \times \mathbf{v} &= \mathbf{m}' + (\operatorname{grad}\mathbf{m})\mathbf{v} - (\operatorname{grad}\mathbf{m})^{\top}\mathbf{v} \\ &= \dot{\mathbf{m}} - (\operatorname{grad}\mathbf{m})^{\top}\mathbf{v} \\ &= \overset{\vartriangle}{\mathbf{m}} - \mathbf{L}^{\top}\mathbf{m} - (\operatorname{grad}\mathbf{m})^{\top}\mathbf{v} \\ &= \overset{\vartriangle}{\mathbf{m}} - \operatorname{grad}(\mathbf{m}\cdot\mathbf{v}) \\ &= \overset{\vartriangle}{\mathbf{m}} + \operatorname{grad}(\mathbf{v}\cdot\operatorname{grad}\varphi - |\mathbf{v}|^2)\end{aligned}$$

and stipulating that the gauge evolve according to

$$\dot{\varphi} = \nu\Delta\varphi - \frac{p}{\rho} + \tfrac{1}{2}|\mathbf{v}|^2 + \beta, \tag{46.35}$$

in which case (46.32) reduces

$$\overset{\vartriangle}{\mathbf{m}} = \nu\Delta\mathbf{m}. \tag{46.36}$$

Recall from $(43.9)_1$ that a spatial vector field convects like a normal if and only if its covariant rate vanishes. Analogous to the statement following the vorticity-transport equation (46.18), the evolution equation (46.36) shows that under the gauge (46.35) the impetus **m** would convect like a normal if not for the presence of viscosity.[225]

EXERCISES

1. Assume that φ evolves according to (46.35). Show that

$$\overline{\boldsymbol{\omega}\cdot\mathbf{m}}^{\boldsymbol{\cdot}} = \nu\Delta(\boldsymbol{\omega}\cdot\mathbf{m}) - 2\nu(\operatorname{grad}\boldsymbol{\omega}):(\operatorname{grad}\mathbf{m}).$$

2. Based on variational considerations, MADDOCKS & PEGO (1995) introduce the gauge

$$\dot{\varphi} = \nu\Delta\varphi - \frac{p}{\rho} + \beta.$$

Show that, for this choice of φ, the impetus evolves according to

$$\dot{\mathbf{m}} = \mathbf{L}^{\top}\operatorname{grad}\varphi + \nu\Delta\mathbf{m}.$$

3. E & LIU (2003) work with the gauge

$$\varphi' = \nu\Delta\varphi - \frac{p}{\rho} + \beta.$$

Show that, for this choice of φ, the impetus evolves according to

$$\dot{\mathbf{m}} + \mathbf{L}\operatorname{grad}\varphi = \nu\Delta\mathbf{m}.$$

46.9 Perfect Fluids

A **perfect fluid** is both incompressible and inviscid. Within our framework, such fluids arise on assuming that the deviatoric (i.e., traceless) part $\mathbf{T}_0$ of the stress in

[225] The gauge (46.35) was introduced by BUTTKE (1993). Cf. RUSSO & SMEREKA (1999), who refer to this choice as the geometric gauge. To our knowledge, it has not previously been noticed that, for the choice (46.35), the evolution equation (46.32) can be expressed simply in terms of the covariant rate.

the general decomposition (46.4) vanishes to leave a pure pressure

$$\mathbf{T} = -p\mathbf{1}. \tag{46.37}$$

The stress power therefore vanishes and, since the specific free energy is constant, the free-energy balance (46.6) reduces to the unconditionally true statement "0 = 0." In this sense, the thermodynamic structure of the theory of perfect fluids is vacuous. Nevertheless, due to the vanishing of the internal (and, hence, external) power and of the dissipation, the kinetic energy balance (46.14) specializes to

$$\overline{\int_{\mathcal{P}_t} \tfrac{1}{2}\rho\,|\mathbf{v}|^2\,dv}^{\centerdot} = -\int_{\partial\mathcal{P}_t} p\mathbf{v}\cdot\mathbf{n}\,da + \int_{\mathcal{P}_t} \mathbf{b}_0\cdot\mathbf{v}\,dv. \tag{46.38}$$

In view of (46.37), momentum balance yields

$$\rho\dot{\mathbf{v}} = -\operatorname{grad} p + \mathbf{b}_0, \tag{46.39}$$

which with the constraint equation

$$\operatorname{div}\mathbf{v} = 0$$

comprise the basic equations for the theory of perfect fluids, the **Euler equations.**

Suppose that the body force is conservative with potential β, in which case the flow equation (46.39) simplifies to

$$\dot{\mathbf{v}} = -\operatorname{grad}\left(\frac{p}{\rho} - \beta\right) \tag{46.40}$$

and it can be shown that, like an elastic fluid, a perfect fluid obeys a simplified version of Bernoulli's theorem.[226] Specifically,

PROPERTIES OF PERFECT FLUIDS *Consider the flow of a perfect fluid under a conservative conventional body force with potential β.*

(i) *If the flow is once irrotational, then it is always irrotational.*
(ii) *The flow preserves circulation.*
(iii) *If the flow is steady, then the field*

$$\frac{p}{\rho} + \tfrac{1}{2}|\mathbf{v}|^2 - \beta \tag{46.41}$$

is constant on streamlines. If, in addition, the flow is irrotational, then (46.41) *is constant everywhere and for all time.*

For a perfect fluid, the vorticity-transport equation can be obtained simply by setting the kinematic viscosity $\nu = 0$ in (46.19) to yield

$$\overset{\diamond}{\boldsymbol{\omega}} = \mathbf{0}. \tag{46.42}$$

The vorticity therefore convects like a tangent in the motion of a perfect fluid.

Similarly, for a perfect fluid, the gauge formulation is obtained by setting $\nu = 0$ in (46.32) to yield

$$\mathbf{m}' + (\operatorname{curl}\mathbf{m})\times\mathbf{v} = -\operatorname{grad}\left(\varphi' + \frac{p}{\rho} + \tfrac{1}{2}|\mathbf{v}|^2 - \beta\right). \tag{46.43}$$

In particular, the specialization of (46.43) arising from the choice

$$\dot{\varphi} = -\frac{p}{\rho} + \tfrac{1}{2}|\mathbf{v}|^2 + \beta \tag{46.44}$$

226 Cf. page 248.

of gauge, which is (46.35) with $\nu = 0$, yields

$$\overset{\triangle}{\mathbf{m}} = \mathbf{0}, \tag{46.45}$$

which is (46.36) with $\nu = 0$. Under the gauge (46.44), the impetus therefore convects like a normal in a perfect fluid. For this reason it seems reasonable to refer to (46.44) as the **normally convected gauge.**

EXERCISES

1. Show that in the flow of a perfect fluid the stress power vanishes.
2. Consider the flow of a perfect fluid in a bounded region R and suppose that

$$\mathbf{v} \cdot \mathbf{n} = 0 \qquad \text{on} \qquad \partial R.$$

Show that

$$\overline{\int_R |\mathbf{v}|^2 \, dv}^{\boldsymbol{\cdot}} = 0,$$

so that the kinetic energy of R is constant.

3. Consider a flow of a perfect fluid in a region R and suppose that

$$\boldsymbol{\omega} \cdot \mathbf{n} = 0 \qquad \text{on} \qquad \partial R.$$

The integrals

$$\int_R \boldsymbol{\omega} \, dv \qquad \text{and} \qquad \int_R \mathbf{v} \cdot \boldsymbol{\omega} \, dv$$

represent, respectively, the net vorticity and net helicity of R. Show that

$$\overline{\int_R \boldsymbol{\omega} \, dv}^{\boldsymbol{\cdot}} = \mathbf{0} \qquad \text{and} \qquad \overline{\int_R \mathbf{v} \cdot \boldsymbol{\omega} \, dv}^{\boldsymbol{\cdot}} = 0.$$

4. Show that the vorticity-transport equation (46.42) for a perfect fluid can be written alternatively as

$$\boldsymbol{\omega}' + \operatorname{curl}(\boldsymbol{\omega} \times \mathbf{v}) = 0.$$

5. Show that, for a perfect fluid,

$$\overline{\mathbf{D}\boldsymbol{\omega}}^{\boldsymbol{\cdot}} = -\left[\operatorname{grad}\operatorname{grad}\left(\frac{P}{\rho} - \beta\right)\right]\boldsymbol{\omega}.$$

6. Show that, if the normally convected gauge (46.44) is imposed, then the vorticity and impetus obey

$$\overline{\boldsymbol{\omega} \cdot \mathbf{m}}^{\boldsymbol{\cdot}} = 0.$$

PART X

MECHANICAL THEORY OF ELASTIC SOLIDS

This chapter discusses the classical theory of elastic solids, neglecting thermal effects. This theory is important not only in its own right but also because it represents a simple context in which to discuss the central steps in the construction of sound constitutive equations for solids.

47 Brief Review

47.1 Kinematical Relations

Basic to our discussion of constitutive equations for elastic solids is the polar decomposition[227]

$$\mathbf{F} = \mathbf{RU} = \mathbf{VR} \tag{47.1}$$

of the deformation gradient

$$\mathbf{F} = \nabla\chi$$

into a *rotation* $\mathbf{R}$, a right stretch tensor $\mathbf{U}$, and a left stretch tensor $\mathbf{V}$, with

$$\mathbf{U} = \sqrt{\mathbf{F}^{\top}\mathbf{F}} \qquad \text{and} \qquad \mathbf{V} = \sqrt{\mathbf{F}\mathbf{F}^{\top}}. \tag{47.2}$$

The tensor fields

$$\mathbf{C} = \mathbf{U}^2 = \mathbf{F}^{\top}\mathbf{F} \qquad \text{and} \qquad \mathbf{B} = \mathbf{V}^2 = \mathbf{F}\mathbf{F}^{\top}, \tag{47.3}$$

respectively, are referred to as the *right* and *left Cauchy–Green tensors*, while

$$\mathbf{E} = \tfrac{1}{2}(\mathbf{C} - \mathbf{1}) \tag{47.4}$$

is the *Green–St. Venant strain tensor*. Important consequences of (47.3) and the polar decomposition are the relations

$$\mathbf{V} = \mathbf{R}\mathbf{U}\mathbf{R}^{\top} \qquad \text{and} \qquad \mathbf{B} = \mathbf{R}\mathbf{C}\mathbf{R}^{\top}. \tag{47.5}$$

47.2 Basic Laws

For solids, it is generally most convenient to use a referential description; we, therefore, begin with the local **momentum balances**[228]

$$\begin{aligned} \rho_{\mathrm{R}}\ddot{\chi} &= \operatorname{Div}\mathbf{T}_{\mathrm{R}} + \mathbf{b}_{0\mathrm{R}}, \\ \mathbf{T}_{\mathrm{R}}\mathbf{F}^{\top} &= \mathbf{F}\mathbf{T}_{\mathrm{R}}^{\top}. \end{aligned} \tag{47.6}$$

[227] Cf. §2.15 and §7.1.

[228] The local referential form (18.9) of mass balance is simply an algebraic equation, $\rho = \rho_{\mathrm{R}}/J$, for the current density. The problem of determining ρ is therefore decoupled from that of determining the deformation χ and of limited interest.

Here, $\mathbf{T}_R$ is the Piola stress defined by

$$\mathbf{T}_R = J\mathbf{T}\mathbf{F}^{-\top}, \tag{47.7}$$

with $\mathbf{T}$ being the Cauchy stress; $\mathbf{T}_R$ and $\mathbf{b}_{0R}$ represent the stress and conventional body force measured per unit area and volume in the reference body.[229] We recall also that the Piola and second Piola stresses are related via[230]

$$\mathbf{T}_R = \mathbf{F}\mathbf{T}_{RR}. \tag{47.8}$$

By $(47.6)_2$ and (47.8), $\mathbf{F}\mathbf{T}_{RR}\mathbf{F}^\top = \mathbf{F}\mathbf{T}_{RR}^\top\mathbf{F}^\top$ and[231]

$$\mathbf{T}_{RR} = \mathbf{T}_{RR}^\top. \tag{47.9}$$

As the theory under consideration is mechanical,[232] we take as our basic thermodynamical law the **free-energy imbalance**

$$\dot{\psi}_R - \mathbf{T}_R : \dot{\mathbf{F}} = -\delta_R \le 0. \tag{47.10}$$

47.3 Transformation Laws Under a Change in Frame

The basic kinematical fields transform as follows under a change in frame:[233]

$$\left.\begin{aligned}
\mathbf{F}^* &= \mathbf{Q}\mathbf{F},\\
\mathbf{R}^* &= \mathbf{Q}\mathbf{R},\\
\mathbf{U}^* &= \mathbf{U},\\
\mathbf{C}^* &= \mathbf{C},\\
\mathbf{E}^* &= \mathbf{E},\\
\mathbf{V}^* &= \mathbf{Q}\mathbf{V}\mathbf{Q}^\top,\\
\mathbf{B}^* &= \mathbf{Q}\mathbf{B}\mathbf{Q}^\top.
\end{aligned}\right\} \tag{47.11}$$

In addition, the Cauchy, Piola, and second Piola stresses transform according to[234]

$$\left.\begin{aligned}
\mathbf{T}^* &= \mathbf{Q}\mathbf{T}\mathbf{Q}^\top,\\
\mathbf{T}_R^* &= \mathbf{Q}\mathbf{T}_R,\\
\mathbf{T}_{RR}^* &= \mathbf{T}_{RR}.
\end{aligned}\right\} \tag{47.12}$$

The transformation law for $\mathbf{T}_R$ follows from $(47.11)_1$ and the transformation law (21.6) for $\mathbf{T}$. Further, granted $(47.12)_2$, the transformation law for $\mathbf{T}_{RR}$ follows from

229 Cf. (24.1).
230 Cf. (25.2).
231 Cf. (25.4).
232 The role of a free-energy imbalance in a mechanical theory is discussed in §29. The local balance (47.10) is derived in the steps leading to (31.24).
233 Cf. (20.16).
234 Cf. (21.6), (25.8), and (25.9).

the definition (25.2) of the second Piola stress and the transformation law for $\mathbf{T}_{\mathrm{R}}$. Notice that

- *the Cauchy stress is frame-indifferent;*
- *the second Piola stress is invariant;*
- *the Piola stress is neither frame-indifferent nor invariant.*

Moreover, these transformation properties are strictly analogous to those of the deformation measures **D**, **F**, and **C** underlying the rates entering the respective power-conjugate pairings $\mathbf{T}:\mathbf{D}$, $\mathbf{T}_{\mathrm{R}}:\dot{\mathbf{F}}$, and $\frac{1}{2}\mathbf{T}_{\mathrm{RR}}:\dot{\mathbf{C}}$.[235]

[235] Cf. (25.6)

48 Constitutive Theory

In classical mechanics, the force and energy within an elastic spring depend only on the change in length of the spring; moreover, the force is independent of the past history of the length as well as the rate at which the length is changing in time. In continuum mechanics, local length changes are characterized by the deformation gradient $\mathbf{F}$; we, therefore, define an **elastic body** through constitutive equations giving the free energy and stress when $\mathbf{F}$ is known:

$$\begin{aligned} \psi_R &= \hat{\psi}_R(\mathbf{F}), \\ \mathbf{T}_R &= \hat{\mathbf{T}}_R(\mathbf{F}). \end{aligned} \tag{48.1}$$

The response functions $\hat{\psi}$ and $\hat{\mathbf{T}}_R$ are defined on the set of all tensors with strictly positive determinant. In view of (47.7) and (47.8), the relation $(48.1)_2$ for the Piola stress determines auxiliary constitutive equations

$$\begin{aligned} \mathbf{T} = \hat{\mathbf{T}}(\mathbf{F}) &= (\det \mathbf{F})^{-1}\hat{\mathbf{T}}_R(\mathbf{F})\mathbf{F}^{\top}, \\ \mathbf{T}_{RR} = \hat{\mathbf{T}}_{RR}(\mathbf{F}) &= \mathbf{F}^{-1}\hat{\mathbf{T}}_R(\mathbf{F}), \end{aligned} \tag{48.2}$$

for the Cauchy and second Piola stresses.

We now explore the consequences of the hypotheses (page 225) stipulating that the constitutive equations be frame-indifferent and compatible with thermodynamics.

48.1 Consequences of Frame-Indifference

The free energy ψ_R, being a scalar field, is invariant under a change in frame: $\psi_R^* = \psi_R$; thus, since the deformation gradient transforms according to $\mathbf{F}^* = \mathbf{QF}$,

$$\begin{aligned} \hat{\psi}_R(\mathbf{F}) &= \hat{\psi}_R(\mathbf{F}^*) \\ &= \hat{\psi}_R(\mathbf{QF}). \end{aligned} \tag{48.3}$$

Further, by $(47.12)_2$, the Piola stress transforms according to $\mathbf{T}_R^* = \mathbf{QT}_R$, so that

$$\begin{aligned} \mathbf{T}_R^* &= \hat{\mathbf{T}}_R(\mathbf{F}^*) \\ &= \hat{\mathbf{T}}_R(\mathbf{QF}). \end{aligned}$$

The response functions $\hat{\psi}_R$ and $\hat{\mathbf{T}}_R$ must therefore satisfy

$$\begin{aligned}\hat{\psi}_R(\mathbf{F}) &= \hat{\psi}_R(\mathbf{QF}),\\ \hat{\mathbf{T}}_R(\mathbf{F}) &= \mathbf{Q}^{\top}\hat{\mathbf{T}}_R(\mathbf{QF}),\end{aligned} \tag{48.4}$$

for all rotations $\mathbf{Q}$ and all $\mathbf{F}$.

Fix $\mathbf{F}$ and consider the polar decomposition $\mathbf{F} = \mathbf{RU}$. Since $\mathbf{Q}$ is arbitrary, we are at liberty to choose $\mathbf{Q} = \mathbf{R}^{\top}$; hence, $\mathbf{QF} = \mathbf{U}$ and (48.4) specializes to

$$\begin{aligned}\hat{\psi}_R(\mathbf{F}) &= \hat{\psi}_R(\mathbf{U}),\\ \hat{\mathbf{T}}_R(\mathbf{F}) &= \mathbf{R}\hat{\mathbf{T}}_R(\mathbf{U}).\end{aligned} \tag{48.5}$$

Replacing $\mathbf{U}$ by $\sqrt{\mathbf{C}}$ in $(48.5)_1$, we may therefore introduce a response function $\bar{\psi}_R$ determining the free energy ψ_R as a function of $\mathbf{C}$ via

$$\bar{\psi}_R(\mathbf{C}) = \hat{\psi}_R\big(\sqrt{\mathbf{C}}\big).$$

Similarly, in view of the auxiliary constitutive equation $(48.2)_2$ for the second Piola stress $\mathbf{T}_{RR}$, we find that

$$\begin{aligned}\mathbf{R}\hat{\mathbf{T}}_R(\mathbf{U}) &= \mathbf{F}\mathbf{U}^{-1}\hat{\mathbf{T}}_R(\mathbf{U})\\ &= \mathbf{F}\hat{\mathbf{T}}_{RR}(\mathbf{F})\\ &= \mathbf{R}\mathbf{U}\hat{\mathbf{T}}_{RR}(\mathbf{F}).\end{aligned}$$

Hence, $\hat{\mathbf{T}}_R(\mathbf{U}) = \mathbf{U}\hat{\mathbf{T}}_{RR}(\mathbf{F})$,

$$\hat{\mathbf{T}}_{RR}(\mathbf{F}) = \mathbf{U}^{-1}\hat{\mathbf{T}}_R(\mathbf{U}),$$

and, replacing $\mathbf{U}$ by $\sqrt{\mathbf{C}}$, we may introduce a response function $\bar{\mathbf{T}}_{RR}$ determining the second Piola stress as a function of $\mathbf{C}$ via

$$\begin{aligned}\mathbf{T}_{RR} &= \mathbf{C}^{-1/2}\hat{\mathbf{T}}_R\big(\sqrt{\mathbf{C}}\big)\\ &= \bar{\mathbf{T}}_{RR}(\mathbf{C}).\end{aligned} \tag{48.6}$$

If the constitutive equations (48.1) are to be frame-indifferent, the foregoing results show that they must reduce to constitutive equations of the specific form

$$\begin{aligned}\psi_R &= \bar{\psi}_R(\mathbf{C}),\\ \mathbf{T}_R &= \mathbf{F}\bar{\mathbf{T}}_{RR}(\mathbf{C}).\end{aligned} \tag{48.7}$$

To prove the converse assertion — that the constitutive equations (48.7) are frame-indifferent — we note that, by $(47.11)_4$, $\mathbf{C} = \mathbf{C}^*$ and, hence that

$$\begin{aligned}\psi_R^* &= \bar{\psi}_R(\mathbf{C}^*)\\ &= \bar{\psi}_R(\mathbf{C})\\ &= \psi_R.\end{aligned}$$

Thus, $(48.7)_1$ is frame-indifferent. Turning to $(48.7)_2$, since $\mathbf{F}^* = \mathbf{QF}$,

$$\begin{aligned}\mathbf{T}_R^* &= \mathbf{F}^*\bar{\mathbf{T}}_{RR}(\mathbf{C}^*)\\ &= \mathbf{Q}[\mathbf{F}\bar{\mathbf{T}}_{RR}(\mathbf{C})]\\ &= \mathbf{QT}_R;\end{aligned}$$

by $(47.12)_2$, $(48.7)_2$ is therefore also frame-indifferent.

Recalling from (47.9) that the second Piola stress is symmetric, an important consequence of the result $(48.7)_2$ is that

$$\begin{aligned}\mathbf{T}_R\mathbf{F}^\top - \mathbf{F}\mathbf{T}_R^\top &= \mathbf{F}\mathbf{T}_{RR}\mathbf{F}^\top - \mathbf{F}\mathbf{T}_{RR}^\top\mathbf{F}^\top\\ &= \mathbf{F}(\mathbf{T}_{RR} - \mathbf{T}_{RR}^\top)\mathbf{F}^\top\\ &= \mathbf{0}.\end{aligned}$$

Hence, the requirement that the constitutive equations (48.1) be frame-indifferent implies satisfaction of the moment balance $(47.6)_2$. Granted frame-indifferent constitutive equations, we may therefore neglect the moment balance from further consideration.

Finally, by (47.7), the auxiliary constitutive equations (48.2) for the Cauchy and second Piola stresses become

$$\left.\begin{aligned}\mathbf{T} &= J^{-1}\mathbf{F}\bar{\mathbf{T}}_{RR}(\mathbf{C})\mathbf{F}^\top,\\ \mathbf{T}_{RR} &= \bar{\mathbf{T}}_{RR}(\mathbf{C}).\end{aligned}\right\} \tag{48.8}$$

In view of $(48.7)_2$ and (48.8), the crucial ingredient for determining the Piola, Cauchy, and second Piola stresses is the response function $\bar{\mathbf{T}}_{RR}$ for the second Piola stress.

To summarize the results of this section, the hypothesis that the constitutive equations (48.1) be frame-indifferent reduces the problem of characterizing an elastic solid to one of determining response functions $\bar{\psi}$ and $\bar{\mathbf{T}}_{RR}$ — both depending on the right Cauchy–Green tensor $\mathbf{C}$ — for the free energy ψ_R and the second Piola stress $\mathbf{T}_{RR}$. In addition, this hypothesis ensures satisfaction of the moment balance $(47.6)_2$ and, as an important consequence, obviates the need for further consideration of that balance.

48.2 Thermodynamic Restrictions

48.2.1 The Stress Relation

The Coleman–Noll procedure, introduced in Part D, represents a paradigm for the derivation of thermodynamically consistent constitutive equations.[236] We now apply this procedure to the frame-indifferent constitutive equations (48.7).

Consider an arbitrary **constitutive process**; that is, consider a motion χ together with fields ψ and $\mathbf{T}_R$ determined by the motion through the constitutive equations (48.7). The local force balance $(47.6)_1$ then provides an explicit relation,

$$\mathbf{b}_{0R} = \rho_R\ddot{\chi} - \mathrm{Div}\,\mathbf{T}_R,$$

[236] Cf. the paragraph in petite type just before equation (39.1).

for the conventional body force $\mathbf{b}_{0R}$ needed to support the process under consideration. As a basic hypothesis of the Coleman–Noll procedure,

- *we assume that the conventional body force is arbitrarily assignable.*

Because of this assumption, the force balance in no way restricts the class of processes that the material may undergo. On the other hand, unless the constitutive equations are suitably restricted, not all constitutive processes will be compatible with the laws of thermodynamics as embodied in the free-energy imbalance (47.10).[237] For that reason, we require that

(‡) *all constitutive processes be consistent with the free-energy imbalance* (47.10).

This requirement has strong consequences. Consider an arbitrary constitutive process. By $(48.7)_1$,

$$\dot{\psi}_R = \frac{\partial \bar{\psi}_R(\mathbf{C})}{\partial \mathbf{C}} : \dot{\mathbf{C}}; \tag{48.9}$$

further, by $(25.6)_3$ and (48.6),

$$\begin{aligned} \mathbf{T}_R : \dot{\mathbf{F}} &= \tfrac{1}{2}\mathbf{T}_{RR} : \dot{\mathbf{C}} \\ &= \tfrac{1}{2}\bar{\mathbf{T}}_{RR}(\mathbf{C}) : \dot{\mathbf{C}}. \end{aligned} \tag{48.10}$$

In view of (48.9) and (48.10), the free-energy imbalance (47.10) is equivalent to the requirement that the inequality

$$\left(2\frac{\partial \bar{\psi}_R(\mathbf{C})}{\partial \mathbf{C}} - \bar{\mathbf{T}}_{RR}(\mathbf{C})\right) : \dot{\mathbf{C}} \le 0 \tag{48.11}$$

be satisfied for all motions of the body. Note that, as a consequence of the symmetry of $\mathbf{C}$, $\partial \bar{\psi}_R(\mathbf{C})/\partial \mathbf{C}$ is symmetric. Recalling from (47.9) that $\mathbf{T}_{RR}$ is symmetric, the difference

$$2\frac{\partial \bar{\psi}_R(\mathbf{C})}{\partial \mathbf{C}} - \bar{\mathbf{T}}_{RR}(\mathbf{C})$$

is thus symmetric. Next,

(†) given any point of the body and any time, it is possible to find a motion such that $\mathbf{C}$ and $\dot{\mathbf{C}}$ have arbitrarily prescribed values at that point and time.

Granted this assertion — which we prove on page 281 — the coefficient of $\dot{\mathbf{C}}$ in (48.11) must vanish, for otherwise $\dot{\mathbf{C}}$ may be chosen to violate (48.11).

We therefore have the following **thermodynamic restriction**:

(‡) *The free energy determines the second Piola stress through the* **stress relation**

$$\boxed{\mathbf{T}_{RR} = 2\frac{\partial \bar{\psi}_R(\mathbf{C})}{\partial \mathbf{C}}, \qquad (\mathbf{T}_{RR})_{ij} = 2\frac{\partial \bar{\psi}_R(\mathbf{C})}{\partial C_{ij}}.} \tag{48.12}$$

237 Cf. the discussion in the first two paragraphs of §29.

48.2.2 Consequences of the Stress Relation

By (47.10) and (48.14), the dissipation δ_R satisfies

$$\delta_R = \mathbf{T}_R : \dot{\mathbf{F}} - \dot{\psi}_R$$
$$= 0; \tag{48.13}$$

the dissipation therefore vanishes in smooth constitutive processes.[238] This property, which distinguishes elastic solids from other materials, reinforces the analogy, raised earlier, between elastic solids in continuum mechanics and elastic springs in classical mechanics.

Next, (47.8) and (48.12) imply that the Piola stress is given by a constitutive equation of the form

$$\boxed{\mathbf{T}_R = 2\mathbf{F}\frac{\partial \bar{\psi}_R(\mathbf{C})}{\partial \mathbf{C}}, \qquad (\mathbf{T}_R)_{ij} = 2F_{ik}\frac{\partial \bar{\psi}_R(\mathbf{C})}{\partial C_{kj}}.} \tag{48.14}$$

Similarly, (47.7) and (48.12) imply that the Cauchy stress is given by a constitutive equation of the form

$$\boxed{\mathbf{T} = 2J^{-1}\mathbf{F}\frac{\partial \bar{\psi}_R(\mathbf{C})}{\partial \mathbf{C}}\mathbf{F}^{\top}, \qquad T_{ij} = 2J^{-2}F_{ik}\frac{\partial \bar{\psi}_R(\mathbf{C})}{\partial C_{kj}}F_{jk}.} \tag{48.15}$$

Materials consistent with (48.15) — or, equivalently (48.12) or (48.14) — are commonly termed *hyperelastic*.

48.2.3 Natural Reference Configuration

We say that the reference configuration is **natural** if

$$\bar{\psi}_R(\mathbf{C}) \ \text{has a local minimum at} \ \mathbf{C} = \mathbf{1}; \tag{48.16}$$

that is, the reference configuration is natural if there is a scalar $\alpha > 0$ such that, for all symmetric, positive-definite tensors $\mathbf{C}$ with $|\mathbf{C} - \mathbf{1}| < \alpha$,

$$\bar{\psi}_R(\mathbf{C}) \geq \bar{\psi}_R(\mathbf{1}).$$

An immediate consequence of (48.16) is that

$$\left.\frac{\partial \bar{\psi}_R(\mathbf{C})}{\partial \mathbf{C}}\right|_{\mathbf{C}=\mathbf{1}} = \mathbf{0}; \tag{48.17}$$

thus, as a consequence of the stress relation (48.14),

(‡) *a natural reference configuration is* **stress-free**; *that is, the Cauchy, Piola, and second Piola stresses vanish in a natural reference configuration*

$$\mathbf{T} = \mathbf{T}_R = \mathbf{T}_{RR} = \mathbf{0} \qquad \text{when} \qquad \mathbf{F} = \mathbf{1}. \tag{48.18}$$

A second consequence of (48.16) is that

$$\left.\frac{\partial^2 \bar{\psi}_R(\mathbf{C})}{\partial \mathbf{C}^2}\right|_{\mathbf{C}=\mathbf{1}} \quad \text{is positive-semidefinite;} \tag{48.19}$$

---

[238] Shock waves in elastic materials generally dissipate energy. Cf. the parenthetical remark in petite type following (29.8).

that is, given any symmetric tensor **A**,

$$\left(\frac{\partial^2\bar{\psi}_R(\mathbf{C})}{\partial\mathbf{C}^2}\bigg|_{\mathbf{C}=\mathbf{1}}\mathbf{A}\right):\mathbf{A}\geq 0, \qquad A_{ij}\frac{\partial^2\bar{\psi}_R(\mathbf{C})}{\partial C_{ij}\partial C_{kl}}\bigg|_{\mathbf{C}=\mathbf{1}}A_{kl}\geq 0.$$

Remark. As noted in the sentence containing (29.9), the free energy of each material point may be additively scaled by a term dependent only on the material point. Therefore, without loss in generality, we may assume that

$$\bar{\psi}_R(\mathbf{1}) = 0. \tag{48.20}$$

48.2.4 Verification of (†)

Choose an arbitrary material point $\mathbf{X}_0$ and time t_0. Let $\mathbf{F}_0$ be an arbitrary constant tensor with $\det\mathbf{F}_0 > 0$, and let **D** be an arbitrary constant symmetric tensor. Then, as noted in §14.1, the motion defined by (14.5) with $\mathbf{L} = \mathbf{D}$,

$$\chi(\mathbf{X}, t) = \mathbf{x}_0 + e^{(t-t_0)\mathbf{D}}\mathbf{F}_0(\mathbf{X}-\mathbf{X}_0), \qquad -\infty < t < \infty, \tag{48.21}$$

is well defined for any reference body B, no matter what the shape of that body may be. Moreover, the velocity gradient of this motion is the constant **D**; by virtue of its symmetry, **D** represents the stretching. A consequence of (14.6) is then that

$$\mathbf{C}(\mathbf{X}_0, t_0) = \mathbf{F}_0^{\top}\mathbf{F}_0,$$

$$\dot{\mathbf{C}}(\mathbf{X}_0, t_0) = 2\mathbf{F}_0^{\top}\mathbf{D}\mathbf{F}_0.$$

Since $\det\mathbf{F}_0$ with $\det\mathbf{F}_0 > 0$ and the symmetric tensor **D** were arbitrarily chosen, the verification of (†) is complete.

49 Summary of Basic Equations. Initial/Boundary-Value Problems

49.1 Basic Field Equations

Granted an inertial frame, the basic field equations describing the motion of an elastic body consist of the kinematical relations (6.1) and $(7.3)_1$ defining the deformation gradient and right Cauchy–Green tensor, the relation (48.14) determining the Piola stress, and the local balance (24.10) of linear momentum[239]

$$\boxed{\begin{aligned} &\mathbf{F} = \nabla\boldsymbol{\chi}, \qquad \mathbf{C} = \mathbf{F}^{\top}\mathbf{F}, \\ &\mathbf{T}_{\mathrm{R}} = 2\mathbf{F}\frac{\partial \bar{\psi}_{\mathrm{R}}(\mathbf{C})}{\partial \mathbf{C}}, \\ &\rho_{\mathrm{R}}\ddot{\boldsymbol{\chi}} = \operatorname{Div}\mathbf{T}_{\mathrm{R}} + \mathbf{b}_{0\mathrm{R}}. \end{aligned}} \tag{49.1}$$

These equations hold on the reference body B.

Equilibrium solutions of these equations are of great importance; as such solutions are independent of time, the momentum balance $(49.1)_4$ is replaced by the **equilibrium balance**

$$\operatorname{Div}\mathbf{T}_{\mathrm{R}} + \mathbf{b}_{0\mathrm{R}} = \mathbf{0}. \tag{49.2}$$

A simple equilibrium solution of the basic equations is constructed as follows. Consider a homogeneous deformation $\boldsymbol{\chi}$.[240] Then — bearing in mind our initial assumption that the body be homogeneous — since the deformation gradient $\mathbf{F}$ is identically constant, the constitutive equation $(49.1)_3$ implies that the Piola stress $\mathbf{T}_{\mathrm{R}}$ is also identically constant and hence trivially satisfies the equilibrium equation without body forces:

$$\operatorname{Div}\mathbf{T}_{\mathrm{R}} = \mathbf{0}.$$

Thus, $\boldsymbol{\chi}$ represents a deformation that can be produced in a body by the application of surface tractions alone; such deformations are referred to as **controllable**. We have shown that

(‡) *all homogeneous deformations of a homogeneous elastic body are controllable.*

[239] Cf. §47.1 and §47.2.
[240] Cf. §6.1.

49.2 A Typical Initial/Boundary-Value Problem

Let S_1 and S_2 be complementary subsurfaces of the boundary ∂B, so that

(i) ∂B is the union of S_1 and S_2;
(ii) S_1 and S_2 intersect at most along their boundaries ∂S_1 and ∂S_2.

Possible boundary conditions might entail specifying the motion on S_1 and the surface traction on S_2:

$$\begin{aligned} \chi &= \hat{\chi} \quad \text{a prescribed function on } S_1 \text{ for all } t \geq 0; \\ \mathbf{T}_R\mathbf{n}_R &= \hat{\mathbf{t}}_R \quad \text{a prescribed function on } S_2 \text{ for all } t \geq 0. \end{aligned} \tag{49.3}$$

Standard initial conditions involve a specification of the initial deformation and the velocity

$$\chi(\mathbf{X}, 0) = \chi_0(\mathbf{X}), \qquad \dot{\chi}(\mathbf{X}, 0) = \mathbf{v}_0(\mathbf{X}), \tag{49.4}$$

with χ_0 and $\mathbf{v}_0$ prescribed functions on B.

The *initial/boundary-value problem* corresponding to the prescribed data

$$\{\mathbf{b}_{0R}, \rho_R, \hat{\chi}, \hat{\mathbf{t}}_R, \chi_0, \mathbf{v}_0\}$$

then consists of finding a motion $\chi(\mathbf{X}, t)$ — defined for $\mathbf{X}$ in B and $t \geq 0$ — that satisfies the field equations (49.1) for $\mathbf{X}$ in B and $t \geq 0$, the boundary conditions (49.3) on ∂B for $t \geq 0$, and the initial conditions (49.4) for $\mathbf{X}$ in B.

Also of interest are boundary-value problems involving equilibrium solutions of the basic field equations and boundary conditions. Since initial conditions are irrelevant to such problems, the data consists of $\{\mathbf{b}_{0R}, \hat{\chi}, \hat{\mathbf{t}}_R\}$ and the corresponding **boundary-value problem** consists of finding a deformation χ satisfying the field equations (49.1) — with the momentum balance replaced by the equilibrium balance (49.2) — and the obvious time-independent analogs of the boundary conditions (49.3).

Remark. Another form of boundary condition arises when one specifies the surface traction $\mathbf{Tn}$ on the deformed surface

$$\mathcal{S}_2(t) = \chi(S_2, t).$$

A simple example of this type of condition arises when one considers the effects of a uniform pressure p_0:

$$\mathbf{Tn} = -p_0\mathbf{n} \quad \text{on } \mathcal{S}_2. \tag{49.5}$$

Using the relation $\mathbf{n}\,da = J\mathbf{F}^{-\top}\mathbf{n}_R\,da_R = \mathbf{F}^{\mathrm{c}}\mathbf{n}_R\,da_R$, (49.5) may be expressed as

$$\mathbf{T}_R\mathbf{n}_R = -p_0\mathbf{F}^{\mathrm{c}}\mathbf{n}_R \quad \text{on } S_2$$

and furnishes an example of a **configuration-dependent** boundary condition.

50 Material Symmetry

Assume, for the moment, that the reference body is an undistorted cubic crystal. Then, a rotation of the reference body by 90° would not affect its response to deformation, nor would any two such rotations (about the same or different axes) applied one after the other, nor would the inverse of any such rotation. In fact, the set of all rotations (including the identity **1**) that leave a cube unaltered forms a group that crystallographers refer to as the cubic point group.

In §20.1 we note that — to discuss the notion of invariance under changes in frame — it is useful to differentiate conceptually between the ambient space for B and the ambient space through which $\mathcal{B}_t$ evolves; in accord with this, we introduce the terminology:

(i) the ambient space through which $\mathcal{B}_t$ evolves is termed the *observed space*;
(ii) the ambient space for the reference body B is termed the *reference space*.

In our discussion in §20.2 of frame-indifferent fields, a tensor **D**, here the stretching measured with respect to a frame $\mathcal{F}$, transforms to the tensor

$$\mathbf{Q}\mathbf{D}\mathbf{Q}^{\top} \tag{50.1}$$

when measured in a new frame $\mathcal{F}^* = \mathbf{Q}\mathcal{F}$. Thus, **Q** is a rotation of spatial vectors and, hence, a rotation of objects in the *observed space*. In contrast,

- discussions of material symmetry involve rotations **Q** of the body within the *reference space*;

but, as we show in §50.2, transformations of the generic form

$$\mathbf{Q}(\text{tensor})\mathbf{Q}^{\top} \tag{50.2}$$

arise in that space also.

50.1 The Notion of a Group. Invariance Under a Group

A set $\mathcal{G}$ *of rotations* forms a **group** if it is closed under multiplication and inversion:

(G1) if $\mathbf{Q}_1$ and $\mathbf{Q}_2$ belong to $\mathcal{G}$, then so also does their product $\mathbf{Q}_1\mathbf{Q}_2$;
(G2) if **Q** belongs to $\mathcal{G}$, then so also does its inverse $\mathbf{Q}^{\top}$.

The set of *all rotations* forms the **proper orthogonal group**

$$\text{Orth}^+ = \{\text{all rotations}\}, \tag{50.3}$$

and the group $\mathcal{G}$ defined above is a **subgroup** of Orth^+.

Each of the functions we consider in this section has as its domain of definition one of the following sets:

$$\text{Lin}^+ = \{\text{all tensors with positive determinant}\},$$

$$\text{Sym} = \{\text{all symmetric tensors}\},$$

$$\text{Psym} = \{\text{all symmetric, postitive-definite tensors}\}.$$

Let $\mathcal{G}$ be a subgroup of Orth^+. It then follows that Lin^+, Sym, and Psym **are invariant under** $\mathcal{G}$ in the sense that — for $\mathcal{A}$ any one of these sets — if $\mathbf{A}$ belongs to $\mathcal{A}$, then so also does $\mathbf{QAQ}^\top$ for all $\mathbf{Q}$ in $\mathcal{G}$.[241] (We leave the proof of this as an exercise.)

The discussion in the paragraph containing (50.2) should help to motivate the following terminology. Let $\mathcal{A}$ be invariant under $\mathcal{G}$. Then:

(i) a scalar function φ with domain $\mathcal{A}$ is **invariant under** $\mathcal{G}$ if given any $\mathbf{A}$ in $\mathcal{A}$,

$$\varphi(\mathbf{QAQ}^\top) = \varphi(\mathbf{A}) \quad \text{for all } \mathbf{Q} \text{ in } \mathcal{G}; \tag{50.4}$$

(ii) a tensor-valued function $\mathbf{\Phi}$ with domain $\mathcal{A}$ is **invariant under** $\mathcal{G}$ if given any $\mathbf{A}$ in $\mathcal{A}$,

$$\mathbf{\Phi}(\mathbf{QAQ}^\top) = \mathbf{Q\Phi}(\mathbf{A})\mathbf{Q}^\top \quad \text{for all } \mathbf{Q} \text{ in } \mathcal{G}. \tag{50.5}$$

EXERCISE

1. Show that if two tensor functions $\mathbf{\Phi}_1$ and $\mathbf{\Phi}_2$ are invariant under $\mathcal{G}$, then so also is the product function $\mathbf{\Phi}_1\mathbf{\Phi}_2$.

50.2 The Symmetry Group $\mathcal{G}$

Within the constitutive framework under discussion, a symmetry transformation is defined as a *rotation of the reference body* that leaves the response to deformation unaltered. Because of the underlying thermodynamic structure, the energetic response to deformation determines the stress response and this result allows us to define material symmetry in terms of the response function for the free energy.

Choose a point $\mathbf{X}$ in B, and let $\mathbf{f_F}$ denote the **homogeneous deformation** from $\mathbf{X}$ with deformation gradient $\mathbf{F}$

$$\mathbf{f_F}(\mathbf{Y}) = \mathbf{X} + \mathbf{F}(\mathbf{Y} - \mathbf{X}) \tag{50.6}$$

for all $\mathbf{Y}$ in B. Then, given a rotation $\mathbf{Q}$, consider the following thought experiments:

- **Experiment** 1. Deform B with the homogeneous deformation $\mathbf{f_F}$. In this experiment, the deformation gradient is $\mathbf{F}$, and the free energy is

$$\begin{aligned}\psi_{\mathrm{R}1} &= \hat{\psi}_{\mathrm{R}}(\mathbf{F}) \\ &= \bar{\psi}_{\mathrm{R}}(\mathbf{C}).\end{aligned} \tag{50.7}$$

- **Experiment** 2. First rotate B with the rotation $\mathbf{Q}$ via the homogeneous deformation

$$\mathbf{f_Q}(\mathbf{Y}) = \mathbf{X} + \mathbf{Q}(\mathbf{Y} - \mathbf{X}),$$

and then deform the rotated body with the same homogeneous deformation $\mathbf{f_F}$ as in experiment 1 (Figure 50.1). In this experiment, the composite deformation

[241] Cf. (50.2).

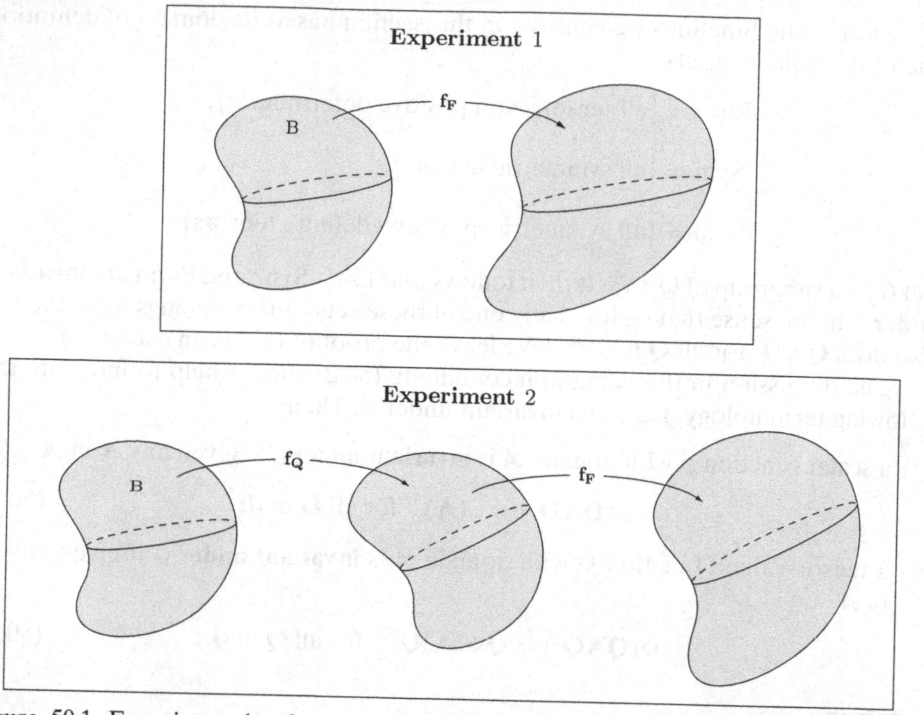

Figure 50.1. Experiment 1: a homogeneous deformation with deformation gradient **F** and corresponding free energy $\psi_{R1} = \hat{\psi}(\mathbf{F})$. Experiment 2: two successive homogeneous deformations with deformation gradients **Q** and **F** resulting in a free energy $\psi_{R2} = \hat{\psi}_R(\mathbf{FQ})$.

gradient relative to the original unrotated reference body is

$$\mathbf{F}_2 = \mathbf{FQ}. \tag{50.8}$$

Then

$$\begin{aligned}\mathbf{C}_2 &= \mathbf{F}_2^\top \mathbf{F}_2 \\ &= \mathbf{Q}^\top \mathbf{F}^\top \mathbf{F}\mathbf{Q} \\ &= \mathbf{Q}^\top \mathbf{C}\mathbf{Q}\end{aligned} \tag{50.9}$$

and the free energy in this second experiment is given by

$$\begin{aligned}\psi_{R2} &= \hat{\psi}_R(\mathbf{FQ}) \\ &= \bar{\psi}_R(\mathbf{Q}^\top \mathbf{C}\mathbf{Q}).\end{aligned} \tag{50.10}$$

In general, we would expect that $\psi_{R1} \neq \psi_{R2}$; however, if for a given rotation $\mathbf{Q}$, $\psi_{R1} = \psi_{R2}$ for *every* $\mathbf{F}$, then $\mathbf{Q}$ is referred to as a symmetry transformation.[242] Thus, by (50.7) and (50.10), a **symmetry transformation** is a rotation $\mathbf{Q}$ such that

$$\hat{\psi}_R(\mathbf{F}) = \hat{\psi}_R(\mathbf{FQ}) \tag{50.11}$$

[242] Cf., e.g., TRUESDELL & NOLL (1965, §§31,85) and GURTIN (1981, §25). Here, to avoid complicated notation, we restrict attention to a homogeneous body, (Cf. the paragraph containing (35.2)) so that the symmetry group does not vary from point to point. For an inhomogeneous body, the symmetry group for a given material point **X** may be defined in a strictly analogous manner; the consequences for **X** are then no different than those derived here for B.

for every deformation gradient $\mathbf{F}$; or equivalently,

$$\bar{\psi}_{\mathrm{R}}(\mathbf{C}) = \bar{\psi}_{\mathrm{R}}(\mathbf{Q}^{\top}\mathbf{C}\mathbf{Q}) \tag{50.12}$$

for every symmetric, positive-definite tensor $\mathbf{C}$. We write $\mathcal{G}$ for the set of *all symmetry transformations* and refer to $\mathcal{G}$ as the **symmetry group** for the body. We show — at the end of this subsection — that $\mathcal{G}$ is indeed a group.

Note that, by (48.3) (which follows from frame-indifference) with $\mathbf{Q}$ replaced by $\mathbf{Q}^{\top}$,

$$\hat{\psi}_{\mathrm{R}}(\mathbf{F}) = \hat{\psi}_{\mathrm{R}}(\mathbf{Q}^{\top}\mathbf{F})$$

for every rotation $\mathbf{Q}$. Thus, (50.11) implies that

$$\hat{\psi}_{\mathrm{R}}(\mathbf{F}) = \hat{\psi}_{\mathrm{R}}(\mathbf{Q}^{\top}\mathbf{F}\mathbf{Q}) \tag{50.13}$$

for every symmetry transformation $\mathbf{Q}$.

Succinctly, (50.12) and (50.13) therefore imply that the response functions $\bar{\psi}_{\mathrm{R}}(\mathbf{C})$ and $\hat{\psi}_{\mathrm{R}}(\mathbf{F})$ for the free energy are each invariant under the symmetry group $\mathcal{G}$.[243]

Remark. It is important to note that the symmetry group depends on the choice of reference. For the body a cubic crystal, a reference in which the lattice is undistorted would have the cubic point group as symmetry group; but this would not be so if we changed reference via a homogeneous deformation $\mathbf{f_H}$ with $\mathbf{H}$ neither a rotation nor a dilatation, for such a deformation would distort the lattice. Specifically, if subjected to such a homogeneous deformation, the body would have an extended symmetry group $\mathcal{G}^{\mathrm{ext}}$, but the elements of $\mathcal{G}^{\mathrm{ext}}$ would not be rotations, as each such element would be of the form $\mathbf{H}\mathbf{Q}\mathbf{H}^{-1}$ with $\mathbf{Q}$ in $\mathcal{G}$.[244] Therefore,

- the assumption that $\mathcal{G}$ is a nontrivial[245] group of rotations carries with it the tacit assumption that the reference configuration is *undistorted.*

In view of (48.14) and (48.15), the basic ingredient in the constitutive relations for the Piola and Cauchy stresses is the relation (48.12) for the second Piola stress. We now determine the symmetry group for this stress. Choose, arbitrarily, a symmetric, positive-definite tensor $\mathbf{C}_0$ and an arbitrary symmetric tensor $\mathbf{A}$. Let $\mathbf{C}(t)$ be a symmetric, positive-definite tensor function of a parameter t consistent with[246]

$$\mathbf{C}(0) = \mathbf{C}_0, \qquad \dot{\mathbf{C}}(0) = \mathbf{A}, \tag{50.14}$$

and let $\psi_{\mathrm{R}}(t)$ be defined by

$$\psi_{\mathrm{R}}(t) = \bar{\psi}_{\mathrm{R}}(\mathbf{C}(t));$$

then, by (48.12) and the chain-rule,

$$2\dot{\psi}_{\mathrm{R}}(t) = \bar{\mathbf{T}}_{\mathrm{RR}}(\mathbf{C}(t)) : \dot{\mathbf{C}}(t)$$

and (50.14) yields

$$2\dot{\psi}_{\mathrm{R}}(0) = \bar{\mathbf{T}}_{\mathrm{RR}}(\mathbf{C}_0) : \mathbf{A}. \tag{50.15}$$

Choose an arbitrary symmetry transformation $\mathbf{Q}$. Then, by (50.12)

$$\psi_{\mathrm{R}}(t) = \bar{\psi}_{\mathrm{R}}(\mathbf{Q}^{\top}\mathbf{C}(t)\mathbf{Q}),$$

243 Because $\mathbf{Q}$ belongs to $\mathcal{G}$ if and only if $\mathbf{Q}^{\top}$ belongs to $\mathcal{G}$, (50.12) could equally well have been written $\bar{\psi}_{\mathrm{R}}(\mathbf{C}) = \bar{\psi}_{\mathrm{R}}(\mathbf{Q}\mathbf{C}\mathbf{Q}^{\top})$, which is the more standard form used on page 285 in the definition of "invariance under $\mathcal{G}$."

244 Cf. equation (33.1) of Truesdell & Noll (1965).

245 $\mathcal{G}$ contains elements other than $\mathbf{1}$.

246 It is helpful to view $\mathbf{C}(t)$ as the right Cauchy–Green tensor and t as the time.

and differentiating this relation with respect to t at $t = 0$ yields, by virtue of (50.14),

$$
\begin{aligned}
2\dot{\psi}_{R}(0) &= \bar{\mathbf{T}}_{RR}(\mathbf{Q}^{\top}\mathbf{C}(0)\mathbf{Q}):(\mathbf{Q}^{\top}\dot{\mathbf{C}}(0)\mathbf{Q}) \\
&= \bar{\mathbf{T}}_{RR}(\mathbf{Q}^{\top}\mathbf{C}_0\mathbf{Q}):(\mathbf{Q}^{\top}\mathbf{A}\mathbf{Q}) \\
&= (\mathbf{Q}\bar{\mathbf{T}}_{RR}(\mathbf{Q}^{\top}\mathbf{C}_0\mathbf{Q})\mathbf{Q}^{\top}):\mathbf{A}.
\end{aligned}
\tag{50.16}
$$

By (50.15) and (50.16),

$$
(\bar{\mathbf{T}}_{RR}(\mathbf{C}_0) - \mathbf{Q}\bar{\mathbf{T}}_{RR}(\mathbf{Q}^{\top}\mathbf{C}_0\mathbf{Q})\mathbf{Q}^{\top}):\mathbf{A} = 0,
$$

and since this must hold for all symmetric tensors $\mathbf{A}$,

$$
\bar{\mathbf{T}}_{RR}(\mathbf{C}_0) = \mathbf{Q}\bar{\mathbf{T}}_{RR}(\mathbf{Q}^{\top}\mathbf{C}_0\mathbf{Q})\mathbf{Q}^{\top}.
$$

Thus, since $\mathbf{C}_0$ was arbitrarily chosen, the response function for the second Piola stress obeys the following transformation law

$$
\mathbf{Q}^{\top}\bar{\mathbf{T}}_{RR}(\mathbf{C})\mathbf{Q} = \bar{\mathbf{T}}_{RR}(\mathbf{Q}^{\top}\mathbf{C}\mathbf{Q}). \tag{50.17}
$$

In summary, the central results of this section are the transformation laws

$$
\boxed{
\begin{aligned}
\bar{\psi}_{R}(\mathbf{C}) &= \bar{\psi}_{R}(\mathbf{Q}^{\top}\mathbf{C}\mathbf{Q}), \\
\mathbf{Q}^{\top}\bar{\mathbf{T}}_{RR}(\mathbf{C})\mathbf{Q} &= \bar{\mathbf{T}}_{RR}(\mathbf{Q}^{\top}\mathbf{C}\mathbf{Q}),
\end{aligned}
}
\tag{50.18}
$$

which are required to hold for all symmetric, positive-definite tensors $\mathbf{C}$ and all symmetry transformations $\mathbf{Q}$. The transformation laws (50.18) are standard for scalar and tensor functions. They assert that

- the response functions $\bar{\psi}_{R}$ and $\bar{\mathbf{T}}_{RR}$ *are invariant under the symmetry group* $\mathcal{G}$.

The argument resulting in (50.17) yields the following result:

(‡) Let φ be a scalar function with domain Sym or Psym and assume that φ is invariant under $\mathcal{G}$. Let $\mathbf{M}$ be the tensor function defined by

$$
\mathbf{M}(\mathbf{A}) = \frac{\partial\varphi(\mathbf{A})}{\partial\mathbf{A}}.
$$

Then $\mathbf{M}$ is also invariant under $\mathcal{G}$.

50.2.1 Proof That $\mathcal{G}$ Is a Group

We now show that $\mathcal{G}$ is a group. To accomplish this, we must show that (G1) and (G2) on page 284 are satisfied. Assume first that $\mathbf{Q}_1$ and $\mathbf{Q}_2$ belong to $\mathcal{G}$. Choose $\mathbf{F}$ arbitrarily. Then, by (50.11) applied twice, first with $\mathbf{Q} = \mathbf{Q}_2$ and then with $\mathbf{Q} = \mathbf{Q}_1$,

$$
\begin{aligned}
\hat{\psi}_{R}(\mathbf{F}) &= \hat{\psi}_{R}(\mathbf{F}\mathbf{Q}_1) \\
&= \hat{\psi}_{R}(\mathbf{F}\mathbf{Q}_1\mathbf{Q}_2);
\end{aligned}
$$

hence, $\mathbf{Q}_1\mathbf{Q}_2$ belongs to $\mathcal{G}$, which is (G1). Assume, next, that $\mathbf{Q}$ belongs to $\mathcal{G}$. Choose $\mathbf{F}$ arbitrarily and let $\tilde{\mathbf{F}} = \mathbf{F}\mathbf{Q}^{\top}$. Then, $\hat{\psi}_{R}(\tilde{\mathbf{F}}\mathbf{Q}) = \hat{\psi}_{R}(\tilde{\mathbf{F}})$ and, therefore, $\hat{\psi}_{R}(\mathbf{F}) = \hat{\psi}_{R}(\mathbf{F}\mathbf{Q}^{\top})$, whereby $\mathbf{Q}^{\top}$ belongs to $\mathcal{G}$, which is (G2). Hence, $\mathcal{G}$ is a group.

50.3 Isotropy

In terms more suggestive than precise, an isotropic body is a body whose properties are the same in all directions. The notion of a symmetry transformation allows us to make this notion precise. We say that the body is **isotropic** if every rotation is a symmetry transformation; thus, for an isotropic body,

$$\mathcal{G} = \mathrm{Orth}^{+}.$$

On the other hand, the body is *anisotropic* if $\mathcal{G}$ is a *proper* subgroup of Orth^{+} (so that $\mathcal{G} \neq \mathrm{Orth}^{+}$).

- We assume, for the remainder of this subsection, that *the body is isotropic*, so that

$$\bar{\psi}_{\mathrm{R}}(\mathbf{Q}^{\top}\mathbf{C}\mathbf{Q}) = \bar{\psi}_{\mathrm{R}}(\mathbf{C}) \tag{50.19}$$

  for all rotations $\mathbf{Q}$ and all symmetric, positive-definite tensors $\mathbf{C}$.[247]

Then, since (50.19) holds for all rotations $\mathbf{Q}$, we may, without loss in generality, choose $\mathbf{Q} = \mathbf{R}^{\top}$ where $\mathbf{R}$ is the rotation in the polar decomposition $\mathbf{F} = \mathbf{R}\mathbf{U} = \mathbf{V}\mathbf{R}$. Recalling the relation

$$\mathbf{B} = \mathbf{R}\mathbf{C}\mathbf{R}^{\top} \tag{50.20}$$

between the right and left Cauchy–Green tensors

$$\mathbf{C} = \mathbf{F}^{\top}\mathbf{F} \qquad \text{and} \qquad \mathbf{B} = \mathbf{F}\mathbf{F}^{\top},$$

we conclude that

$$\bar{\psi}_{\mathrm{R}}(\mathbf{C}) = \bar{\psi}_{\mathrm{R}}(\mathbf{B}), \tag{50.21}$$

and, by (48.7), we may express the free energy in terms of $\mathbf{B}$:

$$\psi_{\mathrm{R}} = \bar{\psi}_{\mathrm{R}}(\mathbf{B}).$$

Choose an arbitrary constitutive process. Then, by (48.13),

$$\dot{\psi}_{\mathrm{R}} = \mathbf{T}_{\mathrm{R}} : \dot{\mathbf{F}},$$

and arguing as in the steps leading to (48.11), we find that, by $(48.1)_2$,

$$\frac{\partial \bar{\psi}_{\mathrm{R}}(\mathbf{B})}{\partial \mathbf{B}} : \dot{\mathbf{B}} = \hat{\mathbf{T}}_{\mathrm{R}}(\mathbf{F}) : \dot{\mathbf{F}}.$$

Further,

$$\begin{aligned}
\dot{\mathbf{B}} &= \overline{\mathbf{F}\mathbf{F}^{\top}}^{\boldsymbol{\cdot}} \\
&= (\dot{\mathbf{F}}\mathbf{F}^{\top} + \mathbf{F}\dot{\mathbf{F}}^{\top}) \\
&= 2\,\mathrm{sym}\,(\dot{\mathbf{F}}\mathbf{F}^{\top}),
\end{aligned}$$

and, since $\partial \bar{\psi}_{\mathrm{R}}(\mathbf{B})/\partial \mathbf{B}$ is symmetric, it follows that

$$\begin{aligned}
\frac{\partial \bar{\psi}_{\mathrm{R}}(\mathbf{B})}{\partial \mathbf{B}} : \dot{\mathbf{B}} &= 2\frac{\partial \bar{\psi}_{\mathrm{R}}(\mathbf{B})}{\partial \mathbf{B}} : (\dot{\mathbf{F}}\mathbf{F}^{\top}) \\
&= 2\left(\frac{\partial \bar{\psi}_{\mathrm{R}}(\mathbf{B})}{\partial \mathbf{B}}\mathbf{F}\right) : \dot{\mathbf{F}};
\end{aligned}$$

therefore

$$\left(2\frac{\partial \bar{\psi}_{\mathrm{R}}(\mathbf{B})}{\partial \mathbf{B}}\mathbf{F} - \hat{\mathbf{T}}_{\mathrm{R}}(\mathbf{F})\right) : \dot{\mathbf{F}} = 0,$$

[247] Note that (50.19) holds trivially when $\mathbf{Q}$ is replaced by $-\mathbf{Q}$, and therefore the phrase "for all rotations" may be replaced by "for all orthogonal tensors." Thus $\bar{\psi}$ is an *isotropic function*: A scalar or tensor function invariant under the full orthogonal group is referred to as an *isotropic function*.

and, appealing to (†) on page 279, we are led to the relation

$$\mathbf{T}_{\mathrm{R}} = 2\frac{\partial \bar{\psi}_{\mathrm{R}}(\mathbf{B})}{\partial \mathbf{B}}\mathbf{F}. \tag{50.22}$$

Finally, since $\mathbf{T} = J^{-1}\mathbf{T}_{\mathrm{R}}\mathbf{F}^{\top}$ and $J = \det \mathbf{F} = \sqrt{\det \mathbf{B}}$, we obtain an expression giving the Cauchy stress as a function

$$\mathbf{T} = \frac{2}{\sqrt{\det \mathbf{B}}}\frac{\partial \bar{\psi}_{\mathrm{R}}(\mathbf{B})}{\partial \mathbf{B}}\mathbf{B} \tag{50.23}$$

of $\mathbf{B}$ only.

An immediate consequence of (50.23) and the symmetry of $\mathbf{T}$, $\mathbf{B}$, and $\partial \bar{\psi}_{\mathrm{R}}(\mathbf{B})/\partial \mathbf{B}$ is that $\mathbf{T}$ *commutes with* $\mathbf{B}$:

$$\mathbf{BT} = \mathbf{TB}. \tag{50.24}$$

Further, by (‡) on page 288, $\partial \bar{\psi}_{\mathrm{R}}(\mathbf{B})/\partial \mathbf{B}$ is an isotropic function of $\mathbf{B}$ and we may therefore conclude from (50.23) that $\bar{\mathbf{T}}$ *is an isotropic function, viz.*

$$\mathbf{Q}^{\top}\bar{\mathbf{T}}(\mathbf{B})\mathbf{Q} = \bar{\mathbf{T}}(\mathbf{Q}^{\top}\mathbf{B}\mathbf{Q}) \tag{50.25}$$

for all $\mathbf{B}$ and all rotations $\mathbf{Q}$.

An important consequence of (50.25) is that $\mathbf{T}$ *commutes with* $\mathbf{C}$:

$$\mathbf{CT} = \mathbf{TC}. \tag{50.26}$$

To verify this note first that, since $\mathbf{B} = \mathbf{RCR}^{\top}$,[248]

$$\begin{aligned}\mathbf{TB} &= \bar{\mathbf{T}}(\mathbf{B})\mathbf{B}\\ &= \bar{\mathbf{T}}(\mathbf{RCR}^{\top})\mathbf{RCR}^{\top}\\ &= \mathbf{R}\bar{\mathbf{T}}(\mathbf{C})\mathbf{R}^{\top}\mathbf{RCR}^{\top}\\ &= \mathbf{R}\bar{\mathbf{T}}(\mathbf{C})\mathbf{CR}^{\top}.\end{aligned}$$

An analogous argument yields $\mathbf{BT} = \mathbf{RC}\bar{\mathbf{T}}(\mathbf{C})\mathbf{R}^{\top}$. Thus, by (50.24),

$$\mathbf{R}\bar{\mathbf{T}}(\mathbf{C})\mathbf{CR}^{\top} = \mathbf{RC}\bar{\mathbf{T}}(\mathbf{C})\mathbf{R}^{\top}$$

and (50.26) follows.

50.3.1 Free Energy Expressed in Terms of Invariants

A classical representation theorem asserts that an isotropic scalar function of a symmetric tensor $\mathbf{B}$ may be expressed as a function of the principal invariants $I_1(\mathbf{B})$, $I_2(\mathbf{B})$, and $I_3(\mathbf{B})$ of $\mathbf{B}$.[249] The constitutive relation for the free energy of an isotropic elastic solid may therefore be written in the form

$$\psi_{\mathrm{R}} = \tilde{\psi}_{\mathrm{R}}(\mathcal{I}_{\mathbf{B}}), \tag{50.27}$$

with[250]

$$\mathcal{I}_{\mathbf{B}} \stackrel{\text{def}}{=} (I_1(\mathbf{B}), I_2(\mathbf{B}), I_3(\mathbf{B})); \tag{50.28}$$

248 Cf. $(47.5)_2$.
249 Cf. Appendix 113.3.
250 Cf. (2.142).

Then, by (50.23),

$$\mathbf{T} = \frac{2}{\sqrt{I_3}}\left(\frac{\partial \tilde{\psi}_{\mathrm{R}}(\mathcal{I}_{\mathbf{B}})}{\partial \mathbf{B}}\right)\mathbf{B}, \tag{50.29}$$

and, therefore, recalling, from (3.35), the identities

$$\left.\begin{aligned} \frac{\partial I_1(\mathbf{B})}{\partial \mathbf{B}} &= \mathbf{1}, \\ \frac{\partial I_2(\mathbf{B})}{\partial \mathbf{B}} &= I_1(\mathbf{B})\mathbf{1} - \mathbf{B}, \\ \frac{\partial I_3(\mathbf{B})}{\partial \mathbf{B}} &= I_3(\mathbf{B})\mathbf{B}^{-1}, \end{aligned}\right\} \tag{50.30}$$

we arrive at the following expression for the Cauchy stress:

$$\mathbf{T} = \frac{2}{\sqrt{I_3}}\left[I_3\frac{\partial \tilde{\psi}_{\mathrm{R}}(\mathcal{I}_{\mathbf{B}})}{\partial I_3}\mathbf{1} + \left(\frac{\partial \tilde{\psi}_{\mathrm{R}}(\mathcal{I}_{\mathbf{B}})}{\partial I_1} + I_1\frac{\partial \tilde{\psi}_{\mathrm{R}}(\mathcal{I}_{\mathbf{B}})}{\partial I_2}\right)\mathbf{B} - \frac{\partial \tilde{\psi}_{\mathrm{R}}(\mathcal{I}_{\mathbf{B}})}{\partial I_2}\mathbf{B}^2\right]. \tag{50.31}$$

An alternative to (50.31) follows from the Cayley–Hamilton equation[251]

$$\mathbf{B}^3 - I_1(\mathbf{B})\mathbf{B}^2 + I_2(\mathbf{B})\mathbf{B} - I_3(\mathbf{B})\mathbf{1} = \mathbf{0}. \tag{50.32}$$

Multiplying (50.32) by $\mathbf{B}^{-1}$ yields

$$\mathbf{B}^2 = I_1\mathbf{B} - I_2\mathbf{1} + I_3\mathbf{B}^{-1}, \tag{50.33}$$

and, using this relation to eliminate $\mathbf{B}^2$ from (50.31), we find that

$$\mathbf{T} = \beta_0(\mathcal{I}_{\mathbf{B}})\mathbf{1} + \beta_1(\mathcal{I}_{\mathbf{B}})\mathbf{B} + \beta_2(\mathcal{I}_{\mathbf{B}})\mathbf{B}^{-1}, \tag{50.34}$$

with

$$\left.\begin{aligned} \beta_0(\mathcal{I}_{\mathbf{B}}) &= \frac{2}{\sqrt{I_3}}\left(I_2\frac{\partial \tilde{\psi}_{\mathrm{R}}(\mathcal{I}_{\mathbf{B}})}{\partial I_2} + I_3\frac{\partial \tilde{\psi}_{\mathrm{R}}(\mathcal{I}_{\mathbf{B}})}{\partial I_3}\right), \\ \beta_1(\mathcal{I}_{\mathbf{B}}) &= \frac{2}{\sqrt{I_3}}\frac{\partial \tilde{\psi}_{\mathrm{R}}(\mathcal{I}_{\mathbf{B}})}{\partial I_1}, \\ \beta_2(\mathcal{I}_{\mathbf{B}}) &= -2\sqrt{I_3}\frac{\partial \tilde{\psi}_{\mathrm{R}}(\mathcal{I}_{\mathbf{B}})}{\partial I_2}. \end{aligned}\right\} \tag{50.35}$$

The stress $\mathbf{T}$ when $\mathbf{F} = \mathbf{1}$ represents the stress in the (undistorted) reference configuration. Since $\mathbf{B} = \mathbf{1}$ when $\mathbf{F} = \mathbf{1}$, $\mathcal{I}_{\mathbf{B}} = \mathcal{I}_{\mathbf{1}} = (3, 3, 1)$ and (50.34) implies that this stress is given by

$$\mathbf{T}|_{\mathbf{B}=\mathbf{1}} = [\beta_0(3,3,1) + \beta_1(3,3,1) + \beta_2(3,3,1)]\mathbf{1}, \tag{50.36}$$

and is therefore hydrostatic. If, in addition, the reference configuration is stress-free, (48.18) and (50.36) imply that

$$\beta_0(3,3,1) + \beta_1(3,3,1) + \beta_2(3,3,1) = 0. \tag{50.37}$$

[251] Cf. (2.144).

50.3.2 Free Energy Expressed in Terms of Principal stretches

In view of (2.143), we recall that the principal invariants $I_1(\mathbf{B})$, $I_2(\mathbf{B})$, and $I_3(\mathbf{B})$ of $\mathbf{B}$ may be expressed in terms of the principal stretches λ_1, λ_2, and λ_3 as[252]

$$\left.\begin{aligned} I_1(\mathbf{B}) &= \lambda_1^2 + \lambda_2^2 + \lambda_3^2, \\ I_2(\mathbf{B}) &= \lambda_1^2\lambda_2^2 + \lambda_2^2\lambda_3^2 + \lambda_3^2\lambda_1^2, \\ I_3(\mathbf{B}) &= \lambda_1^2\lambda_2^2\lambda_3^2. \end{aligned}\right\} \tag{50.38}$$

In writing (50.38) it is tacit that the list $(\lambda_1, \lambda_2, \lambda_3)$ of principal stretches is presumed to have each stretch repeated a number of times equal to its multiplicity as an eigenvalue of $\mathbf{V}$. Next, we may use (50.38) in (50.27) to express the free energy in terms of the principal stretches

$$\begin{aligned} \psi_R &= \tilde{\psi}_R(\mathcal{I}_{\mathbf{B}}) \\ &= \breve{\psi}_R(\lambda_1, \lambda_2, \lambda_3). \end{aligned} \tag{50.39}$$

Since the expressions (50.38) for $I_1(\mathbf{B})$, $I_2(\mathbf{B})$, and $I_3(\mathbf{B})$ in terms of λ_1, λ_2, and λ_3 are invariant under permutations of the integers (1, 2, 3) labeling the principal stretches, so also is $\breve{\psi}_R(\lambda_1, \lambda_2, \lambda_3)$; that is,

$$\breve{\psi}_R(\lambda_1, \lambda_2, \lambda_3) = \breve{\psi}_R(\lambda_1, \lambda_3, \lambda_2) \quad \text{and so forth.}^{253}$$

Next, let

$$\omega_k = \lambda_k^2, \quad k = 1, 2, 3. \tag{50.40}$$

Then, by the chain-rule and (50.29), the Cauchy stress is given by

$$\begin{aligned} \mathbf{T} &= \frac{2}{\lambda_1\lambda_2\lambda_3}\left(\frac{\partial \breve{\psi}_R(\lambda_1, \lambda_2, \lambda_3)}{\partial \mathbf{B}}\right)\mathbf{B} \\ &= \frac{2}{\lambda_1\lambda_2\lambda_3}\left(\sum_{i=1}^{3} \frac{\partial \breve{\psi}_R(\lambda_1, \lambda_2, \lambda_3)}{\partial \lambda_i}\frac{\partial \lambda_i}{\partial \mathbf{B}}\right)\mathbf{B} \\ &= \frac{1}{\lambda_1\lambda_2\lambda_3}\left(\sum_{i=1}^{3} \frac{1}{\lambda_i}\frac{\partial \breve{\psi}_R(\lambda_1, \lambda_2, \lambda_3)}{\partial \lambda_i}\frac{\partial \omega_i}{\partial \mathbf{B}}\right)\mathbf{B}. \end{aligned} \tag{50.41}$$

By (7.28), the spectral representation of the left Cauchy–Green tensor is

$$\mathbf{B} = \sum_{i=1}^{3} \omega_i \mathbf{l}_i \otimes \mathbf{l}_i, \qquad \omega_i = \lambda_i^2. \tag{50.42}$$

Assume that the squared principal stretches ω_i are distinct, so that the ω_i and the principal directions l_i may be considered as functions of $\mathbf{B}$. Then, as we show at the end of this subsection,

$$\frac{\partial \omega_i}{\partial \mathbf{B}} = \mathbf{l}_i \otimes \mathbf{l}_i, \tag{50.43}$$

[252] Cf. (2.143).

[253] Were this not true, (50.39) would *not* define a functional relationship.

and, granted this, (50.42) and (50.41) imply that

$$\mathbf{T} = \frac{1}{\lambda_1\lambda_2\lambda_3}\sum_{i=1}^{3}\lambda_i\frac{\partial\breve{\psi}_{\rm R}(\lambda_1,\lambda_2,\lambda_3)}{\partial\lambda_i}\mathbf{l}_i\otimes\mathbf{l}_i. \tag{50.44}$$

Finally, by $(2.140)_3$,

$$\mathbf{F}^{-\top} = \sum_{i=1}^{3}\lambda_i^{-1}\mathbf{l}_i\otimes\mathbf{r}_i,$$

and the Piola stress $\mathbf{T}_{\rm R} = J\mathbf{T}\mathbf{F}^{-\top}$ is given by

$$\mathbf{T}_{\rm R} = \sum_{i=1}^{3}\frac{\partial\breve{\psi}_{\rm R}(\lambda_1,\lambda_2,\lambda_3)}{\partial\lambda_i}\mathbf{l}_i\otimes\mathbf{r}_i. \tag{50.45}$$

50.3.3 Verification of (50.43)

Consider the tensorial set

$$\text{Psym} = \{\text{all symmetric, positive-definite tensors with distinct principal values}\},$$

which is an *open* set in the space of symmetric tensors. Within this set the principal values ω_i and principal directions $\mathbf{l}_i$ of a tensor $\mathbf{B}$ are themselves smooth functions of $\mathbf{B}$ with derivatives

$$\frac{\partial\omega_i}{\partial\mathbf{B}} \quad\text{and}\quad \frac{\partial\mathbf{l}_i}{\partial\mathbf{B}}.$$

Next, choose an arbitrary smooth curve $\mathbf{B}(s)$ in Psym and let $\omega_i(s)$ and $\mathbf{l}_i(s)$ denote the corresponding principal values and directions. These functions then satisfy[254]

$$\omega_i(s) = \mathbf{l}_i(s)\cdot\mathbf{B}(s)\mathbf{l}_i(s) \qquad \text{(no sum)}.$$

If we differentiate these relations and use the symmetry of $\mathbf{B}$, we find that

$$\frac{\partial\omega_i}{\partial\mathbf{B}}:\frac{d\mathbf{B}}{ds} = (\mathbf{l}_i\otimes\mathbf{l}_i):\frac{d\mathbf{B}}{ds} + 2\frac{d\mathbf{l}_i}{ds}\cdot\mathbf{B}\mathbf{l}_i \qquad \text{(no sum)}.$$

But

$$\mathbf{B}\mathbf{l}_i = \omega_i\mathbf{l}_i \qquad \text{(no sum)}$$

and, since the principal directions are unit vectors,

$$\frac{d\mathbf{l}_i}{ds}\cdot\mathbf{l}_i = 0 \qquad \text{(no sum)}.$$

Thus,

$$\left(\frac{\partial\omega_i}{\partial\mathbf{B}} - \mathbf{l}_i\otimes\mathbf{l}_i\right):\frac{d\mathbf{B}}{ds} = 0 \qquad \text{(no sum)}. \tag{50.46}$$

Choose an arbitrary symmetric, positive-definite tensor $\mathbf{B}_0$ and an arbitrary symmetric tensor $\mathbf{A}$ and consider the curve

$$\mathbf{B}(s) = \mathbf{B}_0 + s\mathbf{A}, \qquad -s_0 < s < s_0,$$

with $s_0 > 0$ small enough that this curve lies in Psym. Then, by (50.46),

$$\mathbf{A}:\left[\frac{\partial\omega_i(\mathbf{B})}{\partial\mathbf{B}} - \mathbf{l}_i(\mathbf{B})\otimes\mathbf{l}_i(\mathbf{B})\right]_{\mathbf{B}=\mathbf{B}_0} = 0 \qquad \text{(no sum)}. \tag{50.47}$$

Since the tensor in parentheses is symmetric and since $\mathbf{A}$ is an arbitrary symmetric tensor, we must have

$$\left[\frac{\partial\omega_i(\mathbf{B})}{\partial\mathbf{B}} - \mathbf{l}_i(\mathbf{B})\otimes\mathbf{l}_i(\mathbf{B})\right]_{\mathbf{B}=\mathbf{B}_0} = \mathbf{0} \qquad \text{(no sum)}.$$

Finally, since $\mathbf{B}_0$ in Psym was chosen arbitrarily, the desired identity (50.43) follows.

[254] Cf. (2.121).

51 Simple Shear of a Homogeneous, Isotropic Elastic Body

In this section we discuss the equilibrium problem of simple shear of a homogeneous, isotropic elastic body, a problem that demonstrates basic differences between the theory of isotropic elasticity presented here — in which the strain may be arbitrarily large — and the classical linear theory in which the strain is infinitesimal.

Let B be a homogeneous, isotropic body in the shape of a cube. Consider a homogeneous deformation $\mathbf{x} = \boldsymbol{\chi}(\mathbf{X})$ defined (in Cartesian components) by

$$x_1 = X_1 + \gamma\, X_2, \qquad x_2 = X_2, \qquad x_3 = X_3, \tag{51.1}$$

where

$$\gamma = \tan\theta \tag{51.2}$$

is the *shear strain* (Figure 51.1). As noted in the paragraph containing (49.2), such a deformation generates a solution of the equilibrium equation

$$\mathrm{Div}\,\mathbf{T}_{\mathrm{R}} = \mathbf{0}$$

(without body forces) and hence represents a deformation that can be produced in a body by the application of surface tractions alone.

The matrix $[\mathbf{F}]$ corresponding to the deformation gradient $\mathbf{F}$ corresponding to (51.1) is

$$[\mathbf{F}] = \begin{bmatrix} 1 & \gamma & 0 \\ 0 & 1 & 0 \\ 0 & 0 & 1 \end{bmatrix}, \tag{51.3}$$

and the matrices of the left Cauchy–Green tensor $\mathbf{B} = \mathbf{F}\mathbf{F}^{\top}$ and its inverse are

$$[\mathbf{B}] = \begin{bmatrix} 1+\gamma^2 & \gamma & 0 \\ \gamma & 1 & 0 \\ 0 & 0 & 1 \end{bmatrix} \quad \text{and} \quad [\mathbf{B}]^{-1} = \begin{bmatrix} 1 & -\gamma & 0 \\ -\gamma & 1+\gamma^2 & 0 \\ 0 & 0 & 1 \end{bmatrix}. \tag{51.4}$$

Also,

$$\det(\mathbf{B} - \omega\mathbf{1}) = -\omega^3 + (3+\gamma^2)\omega^2 - (3+\gamma^2)\omega + 1,$$

and the list of principal invariants hence has the particular form $(3+\gamma^2, 3+\gamma^2, 1)$[255]

$$\mathcal{I}_{\mathbf{B}} = (3+\gamma^2, 3+\gamma^2, 1). \tag{51.5}$$

[255] Cf. (2.142) and (50.28).

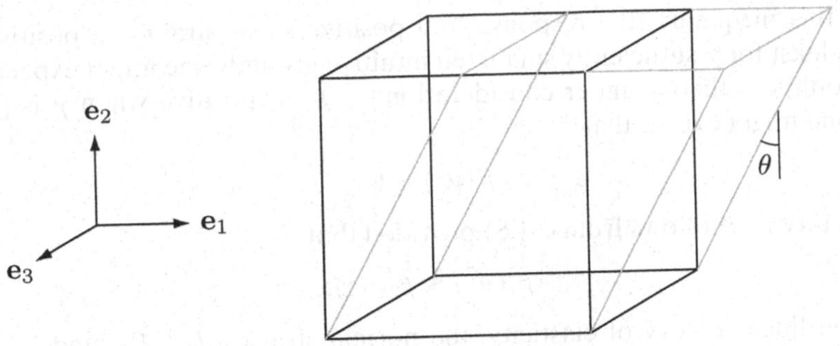

Figure 51.1. Schematic of the simple shear of a rectangular block consistent with (51.1) and (51.2).

The material response functions β_0, β_1, and β_2 are therefore functions of γ^2 alone. For convenience, we therefore write

$$\tilde{\beta}_0(\gamma^2) = \beta_0(3+\gamma^2, 3+\gamma^2, 1),$$

$$\tilde{\beta}_1(\gamma^2) = \beta_1(3+\gamma^2, 3+\gamma^2, 1),$$

$$\tilde{\beta}_2(\gamma^2) = \beta_2(3+\gamma^2, 3+\gamma^2, 1).$$

We may also conclude from (50.34) that

$$\begin{bmatrix} T_{11} & T_{12} & T_{13} \\ T_{21} & T_{22} & T_{23} \\ T_{31} & T_{32} & T_{33} \end{bmatrix} = \tilde{\beta}_0(\gamma^2)\begin{bmatrix} 1 & 0 & 0 \\ 0 & 1 & 0 \\ 0 & 0 & 1 \end{bmatrix} + \tilde{\beta}_1(\gamma^2)\begin{bmatrix} 1+\gamma^2 & \gamma & 0 \\ \gamma & 1 & 0 \\ 0 & 0 & 1 \end{bmatrix}$$

$$+\tilde{\beta}_2(\gamma^2)\begin{bmatrix} 1 & -\gamma & 0 \\ -\gamma & 1+\gamma^2 & 0 \\ 0 & 0 & 1 \end{bmatrix}, \tag{51.6}$$

and, hence, that

$$\left.\begin{aligned} T_{11} &= \tilde{\beta}_0(\gamma^2) + (1+\gamma^2)\tilde{\beta}_1(\gamma^2) + \tilde{\beta}_2(\gamma^2), \\ T_{22} &= \tilde{\beta}_0(\gamma^2) + \tilde{\beta}_1(\gamma^2) + (1+\gamma^2)\tilde{\beta}_2(\gamma^2), \\ T_{33} &= \tilde{\beta}_0(\gamma^2) + \tilde{\beta}_1(\gamma^2) + \tilde{\beta}_2(\gamma^2), \\ T_{12} &= \mu(\gamma^2)\gamma, \\ T_{13} &= T_{23} = 0, \end{aligned}\right\} \tag{51.7}$$

where

$$\mu(\gamma^2) \overset{\text{def}}{=} \tilde{\beta}_1(\gamma^2) - \tilde{\beta}_2(\gamma^2) \tag{51.8}$$

is a shear modulus. As a consequence of $(51.7)_4$ and (51.8), the shear stress T_{12} is an odd function of the shear strain γ: If the direction of shear strain is reversed, the shear stress changes sign.

The value

$$\mu_0 \overset{\text{def}}{=} \mu(0)$$

represents the (infinitesimal) shear modulus of the classical linear theory of isotropic elasticity, and, as such, is known to satisfy

$$\mu_0 > 0;$$

granted this inequality, the response to a positive shear strain is a positive shear stress, at least for γ sufficiently small. On intuitive grounds one might expect that — in the nonlinear theory under consideration — T_{12} is positive when γ is positive; that is, one might expect that

$$\mu(\gamma^2) > 0, \tag{51.9}$$

an inequality that follows from (51.8) provided that

$$\tilde{\beta}_1(\gamma^2) > \tilde{\beta}_2(\gamma^2). \tag{51.10}$$

In the linear theory of elasticity, the normal stresses T_{11}, T_{22}, and T_{33} vanish in simple shear. On the other hand, in the nonlinear theory a shear stress alone does not suffice to determine simple shear: additional normal stresses, $(51.7)_{1\text{–}3}$, are required. Further, because the response functions β_0, β_1, and β_2 are even functions of γ, the normal stresses are unchanged when the sign of shear strain is reversed. For these normal stresses to vanish, both β_1 and β_2 would have to vanish, and this, in turn, would imply that $\mu = 0$. Thus, if

$$\mu(\gamma^2) \neq 0 \quad \text{for} \quad \gamma \neq 0, \tag{51.11}$$

which is a physically reasonable assumption, then

- it is impossible to produce simple shear by applying shear stresses alone.

Next, solving equations $(51.7)_{1\text{–}3}$ for β_0, β_1, and β_2, we obtain

$$\left.\begin{aligned}
\tilde{\beta}_0(\gamma^2) &= \frac{(2+\gamma^2)T_{33} - T_{11} - T_{22}}{\gamma^2},\\
\tilde{\beta}_1(\gamma^2) &= \frac{T_{11} - T_{33}}{\gamma^2},\\
\tilde{\beta}_2(\gamma^2) &= \frac{T_{22} - T_{33}}{\gamma^2},
\end{aligned}\right\} \tag{51.12}$$

and therefore conclude the response functions $\tilde{\beta}_1$ and $\tilde{\beta}_2$ are determined by the normal stress differences $T_{11} - T_{33}$ and $T_{11} - T_{22}$, while $\tilde{\beta}_0$ is determined in terms of T_{11}, T_{22}, and T_{33}.

Further, $(51.7)_{1,2,4}$ imply that

$$T_{11} - T_{22} = \gamma\, T_{12}. \tag{51.13}$$

This relation is independent of the material response function β_0, β_1, and β_2; therefore,

- (51.13) *is satisfied by every isotropic elastic body in simple shear.*

If experimental measurements are inconsistent with this relation, then the material being studied is not an isotropic elastic material.

Finally, a consequence of (51.8) and (51.13) is that

$$\mu(\gamma^2) \neq 0 \quad \text{implies that} \quad T_{11} \neq T_{22}. \tag{51.14}$$

This property of unequal normal stresses in simple shear is generally referred to as the Poynting effect.

52 The Linear Theory of Elasticity

Within the present framework, the linear theory of elasticity may be viewed as an approximation of the general theory appropriate to situations where

SD1: the reference configuration is natural;
SD2: the magnitude of the difference $\mathbf{F} - \mathbf{1}$ between the deformation gradient and the identity tensor is small.

52.1 Small Deformations

The quantity

$$\mathbf{u}(\mathbf{X}, t) = \chi(\mathbf{X}, t) - \mathbf{X}$$

represents the **displacement** of the material point $\mathbf{X}$ at time t, and the deformation gradient $\mathbf{F}$ and the (Green–St. Venant) strain tensor

$$\begin{aligned} \mathbf{E} &= \tfrac{1}{2}(\mathbf{C} - \mathbf{1}) \\ &= \tfrac{1}{2}(\mathbf{F}^{\top}\mathbf{F} - \mathbf{1}), \end{aligned} \tag{52.1}$$

are related to the **displacement gradient**

$$\mathbf{H} = \nabla\mathbf{u}, \qquad H_{ij} = \frac{\partial u_i}{\partial X_j} \tag{52.2}$$

through the relations

$$\mathbf{F} = \mathbf{1} + \nabla\mathbf{u} = \mathbf{1} + \mathbf{H}, \qquad F_{ij} = \delta_{ij} + \frac{\partial u_i}{\partial X_j} = \delta_{ij} + H_{ij}, \tag{52.3}$$

and

$$\begin{aligned} \mathbf{E} &= \tfrac{1}{2}(\nabla\mathbf{u} + (\nabla\mathbf{u})^{\top} + (\nabla\mathbf{u})^{\top}\nabla\mathbf{u}) = \tfrac{1}{2}(\mathbf{H} + \mathbf{H}^{\top} + \mathbf{H}^{\top}\mathbf{H}), \\ E_{ij} &= \tfrac{1}{2}\left(\frac{\partial u_i}{\partial X_j} + \frac{\partial u_j}{\partial X_i}\right) + \frac{\partial u_k}{\partial X_i}\frac{\partial u_k}{\partial X_j} = \tfrac{1}{2}(H_{ij} + H_{ji} + H_{ki}H_{kj}). \end{aligned} \tag{52.4}$$

In what follows, we use the term **small deformations** when discussing theories in which SD1 and SD2 hold.[256] Such theories, although derived formally, are based

[256] In applications the precise meaning of "small" would depend on the scalings used to render the basic fields dimensionless; the asymptotic analysis below is independent of such considerations, since all estimates are in the limit $\mathbf{H} \to \mathbf{0}$.

on precise estimates for the underlying fields in the limit as $\mathbf{H} \to \mathbf{0}$. In particular,

$$\begin{aligned} \mathbf{F} &= \mathbf{1} + o(1), \\ \mathbf{E} &= \tfrac{1}{2}(\nabla\mathbf{u} + (\nabla\mathbf{u})^{\top}) + o(|\mathbf{H}|), \end{aligned} \tag{52.5}$$

as $\mathbf{H} \to \mathbf{0}$.[257] Thus, to within a term of $o(|\mathbf{H}|)$ the strain $\mathbf{E}$ is approximated by

$$\tfrac{1}{2}(\nabla\mathbf{u} + (\nabla\mathbf{u})^{\top}),$$

generally referred to as the **infinitesimal strain**. Further, since the deformation gradient is close to the identity, the deformed and undeformed bodies are approximately coincident, at least up to a constant displacement. It is important to note that, while the choice of reference configuration is arbitrary in the general theory,

- *in theories of small deformations the reference configuration is that configuration from which the small deformations take place.*

52.2 The Stress-Strain Law for Small Deformations

Our next step is to determine constitutive equations that approximate the stress in the limit of small deformations. With this in mind, we begin with the constitutive equation

$$\begin{aligned} \mathbf{T}_{\mathrm{RR}} &= \bar{\mathbf{T}}_{\mathrm{RR}}(\mathbf{C}) \\ &= 2\frac{\partial \bar{\psi}_{\mathrm{R}}(\mathbf{C})}{\partial \mathbf{C}} \end{aligned} \tag{52.6}$$

for the *second Piola stress tensor.*[258] Unlike the relations (48.14) and (48.15) for the stresses $\mathbf{T}_{\mathrm{R}}$ and $\mathbf{T}$, which involve not only $\mathbf{C}$, but also $\mathbf{F}$, (52.6) involves only $\mathbf{C}$; and, as we now show, the derivative $\partial\bar{\mathbf{T}}_{\mathrm{RR}}(\mathbf{C})/\partial\mathbf{C}$ at $\mathbf{C} = \mathbf{1}$ is basic to a discussion of small deformations.

Since the reference configuration is assumed to be natural, it follows from $(47.3)_1$, $(48.8)_2$, and (48.18) that

$$\bar{\mathbf{T}}_{\mathrm{RR}}(\mathbf{1}) = \mathbf{0}. \tag{52.7}$$

52.2.1 The Elasticity Tensor

The derivative

$$\mathbb{C} = 2\frac{\partial \bar{\mathbf{T}}_{\mathrm{RR}}(\mathbf{C})}{\partial \mathbf{C}}\bigg|_{\mathbf{C}=\mathbf{1}}, \qquad C_{ijkl} = 2\frac{\partial (\bar{\mathbf{T}}_{\mathrm{RR}})_{ij}(\mathbf{C})}{\partial C_{kl}}\bigg|_{\mathbf{C}=\mathbf{1}}, \tag{52.8}$$

which we refer to as the **elasticity tensor**, is a linear transformation of symmetric tensors into symmetric tensors, and therefore a *fourth-order tensor*: Specifically, $\mathbb{C}$ associates with each symmetric tensor $\mathbf{A}$ a symmetric tensor

$$\mathbf{B} = \mathbb{C}\mathbf{A}, \qquad B_{ij} = C_{ijkl}A_{kl},$$

and, hence, its components satisfy

$$C_{ijkl} = C_{jikl} = C_{ijlk}. \tag{52.9}$$

---

[257] By definition, $o(1)$ represents a term that goes to zero when $|\mathbf{H}| \to 0$, while $o(|\mathbf{H}|)$ represents a term that goes to zero when $|\mathbf{H}| \to 0$ faster than $\mathbf{H}$ and, hence, faster than the term $\frac{1}{2}(\mathbf{H} + \mathbf{H}^{\top})$.

[258] Cf. (25.2).

Next, by $(52.6)_2$,

$$\mathbb{C} = 4\frac{\partial^2 \bar{\psi}_R(\mathbf{C})}{\partial \mathbf{C}^2}\bigg|_{\mathbf{C}=\mathbf{1}}, \tag{52.10}$$

so that, in components,

$$\begin{aligned} C_{ijkl} &= 4\frac{\partial^2 \bar{\psi}_R(\mathbf{C})}{\partial C_{ij}\partial C_{kl}}\bigg|_{\mathbf{C}=\mathbf{1}} \\ &= 4\frac{\partial^2 \bar{\psi}_R(\mathbf{C})}{\partial C_{kl}\partial C_{ij}}\bigg|_{\mathbf{C}=\mathbf{1}} \\ &= C_{klij}. \end{aligned} \tag{52.11}$$

Thus, *the elasticity tensor is symmetric* in the sense that

$$C_{ijkl} = C_{klij}. \tag{52.12}$$

Because of (52.9) and (52.11), the elasticity tensor $\mathbb{C}$ has at most 21 independent components. The following results follow from (48.19) — which is satisfied because the reference configuration is natural — and (52.10).

Since the reference configuration is assumed to be natural, it follows from (48.19) and (52.8) that the elasticity tensor $\mathbb{C}$ is positive-semidefinite. Thus, $\mathbb{C}$ obeys

$$\mathbf{A}:\mathbb{C}\mathbf{A} \geq 0 \tag{52.13}$$

for all symmetric tensors $\mathbf{A}$.

Our next step is to investigate the symmetry properties of the elasticity tensor. Thus, choose an arbitrary symmetric tensor $\mathbf{A}$. Let $\mathbf{C}(t)$ be a symmetric, positive-definite tensor function of a parameter t consistent with[259]

$$\mathbf{C}(0) = \mathbf{1} \qquad \text{and} \qquad \dot{\mathbf{C}}(0) = \mathbf{A}, \tag{52.14}$$

and let $\mathbf{T}_{RR}(t)$ be defined by

$$\mathbf{T}_{RR}(t) = \bar{\mathbf{T}}_{RR}(\mathbf{C}(t)). \tag{52.15}$$

Then, by the chain-rule,

$$\dot{\mathbf{T}}_{RR} = \frac{\partial \bar{\mathbf{T}}_{RR}(\mathbf{C})}{\partial \mathbf{C}}\dot{\mathbf{C}}, \qquad (\dot{\mathbf{T}}_{RR})_{ij} = \frac{\partial (\bar{\mathbf{T}}_{RR})_{ij}}{\partial C_{kl}}\dot{C}_{kl},$$

and, bearing in mind (52.8) and (52.14),

$$\dot{\mathbf{T}}_{RR}(\mathbf{1}) = \tfrac{1}{2}\mathbb{C}\mathbf{A}. \tag{52.16}$$

Let $\mathbf{Q}$ be a symmetry transformation. Then, by (52.14) and (52.15),

$$\mathbf{Q}^{\top}\mathbf{T}_{RR}\mathbf{Q} = \bar{\mathbf{T}}_{RR}(\mathbf{Q}^{\top}\mathbf{C}\mathbf{Q}),$$

while, by (52.8) and (52.14),

$$\mathbf{Q}^{\top}\dot{\mathbf{T}}_{RR}(\mathbf{1})\mathbf{Q} = \tfrac{1}{2}\mathbf{Q}^{\top}(\mathbb{C}\mathbf{A})\mathbf{Q},$$

so that, by (52.16),

$$\mathbf{Q}^{\top}(\mathbb{C}\mathbf{A})\mathbf{Q} = \mathbb{C}(\mathbf{Q}^{\top}\mathbf{A}\mathbf{Q}). \tag{52.17}$$

This relation, which must be satisfied for every symmetric tensor $\mathbf{A}$ and every symmetry transformation $\mathbf{Q}$, represents the transformation law for the elasticity tensor;

---

[259] The ensuing argument parallels that leading from (50.14) to $(50.18)_2$.

it asserts that *the elasticity tensor is invariant under the symmetry group* $\mathcal{G}$ of the material.

PROPERTIES OF THE ELASTICITY TENSOR

(i) *The elasticity tensor is symmetric*:

$$C_{ijkl} = C_{klij}, \tag{52.18}$$

so that, *for all symmetric tensors* **G** *and* **A**,

$$\mathbf{G}:\mathbb{C}\mathbf{A} = \mathbf{A}:\mathbb{C}\mathbf{G}. \tag{52.19}$$

(ii) *The elasticity tensor is positive-semidefinite*:

$$\mathbf{A}:\mathbb{C}\mathbf{A} \geq 0 \tag{52.20}$$

for all symmetric tensors **A**.

(iii) *The elasticity tensor obeys*

$$\mathbf{Q}^{\top}(\mathbb{C}\mathbf{A})\mathbf{Q} = \mathbb{C}(\mathbf{Q}^{\top}\mathbf{A}\mathbf{Q}) \tag{52.21}$$

for all symmetric tensors **A**.

Remark. The definition (52.8) of $\mathbb{C}$ renders $\mathbb{C}\mathbf{G}$ defined only on tensors **G** that are symmetric. But we may trivially extend this definition to all tensors **A** by

$$\mathbb{C}\mathbf{A} \stackrel{\text{def}}{=} \mathbb{C}(\operatorname{sym}\mathbf{A}). \tag{52.22}$$

52.2.2 The Compliance Tensor

Assume that the elasticity tensor $\mathbb{C}$ is positive-definite. It then follows that there exists a unique fourth-order tensor $\mathbb{K}$, the **compliance tensor**, such that

$$\mathbb{K}(\mathbb{C}\mathbf{A}) = \mathbf{A} \qquad \text{and} \qquad \mathbb{C}(\mathbb{K}\mathbf{A}) = \mathbf{A} \tag{52.23}$$

for every symmetric tensor **A**. As one might expect, the compliance tensor has the exact same material-symmetry properties as the elasticity tensor. That is, the elasticity and compliance tensors are invariant under the same symmetry group.

52.2.3 Estimates for the Stress and Free Energy

We begin with the second Piola stress. By (48.12), (48.18), and (52.8),

$$\begin{aligned} \bar{\mathbf{T}}(\mathbf{C})\big|_{\mathbf{C}=\mathbf{1}} &= \mathbf{0}, \\ \frac{\partial \bar{\mathbf{T}}_{\text{RR}}(\mathbf{C})}{\partial \mathbf{C}}\bigg|_{\mathbf{C}=\mathbf{1}} &= \tfrac{1}{2}\mathbb{C}, \end{aligned} \tag{52.24}$$

and, by (47.4), the Taylor expansion of $\bar{\mathbf{T}}_{\text{RR}}(\mathbf{C})$ about $\mathbf{C} = \mathbf{1}$ has the form

$$\begin{aligned} \bar{\mathbf{T}}_{\text{RR}}(\mathbf{C}) &= \tfrac{1}{2}\mathbb{C}(\mathbf{C}-\mathbf{1}) + o(|\mathbf{C}-\mathbf{1}|) \quad \text{as } \mathbf{C}\to\mathbf{1} \\ &= \mathbb{C}\mathbf{E} + o(|\mathbf{E}|) \qquad \text{as } \mathbf{E}\to\mathbf{0}. \end{aligned} \tag{52.25}$$

Further, by $(52.5)_1$, $J = \det\mathbf{F} = 1 + \mathrm{o}(1)$ and hence $J^{-1} = 1 + \mathrm{o}(1)$; in view of (47.7) and (47.8), we therefore have the following estimate for the Cauchy stress $\mathbf{T}$:

$$\begin{aligned}\mathbf{T} &= J^{-1}\mathbf{F}\mathbf{T}_{\mathrm{RR}}\mathbf{F}^{\top}\\ &= [1 + \mathrm{o}(1)][\mathbf{1} + \mathrm{o}(1)]\mathbf{T}_{\mathrm{RR}}[\mathbf{1} + \mathrm{o}(1)]\\ &= \mathbf{T}_{\mathrm{RR}} + \mathrm{o}(|\mathbf{H}|) \quad \text{as} \quad \mathbf{H} \to \mathbf{0},\end{aligned} \tag{52.26}$$

where we have used $(52.5)_1$. Similarly, by (47.7),

$$\begin{aligned}\mathbf{T}_{\mathrm{R}} &= J\mathbf{T}\mathbf{F}^{-\top}\\ &= [1 + \mathrm{o}(1)]\mathbf{T}[\mathbf{1} + \mathrm{o}(1)]\\ &= \mathbf{T} + \mathrm{o}(|\mathbf{H}|) \quad \text{as} \quad \mathbf{H} \to \mathbf{0}.\end{aligned} \tag{52.27}$$

Hence,

- *to within an error of* $\mathrm{o}(|\mathbf{H}|)$, *the Cauchy and Piola stresses coincide*:

$$\left.\begin{aligned}\mathbf{T} &= \mathbf{T}_{\mathrm{RR}} + \mathrm{o}(|\mathbf{H}|),\\ \mathbf{T} &= \mathbf{T}_{\mathrm{R}} + \mathrm{o}(|\mathbf{H}|).\end{aligned}\right. \tag{52.28}$$

Moreover, by (18.9), the spatial and material forms of the density and conventional body force are related through

$$\rho = [1 + \mathrm{o}(1)]\rho_{\mathrm{R}} \qquad \text{and} \qquad \mathbf{b}_0 = [1 + \mathrm{o}(1)]\mathbf{b}_{0\mathrm{R}}. \tag{52.29}$$

Next, the expansion — to quadratic terms — of $\bar{\psi}(\mathbf{C})$ about $\mathbf{C} = \mathbf{1}$ has the form

$$\begin{aligned}\bar{\psi}_{\mathrm{R}}(\mathbf{C}) = \bar{\psi}_{\mathrm{R}}(\mathbf{1}) &+ (\mathbf{C} - \mathbf{1}) : \left.\frac{\partial\bar{\psi}_{\mathrm{R}}(\mathbf{C})}{\partial\mathbf{C}}\right|_{\mathbf{C}=\mathbf{1}}\\ &+ \tfrac{1}{2}(\mathbf{C} - \mathbf{1}) : \left.\frac{\partial^2\bar{\psi}_{\mathrm{R}}(\mathbf{C})}{\partial\mathbf{C}^2}\right|_{\mathbf{C}=\mathbf{1}}[\mathbf{C} - \mathbf{1}] + \mathrm{o}(|\mathbf{C} - \mathbf{1}|^2) \quad \text{as} \quad \mathbf{C} \to \mathbf{1};\end{aligned}$$

since, by (48.20), (50.23), and (52.10),

$$\left.\begin{aligned}\bar{\psi}_{\mathrm{R}}(\mathbf{C})\big|_{\mathbf{C}=\mathbf{1}} &= 0,\\ \left.\frac{\partial\bar{\psi}_{\mathrm{R}}(\mathbf{C})}{\partial\mathbf{C}}\right|_{\mathbf{C}=\mathbf{1}} &= \mathbf{0},\\ \left.\frac{\partial^2\bar{\psi}_{\mathrm{R}}(\mathbf{C})}{\partial\mathbf{C}^2}\right|_{\mathbf{C}=\mathbf{1}} &= \tfrac{1}{4}\mathbb{C},\end{aligned}\right\} \tag{52.30}$$

this expansion reduces to

$$\begin{aligned}\bar{\psi}_{\mathrm{R}}(\mathbf{C}) &= \tfrac{1}{8}(\mathbf{C} - \mathbf{1}) : \mathbb{C}(\mathbf{C} - \mathbf{1}) + \mathrm{o}(|\mathbf{C} - \mathbf{1}|^2) \quad \text{as} \quad \mathbf{C} \to \mathbf{1}\\ &= \tfrac{1}{2}\mathbf{E} : \mathbb{C}\mathbf{E} + \mathrm{o}(|\mathbf{E}|^2) \quad \text{as} \quad \mathbf{E} \to \mathbf{0}.\end{aligned} \tag{52.31}$$

52.3 Basic Equations of the Linear Theory of Elasticity

The linear theory of elasticity is based on approximate equations obtained when the higher-order terms in (52.5), (52.26), (52.27), (52.29), and (52.31) are neglected.[260] We therefore take $\rho = \rho_R$, $\mathbf{T} = \mathbf{T}_R$, and $\mathbf{b}_0 = \mathbf{b}_{0R}$,[261] and we base the theory on the **strain-displacement relation**

$$\mathbf{E} = \tfrac{1}{2}(\mathbf{H}+\mathbf{H}^{\top}), \qquad E_{ij} = \frac{1}{2}\left(\frac{\partial u_i}{\partial X_j} + \frac{\partial u_j}{\partial X_i}\right), \tag{52.32}$$

the **stress-strain relation**

$$\mathbf{T} = \mathbb{C}\mathbf{E}, \qquad T_{ij} = C_{ijkl}E_{kl}, \tag{52.33}$$

and the **free energy**

$$\psi_R = \tfrac{1}{2}\mathbf{E}:\mathbb{C}\mathbf{E}, \qquad \psi_R = \tfrac{1}{2}C_{ijkl}E_{ij}E_{kl}. \tag{52.34}$$

The stress is, of course, related to the free energy via

$$\mathbf{T} = \frac{\partial \psi_R(\mathbf{E})}{\partial \mathbf{E}}. \tag{52.35}$$

By (52.19) and (52.20), the *elasticity tensor* $\mathbb{C}$ *is symmetric* and positive-semidefinite; here we assume, in addition, that $\mathbb{C}$ is *positive-definite*:

$$\mathbf{A}:\mathbb{C}\mathbf{A} > 0 \quad \text{for all symmetric tensors } \mathbf{A} \neq \mathbf{0}. \tag{52.36}$$

The *basic equations of the linear theory of elasticity* consist of (52.32), (52.33), and the local momentum balance $(47.6)_1$ written in terms of displacement:

$$\rho\ddot{\mathbf{u}} = \operatorname{Div}\mathbf{T} + \mathbf{b}_0, \qquad \rho\ddot{u} = \frac{\partial T_{ij}}{\partial X_j} + b_{0i}. \tag{52.37}$$

To avoid repeated assumptions, we assume throughout this subsection that

- the body is homogeneous,

so that the elasticity tensor $\mathbb{C}$ and the density ρ are constant.

By (52.22),

$$\mathbb{C}\mathbf{E} = \mathbb{C}\nabla\mathbf{u}$$

and (52.32), (52.33), and (52.37) may be combined to form a single partial differential equation for the displacement field $\mathbf{u}$: the **displacement equation of motion**

$$\rho\ddot{\mathbf{u}} = \operatorname{Div}(\mathbb{C}\nabla\mathbf{u}) + \mathbf{b}_0, \qquad \rho\ddot{u}_i = C_{ijkl}\frac{\partial^2 u_k}{\partial X_j \partial X_l} + b_{0i}. \tag{52.38}$$

52.4 Special Forms for the Elasticity Tensor

We here give specific forms for the elasticity tensor for two special cases: when the body is isotropic; when the body is a cubic crystal.[262]

---

260 Precisely, the $o(|\mathbf{H}|)$ terms in (52.5), (52.26), and (52.27), the $o(|\mathbf{E}|^2)$ term in (52.31), and the $o(1)$ terms in (52.29) are neglected.

261 Because $\psi_R = \rho\psi$, we continue to use the symbol ψ_R for the density measured per unit volume.

262 A discussion of the forms taken by the elasticity tensor for the other crystal classes is beyond the scope of this book.

52.4.1 Isotropic Material

Assume now that the body is isotropic. Then $\mathbb{C}\mathbf{E}$ is a (linear) isotropic function of $\mathbf{E}$ and, by the Representation Theorem for Isotropic Linear Tensor Functions in Appendix 113.3,[263]

$$\mathbb{C}\mathbf{E} = 2\mu\mathbf{E} + \lambda(\operatorname{tr}\mathbf{E})\mathbf{1}, \tag{52.39}$$

with μ and λ scalar constitutive moduli.

To determine the inequalities satisfied by these moduli as a consequence of the positive-semidefiniteness of the elasticity tensor, choose an arbitrary symmetric tensor $\mathbf{E}$ and let $\mathbf{E}_0$ denote its deviatoric part:

$$\mathbf{E}_0 = \mathbf{E} - \tfrac{1}{3}(\operatorname{tr}\mathbf{E})\mathbf{1}.$$

Then

$$\operatorname{tr}\mathbf{E}_0 = \mathbf{1}:\mathbf{E}_0 = 0$$

and

$$\begin{aligned}|\mathbf{E}|^2 &= (\mathbf{E}_0 + \tfrac{1}{3}(\operatorname{tr}\mathbf{E})\mathbf{1}):(\mathbf{E}_0 + \tfrac{1}{3}(\operatorname{tr}\mathbf{E})\mathbf{1})\\ &= |\mathbf{E}_0|^2 + \tfrac{1}{3}(\operatorname{tr}\mathbf{E})^2.\end{aligned}$$

Thus, by (52.39) and (52.20),

$$\begin{aligned}0 &\le \mathbf{E}:\mathbb{C}\mathbf{E}\\ &= 2\mu|\mathbf{E}|^2 + \lambda(\operatorname{tr}\mathbf{E})^2\\ &= 2\mu|\mathbf{E}_0|^2 + \kappa(\operatorname{tr}\mathbf{E})^2,\end{aligned} \tag{52.40}$$

with

$$\kappa = \frac{2\mu + 3\lambda}{3}. \tag{52.41}$$

Choosing $\mathbf{E} = \mathbf{1}$, so that $\operatorname{tr}\mathbf{E} = 3$ and $\mathbf{E}_0 = \mathbf{0}$, yields $2\mu + 3\lambda \ge 0$; choosing

$$\mathbf{E} = \mathbf{e}\otimes\mathbf{f} + \mathbf{f}\otimes\mathbf{e}$$

with $\mathbf{e}$ and $\mathbf{f}$ orthonormal (so that $\operatorname{tr}\mathbf{E} = 0$ and $|\mathbf{E}_0|^2 = 2$) yields $\mu \ge 0$. Thus,[264]

$$\mu \ge 0 \qquad \text{and} \qquad 2\mu + 3\lambda \ge 0. \tag{52.42}$$

If $\mathbb{C}$ is invertible, then — since an invertible, positive-semidefinite linear transformation is positive-definite — the inequality (52.40) is strict for $\mathbf{E} \ne \mathbf{0}$ and the argument above implies that

$$\mu > 0 \qquad \text{and} \qquad 2\mu + 3\lambda > 0. \tag{52.43}$$

Conversely, since $\mathbf{E}$ vanishes if and only if both $\mathbf{E}_0$ and $\operatorname{tr}\mathbf{E}$ vanish, (52.40) and (52.43) imply that $\mathbb{C}$ is positive-definite.

263 Cf. (113.9).

264 Cf. the inequalities (45.21) arising in the theory of compressible, linearly viscous fluids. Whereas (45.21) stems from the requirement that the free-energy imbalance hold for all constitutive processes, the inequalities (52.42) reflect the less basic assumption that $\mathbb{C}$ be positive-semidefinite.

52.4.2 Cubic Crystal

As before we let $\mathcal{G}$ — a subgroup of the group of rotations — denote the symmetry group (point group) of the crystal. Let $\mathcal{M}$ denote the space of symmetric tensors and consider a free energy $\psi_{\rm R}(\mathbf{E})$ that is quadratic on $\mathcal{M}$ and invariant with respect to $\mathcal{G}$. Thus, for each symmetric tensor $\mathbf{E}$,

$$\psi_{\rm R}(\mathbf{E}) = \psi_{\rm R}(\mathbf{QEQ}^{\top}) \qquad \text{for all } \mathbf{Q} \in \mathcal{G}. \tag{52.44}$$

Assume that the crystal has cubic symmetry, so that $\mathcal{G}$ is the cubic group, and assume that the unit cube of the crystal is generated by a rectangular Cartesian basis $\{\mathbf{e}_i\}$. With respect to this basis the most general invariant free energy $\psi_{\rm R}(\mathbf{E})$, quadratic in $\mathbf{E}$, is a linear combination of three independent invariants and may be expressed in the form[265]

$$\begin{aligned}\psi_{\rm R}(\mathbf{E}) = \kappa_0(E_{11}^2 + E_{22}^2 + E_{33}^2) + \kappa_1(E_{11}E_{22} + E_{22}E_{33} + E_{11}E_{33})\\ + \kappa_2(E_{12}^2 + E_{13}^2 + E_{23}^2 + E_{21}^2 + E_{31}^2 + E_{32}^2),\end{aligned} \tag{52.45}$$

with κ_0, κ_1, and κ_2 scalar constants. More useful for our purposes is the fact that this free energy can be rewritten in the alternative form

$$\psi_{\rm R}(\mathbf{E}) = \mu|\mathbf{E}|^2 + \tfrac{1}{2}\lambda(\operatorname{tr}\mathbf{E})^2 + \tfrac{1}{2}c\Phi(\mathbf{E}), \tag{52.46}$$

where

$$\Phi(\mathbf{E}) = E_{11}^2 + E_{22}^2 + E_{33}^2. \tag{52.47}$$

Note that the terms in (52.46) with coefficients μ and λ are isotropic invariants, so that the cubic nature of $\psi_{\rm R}$ is characterized solely by $\Phi(\mathbf{E})$.

As in the case of an isotropic body, the requirement that the elasticity tensor be positive-semidefinite yields inequalities involving the moduli μ, λ, and c entering (52.46). To determine the relevant inequalities, choose an arbitrary symmetric tensor $\mathbf{E}$ and note that, trivially, $\mathbf{E}$ admits the decomposition

$$\mathbf{E} = \tilde{\mathbf{E}} + \hat{\mathbf{E}}$$

where, in terms of matrices expressed with respect to the basis $\{\mathbf{e}_i\}$,

$$[\tilde{\mathbf{E}}] = \begin{bmatrix} E_{11} & 0 & 0 \\ 0 & E_{22} & 0 \\ 0 & 0 & E_{33} \end{bmatrix} \qquad \text{and} \qquad [\hat{\mathbf{E}}] = \begin{bmatrix} 0 & E_{12} & E_{13} \\ E_{21} & 0 & E_{23} \\ E_{31} & E_{32} & 0 \end{bmatrix}.$$

Immediate consequences of this decomposition are that $\tilde{\mathbf{E}}$ and $\hat{\mathbf{E}}$ are orthogonal,

$$\tilde{\mathbf{E}}:\hat{\mathbf{E}} = 0, \tag{52.48}$$

and that

$$\Phi(\mathbf{E}) = |\tilde{\mathbf{E}}|^2. \tag{52.49}$$

Letting $\tilde{\mathbf{E}}_0 = \tilde{\mathbf{E}} - \frac{1}{3}(\operatorname{tr}\tilde{\mathbf{E}})\mathbf{1}$ and bearing in mind that

$$\operatorname{tr}\hat{\mathbf{E}} = \hat{\mathbf{E}}:\mathbf{1} = 0, \tag{52.50}$$

[265] Cf. Sirotin & Shaskolskaya (1982, p. 620).

it therefore follows that

$$\begin{aligned}
|\mathbf{E}|^2 &= (\tilde{\mathbf{E}} + \hat{\mathbf{E}}) : (\tilde{\mathbf{E}} + \hat{\mathbf{E}}) \\
&= |\tilde{\mathbf{E}}|^2 + |\hat{\mathbf{E}}|^2 \\
&= (\tilde{\mathbf{E}}_0 + \tfrac{1}{3}(\operatorname{tr}\tilde{\mathbf{E}})\mathbf{1}) : (\tilde{\mathbf{E}}_0 + \tfrac{1}{3}(\operatorname{tr}\tilde{\mathbf{E}})\mathbf{1}) + |\hat{\mathbf{E}}|^2 \\
&= |\tilde{\mathbf{E}}_0|^2 + \tfrac{1}{3}(\operatorname{tr}\tilde{\mathbf{E}})^2 + |\hat{\mathbf{E}}|^2.
\end{aligned} \tag{52.51}$$

Using (52.49) and (52.51) in the free energy (52.46), we obtain

$$\begin{aligned}
\psi_{\mathrm{R}}(\mathbf{E}) &= \mu|\tilde{\mathbf{E}}_0|^2 + \tfrac{1}{3}\mu(\operatorname{tr}\tilde{\mathbf{E}})^2 + \mu|\hat{\mathbf{E}}|^2 + \tfrac{1}{2}\lambda(\operatorname{tr}\tilde{\mathbf{E}})^2 + \tfrac{1}{2}c|\tilde{\mathbf{E}}_0|^2 + \tfrac{1}{6}c(\operatorname{tr}\tilde{\mathbf{E}})^2 \\
&= \mu|\hat{\mathbf{E}}|^2 + (\mu + \tfrac{1}{2}c)|\tilde{\mathbf{E}}_0|^2 + \tfrac{1}{3}(\mu + \tfrac{3}{2}\lambda + \tfrac{1}{2}c)(\operatorname{tr}\tilde{\mathbf{E}})^2.
\end{aligned} \tag{52.52}$$

In view of the orthogonality relations (52.48) and (52.50), $\hat{\mathbf{E}}$, $\tilde{\mathbf{E}}_0$, and $\operatorname{tr}\tilde{\mathbf{E}}$ can be varied independently. To ensure that $\psi_{\mathrm{R}}(\mathbf{E}) \geq 0$, it follows from (52.52) that the coefficients of $|\hat{\mathbf{E}}|^2$, $|\tilde{\mathbf{E}}_0|^2$, and $(\operatorname{tr}\tilde{\mathbf{E}})^2$ must be nonnegative:

$$\mu \geq 0, \qquad \mu + \tfrac{1}{2}c \geq 0, \qquad \text{and} \qquad \mu + \tfrac{3}{2}\lambda + \tfrac{1}{2}c \geq 0. \tag{52.53}$$

We leave it as an exercise to show that the strict versions of the inequalities (52.53) are necessary and sufficient to ensure that the elasticity tensor for a cubic material be positive-definite.

A direct calculation shows that, for Φ as defined in (52.47),

$$\frac{\partial \Phi}{\partial \mathbf{E}} = 2\sum_{i=1}^{3} E_{ii}\mathbf{e}_i \otimes \mathbf{e}_i.$$

On applying (52.35) to (52.46), the constitutive relation for the stress $\mathbf{T}$ therefore has the well-known form

$$\mathbf{T} = \underbrace{2\mu\mathbf{E} + \lambda(\operatorname{tr}\mathbf{E})\mathbf{1}}_{\mathbf{T}^{\mathrm{iso}}} + \mathbf{T}^{\mathrm{cub}} \tag{52.54}$$

in which $\mathbf{T}^{\mathrm{iso}}$ is the isotropic part of $\mathbf{T}$, while $\mathbf{T}^{\mathrm{cub}}$, the cubic part of $\mathbf{T}$, has the form

$$\mathbf{T}^{\mathrm{cub}} = c\sum_{i=1}^{3} E_{ii}\mathbf{e}_i \otimes \mathbf{e}_i, \tag{52.55}$$

and is, hence, represented by a diagonal matrix with respect to the cubic basis.

To make contact with the crystallography literature, we use the Voigt single index notation[266] for the components of $\mathbf{T}$ and $\mathbf{E}$,

$$\begin{pmatrix} T_1 \\ T_2 \\ T_3 \\ T_4 \\ T_5 \\ T_6 \end{pmatrix} \stackrel{\text{def}}{=} \begin{pmatrix} T_{11} \\ T_{22} \\ T_{33} \\ T_{23} \\ T_{13} \\ T_{12} \end{pmatrix}, \qquad \begin{pmatrix} E_1 \\ E_2 \\ E_3 \\ E_4 \\ E_5 \\ E_6 \end{pmatrix} \stackrel{\text{def}}{=} \begin{pmatrix} E_{11} \\ E_{22} \\ E_{33} \\ 2E_{23} \\ 2E_{13} \\ 2E_{12} \end{pmatrix}, \tag{52.56}$$

[266] Cf., e.g., TING (1996, §2.3).

and rewrite the constitutive relation $\mathbf{T} = \mathbb{C}\mathbf{E}$ in the form

$$\begin{pmatrix} T_1 \\ T_2 \\ T_3 \\ T_4 \\ T_5 \\ T_6 \end{pmatrix} = \begin{pmatrix} C_{11} & C_{12} & C_{13} & C_{14} & C_{15} & C_{16} \\ C_{12} & C_{22} & C_{23} & C_{24} & C_{25} & C_{26} \\ C_{13} & C_{23} & C_{33} & C_{34} & C_{35} & C_{36} \\ C_{14} & C_{24} & C_{34} & C_{44} & C_{45} & C_{46} \\ C_{15} & C_{25} & C_{35} & C_{45} & C_{55} & C_{56} \\ C_{16} & C_{26} & C_{36} & C_{46} & C_{56} & C_{66} \end{pmatrix} \begin{pmatrix} E_1 \\ E_2 \\ E_3 \\ E_4 \\ E_5 \\ E_6 \end{pmatrix}, \tag{52.57}$$

with the elastic constants C_{IJ} in the matrix relation above satisfying $C_{IJ} = C_{JI}$. Then, granted the components of $\mathbb{C}$ are expressed with respect to the cubic basis, (52.57) reduces to

$$\begin{pmatrix} T_1 \\ T_2 \\ T_3 \\ T_4 \\ T_5 \\ T_6 \end{pmatrix} = \begin{pmatrix} C_{11} & C_{12} & C_{12} & 0 & 0 & 0 \\ C_{12} & C_{11} & C_{12} & 0 & 0 & 0 \\ C_{12} & C_{12} & C_{11} & 0 & 0 & 0 \\ 0 & 0 & 0 & C_{44} & 0 & 0 \\ 0 & 0 & 0 & 0 & C_{44} & 0 \\ 0 & 0 & 0 & 0 & 0 & C_{44} \end{pmatrix} \begin{pmatrix} E_1 \\ E_2 \\ E_3 \\ E_4 \\ E_5 \\ E_6 \end{pmatrix}, \tag{52.58}$$

with C_{11}, C_{12}, and C_{44} the independent elastic constants. We leave it as an exercise to relate the moduli μ, λ, and c to these constants.

52.5 Basic Equations of the Linear theory of Elasticity for an Isotropic Material

By (52.39), when the body is isotropic the stress-strain law takes the simple form

$$\boxed{\mathbf{T} = 2\mu\mathbf{E} + \lambda(\mathrm{tr}\,\mathbf{E})\mathbf{1}} \tag{52.59}$$

and corresponds to the energy

$$\boxed{\psi_{\mathrm{R}} = \mu|\mathbf{E}|^2 + \tfrac{1}{2}\lambda(\mathrm{tr}\,\mathbf{E})^2.} \tag{52.60}$$

Because $\mathbb{C}$ is positive-definite, the **Lamé moduli** μ and λ satisfy the strict inequalities (52.43); as a consequence the stress-strain relation (52.59) may be inverted to give

$$\mathbf{E} = \frac{1}{2\mu}\left(\mathbf{T} - \frac{\lambda}{2\mu + 3\lambda}(\mathrm{tr}\,\mathbf{T})\mathbf{1}\right), \tag{52.61}$$

an assertion whose proof we leave as an exercise.

Next,

$$\begin{aligned} (2\mathrm{Div}\,\mathbf{E})_i &= \frac{\partial}{\partial X_j}\left(\frac{\partial u_i}{\partial X_j} + \frac{\partial u_j}{\partial X_i}\right) \\ &= \frac{\partial^2 u_i}{\partial X_j \partial X_j} + \frac{\partial^2 u_j}{\partial X_i \partial X_j} \\ &= (\Delta\mathbf{u} + \nabla\mathrm{Div}\,\mathbf{u})_i \end{aligned} \tag{52.62}$$

and the displacement equation of motion for an isotropic body takes the form

$$\boxed{\rho\ddot{\mathbf{u}} = \mu\Delta\mathbf{u} + (\lambda + \mu)\nabla\mathrm{Div}\,\mathbf{u} + \mathbf{b}_0, \qquad \rho\ddot{u}_i = \mu\frac{\partial^2 u_i}{\partial X_j \partial X_j} + (\lambda + \mu)\frac{\partial^2 u_j}{\partial X_i \partial X_j} + b_{0i}.} \tag{52.63}$$

52.5.1 Statical Equations

Statical (i.e., time-independent) solutions of (52.32), (52.33), and the local balance

$$\boxed{\operatorname{Div}\mathbf{T} + \mathbf{b}_0 = \mathbf{0}} \tag{52.64}$$

describe situations in which the body is in equilibrium. Such solutions satisfy the **displacement equation of equilibrium**

$$\operatorname{Div}(\mathbb{C}\nabla\mathbf{u}) + \mathbf{b}_0 = \mathbf{0}, \qquad C_{ijkl}\frac{\partial^2 u_k}{\partial X_j \partial X_l} + b_{0i} = 0, \tag{52.65}$$

an equation that for B isotropic has the form

$$\boxed{\mu\Delta\mathbf{u} + (\lambda+\mu)\nabla\operatorname{Div}\mathbf{u} + \mathbf{b}_0 = \mathbf{0}.} \tag{52.66}$$

EXERCISES

1. Establish the dynamical **energy balance**

$$\overline{\int_{\mathrm{P}} \tfrac{1}{2}(\mathbf{E}:\mathbb{C}\mathbf{E} + \rho|\dot{\mathbf{u}}|^2)\,dv} = \int_{\partial\mathrm{P}} \mathbf{T}\mathbf{n}\cdot\dot{\mathbf{u}}\,da + \int_{\mathrm{P}} \mathbf{b}_0\cdot\dot{\mathbf{u}}\,dv \tag{52.67}$$

for any subregion P of the body. Establish the **energy-conservation theorem:** If $\mathbf{b}_0 = \mathbf{0}$, and if $\mathbf{T}\mathbf{n} = \mathbf{0}$ on a portion of ∂B and $\dot{\mathbf{u}} = \mathbf{0}$ on the remainder of ∂B, then

$$\int_{\mathrm{P}} \tfrac{1}{2}(\mathbf{E}:\mathbb{C}\mathbf{E} + \rho|\dot{\mathbf{u}}|^2)\,dv \quad \text{is independent of time} \tag{52.68}$$

and energy is conserved; if, in addition, $\mathbf{u}$ and $\dot{\mathbf{u}}$ vanish on B at some time, then

$$\mathbf{u} = \mathbf{0} \quad \text{on B for all time.} \tag{52.69}$$

2. Show, as a consequence of the symmetry of $\mathbb{C}$, that

$$\mathbf{T} = \frac{\partial(\tfrac{1}{2}\mathbf{E}:\mathbb{C}\mathbf{E})}{\partial\mathbf{E}}, \tag{52.70}$$

and use this relation to establish the local **free-energy balance**

$$\dot{\psi}_{\mathrm{R}} = \mathbf{T}:\dot{\mathbf{E}}. \tag{52.71}$$

3. Show that the statical equations (52.32), (52.33), and (52.64) imply the **work and energy balance**

$$\int_{\partial\mathrm{P}} \mathbf{T}\mathbf{n}\cdot\mathbf{u}\,da + \int_{\mathrm{P}} \mathbf{b}_0\cdot\mathbf{u}\,dv = \int_{\mathrm{P}} \mathbf{E}:\mathbb{C}\mathbf{E}\,dv \tag{52.72}$$

for any subregion P of the body.[267]

[267] Note that the right side of (52.72) is twice the strain energy.

4. Assume that the material is isotropic and that $\mathbf{b}_0 = \mathbf{0}$. Let

$$\vartheta = \mathrm{Div}\,\mathbf{u}, \qquad \boldsymbol{\omega} = \mathrm{Curl}\,\mathbf{u}. \tag{52.73}$$

Show that

$$\ddot{\vartheta} = V_1 \Delta \vartheta, \qquad V_1 = \sqrt{\frac{2\mu + \lambda}{\rho}},$$
$$\ddot{\boldsymbol{\omega}} = V_2 \Delta \boldsymbol{\omega}, \qquad V_2 = \sqrt{\frac{\mu}{\rho}}. \tag{52.74}$$

Discuss the physical meaning of these equations.

5. Show that the statical counterparts of (52.74) are

$$\Delta \vartheta = 0,$$
$$\Delta \boldsymbol{\omega} = 0,$$

so that ϑ and $\boldsymbol{\omega}$ are harmonic. Show further that the displacement field $\mathbf{u}$ is biharmonic:

$$\Delta\Delta \mathbf{u} = \mathbf{0}, \qquad \frac{\partial^4 u_i}{\partial X_j \partial X_j \partial X_k \partial X_k} = 0.$$

6. Derive the strain-stress relation (52.61).
7. Arguing as in the case of an isotropic body, show that the strict versions of the inequalities (52.53) are necessary and sufficient for the elasticity tensor of a cubic material to be positive-definite.
8. Show that the moduli $\{\mu, \lambda, c\}$ and the cubic constants $\{C_{11}, C_{12}, C_{44}\}$ are related via

$$\mu = C_{44}, \qquad \lambda = C_{12}, \qquad \text{and} \qquad c = C_{11} - C_{12} - 2C_{44}.$$

Thus, conclude that the extent to which a linearly elastic solid with cubic symmetry deviates from isotropic is determined by the extent to which C_{44} differs from $\frac{1}{2}(C_{11} - C_{12})$.

9. Assume that the material is a cubic crystal. Show that

$$\mathrm{Div}\,\mathbf{T} = \mathrm{Div}\,\mathbf{T}^{\mathrm{iso}} + \mathrm{Div}\,\mathbf{T}^{\mathrm{cub}},$$

where

$$\mathrm{Div}\,\mathbf{T}^{\mathrm{iso}} = \mu \Delta \mathbf{u} + (\lambda + \mu)\nabla \mathrm{Div}\,\mathbf{u}, \tag{52.75}$$

and where the cubic term $\mathrm{Div}\,\mathbf{T}^{\mathrm{cub}}$ (which does not lend itself to indicial notation) has the matrix form

$$[\mathrm{Div}\,\mathbf{T}^{\mathrm{cub}}] = c \begin{bmatrix} u_{1,11} \\ u_{2,22} \\ u_{3,33} \end{bmatrix}. \tag{52.76}$$

Show further that the displacement equation of motion for a cubic crystal takes the form

$$\rho \ddot{\mathbf{u}} = \mu \Delta \mathbf{u} + (\lambda + \mu)\nabla \mathrm{Div}\,\mathbf{u} + \mathrm{Div}\,\mathbf{T}^{\mathrm{cub}} + \mathbf{b}_0 \tag{52.77}$$

with $\mathrm{Div}\,\mathbf{T}^{\mathrm{cub}}$ given by (52.76).

52.6 Some Simple Statical Solutions

Any statical displacement field $\mathbf{u}$ with $\nabla\mathbf{u}$ constant generates a solution of the basic equations (52.32), (52.37), and (52.59) with $\mathbf{b}_0 = \mathbf{0}$. We now discuss three such displacement fields.

(i) *Pure shear.* Let

$$\mathbf{u}(\mathbf{X}) = \gamma X_2 \mathbf{e}_1$$

so that, on defining the shear stress

$$\tau = \mu\gamma,$$

the matrices of $\mathbf{E}$ and $\mathbf{T}$ are given by

$$[\mathbf{E}] = \begin{bmatrix} 0 & \gamma & 0 \\ \gamma & 0 & 0 \\ 0 & 0 & 0 \end{bmatrix}, \qquad [\mathbf{T}] = \begin{bmatrix} 0 & \tau & 0 \\ \tau & 0 & 0 \\ 0 & 0 & 0 \end{bmatrix}; \tag{52.78}$$

μ is called the **shear modulus.**

(ii) *Uniform compression or expansion.* Let

$$\mathbf{u}(\mathbf{X}) = \upsilon(\mathbf{X} - \mathbf{0}),$$

with υ constant. In this case,

$$\mathbf{E} = \upsilon\mathbf{1}, \qquad \mathbf{T} = -p\mathbf{1},$$

with

$$p = -3\kappa\upsilon,$$

where κ, as given by (52.41), is referred to as the **bulk modulus** (or **modulus of compression**).

(iii) *Pure tension.* Here, the stress has the form

$$[\mathbf{T}] = \begin{bmatrix} \sigma & 0 & 0 \\ 0 & 0 & 0 \\ 0 & 0 & 0 \end{bmatrix}, \tag{52.79}$$

so that, by (52.61),[268]

$$[\mathbf{E}] = \begin{bmatrix} \varepsilon & 0 & 0 \\ 0 & \ell & 0 \\ 0 & 0 & \ell \end{bmatrix}, \tag{52.80}$$

with

$$\varepsilon = \frac{\sigma}{E}, \qquad \ell = -\nu\varepsilon,$$

and

$$E = \frac{\mu(2\mu + 3\lambda)}{\mu + \lambda}, \qquad \nu = \frac{\lambda}{2(\mu + \lambda)}.$$

Note that the displacement field has the form

$$\mathbf{u}(\mathbf{X}) = \varepsilon X_1 \mathbf{e}_1 + \ell X_2 \mathbf{e}_2 + \ell X_3 \mathbf{e}_3.$$

The material parameter E, known as **Young's modulus**, is obtained by dividing the tensile stress σ by the longitudinal strain ε produced by it; the material

[268] The use of ε for tensile strain is standard. Our use of the same symbol for the specific internal-energy should not cause confusion: ε is used for tensile strain only here and in a similar paragraph in §52.6.

parameter ν, known as **Poisson's ratio**, is the ratio of the lateral contraction to the longitudinal strain of a bar under pure tension. For classical materials, it is expected that an elastic solid should increase its length when pulled, should decrease its volume when acted on by a pressure, and should respond to a positive shearing strain by a positive shearing stress. Consistent with these expectations being so, it follows that

$$E > 0, \qquad \kappa > 0, \qquad \text{and} \qquad \mu > 0.$$

These inequalities (actually $\kappa > 0$ and $\mu > 0$, or $E > 0$ and $-1 < \nu < \frac{1}{2}$), are equivalent to the positive-definiteness of $\mathbb{C}$.

52.7 Boundary-Value Problems

In this section, we establish classical theorems regarding the boundary-value problem of elastostatics and the initial/boundary-value problem of elastodynamics. In this regard, recall our assumption that[269]

- *the elasticity tensor $\mathbb{C}$ is symmetric and positive-definite.*

In the linear theory, the analog of a rigid deformation is an (infinitesimal) **rigid displacement**, which is a field $\mathbf{w}$ of the form

$$\mathbf{w}(\mathbf{X}) = \boldsymbol{\alpha} + \boldsymbol{\lambda} \times (\mathbf{X} - \mathbf{o}), \tag{52.81}$$

with $\boldsymbol{\alpha}$ and $\boldsymbol{\lambda}$ constant vectors.[270] It then follows that a vector field $\mathbf{w}$ is a rigid displacement field if and only if[271]

$$\nabla\mathbf{w} + (\nabla\mathbf{w})^{\top} = \mathbf{0}; \tag{52.82}$$

that is, an infinitesimal displacement field is rigid if and only if the corresponding strain field vanishes.

52.7.1 Elastostatics

Let $\mathcal{S}_1$ and $\mathcal{S}_2$ be complementary subsurfaces of the boundary $\partial\mathrm{B}$ of the body B. We consider **boundary conditions** in which the displacement is specified on $\mathcal{S}_1$ and the surface traction is specified on $\mathcal{S}_2$:

$$\left.\begin{aligned} \mathbf{u} &= \hat{\mathbf{u}} \quad \text{on } \mathcal{S}_1, \\ \mathbf{T}\mathbf{n} &= \hat{\mathbf{t}} \quad \text{on } \mathcal{S}_2, \end{aligned}\right. \tag{52.83}$$

with $\hat{\mathbf{u}}$ and $\hat{\mathbf{t}}$ prescribed functions. The **mixed problem of elastostatics** may then be stated as follows: Given boundary data $\hat{\mathbf{u}}$ and $\hat{\mathbf{t}}$, find a displacement field $\mathbf{u}$, a strain field $\mathbf{E}$, and a stress field $\mathbf{T}$ that satisfy the field equations

$$\left.\begin{aligned} \mathbf{E} &= \tfrac{1}{2}(\nabla\mathbf{u} + (\nabla\mathbf{u})^{\top}), \\ \mathbf{T} &= \mathbb{C}\mathbf{E}, \\ \operatorname{Div}\mathbf{T} + \mathbf{b}_0 &= \mathbf{0}, \end{aligned}\right\} \quad \text{on } \mathrm{B} \tag{52.84}$$

and the boundary conditions (52.83).[272]

---

[269] Cf. (52.19) and (52.36).

[270] Cf. (10.5).

[271] Cf. GURTIN (1981, p. 56).

[272] Here, it is assumed that the elasticity tensor $\mathbb{C}$ and the conventional body force $\mathbf{b}_0$ are known.

Uniqueness Theorem *The mixed problem of elastostatics has at most one solution up to a rigid displacement of the body; that is, any two solutions must have the same stress and strain fields and the displacement fields of the two solutions may differ at most by a rigid displacement.*

To establish this theorem, let $\mathbf{u}_1$ and $\mathbf{u}_2$ (and corresponding strains and stresses) represent two solutions of the mixed problem of elastostatics. The displacement difference $\mathbf{u} = \mathbf{u}_1 - \mathbf{u}_2$ then represents a solution of a mixed problem in which the body force vanishes, as do the prescribed surface displacement and surface traction, so that

$$\int_{\mathrm{B}} \mathbf{b}_0 \cdot \mathbf{u}\, dv = 0$$

and since, by (52.83), $\mathbf{u} = \mathbf{u}_1 - \mathbf{u}_2 = \mathbf{0}$ on $\mathcal{S}_1$ and $\mathbf{Tn} = \mathbf{T}_1\mathbf{n} - \mathbf{T}_2\mathbf{n} = \mathbf{0}$ on $\mathcal{S}_2$,

$$\begin{aligned}\int_{\partial\mathrm{B}} \mathbf{Tn} \cdot \mathbf{u}\, da &= \int_{\mathcal{S}_1} \mathbf{Tn} \cdot \mathbf{u}\, da + \int_{\mathcal{S}_2} \mathbf{Tn} \cdot \mathbf{u}\, da \\ &= 0.\end{aligned}$$

Hence, by the work and energy balance (52.72) applied to the body B,

$$\int_{\mathrm{B}} \mathbf{E}\colon\mathbb{C}\mathbf{E}\, dv = 0, \qquad (52.85)$$

and, since the integrand is nonnegative, it must vanish: $\mathbf{E}\colon\mathbb{C}\mathbf{E} = 0$. In view of the assumed positive-definiteness of $\mathbb{C}$, we must therefore have $\mathbf{E} = \mathbf{T} = \mathbf{0}$. In particular, the vanishing of the strain $\mathbf{E}$ renders the displacement $\mathbf{u}$ rigid.[273]

Our next step is to state and prove the principle of minimum potential energy within the context of elastostatics. By a **kinematically admissible displacement field**, we mean an arbitrary field $\tilde{\mathbf{u}}$ on B that satisfies the displacement boundary condition (52.83):

$$\tilde{\mathbf{u}} = \hat{\mathbf{u}} \quad \text{on } \mathcal{S}_1. \qquad (52.86)$$

Let $\mathcal{F}$ denote the net free-energy

$$\mathcal{F}(\mathbf{E}) = \tfrac{1}{2} \int_{\mathrm{B}} \mathbf{E}\colon\mathbb{C}\mathbf{E}\, dv \qquad (52.87)$$

corresponding to any strain field $\mathbf{E}$, and let Φ — the potential energy functional — be defined on the set of kinematically admissible displacement fields by

$$\begin{aligned}\Phi(\tilde{\mathbf{u}}) &= \mathcal{F}(\tilde{\mathbf{E}}) - \int_{\partial\mathrm{B}} \mathbf{Tn} \cdot \tilde{\mathbf{u}}\, da - \int_{\mathrm{B}} \mathbf{b}_0 \cdot \tilde{\mathbf{u}}\, dv, \\ \tilde{\mathbf{E}} &= \tfrac{1}{2}(\nabla\tilde{\mathbf{u}} + (\nabla\tilde{\mathbf{u}})^{\mathsf{T}}).\end{aligned} \qquad (52.88)$$

Principle of Minimum Potential Energy *Let* $\mathbf{u}$*,* $\mathbf{E}$*, and* $\mathbf{T}$ *define a solution of the mixed problem of elastostatics. Then*

$$\Phi(\mathbf{u}) \le \Phi(\tilde{\mathbf{u}}) \qquad (52.89)$$

for every kinematically admissible displacement field $\tilde{\mathbf{u}}$*, with equality holding only if* $\mathbf{u}$ *and* $\tilde{\mathbf{u}}$ *differ by a rigid displacement field.*

---

273 Cf. the sentence containing (52.82).

To establish this principle, choose a kinematically admissible displacement field $\tilde{\mathbf{u}}$ and let

$$\mathbf{w} = \tilde{\mathbf{u}} - \mathbf{u},$$

$$\bar{\mathbf{E}} = \tilde{\mathbf{E}} - \mathbf{E}.$$

Then, since $\mathbf{u}$ is a solution and $\tilde{\mathbf{u}}$ is kinematically admissible,

$$\begin{aligned} \mathbf{w} &= \mathbf{0} \quad \text{on } \mathcal{S}_1, \\ \bar{\mathbf{E}} &= \tfrac{1}{2}(\nabla\mathbf{w} + (\nabla\mathbf{w})^{\top}). \end{aligned} \tag{52.90}$$

Further, since $\mathbb{C}$ is symmetric and $\mathbf{T} = \mathbb{C}\mathbf{E}$,

$$\begin{aligned} \tilde{\mathbf{E}}:\mathbb{C}\tilde{\mathbf{E}} &= \mathbf{E}:\mathbb{C}\mathbf{E} + \bar{\mathbf{E}}:\mathbb{C}\bar{\mathbf{E}} + \mathbf{E}:\mathbb{C}\bar{\mathbf{E}} + \bar{\mathbf{E}}:\mathbb{C}\mathbf{E} \\ &= \mathbf{E}:\mathbb{C}\mathbf{E} + \bar{\mathbf{E}}:\mathbb{C}\bar{\mathbf{E}} + 2\mathbf{T}:\bar{\mathbf{E}}; \end{aligned} \tag{52.91}$$

hence, by (52.87),

$$\mathcal{F}(\tilde{\mathbf{E}}) - \mathcal{F}(\mathbf{E}) = \mathcal{F}(\bar{\mathbf{E}}) + \int_{\mathrm{B}} \mathbf{T}:\bar{\mathbf{E}}\,dv. \tag{52.92}$$

On the other hand, by $(52.90)_{1,2}$, $(52.84)_3$, the symmetry of $\mathbf{T}$, and the divergence theorem,

$$\begin{aligned} \int_{\mathrm{B}} \mathbf{T}:\bar{\mathbf{E}}\,dv &= \int_{\mathrm{B}} \mathbf{T}:\nabla\mathbf{w}\,dv \\ &= \int_{\partial\mathrm{B}} \mathbf{T}\mathbf{n}\cdot\mathbf{w}\,da - \int_{\mathrm{B}} \operatorname{Div}\mathbf{T}\cdot\mathbf{w}\,dv \\ &= \int_{\partial\mathrm{B}} \mathbf{T}\mathbf{n}\cdot\mathbf{w}\,da + \int_{\mathrm{B}} \mathbf{b}_0\cdot\mathbf{w}\,dv \\ &= \int_{\mathcal{S}_2} \mathbf{T}\mathbf{n}\cdot\mathbf{w}\,da + \int_{\mathrm{B}} \mathbf{b}_0\cdot\mathbf{w}\,dv. \end{aligned} \tag{52.93}$$

In view of (52.88), (52.92), and (52.93),

$$\Phi(\tilde{\mathbf{u}}) - \Phi(\mathbf{u}) = \mathcal{F}(\bar{\mathbf{E}}).$$

Thus, since $\mathbb{C}$ is positive-definite,

$$\Phi(\mathbf{u}) \le \Phi(\tilde{\mathbf{u}})$$

and, by the argument following (52.85),

$$\Phi(\mathbf{u}) = \Phi(\tilde{\mathbf{u}})$$

only when $\bar{\mathbf{E}} = \mathbf{0}$ and, hence, only when $\mathbf{w} = \tilde{\mathbf{u}} - \mathbf{u}$ is a rigid displacement. This completes the proof of the principle of minimum potential energy.

52.7.2 Elastodynamics

As before, we consider **boundary conditions** in which the displacement is specified on $\mathcal{S}_1$ and the surface traction is specified on $\mathcal{S}_2$:

$$\begin{aligned} \mathbf{u} &= \hat{\mathbf{u}} \quad \text{on } \mathcal{S}_2 \text{ for all time } (\geq 0), \\ \mathbf{Tn} &= \hat{\mathbf{t}} \quad \text{on } \mathcal{S}_1 \text{ for all time } (\geq 0), \end{aligned} \tag{52.94}$$

with $\hat{\mathbf{u}}$ and $\hat{\mathbf{t}}$ prescribed functions. In dynamics, these conditions are supplemented by **initial conditions** in which the displacement $\mathbf{u}$ and the velocity $\dot{\mathbf{u}}$ are specified initially:

$$\mathbf{u}(\mathbf{X}, 0) = \mathbf{u}_0(\mathbf{X}) \qquad \text{and} \qquad \dot{\mathbf{u}}(\mathbf{X}, 0) = \mathbf{v}_0(\mathbf{X}) \qquad \text{for all } \mathbf{X} \text{ in B}, \tag{52.95}$$

with $\mathbf{u}_0$ and $\mathbf{v}_0$ prescribed functions. The **mixed problem of elastodynamics** may then be stated as follows: Given boundary data $\hat{\mathbf{u}}$ and $\hat{\mathbf{t}}$ and initial data $\mathbf{u}_0$ and $\mathbf{v}_0$, find a displacement field $\mathbf{u}$, a strain field $\mathbf{E}$, and a stress field $\mathbf{T}$ that satisfy the field equations

$$\left.\begin{aligned} \mathbf{E} &= \tfrac{1}{2}(\nabla\mathbf{u} + (\nabla\mathbf{u})^{\top}), \\ \mathbf{T} &= \mathbb{C}\mathbf{E}, \\ \rho\ddot{\mathbf{u}} &= \operatorname{Div}\mathbf{T} + \mathbf{b}_0, \end{aligned}\right\} \quad \text{on B for all time } (\geq 0), \tag{52.96}$$

the boundary conditions (52.94), and the the initial conditions (52.95).[274]

UNIQUENESS THEOREM *The mixed problem of elastodynamics has at most one solution.*

This theorem is a consequence of the energy-conservation theorem on page 307. Indeed, the difference between two solutions represents a solution of a mixed problem in which the body force vanishes, as do the prescribed surface displacement, surface traction, initial displacement, and initial velocity; by (52.69), the two solutions must therefore coincide.

52.8 Sinusoidal Progressive Waves

Sinusoidal progressive waves form an important class of solutions to the equations of linear elastodynamics. We consider these waves for isotropic media and assume that the conventional body force $\mathbf{b}_0$ vanishes. The underlying field equation is then the displacement equation of motion

$$\rho\ddot{\mathbf{u}} = \mu\Delta\mathbf{u} + (\lambda + \mu)\nabla\operatorname{Div}\mathbf{u}. \tag{52.97}$$

A displacement field $\mathbf{u}$ of the form

$$\mathbf{u}(\mathbf{X}, t) = \mathbf{a}\sin(\mathbf{r}\cdot\boldsymbol{\nu} - ct), \qquad |\boldsymbol{\nu}| = 1, \qquad \mathbf{r} = \mathbf{X} - \mathbf{o}, \tag{52.98}$$

is called a sinusoidal progressive wave with amplitude $\mathbf{a}$, direction $\boldsymbol{\nu}$, and velocity c. Such a wave is longitudinal if $\mathbf{a}$ and $\boldsymbol{\nu}$ are parallel or transverse if $\mathbf{a}$ and $\boldsymbol{\nu}$ are perpendicular.

We now determine conditions necessary and sufficient for (52.98) to satisfy (52.97). Applying the chain-rule to (52.98) and writing $\varphi(\mathbf{X}, t) = \mathbf{r}\cdot\boldsymbol{\nu} - ct$, we

274 Here, it is assumed that, in addition to the elasticity tensor $\mathbb{C}$ and the conventional body force $\mathbf{b}_0$, the density ρ is known.

find that

$$\begin{aligned} \nabla \mathbf{u} &= \mathbf{a} \otimes \boldsymbol{\nu} \cos \varphi, \\ \Delta \mathbf{u} &= -\mathbf{a} \sin \varphi, \\ \nabla \operatorname{Div} \mathbf{u} &= -(\mathbf{a} \cdot \boldsymbol{\nu}) \boldsymbol{\nu} \sin \varphi, \\ \ddot{\mathbf{u}} &= -c^2 \mathbf{a} \sin \varphi. \end{aligned} \tag{52.99}$$

By $(52.99)_1$,

$$\operatorname{Div} \mathbf{u} = \mathbf{a} \cdot \boldsymbol{\nu} \cos \varphi,$$

$$\operatorname{Curl} \mathbf{u} = -\mathbf{a} \times \boldsymbol{\nu} \cos \varphi,$$

and it follows that the wave is longitudinal if and only if $\operatorname{Curl} \mathbf{u} = \mathbf{0}$ and transverse if and only if $\operatorname{Div} \mathbf{u} = 0$.

Next, by $(52.99)_{2\text{–}4}$, $\mathbf{u}$ as defined in (52.98) satisfies (52.97) if and only if

$$\rho c^2 \mathbf{a} = \mu \mathbf{a} + (\lambda + \mu)(\mathbf{a} \cdot \boldsymbol{\nu}) \boldsymbol{\nu}. \tag{52.100}$$

Defining the acoustic tensor $\mathbf{A}(\boldsymbol{\nu})$ via

$$\rho \mathbf{A}(\boldsymbol{\nu}) = \mu \mathbf{1} + (\lambda + \mu) \boldsymbol{\nu} \otimes \boldsymbol{\nu}$$

allows us to rewrite (52.100) in the form of a propagation condition

$$\mathbf{A}(\boldsymbol{\nu}) \mathbf{a} = c^2 \mathbf{a}. \tag{52.101}$$

A condition necessary and sufficient for $\mathbf{u}$ to satisfy (52.97) is therefore that $\mathbf{a}$ be an eigenvector and that c^2 be a corresponding eigenvalue of the acoustic tensor $\mathbf{A}(\boldsymbol{\nu})$.

A direct calculation shows that

$$\mathbf{A}(\boldsymbol{\nu}) = \frac{\mu}{\rho}(\mathbf{1} - \boldsymbol{\nu} \otimes \boldsymbol{\nu}) + \frac{\lambda + 2\mu}{\rho} \boldsymbol{\nu} \otimes \boldsymbol{\nu}.$$

But, this is simply the spectral decomposition of $\mathbf{A}(\boldsymbol{\nu})$ and we may conclude from (ii) of the Spectral Theorem on page 28 that $\rho^{-1}\mu$ and $\rho^{-1}(\lambda + 2\mu)$ are the eigenvalues of $\mathbf{A}(\boldsymbol{\nu})$, while the line spanned by $\boldsymbol{\nu}$ and the plane perpendicular to $\boldsymbol{\nu}$ (containing the origin) are the corresponding characteristic spaces. A sinusoidal progressive wave with amplitude $\mathbf{a}$, direction $\boldsymbol{\nu}$, and velocity c will therefore be a solution of the displacement equation of motion (52.97) if and only if either

- $\rho c^2 = \lambda + 2\mu$ and the wave is longitudinal

or

- $\rho c^2 = \mu$ and the wave is transverse.

This result shows that for an isotropic medium, only two types of sinusoidal progressive waves are possible: longitudinal and transverse. The corresponding wave speeds $\sqrt{(\lambda + 2\mu)/\rho}$ and $\sqrt{\mu/\rho}$ are called, respectively, the **longitudinal sound speed** and the **transverse sound speed** of the medium. Importantly, these speeds are real when the elasticity tensor is positive-semidefinite. For an anisotropic medium the situation is far more complicated. A propagation condition of the form (52.101) can be obtained. However, the waves determined by that condition are generally neither longitudinal nor transverse and those waves may propagate with different speeds in different directions.

EXERCISE

1. For an anisotropic medium subject to no conventional body force, the displacement equation of motion can be written in the form

$$\rho\ddot{\mathbf{u}} = \operatorname{Div}(\mathbb{C}\nabla\mathbf{u}). \tag{52.102}$$

Show that

a. A sinusoidal progressive wave of the form (52.98) satisfies (52.102) if and only if the propagation condition (52.101) holds, where now $\mathbf{A}(\boldsymbol{\nu})$ is defined by

$$\rho\mathbf{A}(\boldsymbol{\nu})\mathbf{k} = [\mathbb{C}(\mathbf{k}\otimes\boldsymbol{\nu})]\boldsymbol{\nu}$$

for every vector **k**.

b. The acoustic tensor $\mathbf{A}(\boldsymbol{\nu})$ is positive-semidefinite if and only if the elasticity tensor $\mathbb{C}$ obeys

$$(\mathbf{e}\otimes\mathbf{f})\cdot\mathbb{C}(\mathbf{e}\otimes\mathbf{f}) \geq 0$$

for all vectors **e** and **f** — in which case $\mathbb{C}$ is said to be *elliptic*.

c. The elasticity tensor $\mathbb{C}$ for an isotropic medium is elliptic if and only if $\mu \geq 0$ and $\lambda + 2\mu \geq 0$ (in which case the longitudinal and transverse wave speeds are real).

53 Digression: Incompressibility

In preparation for the next section, which treats incompressible elastic solids, we now introduce the notion of incompressibility and discuss several of its most important consequences.

53.1 Kinematics of Incompressibility

Constraints are constitutive assumptions that limit the class of constitutive processes a body may undergo. Here, we consider only the constraint of incompressibility.[275]

An **incompressible body** is one for which only isochoric motions are possible, a requirement manifested in the **constraint**[276]

$$\operatorname{div}\mathbf{v} = 0. \tag{53.1}$$

In view of the proposition labeled (‡) on page 114, either of the following equivalent conditions may also be used to characterize incompressibility:[277]

$$\begin{aligned} \overline{\det \mathbf{F}}^{\boldsymbol{\cdot}} &= 0, \\ \operatorname{tr}\mathbf{L} &= 0. \end{aligned} \tag{53.2}$$

Moreover, since the stretching $\mathbf{D}$ is the symmetric part of the velocity gradient $\mathbf{L}$, the incompressibility condition $(53.2)_2$ is equivalent to the requirement that

$$\operatorname{tr}\mathbf{D} = 0. \tag{53.3}$$

By $(53.2)_1$, we may, without loss in generality, assume that

$$J = \det \mathbf{F} \equiv 1 \tag{53.4}$$

in all motions of an incompressible body, so that at any instant t of any such motion, each part P deforms to a spatial region $\mathcal{P}_t$ with the same volume. An immediate consequence of (47.2) and (47.3) is then that

$$\begin{aligned} \det \mathbf{C} &\equiv 1, \\ \det \mathbf{B} &\equiv 1. \end{aligned} \tag{53.5}$$

275 Cf. TRUESDELL and NOLL (1965, §30) for a general discussion of constraints.

276 Cf. the results concerning isochoric motions in §16.

277 The second of these conditions follows from the first and the identity $\operatorname{tr}\mathbf{A} = \operatorname{tr}(\mathbf{R}\mathbf{A}\mathbf{R}^{\top})$, which is valid for any tensor $\mathbf{A}$ and any orthogonal tensor $\mathbf{R}$.

53.2 Indeterminacy of the Pressure. Free-Energy Imbalance

Assume throughout this subsection that the body is incompressible. Then, assuming that the referential density ρ_R is constant, we may conclude from (53.4) and the referential form (18.9) of mass balance that

$$\begin{aligned}\rho &= J\rho_R \\ &= \rho_R \\ &\equiv \text{constant.}\end{aligned} \tag{53.6}$$

The essential change induced by the constraint of incompressibility lies with the stress power $\mathbf{T}:\mathbf{D}$. Indeed, defining the **extra stress S** and the **pressure** p through

$$\mathbf{S} = \mathbf{T} + p\mathbf{1} \qquad \text{and} \qquad p = -\tfrac{1}{3}\mathrm{tr}\,\mathbf{T}, \tag{53.7}$$

we see that

$$\mathrm{tr}\,\mathbf{S} = \mathbf{0}. \tag{53.8}$$

Thus, since

$$\begin{aligned}\mathbf{1}:\mathbf{D} &= \mathrm{tr}\,\mathbf{D} \\ &= \mathrm{div}\,\mathbf{v} \\ &= 0,\end{aligned}$$

it follows that

$$\mathbf{T}:\mathbf{D} = \mathbf{S}:\mathbf{D}. \tag{53.9}$$

We therefore have a central consequence of incompressibility:

(‡) *In the motion of an incompressible body, the pressure expends no power and hence performs no work.*

Next, by (24.1) and (53.5), the Piola stress has the form

$$\mathbf{T}_R = \mathbf{T}\mathbf{F}^{-\top} \tag{53.10}$$

and, therefore, $(53.7)_1$ yields

$$\mathbf{T}_R = -p\mathbf{F}^{-\top} + \mathbf{S}_R, \tag{53.11}$$

where, bearing in mind (53.4),

$$\mathbf{S}_R \stackrel{\text{def}}{=} \mathbf{S}\mathbf{F}^{-\top} \tag{53.12}$$

is the **extra Piola stress**. Thus, by (53.9), the stress power takes the form

$$\begin{aligned}\mathbf{S}:\mathbf{D} &= \mathbf{S}:\mathbf{L} \\ &= \mathbf{S}:(\dot{\mathbf{F}}\mathbf{F}^{-1}) \\ &= (\mathbf{S}\mathbf{F}^{-\top}):\dot{\mathbf{F}} \\ &= \mathbf{S}_R:\dot{\mathbf{F}}.\end{aligned} \tag{53.13}$$

We may use (53.13) to write the referential free-energy imbalance (31.24) in the form

$$\dot{\psi}_R - \mathbf{S}_R:\dot{\mathbf{F}} = -\delta_R \le 0. \tag{53.14}$$

53.3 Changes in Frame

By (21.6) and (53.7), under a change in frame, since $p^* = p$,

$$\underbrace{-p^*\mathbf{1} + \mathbf{S}^*}_{\mathbf{T}^*} = \mathbf{Q}\,\underbrace{(-p\mathbf{1} + \mathbf{S})}_{\mathbf{T}}\,\mathbf{Q}^{\top}$$

$$= -p\mathbf{1} + \mathbf{Q}\mathbf{S}\mathbf{Q}^{\top},$$

and the extra stress $\mathbf{S}$ must be frame-indifferent:

$$\mathbf{S}^* = \mathbf{Q}\mathbf{S}\mathbf{Q}^{\top}. \tag{53.15}$$

Similarly, for the extra Piola stress $\mathbf{S}_{\mathrm{R}}$, by (53.12), since $\mathbf{F}^* = \mathbf{Q}\mathbf{F}$ implies that $(\mathbf{F}^{-\top})^* = (\mathbf{F}^*)^{-\top} = (\mathbf{Q}\mathbf{F})^{-\top} = (\mathbf{F}^{-1}\mathbf{Q}^{-1})^{\top} = (\mathbf{F}^{-1}\mathbf{Q}^{\top})^{\top} = \mathbf{Q}\mathbf{F}^{-\top}$,

$$\mathbf{S}^*_{\mathrm{R}} = \mathbf{S}^*(\mathbf{F}^{-\top})^*$$

$$= \mathbf{Q}\mathbf{S}\mathbf{F}^{-\top}$$

$$= \mathbf{Q}\mathbf{S}_{\mathrm{R}}. \tag{53.16}$$

54 Incompressible Elastic Materials

54.1 Constitutive Theory

The discussion of compressible elastic solids in §48 is based, in part, on a constitutive equation

$$\mathbf{T}_R = \hat{\mathbf{T}}_R(\mathbf{F}) \tag{54.1}$$

giving the Piola stress when the motion (and hence $\mathbf{F}$) is known, and one might ask: Would this be a valid constitutive relation for an incompressible body? To answer this question, consider the following thought experiment involving a homogeneous, incompressible elastic body in the shape of a ball of diameter d. Assume that there are no body forces and that the ball is in equilibrium under a uniform pressure p_0. A second experiment in which the ball is in equilibrium under a pressure different from p_0 would then leave the body unchanged: Since the ball is incompressible it would remain a ball of diameter d. In view of this thought experiment,

- *it would seem unreasonable to allow a constitutive relation for an incompressible elastic body to involve the pressure.*

We should therefore replace the constitutive relation (54.1), which involves the full Piola stress tensor $\mathbf{T}_R$, with one involving only its "pressureless part" $\mathbf{S}_R$. We thus take — as appropriate *incompressible* counterparts of the constitutive equations (48.1) — constitutive equations of the form

$$\left.\begin{aligned} \psi_R &= \hat{\psi}_R(\mathbf{F}), \\ \mathbf{S}_R &= \hat{\mathbf{S}}_R(\mathbf{F}), \end{aligned}\right\} \tag{54.2}$$

with $\mathbf{F}$ constrained to satisfy

$$\det \mathbf{F} = 1.$$

Because the constitutive equations (54.2), by themselves, cannot determine the pressure, the pressure is usually referred to as *indeterminate*.[278]

54.1.1 Consequences of Frame-Indifference

The relation (53.16) and an argument identical to that surrounding (48.7) here imply that the constitutive relations (54.2) are frame-indifferent if and only if they reduce

[278] By (‡) on page 317, the pressure in an incompressible body performs no work; our use of the term "indeterminate" is therefore consistent with classical mechanics, where a force is indeterminate if it performs no work.

to constitutive relations of the specific form

$$\begin{aligned}\psi_R &= \bar{\psi}_R(\mathbf{C}),\\ \mathbf{S}_R &= \mathbf{F}\bar{\mathbf{S}}_{RR}(\mathbf{C}),\end{aligned}\tag{54.3}$$

where

$$\mathbf{S}_{RR} = \bar{\mathbf{S}}_{RR}(\mathbf{C})$$

denotes the pressureless part of the second Piola stress. By (53.10) and (53.11), we may rewrite the constitutive equation $(54.3)_2$ in either of the equivalent forms[279]

$$\begin{aligned}\mathbf{T}_R &= -p\mathbf{F}^{-\top} + \mathbf{F}\bar{\mathbf{S}}_{RR}(\mathbf{C}),\\ \mathbf{T} &= -p\mathbf{1} + \mathbf{F}\bar{\mathbf{S}}_{RR}(\mathbf{C})\mathbf{F}^{\top}.\end{aligned}\tag{54.4}$$

54.1.2 Domain of Definition of the Response Functions

The common domain of the response functions $\bar{\psi}(\mathbf{C})$ and $\bar{\mathbf{S}}_{RR}(\mathbf{C})$ is not the set of all symmetric and positive-definite tensors $\mathbf{C}$ but instead

the set of all symmetric, positive-definite tensors $\mathbf{C}$ with $\det\mathbf{C} = 1$; (54.5)

because of the constraint $\det\mathbf{C} = 1$, care must be taken in ascribing a meaning to the derivative $\partial\bar{\psi}(\mathbf{C})/\partial\mathbf{C}$. We can, however, bypass this difficulty by noting that any function $f(\mathbf{C})$ defined on the constrained set (54.5) may be extended to a function $f_{\mathrm{ext}}(\mathbf{C})$ defined on *all symmetric, positive-definite tensors* $\mathbf{C}$, even those with $\det\mathbf{C} \neq 1$, as follows:[280]

$$f_{\mathrm{ext}}(\mathbf{C}) = f((\det\mathbf{C})^{-1/3}\mathbf{C}).\tag{54.6}$$

We may therefore assume, without loss in generality, that $\bar{\psi}_R(\mathbf{C})$ is well-defined on *all symmetric, positive-definite tensors* $\mathbf{C}$.[281] Then, as in the compressible theory,

$$\frac{\partial\bar{\psi}_R(\mathbf{C})}{\partial\mathbf{C}} \text{ is symmetric,}\tag{54.7}$$

with $\partial\bar{\psi}_R(\mathbf{C})/\partial\mathbf{C}$ a conventional derivative on the space of symmetric tensors. Even so, when we write $\bar{\psi}_R(\mathbf{C})$ or $\partial\bar{\psi}_R(\mathbf{C})/\partial\mathbf{C}$ it should be understood that $\det\mathbf{C} = 1$, even though the derivative represents that of the extended function.

We find it neither necessary nor helpful to extend the domain of $\bar{\mathbf{S}}_{RR}(\mathbf{C})$. Since

$$\mathbf{S} = \mathbf{S}_R\mathbf{F}^{\top},$$

(54.3) yields

$$\mathbf{S} = \mathbf{F}\bar{\mathbf{S}}_{RR}(\mathbf{C})\mathbf{F}^{\top};$$

thus, since $\mathbf{S}$ is symmetric and deviatoric,

$$\mathbf{F}\bar{\mathbf{S}}_{RR}(\mathbf{C})\mathbf{F}^{\top} \text{ is symmetric and deviatoric.}\tag{54.8}$$

For the purist, a definition of $\partial\bar{\psi}_R(\mathbf{C})/\partial\mathbf{C}$ that does not involve extending the function ψ_R follows from the discussion of derivatives in §3.4: $\partial\bar{\psi}_R(\mathbf{C})/\partial\mathbf{C}$ is defined via the requirement that, for any function

---

[279] Cf. (48.14) and (48.15).

[280] The right side of (54.6) is well defined, because $(\det\mathbf{C})^{-1/3}\mathbf{C}$ has unit determinant.

[281] We do not require that the particular extension (54.6) be used: For example, one might choose to use a free energy of the compressible theory, suitably tailored.

$\mathbf{C}(t)$ with values in the incompressibility set (54.5), and for $\psi_R(t) = \bar{\psi}_R(\mathbf{C}(t))$,

$$\dot{\psi}_R = \frac{\partial \bar{\psi}_R(\mathbf{C})}{\partial \mathbf{C}} : \dot{\mathbf{C}}. \tag{54.9}$$

Granted this, for any $\mathbf{C}$,

$$\text{the derivative } \frac{\partial \bar{\psi}_R(\mathbf{C})}{\partial \mathbf{C}} \text{ must belong to the same tensorial set as } \dot{\mathbf{C}}.$$

Next, by $(3.3)_1$, $\overline{\det \mathbf{C}}^{\displaystyle\cdot} = (\det \mathbf{C})\mathrm{tr}\,(\dot{\mathbf{C}}\mathbf{C}^{-1})$, and we may use $(53.5)_1$ to conclude that

$$\mathrm{tr}\,(\dot{\mathbf{C}}\mathbf{C}^{-1}) = 0 \qquad \text{or, equivalently,} \qquad \dot{\mathbf{C}} : \mathbf{C}^{-1} = 0;$$

hence, in the space of symmetric tensors,

$$\dot{\mathbf{C}} \qquad \text{is orthogonal to} \qquad \mathbf{C}^{-1}.$$

Thus,

$$\frac{\partial \bar{\psi}_R(\mathbf{C})}{\partial \mathbf{C}} \qquad \text{must be orthogonal to} \qquad \mathbf{C}^{-1}.$$

54.1.3 Thermodynamic Restrictions

As in the steps leading to (48.11), the Coleman–Noll procedure — here based on the definition of $\partial \bar{\psi}_R(\mathbf{C})/\partial \mathbf{C}$ using an appropriate extension of $\bar{\psi}$ together with the free-energy imbalance (53.14) applied to the constitutive equations (54.3) — yields the requirement that the inequality

$$\underbrace{\left(2\frac{\partial \bar{\psi}_R(\mathbf{C})}{\partial \mathbf{C}} - \bar{\mathbf{S}}_{RR}(\mathbf{C})\right)}_{\mathbf{M}} : \dot{\mathbf{C}} \le 0 \tag{54.10}$$

be satisfied in all motions of the body. Equivalently, noting that, by (11.10) and the fact that $\mathbf{C} = 2\mathbf{E} + \mathbf{1}$,

$$2\mathbf{F}^{\top}\mathbf{D}\mathbf{F} = \dot{\mathbf{C}}, \tag{54.11}$$

and hence that, for $\mathbf{M}$ as defined in (54.10),

$$\begin{aligned} \mathbf{M} : \dot{\mathbf{C}} &= 2\mathbf{M} : (\mathbf{F}^{\top}\mathbf{D}\mathbf{F}) \\ &= 2(\mathbf{F}\mathbf{M}\mathbf{F}^{\top}) : \mathbf{D}, \end{aligned}$$

all motions must be consistent with

$$\left[\mathbf{F}\left(2\frac{\partial \bar{\psi}_R(\mathbf{C})}{\partial \mathbf{C}} - \bar{\mathbf{S}}_{RR}(\mathbf{C})\right)\mathbf{F}^{\top}\right] : \mathbf{D} \le 0. \tag{54.12}$$

The following proposition is useful in determining the consequences of (54.12):

(†) Given any point of the body and any time, it is possible to find a motion such that[282]
  (a) $\det \mathbf{C} \equiv 1$ (so that $\mathrm{tr}\,\mathbf{D} \equiv 0$);
  (b) $\mathbf{C}$ and $\mathbf{D}$ have arbitrarily prescribed values — consistent with (a) — at that point and time.

Granted this assertion — which we prove on page 322 — the symmetric-deviatoric part of the "coefficient" of $\mathbf{D}$ in (54.12), namely

$$\mathbf{F}\left(2\frac{\partial \bar{\psi}_R(\mathbf{C})}{\partial \mathbf{C}} - \bar{\mathbf{S}}_{RR}(\mathbf{C})\right)\mathbf{F}^{\top},$$

[282] Cf. (†) on page 279.

must vanish, for otherwise it is possible to choose $\mathbf{D}$ so that (54.12) is violated. Thus, bearing in mind that, by (54.7) and (54.8),

(i) $\dfrac{\partial \bar{\psi}_{\rm RR}(\mathbf{C})}{\partial \mathbf{C}}$ is symmetric, but not necessarily deviatoric,

(ii) $\mathbf{D}$ and $\mathbf{F}\bar{\mathbf{S}}_{\rm RR}(\mathbf{C})\mathbf{F}^{\scriptscriptstyle\top}$ are symmetric and deviatoric,

and it follows that[283]

$$\mathbf{F}\bar{\mathbf{S}}_{\rm RR}(\mathbf{C})\mathbf{F}^{\scriptscriptstyle\top} = 2\,\mathrm{dev}\left(\mathbf{F}\frac{\partial \bar{\psi}_{\rm R}(\mathbf{C})}{\partial \mathbf{C}}\mathbf{F}^{\scriptscriptstyle\top}\right),$$

and hence, by (54.4), that

$$\mathbf{T} = -p\mathbf{1} + 2\,\mathrm{dev}\left(\mathbf{F}\frac{\partial \bar{\psi}_{\rm R}(\mathbf{C})}{\partial \mathbf{C}}\mathbf{F}^{\scriptscriptstyle\top}\right). \tag{54.13}$$

The presence of the operation dev in (54.13) is necessary if p is to represent the pressure, $p = -\mathrm{tr}\,\mathbf{T}$. On the other hand, if we replace p in (54.13) by

$$P = p + \tfrac{2}{3}\,\mathrm{tr}\left(\mathbf{F}\frac{\partial \bar{\psi}_{\rm R}(\mathbf{C})}{\partial \mathbf{C}}\mathbf{F}^{\scriptscriptstyle\top}\right), \tag{54.14}$$

then we may remove the operation dev in (54.13); thus, by (53.10), we have the following **thermodynamic restriction:**

(‡) *The free energy determines the Piola and Cauchy stresses through the equivalent* **stress relations**

$$\begin{aligned}\mathbf{T}_{\rm R} &= -P\mathbf{F}^{-\scriptscriptstyle\top} + 2\mathbf{F}\frac{\partial \bar{\psi}_{\rm R}(\mathbf{C})}{\partial \mathbf{C}},\\ \mathbf{T} &= -P\mathbf{1} + 2\mathbf{F}\frac{\partial \bar{\psi}_{\rm R}(\mathbf{C})}{\partial \mathbf{C}}\mathbf{F}^{\scriptscriptstyle\top},\end{aligned} \tag{54.15}$$

in which P is an arbitrary scalar field, a field that is generally distinct from the pressure.[284]

We refer to P as the *effective pressure*. For any initial-boundary-value problem involving applied pressure on any portion of the boundary, it is the actual pressure p as opposed to the effective pressure P that is of primary physical importance. However, given χ and P, the relation (54.14) determines p via

$$p = P - \tfrac{2}{3}\,\mathrm{tr}\left(\mathbf{F}\frac{\partial \bar{\psi}_{\rm R}(\mathbf{C})}{\partial \mathbf{C}}\mathbf{F}^{\scriptscriptstyle\top}\right).$$

54.1.4 Verification of (†)

Choose an arbitrary material point $\mathbf{X}_0$ and time t_0. Let $\mathbf{F}_0$ be an arbitrary constant tensor with

$$\det \mathbf{F}_0 = 1, \tag{54.16}$$

and let $\mathbf{D}$ be an arbitrary constant symmetric tensor with

$$\mathrm{tr}\,\mathbf{D} = 0. \tag{54.17}$$

The motion (48.21) is then, as before, well defined for any reference body B, no matter the shape. Moreover, the constant $\mathbf{D}$ represents the stretching, so that, by (54.17), (48.21) represents the motion of an incompressible body. A consequence of (14.6) is then that

$$\mathbf{C}(\mathbf{X}_0, t_0) = \mathbf{F}_0^{\scriptscriptstyle\top}\mathbf{F}_0.$$

Since $\mathbf{F}_0$ with unit determinant and the symmetric and deviatoric tensor $\mathbf{D}$ were arbitrarily chosen, this completes the verification of (†).

---

283 Recall that $\mathrm{dev}\mathbf{A} = \mathbf{A} - \frac{1}{3}(\mathrm{tr}\,\mathbf{A})\mathbf{1}$ denotes the deviatoric part of $\mathbf{A}$.

284 The effective P is arbitrary because the pressure p is arbitrary.

54.2 Incompressible Isotropic Elastic Bodies

Assume that the body is both incompressible and isotropic. Then, bearing in mind (53.4), we have

$$\bar{\psi}_{\mathrm{R}}(\mathbf{B}) = \tilde{\psi}_{\mathrm{R}}(\tilde{\mathcal{I}}_{\mathbf{B}}), \tag{54.18}$$

where

$$\tilde{\mathcal{I}}_{\mathbf{B}} \stackrel{\text{def}}{=} (I_1(\mathbf{B}), I_2(\mathbf{B})) \tag{54.19}$$

denotes the list of nontrivial principal invariants. Under these circumstances, the counterpart of the stress relation (50.34) is

$$\mathbf{T} = -P\mathbf{1} + \beta_1(\tilde{\mathcal{I}}_{\mathbf{B}})\mathbf{B} + \beta_2(\tilde{\mathcal{I}}_{\mathbf{B}})\mathbf{B}^{-1}, \tag{54.20}$$

with

$$\begin{aligned} \beta_1(\tilde{\mathcal{I}}_{\mathbf{B}}) &= 2\frac{\partial\tilde{\psi}_{\mathrm{R}}(\tilde{\mathcal{I}}_{\mathbf{B}})}{\partial I_1}, \\ \beta_2(\tilde{\mathcal{I}}_{\mathbf{B}}) &= -2\frac{\partial\tilde{\psi}_{\mathrm{R}}(\tilde{\mathcal{I}}_{\mathbf{B}})}{\partial I_2}. \end{aligned} \tag{54.21}$$

When the free energy is expressed in terms of principal stretches,

$$\psi_{\mathrm{R}} = \check{\psi}_{\mathrm{R}}(\lambda_1, \lambda_2, \lambda_3),$$

the constraint of incompressibility reads

$$\lambda_1\lambda_2\lambda_3 = 1,$$

and, for λ_i distinct, the counterparts of the expressions (50.44) and (50.45) are

$$\begin{aligned} \mathbf{T}_{\mathrm{R}} &= \sum_{i=1}^{3}\left(\frac{\partial\check{\psi}_{\mathrm{R}}(\lambda_1, \lambda_2, \lambda_3)}{\partial\lambda_i} - \frac{P}{\lambda_i}\right)\mathbf{l}_i \otimes \mathbf{r}_i, \\ \mathbf{T} &= \sum_{i=1}^{3}\left(\lambda_i\frac{\partial\check{\psi}_{\mathrm{R}}(\lambda_1, \lambda_2, \lambda_3)}{\partial\lambda_i} - P\right)\mathbf{l}_i \otimes \mathbf{l}_i, \end{aligned} \tag{54.22}$$

with $\mathbf{l}_i$ and $\mathbf{r}_i$ being the left and right principal directions.

A consequence of $(54.22)_2$ is that the principal values of the Cauchy stress are given by

$$\sigma_i = \lambda_i\frac{\partial\check{\psi}_{\mathrm{R}}(\lambda_1, \lambda_2, \lambda_3)}{\partial\lambda_i} - P \quad \text{(no sum)}. \tag{54.23}$$

Further, granted that $\mathbf{R} \equiv \mathbf{1}$, so that the deformation is a pure homogeneous strain, the Piola stress $\mathbf{T}_{\mathrm{R}}$ is symmetric with principal values given by

$$s_i = \frac{\partial\check{\psi}_{\mathrm{R}}(\lambda_1, \lambda_2, \lambda_3)}{\partial\lambda_i} - \frac{P}{\lambda_i}. \tag{54.24}$$

The equations (54.23) and (54.24) in terms of principal stresses are widely used in comparing theory with experiment for isotropic materials with negligible compressibility. Since, for incompressible materials only two stretches may be varied independently, *biaxial* tests alone are sufficient to determine the form of the response function $\check{\psi}$ determining the free energy.

54.3 Simple Shear of a Homogeneous, Isotropic, Incompressible Elastic Body

Using (54.20) and (54.21), and the kinematical relations discussed in the case of simple shear for a homogeneous, isotropic, compressible elastic body,[285] we find that for an isotropic, incompressible elastic body

$$\begin{bmatrix} T_{11} & T_{12} & T_{13} \\ T_{21} & T_{22} & T_{23} \\ T_{31} & T_{32} & T_{33} \end{bmatrix} = -P \begin{bmatrix} 1 & 0 & 0 \\ 0 & 1 & 0 \\ 0 & 0 & 1 \end{bmatrix} + \tilde{\beta}_1(\gamma^2) \begin{bmatrix} 1+\gamma^2 & \gamma & 0 \\ \gamma & 1 & 0 \\ 0 & 0 & 1 \end{bmatrix} + \tilde{\beta}_2(\gamma^2) \begin{bmatrix} 1 & -\gamma & 0 \\ -\gamma & 1+\gamma^2 & 0 \\ 0 & 0 & 1 \end{bmatrix}, \tag{54.25}$$

with

$$\tilde{\beta}_1(\gamma^2) = \beta_1(3+\gamma^2, 3+\gamma^2) \quad \text{and} \quad \tilde{\beta}_2(\gamma^2) = \beta_2(3+\gamma^2, 3+\gamma^2).$$

Thus,

$$\left.\begin{aligned} T_{11} &= -P + (1+\gamma^2)\tilde{\beta}_1(\gamma^2) + \tilde{\beta}_2(\gamma^2), \\ T_{22} &= -P + \tilde{\beta}_1(\gamma^2) + (1+\gamma^2)\tilde{\beta}_2(\gamma^2), \\ T_{33} &= -P + \tilde{\beta}_1(\gamma^2) + \tilde{\beta}_2(\gamma^2), \\ T_{12} &= \mu(\gamma^2)\gamma, \\ T_{13} &= T_{23} = 0, \end{aligned}\right\} \tag{54.26}$$

where the modulus $\mu(\gamma^2)$ entering the shear stress T_{12} is defined by

$$\mu(\gamma^2) = \tilde{\beta}_1(\gamma^2) - \tilde{\beta}_2(\gamma^2). \tag{54.27}$$

In contrast to the compressible case, one can set $T_{33} = 0$ by selecting the arbitrary effective pressure P as

$$P = \tilde{\beta}_1(\gamma^2) + \tilde{\beta}_2(\gamma^2). \tag{54.28}$$

With this choice, we obtain

$$\left.\begin{aligned} T_{11} &= \tilde{\beta}_1(\gamma^2)\gamma^2, \\ T_{22} &= \tilde{\beta}_2(\gamma^2)\gamma^2, \\ T_{12} &= \mu(\gamma^2)\gamma, \end{aligned}\right\} \tag{54.29}$$

and

$$T_{13} = T_{23} = T_{33} = 0. \tag{54.30}$$

Granted that P is chosen consistent with (54.28), the response functions β_1 and β_2 are therefore determined directly in terms of the nonvanishing normal stresses. Further, if

$$\beta_1(3+\gamma^2, 3+\gamma^2) > \beta_2(3+\gamma^2, 3+\gamma^2), \tag{54.31}$$

then $\mu(\gamma^2) > 0$.[286]

---

[285] Cf. §51.

[286] Cf. (51.10).

EXERCISES

1. (Project) Develop a linear theory for incompressible elastic solids.
2. A simple phenomenological model for the nonlinearly elastic response of a rubber-like material is provided by the MOONEY (1940) free-energy function

$$\psi_R = c_1(I_1 - 3) + c_2(I_2 - 3),$$

with $c_1 \geq 0$ and $c_2 \geq 0$ constant. We refer to an incompressible, isotropic, hyperelastic body with free energy of this form as a "Mooney material." For $c_2 = 0$, the Mooney material reduces to the neo-Hookean material arising from the molecular-statistical Gaussian theory of rubber elasticity (GUTH & MARK 1934; KUHN 1934; WALL 1942; TRELOAR 1943; FLORY 1944).

a. Using (54.20), show that the Cauchy stress for a Mooney material has the specific form

$$\mathbf{T} = -P\mathbf{1} + 2c_1\mathbf{B} - 2c_2\mathbf{B}^{-1}.$$

b. Using (54.27), show that the shear modulus for a Mooney material is given by

$$\mu = 2(c_1 + c_2).$$

3. A phenomenological model that incorporates the finite extensibility of the polymer chains comprising a rubber network was provided by GENT (1996) in the form of the free-energy function

$$\psi_R = -c_1 I_1^m \ln\left(1 - \frac{I_1 - 3}{I_1^m}\right) + c_2 \ln\left(\frac{I_2}{3}\right),$$

with $c_1 \geq 0$, $c_2 \geq 0$, and $I_1^m > 3$ constant. In particular, I_1^m denotes the maximum possible value of $(I_1 - 3)$. We refer to an incompressible, isotropic, hyperelastic body with free energy of this form as a "Gent material."

a. Using (54.20), show that the Cauchy stress for a Gent material has the specific form

$$\mathbf{T} = -P\mathbf{1} + 2c_1\left(1 - \frac{I_1 - 3}{I_1^m}\right)^{-1}\mathbf{B} - 2c_2 I_2^{-1}\mathbf{B}^{-1}.$$

b. Using (54.27), show that the shear modulus for a Gent material is given by

$$\mu = 2\left(c_1\left(1 - \frac{\gamma^2}{I_1^m}\right)^{-1} + \frac{c_2}{3 + \gamma^2}\right).$$

55 Approximately Incompressible Elastic Materials

A theory for approximately incompressible elastic materials can be developed based on a multiplicative decomposition

$$\mathbf{F} = \mathbf{F}^v\mathbf{F}^i \tag{55.1}$$

of the deformation gradient $\mathbf{F}$ into volumetric and isochoric factors $\mathbf{F}^v$ and $\mathbf{F}^i$. The particular expressions for $\mathbf{F}^v$ and $\mathbf{F}^i$ are easily determined. To be volumetric, $\mathbf{F}^v$ must have the form

$$\mathbf{F}^v = \alpha\mathbf{1} \tag{55.2}$$

for some $\alpha > 0$. Further, to be isochoric, $\mathbf{F}^i$ must obey

$$\det \mathbf{F}^i = 1. \tag{55.3}$$

Then, by $(2.82)_2$ and (55.1)–(55.3),

$$\begin{aligned} J &= \det \mathbf{F} \\ &= (\det \mathbf{F}^v)(\det \mathbf{F}^i) \\ &= \alpha^3; \end{aligned}$$

hence,

$$\mathbf{F}^v = J^{1/3}\mathbf{1} \qquad \text{and} \qquad \mathbf{F}^i = J^{-1/3}\mathbf{F}. \tag{55.4}$$

The decomposition (55.1) and the representations (55.4) lead to a multiplicative decomposition of the left Cauchy–Green tensor in the form

$$\left.\begin{aligned} \mathbf{C} &= \mathbf{C}^v\mathbf{C}^i, \\ \mathbf{C}^v &= J^{2/3}\mathbf{1}, \\ \mathbf{C}^i &= J^{-2/3}\mathbf{C}. \end{aligned}\right\} \tag{55.5}$$

Additionally, the frame-indifferent constitutive equation $(48.7)_1$ determining the free energy of an elastic solid can be expressed in the alternative form

$$\begin{aligned} \psi_{\mathrm{R}} &= \bar{\psi}_{\mathrm{R}}(\mathbf{C}) \\ &= \tilde{\psi}_{\mathrm{R}}(\mathbf{C}^i, J). \end{aligned} \tag{55.6}$$

We may now use the relations (48.14) and (48.15) to determine expressions for the Piola and Cauchy stresses in an elastic solid with free energy $\psi = \tilde{\psi}_{\mathrm{R}}(\mathbf{C}^i, J)$. To achieve this, we first note that, by (3.34), the chain-rule, and the symmetry of $\mathbf{C}$,

$$\begin{aligned}\frac{\partial J}{\partial \mathbf{C}} &= \frac{\partial\sqrt{\det\mathbf{C}}}{\partial\mathbf{C}}\\ &= \frac{1}{2\sqrt{\det\mathbf{C}}}\frac{\partial(\det\mathbf{C})}{\partial\mathbf{C}}\\ &= \frac{\det\mathbf{C}}{2\sqrt{\det\mathbf{C}}}\mathbf{C}^{-1}\\ &= \tfrac{1}{2}J\mathbf{C}^{-1};\end{aligned}$$

similarly, by (55.5)$_3$,

$$\begin{aligned}\frac{\partial\mathbf{C}^i}{\partial\mathbf{C}} &= \frac{\partial(J^{-2/3}\mathbf{C})}{\partial\mathbf{C}}\\ &= J^{-2/3}\frac{\partial\mathbf{C}}{\partial\mathbf{C}} + \mathbf{C}\otimes\frac{\partial(J^{-2/3})}{\partial\mathbf{C}}\\ &= J^{-2/3}\mathbb{I} - \tfrac{1}{3}J^{-5/3}\mathbf{C}\otimes\frac{\partial(\det\mathbf{C})}{\partial\mathbf{C}}\\ &= J^{-2/3}(\mathbb{I} - \tfrac{1}{3}\mathbf{C}\otimes\mathbf{C}^{-1})\\ &= J^{-2/3}(\mathbb{I} - \tfrac{1}{3}J^{2/3}\mathbf{C}^i\otimes\mathbf{C}^{-1})\\ &= J^{-2/3}(\mathbb{I} - \tfrac{1}{3}\mathbf{C}^i\otimes(J^{-2/3}\mathbf{C})^{-1})\\ &= J^{-2/3}(\mathbb{I} - \tfrac{1}{3}\mathbf{C}^i\otimes\mathbf{C}^{i-1}),\end{aligned}$$

where $\mathbb{I}$ denotes the fourth-order identity tensor, and the tensor product $\mathbf{A}\otimes\mathbf{G}$ of two second-order tensors $\mathbf{A}$ and $\mathbf{G}$ is the fourth-order tensor defined by

$$(\mathbf{A}\otimes\mathbf{G})\mathbf{K} = (\mathbf{G}:\mathbf{K})\mathbf{A}, \qquad (A_{ij}G_{kl})K_{kl} = (G_{kl}K_{kl})A_{ij},$$

for any second-order tensor $\mathbf{K}$. Thus, by (47.9) and (55.6), the second Piola stress has the form

$$\begin{aligned}\mathbf{T}_{\mathrm{RR}} &= 2\frac{\partial\bar{\psi}_{\mathrm{R}}(\mathbf{C})}{\partial\mathbf{C}}\\ &= 2J^{-2/3}(\mathbb{I} - \tfrac{1}{3}\mathbf{C}^{i-1}\otimes\mathbf{C}^i)\frac{\partial\tilde{\psi}_{\mathrm{R}}(\mathbf{C}^i, J)}{\partial\mathbf{C}^i} + J\frac{\partial\tilde{\psi}_{\mathrm{R}}(\mathbf{C}^i, J)}{\partial J}\mathbf{C}^{-1}\\ &= \frac{2}{J^{2/3}}\left[\frac{\partial\tilde{\psi}_{\mathrm{R}}(\mathbf{C}^i, J)}{\partial\mathbf{C}^i} - \frac{1}{3}\left(\mathbf{C}^i : \frac{\partial\tilde{\psi}_{\mathrm{R}}(\mathbf{C}^i, J)}{\partial\mathbf{C}^i}\right)\mathbf{C}^{i-1}\right] + J\frac{\partial\tilde{\psi}_{\mathrm{R}}(\mathbf{C}^i, J)}{\partial J}\mathbf{C}^{-1}.\end{aligned} \tag{55.7}$$

Next, using (55.7) in the expression (47.8) relating the Piola stresses, we obtain

$$\mathbf{T}_{\mathrm{R}} = \frac{2}{J^{1/3}}\left[\mathbf{F}^i\frac{\partial\tilde{\psi}_{\mathrm{R}}(\mathbf{C}^i, J)}{\partial\mathbf{C}^i} - \frac{1}{3}\left(\mathbf{C}^i : \frac{\partial\tilde{\psi}_{\mathrm{R}}(\mathbf{C}^i, J)}{\partial\mathbf{C}^i}\right)\mathbf{F}^{i-\top}\right] + J\frac{\partial\tilde{\psi}_{\mathrm{R}}(\mathbf{C}^i, J)}{\partial J}\mathbf{F}^{-\top}; \tag{55.8}$$

further, by (47.7), (55.4), and (55.8),

$$\mathbf{T} = \frac{2}{J}\left[\mathbf{F}^i\frac{\partial\tilde{\psi}_{\mathrm{R}}(\mathbf{C}^i, J)}{\partial\mathbf{C}^i}\mathbf{F}^{i\top} - \frac{1}{3}\left(\mathbf{C}^i : \frac{\partial\tilde{\psi}_{\mathrm{R}}(\mathbf{C}^i, J)}{\partial\mathbf{C}^i}\right)\mathbf{1}\right] + \frac{\partial\tilde{\psi}_{\mathrm{R}}(\mathbf{C}^i, J)}{\partial J}\mathbf{1}. \tag{55.9}$$

Computing the trace of $\mathbf{T}$ as given by (55.9), we find that

$$\begin{aligned}
\operatorname{tr}\mathbf{T} &= \frac{2}{J}\operatorname{tr}\left[\mathbf{F}^{i}\frac{\partial\tilde{\psi}_{\mathrm{R}}(\mathbf{C}^{i},J)}{\partial\mathbf{C}^{i}}\mathbf{F}^{i\top}-\frac{1}{3}\left(\mathbf{C}^{i}:\frac{\partial\tilde{\psi}_{\mathrm{R}}(\mathbf{C}^{i},J)}{\partial\mathbf{C}^{i}}\right)\mathbf{1}\right]+3\frac{\partial\tilde{\psi}_{\mathrm{R}}(\mathbf{C}^{i},J)}{\partial J}\\
&= \frac{2}{J}\left[\mathbf{F}^{i\top}:\frac{\partial\tilde{\psi}_{\mathrm{R}}(\mathbf{C}^{i},J)}{\partial\mathbf{C}^{i}}\mathbf{F}^{i\top}-\mathbf{C}^{i}:\frac{\partial\tilde{\psi}_{\mathrm{R}}(\mathbf{C}^{i},J)}{\partial\mathbf{C}^{i}}\right]+3\frac{\partial\tilde{\psi}_{\mathrm{R}}(\mathbf{C}^{i},J)}{\partial J}\\
&= \frac{2}{J}\left[\mathbf{F}^{i\top}\mathbf{F}^{i}:\frac{\partial\tilde{\psi}_{\mathrm{R}}(\mathbf{C}^{i},J)}{\partial\mathbf{C}^{i}}-\mathbf{C}^{i}:\frac{\partial\tilde{\psi}_{\mathrm{R}}(\mathbf{C}^{i},J)}{\partial\mathbf{C}^{i}}\right]+3\frac{\partial\tilde{\psi}_{\mathrm{R}}(\mathbf{C}^{i},J)}{\partial J}\\
&= \frac{2}{J}\left[\mathbf{C}^{i}:\frac{\partial\tilde{\psi}_{\mathrm{R}}(\mathbf{C}^{i},J)}{\partial\mathbf{C}^{i}}-\mathbf{C}^{i}:\frac{\partial\tilde{\psi}_{\mathrm{R}}(\mathbf{C}^{i},J)}{\partial\mathbf{C}^{i}}\right]+3\frac{\partial\tilde{\psi}_{\mathrm{R}}(\mathbf{C}^{i},J)}{\partial J}\\
&= 3\frac{\partial\tilde{\psi}_{\mathrm{R}}(\mathbf{C}^{i},J)}{\partial J}.
\end{aligned}$$

Along with the multiplicative decomposition (55.1) of the deformation gradient $\mathbf{F}$ into volumetric and isochoric factors $\mathbf{F}^{v}$ and $\mathbf{F}^{i}$ and the representations (55.4) $\mathbf{F}^{v}$ and $\mathbf{F}^{i}$, the alternative constitutive equation (55.6) determining the free energy ψ_{R} as a function $\tilde{\psi}_{\mathrm{R}}$ of the isochoric component $\mathbf{C}^{i}$ of the right Cauchy–Green tensor $\mathbf{C}$ and the volumetric determinant J leads naturally to a decomposition of the Cauchy stress tensor $\mathbf{T}$ into a sum

$$\mathbf{T}=\mathbf{T}_{0}-p\mathbf{1} \tag{55.10}$$

of a deviatoric component

$$\mathbf{T}_{0}=\frac{2}{J}\left[\mathbf{F}^{i}\frac{\partial\tilde{\psi}_{\mathrm{R}}(\mathbf{C}^{i},J)}{\partial\mathbf{C}^{i}}\mathbf{F}^{i\top}-\frac{1}{3}\left(\frac{\partial\tilde{\psi}_{\mathrm{R}}(\mathbf{C}^{i},J)}{\partial\mathbf{C}^{i}}:\mathbf{C}^{i}\right)\mathbf{1}\right] \tag{55.11}$$

and a spherical component with pressure

$$p=-\frac{\partial\tilde{\psi}_{\mathrm{R}}(\mathbf{C}^{i},J)}{\partial J}. \tag{55.12}$$

For an approximately incompressible material, it seems reasonable to expect that the deviatoric component $\mathbf{T}_{0}$ of the Cauchy stress be inversely proportional to the volumetric Jacobian J (thereby accounting for the change in volume between the reference and spatial configurations) and that the pressure p be independent of the isochoric factor $\mathbf{C}^{i}$ of the right Cauchy–Green tensor $\mathbf{C}$. Necessary and sufficient for the satisfaction of these requirements is the assumption that $\tilde{\psi}$ have the separable form

$$\tilde{\psi}_{\mathrm{R}}(\mathbf{C}^{i},J)=\psi_{\mathrm{R}}^{i}(\mathbf{C}^{i})+\psi_{\mathrm{R}}^{v}(J). \tag{55.13}$$

This being the case, (55.8) and (55.9) simplify somewhat to

$$\mathbf{T}_{\mathrm{R}}=\frac{2}{J^{1/3}}\left[\mathbf{F}^{i}\frac{\partial\psi_{\mathrm{R}}^{i}(\mathbf{C}^{i})}{\partial\mathbf{C}^{i}}-\frac{1}{3}\left(\mathbf{C}^{i}:\frac{\partial\psi_{\mathrm{R}}^{i}(\mathbf{C}^{i})}{\partial\mathbf{C}^{i}}\right)\mathbf{F}^{i-\top}\right]+J\frac{\partial\psi_{\mathrm{R}}^{v}(J)}{\partial J}\mathbf{F}^{-\top} \tag{55.14}$$

and

$$\mathbf{T}=\frac{2}{J}\left[\mathbf{F}^{i}\frac{\partial\psi_{\mathrm{R}}^{i}(\mathbf{C}^{i})}{\partial\mathbf{C}^{i}}\mathbf{F}^{i\top}-\frac{1}{3}\left(\mathbf{C}^{i}:\frac{\partial\psi_{\mathrm{R}}^{i}(\mathbf{C}^{i})}{\partial\mathbf{C}^{i}}\right)\mathbf{1}\right]+\frac{\partial\psi_{\mathrm{R}}^{v}(J)}{\partial J}\mathbf{1}. \tag{55.15}$$

Volume changes accompanying the deformation of elastomeric materials under ambient pressures are such that $J=\det\mathbf{F}$ differs from 1 by about 10^{-4}. For conventional applications, the assumption that the material is incompressible is therefore usually a good approximation. However, for situations involving high pressures, and in cases where the material is confined (as in applications involving gaskets and o-rings), it is important to include the effects of the slight compressibility of such materials. Specializing the separable expression (55.13) to the case of an isotropic material, we obtain

$$\psi_R = \psi_R^i(I_1(\mathbf{B}^i), I_2(\mathbf{B}^i)) + \psi_R^v(J), \tag{55.16}$$

with

$$\mathbf{B}^i = \mathbf{F}^i\mathbf{F}^{i\top}. \tag{55.17}$$

Calculations analogous to those performed in the anisotropic case then yield the identities

$$\frac{\partial J}{\partial \mathbf{B}} = J^2\mathbf{B}^{-1}$$

and

$$\frac{\partial \mathbf{B}^i}{\partial \mathbf{B}} = J^{-2/3}(\mathbb{I} - \tfrac{1}{3}\mathbf{B}^i \otimes \mathbf{B}^{i-1});$$

thus, since

$$\frac{\partial I_1(\mathbf{B}^i)}{\partial \mathbf{B}^i} = \mathbf{1} \quad \text{and} \quad \frac{\partial I_2(\mathbf{B}^i)}{\partial \mathbf{B}^i} = I_1(\mathbf{B}^i)\mathbf{1} - \mathbf{B}^i,$$

we find that

$$\mathbf{T}_R = 2J^{-1/3}\mathrm{dev}\left(\frac{\partial \psi_R^i(I_1, I_2)}{\partial I_1}\mathbf{F}^i + \frac{\partial \psi^i(I_1, I_2)}{\partial I_2}(I_1\mathbf{1} - \mathbf{B}^i)\mathbf{F}^i\right) + J\frac{\partial \tilde{\psi}_R^v(J)}{\partial J}\mathbf{F}^{-\top} \tag{55.18}$$

and

$$\mathbf{T} = 2J^{-1}\mathrm{dev}\left(\frac{\partial \psi_R^i(I_1, I_2)}{\partial I_1}\mathbf{B}^i + \frac{\partial \psi_R^i(I_1, I_2)}{\partial I_2}(I_1\mathbf{1} - \mathbf{B}^i)\mathbf{B}^i\right) + \frac{\partial \tilde{\psi}_R^v(J)}{\partial J}\mathbf{1}. \tag{55.19}$$

EXERCISES

1. To model the compressible elastic response of rubber-like materials under high pressures, consider a variation of the Gent free energy of the form

$$\psi_R = -\tfrac{1}{2}\mu\, I_1^m \ln\left(1 - \frac{I_1 - 3}{I_1^m}\right) + \psi_R^v(J),$$

where $I_1 = I_1(\mathbf{B}^i)$ is the first invariant of $\mathbf{B}^i = J^{-2/3}\mathbf{B}$, $\mu \geq 0$ and $I_1^m > 3$ are constants, and ψ_R^v is the volumetric contribution to the free energy.

(a) Using (55.19), show that the Cauchy stress for such a material has the specific form

$$\mathbf{T} = J^{-1}\mu\left(1 - \frac{I_1 - 3}{I_1^m}\right)^{-1}\mathbf{B}_0^i + \frac{\partial \psi_R^v(J)}{\partial J}\mathbf{1}.$$

(b) Supposing that $\psi_R^v(J) = \tfrac{1}{2}\kappa(\ln J)^2$, $\kappa > 0$, show that the Kirchhoff stress $\mathbf{T}_K = J\mathbf{T}$ for such a material has the specific form

$$\mathbf{T}_K = \mu\left(1 - \frac{I_1 - 3}{I_1^m}\right)^{-1}\mathbf{B}_0^i + \kappa(\ln J)\mathbf{1}.$$

2. Motivated by the simple form of the expression (52.60) for the strain energy of an infinitesimally strained isotropic elastic body, one might ask whether an analogous expression, in which dependence upon the infinitesimal strain measure is replaced by dependence upon a finite strain, is capable of describing the

behavior of a moderately strained isotropic, compressible elastic body. A model of this type, introduced by HENCKY (1928, 1933), has the form

$$\psi_{\mathrm{R}} = \mu|\mathbf{E}_0^{\mathrm{H}}|^2 + \tfrac{1}{2}\kappa(\operatorname{tr}\mathbf{E}^{\mathrm{H}})^2, \tag{55.20}$$

where μ and κ are the shear and bulk moduli from the linear theory,

$$\mathbf{E}^{\mathrm{H}} = \ln \mathbf{V} \tag{55.21}$$

is the Hencky strain introduced previously in $(7.30)_2$, and

$$\mathbf{E}_0^{\mathrm{H}} = \mathbf{E}^{\mathrm{H}} - \tfrac{1}{3}(\operatorname{tr}\mathbf{E}^{\mathrm{H}})\mathbf{1}$$

is the deviatoric part of $\mathbf{E}^{\mathrm{H}}$.

(a) Using the definition of $\mathbf{E}^{\mathrm{H}}$, show that

$$\operatorname{tr}\mathbf{E}^{\mathrm{H}} = \ln J.$$

(b) Noting that $\mathbf{V} = (J^{-1/3}\mathbf{V})(J^{1/3}\mathbf{1})$, show that

$$\mathbf{E}_0^{\mathrm{H}} = \ln(J^{-1/3}\mathbf{V}).$$

(c) Show that the Kirchhoff stress $\mathbf{T}_{\mathrm{K}} = J\mathbf{T}$ for a model of this type can be expressed as

$$\mathbf{T}_{\mathrm{K}} = 2\mu\mathbf{E}_0^{\mathrm{H}} + \kappa(\operatorname{tr}\mathbf{E}^{\mathrm{H}})\mathbf{1}, \tag{55.22}$$

which is formally analogous to the expression (52.59) for the stress of an infinitesimally strained isotropic elastic body.[287]

(d) Show that for simple shear (cf. §51) the constitutive equation (55.22) yields the following nonzero components of the Cauchy stress:

$$T_{12} = \frac{\mu \ln\left(1 + \frac{1}{2}\gamma^2 + \gamma\sqrt{1 + \frac{1}{4}\gamma^2}\right)}{\sqrt{1 + \frac{1}{4}\gamma^2}}, \qquad T_{11} = -T_{22} = \tfrac{1}{2}\gamma T_{12}.$$

[287] ANAND (1979, 1986) shows that the quadratic free-energy function $\psi_{\mathrm{R}} = \mu|\mathbf{E}_0^{\mathrm{H}}|^2 + \frac{1}{2}\kappa(\operatorname{tr}\mathbf{E}^{\mathrm{H}})^2$ and the corresponding stress relation $\mathbf{T}_{\mathrm{K}} = 2\mu\mathbf{E}_0^{\mathrm{H}} + \kappa(\operatorname{tr}\mathbf{E}^{\mathrm{H}})\mathbf{1}$ are in good agreement with experiments on a wide class of materials for principal stretches ranging between 0.7 and 1.3. Importantly, since the material constants μ and κ are the classical elastic constants, they may be determined from experimental data at infinitesimal strains. As a consequence of these results, it appears that all moderate strain nonlinearities are incorporated in the logarithmic strain measure. Indeed, for this reasonably large range of stretches, all other commonly used strain measures (including those of Green, Almansi, Swainger, Biot) when used to generalize (55.20) (using the values of μ and κ determined from experimental data at infinitesimal strains), give predictions (for the elastic stress response of materials) that show only poor agreement with experiments.

PART XI

THERMOELASTICITY

It is well known that heating or cooling an unconfined solid specimen generally leads to dimensional changes of the specimen. For confined specimens, the deformation produced by heating may generate complex stress distributions, and the peak magnitudes of such thermally induced stresses are often substantial. Conversely, temperature changes and distributions generated in the mechanical loading of metals may also be important. We now present a framework for the coupled thermal and mechanical response of solids, restricting our attention to situations in which the deformation is elastic. The general framework — known as the theory of thermoelasticity — is broad enough to describe both metals and rubber-like elastomeric materials, including the anomalous contraction of a stressed rubber-like material on heating, known as a Gough–Joule effect.[288]

[288] Cf. GOUGH (1805) and JOULE (1859).

56 Brief Review

56.1 Kinematical Relations

Our discussion of thermoelastic solids makes use of the following kinematical results:[289]

$$\left.\begin{aligned} \mathbf{F} &= \mathbf{R}\mathbf{U}, \\ \mathbf{C} &= \mathbf{U}^2 = \mathbf{F}^{\top}\mathbf{F}, \\ \mathbf{E} &= \tfrac{1}{2}(\mathbf{C}-\mathbf{1}). \end{aligned}\right\} \tag{56.1}$$

Here, $(56.1)_1$ is the right polar decomposition of the deformation gradient, so that $\mathbf{R}$ is a rotation and $\mathbf{U}$ is the right stretch tensor, and $\mathbf{C}$ is the right Cauchy–Green tensor and $\mathbf{E}$ is the Green–St. Venant strain tensor.

56.2 Basic Laws

For solids, it is generally most convenient to use a referential description; we, therefore, recall the local forms of the momentum and energy balances and the free-energy imbalance[290]

$$\begin{gathered} \rho_{\mathrm{R}}\ddot{\boldsymbol{\chi}} = \mathrm{Div}\,\mathbf{T}_{\mathrm{R}} + \mathbf{b}_{0\mathrm{R}}, \\ \mathbf{T}_{\mathrm{R}}\mathbf{F}^{\top} = \mathbf{F}\mathbf{T}_{\mathrm{R}}^{\top}, \\ \dot{\varepsilon}_{\mathrm{R}} = \mathbf{T}_{\mathrm{R}} : \dot{\mathbf{F}} - \mathrm{Div}\,\mathbf{q}_{\mathrm{R}} + q_{\mathrm{R}}, \\ \dot{\psi}_{\mathrm{R}} + \eta_{\mathrm{R}}\dot{\vartheta} - \mathbf{T}_{\mathrm{R}} : \dot{\mathbf{F}} + \frac{1}{\vartheta}\mathbf{q}_{\mathrm{R}} \cdot \nabla\vartheta = -\vartheta\Gamma_{\mathrm{R}} \le 0. \end{gathered} \tag{56.2}$$

Here, ρ_{R}, $\mathbf{b}_{0\mathrm{R}}$, ε_{R}, q_{R},

$$\psi_{\mathrm{R}} = \varepsilon_{\mathrm{R}} - \vartheta\eta_{\mathrm{R}}, \tag{56.3}$$

η_{R}, and Γ_{R} denote the density, conventional body force, internal energy, heat supply, free energy, entropy, and dissipation; $\mathbf{T}_{\mathrm{R}}$ and $\mathbf{q}_{\mathrm{R}}$ denote the Piola stress and the referential heat flux; and ϑ denotes the absolute temperature.

---

[289] Cf. §7.1.

[290] Cf. (24.10), (24.11), (31.10), (31.12), and (31.13).

The Piola stress $\mathbf{T}_R$ is related to the Cauchy stress $\mathbf{T}$ and the second Piola stress $\mathbf{T}_{RR}$ (which is symmetric) by[291]

$$\mathbf{T}_R = J\mathbf{T}\mathbf{F}^{-\top} \tag{56.4}$$

$$= \mathbf{F}\mathbf{T}_{RR}; \tag{56.5}$$

also, the stresses powers of the Piola and second Piola stresses are related by

$$\mathbf{T}_R : \dot{\mathbf{F}} = \tfrac{1}{2}\mathbf{T}_{RR} : \dot{\mathbf{C}}, \tag{56.6}$$

which allows us to rewrite the free-energy imbalance $(56.2)_4$ in the form

$$\dot{\psi}_R + \eta_R\dot{\vartheta} - \tfrac{1}{2}\mathbf{T}_{RR} : \dot{\mathbf{C}} + \frac{1}{\vartheta}\mathbf{q}_R \cdot \nabla\vartheta = -\vartheta\Gamma_R \leq 0. \tag{56.7}$$

Finally, the referential heat flux $\mathbf{q}_R$ and temperature gradient $\nabla\vartheta$ are related to their spatial counterparts by[292]

$$\mathbf{q}_R = J\mathbf{F}^{-1}\mathbf{q} \qquad \text{and} \qquad \nabla\vartheta = \mathbf{F}^{\top}\operatorname{grad}\vartheta. \tag{56.8}$$

291 Cf. (24.1), (25.2), and (25.6).
292 Cf. $(9.2)_1$ and (31.5).

57 Constitutive Theory

Guided by the free-energy imbalance $(56.2)_4$, we assume that the free energy ψ_{R}, the Piola stress $\mathbf{T}_{\mathrm{R}}$, the entropy η_{R}, and the heat flux $\mathbf{q}_{\mathrm{R}}$ are determined by constitutive equations of the form

$$
\begin{aligned}
\psi_{\mathrm{R}} &= \hat{\psi}_{\mathrm{R}}(\mathbf{F}, \vartheta, \nabla\vartheta),\\
\mathbf{T}_{\mathrm{R}} &= \hat{\mathbf{T}}_{\mathrm{R}}(\mathbf{F}, \vartheta, \nabla\vartheta),\\
\eta_{\mathrm{R}} &= \hat{\eta}_{\mathrm{R}}(\mathbf{F}, \vartheta, \nabla\vartheta),\\
\mathbf{q}_{\mathrm{R}} &= \hat{\mathbf{q}}_{\mathrm{R}}(\mathbf{F}, \vartheta, \nabla\vartheta).
\end{aligned}
\tag{57.1}
$$

57.1 Consequences of Frame-Indifference

Consider a change in frame with frame-rotation $\mathbf{Q}$. By $(20.16)_1$ and (25.8), the transformation laws for $\mathbf{F}$ and $\mathbf{T}_{\mathrm{R}}$ are

$$
\mathbf{F}^* = \mathbf{QF} \qquad \text{and} \qquad \mathbf{T}^*_{\mathrm{R}} = \mathbf{QT}_{\mathrm{R}},
$$

while $\mathbf{q}_{\mathrm{R}}$ and $\nabla\vartheta$, being material vector fields, are invariant:[293]

$$
\mathbf{q}^*_{\mathrm{R}} = \mathbf{q}_{\mathrm{R}} \qquad \text{and} \qquad (\nabla\vartheta)^* = \nabla\vartheta.
$$

Frame-indifference therefore requires that the response functions $\hat{\psi}_{\mathrm{R}}$, $\hat{\mathbf{T}}_{\mathrm{R}}$, $\hat{\eta}_{\mathrm{R}}$, and $\hat{\mathbf{q}}_{\mathrm{R}}$ satisfy

$$
\begin{aligned}
\hat{\psi}_{\mathrm{R}}(\mathbf{F}, \vartheta, \nabla\vartheta) &= \hat{\psi}_{\mathrm{R}}(\mathbf{QF}, \vartheta, \nabla\vartheta),\\
\hat{\mathbf{T}}_{\mathrm{R}}(\mathbf{F}, \vartheta, \nabla\vartheta) &= \mathbf{Q}^{\top}\hat{\mathbf{T}}_{\mathrm{R}}(\mathbf{QF}, \vartheta, \nabla\vartheta),\\
\hat{\eta}_{\mathrm{R}}(\mathbf{F}, \vartheta, \nabla\vartheta) &= \hat{\eta}_{\mathrm{R}}(\mathbf{QF}, \vartheta, \nabla\vartheta),\\
\hat{\mathbf{q}}_{\mathrm{R}}(\mathbf{F}, \vartheta, \nabla\vartheta) &= \hat{\mathbf{q}}_{\mathrm{R}}(\mathbf{QF}, \vartheta, \nabla\vartheta)
\end{aligned}
\tag{57.2}
$$

for all rotations $\mathbf{Q}$ and all $(\mathbf{F}, \vartheta, \nabla\vartheta)$.

Arguing as in §48.1, we take $\mathbf{Q} = \mathbf{R}^{\top}$ in (57.2). Then, since, by $(56.1)_{1,2}$, $\mathbf{QF} = \mathbf{U}$ and $\mathbf{U} = \sqrt{\mathbf{C}}$, if we use (56.5) we see that there are new response functions $\bar{\psi}_{\mathrm{R}}$, $\bar{\mathbf{T}}_{\mathrm{RR}}$,

[293] Cf. the bullet on page 147.

$\bar{\eta}_R$, and $\bar{\mathbf{q}}_R$ such that[294]

$$
\begin{aligned}
\psi_R &= \bar{\psi}_R(\mathbf{C}, \vartheta, \nabla\vartheta),\\
\mathbf{T}_R &= \mathbf{F}\bar{\mathbf{T}}_{RR}(\mathbf{C}, \vartheta, \nabla\vartheta),\\
\eta_R &= \bar{\eta}_R(\mathbf{C}, \vartheta, \nabla\vartheta),\\
\mathbf{q}_R &= \bar{\mathbf{q}}_R(\mathbf{C}, \vartheta, \nabla\vartheta).
\end{aligned}
\tag{57.3}
$$

As in the theory of isothermal elasticity, (56.4), $(57.3)_2$, and the symmetry of $\mathbf{T}$ combine to yield the symmetry of $\mathbf{T}_R\mathbf{F}^\top$ as required by the angular momentum balance $(56.2)_2$.

EXERCISE

1. Show that the constitutive relations (57.3) are frame-indifferent.

57.2 Thermodynamic Restrictions

We now apply the Coleman–Noll procedure to the frame-indifferent constitutive equations (57.3). In the setting at hand, a **constitutive process** consists of a motion χ and a temperature field ϑ together with the fields ψ_R, $\mathbf{T}_R$, η_R, and $\mathbf{q}_R$ determined through the constitutive equations (57.3).

Consider an arbitrary constitutive process. The linear momentum balance $(56.2)_1$ and the energy balance $(56.2)_3$ then provide explicit relations

$$\mathbf{b}_{0R} = \rho_R\ddot{\chi} - \mathrm{Div}\,\mathbf{T}_R$$

and

$$q_R = \dot{\varepsilon}_R - \mathbf{T}_R : \dot{\mathbf{F}} + \mathrm{Div}\,\mathbf{q}_R$$

for the conventional body force $\mathbf{b}_{0R}$ and the external heat supply q_R needed to support the process. As a basic hypothesis of the Coleman–Noll procedure, *we assume that* $\mathbf{b}_{0R}$ *and* q_R *are arbitrarily assignable*. Because of this assumption, the linear momentum and energy balances in no way restrict the class of processes that the material may undergo. On the other hand, unless the constitutive equations (57.3) are suitably restricted, not all constitutive processes will be compatible with the laws of thermodynamics as embodied in the free-energy imbalance $(56.2)_4$.[295] For that reason we require that *all constitutive processes be consistent with the free-energy imbalance* $(56.2)_4$.

As we saw in our treatment of the mechanical theory of elastic solids, this requirement has strong consequences. Consider an arbitrary constitutive process. If we differentiate the constitutive relation $(57.3)_1$ for the free energy with respect to time, we find, upon writing

$$\mathbf{g} = \nabla\vartheta \tag{57.4}$$

for the temperature gradient, that

$$\dot{\psi} = \frac{\partial\bar{\psi}_R(\mathbf{C},\vartheta,\mathbf{g})}{\partial\mathbf{C}} : \dot{\mathbf{C}} + \frac{\partial\bar{\psi}_R(\mathbf{C},\vartheta,\mathbf{g})}{\partial\vartheta}\dot{\vartheta} + \frac{\partial\bar{\psi}_R(\mathbf{C},\vartheta,\mathbf{g})}{\partial\mathbf{g}}\cdot\dot{\mathbf{g}}; \tag{57.5}$$

[294] Cf. (48.7).

[295] Cf. the discussion in the first two paragraphs of §29.

further, by $(57.3)_2$,

$$\mathbf{T}_{RR}:\dot{\mathbf{C}} = \bar{\mathbf{T}}_{RR}(\mathbf{C},\vartheta,\mathbf{g}):\dot{\mathbf{C}},$$

while $(57.3)_4$ and (57.4) yield

$$\mathbf{q}_R\cdot\nabla\vartheta = \bar{\mathbf{q}}_R(\mathbf{C},\vartheta,\mathbf{g})\cdot\mathbf{g}.$$

A consequence of the last three relations is that the free-energy imbalance (56.7) is equivalent to the requirement that the inequality

$$\left(\frac{\partial\bar{\psi}_R(\mathbf{C},\vartheta,\mathbf{g})}{\partial\mathbf{C}} - \tfrac{1}{2}\bar{\mathbf{T}}_{RR}(\mathbf{C},\vartheta,\mathbf{g})\right):\dot{\mathbf{C}} + \left(\frac{\partial\bar{\psi}_R(\mathbf{C},\vartheta,\mathbf{g})}{\partial\vartheta} + \bar{\eta}_R(\mathbf{C},\vartheta,\mathbf{g})\right)\dot{\vartheta}$$
$$+\frac{\partial\bar{\psi}_R(\mathbf{C},\vartheta,\mathbf{g})}{\partial\mathbf{g}}\cdot\dot{\mathbf{g}} + \frac{1}{\vartheta}\bar{\mathbf{q}}_R(\mathbf{C},\vartheta,\mathbf{g})\cdot\mathbf{g}\le 0 \qquad (57.6)$$

be satisfied in all constitutive processes.

The essential step in using (57.6) to obtain thermodynamically based constitutive restrictions is the observation that

(†) given any point of the body and any time, it is possible to find a motion and a temperature field ϑ such that $\mathbf{C}, \vartheta, \mathbf{g} = \nabla\vartheta$, and their time derivatives $\dot{\mathbf{C}}, \dot{\vartheta}$, and $\dot{\mathbf{g}}$ have arbitrarily prescribed values at that point and time.

Granted this assertion, which we prove on page 338, the coefficients of $\dot{\mathbf{C}}$, $\dot{\vartheta}$, and $\dot{\mathbf{g}}$ must vanish, for otherwise these rates may be chosen to violate the inequality (57.6). We therefore have the **thermodynamic restrictions**:

(i) *the free energy, second Piola stress, and entropy are independent of the temperature gradient*;

(ii) *the free energy determines the second Piola stress and the entropy through the* **stress** and **entropy relations**

$$\mathbf{T}_{RR} = \bar{\mathbf{T}}_{RR}(\mathbf{C},\vartheta) = 2\frac{\partial\bar{\psi}_R(\mathbf{C},\vartheta)}{\partial\mathbf{C}} \qquad (57.7)$$

and

$$\eta_R = \bar{\eta}_R(\mathbf{C},\vartheta) = -\frac{\partial\bar{\psi}_R(\mathbf{C},\vartheta)}{\partial\vartheta}; \qquad (57.8)$$

(iii) *the heat flux satisfies the* **heat-conduction inequality**

$$\bar{\mathbf{q}}_R(\mathbf{C},\vartheta,\nabla\vartheta)\cdot\nabla\vartheta\le 0 \qquad (57.9)$$

for all $(\mathbf{C},\vartheta,\nabla\vartheta)$.

We refer to (57.7) and (57.8) as **state relations.**

Remarks.

(a) A consequence of (ii) and the free-energy imbalance $(56.2)_4$ is that the entropy production in any constitutive process is given by

$$\Gamma_R = -\frac{1}{\vartheta^2}\mathbf{q}_R\cdot\nabla\vartheta\ge 0.$$

(b) By (ii), the state relations can be determined by experiments in which the underlying fields $\mathbf{C}$ and ϑ are homogeneous.

(c) A consequence of (56.5) and (57.7) is that the Piola stress $\mathbf{T}_R$ satisfies

$$\begin{aligned}\mathbf{T}_R &= \mathbf{F}\bar{\mathbf{T}}_{RR}(\mathbf{C}, \vartheta)\\ &= 2\mathbf{F}\frac{\partial \bar{\psi}_R(\mathbf{C}, \vartheta)}{\partial \mathbf{C}}.\end{aligned} \tag{57.10}$$

(d) The thermomechanical response of the material is determined by the response functions $\bar{\psi}_R$ and $\bar{\mathbf{q}}_R$ for the free energy and heat flux.

(e) In view of (i), the identity (56.3) yields an auxiliary constitutive equation

$$\begin{aligned}\varepsilon_R &= \bar{\varepsilon}_R(\mathbf{C}, \vartheta)\\ &= \bar{\psi}_R(\mathbf{C}, \vartheta) + \vartheta\bar{\eta}_R(\mathbf{C}, \vartheta)\end{aligned} \tag{57.11}$$

for the internal energy.

We say that the material is **strictly dissipative** if

$$\bar{\mathbf{q}}_R(\mathbf{C}, \vartheta, \mathbf{g}) \cdot \mathbf{g} < 0 \qquad \text{whenever } \mathbf{g} \neq \mathbf{0}. \tag{57.12}$$

57.2.1 Verification of (†)

Choose an arbitrary material point $\mathbf{X}_0$ and time t_0. The assertions regarding $\mathbf{C}$ and $\dot{\mathbf{C}}$ are verified on page 281. To establish the assertions regarding ϑ, $\dot{\vartheta}$, $\mathbf{g} = \nabla\vartheta$, and $\dot{\mathbf{g}} = \nabla\dot{\vartheta}$, choose scalars $\vartheta_0 > 0$ and β and vectors $\mathbf{a}$ and $\mathbf{b}$ and consider the temperature field defined by

$$\vartheta(\mathbf{X}, t) = \vartheta_0 e^{\phi(\mathbf{X}, t)}$$

with

$$\phi(\mathbf{X}, t) = \frac{(t - t_0)\beta + \mathbf{a} \cdot (\mathbf{X} - \mathbf{X}_0) + (t - t_0)\mathbf{b} \cdot (\mathbf{X} - \mathbf{X}_0)}{\vartheta_0}.$$

Then,

$$\vartheta(\mathbf{X}_0, t_0) = \vartheta_0, \qquad \dot{\vartheta}(\mathbf{X}_0, t_0) = \beta, \qquad \nabla\vartheta(\mathbf{X}_0, t_0) = \mathbf{a}, \qquad \nabla\dot{\vartheta}(\mathbf{X}_0, t_0) = \mathbf{b} + \beta\mathbf{a}.$$

Since $\vartheta_0 > 0$, β, and $\mathbf{a}$ are arbitrary, ϑ, $\dot{\vartheta}$, and $\nabla\vartheta$ have arbitrary values at $(\mathbf{X}_0, t_0)$, and similarly for $\nabla\dot{\vartheta}$, since $\mathbf{b}$ is arbitrary. This completes the verification of (†).

57.3 Consequences of the Thermodynamic Restrictions

57.3.1 Consequences of the State Relations

By (57.7) and (57.8), we have the **Gibbs relation**

$$\dot{\psi}_R = \tfrac{1}{2}\mathbf{T}_{RR} : \dot{\mathbf{C}} - \eta_R\dot{\vartheta}; \tag{57.13}$$

further, since, by (56.3),

$$\dot{\psi}_R + \eta_R\dot{\vartheta} = \dot{\varepsilon}_R - \vartheta\dot{\eta}_R,$$

(57.13) yields a second Gibbs relation

$$\dot{\varepsilon}_R = \tfrac{1}{2}\mathbf{T}_{RR} : \dot{\mathbf{C}} + \vartheta\dot{\eta}_R. \tag{57.14}$$

Next, (56.6), (57.14), and the energy balance $(56.2)_3$ imply that

$$\begin{aligned}\vartheta\dot{\eta}_R &= \dot{\varepsilon}_R - \mathbf{T}_R : \dot{\mathbf{F}}\\ &= (\mathbf{T}_R : \dot{\mathbf{F}} - \operatorname{Div}\mathbf{q}_R + q_R) - \mathbf{T}_R : \dot{\mathbf{F}}\\ &= -\operatorname{Div}\mathbf{q}_R + q_R.\end{aligned}$$

Hence, an important consequence of the Gibbs relations is that the energy balance reduces to an **entropy balance**

$$\dot{\eta}_R = -\frac{1}{\vartheta}\mathrm{Div}\,\mathbf{q}_R + \frac{q_R}{\vartheta}. \tag{57.15}$$

Next, (57.7) and (57.8) imply that

$$\begin{aligned}\frac{\partial \bar{\mathbf{T}}_{RR}(\mathbf{C},\vartheta)}{\partial \vartheta} &= 2\frac{\partial^2 \bar{\psi}_R(\mathbf{C},\vartheta)}{\partial \mathbf{C}\partial \vartheta} \\ &= -2\frac{\partial \bar{\eta}_R(\mathbf{C},\vartheta)}{\partial \mathbf{C}},\end{aligned}$$

and we have the **Maxwell relation**

$$\frac{\partial \bar{\mathbf{T}}_{RR}(\mathbf{C},\vartheta)}{\partial \vartheta} = -2\frac{\partial \bar{\eta}_R(\mathbf{C},\vartheta)}{\partial \mathbf{C}}. \tag{57.16}$$

57.3.2 Consequences of the Heat-Conduction Inequality

As we now show, the heat-conduction inequality (57.9), while seemingly innocuous, has deep physical consequences of great importance.

Assume that at some material point $\mathbf{X}_0$ and time (which we suppress as an argument)

$$\nabla\vartheta(\mathbf{X}_0) \neq \mathbf{0}$$

and define

$$\mathbf{e} = \frac{\nabla\vartheta(\mathbf{X}_0)}{|\nabla\vartheta(\mathbf{X}_0)|}. \tag{57.17}$$

Then, $\mathbf{e}\cdot\nabla\vartheta(\mathbf{X}_0) = |\nabla\vartheta(\mathbf{X}_0)|$ so that $\vartheta(\mathbf{X}_0 + h\mathbf{e}) = \vartheta(\mathbf{X}_0) + h|\nabla\vartheta(\mathbf{X}_0)| + o(h)$ and, for all sufficiently small $h > 0$,

$$\vartheta(\mathbf{X}_0 + h\mathbf{e}) > \vartheta(\mathbf{X}_0);$$

the point $\mathbf{X}_0 + h\mathbf{e}$ is therefore hotter than the point $\mathbf{X}_0$. Further, for $\mathbf{q}_R(\mathbf{X})$ the heat flux corresponding to the fields $\mathbf{C}(\mathbf{X})$ and $\vartheta(\mathbf{X})$, the heat-conduction inequality and (57.17) imply that

$$\begin{aligned}0 &\geq \mathbf{q}_R(\mathbf{X}_0)\cdot\nabla\vartheta(\mathbf{X}_0) \\ &= \mathbf{q}_R(\mathbf{X}_0)\cdot(|\nabla\vartheta(\mathbf{X}_0)|\mathbf{e}) \\ &= (\mathbf{q}_R(\mathbf{X}_0)\cdot\mathbf{e})\underbrace{|\nabla\vartheta(\mathbf{X}_0)|}_{>0},\end{aligned}$$

so that $\mathbf{q}_R(\mathbf{X}_0)\cdot\mathbf{e} \leq 0$.[296] Thus, the component of $\mathbf{q}_R(\mathbf{X}_0)$ in the direction $-\mathbf{e}$, which is the unit vector that represents the direction from the hotter point $\mathbf{X}_0 + h\mathbf{e}$ to the colder point $\mathbf{X}_0$, must be nonnegative. In this sense, heat flows from the hotter point to the colder point; we, therefore, have the classical result

- *heat flows from hot to cold.*

Next, let

$$\varphi(\mathbf{C},\vartheta,\mathbf{g}) = \bar{\mathbf{q}}_R(\mathbf{C},\vartheta,\mathbf{g})\cdot\mathbf{g};$$

[296] In fact, if the material is strictly dissipative then, by (57.12), $\mathbf{q}_R(\mathbf{X}_0)\cdot\mathbf{e} < 0$.

then, since $\varphi(\mathbf{C}, \vartheta, \mathbf{g}) \le 0$ and $\varphi(\mathbf{C}, \vartheta, \mathbf{0}) = 0$,

$$\varphi(\mathbf{C}, \vartheta, \mathbf{g}), \text{ as a function of } \mathbf{g}, \text{ has a maximum at } \mathbf{g} = \mathbf{0} \tag{57.18}$$

and, hence,

$$\left.\frac{\partial \varphi(\mathbf{C}, \vartheta, \mathbf{g})}{\partial \mathbf{g}}\right|_{\mathbf{g}=\mathbf{0}} = \mathbf{0}. \tag{57.19}$$

If we evaluate the relation[297]

$$\frac{\partial \varphi(\mathbf{C}, \vartheta, \mathbf{g})}{\partial \mathbf{g}} = \bar{\mathbf{q}}_{\mathrm{R}}(\mathbf{C}, \vartheta, \mathbf{g}) + \left(\frac{\partial \bar{\mathbf{q}}_{\mathrm{R}}(\mathbf{C}, \vartheta, \mathbf{g})}{\partial \mathbf{g}}\right)^{\top} \mathbf{g} \tag{57.20}$$

at $\mathbf{g} = \mathbf{0}$ and use (57.19), we therefore find that

$$\bar{\mathbf{q}}_{\mathrm{R}}(\mathbf{C}, \vartheta, \mathbf{0}) = \mathbf{0} \tag{57.21}$$

or, equivalently, that

$$\mathbf{q}_{\mathrm{R}} = \mathbf{0} \qquad \text{whenever} \qquad \nabla\vartheta = \mathbf{0}. \tag{57.22}$$

In words,

- *the heat flux vanishes when the temperature gradient vanishes, independent of the values of the deformation gradient and temperature.*

This result, known as the **absence of a piezo-caloric effect**, implies that *a deformation, no matter how large, cannot induce a flow of heat in the absence of a thermal gradient.*

The heat-conduction inequality has yet another important physical consequence. Choose a right Cauchy–Green tensor $\mathbf{C}_0$ and a temperature ϑ_0 and, for any function $\Phi(\mathbf{C}, \vartheta, \mathbf{g})$, write

$$\Phi\big|_0 = \Phi(\mathbf{C}, \vartheta, \mathbf{g})\big|_{(\mathbf{C},\vartheta,\mathbf{g})=(\mathbf{C}_0,\vartheta_0,\mathbf{0})}.$$

Then (57.21) implies that

$$\bar{\mathbf{q}}_{\mathrm{R}}\big|_0 = \mathbf{0}, \qquad \left.\frac{\partial \bar{\mathbf{q}}_{\mathrm{R}}}{\partial \mathbf{C}}\right|_0 = \mathbf{0}, \qquad \text{and} \qquad \left.\frac{\partial \bar{\mathbf{q}}_{\mathrm{R}}}{\partial \vartheta}\right|_0 = \mathbf{0}. \tag{57.23}$$

Consistent with standard terminology, we refer to the tensor

$$\mathbf{K}_0 = -\left.\frac{\partial \bar{\mathbf{q}}_{\mathrm{R}}}{\partial \mathbf{g}}\right|_0 \tag{57.24}$$

as the **conductivity tensor** at $(\mathbf{C}_0, \vartheta_0)$. Let ϵ denote the dimensionless norm

$$\epsilon = \sqrt{|\mathbf{C} - \mathbf{C}_0|^2 + \frac{|\vartheta - \vartheta_0|^2}{\vartheta_0^2} + \frac{L^2|\mathbf{g}|^2}{\vartheta_0^2}},$$

with L a characteristic length associated with the reference body B. Expanding $\bar{\mathbf{q}}_{\mathrm{R}}(\mathbf{C}, \vartheta, \mathbf{g})$ in a Taylor series about $(\mathbf{C}_0, \vartheta_0, \mathbf{0})$, we then find using (57.23) and (57.24) that the heat flux $\mathbf{q}_{\mathrm{R}} = \bar{\mathbf{q}}_{\mathrm{R}}(\mathbf{C}, \vartheta, \mathbf{g})$ obeys the estimate

$$\underbrace{\mathbf{q}_{\mathrm{R}} = -\mathbf{K}_0 \nabla\vartheta}_{\text{Fourier's law}} + o(\epsilon) \quad \text{as } \epsilon \to 0. \tag{57.25}$$

[297] Recall that $\nabla(\mathbf{v} \cdot \mathbf{w}) = (\nabla\mathbf{w})^{\top}\mathbf{v} + (\nabla\mathbf{v})^{\top}\mathbf{w}$; cf. $(3.20)_2$.

In words, *Fourier's law approximates the general constitutive relation for the heat flux to within terms of order* $o(\varepsilon)$. Further, by (57.20) and (57.24),

$$\left.\frac{\partial^2 \varphi}{\partial \mathbf{g}^2}\right|_0 = -(\mathbf{K}_0 + \mathbf{K}_0^{\top}). \tag{57.26}$$

By (57.18), the left side of (57.26) is positive-semidefinite; thus, the symmetric part of $\mathbf{K}_0$ is positive-semidefinite. But a tensor is positive-semidefinite if and only if its symmetric part is positive-semidefinite;[298] we may hence conclude that

- *the conductivity tensor* $\mathbf{K}_0$ *is positive-semidefinite.*

57.4 Elasticity Tensor. Stress-Temperature Modulus. Heat Capacity

Let $\mathbf{C}(t)$ be a time-dependent right Cauchy–Green tensor, and let $\vartheta(t)$ be a time-dependent temperature field. Then by (57.7) and the chain-rule,

$$\dot{\mathbf{T}}_{\mathrm{RR}} = \frac{\partial \bar{\mathbf{T}}_{\mathrm{RR}}(\mathbf{C}, \vartheta)}{\partial \mathbf{C}}\dot{\mathbf{C}} + \frac{\partial \bar{\mathbf{T}}_{\mathrm{RR}}(\mathbf{C}, \vartheta)}{\partial \vartheta}\dot{\vartheta},$$

suggesting the introduction of two constitutive moduli: the **elasticity tensor**

$$\mathbb{C}(\mathbf{C}, \vartheta) = 2\frac{\partial \bar{\mathbf{T}}_{\mathrm{RR}}(\mathbf{C}, \vartheta)}{\partial \mathbf{C}} \tag{57.27}$$

(at fixed temperature) and the **stress-temperature modulus**

$$\mathbf{M}(\mathbf{C}, \vartheta) = \frac{\partial \bar{\mathbf{T}}_{\mathrm{RR}}(\mathbf{C}, \vartheta)}{\partial \vartheta} \tag{57.28}$$

(at fixed strain).[299] In view of the stress relation (57.7),

$$\mathbb{C}(\mathbf{C}, \vartheta) = 4\frac{\partial^2 \bar{\psi}_{\mathrm{R}}(\mathbf{C}, \vartheta)}{\partial \mathbf{C}^2}, \tag{57.29}$$

while the Maxwell relation (57.16) implies that

$$\mathbf{M}(\mathbf{C}, \vartheta) = -2\frac{\partial \bar{\eta}_{\mathrm{R}}(\mathbf{C}, \vartheta)}{\partial \mathbf{C}}. \tag{57.30}$$

For each $(\mathbf{C}, \vartheta)$, the elasticity tensor $\mathbb{C}(\mathbf{C}, \vartheta)$ — a linear transformation that maps symmetric tensors to symmetric tensors[300] — has symmetry properties strictly analogous to those spelled out in (52.9), (52.18), and (52.19); in particular, since $\mathbb{C}(\mathbf{C}, \vartheta)$ is symmetric,

$$\mathbf{G} : \mathbb{C}(\mathbf{C}, \vartheta)\mathbf{A} = \mathbf{A} : \mathbb{C}(\mathbf{C}, \vartheta)\mathbf{G} \tag{57.31}$$

for all symmetric tensors $\mathbf{G}$ and $\mathbf{A}$ — so that

$$C_{ijkl} = C_{klij}. \tag{57.32}$$

Since the right Cauchy–Green tensor $\mathbf{C}$ is symmetric, we may conclude from (57.30) that the stress-temperature modulus $\mathbf{M}(\mathbf{C}, \vartheta)$ is a symmetric tensor; this tensor measures the marginal change in stress due to a change in temperature holding the strain fixed.

Another important modulus is the **heat capacity**

$$c(\mathbf{C}, \vartheta) = \frac{\partial \bar{\varepsilon}_{\mathrm{R}}(\mathbf{C}, \vartheta)}{\partial \vartheta} \tag{57.33}$$

---

298 Since $\mathbf{a} \cdot \mathbf{A}\mathbf{a} = \mathbf{a} \cdot (\operatorname{sym} \mathbf{A})\mathbf{a}$ for every tensor $\mathbf{A}$ and every vector $\mathbf{a}$.

299 By (56.1), a derivative holding $\mathbf{E}$ fixed is equivalent to a derivative holding $\mathbf{C}$ fixed.

300 In the mechanical theory the elasticity tensor is defined by (52.8) only at $\mathbf{C}$.

(at fixed strain). By (57.8), (57.11), and (57.33),

$$c(\mathbf{C},\vartheta)=\frac{\partial\bar{\varepsilon}_{\rm R}(\mathbf{C},\vartheta)}{\partial\vartheta}=\frac{\partial\bar{\psi}_{\rm R}(\mathbf{C},\vartheta)}{\partial\vartheta}+\bar{\eta}_{\rm R}(\mathbf{C},\vartheta)+\vartheta\frac{\partial\bar{\eta}_{\rm R}(\mathbf{C},\vartheta)}{\partial\vartheta}$$

$$=\vartheta\frac{\partial\bar{\eta}_{\rm R}(\mathbf{C},\vartheta)}{\partial\vartheta} \tag{57.34}$$

$$=-\vartheta\frac{\partial^2\bar{\psi}_{\rm R}(\mathbf{C},\vartheta)}{\partial\vartheta^2}. \tag{57.35}$$

An important consequence of (57.35) and the positivity of temperature is that the following three assertions are equivalent:

(i) The heat capacity $c(\mathbf{C},\vartheta)$ is strictly positive for all ϑ.
(ii) The entropy $\bar{\eta}_{\rm R}(\mathbf{C},\vartheta)$ is a strictly increasing function of ϑ.
(iii) $\partial^2\bar{\psi}_{\rm R}(\mathbf{C},\vartheta)/\partial\vartheta^2<0$ for all ϑ, so that $\bar{\psi}_{\rm R}(\mathbf{C},\vartheta)$ is strictly concave in ϑ.

Next, differentiating the entropy with respect to time we find, using (57.30) and (57.34), that

$$\vartheta\dot{\eta}_{\rm R}=\vartheta\frac{\partial\bar{\eta}_{\rm R}(\mathbf{C},\vartheta)}{\partial\mathbf{C}}:\dot{\mathbf{C}}+\vartheta\frac{\partial\bar{\eta}_{\rm R}(\mathbf{C},\vartheta)}{\partial\vartheta}\dot{\vartheta}$$

$$=-\tfrac{1}{2}\vartheta\,\mathbf{M}(\mathbf{C},\vartheta):\dot{\mathbf{C}}+c(\mathbf{C},\vartheta)\dot{\vartheta}.$$

This identity, the constitutive equation $(57.3)_4$ for the heat flux, and the entropy balance (57.15) yield the evolution equation

$$c(\mathbf{C},\vartheta)\dot{\vartheta}=-\mathrm{Div}\,\bar{\mathbf{q}}_{\rm R}(\mathbf{C},\vartheta,\nabla\vartheta)+\tfrac{1}{2}\vartheta\,\mathbf{M}(\mathbf{C},\vartheta):\dot{\mathbf{C}}+q_{\rm R}. \tag{57.36}$$

This equation is generic: Its derivation requires no assumptions other than frame-indifference and consistency with thermodynamics. Classical simple models of heat conduction assume that c is constant and that the heat flux is given by Fourier's law, which is (57.25) with the term of order $o(\varepsilon)$ neglected. Then, since $\mathbf{K}_0$ is constant, (57.36) takes the form

$$c\dot{\vartheta}=\mathbf{K}_0:\nabla\nabla\vartheta+\tfrac{1}{2}\vartheta\,\mathbf{M}(\mathbf{C},\vartheta):\dot{\mathbf{C}}+q_{\rm R}, \tag{57.37}$$

which is the classical anisotropic heat equation augmented by a term $\frac{1}{2}\vartheta\,\mathbf{M}(\mathbf{C},\vartheta):\dot{\mathbf{C}}$ representing a local expenditure of stress power.

57.5 The Basic Thermoelastic Field Equations

The basic thermoelastic field equations consist of the **kinematical equations**

$$\begin{aligned}\mathbf{F}&=\nabla\chi,\\ \mathbf{C}&=\mathbf{F}^{\top}\mathbf{F};\end{aligned} \tag{57.38}$$

the **constitutive equations**

$$\begin{aligned}\psi_{\rm R}&=\bar{\psi}_{\rm R}(\mathbf{C},\vartheta),\\ \mathbf{T}_{\rm R}&=2\mathbf{F}\frac{\partial\bar{\psi}_{\rm R}(\mathbf{C},\vartheta)}{\partial\mathbf{C}},\\ \eta_{\rm R}&=-\frac{\partial\bar{\psi}_{\rm R}(\mathbf{C},\vartheta)}{\partial\vartheta},\\ \mathbf{q}_{\rm R}&=\bar{\mathbf{q}}_{\rm R}(\mathbf{C},\vartheta,\nabla\vartheta);\end{aligned} \tag{57.39}$$

balance of linear momentum

$$\rho_R \ddot{\chi} = \operatorname{Div} \mathbf{T}_R + \mathbf{b}_{0R}; \tag{57.40}$$

balance of energy (written as an entropy balance)

$$\vartheta \dot{\eta}_R = -\operatorname{Div} \mathbf{q}_R + q_R. \tag{57.41}$$

Balance of energy (57.41) is equivalent to the relation

$$c\dot{\vartheta} = -\operatorname{Div} \mathbf{q}_R + \tfrac{1}{2}\vartheta \mathbf{M}:\dot{\mathbf{C}} + q_R, \tag{57.42}$$

with

$$c(\mathbf{C}, \vartheta) = -\vartheta \frac{\partial^2 \bar{\psi}_R(\mathbf{C}, \vartheta)}{\partial \vartheta^2} \qquad \text{and} \qquad \mathbf{M}(\mathbf{C}, \vartheta) = 2\frac{\partial^2 \bar{\psi}_R(\mathbf{C}, \vartheta)}{\partial \mathbf{C} \partial \vartheta}. \tag{57.43}$$

57.6 Entropy as Independent Variable. Nonconductors

For problems involving heat conduction, the right Cauchy–Green tensor $\mathbf{C}$ and the temperature ϑ are the natural choice of independent constitutive variables. However, for processes that occur over time scales so short that heat conduction is negligible, it is often preferable to replace constitutive dependence upon ϑ by constitutive dependence upon the entropy η_R.

Unless specified otherwise, we assume that the specific heat is strictly positive, so that

$$c(\mathbf{C}, \vartheta) > 0$$

for all $(\mathbf{C}, \vartheta)$. Since $\vartheta > 0$, it follows from (57.34) that

$$\frac{\partial \bar{\eta}_R(\mathbf{C}, \vartheta)}{\partial \vartheta} = \frac{c(\mathbf{C}, \vartheta)}{\vartheta} > 0. \tag{57.44}$$

This allows us to conclude that, for each fixed $\mathbf{C}$, the relation

$$\eta_R = \bar{\eta}_R(\mathbf{C}, \vartheta) \tag{57.45}$$

is smoothly invertible in ϑ, so that

$$\vartheta = \breve{\vartheta}(\mathbf{C}, \eta_R), \tag{57.46}$$

where, by (57.34),

$$\begin{aligned} \frac{\partial \breve{\vartheta}(\mathbf{C}, \eta_R)}{\partial \eta_R} &= \left(\frac{\partial \bar{\eta}_R(\mathbf{C}, \vartheta)}{\partial \vartheta} \right)^{-1} \\ &= \frac{\vartheta}{c(\mathbf{C}, \vartheta)} \end{aligned} \tag{57.47}$$

for $\vartheta = \breve{\vartheta}(\mathbf{C}, \eta_R)$. Then, by (57.11),

$$\begin{aligned} \varepsilon_R &= \breve{\varepsilon}_R(\mathbf{C}, \eta_R) \\ &= \bar{\psi}_R(\mathbf{C}, \breve{\vartheta}(\mathbf{C}, \eta_R)) + \breve{\vartheta}(\mathbf{C}, \eta_R)\eta_R, \end{aligned} \tag{57.48}$$

while (57.7) yields

$$\begin{aligned} \mathbf{T}_{RR} &= \breve{\mathbf{T}}_{RR}(\mathbf{C}, \eta_R) \\ &= \bar{\mathbf{T}}_{RR}(\mathbf{C}, \breve{\vartheta}(\mathbf{C}, \eta_R)). \end{aligned} \tag{57.49}$$

Thus, bearing in mind that a "breve" denotes a function of $(\mathbf{C}, \eta_R)$ while a "bar" denotes a function of $(\mathbf{C}, \vartheta)$, we find, using (57.7) and (57.8), that

$$\frac{\partial \breve{\varepsilon}_R}{\partial \mathbf{C}} = \frac{\partial \bar{\psi}_R}{\partial \mathbf{C}} + \underbrace{\left(\frac{\partial \bar{\psi}_R}{\partial \vartheta} + \eta_R\right)}_{=0} \frac{\partial \breve{\vartheta}}{\partial \mathbf{C}}$$

$$= \tfrac{1}{2}\breve{\mathbf{T}}_{RR}$$

and

$$\frac{\partial \breve{\varepsilon}_R}{\partial \eta_R} = \underbrace{\left(\frac{\partial \bar{\psi}_R}{\partial \vartheta} + \eta_R\right)}_{=0} \frac{\partial \breve{\vartheta}}{\partial \vartheta} + \breve{\vartheta}$$

$$= \breve{\vartheta}.$$

The second Piola stress and the temperature are therefore determined by the response function $\breve{\varepsilon}_R$ for the internal energy via the relations

$$\left.\begin{aligned} \mathbf{T}_{RR} &= 2\frac{\partial \breve{\varepsilon}_R(\mathbf{C}, \eta_R)}{\partial \mathbf{C}}, \\ \vartheta &= \frac{\partial \breve{\varepsilon}_R(\mathbf{C}, \eta_R)}{\partial \eta_R}; \end{aligned}\right\} \tag{57.50}$$

further, by (56.5) and $(57.50)_1$, the Piola stress is given by

$$\mathbf{T}_R = 2\mathbf{F}\frac{\partial \breve{\varepsilon}_R(\mathbf{C}, \eta_R)}{\partial \mathbf{C}}. \tag{57.51}$$

An immediate consequence of (57.50) is the **Maxwell relation**

$$\frac{\partial \breve{\mathbf{T}}_{RR}(\mathbf{C}, \eta_R)}{\partial \eta_R} = 2\frac{\partial \breve{\vartheta}(\mathbf{C}, \eta_R)}{\partial \mathbf{C}}. \tag{57.52}$$

The elasticity and stress-temperature tensors $\mathbb{C}(\mathbf{C}, \vartheta)$ and $\mathbf{M}(\mathbf{C}, \vartheta)$ have natural counterparts in the theory with entropy as independent variable; they are the **elasticity tensor**

$$\mathbb{C}^{\text{ent}}(\mathbf{C}, \eta_R) = 2\frac{\partial \breve{\mathbf{T}}_{RR}(\mathbf{C}, \eta_R)}{\partial \mathbf{C}} \tag{57.53}$$

(at fixed entropy) and the **stress-entropy modulus**

$$\mathbf{M}^{\text{ent}}(\mathbf{C}, \eta_R) = \frac{\partial \breve{\mathbf{T}}_{RR}(\mathbf{C}, \eta_R)}{\partial \eta_R} \tag{57.54}$$

(at fixed strain). By $(57.50)_1$,

$$\mathbb{C}^{\text{ent}}(\mathbf{C}, \eta_R) = 4\frac{\partial^2 \breve{\varepsilon}(\mathbf{C}, \eta_R)}{\partial \mathbf{C}^2}, \tag{57.55}$$

while the Maxwell relation (57.52) implies that

$$\mathbf{M}^{\text{ent}}(\mathbf{C}, \eta_R) = 2\frac{\partial \breve{\vartheta}(\mathbf{C}, \eta_R)}{\partial \mathbf{C}}. \tag{57.56}$$

We now determine relations between these various material functions. Toward this, we note, by (57.46), that we may relate the alternative descriptions of the second Piola stress in terms of $(\mathbf{C}, \vartheta)$ and $(\mathbf{C}, \eta_R)$ via

$$\breve{\mathbf{T}}_{RR}(\mathbf{C}, \eta_R) = \bar{\mathbf{T}}_{RR}(\mathbf{C}, \breve{\vartheta}(\mathbf{C}, \eta_R)).$$

Thus, by (57.54),

$$\mathbf{M}^{\text{ent}} = \frac{\partial \breve{\mathbf{T}}_{\text{RR}}}{\partial \eta_{\text{R}}} = \frac{\partial \bar{\mathbf{T}}_{\text{RR}}}{\partial \vartheta}\frac{\partial \breve{\vartheta}}{\partial \eta_{\text{R}}}$$

and, using (57.28) and (57.47), we conclude that the stress-entropy and stress-temperature moduli are related via

$$\mathbf{M}^{\text{ent}}(\mathbf{C}, \eta_{\text{R}}) = \frac{\vartheta}{c(\mathbf{C}, \vartheta)}\mathbf{M}(\mathbf{C}, \vartheta) \tag{57.57}$$

for $\vartheta = \breve{\vartheta}(\mathbf{C}, \eta_{\text{R}})$.[301]

The relation between the elasticity tensor $\mathbb{C}^{\text{ent}}$ at fixed entropy with the elasticity tensor $\mathbb{C}$ at fixed temperature is based on computing the partial derivative

$$\mathbb{C}^{\text{ent}}(\mathbf{C}, \eta_{\text{R}}) = 2\frac{\partial \breve{\mathbf{T}}_{\text{RR}}(\mathbf{C}, \eta_{\text{R}})}{\partial \mathbf{C}} = 2\frac{\partial}{\partial \mathbf{C}}\Big(\bar{\mathbf{T}}_{\text{RR}}(\mathbf{C}, \breve{\vartheta}(\mathbf{C}, \eta_{\text{R}}))\Big)$$

with respect to $\mathbf{C}$ holding η_{R} fixed. Suppressing arguments and using components, this Derivative of $\bar{\mathbf{T}}_{\text{RR}}(\mathbf{C}, \breve{\vartheta}(\mathbf{C}, \eta_{\text{R}}))$ is given by

$$\frac{\partial(\bar{\mathbf{T}}_{\text{RR}})_{ij}}{\partial C_{kl}} + \frac{\partial(\bar{\mathbf{T}}_{\text{RR}})_{ij}}{\partial \vartheta}\frac{\partial \breve{\vartheta}}{\partial C_{kl}}.$$

Thus, since the term

$$\frac{\partial(\bar{\mathbf{T}}_{\text{RR}})_{ij}}{\partial \vartheta}\frac{\partial \breve{\vartheta}}{\partial C_{kl}} \quad \text{is the component form of} \quad \frac{\partial \bar{\mathbf{T}}_{\text{RR}}}{\partial \vartheta} \otimes \frac{\partial \breve{\vartheta}}{\partial \mathbf{C}}$$

we find, with the aid of (57.28), (57.56), and (57.57), that

$$\mathbb{C}^{\text{ent}}(\mathbf{C}, \eta_{\text{R}}) = \mathbb{C}(\mathbf{C}, \vartheta) + \frac{\vartheta}{c(\mathbf{C}, \vartheta)}\mathbf{M}(\mathbf{C}, \vartheta) \otimes \mathbf{M}(\mathbf{C}, \vartheta) \tag{57.58}$$

for $\vartheta = \breve{\vartheta}(\mathbf{C}, \eta_{\text{R}})$; equivalently, in components, suppressing arguments,

$$C^{\text{ent}}_{ijkl} = C_{ijkl} + \frac{\vartheta}{c}M_{ij}M_{kl}. \tag{57.59}$$

The identity (57.58) has two important consequences. First, given symmetric tensors $\mathbf{A}$ and $\mathbf{G}$,

$$\mathbf{A}:(\mathbf{M} \otimes \mathbf{M})\mathbf{G} = (\mathbf{A}:\mathbf{M})(\mathbf{M}:\mathbf{G}) \tag{57.60}$$

$$= \mathbf{G}:(\mathbf{M} \otimes \mathbf{M})\mathbf{A}, \tag{57.61}$$

---

301 An alternate measure of thermomechanical coupling related to the stress-entropy modulus is the Grüneisen tensor defined by

$$\mathbf{G}(\mathbf{C}, \eta_{\text{R}}) \overset{\text{def}}{=} -\frac{1}{\vartheta(\mathbf{C}, \eta_{\text{R}})}\mathbf{M}^{\text{ent}}(\mathbf{C}, \eta_{\text{R}}).$$

Using (57.57), the Grüneisen tensor may be expressed in terms of the heat capacity and the stress-temperature modulus as

$$\mathbf{G}(\mathbf{C}, \eta_{\text{R}}) = -\frac{1}{c(\mathbf{C}, \vartheta)}\mathbf{M}(\mathbf{C}, \vartheta).$$

so that the fourth-order tensor $\mathbf{M}\otimes\mathbf{M}$ is symmetric. Thus, since $\mathbb{C}$ is symmetric, the tensor $\mathbb{C}^{\rm ent}$ is symmetric.[302]

Further, by (57.58) and (57.60),

$$\mathbf{A}:\mathbb{C}^{\rm ent}(\mathbf{C},\eta)\mathbf{A}-\mathbf{A}:\mathbb{C}(\mathbf{C},\vartheta)\mathbf{A}=\frac{\vartheta}{c}(\mathbf{A}:\mathbf{M})^2$$

for any tensor $\mathbf{A}$, and, since ϑ and c are both positive,

$$\mathbf{A}:\mathbb{C}^{\rm ent}(\mathbf{C},\eta)\mathbf{A}\geq\mathbf{A}:\mathbb{C}(\mathbf{C},\vartheta)\mathbf{A};$$

$\mathbb{C}^{\rm ent}$ *is thus positive-definite whenever* $\mathbb{C}$ *is positive-definite.*

57.7 Nonconductors

A body is referred to as a **nonconductor** if the constitutive response function for the heat flux vanishes identically, so that

$$\mathbf{q}_{\rm R}\equiv\mathbf{0}$$

in all constitutive processes. For a nonconductor, (57.15) shows that, if the heat supply $q_{\rm R}$ vanishes, then in any smooth constitutive process the entropy of each material point is constant in time:[303]

$$\dot{\eta}_{\rm R}=0.$$

Processes that satisfy $\dot{\eta}_{\rm R}=0$ are termed *isentropic*. This result shows why entropy is the variable of choice for nonconductors.

57.8 Material Symmetry

Because the response — to deformation and temperature — of the free energy determines the response of the stress and entropy, but not that of the heat flux, a discussion of material symmetry must account not only for the constitutive behavior of the free energy but also the constitutive behavior of the heat flux.

In a thermomechanical setting, symmetry transformations represent rigid transformations of the reference body that leave the response to deformation and temperature unaltered. Given a rotation $\mathbf{Q}$, consider the following generalization of the two experiments discussed in §50:

- *Experiment* 1. In this experiment the deformation gradient is $\mathbf{F}$ and the temperature *field* is ϑ.
- *Experiment* 2. In this experiment the deformation gradient is $\mathbf{FQ}$, but the temperature field remains ϑ.

Suppose that $\mathbf{Q}$ is a symmetry transformation. Arguing as in the mechanical theory, we should then have[304]

$$\bar{\psi}_{\rm R}(\mathbf{Q}^{\top}\mathbf{C}\mathbf{Q},\vartheta)=\bar{\psi}_{\rm R}(\mathbf{C},\vartheta). \tag{57.62}$$

Next, for the experiments 1 and 2, respectively, let $\mathbf{q}_1$ and $\mathbf{q}_2$ denote the heat-flux fields measured in the deformed body. Because the deformation gradient $\mathbf{F}$ applied to the rotated body as well as the spatial temperature field in the second experiment are the same as the deformation gradient and the spatial temperature field in the first

[302] This conclusion also follows from (57.59).

[303] The adjective "smooth" in this assertion is essential. In the presence of a shock wave the material away from the shock is isentropic on either side, but the values of the entropy on the two sides of the shock need not be the same; cf. $(33.24)_2$.

[304] Cf. (50.12).

experiment, and because we have assumed that $\mathbf{Q}$ is a symmetry transformation, the corresponding heat-flux fields $\mathbf{q}_1$ and $\mathbf{q}_2$ — *measured in the deformed body* — should be the same:

$$\mathbf{q}_1 = \mathbf{q}_2 \equiv \mathbf{q}. \tag{57.63}$$

We would not, however, expect the corresponding referential fluxes $\mathbf{q}_{R1}$ and $\mathbf{q}_{R2}$ to coincide, since the reference body has been rotated; in fact, by $(56.8)_1$, these fluxes must satisfy

$$\mathbf{F}\mathbf{q}_{R1} = \mathbf{F}\mathbf{Q}\mathbf{q}_{R2} = J\mathbf{q}. \tag{57.64}$$

Since the two experiments are associated with the same spatial temperature field, $\operatorname{grad}\vartheta$ should coincide in the two experiments. Thus, by $(56.8)_2$, the referential temperature gradients $\mathbf{g}_1$ and $\mathbf{g}_2$ measured in the two experiments must satisfy

$$\mathbf{F}^{-\top}\mathbf{g}_1 = \mathbf{F}^{-\top}\mathbf{Q}\mathbf{g}_2 = \operatorname{grad}\vartheta. \tag{57.65}$$

By (57.64), (57.65), and (50.9),

$$\mathbf{g}_2 = \mathbf{Q}^{\top}\mathbf{g}_1, \qquad \mathbf{q}_{R2} = \mathbf{Q}^{\top}\mathbf{q}_{R1}, \qquad \mathbf{C}_2 = \mathbf{Q}^{\top}\mathbf{C}_1\mathbf{Q}.$$

Thus, since

$$\mathbf{q}_{R1} = \bar{\mathbf{q}}_R(\mathbf{C}_1, \vartheta, \mathbf{g}_1) \qquad \text{and} \qquad \mathbf{q}_{R2} = \bar{\mathbf{q}}_R(\mathbf{C}_2, \vartheta, \mathbf{g}_2),$$

it follows that

$$\mathbf{Q}^{\top}\bar{\mathbf{q}}_R(\mathbf{C}, \vartheta, \mathbf{g}) = \bar{\mathbf{q}}_R(\mathbf{Q}^{\top}\mathbf{C}\mathbf{Q}, \vartheta, \mathbf{Q}^{\top}\mathbf{g}). \tag{57.66}$$

Summarizing: A rotation $\mathbf{Q}$ is a **symmetry transformation** if the symmetry relations (57.62) and (57.66) are satisfied for all $(\mathbf{C}, \vartheta, \mathbf{g})$.

58 Natural Reference Configuration for a Given Temperature

Roughly speaking, the reference configuration is natural for a temperature ϑ_0 if — in the absence of external loads — the configuration at that temperature is stable relative to minor perturbations in strain and temperature. We now make this notion precise and establish its local consequences.

58.1 Asymptotic Stability and its Consequences. The Gibbs Function

To be more specific, assume that at time $t = 0$ the body B is perturbed slightly — but in a homogeneous manner — from an undeformed state at temperature ϑ_0; more precisely, consider initial conditions for the right Cauchy–Green tensor $\mathbf{C}_*$ and the temperature ϑ_* of the form

$$\mathbf{C}(\mathbf{X}, 0) = \mathbf{C}_*, \qquad \vartheta(\mathbf{X}, 0) = \vartheta_*, \tag{58.1}$$

where $\mathbf{C}_*$ and ϑ_* are constant fields with

(i) $\mathbf{C}_*$ close to $\mathbf{1}$, so that the Green–St. Venant strain

$$\mathbf{E}_* = \tfrac{1}{2}(\mathbf{C}_* - \mathbf{1})$$

is small;

(ii) ϑ_* close to a constant temperature ϑ_0.

Assume, in addition, that

$$\dot{\chi}(\mathbf{X}, \mathbf{0}) = \mathbf{0},$$

and that the boundary ∂B is free and in thermal equilibrium at temperature ϑ_0:

$$\mathbf{T}_{\mathrm{R}}\mathbf{n}_{\mathrm{R}} = \mathbf{0} \quad \text{and} \quad \vartheta = \vartheta_0 \quad \text{on } \partial\mathrm{B}.$$

Then, by the global decay relation (31.20) (with $\mathbf{S}_0 = \mathbf{0}$),

$$\overline{\int_{\mathrm{B}} (\psi_{\mathrm{R}} + (\vartheta - \vartheta_0)\eta_{\mathrm{R}} + \tfrac{1}{2}\rho_{\mathrm{R}}|\dot{\chi}|^2)\, dv_{\mathrm{R}}}^{\,\boldsymbol{\cdot}} \leq 0,$$

so that, for $t = T > 0$,

$$\int_B (\psi_R + (\vartheta - \vartheta_0)\eta_R)\,dv_R\bigg|_{t=0} \geq \int_B (\psi_R + (\vartheta - \vartheta_0)\eta_R + \underbrace{\tfrac{1}{2}\rho_R|\dot{\chi}|^2}_{\geq 0})\,dv_R\bigg|_{t=T}$$

$$\geq \int_B (\psi_R + (\vartheta - \vartheta_0)\eta_R)\,dv_R\bigg|_{t=T};$$

thus, in view of the constitutive equations and the initial conditions (58.1),

$$\int_B [\bar{\psi}_R(\mathbf{C}_*, \vartheta_*) + (\vartheta_* - \vartheta_0)\bar{\eta}_R(\mathbf{C}_*, \vartheta_*)]\,dv_R$$

$$\geq \int_B [\bar{\psi}_R(\mathbf{C}, \vartheta) + (\vartheta - \vartheta_0)\bar{\eta}_R(\mathbf{C}, \vartheta_0)]\,dv_R\bigg|_{t=T}. \tag{58.2}$$

If the reference configuration at the temperature ϑ_0 is asymptotically stable against all such perturbations, then the body should ultimately relax to an undeformed state at temperature ϑ_0; granted this,

$$\mathbf{C}(\mathbf{X}, t) \to \mathbf{1} \qquad \text{and} \qquad \vartheta(\mathbf{X}, t) \to \vartheta_0$$

as $t \to \infty$, and passing to the limit as $T \to \infty$ in (58.2) gives

$$\int_B \underbrace{[\bar{\psi}_R(\mathbf{C}_*, \vartheta_*) + (\vartheta_* - \vartheta_0)\bar{\eta}_R(\mathbf{C}_*, \vartheta_*)]}_{\text{constant}}\,dv_R \geq \int_B \underbrace{\bar{\psi}_R(\mathbf{1}, \vartheta_0)}_{\text{constant}}\,dv_R,$$

or equivalently, since both integrands are constant,

$$\bar{\psi}_R(\mathbf{C}_*, \vartheta_*) + (\vartheta_* - \vartheta_0)\bar{\eta}_R(\mathbf{C}_*, \vartheta_*) \geq \bar{\psi}_R(\mathbf{1}, \vartheta_0).$$

Guided by this discussion, we introduce a **Gibbs function**

$$\omega(\mathbf{C}, \vartheta, \vartheta_0) = \bar{\psi}_R(\mathbf{C}, \vartheta) + (\vartheta - \vartheta_0)\bar{\eta}_R(\mathbf{C}, \vartheta) \tag{58.3}$$

and refer to the reference configuration as **natural for the temperature** ϑ_0 if

$$\omega(\mathbf{C}, \vartheta, \vartheta_0) \text{ has a local minimum at } (\mathbf{C}, \vartheta) = (\mathbf{1}, \vartheta_0). \tag{58.4}$$

58.2 Local Relations at a Reference Configuration that is Natural for a Temperature ϑ_0

Assume that the reference configuration is natural for the temperature ϑ_0. Then (58.4) implies that, at $(\mathbf{C}, \vartheta) = (\mathbf{1}, \vartheta_0)$, the derivatives of the Gibbs function ω with respect to $\mathbf{C}$ and ϑ vanish, while the matrix of second derivatives is positive-semidefinite. Thus, introducing the notation

$$\Phi\big|_0 = \Phi(\mathbf{C}, \vartheta)\big|_{(\mathbf{C},\vartheta)=(\mathbf{1},\vartheta_0)},$$

it follows that

$$\frac{\partial \omega}{\partial \mathbf{C}}\bigg|_0 = \mathbf{0}, \qquad \frac{\partial \omega}{\partial \vartheta}\bigg|_0 = 0, \tag{58.5}$$

and, for any symmetric tensor $\mathbf{A}$ and any scalar α,

$$\mathbf{A}\colon \frac{\partial^2 \omega}{\partial \mathbf{C}^2}\bigg|_0 \mathbf{A} + 2\mathbf{A}\colon \frac{\partial^2 \omega}{\partial \mathbf{C}\partial \vartheta}\bigg|_0 \alpha + \alpha^2 \frac{\partial^2 \omega}{\partial \vartheta^2}\bigg|_0 \geq 0, \tag{58.6}$$

or, in components,

$$\frac{\partial^2\omega}{\partial C_{ij}\partial C_{kl}}\bigg|_0 A_{ij}A_{kl} + 2\frac{\partial^2\omega}{\partial C_{ij}\partial\vartheta}\bigg|_0 \alpha A_{ij} + \frac{\partial^2\omega}{\partial\vartheta^2}\bigg|_0 \alpha^2 \geq 0.$$

By (58.5) and (58.6) together with the stress and entropy relations (57.7) and (57.8),

$$\left.\begin{aligned}
\frac{\partial\omega}{\partial\mathbf{C}}\bigg|_0 &= \left[\frac{\partial\bar{\psi}_{\mathrm{R}}}{\partial\mathbf{C}} + (\vartheta-\vartheta_0)\frac{\partial\bar{\eta}_{\mathrm{R}}}{\partial\mathbf{C}}\right]_0 = \frac{\partial\bar{\psi}_{\mathrm{R}}}{\partial\mathbf{C}}\bigg|_0 = \tfrac{1}{2}\bar{\mathbf{T}}_{\mathrm{RR}}\big|_0 = \mathbf{0},\\
\frac{\partial\omega}{\partial\vartheta}\bigg|_0 &= \left[\frac{\partial\bar{\psi}_{\mathrm{R}}}{\partial\vartheta} + \bar{\eta}_{\mathrm{R}} + (\vartheta-\vartheta_0)\frac{\partial\bar{\eta}_{\mathrm{R}}}{\partial\vartheta}\right]_0 = \frac{\partial\bar{\psi}_{\mathrm{R}}}{\partial\vartheta}\bigg|_0 + \bar{\eta}_{\mathrm{R}}\big|_0 = 0
\end{aligned}\right. \tag{58.7}$$

and using, in addition, (57.29) and (57.34)

$$\left.\begin{aligned}
\frac{\partial^2\omega}{\partial\mathbf{C}^2}\bigg|_0 &= \left[\frac{\partial^2\bar{\psi}_{\mathrm{R}}}{\partial\mathbf{C}^2} + (\vartheta-\vartheta_0)\frac{\partial^2\bar{\eta}_{\mathrm{R}}}{\partial\mathbf{C}^2}\right]_0 = \frac{\partial^2\bar{\psi}_{\mathrm{R}}}{\partial\mathbf{C}^2}\bigg|_0 = \tfrac{1}{4}\mathbb{C}\big|_0,\\
\frac{\partial^2\omega}{\partial\mathbf{C}\partial\vartheta}\bigg|_0 &= \left[\frac{\partial\bar{\psi}_{\mathrm{R}}}{\partial\mathbf{C}\partial\vartheta} + \frac{\partial\bar{\eta}_{\mathrm{R}}}{\partial\mathbf{C}} + (\vartheta-\vartheta_0)\frac{\partial^2\bar{\eta}_{\mathrm{R}}}{\partial\mathbf{C}\partial\vartheta}\right]_0 = \mathbf{0},\\
\frac{\partial^2\omega}{\partial\vartheta^2}\bigg|_0 &= \left[\frac{\partial^2\bar{\psi}_{\mathrm{R}}}{\partial\vartheta^2} + 2\frac{\partial\bar{\eta}_{\mathrm{R}}}{\partial\vartheta} + (\vartheta-\vartheta_0)\frac{\partial^2\bar{\eta}_{\mathrm{R}}}{\partial\vartheta^2}\right]_0 = 2\frac{\partial\bar{\eta}_{\mathrm{R}}}{\partial\vartheta}\bigg|_0 = \frac{c}{\vartheta}\bigg|_0.
\end{aligned}\right\} \tag{58.8}$$

Further, (58.6) and (58.8) imply that

$$\mathbf{A}:\mathbb{C}\big|_0\mathbf{A} \geq 0 \tag{58.9}$$

for every symmetric tensor **A**, and, in addition, that

$$c|_0 \geq 0. \tag{58.10}$$

Summarizing, we have shown that

(‡) *if the reference configuration is natural for the temperature* ϑ_0, *then*

- (i) *the residual stress* $\mathbf{T}_{RR}|_0$ *vanishes;*
- (ii) *the elasticity tensor* $\mathbb{C}|_0$ — *aside from being symmetric* — *is positive-semidefinite;*
- (iii) *the heat capacity* $c|_0$ *is nonnegative.*

EXERCISES

1. Consider a general thermoelastic material with internal energy, entropy, and free energy given by

$$\varepsilon_{\mathrm{R}} = \bar{\varepsilon}_{\mathrm{R}}(\mathbf{C},\vartheta), \qquad \eta_{\mathrm{R}} = \bar{\eta}_{\mathrm{R}}(\mathbf{C},\vartheta), \qquad \text{and} \qquad \psi_{\mathrm{R}} = \bar{\varepsilon}_{\mathrm{R}}(\mathbf{C},\vartheta) - \vartheta\bar{\eta}_{\mathrm{R}}(\mathbf{C},\vartheta) = \bar{\psi}_{\mathrm{R}}(\mathbf{C},\vartheta).$$

For rubber-like elastomeric materials, experiments show that the internal energy ε_{R} is *essentially independent of* **C**:

$$\bar{\varepsilon}_{\mathrm{R}}(\mathbf{C},\vartheta) = \bar{\varepsilon}_{\mathrm{R}}(\vartheta).$$

In this case, the heat capacity is also independent of **C**,

$$c(\vartheta) = \frac{d\bar{\varepsilon}_{\mathrm{R}}(\vartheta)}{d\vartheta}.$$

Using (57.34), show that the entropy $\eta_{\mathrm{R}}(\mathbf{C},\vartheta)$ must then have the separable form

$$\bar{\eta}_{\mathrm{R}}(\mathbf{C},\vartheta) = f(\vartheta) + g(\mathbf{C}).$$

Further, using (57.39), show that

$$\mathbf{T}_R = -2\vartheta \mathbf{F}\frac{dg(\mathbf{C})}{d\mathbf{C}}.$$

Therefore, at constant temperature, the stress in a rubber-like material is related to the change of the mechanical contribution g to the entropy with $\mathbf{C}$. Response of this kind is called **entropic elasticity** and approximately satisfied by many elastomeric materials.

2. Consider the general theory of thermoelastic materials in §57. Show, using the symmetry considerations in §57.8 and §50.3, that the constitutive equations of state (57.39) for the free energy, Piola stress, and entropy for an isotropic thermoelastic materials reduce to

$$\psi_R = \bar{\psi}_R(\mathbf{B}, \vartheta), \qquad \mathbf{T}_R = 2\frac{\partial \bar{\psi}_R(\mathbf{B}, \vartheta)}{\partial \mathbf{B}}\mathbf{F}, \qquad \eta_R = -\frac{\partial \bar{\psi}_R(\mathbf{B}, \vartheta)}{\partial \vartheta},$$

and since $\mathbf{T} = J^{-1}\mathbf{T}_R\mathbf{F}^\top$ the Cauchy stress is given by

$$\mathbf{T} = 2J^{-1}\frac{\partial \bar{\psi}_R(\mathbf{B}, \vartheta)}{\partial \mathbf{B}}\mathbf{B}.$$

Further, following the arguments of §50.3, show that the free energy for an isotropic thermoelastic material may be expressed in terms of the principal stretches λ_1, λ_2, and λ_3 as

$$\psi_R = \check{\psi}_R(\lambda_1, \lambda_2, \lambda_3, \vartheta),$$

and in this case the entropy is given by

$$\eta_R = -\frac{\partial \check{\psi}_R(\lambda_1, \lambda_2, \lambda_3, \vartheta)}{\partial \vartheta},$$

and, if the principal stretches are distinct, then

$$\begin{aligned} \mathbf{T}_R &= \sum_{i=1}^{3} \frac{\partial \check{\psi}_R(\lambda_1, \lambda_2, \lambda_3, \vartheta)}{\partial \lambda_i}\mathbf{l}_i \otimes \mathbf{r}_i, \\ \mathbf{T} &= \frac{1}{\lambda_1\lambda_2\lambda_3}\sum_{i=1}^{3} \lambda_i \frac{\partial \check{\psi}_R(\lambda_1, \lambda_2, \lambda_3, \vartheta)}{\partial \lambda_i}\mathbf{l}_i \otimes \mathbf{l}_i, \end{aligned} \tag{58.11}$$

where $\{\mathbf{r}_i\}$ and $\{\mathbf{l}_i\}$ are the right and left principal directions, respectively.

3. Following the arguments of §54, show that for an incompressible material, for which the principal stretches satisfy the constraint

$$\lambda_1\lambda_2\lambda_3 = 1,$$

the counterparts of (58.11) are

$$\mathbf{T}_R = \sum_{i=1}^{3}\left(\frac{\partial \check{\psi}_R(\lambda_1, \lambda_2, \lambda_3, \vartheta)}{\partial \lambda_i} - \frac{P}{\lambda_i}\right)\mathbf{l}_i \otimes \mathbf{r}_i,$$

$$\mathbf{T} = \sum_{i=1}^{3}\left(\lambda_i\frac{\partial \check{\psi}_R(\lambda_1, \lambda_2, \lambda_3, \vartheta)}{\partial \lambda_i} - P\right)\mathbf{l}_i \otimes \mathbf{l}_i,$$

where P is an arbitrary scalar field. Thus, for an isotropic, incompressible, thermoelastic material in the absence of temperature gradients, with free energy expressed in terms of principal stretches and temperature, the principal values of the Cauchy stress and the entropy are given by

$$\sigma_i = \lambda_i \frac{\partial \breve{\psi}_R(\lambda_1, \lambda_2, \lambda_3, \vartheta)}{\partial \lambda_i} - P \quad \text{(no sum)}, \qquad \eta = -\frac{\partial \breve{\psi}_R(\lambda_1, \lambda_2, \lambda_3, \vartheta)}{\partial \vartheta},$$

and granted that $\mathbf{R} \equiv \mathbf{1}$, so that the deformation is a pure homogeneous strain, the Piola stress $\mathbf{T}_R$ is symmetric with principal values given by

$$s_i = \frac{\partial \breve{\psi}_R(\lambda_1, \lambda_2, \lambda_3, \vartheta)}{\partial \lambda_i} - P\lambda_i^{-1}.$$

4. Define a scalar effective stretch by

$$\bar{\lambda} \stackrel{\text{def}}{=} \sqrt{\tfrac{1}{3}(\lambda_1^2 + \lambda_2^2 + \lambda_3^2)} \equiv \sqrt{\tfrac{1}{3}\mathrm{tr}\,\mathbf{B}},$$

and consider the following special free energy for rubber-like materials[305]

$$\psi_R = \bar{\psi}_R(\bar{\lambda}, \vartheta).$$

(a) Show that, for such a free energy, the principal values of the Cauchy stress and the first Piola stress (in a pure homogeneous strain) for an incompressible material are given by

$$\sigma_i = \mu\lambda_i^2 - P \quad \text{(no sum)}, \qquad s_i = \mu\lambda_i - P\lambda_i^{-1},$$

where

$$\mu \stackrel{\text{def}}{=} \frac{1}{3\bar{\lambda}} \frac{\partial \bar{\psi}_R(\bar{\lambda}, \vartheta)}{\partial \bar{\lambda}}.$$

(b) Show that the shear stress in simple shear is given by

$$T_{12} = \mu\gamma,$$

where γ is the amount of shear.[306] Thus μ represents a generalized shear modulus; assume that $\mu > 0$.

(c) For rubber-like materials (cf. Exercise 1 above), which satisfy

$$\bar{\varepsilon}_R(\bar{\lambda}, \vartheta) = \bar{\varepsilon}_R(\vartheta),$$

with a corresponding separable entropy

$$\bar{\eta}_R(\bar{\lambda}, \vartheta) = f(\vartheta) + g(\bar{\lambda}),$$

show that the generalized shear modulus is given by

$$\mu = -\vartheta\left(\frac{1}{3\bar{\lambda}} \frac{dg(\bar{\lambda})}{d\bar{\lambda}}\right) > 0.$$

(d) Consider a cylindrical bar with traction-free lateral surfaces subject to a simple extension, so that

$$\lambda_1 = \lambda, \quad \lambda_2 = \lambda_3 = \lambda^{-1/2}, \quad \bar{\lambda} = \sqrt{\tfrac{1}{3}(\lambda^2 + 2\lambda^{-1})}, \quad \sigma_1 = \sigma, \quad \sigma_2 = \sigma_3 = 0.$$

Show that the Piola stress $s \equiv s_1$ for rubber-like materials is given by

$$s = \underbrace{\vartheta\left(-\frac{1}{3\bar{\lambda}} \frac{dg(\bar{\lambda})}{d\bar{\lambda}}\right)}_{\mu}\left(\lambda - \frac{1}{\lambda^2}\right).$$

[305] Cf. Anand (1996).

[306] Cf. §51.

This demonstrates that, since $\mu > 0$, at a fixed stretch λ the stress s increases linearly as the temperature increases. Alternatively, if ϑ is increased at constant s, then λ must decrease. This is the first of Gough–Joule thermoelastic effects — *a rubber under a constant uniaxial stress contracts on heating.*[307]

(e) Consider a part P of a homogeneous rubber-like body in the form of a cylindrical bar of volume v_R at a uniform temperature ϑ. Show that the net internal energy $\mathcal{E}(\mathrm{P})$ of the bar and its time rate of change are

$$\mathcal{E}(\mathrm{P}) = v_R \bar{\epsilon}(\vartheta), \qquad \dot{\overline{\mathcal{E}(\mathrm{P})}} = v_R c(\vartheta)\dot{\vartheta},$$

where $c(\vartheta) = d\bar{\varepsilon}(\vartheta)/d\vartheta > 0$ is the heat capacity.[308]

(f) Show that, on neglecting body forces and kinetic energy, the power $\mathcal{W}(\mathrm{P})$, as defined in (24.19), exerted on the bar in simple extension is

$$\mathcal{W}(\mathrm{P}) = v_R \vartheta \left(-\frac{1}{3\bar{\lambda}} \frac{\mathrm{d}g(\bar{\lambda})}{\mathrm{d}\bar{\lambda}} \right) \left(\lambda - \frac{1}{\lambda^2} \right) \dot{\lambda}.$$

(g) Show that, under isothermal conditions,

$$\mathcal{Q}(\mathrm{P}) = -v_R \vartheta \left(-\frac{1}{3\bar{\lambda}} \frac{dg(\bar{\lambda})}{\partial \bar{\lambda}} \right) \left(\lambda - \frac{1}{\lambda^2} \right) \dot{\lambda} \tag{58.12}$$

in simple extension, and, hence, show that to maintain isothermal conditions, a rubber-like material must give out heat when stretched; this is the second of the Gough–Joule thermoelastic effects.

(h) Finally, show that, under adiabatic conditions — that is, when $\mathcal{Q}(\mathrm{P}) = 0$ — in simple extension

$$\dot{\vartheta} = \frac{\vartheta}{c(\vartheta)} \left(-\frac{1}{3\bar{\lambda}} \frac{\mathrm{d}g(\bar{\lambda})}{\mathrm{d}\bar{\lambda}} \right) (\lambda - \lambda^{-2})\dot{\lambda}, \tag{58.13}$$

and, hence, show that the temperature increases as the stretch increases. This represents the third of the classical Gough–Joule thermoelastic effects — *under adiabatic conditions a rubber-like material increases in temperature when stretched.*

[307] Cf. Gough (1805) and Joule (1859).
[308] Cf. §26.

59 Linear Thermoelasticity

Our derivation of the linear theory is based on the following hypotheses in which ϑ_0 is a prescribed (constant value) of the temperature and ℓ is a characteristic length associated with the reference body B:

- The reference configuration is natural for the temperature ϑ_0.
- The temperature ϑ is everywhere close to ϑ_0.
- The magnitudes of the displacement gradient $\mathbf{H} = \nabla\mathbf{u}$ and the scaled temperature gradient $\ell\mathbf{g}/\vartheta_0 = \ell\nabla\vartheta/\vartheta_0$ are everywhere small.

We derive asymptotic forms of the governing equations appropriate to the limit as

$$\epsilon = \sqrt{|\mathbf{H}|^2 + \frac{|\vartheta - \vartheta_0|^2}{\vartheta_0^2} + \frac{\ell^2|\mathbf{g}|^2}{\vartheta_0^2}}$$

tends to zero.[309] In this regard, given a function $\Phi(\mathbf{C}, \vartheta, \mathbf{g})$, we write

$$\Phi_0 = \Phi|_0 \text{ for } \Phi \text{ evaluated at } \mathbf{H} = \mathbf{0},\ \mathbf{g} = \mathbf{0},\ \vartheta = \vartheta_0. \tag{59.1}$$

Bearing in mind that $\mathbf{C} = \mathbf{F} = \mathbf{1}$ when $\mathbf{H} = \mathbf{0}$, this notation makes sense for functions $\Phi(\mathbf{C}, \vartheta, \mathbf{g})$ ($\Phi_0 = \Phi(\mathbf{1}, \vartheta_0, \mathbf{0})$), for functions $\Phi(\mathbf{F}, \vartheta, \mathbf{g})$ ($\Phi_0 = \Phi(\mathbf{1}, \vartheta_0, \mathbf{0})$), and trivially for functions $\Phi(\mathbf{F}, \vartheta)$ and $\Phi(\mathbf{C}, \vartheta)$.

59.1 Approximate Constitutive Equations for the Stress and Entropy

We now determine asymptotic constitutive equations for the stress and entropy appropriate to small departures from a reference configuration that is natural at the temperature ϑ_0.

Expanding $\bar{\mathbf{T}}_{\mathrm{RR}}(\mathbf{C}, \vartheta)$ and $\bar{\eta}_{\mathrm{R}}(\mathbf{C}, \vartheta)$ about $\mathbf{C} = \mathbf{1}$ and $\vartheta = \vartheta_0$, we obtain

$$\bar{\mathbf{T}}_{\mathrm{RR}}(\mathbf{C}, \vartheta) = \bar{\mathbf{T}}_{\mathrm{RR}}|_0 + \frac{\partial \bar{\mathbf{T}}_{\mathrm{RR}}}{\partial \mathbf{C}}\bigg|_0 (\mathbf{C} - \mathbf{1}) + \frac{\partial \bar{\mathbf{T}}_{\mathrm{RR}}}{\partial \vartheta}\bigg|_0 (\vartheta - \vartheta_0) + \mathrm{o}(\epsilon),$$

$$\bar{\eta}_{\mathrm{R}}(\mathbf{C}, \vartheta) = \bar{\eta}_{\mathrm{R}}|_0 + \frac{\partial \bar{\eta}_{\mathrm{R}}}{\partial \mathbf{C}}\bigg|_0 : (\mathbf{C} - \mathbf{1}) + \frac{\partial \bar{\eta}_{\mathrm{R}}}{\partial \vartheta}\bigg|_0 (\vartheta - \vartheta_0) + \mathrm{o}(\epsilon),$$

[309] This makes precise the sense of the term "small" in the second and third bullets above.

so that, by (57.27), (57.28), (57.30), and (57.34),

$$\left.\frac{\partial \bar{\mathbf{T}}_{\mathrm{RR}}}{\partial \mathbf{C}}\right|_0 = \tfrac{1}{2}\mathbb{C}_0,$$

$$\left.\frac{\partial \bar{\mathbf{T}}_{\mathrm{RR}}}{\partial \vartheta}\right|_0 = -2\left.\frac{\partial \bar{\eta}_{\mathrm{R}}}{\partial \mathbf{C}}\right|_0 = \mathbf{M}_0,$$

$$\left.\frac{\partial \bar{\eta}_{\mathrm{R}}}{\partial \vartheta}\right|_0 = \frac{c_0}{\vartheta_0},$$

and, by (58.7),

$$\bar{\mathbf{T}}_{\mathrm{RR}}|_0 = \mathbf{0}.$$

Assuming, without loss of generality, that[310]

$$\bar{\eta}_{\mathrm{R}}|_0 = 0,$$

introducing the shorthand notation

$$\mathbb{C} = \mathbb{C}_0, \qquad \mathbf{M} = \mathbf{M}_0, \qquad c = c_0,$$

and using the relation

$$\mathbf{E} = \tfrac{1}{2}(\mathbf{C} - \mathbf{1})$$

giving the Green–St. Venant strain in terms of the right Cauchy–Green tensor, we therefore obtain the following estimates for the second Piola stress and the entropy:

$$\begin{aligned} \mathbf{T}_{\mathrm{RR}} &= \mathbb{C}\mathbf{E} + \mathbf{M}(\vartheta - \vartheta_0) + \mathrm{o}(\epsilon), \\ \eta_{\mathrm{R}} &= -\mathbf{M}\colon\mathbf{E} + \frac{c_0}{\vartheta_0}(\vartheta - \vartheta_0) + \mathrm{o}(\epsilon), \end{aligned} \tag{59.2}$$

as $\epsilon \to 0$.

Further, arguing as in the derivations of (52.26) and (52.27), we find that, as $\epsilon \to 0$,

$$\begin{aligned} \mathbf{T}_{\mathrm{R}} &= \mathbb{C}\mathbf{E} + \mathbf{M}(\vartheta - \vartheta_0) + \mathrm{o}(\epsilon), \\ \mathbf{T} &= \mathbb{C}\mathbf{E} + \mathbf{M}(\vartheta - \vartheta_0) + \mathrm{o}(\epsilon), \end{aligned} \tag{59.3}$$

so that, as in the mechanical theory, to within a small error the Cauchy and Piola stresses are symmetric and coincident. Similarly, arguing as is the derivation of (52.31), we may show that the free-energy function $\bar{\psi}_{\mathrm{R}}(\mathbf{C}, \vartheta)$ admits the estimate

$$\psi_{\mathrm{R}} = \frac{1}{2}\mathbf{E}\colon\mathbb{C}\mathbf{E} + (\vartheta - \vartheta_0)\mathbf{M}\colon\mathbf{E} - \frac{c}{2\vartheta_0}(\vartheta - \vartheta_0)^2 + \mathrm{o}(\epsilon^2) \tag{59.4}$$

as $\epsilon \to 0$. Moreover, by (18.9), the spatial and material forms of the density, conventional body force, and external heat supply are related through

$$\rho = [1 + \mathrm{o}(1)]\rho_{\mathrm{R}}, \qquad \mathbf{b}_0 = [1 + \mathrm{o}(1)]\mathbf{b}_{0\mathrm{R}}, \qquad \text{and} \qquad q = [1 + \mathrm{o}(1)]q_{\mathrm{R}}. \tag{59.5}$$

EXERCISE

1. Verify (59.3)–(59.5).

---

310 Cf. (31.9).

59.2 Basic Field Equations of Linear Thermoelasticity

The linear theory of thermoelasticity is based on approximate equations obtained when the higher-order terms in

$$\mathbf{E} = \tfrac{1}{2}(\mathbf{H} + \mathbf{H}^{\top}) + o(|\mathbf{H}|^2), \tag{59.6}$$

(57.25), (59.2), (59.3), (59.4), and (59.5) are neglected.[311] We therefore take $\rho = \rho_{\mathrm{R}}$, $\mathbf{T} = \mathbf{T}_{\mathrm{R}}$, $\mathbf{b}_0 = \mathbf{b}_{0\mathrm{R}}$, $\mathbf{q} = \mathbf{q}_{\mathrm{R}}$, and $q = q_{\mathrm{R}}$ and base the theory on the **strain-displacement relation**

$$\mathbf{E} = \tfrac{1}{2}(\nabla\mathbf{u} + (\nabla\mathbf{u})^{\top}) \tag{59.7}$$

and the **constitutive equations**

$$\begin{aligned}
\psi_{\mathrm{R}} &= \frac{1}{2}\mathbf{E}:\mathbb{C}\mathbf{E} + (\vartheta - \vartheta_0)\mathbf{M}:\mathbf{E} - \frac{c}{2\vartheta_0}(\vartheta - \vartheta_0)^2,\\
\mathbf{T} &= \mathbb{C}\mathbf{E} + \mathbf{M}(\vartheta - \vartheta_0),\\
\eta_{\mathrm{R}} &= -\mathbf{M}:\mathbf{E} + \frac{c}{\vartheta_0}(\vartheta - \vartheta_0),\\
\mathbf{q} &= -\mathbf{K}\nabla\vartheta,
\end{aligned} \tag{59.8}$$

where $\mathbb{C}$, $\mathbf{M}$, c, and $\mathbf{K} = \mathbf{K}_0$ are, respectively, the elasticity tensor, the stress-temperature modulus, the heat capacity, and the thermal conductivity tensor at the reference temperature ϑ_0.

By (58.9), $\mathbb{C}$ is symmetric and positive-semidefinite; here we assume, in addition, that $\mathbb{C}$ is *positive-definite*:[312]

$$\mathbf{A}:\mathbb{C}\mathbf{A} > 0 \quad \text{for all symmetric tensors } \mathbf{A} \neq \mathbf{0}. \tag{59.9}$$

Similarly, by (58.10), c is nonnegative and we strengthen this by requiring that c be positive

$$c > 0. \tag{59.10}$$

Finally, by the bullet on page 341, $\mathbf{K}$ is positive-semidefinite and we strengthen this by requiring that $\mathbf{K}$ be positive-definite:

$$\mathbf{a}\cdot\mathbf{K}\mathbf{a} > 0 \quad \text{for all vectors } \mathbf{a} \neq \mathbf{0}. \tag{59.11}$$

The *basic equations of the linear theory of thermoelasticity* consist of (59.7), (59.8), the local momentum balance

$$\rho\ddot{\mathbf{u}} = \mathrm{Div}[\mathbb{C}\mathbf{E} + \mathbf{M}(\vartheta - \vartheta_0)] + \mathbf{b}_0 \tag{59.12}$$

and the local energy balance

$$c\dot{\vartheta} = \mathrm{Div}(\mathbf{K}\nabla\vartheta) + \vartheta_0\mathbf{M}:\dot{\mathbf{E}} + q. \tag{59.13}$$

59.3 Isotropic Linear Thermoelasticity

When one uses $\mathbf{E}$ instead of $\mathbf{C}$ to express the constitutive equations for thermoelasticity, the transformation rules (57.62) and (57.66) under a symmetry transformation

311 Precisely, the $o(\epsilon)$ terms in (59.2) and (59.3), the $o(\epsilon^2)$ term in (59.4), and the $o(1)$ terms in (59.5) are neglected.

312 A condition sufficient to ensure (59.9) is that $\mathbb{C}$ be invertible, since an invertible, positive-semidefinite linear transformation is positive-definite.

$\mathbf{Q}$, respectively, become

$$\tilde{\psi}_{\mathrm{R}}(\mathbf{Q}^{\top}\mathbf{E}\mathbf{Q}, \vartheta) = \tilde{\psi}_{\mathrm{R}}(\mathbf{E}, \vartheta) \qquad \text{and} \qquad \mathbf{Q}^{\top}\tilde{\mathbf{q}}_{\mathrm{R}}(\mathbf{E}, \vartheta, \mathbf{g}) = \tilde{\mathbf{q}}_{\mathrm{R}}(\mathbf{Q}^{\top}\mathbf{E}\mathbf{Q}, \vartheta, \mathbf{Q}^{\top}\mathbf{g}). \quad (59.14)$$

For the free-energy function $(59.8)_1$ and the Fourier law $(59.8)_3$, these rules immediately imply that

$$\mathbf{Q}^{\top}(\mathbb{C}\mathbf{E})\mathbf{Q} = \mathbb{C}(\mathbf{Q}^{\top}\mathbf{E}\mathbf{Q}), \qquad \mathbf{Q}^{\top}\mathbf{M}\mathbf{Q} = \mathbf{M}, \qquad \text{and} \qquad \mathbf{Q}^{\top}\mathbf{K}\mathbf{Q} = \mathbf{K}, \qquad (59.15)$$

for a symmetry transformation $\mathbf{Q}$ and all symmetric tensors $\mathbf{E}$. If the body is isotropic, then $\mathbb{C}\mathbf{E}$, $\mathbf{M}$, and $\mathbf{K}$ have the specific forms

$$\left.\begin{aligned} \mathbb{C}\mathbf{E} &= 2\mu\mathbf{E} + \lambda(\operatorname{tr}\mathbf{E})\mathbf{1}, \\ \mathbf{M} &= \beta\mathbf{1}, \\ \mathbf{K} &= k\mathbf{1}, \end{aligned}\right\} \qquad (59.16)$$

with μ and λ **elastic moduli**, β the **stress-temperature modulus**, and k the **thermal conductivity**.[313]

Next, we determine the restrictions placed on the moduli μ and λ by the requirement (59.9) that the elasticity tensor $\mathbb{C}$ be positive-definite. Choose an arbitrary symmetric tensor $\mathbf{A}$ and let $\mathbf{A}_0$ denote its deviatoric part:

$$\mathbf{A}_0 = \mathbf{A} - \tfrac{1}{3}(\operatorname{tr}\mathbf{A})\mathbf{1}.$$

Then $\operatorname{tr}\mathbf{A}_0 = 0$ and

$$\begin{aligned} |\mathbf{A}|^2 &= (\mathbf{A}_0 + \tfrac{1}{3}(\operatorname{tr}\mathbf{A})\mathbf{1}):(\mathbf{A}_0 + \tfrac{1}{3}(\operatorname{tr}\mathbf{A})\mathbf{1}) \\ &= |\mathbf{A}_0|^2 + \tfrac{1}{3}(\operatorname{tr}\mathbf{A})^2. \end{aligned}$$

Thus, by (59.9) and $(59.16)_1$,

$$\begin{aligned} 0 &< \mathbf{A}:\mathbb{C}\mathbf{A} \\ &= 2\mu|\mathbf{A}|^2 + \lambda(\operatorname{tr}\mathbf{A})^2 \\ &= 2\mu|\mathbf{A}_0|^2 + \kappa(\operatorname{tr}\mathbf{A})^2, \end{aligned} \qquad (59.17)$$

with

$$\kappa = \lambda + \tfrac{2}{3}\mu. \qquad (59.18)$$

Choosing $\mathbf{A} = \mathbf{1}$ (so that $\operatorname{tr}\mathbf{A} = 3$ and $\mathbf{A}_0 = \mathbf{0}$) yields $\kappa > 0$; choosing $\mathbf{A} = \mathbf{e}\otimes\mathbf{f} + \mathbf{f}\otimes\mathbf{e}$ with $\mathbf{e}$ and $\mathbf{f}$ orthonormal (so that $\operatorname{tr}\mathbf{A} = 0$ and $|\mathbf{A}_0|^2 = 2$) yields $\mu > 0$. Thus, the elastic moduli μ and λ satisfy

$$\mu > 0, \qquad \lambda + \tfrac{2}{3}\mu > 0. \qquad (59.19)$$

The scalars μ and λ are generally referred to as **Lamé moduli**. In view of (59.18), the relation $(59.16)_1$ may alternatively be written in terms of the scalars μ and κ as

$$\mathbb{C}\mathbf{E} = 2\mu\mathbf{E}_0 + \kappa(\operatorname{tr}\mathbf{E})\mathbf{1}, \qquad (59.20)$$

where $\mathbf{E}_0 = \mathbf{E} - \frac{1}{3}(\operatorname{tr}\mathbf{E})\mathbf{1}$ denotes the deviatoric part of $\mathbf{E}$. In view of (59.20), the Lame modulus μ is also known as the **isothermal shear modulus**, while κ is called the **isothermal bulk modulus**.

[313] These results follow from standard representation theorems for isotropic functions. Cf. the Appendix (Part 112.7).

Further, since, by (59.11), $\mathbf{K}$ is positive-definite, the thermal conductivity k must be positive

$$k > 0. \tag{59.21}$$

By (59.16), the defining constitutive equations for an isotropic, linear thermoelastic solid are

$$\begin{aligned}
\psi_{\mathrm{R}} &= \mu|\mathbf{E}|^2 + \frac{\lambda}{2}(\operatorname{tr}\mathbf{E})^2 + \beta(\vartheta - \vartheta_0)\operatorname{tr}\mathbf{E} - \frac{c}{2\vartheta_0}(\vartheta - \vartheta_0)^2,\\
\mathbf{T} &= 2\mu\mathbf{E} + \lambda(\operatorname{tr}\mathbf{E})\mathbf{1} + \beta(\vartheta - \vartheta_0)\,\mathbf{1},\\
\eta_{\mathrm{R}} &= -\beta\operatorname{tr}\mathbf{E} + \frac{c}{\vartheta_0}(\vartheta - \vartheta_0),\\
\mathbf{q} &= -k\nabla\vartheta.
\end{aligned} \tag{59.22}$$

Granted (59.9), this stress-strain relation $(81.53)_2$ may be inverted to give

$$\mathbf{E} = \frac{1}{2\mu}\left(\mathbf{T} - \frac{\lambda}{2\mu + 3\lambda}(\operatorname{tr}\mathbf{T})\mathbf{1}\right) + \alpha(\vartheta - \vartheta_0)\mathbf{1}, \tag{59.23}$$

where

$$\alpha \stackrel{\text{def}}{=} -\frac{\beta}{2\mu + 3\lambda} \tag{59.24}$$

is the *coefficient of thermal expansion*. Thus, using (59.18),

$$\beta = -3\kappa\alpha. \tag{59.25}$$

Recall from (57.58) that the isentropic elasticity tensor $\mathbb{C}^{\mathrm{ent}}$ is generally related to the isothermal elastic tensor $\mathbb{C}$ by

$$\mathbb{C}^{\mathrm{ent}}(\mathbf{C}, \eta) = \mathbb{C}(\mathbf{C}, \vartheta) + \frac{\vartheta}{c(\mathbf{C}, \eta)}\mathbf{M}(\mathbf{C}, \vartheta) \otimes \mathbf{M}(\mathbf{C}, \vartheta). \tag{59.26}$$

Thus, using (59.20) and (59.25),

$$\mathbb{C}^{\mathrm{ent}}\mathbf{E} = 2\mu\mathbf{E}_0 + \kappa^{\mathrm{ent}}(\operatorname{tr}\mathbf{E})\mathbf{1}, \tag{59.27}$$

where

$$\kappa^{\mathrm{ent}} = \kappa + \frac{\vartheta_0\beta^2}{c} \tag{59.28}$$

$$= \kappa\left(1 + \frac{9\vartheta_0\kappa\alpha^2}{c}\right). \tag{59.29}$$

The isothermal and isentropic shear moduli are therefore identical, while the isentropic bulk modulus κ^{ent} is related to the isothermal bulk modulus κ through (59.29).

Next, when B is homogeneous and isotropic, then ρ, μ, λ, β, and k are constants. In this case, since

$$\begin{aligned}
2\operatorname{Div}\mathbf{E} &= \operatorname{Div}(\nabla\mathbf{u} + (\nabla\mathbf{u})^{\top})\\
&= \Delta\mathbf{u} + \nabla\operatorname{Div}\mathbf{u},
\end{aligned}$$

and

$$\begin{aligned}
\operatorname{Div}[(\operatorname{tr}\mathbf{E})\mathbf{1}] &= \operatorname{Div}[(\operatorname{Div}\mathbf{u})\mathbf{1}]\\
&= \nabla\operatorname{Div}\mathbf{u},
\end{aligned}$$

the momentum balance (59.12) yields

$$\rho_R \ddot{\mathbf{u}} = \mu \Delta \mathbf{u} + (\lambda + \mu)\nabla \mathrm{Div}\,\mathbf{u} + \beta \nabla \vartheta + \mathbf{b}_0. \tag{59.30}$$

Further, the balance of energy (59.13), yields

$$c\dot{\vartheta} = k\Delta\vartheta + \beta\vartheta_0 \mathrm{Div}\,\dot{\mathbf{u}} + q. \tag{59.31}$$

In practice, a simplifying approximation that is often imposed to facilitate the solution of actual problems is to neglect the small coupling term $\beta\vartheta_0\mathrm{Div}\,\dot{\mathbf{u}}$ in the partial-differential equation (59.31). Under this approximation, the resulting theory is referred to as the weakly coupled theory of isotropic linear thermoelasticity.

PART XII

SPECIES DIFFUSION COUPLED TO ELASTICITY

This chapter presents a purely mechanical theory for coupled species transport and elastic deformation. The species in question may be ionic, atomic, molecular, or chemical. Underlying our approach is the notion of a structure through or on which the various species diffuse. Examples of such structures include the fissures and voids of a porous medium, the interstices of a polymer network, and the lattice of crystalline solid. The most unfamiliar and conceptually challenging features of the theory are associated with the need to account for the energy flow due to species transport. To convey these, we therefore begin by developing the theory for a single species. After extending the theory to allow for the presence of $N \geq 1$ unconstrained species, we illustrate the impact of a constraint by developing the theory for a substitional alloy.

60 Balance Laws for Forces, Moments, and the Conventional External Power

We have in mind applications where the time scales associated with species diffusion are considerably longer than those associated with wave propagation and, for that reason, we neglect all inertial effects.[314] In particular, when the inertial body force ι is neglected, it follows from (19.15) that the generalized body force $\mathbf{b}$ coincides with the conventional body force $\mathbf{b}_0$:

$$\mathbf{b} \equiv \mathbf{b}_0. \tag{60.1}$$

In view of (60.1), the balances $(19.29)_1$ and $(19.29)_2$ for forces and moments reduce to

$$\begin{aligned} \int\limits_{\partial\mathcal{P}_t} \mathbf{Tn}\, da + \int\limits_{\mathcal{P}_t} \mathbf{b}_0\, dv &= \mathbf{0}, \\ \int\limits_{\partial\mathcal{P}_t} \mathbf{r} \times \mathbf{Tn}\, da + \int\limits_{\mathcal{P}_t} \mathbf{r} \times \mathbf{b}_0\, dv &= \mathbf{0}. \end{aligned} \tag{60.2}$$

Proceeding as in the derivations of (19.30) and (19.34), we are thus led to the local force and moment balances

$$\begin{aligned} \operatorname{div}\mathbf{T} + \mathbf{b}_0 &= \mathbf{0}, \\ \mathbf{T} &= \mathbf{T}^{\top}. \end{aligned} \tag{60.3}$$

Next, by $(4.11)_5$, (19.40), and $(60.3)_1$,

$$\begin{aligned} \int\limits_{\partial\mathcal{P}_t} \mathbf{Tn} \cdot \mathbf{v}\, da &= \int\limits_{\mathcal{P}_t} (\mathbf{v} \cdot \operatorname{div}\mathbf{T} + \mathbf{T} : \operatorname{grad}\mathbf{v})\, dv \\ &= \int\limits_{\mathcal{P}_t} (\mathbf{T} : \mathbf{D} - \mathbf{b}_0 \cdot \mathbf{v})\, dv, \end{aligned} \tag{60.4}$$

and we arrive at a version

$$\underbrace{\int\limits_{\partial\mathcal{P}_t} \mathbf{Tn} \cdot \mathbf{v}\, da + \int\limits_{\mathcal{P}_t} \mathbf{b}_0 \cdot \mathbf{v}\, dv}_{\mathcal{W}_0(\mathcal{P}_t)} = \underbrace{\int\limits_{\mathcal{P}_t} \mathbf{T} : \mathbf{D}\, dv}_{\mathcal{I}(\mathcal{P}_t)} \tag{60.5}$$

of the power balance valid in the absence of inertia.[315]

---

[314] Cf. §19.7.4.

[315] Cf. $(19.50)_5$.

61 Mass Balance for a Single Diffusing Species

We consider a single diffusing species and write $n \geq 0$ for its **mass fraction**,[316] so that ρn represents the **species density**. For $\mathcal{P}_t$ a spatial region convecting with the body, the integral

$$\int_{\mathcal{P}_t} \rho n \, dv \tag{61.1}$$

represents the net mass of the diffusing species in $\mathcal{P}_t$ (at time t). To characterize **species transport** to $\mathcal{P}_t$, we mimic our treatment of the heat flow $\mathcal{Q}(\mathcal{P}_t)$ encountered first in (26.5). Introducing a vectorial **species flux h** and a scalar **species supply** h, we write

$$-\int_{\partial\mathcal{P}_t} \mathbf{h}\cdot\mathbf{n}\, da + \int_{\mathcal{P}_t} h\, dv \tag{61.2}$$

for the net rate of species transported into $\mathcal{P}_t$. The first term in (61.2) gives the rate at which the species is transported to $\mathcal{P}_t$ by diffusion across $\partial\mathcal{P}_t$; because $\mathbf{n}$ is the outward unit normal to $\partial\mathcal{P}_t$, the minus sign renders this term nonnegative when the flux $\mathbf{h}$ points into $\mathcal{P}_t$. Since $\mathbf{n}$ is spatial, $\mathbf{h}$ is a spatial vector field. The second term in (61.2) represents the rate of transport to $\mathcal{P}_t$ by agencies external to the body.

In view of (61.1) and (61.2), **species mass balance** is the requirement that

$$\boxed{\overline{\int_{\mathcal{P}_t} \rho n\, dv}^{\bullet} = -\int_{\partial\mathcal{P}_t} \mathbf{h}\cdot\mathbf{n}\, da + \int_{\mathcal{P}_t} h\, dv} \tag{61.3}$$

for every convecting region $\mathcal{P}_t$. To localize this balance, we note that, by (18.14),

$$\overline{\int_{\mathcal{P}_t} \rho n\, dv}^{\bullet} = \int_{\mathcal{P}_t} \rho \dot{n}\, dv;$$

thus, applying the divergence theorem to the term involving the species flux $\mathbf{h}$, we find that

$$\int_{\mathcal{P}_t} (\rho\dot{n} + \operatorname{div}\mathbf{h} - h)\, dv = 0. \tag{61.4}$$

[316] The mass fraction of a species is the species density divided by the net density of all species.

Since (61.4) holds for every convecting region $\mathcal{P}_t$, we have the **local species mass-balance**

$$\rho\dot{n} = -\mathrm{div}\mathbf{h} + h. \tag{61.5}$$

EXERCISES

1. Using (18.12), show that the local species mass balance can be written equivalently as

$$(\rho n)' = -\mathrm{div}(\mathbf{h} + \rho n\mathbf{v}) + h$$

and provide a physical interpretation of the quantity $\mathbf{h} + \rho n\mathbf{v}$.

2. Show that, for a spatial control volume R, the species mass balance has the form

$$\overline{\int_R \rho n\, dv}^{\,\bullet} + \int_{\partial R} \rho n\mathbf{v}\cdot\mathbf{n}\, da = -\int_{\partial R} \mathbf{h}\cdot\mathbf{n}\, da + \int_R h\, dv$$

or, equivalently,

$$\overline{\int_R \rho n\, dv}^{\,\bullet} = -\int_{\partial R} (\mathbf{h} + \rho n\mathbf{v})\cdot\mathbf{n}\, da + \int_R h\, dv.$$

62 Free-Energy Imbalance Revisited. Chemical Potential

Let $\mathcal{P}_t$ be an arbitrary spatial region convecting with the body. Consistent with our neglect of inertia, we also neglect kinetic energy.[317] As in the mechanical theory discussed in §29, changes in the net free-energy of $\mathcal{P}_t$ are influenced by the conventional power expended on $\mathcal{P}_t$, but we must now account also for energy carried into $\mathcal{P}_t$ by species transport; we, therefore, seek a free-energy imbalance of the form[318]

$$\overline{\int_{\mathcal{P}_t} \rho\psi\, dv} \leq \mathcal{W}_0(\mathcal{P}_t) + \mathcal{T}(\mathcal{P}_t),$$

where the term $\mathcal{T}(\mathcal{P}_t)$ represents energy flow due to species transport. We characterize this flow through the **chemical potential** μ; specifically, we assume that the species flux $\mathbf{h}$ and species supply h carry with them a flux and supply of energy described by $\mu\mathbf{h}$ and μh, so that[319]

$$\mathcal{T}(\mathcal{P}_t) = -\int_{\partial\mathcal{P}_t} \mu\mathbf{h}\cdot\mathbf{n}\, da + \int_{\mathcal{P}_t} \mu h\, dv. \tag{62.1}$$

Thus, trivially, if there is no species flux at some point, then there is no associated flux of energy at that point, and similarly for the supply h; our treatment of energy flow due to species transport therefore bears some resemblance to our discussion of entropy flow due to heat flow.[320] In this regard, the chemical potential μ can be viewed as a quantity roughly analogous to the reciprocal of the absolute temperature ϑ, the most significant differences being that μ is unsigned and $\mu\mathbf{h}$ and μh describe energy exchanges induced by species transport rather than heat transport (as described by the heat flux $\mathbf{q}$ and heat supply q).

We are therefore led to the inequality

$$\boxed{\overline{\int_{\mathcal{P}_t} \rho\psi\, dv} \leq \int_{\partial\mathcal{P}_t} \mathbf{Tn}\cdot\mathbf{v}\, da + \int_{\mathcal{P}_t} \mathbf{b}_0\cdot\mathbf{v}\, dv - \int_{\partial\mathcal{P}_t} \mu\mathbf{h}\cdot\mathbf{n}\, da + \int_{\mathcal{P}_t} \mu h\, dv,} \tag{62.2}$$

[317] Cf. $(19.50)_{2,3}$.

[318] Cf. (29.8).

[319] ECKART (1940), in his discussion of fluid mixtures, notes that chemical potentials should enter the energy balance through terms of this form. (JAUMANN (1911) and LOHR (1917) seem also to have this view, but we are unable to fully comprehend their work.) While Eckart employs constitutive equations, their use is unnecessary. Related works are MEIXNER & REIK (1959), MÜLLER (1968), GURTIN & VARGAS (1971), DAVÌ & GURTIN (1990), FRIED & SELLERS (2000), GURTIN (1991), and PODIO-GUIDUGLI (2006).

[320] Cf. §27.1.

or equivalently, defining the dissipation $\mathcal{D}(\mathcal{P}_t)$ to be the left side of this relation minus the right, to the global **free-energy imbalance**:[321]

$$\mathcal{D}(\mathcal{P}_t) = -\int\limits_{\partial\mathcal{P}_t} \mu\mathbf{h}\cdot\mathbf{n}\,da + \int\limits_{\mathcal{P}_t} \mu h\,dv + \int\limits_{\partial\mathcal{P}_t} \mathbf{Tn}\cdot\mathbf{v}\,da + \int\limits_{\mathcal{P}_t} \mathbf{b}_0\cdot\mathbf{v}\,dv - \overline{\int\limits_{\mathcal{P}_t} \rho\psi\,dv}^{\boldsymbol{\cdot}} \geq 0. \tag{62.3}$$

We view chemical potentials as *primitive quantities* introduced as a means of characterizing energy exchanges induced by species transport. This contrasts sharply with what is done in the materials science literature, where chemical potentials are *defined* as derivatives of the free energy with respect to the species densities, or introduced variationally — via an assumption of equilibrium — as Lagrange multipliers corresponding to a mass constraint; in either case, the chemical potentials require a constitutive structure.

- To the contrary, in our framework it is the free-energy imbalance and, hence, the characterization of energy transport that are basic.

As in the setting without species transport, we may determine an expression for the density of the net dissipation valid under appropriate smoothness hypotheses. To do so, we note first that, by (18.14),

$$\overline{\int\limits_{\mathcal{P}_t} \rho\psi\,dv}^{\boldsymbol{\cdot}} = \int\limits_{\mathcal{P}_t} \rho\dot{\psi}\,dv. \tag{62.4}$$

Further, by the divergence theorem and the species mass balance (61.5),

$$\begin{aligned}
-\int\limits_{\partial\mathcal{P}_t} \mu\mathbf{h}\cdot\mathbf{n}\,da &= -\int\limits_{\mathcal{P}_t} (\mu\operatorname{div}\mathbf{h} + \mathbf{h}\cdot\operatorname{grad}\mu)\,dv \\
&= \int\limits_{\mathcal{P}_t} (\rho\mu\dot{n} - \mathbf{h}\cdot\operatorname{grad}\mu - \mu h)\,dv
\end{aligned}$$

and, therefore,

$$-\int\limits_{\partial\mathcal{P}_t} \mu\mathbf{h}\cdot\mathbf{n}\,da + \int\limits_{\mathcal{P}_t} \mu h\,dv = -\int\limits_{\mathcal{P}_t} (\mathbf{h}\cdot\operatorname{grad}\mu - \rho\mu\dot{n})\,dv. \tag{62.5}$$

Using (60.5) and (62.4) in (62.3), we thus find that

$$\mathcal{D}(\mathcal{P}_t) = \int\limits_{\mathcal{P}_t} [\mathbf{T}:\mathbf{D} - \mathbf{h}\cdot\operatorname{grad}\mu - \rho(\dot{\psi} - \mu\dot{n})]\,dv.$$

Granted sufficient smoothness,[322] the net dissipation therefore has a density δ, measured per unit volume in the deformed body, such that

$$\mathcal{D}(\mathcal{P}_t) = \int\limits_{\mathcal{P}_t} \delta\,dv, \qquad \delta = \mathbf{T}:\mathbf{D} - \mathbf{h}\cdot\operatorname{grad}\mu - \rho(\dot{\psi} - \mu\dot{n}), \tag{62.6}$$

and, hence, since $\mathcal{D}(\mathcal{P}_t) \geq 0$ and since the convecting region $\mathcal{P}_t$ is arbitrary, such that

$$\delta \geq 0.$$

---

[321] Cf. §29.1.

[322] Cf. (‡) on page 187.

Thus, (62.6) yields the local **free-energy imbalance**

$$\boxed{\rho(\dot{\psi} - \mu\dot{n}) - \mathbf{T}:\mathbf{D} + \mathbf{h}\cdot\operatorname{grad}\mu = -\delta \leq 0.} \tag{62.7}$$

EXERCISES

1. Show that the species mass balance (61.5) and the free-energy imbalance (62.3) are invariant under transformations of the form

$$n \to n + n_0, \qquad \dot{n}_0 = 0,$$

$$\psi \to \psi + \psi_0, \qquad \dot{\psi}_0 = 0,$$

$$\mathbf{h} \to \mathbf{h} + \boldsymbol{\lambda} \times \operatorname{grad}\mu, \qquad \operatorname{grad}\boldsymbol{\lambda} = \mathbf{0}.$$

2. Show that, for a spatial control volume R, the free-energy imbalance has the form

$$\overline{\int_R \rho\psi\, dv}^{\,\boldsymbol{\cdot}} + \int_{\partial R} \rho\psi\mathbf{v}\cdot\mathbf{n}\, da \leq \int_{\partial R} \mathbf{Tn}\cdot\mathbf{v}\, da + \int_R \mathbf{b}\cdot\mathbf{v}\, dv - \int_{\partial R} \mu\mathbf{h}\cdot\mathbf{n}\, da + \int_R \mu h\, dv$$

or, equivalently,

$$\overline{\int_R \rho\psi\, dv}^{\,\boldsymbol{\cdot}} \leq \int_{\partial R} \mathbf{Tn}\cdot\mathbf{v}\, da + \int_R \mathbf{b}\cdot\mathbf{v}\, dv - \int_{\partial R} (\mu\mathbf{h} + \rho\psi\mathbf{v})\cdot\mathbf{n}\, da + \int_R \mu h\, dv$$

and provide a physical interpretation of the quantity $\mu\mathbf{h} + \rho\psi\mathbf{v}$.

3. Introducing the specific **grand-canonical energy**

$$\omega = \psi - n\mu, \tag{62.8}$$

show that the free-energy imbalance (62.7) is equivalent to the **grand-canonical energy imbalance**

$$\rho(\dot{\omega} + n\dot{\mu}) - \mathbf{T}:\mathbf{D} + \mathbf{h}\cdot\operatorname{grad}\mu = -\delta \leq 0. \tag{62.9}$$

63 Multiple Species

When considering a multiplicity of species, we

- use lowercase Greek superscripts to denote species labels; these range over the integers $1, 2, \dots, N$;
- do *not* use the summation convention for species labels;
- use the shorthand

$$\sum_{\alpha} = \sum_{\alpha=1}^{N}, \qquad \sum_{\alpha,\beta} = \sum_{\alpha=1}^{N}\sum_{\beta=1}^{N}, \qquad \text{and} \qquad \sum_{\alpha\neq\beta} = \sum_{\substack{\alpha=1 \\ \alpha\neq\beta}}^{N}.$$

63.1 Species Mass Balances

To generalize the foregoing framework to account for $N \geq 1$ diffusing species, we introduce a mass fraction n^α, flux $\mathbf{h}^\alpha$, and supply h^α for each species $\alpha = 1, 2, \dots, N$. The global balance (61.3) is then replaced by the requirement that

$$\overline{\int_{\mathcal{P}_t} \rho n^\alpha \, dv}^{\,\boldsymbol{\cdot}} = -\int_{\partial\mathcal{P}_t} \mathbf{h}^\alpha \cdot \mathbf{n} \, da + \int_{\mathcal{P}_t} h^\alpha \, dv \tag{63.1}$$

for each species α. Similarly, the local balance (61.5) is replaced by N balances of the form

$$\rho \dot{n}^\alpha = -\mathrm{div}\mathbf{h}^\alpha + h^\alpha. \tag{63.2}$$

When diffusion occurs through a network within the solid, the **species densities** ρn^α do not generally sum to the mass density ρ of that solid. However, when the solid is comprised entirely of the diffusing species, as occurs for a substitutional alloy,[323] the species densities must sum to the density of the solid, and it follows that the mass fractions must obey

$$\sum_\alpha n^\alpha = 1.$$

In this event,

$$\sum_\alpha \dot{n}^\alpha = 0,$$

and summing the species mass balances (63.2) over α shows that the **net species flux** $\mathbf{h}_{\mathrm{net}} \equiv \sum_\alpha \mathbf{h}^\alpha$ and the **net species supply** $h_{\mathrm{net}} \equiv \sum_\alpha h^\alpha$ must obey the auxiliary constraint

$$\mathrm{div}\mathbf{h}_{\mathrm{net}} = h_{\mathrm{net}}.$$

[323] Cf. §72.

63.2 Free-Energy Imbalance

To properly characterize the energy flow due to the transport of $N \geq 1$ species, we introduce a chemical potential μ^α for each species α. Then

$$-\int_{\partial \mathcal{P}_t} \mu^\alpha \mathbf{h}^\alpha \cdot \mathbf{n}\, da + \int_{\mathcal{P}_t} \mu^\alpha h^\alpha \, dv$$

represents the energy flow due to the transport of species α, and the *net* energy-flow due to species transport has the form

$$\mathcal{T}(\mathcal{P}_t) = \sum_\alpha \Bigg(-\int_{\partial \mathcal{P}_t} \mu^\alpha \mathbf{h}^\alpha \cdot \mathbf{n}\, da + \int_{\mathcal{P}_t} \mu^\alpha h^\alpha \, dv \Bigg). \tag{63.3}$$

As a consequence of (63.3), the global and local statements (62.3) and (62.7) of free-energy imbalance are replaced by

$$\mathcal{D}(\mathcal{P}_t) = \sum_\alpha \Bigg(-\int_{\partial \mathcal{P}_t} \mu^\alpha \mathbf{h}^\alpha \cdot \mathbf{n}\, da + \int_{\mathcal{P}_t} \mu^\alpha h^\alpha \, dv \Bigg)$$
$$+ \int_{\partial \mathcal{P}_t} \mathbf{Tn} \cdot \mathbf{v}\, da + \int_{\mathcal{P}_t} \mathbf{b}_0 \cdot \mathbf{v}\, dv - \overline{\int_{\mathcal{P}_t} \rho \psi \, dv} \geq 0 \tag{63.4}$$

and

$$\rho \dot{\psi} - \mathbf{T}\colon\mathbf{D} - \sum_\alpha (\rho \mu^\alpha \dot{n}^\alpha - \mathbf{h}^\alpha \cdot \operatorname{grad} \mu^\alpha) = -\delta \leq 0, \tag{63.5}$$

where the net density δ of the dissipation now has the form

$$\delta = \mathbf{T}\colon\mathbf{D} + \sum_\alpha (\rho \mu^\alpha \dot{n}^\alpha - \mathbf{h}^\alpha \cdot \operatorname{grad} \mu^\alpha) - \rho \dot{\psi}. \tag{63.6}$$

In classical thermodynamics, the specific free-energy of a multicomponent system is commonly viewed as a function of the specific volume, the absolute temperature, and the densities of the component species. The derivative of the specific free-energy with respect to the species density of a given component then defines the "chemical potential" of that species. An intuitively appealing interpretation of the term "chemical potential" is evoked by that definition: Namely, the "chemical potential" of a given species can be viewed as a measure of how much the free energy of a system changes with the addition of a unit measure of that species while holding fixed the specific volume, the absolute temperature, and the densities of all remaining species.

EXERCISES

1. Determine transformation rules that leave the mass balance (63.1) for each species α and the free-energy imbalance (63.4) invariant.
2. Develop the forms for the mass balance for each species α and the free-energy imbalance for a spatial control volume R.

64 Digression: The Thermodynamic Laws in the Presence of Species Transport

We now show how the ideas expressed above fit within a general thermodynamical structure in which thermal influences are taken into explicit consideration.

Consider a spatial region $\mathcal{P}_t$ convecting with the body. The balance law for energy is then the balance (26.10) of the theory without species transport augmented by (63.3) but — consistent with our omission of inertial effects — neglecting kinetic energy:

$$\overline{\int_{\mathcal{P}_t} \rho\varepsilon\, dv}^{\boldsymbol{\cdot}} = -\int_{\partial\mathcal{P}_t} \mathbf{q}\cdot\mathbf{n}\, da + \int_{\mathcal{P}_t} q\, dv + \int_{\partial\mathcal{P}_t} \mathbf{T}\mathbf{n}\cdot\mathbf{v}\, da + \int_{\mathcal{P}_t} \mathbf{b}_0\cdot\mathbf{v}\, dv$$
$$+\sum_\alpha\bigg(-\int_{\partial\mathcal{P}_t} \mu^\alpha\mathbf{h}^\alpha\cdot\mathbf{n}\, da + \int_{\mathcal{P}_t} \mu^\alpha h^\alpha\, dv\bigg). \tag{64.1}$$

Using (18.14), (60.5), the straightforward generalization

$$-\int_{\partial\mathcal{P}_t} \mu^\alpha\mathbf{h}^\alpha\cdot\mathbf{n}\, da + \int_{\mathcal{P}_t} \mu^\alpha h^\alpha\, dv = \int_{\mathcal{P}_t} (\rho\mu^\alpha\dot{n}^\alpha - \mathbf{h}^\alpha\cdot\operatorname{grad}\mu^\alpha)\, dv$$

of (62.5), and applying the divergence theorem to the term involving the heat flux, we find, as a consequence of (64.1), that

$$\int_{\mathcal{P}_t}\Big(\rho\dot{\varepsilon} - \mathbf{T}:\mathbf{D} - \sum_\alpha(\rho\mu^\alpha\dot{n}^\alpha - \mathbf{h}^\alpha\cdot\operatorname{grad}\mu^\alpha) + \operatorname{div}\mathbf{q} - q\Big)\, dv = 0; \tag{64.2}$$

since (64.2) must hold for all convecting regions $\mathcal{P}_t$, we have the local **energy balance**[324]

$$\rho\dot{\varepsilon} = \mathbf{T}:\mathbf{D} + \sum_\alpha(\rho\mu^\alpha\dot{n}^\alpha - \mathbf{h}^\alpha\cdot\operatorname{grad}\mu^\alpha) - \operatorname{div}\mathbf{q} + q. \tag{64.3}$$

As in our treatment of thermodynamics without species transport, we impose the second law via the entropy imbalance[325]

$$\overline{\int_{\mathcal{P}_t} \rho\eta\, dv}^{\boldsymbol{\cdot}} = -\int_{\partial\mathcal{P}_t} \frac{\mathbf{q}}{\vartheta}\cdot\mathbf{n}\, da + \int_{\mathcal{P}_t} \frac{q}{\vartheta}\, dv + \int_{\mathcal{P}_t} \Gamma\, dv, \qquad \Gamma \geq 0, \tag{64.4}$$

[324] Cf. (26.8).
[325] Cf. (27.11) and (27.12).

in which case the local **entropy imbalance** has the form[326]

$$\Gamma = \rho\dot{\eta} + \mathrm{div}\left(\frac{\mathbf{q}}{\vartheta}\right) - \frac{q}{\vartheta} \geq 0. \tag{64.5}$$

Then, in view of the local energy balance (64.3),

$$\begin{aligned}-\mathrm{div}\left(\frac{\mathbf{q}}{\vartheta}\right) + \frac{q}{\vartheta} &= \frac{1}{\vartheta}\Big(- \mathrm{div}\mathbf{q} + q\Big) + \frac{1}{\vartheta^2}\mathbf{q}\cdot\mathrm{grad}\,\vartheta \\ &= \frac{1}{\vartheta}\Big(- \mathbf{T}\!:\!\mathbf{D} - \sum_\alpha(\rho\mu^\alpha\dot{n}^\alpha - \mathbf{h}^\alpha\cdot\mathrm{grad}\,\mu^\alpha) + \frac{1}{\vartheta}\mathbf{q}\cdot\mathrm{grad}\,\vartheta\Big),\end{aligned}$$

and this with the local entropy imbalance (64.5) implies that[327]

$$\rho(\dot{\varepsilon} - \vartheta\dot{\eta}) - \mathbf{T}\!:\!\mathbf{D} - \sum_\alpha(\rho\mu^\alpha\dot{n}^\alpha - \mathbf{h}^\alpha\cdot\mathrm{grad}\,\mu^\alpha) + \frac{1}{\vartheta}\mathbf{q}\cdot\mathrm{grad}\,\vartheta = -\vartheta\Gamma \leq 0. \tag{64.6}$$

Finally, using the specific free energy $\psi = \varepsilon - \vartheta\eta$ introduced in (81.8) we obtain from (64.6) the local **free-energy imbalance**[328]

$$\rho(\dot{\psi} + \eta\dot{\vartheta}) - \mathbf{T}\!:\!\mathbf{D} - \sum_\alpha(\rho\mu^\alpha\dot{n}^\alpha - \mathbf{h}^\alpha\cdot\mathrm{grad}\,\mu^\alpha) + \frac{1}{\vartheta}\mathbf{q}\cdot\mathrm{grad}\,\vartheta = -\vartheta\Gamma \leq 0. \tag{64.7}$$

EXERCISES

1. Integrate (64.7) over $\mathcal{P}_t$ with the aid of (18.14) and (60.5) to show that

$$\begin{aligned}\underbrace{\int_{\mathcal{P}_t}\vartheta\Gamma\,dv}_{\substack{\text{dissipation}\\ \geq 0}} &= \underbrace{\int_{\partial\mathcal{P}_t}\mathbf{Tn}\cdot\mathbf{v}\,da + \int_{\mathcal{P}_t}\mathbf{b}_0\cdot\mathbf{v}\,dv}_{\substack{\text{conventional external}\\ \text{power expenditure}}} - \underbrace{\overline{\int_{\mathcal{P}_t}\rho\psi\,dv}^{\,\bullet}}_{\substack{\text{rate of}\\ \text{free energy}}} \\ &\quad \underbrace{+\sum_\alpha\int_{\mathcal{P}_t}(\rho\mu^\alpha\dot{n}^\alpha - \mathbf{h}^\alpha\cdot\mathrm{grad}\,\mu^\alpha)\,dv}_{\substack{\text{species production}\\ \text{of energy}}} \underbrace{- \int_{\mathcal{P}_t}\left(\rho\eta\dot{\vartheta} + \frac{1}{\vartheta}\mathbf{q}\cdot\mathrm{grad}\,\vartheta\right)dv.}_{\substack{\text{thermal production}\\ \text{of energy}}}\end{aligned} \tag{64.8}$$

2. Derive the free-energy imbalance (63.4) by specializing the general free-energy imbalance (64.8) to an isothermal process.
3. Assume that the external body force, heat supply, and species supplies vanish: $\mathbf{b}_0 \equiv \mathbf{0}$, $q \equiv 0$, $h^\alpha \equiv 0$ ($\alpha = 1, 2, \ldots, N$). Assume further that the body is isolated in the sense that, at each time
   (i) $\mathbf{Tn} = \mathbf{0}$ on a portion of $\partial\mathcal{B}_t$ and $\mathbf{v} = \mathbf{0}$ on the remainder of $\partial\mathcal{B}_t$;
   (ii) $\mathbf{q}\cdot\mathbf{n} = 0$ on $\partial\mathcal{B}_t$ and $\mathbf{h}^\alpha\cdot\mathbf{n} = 0$ on $\partial\mathcal{B}_t$ for each species α.
   Show that the net amount of each species α and the net energy are constant, while the net entropy cannot decrease.
4. Assume that the external body force, heat supply, and species supplies vanish: $\mathbf{b}_0 \equiv \mathbf{0}$, $q \equiv 0$, $h^\alpha \equiv 0$ ($\alpha = 1, 2, \ldots, N$). Assume also that there are a constant pressure p_0, a constant temperature $\vartheta_0 > 0$, and — for each species α — a

[326] Cf. (27.14).
[327] Cf. (27.16).
[328] Cf. (27.18).

constant chemical potential μ_0^α such that:
(i) $\mathbf{Tn} = -p_0\mathbf{n}$ on a portion of $\partial\mathcal{B}_t$ and $\mathbf{v} = \mathbf{0}$ on the remainder of $\partial\mathcal{B}_t$;
(ii) $\vartheta = \vartheta_0$ on a portion of $\partial\mathcal{B}_t$ and $\mathbf{q}\cdot\mathbf{n} = 0$ on the remainder of $\partial\mathcal{B}_t$;
(iii) $\mu^\alpha = \mu_0^\alpha$ on a portion of $\partial\mathcal{B}_t$ and $\mathbf{h}^\alpha\cdot\mathbf{n} = 0$ on the remainder of $\partial\mathcal{B}_t$.
Show, in addition, that the first and second laws reduce to the decay inequality

$$\overline{\int\limits_{\mathcal{B}_t} \rho\left(\varepsilon - \vartheta_0\eta + p_0 \upsilon - \sum_\alpha \mu_0^\alpha \dot{n}^\alpha\right) dv}^{\;\boldsymbol{\cdot}} \leq 0$$

and provide an interpretation of the contribution $-\rho\mu_0^\alpha\dot{n}^\alpha$ to the integrand.

65 Referential Laws

65.1 Single Species

In this section, we determine referential counterparts of the species balance, free-energy imbalance, and associated relations discussed in §61 and §62.

We define the species density n_{R},[329] referential flux $\mathbf{h}_{\mathrm{R}}$, and referential supply h_{R} by

$$n_{\mathrm{R}} = \rho_{\mathrm{R}} n, \qquad \mathbf{h}_{\mathrm{R}} = J\mathbf{F}^{-1}\mathbf{h}, \qquad \text{and} \qquad h_{\mathrm{R}} = Jh, \tag{65.1}$$

so that — for P a fixed subregion of the undeformed body B and $\mathcal{P}_t = \boldsymbol{\chi}_t(\mathrm{P})$ the corresponding convecting subregion of the deformed body $\mathcal{B}_t$ — (31.4) yields

$$\left.\begin{aligned}
\int\limits_{\mathcal{P}_t} \rho n\, dv &= \int\limits_{\mathrm{P}} n_{\mathrm{R}}\, dv_{\mathrm{R}},\\
\int\limits_{\partial\mathcal{P}_t} \mathbf{h}\cdot\mathbf{n}\, da &= \int\limits_{\partial\mathrm{P}} \mathbf{h}_{\mathrm{R}}\cdot\mathbf{n}_{\mathrm{R}}\, da_{\mathrm{R}},\\
\int\limits_{\mathcal{P}_t} h\, dv &= \int\limits_{\mathrm{P}} h_{\mathrm{R}}\, dv_{\mathrm{R}},\\
\int\limits_{\partial\mathcal{P}_t} \mu\mathbf{h}\cdot\mathbf{n}\, da &= \int\limits_{\partial\mathrm{P}} \mu\mathbf{h}_{\mathrm{R}}\cdot\mathbf{n}_{\mathrm{R}}\, da_{\mathrm{R}},\\
\int\limits_{\mathcal{P}_t} \mu h\, dv &= \int\limits_{\mathrm{P}} \mu h_{\mathrm{R}}\, dv_{\mathrm{R}}.
\end{aligned}\right\} \tag{65.2}$$

The species density n_{R} and the species supply h_{R} are therefore measured per unit volume in the reference body B, while the species flux $\mathbf{h}_{\mathrm{R}}$ is measured per unit area, also in B. Because $\mathbf{h}_{\mathrm{R}}$ arises via an inner product $\mathbf{h}_{\mathrm{R}}\cdot\mathbf{n}_{\mathrm{R}}$ with the material vector field $\mathbf{n}_{\mathrm{R}}$, $\mathbf{h}_{\mathrm{R}}$ is itself a material vector field. Note that the chemical potential μ, not being a density, is invariant under the spatial to material transformations $(65.2)_{4,5}$.[330]

[329] Note that n_{R} is actually a density rather than a mass fraction.
[330] Cf. our treatment of the temperature ϑ in §31.

Recalling (24.2) and (24.5), we may express the force balance $(60.2)_1$ referentially as

$$\int_{\partial P} \mathbf{T}_R \mathbf{n}_R \, da_R + \int_P \mathbf{b}_{0R} \, dv_R = \mathbf{0} \tag{65.3}$$

and, since the material region P is arbitrary, we have the referential force balance

$$\mathrm{Div}\,\mathbf{T}_R + \mathbf{b}_{0R} = \mathbf{0}. \tag{65.4}$$

Moreover, by (24.1) and $(60.3)_2$, the moment balance shows that, as is the case when inertia is taken into account,

$$\mathbf{T}_R \mathbf{F}^\top = \mathbf{F} \mathbf{T}_R^\top. \tag{65.5}$$

Next, using (65.2) we may rewrite the species mass balance (61.3) as

$$\overline{\int_P n_R \, dv_R}^{\boldsymbol{\cdot}} = -\int_{\partial P} \mathbf{h}_R \cdot \mathbf{n}_R \, da_R + \int_P h_R \, dv_R \tag{65.6}$$

and the free-energy imbalance (62.3) as[331]

$$\overline{\int_P \psi_R \, dv_R}^{\boldsymbol{\cdot}} \le \int_{\partial P} \mathbf{T}_R \mathbf{n}_R \cdot \dot{\boldsymbol{\chi}} \, da_R + \int_P \mathbf{b}_{0R} \cdot \dot{\boldsymbol{\chi}} \, dv_R - \int_{\partial P} \mu \mathbf{h}_R \cdot \mathbf{n}_R \, da_R + \int_P \mu h_R \, dv_R. \tag{65.7}$$

Next,

$$\overline{\int_P n_R \, dv_R}^{\boldsymbol{\cdot}} = \int_P \dot{n}_R \, dv_R \qquad \text{and} \qquad \overline{\int_P \psi_R \, dv_R}^{\boldsymbol{\cdot}} = \int_P \dot{\psi}_R \, dv_R$$

for all material regions P. Thus, if in (65.6) we apply the divergence theorem to the term involving the species flux, we obtain the local **species mass balance**

$$\boxed{\dot{n}_R = -\mathrm{Div}\,\mathbf{h}_R + h_R;} \tag{65.8}$$

further, the divergence theorem and (65.8) yield

$$\begin{aligned} -\int_{\partial P} \mu \mathbf{h}_R \cdot \mathbf{n} \, da_R &= -\int_P (\mu \mathrm{Div}\,\mathbf{h}_R + \mathbf{h}_R \cdot \nabla\mu) \, dv_R \\ &= \int_P (\mu \dot{n}_R - \mathbf{h}_R \cdot \nabla\mu - \mu h_R) \, dv_R, \end{aligned}$$

so that, by (65.7), we have the local **free-energy imbalance**[332]

$$\boxed{\dot{\psi}_R - \mu \dot{n}_R - \mathbf{T}_R : \dot{\mathbf{F}} + \mathbf{h}_R \cdot \nabla\mu = -\delta_R \le 0.} \tag{65.9}$$

We recall that, by (9.2), the referential chemical-potential gradient $\nabla\mu$ appearing in (65.9) is related to its spatial counterpart $\mathrm{grad}\,\mu$ via

$$\nabla\mu = \mathbf{F}^\top \mathrm{grad}\,\mu. \tag{65.10}$$

331 Cf. (31.7).
332 Cf. (31.24).

65.2 Multiple Species

For multiple species, definitions analogous to (65.1) are introduced for each species label $\alpha = 1, 2, \dots, N$:

$$n_R^\alpha = \rho_R n^\alpha, \qquad \mathbf{h}_R^\alpha = J\mathbf{F}^{-1}\mathbf{h}^\alpha, \qquad \text{and} \qquad h_R^\alpha = J h^\alpha. \tag{65.11}$$

Proceeding as in the development of (65.6) and (65.7), we then have the mass balance

$$\overline{\int_P n_R^\alpha \, dv_R}^{\,\cdot} = -\int_{\partial P} \mathbf{h}_R^\alpha \cdot \mathbf{n}_R \, da_R + \int_P h_R^\alpha \, dv_R \tag{65.12}$$

for species α along with the free-energy imbalance

$$\overline{\int_P \psi_R \, dv_R}^{\,\cdot} \le \int_{\partial P} \mathbf{T}_R \mathbf{n}_R \cdot \dot{\boldsymbol{\chi}} \, da_R + \int_P \mathbf{b}_{0R} \cdot \dot{\boldsymbol{\chi}} \, dv_R + \sum_\alpha \left(-\int_{\partial P} \mu^\alpha \mathbf{h}_R^\alpha \cdot \mathbf{n}_R \, da_R + \int_P \mu^\alpha h_R^\alpha \, dv_R \right), \tag{65.13}$$

the corresponding local versions of which are

$$\boxed{\dot{n}_R^\alpha = -\mathrm{Div}\, \mathbf{h}_R^\alpha + h_R^\alpha} \tag{65.14}$$

and

$$\boxed{\dot{\psi}_R - \mathbf{T}_R : \dot{\mathbf{F}} - \sum_\alpha (\mu^\alpha \dot{n}_R^\alpha - \mathbf{h}_R^\alpha \cdot \nabla \mu^\alpha) = -\delta_R \le 0.} \tag{65.15}$$

EXERCISES

1. Using (62.8) along with the definitions of ψ_R and n_R to define the **grand-canonical energy**
$$\omega_R = \psi_R - \mu n_R, \tag{65.16}$$
derive the referential counterpart
$$\dot{\omega}_R + n_R \dot{\mu} - \mathbf{T}_R : \dot{\mathbf{F}} + \mathbf{h}_R \cdot \nabla \mu = -\delta_R \le 0 \tag{65.17}$$
of the **grand-canonical energy imbalance** (62.9).
2. Generalize (65.17) to the case of N species.

66 Constitutive Theory for a Single Species

Guided by (65.15), we assume that the free energy ψ_R, the Piola stress $\mathbf{T}_R$, and the chemical potential μ are determined by constitutive equations of the form

$$\left.\begin{aligned} \psi_R &= \hat{\psi}_R(\mathbf{F}, n_R), \\ \mathbf{T}_R &= \hat{\mathbf{T}}_R(\mathbf{F}, n_R), \\ \mu &= \hat{\mu}(\mathbf{F}, n_R). \end{aligned}\right\} \tag{66.1}$$

Further, we assume that the species flux $\mathbf{h}$ is given by a constitutive equation of the form

$$\mathbf{h}_R = \hat{\mathbf{h}}_R(\mathbf{F}, n_R, \nabla\mu). \tag{66.2}$$

In view of $(66.1)_3$ the chemical-potential gradient $\nabla\mu$ depends on the gradients $\nabla\mathbf{F}$ and ∇n_R. Hence, by (66.2), the species flux $\mathbf{h}_R$ depends not only on $\mathbf{F}$ and n_R but also on the gradients $\nabla\mathbf{F}$ and ∇n_R. The constitutive equations (66.1) and (66.2) therefore violate the principle of equipresence discussed in the paragraph in petite type on page 231.

The decision to violate equipresence is not capricious. The pairing of $\mathbf{h}_R$ with $\nabla\mu$ in the imbalance (65.9) renders $\nabla\mu$ the variable of primary importance in a constitutive equation for $\mathbf{h}_R$.[333]

An alternative approach consistent with equipresence involves rewriting the free-energy imbalance using the grand-canonical energy $\omega_R = \psi_R - \mu n_R$ as in (65.17) and choosing as independent variables the list $(\mathbf{F}, \mu, \nabla\mu)$, as discussed in the exercise on page 390. However, while most theories of the type considered here are consistent with a dependence of μ on n_R, there are situations of great importance involving phase transformations for which n_R *cannot be expressed as a function of* μ.[334]

66.1 Consequences of Frame-Indifference

First of all, recall that the Cauchy and Piola stresses $\mathbf{T}$ and $\mathbf{T}_R$ and the second Piola stress $\mathbf{T}_{RR}$ are related by[335]

$$\begin{aligned} \mathbf{T} &= J^{-1}\mathbf{F}\mathbf{T}_{RR}\mathbf{F}^{\top}, \\ \mathbf{T}_R &= \mathbf{F}\mathbf{T}_{RR}. \end{aligned} \tag{66.3}$$

Being scalar fields, the species density n_R and chemical potential μ are, like the free energy ψ_R, invariant under changes of frame:

$$\psi_R^* = \psi_R, \qquad n_R^* = n_R, \qquad \mu^* = \mu.$$

333 Cf. the bullet on page 382.
334 Cf. Fried & Gurtin (1999, 2004).
335 Cf. (47.7) and (56.5).

Further, as is clear from the bullet on page 147, the species flux $\mathbf{h}_R$ and chemical-potential gradient $\nabla\mu$, being material vector fields, are also invariant under a change of frame:

$$\mathbf{h}_R^* = \mathbf{h}_R, \qquad \nabla\mu^* = \nabla\mu.$$

Thus, since by $(47.11)_1$ and (47.12) the deformation gradient and Piola stress transform according to

$$\mathbf{F}^* = \mathbf{QF} \qquad \text{and} \qquad \mathbf{T}_R^* = \mathbf{QT}_R,$$

the response functions $\hat{\psi}_R$, $\hat{\mathbf{T}}_R$, $\hat{\mu}$, and $\hat{\mathbf{h}}_R$ must satisfy

$$\begin{aligned}\hat{\psi}_R(\mathbf{F}, n_R) &= \hat{\psi}_R(\mathbf{QF}, n_R),\\ \hat{\mathbf{T}}_R(\mathbf{F}, n_R) &= \mathbf{Q}^\top\hat{\mathbf{T}}_R(\mathbf{QF}, n_R),\\ \hat{\mu}(\mathbf{F}, n_R) &= \hat{\mu}(\mathbf{QF}, n_R),\\ \hat{\mathbf{h}}_R(\mathbf{QF}, n_R, \nabla\mu) &= \hat{\mathbf{h}}_R(\mathbf{F}, n_R, \nabla\mu),\end{aligned} \tag{66.4}$$

for all rotations $\mathbf{Q}$ and all $\mathbf{F}$ and n_R.

Arguing as in §48.1, we conclude from (66.4) that there are response functions $\bar{\psi}_R$, $\bar{\mathbf{T}}_{RR}$, $\bar{\mu}$, and $\bar{\mathbf{h}}_R$ such that[336]

$$\begin{aligned}\psi_R &= \bar{\psi}_R(\mathbf{C}, n_R),\\ \mathbf{T}_R &= \mathbf{F}\bar{\mathbf{T}}_{RR}(\mathbf{C}, n_R),\\ \mu &= \bar{\mu}^\alpha(\mathbf{C}, n_R),\\ \mathbf{h}_R &= \bar{\mathbf{h}}_R(\mathbf{C}, n_R, \nabla\mu),\end{aligned} \tag{66.5}$$

so that, by $(66.3)_2$ and $(66.5)_2$,

$$\mathbf{T}_{RR} = \bar{\mathbf{T}}_{RR}(\mathbf{C}, n_R). \tag{66.6}$$

66.2 Thermodynamic Restrictions

We now apply the Coleman–Noll procedure to the frame-indifferent constitutive equations (66.5). For the setting at hand, a constitutive process consists of a motion χ and a species density n_R along with the fields ψ_R, $\mathbf{T}_R$, μ, and $\mathbf{h}_R$ determined through the constitutive equations (66.5). Arguing as in §46.9 and bearing in mind that inertia is neglected, we assume that the conventional body force

$$\mathbf{b}_{0R} = -\mathrm{Div}\,\mathbf{T}_R$$

and the external species supply

$$h_R = \dot{n}_R + \mathrm{Div}\,\mathbf{h}_R$$

needed to support the process are arbitrarily assignable; hence, the linear momentum and species mass balances in no way restrict the class of constitutive processes. Arguing as in §48.2, we assume that the supplies are arbitrary and therefore require that all constitutive processes be consistent with (65.15).

Consider an arbitrary constitutive process. By $(66.5)_1$,

$$\dot{\psi}_R = \frac{\partial\bar{\psi}_R(\mathbf{C}, n_R)}{\partial\mathbf{C}}:\dot{\mathbf{C}} + \frac{\partial\bar{\psi}_R(\mathbf{C}, n_R)}{\partial n_R}\dot{n}_R; \tag{66.7}$$

[336] Cf. (48.7).

further, by (25.6)$_3$ and (66.6),

$$\mathbf{T}_R:\dot{\mathbf{F}} = \tfrac{1}{2}\mathbf{T}_{RR}:\dot{\mathbf{C}}$$
$$= \tfrac{1}{2}\bar{\mathbf{T}}_{RR}(\mathbf{C}, n_R):\dot{\mathbf{C}}. \qquad (66.8)$$

In view of (66.7) and (66.8), the free-energy imbalance (65.9) is equivalent to the requirement that the inequality

$$\left(\frac{\partial\bar{\psi}_R(\mathbf{C}, n_R)}{\partial\mathbf{C}} - \tfrac{1}{2}\bar{\mathbf{T}}_{RR}(\mathbf{C}, n_R)\right):\dot{\mathbf{C}} + \left(\frac{\partial\bar{\psi}_R(\mathbf{C}, n_R)}{\partial n_R} - \bar{\mu}(\mathbf{C}, n_R)\right)\dot{n}_R$$
$$+ \bar{\mathbf{h}}_R(\mathbf{C}, n_R, \nabla\mu)\cdot\nabla\mu \le 0 \qquad (66.9)$$

be satisfied for all constitutive processes. If, for the moment, we restrict attention to processes in which $\mathbf{C}$ and n_R are spatially uniform, but time-dependent, then the inequality (66.9) reduces to

$$\underbrace{\left(\frac{\partial\bar{\psi}_R(\mathbf{C}, n_R)}{\partial\mathbf{C}} - \tfrac{1}{2}\bar{\mathbf{T}}_{RR}(\mathbf{C}, n_R)\right)}_{(*)}:\dot{\mathbf{C}} + \underbrace{\left(\frac{\partial\bar{\psi}_R(\mathbf{C}, n_R)}{\partial n_R} - \bar{\mu}(\mathbf{C}, n_R)\right)}_{(**)}\dot{n}_R \le 0 \qquad (66.10)$$

and must hold for all $\mathbf{C}(t)$ and $n_R(t)$. Proceeding as in the verification of (†) on p. 279, one can show that[337]

(‡) given any point of the body and any time, it is possible to find a constitutive process such that $\mathbf{C}$, $\dot{\mathbf{C}}$, n_R, and $\dot{n}_R$ have arbitrarily prescribed values at that point and time.

Granted this assertion the coefficients (∗) and (∗∗) of $\dot{\mathbf{C}}$ and $\dot{n}_R$ in (66.10) must vanish, for otherwise $\dot{\mathbf{C}}$ and $\dot{n}$ may be chosen to violate (66.10). Hence, it follows that $\mathbf{T}_{RR} = 2\partial\bar{\psi}_R/\partial\mathbf{C}$ and $\mu = \partial\bar{\psi}_R/\partial n_R$. Moreover, these results reduce (66.9) to the residual inequality

$$\bar{\mathbf{h}}_R(\mathbf{C}, n_R, \nabla\mu)\cdot\nabla\mu \le 0.$$

We therefore have the following **thermodynamic restrictions**:

(i) *The free energy determines the second Piola stress and the chemical potential through the* **stress relation**

$$\mathbf{T}_{RR} = \bar{\mathbf{T}}_{RR}(\mathbf{C}, n_R)$$
$$= 2\frac{\partial\bar{\psi}_R(\mathbf{C}, n_R)}{\partial\mathbf{C}} \qquad (66.11)$$

and the **chemical-potential relation**

$$\mu = \bar{\mu}(\mathbf{C}, n_R)$$
$$= \frac{\partial\bar{\psi}_R(\mathbf{C}, n_R)}{\partial n_R}. \qquad (66.12)$$

(ii) *The species flux satisfies the* **species-transport inequality**

$$\bar{\mathbf{h}}_R(\mathbf{C}, n_R, \nabla\mu)\cdot\nabla\mu \le 0 \qquad (66.13)$$

for all $(\mathbf{C}, n_R, \nabla\mu)$.

[337] The assertions regarding $\mathbf{C}$ and $\dot{\mathbf{C}}$ are verified on page 281; those regarding n_R and $\dot{n}_R$ are left as an exercise.

66.3 Consequences of the Thermodynamic Restrictions

By (66.7), (66.11), and (66.12), we have the **Gibbs relation**

$$\dot{\psi}_{\rm R} = \tfrac{1}{2}\mathbf{T}_{\rm RR} : \dot{\mathbf{C}} + \mu \dot{n}_{\rm R}, \tag{66.14}$$

while (66.11) and (66.12) yield the **Maxwell relation**

$$\frac{\partial \bar{\mathbf{T}}_{\rm RR}(\mathbf{C}, n_{\rm R})}{\partial n_{\rm R}} = 2\frac{\partial \bar{\mu}(\mathbf{C}, n_{\rm R})}{\partial \mathbf{C}}. \tag{66.15}$$

Let $\mathbf{C}(t)$ be a time-dependent right Cauchy–Green tensor, let $n_{\rm R}(t)$ be a time-dependent species density, and write

$$\begin{aligned} \mathbf{T}_{\rm RR}(t) &= \bar{\mathbf{T}}_{\rm RR}(\mathbf{C}(t), n_{\rm R}(t)) \\ &= 2\frac{\partial \bar{\psi}_{\rm R}(\mathbf{G}, n_{\rm R}(t))}{\partial \mathbf{G}}\bigg|_{\mathbf{G}=\mathbf{C}(t)}. \end{aligned}$$

The chain-rule then yields a relation,

$$\dot{\mathbf{T}}_{\rm RR} = 2\frac{\partial \bar{\mathbf{T}}_{\rm RR}(\mathbf{C}, n_{\rm R})}{\partial \mathbf{C}}\dot{\mathbf{C}} + \frac{\partial \bar{\mathbf{T}}_{\rm RR}(\mathbf{C}, n_{\rm R})}{\partial n_{\rm R}}\dot{n}_{\rm R},$$

suggesting the introduction of two constitutive moduli: the **elasticity tensor** $\mathbb{C}(\mathbf{C}, n_{\rm R})$ defined by

$$\begin{aligned} \mathbb{C}(\mathbf{C}, n_{\rm R}) &= 2\frac{\partial \bar{\mathbf{T}}_{\rm RR}(\mathbf{C}, n_{\rm R})}{\partial \mathbf{C}} \\ &= 4\frac{\partial^2 \bar{\psi}_{\rm R}(\mathbf{C}, n_{\rm R})}{\partial \mathbf{C}^2}; \end{aligned} \tag{66.16}$$

the **chemistry-strain tensor** $\mathbf{S}(\mathbf{C}, n_{\rm R})$ defined by

$$\begin{aligned} \mathbf{S}(\mathbf{C}, n_{\rm R}) &= \frac{\partial \bar{\mathbf{T}}_{\rm RR}(\mathbf{C}, n_{\rm R})}{\partial n_{\rm R}} \\ &= 2\frac{\partial \bar{\mu}(\mathbf{C}, n_{\rm R})}{\partial \mathbf{C}} \\ &= 2\frac{\partial^2 \bar{\psi}_{\rm R}(\mathbf{C}, n_{\rm R})}{\partial \mathbf{C}\partial n_{\rm R}}. \end{aligned} \tag{66.17}$$

The elasticity tensor $\mathbb{C}(\mathbf{C}, n_{\rm R})$ is a linear transformation of symmetric tensors to symmetric tensors and, hence, a fourth-order tensor; that is, $\mathbb{C}(\mathbf{C}, n_{\rm R})$ associates with each symmetric tensor $\mathbf{G}$ a symmetric tensor $\mathbb{C}(\mathbf{C}, n_{\rm R})\mathbf{G}$. Moreover, the chemistry-strain tensor $\mathbf{S}(\mathbf{C}, n_{\rm R})$ is a symmetric tensor that measures the marginal change in stress due to an increment of the species density $n_{\rm R}$ with $\mathbf{C}$ held fixed.

Proceeding as above, let

$$\begin{aligned} \mu(t) &= \bar{\mu}^{\alpha}(\mathbf{C}(t), n_{\rm R}(t)) \\ &= \frac{\partial \bar{\psi}_{\rm R}(\mathbf{G}, n_{\rm R}(t))}{\partial \mathbf{G}}\bigg|_{\mathbf{G}=\mathbf{C}(t)}. \end{aligned}$$

Then, by the chain-rule, (66.15), and (66.17),

$$\dot{\mu} = \frac{\partial \bar{\mu}(\mathbf{C}, n_{\mathrm{R}})}{\partial \mathbf{C}} : \dot{\mathbf{C}} + \frac{\partial \bar{\mu}^{\alpha}(\mathbf{C}, n_{\mathrm{R}})}{\partial n_{\mathrm{R}}} \dot{n}_{\mathrm{R}}$$

$$= \tfrac{1}{2}\mathbf{S}(\mathbf{C}, n_{\mathrm{R}}) : \dot{\mathbf{C}} + \frac{\partial^2 \bar{\psi}_{\mathrm{R}}(\mathbf{C}, n_{\mathrm{R}})}{\partial n_{\mathrm{R}}^2} \dot{n}_{\mathrm{R}}.$$

This suggests the introduction of a **chemistry modulus** $\Lambda(\mathbf{C}, n_{\mathrm{R}})$ defined by

$$\Lambda(\mathbf{C}, n_{\mathrm{R}}) = \frac{\partial^2 \bar{\psi}_{\mathrm{R}}(\mathbf{C}, n_{\mathrm{R}})}{\partial n_{\mathrm{R}}^2}. \tag{66.18}$$

In the theory of thermoelasticity, we found that important consequences the heat conduction inequality (57.9) are that (i) heat flows from hot to cold and (ii) a deformation, no matter how large, cannot induce a flow of heat in the absence of a thermal gradient. We now show that the species-transport inequality (66.13) has analogous consequences.

First, to arrive at the analog of (i), assume that at some material point $\mathbf{X}_0$ and time (which we suppress as an argument)

$$\nabla\mu(\mathbf{X}_0) \neq \mathbf{0}$$

and define

$$\mathbf{e} = \frac{\nabla\mu(\mathbf{X}_0)}{|\nabla\mu(\mathbf{X}_0)|}. \tag{66.19}$$

Then $\mathbf{e} \cdot \nabla\mu(\mathbf{X}_0) = |\nabla\mu(\mathbf{X}_0)|$ so that $\mu(\mathbf{X}_0 + \ell\mathbf{e}) = \mu(\mathbf{X}_0) + \ell|\nabla\mu(\mathbf{X}_0)| + o(\ell)$ and, for all sufficiently small $\ell > 0$,

$$\mu(\mathbf{X}_0 + \ell\mathbf{e}) > \mu(\mathbf{X}_0);$$

the point $\mathbf{X}_0 + \ell\mathbf{e}$ therefore has higher chemical potential than the point $\mathbf{X}_0$. Further, for $\mathbf{h}_{\mathrm{R}}(\mathbf{X})$ the species flux corresponding to the fields $\mathbf{C}(\mathbf{X})$ and $n_{\mathrm{R}}(\mathbf{X})$, the species-transport inequality and (66.19) imply that

$$\begin{aligned} 0 &\geq \mathbf{h}_{\mathrm{R}}(\mathbf{X}_0) \cdot \nabla\mu(\mathbf{X}_0) \\ &= (\mathbf{h}_{\mathrm{R}}(\mathbf{X}_0) \cdot \mathbf{e})\underbrace{(\nabla\mu(\mathbf{X}_0) \cdot \mathbf{e})}_{>0}, \end{aligned}$$

so that $\mathbf{h}_{\mathrm{R}}(\mathbf{X}_0) \cdot \mathbf{e} \leq 0$. The component of $\mathbf{h}_{\mathrm{R}}(\mathbf{X}_0)$ in the direction $-\mathbf{e}$, which is the unit vector that represents the direction from the point $\mathbf{X}_0 + \ell\mathbf{e}$ of higher chemical potential to the point $\mathbf{X}_0$ of lower chemical potential, must therefore be nonnegative. In this sense, species flow occurs from the point with higher chemical potential to the point with lower chemical potential; thus, we have the classical result

- *for a single-component system, species transport occurs down a chemical-potential gradient.*

Next, to arrive at the analog of (ii), define

$$\varphi(\mathbf{C}, n_{\mathrm{R}}, \mathbf{p}) = \bar{\mathbf{h}}_{\mathrm{R}}(\mathbf{C}, n_{\mathrm{R}}, \mathbf{p}) \cdot \mathbf{p};$$

then, since by (66.13) $\varphi(\mathbf{C}, n_{\mathrm{R}}, \mathbf{p}) \leq 0$ and $\varphi(\mathbf{C}, n_{\mathrm{R}}, \mathbf{0}) = 0$,

$$\varphi(\mathbf{C}, n_{\mathrm{R}}, \mathbf{p}) \text{ has a maximum at } \mathbf{p} = \mathbf{0}. \tag{66.20}$$

Thus,

$$\frac{\partial \varphi(\mathbf{C}, n_{\mathrm{R}}, \mathbf{p})}{\partial \mathbf{p}} = \bar{\mathbf{h}}_{\mathrm{R}}(\mathbf{C}, n_{\mathrm{R}}, \mathbf{p}) + \left(\frac{\partial \bar{\mathbf{h}}_{\mathrm{R}}(\mathbf{C}, n_{\mathrm{R}}, \mathbf{p})}{\partial \mathbf{p}}\right)^{\!\top} \mathbf{p} \tag{66.21}$$

vanishes at $\mathbf{p} = \mathbf{0}$; hence, the species flux vanishes when the chemical-potential gradient $\mathbf{p} = \nabla\mu$ vanishes, independent of the values of the right Cauchy–Green tensor $\mathbf{C}$ and the species density n_R:

$$\mathbf{h}_R = \mathbf{0} \qquad \text{whenever} \qquad \nabla\mu = \mathbf{0}. \tag{66.22}$$

We, therefore, conclude that

- *for a single-component system, a deformation, no matter how large, cannot induce a flow of species in the absence of a chemical-potential gradient,*

a result that might be referred to as the **absence of a piezo-diffusive effect.**

66.4 Fick's Law

Within the present context, **Fick's law** is the assertion that the species flux $\mathbf{h}_R$ depends linearly on the chemical-potential gradient $\nabla\mu$:

$$\mathbf{h}_R = -\mathbf{M}(\mathbf{C}, n_R)\nabla\mu, \tag{66.23}$$

with $\mathbf{M}(\mathbf{C}, n_R)$ the **mobility tensor**;[338] a consequence of the species-transport inequality (66.13) is then that the mobility tensor is positive-semidefinite:

$$\mathbf{p}\cdot\mathbf{M}(\mathbf{C}, n_R)\mathbf{p} \geq 0 \tag{66.24}$$

for all $\mathbf{p}$.

We assume henceforth that the species flux is determined via Fick's law.

The dependence of $\mathbf{h}_R$ on $\nabla\mu$ renders $\mathbf{h}_R$ a linear function of the gradients of $\mathbf{C}$ and n_R; we now determine the explicit form of these dependencies. By (66.12), (66.17), and (66.18),

$$\begin{aligned}
\frac{\partial\mu}{\partial X_j} &= \frac{\partial}{\partial X_j}\left(\frac{\partial\bar{\psi}_R(\mathbf{C}, n_R)}{\partial n_R}\right) \\
&= \frac{\partial^2\bar{\psi}_R(\mathbf{C}, n_R)}{\partial C_{kl}\partial n_R}\frac{\partial C_{kl}}{\partial X_j} + \frac{\partial^2\bar{\psi}_R(\mathbf{C}, n_R)}{\partial n_R^2}\frac{\partial n_R}{\partial X_j} \\
&= \tfrac{1}{2}S_{kl}(\mathbf{C}, n_R)\frac{\partial C_{kl}}{\partial X_j} + \frac{\partial^2\bar{\psi}_R(\mathbf{C}, n_R)}{\partial n_R^2}\frac{\partial n_R}{\partial X_j} \\
&= (\tfrac{1}{2}\mathbf{S}(\mathbf{C}, n_R)\colon\nabla\mathbf{C} + \Lambda(\mathbf{C}, n_R)\nabla n_R)_j,
\end{aligned}$$

where, for $\mathbf{A}$ a second-order tensor with components A_{ij} and $\mathbf{K}$ a third-order tensor with components K_{ijk}, $\mathbf{A}\colon\mathbf{K}$ is the vector with k-th component

$$(\mathbf{A}\colon\mathbf{K})_k = A_{ij}K_{ijk}. \tag{66.25}$$

Fick's law (70.14) for the flux of species α therefore becomes

$$\mathbf{h}_R = -\mathbf{M}(\mathbf{C}, n_R)(\Lambda(\mathbf{C}, n_R)\nabla n_R + \tfrac{1}{2}\mathbf{S}(\mathbf{C}, n_R)\colon\nabla\mathbf{C}) \tag{66.26}$$

or, using components,

$$(\mathbf{h}_R)_i = -M_{ij}(\mathbf{C}, n_R)\left(\Lambda(\mathbf{C}, n_R)\frac{\partial n_R}{\partial X_j} + \tfrac{1}{2}S^{\beta}_{kl}(\mathbf{C}, n_R)\frac{\partial C_{kl}}{\partial X_j}\right). \tag{66.27}$$

[338] Actually one can show that the most general relation of the form $\mathbf{h} = \bar{\mathbf{h}}_R(\mathbf{C}, n_R, \nabla\mu)$ consistent with the species-transport inequality (66.13) must necessarily have the specific form

$$\mathbf{h} = -\mathbf{M}(\mathbf{C}, n_R, \nabla\mu)\nabla\mu;$$

cf. the equation before (9-23) of GURTIN (2000a, p. 8).

Species diffusion may therefore be driven not only by spatial variations of the species density but also by spatial variations of the right Cauchy–Green tensor $\mathbf{C}$.

EXERCISES

1. Use (66.3) and (66.6) to derive an auxiliary constitutive equation for the Cauchy stress.
2. Combining $(65.11)_2$ and the invariance of $\mathbf{h}_R$ under a change of frame, show that the spatial species flux $\mathbf{h}$ is frame-indifferent:

$$\mathbf{h}^* = \mathbf{Q}\mathbf{h}$$

for all rotations $\mathbf{Q}$.
3. Assume that the grand-canonical energy ω_R, the stress $\mathbf{T}_{RR}$, the species density n_R, and the species flux $\mathbf{h}_R$ are given constitutive equations of the form

$$\begin{aligned} \omega_R &= \breve{\omega}_R(\mathbf{C}, \mu, \nabla\mu), \\ \mathbf{T}_{RR} &= \breve{\mathbf{T}}_{RR}(\mathbf{C}, \mu, \nabla\mu), \\ n_R &= \breve{n}_R(\mathbf{C}, \mu, \nabla\mu), \\ \mathbf{h}_R &= \breve{\mathbf{h}}_R(\mathbf{C}, \mu, \nabla\mu), \end{aligned} \tag{66.28}$$

and use the grand-canonical energy imbalance (65.17) to show that

$$\left.\begin{aligned} \omega_R &= \breve{\omega}_R(\mathbf{C}, \mu), \\ \mathbf{T}_{RR} = \breve{\mathbf{T}}_{RR}(\mathbf{C}, \mu) &= 2\frac{\partial \breve{\omega}_R(\mathbf{C}, \mu)}{\partial \mathbf{C}}, \\ n_R = \breve{n}_R(\mathbf{C}, \mu) &= -\frac{\partial \breve{\omega}_R(\mathbf{C}, \mu)}{\partial \mu}, \end{aligned}\right\} \tag{66.29}$$

and

$$\breve{\mathbf{h}}_R(\mathbf{C}, \mu, \nabla\mu) \cdot \nabla\mu \le 0. \tag{66.30}$$

4. Verify the assertions regarding n_R and $\dot{n}_R$ in (‡) on page 379.
5. Derive the **gradient Gibbs relation**

$$\nabla\psi_R = \tfrac{1}{2}\mathbf{T}_{RR} : \nabla\mathbf{C} + \mu\nabla n_R,$$

where $\mathbf{T}_{RR} : \nabla\mathbf{C}$ is the vector with components $(\mathbf{T}_{RR})_{jk}\partial C_{jk}/\partial X_i$.
6. Establish the Maxwell relation (70.10) showing all steps.
7. Show that the Maxwell relations (70.10) can be expressed alternatively as

$$\frac{\partial \tilde{\mathbf{T}}_{RR}(\mathbf{E}, n_R)}{\partial n_R} = \frac{\partial \tilde{\mu}(\mathbf{E}, n_R)}{\partial \mathbf{E}},$$

where $\tilde{\mathbf{T}}_{RR}(\mathbf{E}, n_R)$ and $\tilde{\mu}(\mathbf{E}, n_R)$ are determined by $\tilde{\psi}_R(\mathbf{E}, n_R) = \bar{\psi}_R(\mathbf{1} + 2\mathbf{E}, n_R)$ via

$$\tilde{\mathbf{T}}_{RR}(\mathbf{E}, n_R) = \frac{\partial \tilde{\psi}_R(\mathbf{E}, n_R)}{\partial \mathbf{E}} \quad \text{and} \quad \tilde{\mu}(\mathbf{E}, n_R) = \frac{\partial \tilde{\psi}_R(\mathbf{E}, n_R)}{\partial n_R}.$$

8. Using the relation $\tilde{\psi}_R(\mathbf{E}, n_R) = \bar{\psi}_R(\mathbf{1} + 2\mathbf{E}, n_R)$ show that, when expressed alternatively as functions of $\mathbf{E}$ and n_R, the elasticity tensor $\mathbb{C}$ and chemistry-strain tensor $\mathbf{S}$ have the forms

$$\mathbb{C}(\mathbf{E}, n_R) = \frac{\partial^2 \tilde{\psi}_R(\mathbf{E}, n_R)}{\partial \mathbf{E}^2} \quad \text{and} \quad \mathbf{S}(\mathbf{E}, n_R) = \frac{\partial^2 \tilde{\psi}_R(\mathbf{E}, n_R)}{\partial \mathbf{E}\partial n_R^\alpha}.$$

9. Use the relations $(66.29)_{2,3}$ to develop counterparts

$$\dot{\omega}_{\mathrm{R}} = \tfrac{1}{2}\mathbf{T}_{\mathrm{RR}}:\dot{\mathbf{C}} - n_{\mathrm{R}}\dot{\mu} \qquad \text{and} \qquad \frac{\partial \check{\mathbf{T}}_{\mathrm{RR}}(\mathbf{C},\mu)}{\partial \mu} = -2\frac{\partial \check{n}_{\mathrm{R}}(\mathbf{C},\mu)}{\partial \mathbf{C}} \tag{66.31}$$

of the Gibbs and Maxwell relations (66.14) and (66.15).

10. Use the Maxwell relation to motivate the introduction of an elasticity tensor

$$\begin{aligned}\mathbb{C}(\mathbf{C},\mu) &= 2\frac{\partial \check{\mathbf{T}}_{\mathrm{RR}}(\mathbf{C},\mu)}{\partial \mathbf{C}}\\ &= 4\frac{\partial^2 \check{\omega}_{\mathrm{R}}(\mathbf{C},\mu)}{\partial \mathbf{C}^2},\end{aligned} \tag{66.32}$$

a **chemistry-stress tensor** $\mathbf{A}(\mathbf{C},\mu)$ defined by

$$\begin{aligned}\mathbf{A}(\mathbf{C},\mu) &= \frac{\partial \check{\mathbf{T}}_{\mathrm{RR}}(\mathbf{C},\mu)}{\partial \mu}\\ &= -2\frac{\partial \check{\mathbf{n}}_{\mathrm{R}}(\mathbf{C},\mu)}{\partial \mathbf{C}}\\ &= 2\frac{\partial^2 \check{\omega}_{\mathrm{R}}(\mathbf{C},\mu)}{\partial \mathbf{C}\partial \mu},\end{aligned} \tag{66.33}$$

and a **density modulus**

$$\begin{aligned}K(\mathbf{C},\mu) &= \frac{\partial \check{n}_{\mathrm{R}}(\mathbf{C},\mu)}{\partial \mu}\\ &= -\frac{\partial \check{\omega}_{\mathrm{R}}(\mathbf{C},\mu)}{\partial \mu^2}.\end{aligned} \tag{66.34}$$

11. Assuming that $\mathbf{h}_{\mathrm{R}} = \acute{\mathbf{h}}_{\mathrm{R}}(\mathbf{C},\mu,\nabla\mu)$,[339] emulate the argument leading to Fourier's law (57.25) for the heat flux to arrive at the linearized version

$$\underbrace{\mathbf{h}_{\mathrm{R}} = -\mathbf{M}_0\nabla\mu}_{\text{linearized Fick's law}} + o(\epsilon) \quad \text{as } \epsilon \to 0 \tag{66.35}$$

of Fick's law, with

$$\mathbf{M}_0 = -\frac{\partial \check{\mathbf{h}}_{\mathrm{R}}(\mathbf{C},\mu,\mathbf{p})}{\partial \mathbf{p}}\bigg|_{(\mathbf{C},\mu,\mathbf{p})=(\mathbf{C}_0,\mu_0,\mathbf{0})}$$

for given values $\mathbf{C}_0$ and μ_0 of the right Cauchy–Green tensor and the chemical potential and

$$\epsilon = \sqrt{|\mathbf{C}-\mathbf{C}_0|^2 + \frac{|\mu-\mu_0|^2}{\mu_0^2} + \frac{\ell^2|\mathbf{p}|^2}{\mu_0^2}},$$

with ℓ a characteristic length associated with the reference body B.

12. Show that the *linearized mobility* $\mathbf{M}_0$ entering (66.35) is positive-semidefinite.

339 Cf. $(66.28)_4$.

67 Material Symmetry

Because the response of the free energy to deformation and variations of the species density determines the response of the stress and chemical potentials but not the response of the species flux, a discussion of material symmetry must account not only for the constitutive behavior free energy but also the constitutive behavior of the species flux. We therefore begin with the constitutive equations $(66.5)_1$ and (66.23):

$$\begin{aligned} \psi_R &= \bar{\psi}_R(\mathbf{C}, n_R), \\ \mathbf{h}_R &= -\mathbf{M}(\mathbf{C}, n_R)\nabla\mu. \end{aligned} \tag{67.1}$$

As defined roughly in §50, a symmetry transformation is a *rotation of the reference body* B that leaves its response to deformation unaltered. Here — because the constitutive relations (67.1) involve also the species densities and the chemical-potential gradients — it is necessary to be more specific.

Consider the following generalization of the two experiments discussed in §50:

- *Experiment* 1. In this experiment the deformation gradient is $\mathbf{F}$ and the species density and spatial chemical-potential gradient *fields* are n_R and $\operatorname{grad}\mu$.
- *Experiment* 2. In this experiment the deformation gradient is $\mathbf{FQ}$, the species density and spatial chemical-potential gradient *fields* remain n_R and $\operatorname{grad}\mu$.

Suppose that $\mathbf{Q}$ is a symmetry transformation. Then, arguing as in the theory of elasticity,[340] we should have

$$\bar{\psi}(\mathbf{Q}^{\top}\mathbf{C}\mathbf{Q}, n_R) = \bar{\psi}(\mathbf{C}, n_R). \tag{67.2}$$

Next, for the experiments 1 and 2, respectively, let $\mathbf{h}_1$ and $\mathbf{h}_2$ denote the species-flux fields measured in the deformed body. Because the deformation gradient $\mathbf{F}$ applied to the rotated body as well as the species density n_R in the second experiment are the same as the deformation gradient and the species density in the first experiment, and because we have assumed that $\mathbf{Q}$ is a symmetry transformation, the corresponding species-flux fields $\mathbf{h}_1$ and $\mathbf{h}_2$ — measured in the deformed body — should be the same:

$$\mathbf{h}_1 = \mathbf{h}_2 \equiv \mathbf{h}. \tag{67.3}$$

We would not, however, expect the corresponding referential fluxes $\mathbf{h}_{R1}$ and $\mathbf{h}_{R2}$ to coincide, since the reference body has been rotated; in fact, by $(65.11)_2$, these fluxes must satisfy

$$\mathbf{F}\mathbf{h}_{R1} = \mathbf{FQ}\mathbf{h}_{R2} = J\mathbf{h}. \tag{67.4}$$

[340] Cf. §50.

Since the two experiments are associated with the same spatial chemical-potential gradients, grad μ should coincide in the two experiments. Thus, by $(9.2)_1$, the referential chemical-potential gradients $\mathbf{p}_1 = (\nabla\mu)_1$ and $\mathbf{p}_2 = (\nabla\mu)_2$ measured in the two experiments must satisfy

$$\mathbf{F}^{-\top}\mathbf{p}_1 = \mathbf{F}^{-\top}\mathbf{Q}\mathbf{p}_2 = \operatorname{grad}\mu. \tag{67.5}$$

By (67.3) and (67.5),

$$\mathbf{p}_2 = \mathbf{Q}^{\top}\mathbf{p}_1, \qquad \mathbf{h}_{R2} = \mathbf{Q}^{\top}\mathbf{h}_{R1}.$$

Thus, since $\mathbf{C}_2 = \mathbf{Q}^{\top}\mathbf{C}_1\mathbf{Q}$,

$$\mathbf{h}_{R1} = -\mathbf{M}(\mathbf{C}_1, n_R)\mathbf{p}_1 \qquad \text{and} \qquad \mathbf{h}_{R2} = -\mathbf{Q}\mathbf{M}(\mathbf{C}_2, n_R)\mathbf{p}_2,$$

and it follows that

$$\mathbf{F}\mathbf{Q}\mathbf{M}(\mathbf{Q}^{\top}\mathbf{C}\mathbf{Q}, n_R)\mathbf{Q}^{\top}\nabla\mu = \mathbf{F}\mathbf{M}(\mathbf{C}, n_R)\nabla\mu$$

for all $\mathbf{F}$, n_R, and $\nabla\mu$; or, equivalently, since $\mathbf{F}$ is invertible, if, for all $\mathbf{C}$ and n_R,

$$[\mathbf{Q}\mathbf{M}(\mathbf{Q}^{\top}\mathbf{C}\mathbf{Q}, n_R)\mathbf{Q}^{\top} - \mathbf{M}(\mathbf{C}, n_R)]\nabla\mu = \mathbf{0} \tag{67.6}$$

for all $\nabla\mu$. We show at the end of this section that

(‡) without loss in generality the referential chemical-potential gradient $\nabla\mu$ may be arbitrarily chosen.

Granted (‡) we have the following result: A rotation $\mathbf{Q}$ is a symmetry transformation for species transport if, in addition to (67.2),

$$\mathbf{Q}^{\top}\mathbf{M}(\mathbf{C}, n_R)\mathbf{Q} = \mathbf{M}(\mathbf{Q}^{\top}\mathbf{C}\mathbf{Q}, n_R) \tag{67.7}$$

for all $\mathbf{C}$ and n_R.

Thus — based on our individual conclusions regarding symmetry transformations for free energy and species transport — we say that a rotation $\mathbf{Q}$ is a **symmetry transformation** for the material if the relations (67.2) and (67.7) hold for all $\mathbf{C}$ and n_R.

67.1 Verification of (‡)

A consequence of the constitutive equation (66.12) is that the chemical-potential gradient must obey the subsidiary constitutive relation

$$\nabla\mu = \frac{\partial\bar{\mu}(\mathbf{C}, n_R)}{\partial n_R}\nabla n_R + \frac{\partial\bar{\mu}(\mathbf{C}, n_R)}{\partial C_{ij}}\nabla C_{ij}. \tag{67.8}$$

Fix the argument $(\mathbf{C}, n_R)$ and suppress it in what follows. Then, rewriting (67.8),

$$\nabla\mu = \frac{\partial\bar{\mu}}{\partial n_R}\nabla n_R + \frac{\partial\bar{\mu}}{\partial C_{ij}}\nabla C_{ij} \stackrel{\text{def}}{=} \mathcal{L}(\nabla n_R, \nabla\mathbf{C}) \tag{67.9}$$

with $\mathcal{L}(\nabla n_R, \nabla\mathbf{C})$ a linear function of ∇n_R and $\nabla\mathbf{C}$. Let $\operatorname{rng}(\mathcal{L})$ denote the range of the linear operator $\mathcal{L}$ (the set of vectors $\nabla\mu$ such that $\nabla\mu = \mathcal{L}(\nabla n_R, \nabla\mathbf{C})$ for some choice of the argument $(\nabla n_R, \nabla\mathbf{C})$). Since $\mathcal{L}$ is linear, if $\nabla\mu$ belongs to $\operatorname{rng}(\mathcal{L})$ then so also does $\lambda\nabla\mu$ for every scalar λ. Further, for $\mathbf{Q}$ an arbitrary rotation, (67.9) implies that

$$\mathbf{Q}\nabla\mu = \frac{\partial\bar{\mu}}{\partial n_R}\mathbf{Q}\nabla n_R + \frac{\partial\bar{\mu}}{\partial C_{ij}}\mathbf{Q}\nabla C_{ij};$$

thus if $\nabla\mu$ belongs to $\operatorname{rng}(\mathcal{L})$ then so also does $\mathbf{Q}\nabla\mu$ for every rotation $\mathbf{Q}$. Thus either:

(i) $\operatorname{rng}(\mathcal{L})$ is the entire three-dimensional space of material vectors; or
(ii) $\operatorname{rng}(\mathcal{L})$ contains only the zero-vector $\mathbf{0}$.

In case (i) (67.6) leads to the desired result (67.7). In case (ii) we must have $\nabla\mu = \mathbf{0}$ and so the constitutive equation $(67.1)_2$ is insensitive to the particular value of $\mathbf{M}$. In particular, we may without loss in generality take $\mathbf{M} = \mathbf{0}$, in which case (67.6) again implies (67.7). This completes the verification of (‡).

EXERCISE

1. Granted that the mobility tensor is independent of $\mathbf{C}$, determine the form for the constitutive equation $(67.1)_2$ for an isotropic material.

68 Natural Reference Configuration

In the theory of elasticity, a reference configuration is identified with a state in which the deformation is trivial, in which case $\mathbf{F} = \mathbf{C} = \mathbf{1}$. When the transport of a single species is taken into account, we must extend this notion to include dependence upon the density of that species.

Emulating our treatment of elasticity, we say that a reference configuration is **natural** for a species density n_{0R} if[341]

$$\bar{\psi}_R(\mathbf{C}, n_R) \text{ has a local minimum at } (\mathbf{C}, n_R) = (\mathbf{1}, n_{0R}). \tag{68.1}$$

Thus, introducing the notation

$$\Phi\big|_0 = \Phi(\mathbf{C}, n_R)\big|_{(\mathbf{C},n_R)=(\mathbf{1},n_{0R})},$$

it follows that

$$\left.\frac{\partial\bar{\psi}_R}{\partial\mathbf{C}}\right|_0 = \mathbf{0}, \qquad \left.\frac{\partial\bar{\psi}_R}{\partial n_R}\right|_0 = 0, \tag{68.2}$$

and, for any symmetric tensor $\mathbf{A}$ and any scalar κ,

$$\mathbf{A}\colon \left.\frac{\partial^2\bar{\psi}_R}{\partial\mathbf{C}^2}\right|_0 \mathbf{A} + 2\kappa \left.\frac{\partial^2\bar{\psi}_R}{\partial\mathbf{C}\partial n_R}\right|_0 \colon\mathbf{A} + \kappa^2 \left.\frac{\partial^2\bar{\psi}_R}{\partial n_R^2}\right|_0 \geq 0, \tag{68.3}$$

or, in components,

$$\left.\frac{\partial^2\bar{\psi}_R}{\partial C_{ij}\partial C_{kl}}\right|_0 A_{ij}A_{kl} + 2\kappa \left.\frac{\partial^2\bar{\psi}_R}{\partial C_{ij}\partial n_R}\right|_0 A_{ij} + \kappa^2 \frac{\partial^2\bar{\psi}_R}{\partial n_R^2} \geq 0.$$

By (68.2) and (68.3) together with the stress and chemical-potential relations (66.11) and (66.12),

$$\left.\begin{aligned} \left.\frac{\partial\bar{\psi}_R}{\partial\mathbf{C}}\right|_0 &= \tfrac{1}{2}\bar{\mathbf{T}}_{RR}\big|_0 = \mathbf{0}, \\ \left.\frac{\partial\bar{\psi}_R}{\partial n_R}\right|_0 &= \bar{\mu}|_0 = 0, \end{aligned}\right\} \tag{68.4}$$

[341] Cf. (48.16).

and using, in addition, (66.16), (66.17), and (66.18)

$$\left.\begin{aligned}\left.\frac{\partial^2\bar{\psi}_{\mathrm{R}}}{\partial\mathbf{C}^2}\right|_0 &= \mathbb{C}|_0,\\ \left.\frac{\partial^2\bar{\psi}_{\mathrm{R}}}{\partial\mathbf{C}\partial n_{\mathrm{R}}}\right|_0 &= \mathbf{S}|_0 = \mathbf{0},\\ \left.\frac{\partial^2\bar{\psi}_{\mathrm{R}}}{\partial n_{\mathrm{R}}^2}\right|_0 &= \Lambda|_0.\end{aligned}\right\} \tag{68.5}$$

Further, (68.3) and (68.5) imply that

$$\mathbf{A}\cdot\mathbb{C}|_0\mathbf{A} + 2\kappa\mathbf{S}|_0\cdot\mathbf{A} + \kappa^2\Lambda|_0 \geq 0, \tag{68.6}$$

from which we conclude that

$$\mathbf{A}\colon\mathbb{C}|_0\mathbf{A} \geq 0 \tag{68.7}$$

for every symmetric tensor **A**, and, in addition, that

$$\Lambda|_0 \geq 0. \tag{68.8}$$

Summarizing, we have shown that

(‡) *if the reference configuration is natural for the species density n_{0R}, then:*

(i) *the residual stress $\mathbf{T}_{RR}|_0$ vanishes;*
(ii) *the residual chemical potential $\mu|_0$ vanishes;*
(iii) *the chemistry-strain tensor $\mathbf{S}|_0$ vanishes;*
(iv) *the elasticity tensor $\mathbb{C}|_0$ — aside from being symmetric — is positive-semidefinite;*
(v) *the chemistry modulus $\Lambda|_0$ is nonnegative.*

As an important consequence of (i), the Piola and Cauchy stresses also vanish when the reference configuration is natural for the species density n_{0R}.

69 Summary of Basic Equations for a Single Species

The basic field equations describing the coupling between the transport of a single species and elastic deformation consist of the kinematical relations (6.1) and $(7.3)_1$ defining the deformation gradient and right Cauchy–Green tensor, the relation (66.11) determining the Piola stress, the relation (66.12) determining the chemical potential, Fick's law (66.13) for the species flux, the local force balance (65.4), and the local species mass balance (63.2):

$$\left.\begin{aligned} &\mathbf{F} = \nabla\boldsymbol{\chi}, \qquad \mathbf{C} = \mathbf{F}^{\top}\mathbf{F}, \\ &\mathbf{T}_{\mathrm{R}} = 2\mathbf{F}\frac{\partial\bar{\psi}_{\mathrm{R}}(\mathbf{C}, n_{\mathrm{R}})}{\partial\mathbf{C}}, \quad \mu = \frac{\partial\bar{\psi}_{\mathrm{R}}(\mathbf{C}, n_{\mathrm{R}})}{\partial n_{\mathrm{R}}}, \quad \mathbf{h}_{\mathrm{R}} = -\mathbf{M}(\mathbf{C}, n_{\mathrm{R}})\nabla\mu, \\ &\mathrm{S}\mathrm{Div}\,\mathbf{T}_{\mathrm{R}} + \mathbf{b}_{0\mathrm{R}} = \mathbf{0}, \qquad \dot{n}_{\mathrm{R}} = -\mathrm{Div}\,\mathbf{h}_{\mathrm{R}} + h_{\mathrm{R}}. \end{aligned}\right\} \tag{69.1}$$

These equations hold on the reference body B.

EXERCISE

1. Using (66.29), (66.33), and (66.34), and assuming that the species flux is given via Fick's law in the form

$$\mathbf{h}_{\mathrm{R}} = -\check{\mathbf{M}}(\mathbf{C}, \mu)\nabla\mu,$$

develop the basic equations for the alternative theory in which $\mathbf{C}$, μ, and $\nabla\mu$ are independent constitutive variables. Note that, for this theory, the final evolution equations are for $\boldsymbol{\chi}$ and μ as opposed to $\boldsymbol{\chi}$ and n_{R}.

70 Constitutive Theory for Multiple Species

The constitutive theory for *N independent* species is largely identical to that for a single species, the major salient differences being related to the treatment of the species fluxes and the conditions satisfied by the mobilities entering the relevant generalization of Fick's law. For that reason, we provide only an abbreviated exposition.

70.1 Consequences of Frame-Indifference and Thermodynamics

Guided by (65.15) and emulating our approach to the theory for a single independent species, we assume that the free energy ψ_R, the Piola stress $\mathbf{T}_R$, and the chemical potential μ^α of each species α are determined by constitutive equations depending on $\mathbf{F}$ and the list

$$\vec{n}_R = (n_R^1, n_R^2, \dots, n_R^N) \tag{70.1}$$

of species densities and that the flux $\mathbf{h}_R^\alpha$ of each species α is determined by a constitutive equation depending on $\mathbf{F}$, $\vec{n}_R$, and the list

$$\nabla\vec{\mu} = (\nabla\mu^1, \nabla\mu^2, \dots, \nabla\mu^N) \tag{70.2}$$

of chemical-potential gradients.

Frame-indifference then requires that there be response functions $\bar{\psi}_R$, $\bar{\mathbf{T}}_{RR}$, $\bar{\mu}^\alpha$, and $\bar{\mathbf{h}}_R^\alpha$ such that[342]

$$\left.\begin{aligned}\psi_R &= \bar{\psi}_R(\mathbf{C}, \vec{n}_R),\\ \mathbf{T}_R &= \mathbf{F}\bar{\mathbf{T}}_{RR}(\mathbf{C}, \vec{n}_R),\\ \mu^\alpha &= \bar{\mu}^\alpha(\mathbf{C}, \vec{n}_R),\\ \mathbf{h}_R^\alpha &= \bar{\mathbf{h}}_R^\alpha(\mathbf{C}, \vec{n}_R, \nabla\vec{\mu});\end{aligned}\right\} \tag{70.3}$$

thus, by $(66.3)_2$ and $(70.3)_2$,

$$\mathbf{T}_{RR} = \bar{\mathbf{T}}_{RR}(\mathbf{C}, \vec{n}_R). \tag{70.4}$$

Next, using the Coleman–Noll procedure in conjunction with the free-energy imbalance (65.15) leads to the following **thermodynamic restrictions:**[343]

[342] Cf. (66.5).
[343] Cf. (66.11), (66.12), and (66.13).

(i) *The free energy determines the second Piola stress and the chemical potential of species α through the* **stress** and **chemical-potential relations**

$$\mathbf{T}_{\mathrm{RR}} = \bar{\mathbf{T}}_{\mathrm{RR}}(\mathbf{C}, \vec{n}_{\mathrm{R}}) = 2\frac{\partial \bar{\psi}_{\mathrm{R}}(\mathbf{C}, \vec{n}_{\mathrm{R}})}{\partial \mathbf{C}}, \tag{70.5}$$

and

$$\mu^{\alpha} = \bar{\mu}^{\alpha}(\mathbf{C}, \vec{n}_{\mathrm{R}}) = \frac{\partial \bar{\psi}_{\mathrm{R}}(\mathbf{C}, \vec{n}_{\mathrm{R}})}{\partial n_{\mathrm{R}}^{\alpha}}. \tag{70.6}$$

(ii) *The species fluxes satisfy the* **species-transport inequality**

$$\sum_{\alpha} \bar{\mathbf{h}}_{\mathrm{R}}^{\alpha}(\mathbf{C}, \vec{n}_{\mathrm{R}}, \nabla\vec{\mu}) \cdot \nabla\mu^{\alpha} \leq 0 \tag{70.7}$$

for all $(\mathbf{C}, \vec{n}_{\mathrm{R}}, \nabla\vec{\mu})$.

As in the theory for a single species, we may use (70.5) and (70.6) to establish the **Gibbs relation**

$$\dot{\psi}_{\mathrm{R}} = \tfrac{1}{2}\mathbf{T}_{\mathrm{RR}} : \dot{\mathbf{C}} + \sum_{\alpha} \mu^{\alpha}\dot{n}_{\mathrm{R}}^{\alpha}, \tag{70.8}$$

the **gradient Gibbs relation**

$$\nabla\psi = \tfrac{1}{2}\mathbf{T}_{\mathrm{RR}} : \nabla\mathbf{C} + \sum_{\alpha} \mu^{\alpha}\nabla n_{\mathrm{R}}^{\alpha}, \tag{70.9}$$

and, for each species α, the **Maxwell relations**

$$\frac{\partial \bar{\mathbf{T}}_{\mathrm{RR}}(\mathbf{C}, \vec{n}_{\mathrm{R}})}{\partial n_{\mathrm{R}}^{\alpha}} = 2\frac{\partial \bar{\mu}^{\alpha}(\mathbf{C}, \vec{n}_{\mathrm{R}})}{\partial \mathbf{C}}. \tag{70.10}$$

Further, we may use the chain-rule to motivate the introduction of the **elasticity tensor**

$$\begin{aligned}\mathbb{C}(\mathbf{C}, \vec{n}_{\mathrm{R}}) &= 2\frac{\partial \bar{\mathbf{T}}_{\mathrm{RR}}(\mathbf{C}, \vec{n}_{\mathrm{R}})}{\partial \mathbf{C}} \\ &= 4\frac{\partial^2 \bar{\psi}_{\mathrm{R}}(\mathbf{C}, \vec{n}_{\mathrm{R}})}{\partial \mathbf{C}^2},\end{aligned} \tag{70.11}$$

a **chemistry-strain tensor**

$$\begin{aligned}\mathbf{S}^{\alpha}(\mathbf{C}, \vec{n}_{\mathrm{R}}) &= \frac{\partial \bar{\mathbf{T}}_{\mathrm{RR}}(\mathbf{C}, \vec{n}_{\mathrm{R}})}{\partial n_{\mathrm{R}}^{\alpha}} \\ &= 2\frac{\partial \bar{\mu}^{\alpha}(\mathbf{C}, \vec{n}_{\mathrm{R}})}{\partial \mathbf{C}} \\ &= 2\frac{\partial^2 \bar{\psi}_{\mathrm{R}}(\mathbf{C}, \vec{n}_{\mathrm{R}})}{\partial \mathbf{C}\,\partial n_{\mathrm{R}}^{\alpha}}\end{aligned} \tag{70.12}$$

for each species α, and a **chemistry modulus**

$$\Lambda^{\alpha\beta}(\mathbf{C},\vec{n}_{\mathrm{R}}) = \frac{\partial\bar{\mu}^{\alpha}(\mathbf{C},\vec{n}_{\mathrm{R}})}{\partial n_{\mathrm{R}}^{\beta}} = \frac{\partial\bar{\mu}^{\beta}(\mathbf{C},\vec{n}_{\mathrm{R}})}{\partial n_{\mathrm{R}}^{\alpha}} = \frac{\partial^2\bar{\psi}_{\mathrm{R}}(\mathbf{C},\vec{n}_{\mathrm{R}})}{\partial n_{\mathrm{R}}^{\alpha}\partial n_{\mathrm{R}}^{\beta}} \tag{70.13}$$

for each pair of species α and β.

70.2 Fick's Law

For N species, **Fick's law** is the assertion that the species flux $\mathbf{h}_{\mathrm{R}}^{\alpha}$ of each species α depends *linearly* on the list $\nabla\vec{\mu}$ chemical-potential gradients

$$\mathbf{h}_{\mathrm{R}}^{\alpha} = -\sum_{\beta}\mathbf{M}^{\alpha\beta}(\mathbf{C},\vec{n}_{\mathrm{R}})\nabla\mu^{\beta}, \tag{70.14}$$

with $\mathbf{M}^{\alpha\beta}(\mathbf{C},\vec{n}_{\mathrm{R}})$ the **mobility tensor** for species α with respect to species β;[344] a consequence of the species-transport inequality (70.7) is then that the $N\times N$ matrix

$$\begin{bmatrix} \mathbf{M}^{11}(\mathbf{C},\vec{n}_{\mathrm{R}}) & \mathbf{M}^{12}(\mathbf{C},\vec{n}_{\mathrm{R}}) & \cdots & \mathbf{M}^{1N}(\mathbf{C},\vec{n}_{\mathrm{R}}) \\ \mathbf{M}^{21}(\mathbf{C},\vec{n}_{\mathrm{R}}) & \mathbf{M}^{22}(\mathbf{C},\vec{n}_{\mathrm{R}}) & \cdots & \mathbf{M}^{2N}(\mathbf{C},\vec{n}_{\mathrm{R}}) \\ \vdots & \vdots & \ddots & \vdots \\ \mathbf{M}^{N1}(\mathbf{C},\vec{n}_{\mathrm{R}}) & \mathbf{M}^{N2}(\mathbf{C},\vec{n}_{\mathrm{R}}) & \cdots & \mathbf{M}^{NN}(\mathbf{C},\vec{n}_{\mathrm{R}}) \end{bmatrix} \tag{70.15}$$

of mobility tensors is positive-semidefinite:

$$\sum_{\alpha\beta}\mathbf{p}^{\alpha}\cdot\mathbf{M}^{\alpha\beta}(\mathbf{C},\vec{n}_{\mathrm{R}})\mathbf{p}^{\beta} \geq 0 \tag{70.16}$$

for all $\vec{\mathbf{p}}$. As in the theory for a single species, we assume henceforth that the species fluxes are determined via Fick's law.

70.3 Natural Reference Configuration

As in the theory for a single species, we say that a reference configuration is **natural** for a density list $\vec{n}_{0\mathrm{R}}$ if[345]

$$\bar{\psi}_{\mathrm{R}}(\mathbf{C},\vec{n}_{\mathrm{R}}) \text{ has a local minimum at } (\mathbf{C},\vec{n}_{\mathrm{R}}) = (\mathbf{1},\vec{n}_{0\mathrm{R}}). \tag{70.17}$$

344 Just as in the theory for a single species, one can show that the most general relation of the form $\bar{\mathbf{h}}_{\mathrm{R}}^{\alpha}(\mathbf{C},\vec{n}_{\mathrm{R}},\nabla\vec{\mu})$ consistent with the species-transport inequality (70.7) must necessarily have the specific form

$$\mathbf{h}_{\mathrm{R}}^{\alpha} = -\sum_{\beta}\mathbf{M}^{\alpha\beta}(\mathbf{C},\vec{n}_{\mathrm{R}},\nabla\vec{\mu})\nabla\mu^{\beta}.$$

345 Cf. (68.1).

Thus, introducing the notation

$$\Phi\big|_0 = \Phi(\mathbf{C}, \vec{n}_{\mathrm{R}})\big|_{(\mathbf{C},\vec{n}_{\mathrm{R}})=(\mathbf{1},\vec{n}_{0\mathrm{R}})},$$

it follows that

$$\frac{\partial\bar{\psi}_{\mathrm{R}}}{\partial\mathbf{C}}\bigg|_0 = \mathbf{0}, \qquad \frac{\partial\bar{\psi}_{\mathrm{R}}}{\partial n^\alpha_{\mathrm{R}}}\bigg|_0 = 0 \qquad (\alpha = 1, 2, \ldots, N), \tag{70.18}$$

and, for any symmetric tensor $\mathbf{A}$ and any list $\vec{\kappa} = (\kappa^1, \kappa^2, \ldots, \kappa^N)$ of N scalars,

$$\mathbf{A}\colon \frac{\partial^2\bar{\psi}_{\mathrm{R}}}{\partial\mathbf{C}^2}\bigg|_0 \mathbf{A} + 2\bigg(\sum_\alpha \kappa^\alpha \frac{\partial^2\bar{\psi}_{\mathrm{R}}}{\partial\mathbf{C}\partial n^\alpha_{\mathrm{R}}}\bigg|_0\bigg)\colon\mathbf{A} + \sum_{\alpha,\beta}\kappa^\alpha\kappa^\beta \frac{\partial^2\bar{\psi}_{\mathrm{R}}}{\partial n^\alpha_{\mathrm{R}}\partial n^\beta_{\mathrm{R}}}\bigg|_0 \geq 0, \tag{70.19}$$

or, in components,

$$\frac{\partial^2\bar{\psi}_{\mathrm{R}}}{\partial C_{ij}\partial C_{kl}}\bigg|_0 A_{ij}A_{kl} + 2\sum_\alpha \frac{\partial^2\bar{\psi}_{\mathrm{R}}}{\partial C_{ij}\partial n^\alpha_{\mathrm{R}}}\bigg|_0 \kappa^\alpha A_{ij} + \sum_{\alpha,\beta}\frac{\partial^2 g}{\partial n^\alpha_{\mathrm{R}}\partial n^\beta_{\mathrm{R}}}\bigg|_0 \kappa^\alpha\kappa^\beta \geq 0.$$

By (70.18) and (70.19) together with the stress and chemical-potential relations (70.5) and (70.6),

$$\left.\begin{aligned} \frac{\partial\bar{\psi}_{\mathrm{R}}}{\partial\mathbf{C}}\bigg|_0 &= \tfrac{1}{2}\mathbf{T}_{\mathrm{RR}}\big|_0 = \mathbf{0},\\ \frac{\partial\bar{\psi}_{\mathrm{R}}}{\partial n^\alpha_{\mathrm{R}}}\bigg|_0 &= \bar{\mu}^\alpha\big|_0 = 0 \end{aligned}\right. \tag{70.20}$$

and using, in addition, (70.11), (70.12), and (70.13)

$$\left.\begin{aligned} \frac{\partial^2\bar{\psi}_{\mathrm{R}}}{\partial\mathbf{C}^2}\bigg|_0 &= \mathbb{C}\big|_0,\\ \frac{\partial^2\bar{\psi}_{\mathrm{R}}}{\partial\mathbf{C}\partial n^\alpha_{\mathrm{R}}}\bigg|_0 &= \mathbf{S}^\alpha\big|_0,\\ \frac{\partial^2\bar{\psi}_{\mathrm{R}}}{\partial n^\alpha_{\mathrm{R}}\partial n^\beta_{\mathrm{R}}}\bigg|_0 &= \Lambda^{\alpha\beta}\big|_0. \end{aligned}\right\} \tag{70.21}$$

Further, (70.19) and (70.21) imply that

$$\mathbf{A}\cdot\mathbb{C}|_0\mathbf{A} + 2\bigg(\sum_\alpha \kappa^\alpha\mathbf{A}^\alpha|_0\bigg)\cdot\mathbf{A} + \sum_{\alpha,\beta}\kappa^\alpha\kappa^\beta\Lambda^{\alpha\beta}\big|_0 \geq 0, \tag{70.22}$$

from which we conclude that

$$\mathbf{A}\colon\mathbb{C}\big|_0\mathbf{A} \geq 0 \tag{70.23}$$

for every symmetric tensor $\mathbf{A}$, and, in addition, that

$$\sum_{\alpha,\beta}\kappa^\alpha\kappa^\beta\Lambda^{\alpha\beta}\big|_0 \geq 0 \tag{70.24}$$

for all lists $\vec{\kappa}$.

Summarizing, we have shown that

(‡) *if the reference configuration is natural for the density list $\vec{n}_{0R}$, then:*

(i) *the residual stress $\mathbf{T}_{RR}\big|_0$ vanishes;*
(ii) *the residual chemical potential $\mu^\alpha\big|_0$ for each species α vanishes;*

(iii) *the elasticity tensor* $\mathbb{C}|_0$ — *aside from being symmetric — is positive–semidefinite;*
(iv) *the matrix of chemistry moduli* $\Lambda^{\alpha\beta}|_0$ — *aside from being symmetric — is positive-semidefinite.*

An important consequence of (i) is that, analogous to what occurs in the theory for a single diffusing species, the Piola and Cauchy stresses vanish, along with the second Piola stress, when the reference configuration is natural for the density list $\vec{n}_{0R}$.[346]

[346] Cf. the statement concluding §68.

71 Summary of Basic Equations for N Independent Species

The basic field equations describing the coupling between the transport of N independent species and elastic deformation are arrived at essentially as in the theory for a single species. These equations consist of the kinematical relations (6.1) and $(7.3)_1$ defining the deformation gradient and right Cauchy–Green tensor, the relation (70.5) determining the Piola stress, the relation (70.6) determining the chemical potential of each species α, Fick's law (70.7) for the flux of each species α, the local force balance (i.e., the local balance (24.10) of linear momentum but with inertia neglected), and the local balance (63.2) for each species α:

$$\left.\begin{aligned} &\mathbf{F} = \nabla\boldsymbol{\chi}, \qquad \mathbf{C} = \mathbf{F}^{\top}\mathbf{F}, \\ &\mathbf{T}_{\mathrm{R}} = 2\mathbf{F}\frac{\partial\bar{\psi}_{\mathrm{R}}(\mathbf{C},\vec{n}_{\mathrm{R}})}{\partial\mathbf{C}}, \quad \mu^{\alpha} = \frac{\partial\bar{\psi}_{\mathrm{R}}(\mathbf{C},\vec{n}_{\mathrm{R}})}{\partial n^{\alpha}_{\mathrm{R}}}, \quad \mathbf{h}^{\alpha}_{\mathrm{R}} = -\sum_{\beta}\mathbf{M}^{\alpha\beta}(\mathbf{C},\vec{n}_{\mathrm{R}})\nabla\mu^{\beta}, \\ &\mathrm{Div}\,\mathbf{T}_{\mathrm{R}} + \mathbf{b}_{0\mathrm{R}} = \mathbf{0}, \qquad \dot{n}^{\alpha}_{\mathrm{R}} = -\mathrm{Div}\,\mathbf{h}^{\alpha}_{\mathrm{R}} + h^{\alpha}_{\mathrm{R}}. \end{aligned}\right\} \tag{71.1}$$

EXERCISES

1. Use (70.5) and (70.6) to establish the Gibbs relation (70.8), the gradient Gibbs relation (70.9), and the Maxwell relations (70.10).
2. Use the species-transport inequality (70.7) to extend the result (66.22) to the case of N independent species and, thus, to show that a piezo-diffusive effect remains absent.
3. Show that the dependence of $\mathbf{h}^{\alpha}_{\mathrm{R}}$ on $\nabla\mu^{\beta}$ renders $\mathbf{h}^{\alpha}_{\mathrm{R}}$ a linear function of the gradients of $\mathbf{C}$ and n^{β}_{R} with the explicit form

$$\mathbf{h}^{\alpha}_{\mathrm{R}} = -\sum_{\beta}\mathbf{M}^{\alpha\beta}(\mathbf{C},\vec{n}_{\mathrm{R}})\Big(\sum_{\gamma}\lambda^{\beta\gamma}(\mathbf{C},\vec{n}_{\mathrm{R}})\nabla n^{\gamma}_{\mathrm{R}} + \tfrac{1}{2}\mathbf{S}^{\beta}(\mathbf{C},\vec{n}_{\mathrm{R}}):\nabla\mathbf{C}\Big), \tag{71.2}$$

or, using components,[347]

$$(\mathbf{h}^{\alpha}_{\mathrm{R}})_i = -\sum_{\beta}M^{\alpha\beta}_{ij}(\mathbf{C},\vec{n}_{\mathrm{R}})\Big(\sum_{\gamma}\lambda^{\beta\gamma}(\mathbf{C},\vec{n}_{\mathrm{R}})\frac{\partial n^{\gamma}_{\mathrm{R}}}{\partial x_j} + \tfrac{1}{2}S^{\beta}_{kl}(\mathbf{C},\vec{n}_{\mathrm{R}})\frac{\partial C_{kl}}{\partial x_j}\Big), \tag{71.3}$$

and conclude that species diffusion may be driven not only by spatial variations of the species densities but also by spatial variations of the right Cauchy–Green tensor $\mathbf{C}$.

[347] Cf. (66.25).

4. Arguing as in the theory for a single species, show that a rotation $\mathbf{Q}$ is a symmetry transformation for the material defined by $\bar{\psi}$ and $\mathbf{M}^{\alpha\beta}$ ($\alpha, \beta = 1, 2, \dots, N$) if, for all $\mathbf{C}$ and $\vec{n}_R$,

$$\bar{\psi}(\mathbf{Q}^{\mathsf{T}}\mathbf{C}\mathbf{Q}, \vec{n}_R) = \bar{\psi}(\mathbf{C}, \vec{n}_R) \tag{71.4}$$

and, for all species α and β,

$$\mathbf{M}^{\alpha\beta}(\mathbf{Q}^{\mathsf{T}}\mathbf{C}\mathbf{Q}, \vec{n}_R) = \mathbf{Q}^{\mathsf{T}}\mathbf{M}^{\alpha\beta}(\mathbf{C}, \vec{n}_R)\mathbf{Q}. \tag{71.5}$$

72 Substitutional Alloys

A crystalline solid can be usefully conceived of as a deformable lattice upon and through which various atomic species may diffuse. Generally, atoms may be substitutional or interstitial. Whereas substitutional atoms occupy lattice sites, interstitial atoms are found between lattice sites. When substitutional exchanges prevail, a constraint must be introduced to account for the fixed number of lattice sites in any sample and such a constraint requires a modification of the theory developed above. We now discuss the essential features needed to describe the coupling between diffusion and deformation in an alloy comprised solely by substitutional atoms.

For a crystalline solid, the density n_R^α of species α measures the number of atoms of that species per unit volume. We therefore refer to n_R^α, $\mathbf{h}_R^\alpha$, and h_R^α respectively as the **atomic density**, **atomic flux**, and **atomic supply** for species α. Moreover, we refer to the balance (65.14) as the **balance of atoms** for species α.

In the absence of plasticity, the displacement gradients in applications involving atomic diffusion are typically small. For that reason, a theory for small strains, but possibly large rotations, would seem most relevant. Within our general setting, such a theory may be achieved by positing that the free energy ψ_R be given by a response function of the generic form

$$\tilde{\psi}_R(\mathbf{E}, \vec{n}_R) = \tfrac{1}{2}\mathbf{E}:\mathbb{C}\mathbf{E} + \sum_\alpha \mathbf{S}^\alpha(\vec{n}_R):\mathbf{E} + \psi_0(\vec{n}_R) \tag{72.1}$$

and assuming as well that the mobilities $\mathbf{M}^{\alpha\beta}$ depend at most on the atomic densities. To maintain the connection with our previous results, we nevertheless continue to work with general response functions depending on $\mathbf{C}$. The specialization of our results to the particular free energy (72.1) and mobilities independent of $\mathbf{C}$ is straightforward.

72.1 Lattice Constraint

For a substitutional alloy, atoms are constrained to lie on lattice sites, and a scalar constant, n_R^{sites}, represents the density of substitutional sites, per unit volume, available for occupation by atoms. The atomic densities n_R^α, $\alpha = 1, 2, \ldots, N$, for a substitutional alloy are then required to satisfy the **lattice constraint**

$$\sum_\alpha n_R^\alpha = n_R^{\text{sites}}. \tag{72.2}$$

The constraint (72.2) is *inconsistent* with the notion of an interstitial defect, which is a substitutional atom forced into a position between lattice sites. However, if one of the species is associated with unoccupied lattice sites then the lattice constraint does allow for *vacancies*.

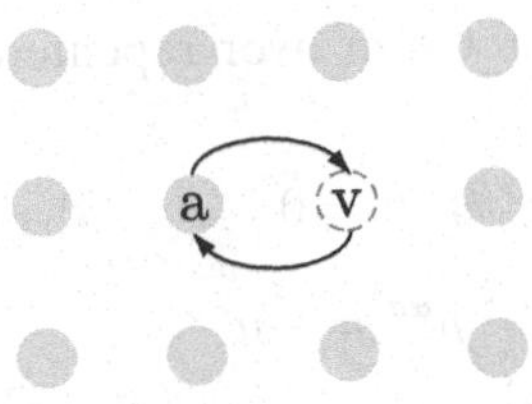

Figure 72.1. Schematic of an atom-vacancy exchange.

A simple but important consequence of the lattice constraint is the **conservation of substitutional atoms,**

$$\sum_\alpha \dot{n}^\alpha_{\mathrm{R}} = 0, \tag{72.3}$$

a condition that, by virtue of the local atomic balance $(71.1)_7$, is equivalent to the **flux-supply constraint**

$$\mathrm{Div}\Big(\sum_\alpha \mathbf{h}^\alpha_{\mathrm{R}}\Big) = \sum_\alpha h^\alpha_{\mathrm{R}}. \tag{72.4}$$

72.2 Substitutional Flux Constraint

On a basis of the view that atomic transport, as represented by the atomic fluxes, arises microscopically from exhanges of atoms or exchanges of atoms and vacancies (Figure 72.1), ÅGREN (1982) and CAHN & LARCHÉ (1983) argue that the atomic fluxes should obey the **substitutional-flux constraint**

$$\sum_\alpha \mathbf{h}^\alpha_{\mathrm{R}} = \mathbf{0}. \tag{72.5}$$

This restriction is considerably stronger than the flux-supply constraint (72.4). Indeed, when applied to (72.4), (72.5) implies that the atomic supplies must satisfy the intuitively appealing condition

$$\sum_\alpha h^\alpha_{\mathrm{R}} = 0, \tag{72.6}$$

which requires that if an external agency adds or deletes atoms of a given species (vacancies included) to the lattice that same agency must delete or add an equal amount of atoms of the remaining species. Henceforth, we assume that

(†) *the substitutional-flux constraint* (72.5) *is satisfied.*

72.3 Relative Chemical Potentials. Free-Energy Imbalance

We now consider the impact of conservation of substitutional atoms (72.3) and the substitutional-flux constraint (72.5) on the free-energy imbalance. Of central importance in this discussion are the **relative chemical-potentials**[348]

$$\mu^{\alpha\sigma} = \mu^\alpha - \mu^\sigma. \tag{72.7}$$

[348] Strictly speaking, $\mu^{\alpha\beta}$ is the chemical potential of species α measured relative to the chemical potential of species β.

Bearing in mind that we do not sum over repeated species labels, the definition (72.7) implies that

$$\begin{aligned}\mu^{\alpha\alpha} &= 0,\\ \mu^{\alpha\sigma} &= -\mu^{\sigma\alpha},\\ \mu^{\alpha\sigma} &= \mu^{\alpha\beta} - \mu^{\sigma\beta}\end{aligned} \tag{72.8}$$

for all relevant choices of the various species labels.

Select and fix an atomic species σ. For a substitutional alloy, the conservation condition (72.3) and the flux constraint (72.5) then give

$$\begin{aligned}\sum_{\alpha}(\mu^{\alpha\sigma}\dot{n}_{\mathrm{R}}^{\alpha} - \mathbf{h}_{\mathrm{R}}^{\alpha}\cdot\nabla\mu^{\alpha\sigma}) &= \sum_{\alpha}(\mu^{\alpha}\dot{n}_{\mathrm{R}}^{\alpha} - \mathbf{h}_{\mathrm{R}}^{\alpha}\cdot\nabla\mu^{\alpha}) - \mu^{\sigma}\sum_{\alpha}\dot{n}_{\mathrm{R}}^{\alpha} + \nabla\mu^{\sigma}\cdot\sum_{\alpha}\mathbf{h}_{\mathrm{R}}^{\alpha}\\ &= \sum_{\alpha}(\mu^{\alpha}\dot{n}_{\mathrm{R}}^{\alpha} - \mathbf{h}_{\mathrm{R}}^{\alpha}\cdot\nabla\mu^{\alpha}),\end{aligned}$$

which, when used in the local free-energy imbalance (65.15), yields

$$\dot{\psi}_{\mathrm{R}} - \mathbf{T}_{\mathrm{R}}\cdot\dot{\mathbf{F}} - \sum_{\alpha}(\mu^{\alpha\sigma}\dot{n}_{\mathrm{R}}^{\alpha} - \mathbf{h}_{\mathrm{R}}^{\alpha}\cdot\nabla\mu^{\alpha\sigma}) \le 0. \tag{72.9}$$

We emphasize that the imbalance (72.9) must hold for any choice of σ.

Remarks.

- LARCHÉ & CAHN (1973, 1985) were apparently the first to emphasize the importance of the relative chemical-potentials when discussing substitutional alloys: LARCHÉ & CAHN (1973) consider a variational problem that, within our framework, consists of minimizing a body's free energy under a mass constraint for each atomic species; LARCHÉ & CAHN define the chemical potentials μ^{α}, $\alpha = 1, 2, \ldots, N$, to be the Lagrange multipliers associated with the mass constraints and show that only the relative chemical-potentials $\mu^{\alpha} - \mu^{\beta}$ enter the corresponding equilibrium conditions.
- Of the basic laws, it is only the free-energy imbalance that involves chemical potentials. We may, therefore, conclude from the foregoing discussion that the individual chemical potentials are irrelevant to the theory in bulk. At external or internal boundaries, however, it is often the individual chemical potentials that are needed, with specific example, being solid-vapor interfaces and grain boundaries.[349]
- The local free-energy imbalance (65.15) may be written instead in the form (72.9) involving only chemical potentials expressed relative to the chemical potential of any arbitrarily chosen species σ, in which case (72.9) is independent of n_{R}^{σ} and $\mathbf{h}_{\mathrm{R}}^{\sigma}$. Thus, like the pressure in an incompressible body, the individual chemical potentials are indeterminate in bulk.

72.4 Elimination of the Lattice Constraint. Larché–Cahn Differentiation

The lattice constraint (72.2) renders the constitutive theory for a substitutional alloy more involved than that for an unconstrained material. In many respects, however, the substitutional theory is identical to that for unconstrained materials; in particular, the theory is based on constitutive equations in which the list $\vec{n}_{\mathrm{R}} = (n_{\mathrm{R}}^{1}, n_{\mathrm{R}}^{2}, \ldots, n_{\mathrm{R}}^{N})$ of atomic densities appears as an independent variable. Difficulties

349 Cf. LARCHÉ & CAHN (1985).

arise because each such list must be admissible in the sense that it must satisfy the lattice constraint (72.2). Because of the lattice constraint, the set of all admissible lists is not open in the N-dimensional space $\mathbb{R}^N$. Thus, since varying one of the densities while holding the remaining densities fixed violates the lattice constraint, standard partial differentiation of the constitutive response functions with respect to the atomic densities is not well-defined. To address this difficulty, we rely on *Larché–Cahn differentiation.*

Let $f(\vec{n}_R)$ be defined on the set of admissible density lists. As noted above, the standard partial derivative of f with respect to the density n_R^α of a given species α is not defined. To free f of the lattice constraint, choose a species σ as reference, use the lattice constraint in the form

$$n_R^\sigma = n_R^{\text{sites}} - \sum_{\alpha\neq\sigma} n_R^\alpha \tag{72.10}$$

to express n_R^σ as a function of the list $(n_R^1, n_R^2, \ldots, n_R^{\sigma-1}, n_R^{\sigma+1}, \ldots, n_R^N)$ of the remaining atomic densities, and consider f a function $f^{(\sigma)}$ of that reduced list by defining

$$f^{(\sigma)}\underbrace{(n_R^1, n_R^2, \ldots, n_R^{\sigma-1}, n_R^{\sigma+1}, \ldots, n_R^N)}_{n^\sigma\ \text{missing}} = f(\vec{n}_R)\Big|_{n_R^\sigma = n_R^{\text{sites}} - \sum_{\alpha\neq\sigma} n_R^\alpha}. \tag{72.11}$$

The domain of $f^{(\sigma)}$ is then an open set in $\mathbb{R}^{N-1}$, since its arguments may be varied without violating the lattice constraint. Most importantly, the partial derivatives

$$\frac{\partial f^{(\sigma)}}{\partial n_R^\alpha} \qquad \text{and} \qquad \frac{\partial^2 f^{(\sigma)}}{\partial n_R^\alpha \partial n_R^\beta}$$

are well defined. Note that when α is equal to σ the left side of (72.11) is independent of n_R^σ, so that, trivially,

$$\frac{\partial f^{(\sigma)}}{\partial n_R^\sigma} = 0. \tag{72.12}$$

We refer to $f^{(\sigma)}$ as the description of f relative to species σ.

An alternative treatment of differentiation that respects the lattice constraint may be developed as follows. Choose species α and σ. If the list $\vec{n}_R = (n_R^1, n_R^2, \ldots, n_R^N)$ is consistent with the lattice constraint, then so also is the list

$$(n_R^1, \ldots, n_R^\alpha + \epsilon, \ldots, n_R^\sigma - \epsilon, \ldots, n_R^N)$$

obtained by increasing the atomic density of species α by an amount ϵ and decreasing the density of species σ by an equal amount (while holding the remaining atomic densities fixed). Bearing this in mind, we define the **Larché–Cahn derivative** $\partial^{(\sigma)}/\partial n_R^\alpha$ by

$$\frac{\partial^{(\sigma)} f(\vec{n}_R)}{\partial n_R^\alpha} = \left[\frac{d}{d\epsilon} f(n_R^1, \ldots, n_R^\alpha + \epsilon, \ldots, n_R^\sigma - \epsilon, \ldots, n_R^N)\right]_{\epsilon=0}; \tag{72.13}$$

$\partial^{(\sigma)} f(\vec{\rho}\,)/\partial\rho^\alpha$ represents the change in $f(\vec{\rho}\,)$ due to a unit increase in the density of α-atoms and an equal decrease in the density of σ-atoms.[350] Second Larché–Cahn

[350] Larché & Cahn (1985, eq. 3.7) use notation consistent with replacing $\partial^{(\sigma)}/\partial n_R^\alpha$ by $\partial/\partial n_R^{\alpha\sigma}$.

derivatives are defined similarly:

$$\frac{\partial^{2(\sigma)} f(\vec{n}_{\mathrm{R}})}{\partial n_{\mathrm{R}}^{\alpha} \partial n_{\mathrm{R}}^{\beta}} = \left[\frac{d^2}{d\epsilon_1\, d\epsilon_2} f(n_{\mathrm{R}}^1, \dots, n_{\mathrm{R}}^{\alpha} + \epsilon_1, \dots, n_{\mathrm{R}}^{\beta} + \epsilon_2, \dots, n_{\mathrm{R}}^{\sigma} - \epsilon_1 - \epsilon_2, \dots, n_{\mathrm{R}}^N)\right]_{\epsilon_1=\epsilon_2=0}. \tag{72.14}$$

For convenience, we define

$$\frac{\partial^{(\sigma)} f}{\partial n_{\mathrm{R}}^{\sigma}} = 0. \tag{72.15}$$

A direct consequence of (72.13) is then the *skew-symmetry relation*

$$\frac{\partial^{(\sigma)} f}{\partial n_{\mathrm{R}}^{\alpha}} = -\frac{\partial^{(\alpha)} f}{\partial n_{\mathrm{R}}^{\sigma}}, \tag{72.16}$$

valid for all species α and σ. Thus,

$$\sum_{\alpha,\sigma} \frac{\partial^{(\sigma)} f}{\partial n_{\mathrm{R}}^{\alpha}} = -\sum_{\alpha,\sigma}^{N} \frac{\partial^{(\alpha)} f}{\partial n_{\mathrm{R}}^{\sigma}} = -\sum_{\alpha,\sigma}^{N} \frac{\partial^{(\sigma)} f}{\partial n_{\mathrm{R}}^{\alpha}}$$

and we have

$$\sum_{\alpha,\sigma} \frac{\partial^{(\sigma)} f}{\partial n_{\mathrm{R}}^{\alpha}} = 0. \tag{72.17}$$

Using the description $f^{(\sigma)}$ of f relative to σ, the Larché–Cahn derivative may be given an alternative representation that is convenient in calculations. Increasing an argument n_{R}^{α} by an amount ϵ (while holding the other arguments of $f^{(\sigma)}$ fixed) corresponds, via the definition (72.11), to decreasing the argument n_{R}^{σ} by ϵ. Therefore, as a consequence of (72.13), the Larché–Cahn derivative $\partial f^{(\sigma)}/\partial n_{\mathrm{R}}^{\alpha}$ is simply the derivative of f with respect to n_{R}^{α} taken with the density n^{σ} eliminated via the lattice constraint; thus, by (72.12) and (72.15),

$$\frac{\partial^{(\sigma)} f}{\partial n_{\mathrm{R}}^{\alpha}} = \frac{\partial f^{(\sigma)}}{\partial n^{\alpha}}, \tag{72.18}$$

and similarly for second derivatives,

$$\frac{\partial^{2(\sigma)} f}{\partial n_{\mathrm{R}}^{\alpha} \partial n_{\mathrm{R}}^{\beta}} = \frac{\partial^2 f^{(\sigma)}}{\partial n_{\mathrm{R}}^{\alpha} \partial n_{\mathrm{R}}^{\beta}}. \tag{72.19}$$

Note that (72.18) and (72.19) are meaningful even though their left sides are functions of the complete list $\vec{n}_{\mathrm{R}} = (n_{\mathrm{R}}^1, n_{\mathrm{R}}^2, \dots, n_{\mathrm{R}}^{\sigma}, \dots, n_{\mathrm{R}}^N)$, while their right sides are functions of the list

$$\underbrace{(n_{\mathrm{R}}^1, n_{\mathrm{R}}^2, \dots, n_{\mathrm{R}}^{\sigma-1}, n_{\mathrm{R}}^{\sigma+1}, \dots, n_{\mathrm{R}}^N)}_{n_{\mathrm{R}}^{\sigma} \text{ missing}};$$

indeed, the left sides are defined only for those arguments $\vec{n}_{\mathrm{R}}$ consistent with the lattice constraint, a constraint that renders n_{R}^{σ} known when the other densities are known.[351]

351 Cf. (72.11).

It may happen that $f(\vec{n}_R)$ may be extended smoothly to an open region of $\mathbb{R}^N$. In that case, the Larché–Cahn derivative may be computed as the difference

$$\frac{\partial^{(\sigma)} f}{\partial n_R^\alpha} = \frac{\partial f}{\partial n_R^\alpha} - \frac{\partial f}{\partial n_R^\sigma}; \tag{72.20}$$

for instance, for the function f defined on the set of admissible density lists by

$$f(\vec{n}_R) = \sum_\alpha \lambda^\alpha n_R^\alpha$$

with $\lambda_1, \lambda_2, \ldots, \lambda_N$ constant,

$$\frac{\partial^{(\sigma)} f}{\partial n_R^\alpha} = \lambda^\alpha - \lambda^\sigma. \tag{72.21}$$

Next, select a reference species σ and bear in mind that $\partial f^{(\sigma)}/\partial n_R^\alpha$ is a standard partial derivative. Then, for $\vec{n}_R(t)$ an admissible, time-dependent density list and

$$\varphi(t) = f(\vec{n}_R(t)),$$

usng the chain-rule in conjunction with the definition (72.11) while invoking (72.12) gives

$$\begin{aligned} \dot{\varphi} &= \sum_\alpha \frac{\partial f^{(\sigma)}}{\partial n_R^\alpha} \dot{n}_R^\alpha \\ &= \sum_\alpha \frac{\partial^{(\sigma)} f}{\partial n_R^\alpha} \dot{n}_R^\alpha. \end{aligned} \tag{72.22}$$

72.5 General Constitutive Equations

Using (25.6)$_3$, we rewrite the free-energy imbalance (72.9) as

$$\dot{\psi}_R - \tfrac{1}{2}\mathbf{T}_{RR} \cdot \dot{\mathbf{C}} - \sum_\alpha (\mu^{\alpha\sigma} \dot{n}_R^\alpha - \mathbf{h}_R^\alpha \cdot \nabla \mu^{\alpha\sigma}) \leq 0. \tag{72.23}$$

Holding the reference species σ fixed and guided by (72.23), we base the theory on constitutive equations

$$\left.\begin{aligned} \psi_R &= \bar{\psi}_R(\mathbf{C}, \vec{n}_R), \\ \mathbf{T}_{RR} &= \bar{\mathbf{T}}_{RR}(\mathbf{C}, \vec{n}_R), \\ \mu^{\alpha\beta} &= \bar{\mu}^{\alpha\beta}(\mathbf{C}, \vec{n}_R), \end{aligned}\right\} \tag{72.24}$$

for the free energy, stress, and relative chemical-potentials, and, guided by the theory for N unconstrained species, on Fick's law

$$\mathbf{h}_R^\alpha = -\sum_\beta \mathbf{M}^{\alpha\beta}(\mathbf{C}, \vec{n}_R) \nabla \mu^{\beta\sigma}, \tag{72.25}$$

with σ arbitrary, for the atomic fluxes.

The constitutive equations (72.24)$_3$, which are prescribed for *all* relative chemical-potentials, are presumed to be consistent with the identities (72.8); more pragmatically, we need only assume that the response functions $\bar{\mu}^{\alpha\sigma}$ are prescribed for all α and some fixed choice of reference species σ, for then the response functions relative to any other species β may be defined by

$$\bar{\mu}^{\alpha\beta} = \bar{\mu}^{\alpha\sigma} - \bar{\mu}^{\beta\sigma}, \tag{72.26}$$

and, granted this, the *skew-symmetry relation*

$$\bar{\mu}^{\alpha\beta} = -\bar{\mu}^{\beta\alpha} \tag{72.27}$$

is satisfied for each pair of species, so that, in particular,

$$\bar{\mu}^{\alpha\alpha} = 0. \tag{72.28}$$

We require that the mobility tensors be consistent with the substitutional-flux constraint (72.5) and render Fick's law (72.25) independent of the choice of reference species σ. To discuss the implications of these requirements, we suppress the arguments $\mathbf{C}$ and $\vec{n}_R$, which are irrelevant to the following discussion. For Fick's law to be independent of the choice of reference species σ, it is sufficient that

$$\sum_\beta \mathbf{M}^{\alpha\beta}\nabla\mu^{\beta\sigma} = \sum_\beta \mathbf{M}^{\alpha\beta}\nabla\mu^{\beta\gamma} \tag{72.29}$$

for all choices of σ and γ and all α. By 72.8$_3$, the relative chemical-potentials necessarily satisfy $\mu^{\beta\sigma} = \mu^{\beta\gamma} - \mu^{\gamma\sigma}$ for all choices of σ and γ and all β; therefore, (72.29) will be satisfied provided that

$$\sum_\beta \mathbf{M}^{\alpha\beta}\nabla\mu^{\gamma\sigma} = \mathbf{0}$$

for all choices of σ and γ and for all α, and, hence, if

$$\sum_\beta \mathbf{M}^{\alpha\beta} = \mathbf{0} \tag{72.30}$$

for all α.

Next, the stipulation that the mobility tensors be consistent with the substitutional-flux constraint (72.5) requires that

$$\begin{aligned}\sum_\alpha \mathbf{h}_R^\alpha &= -\sum_{\alpha,\beta} \mathbf{M}^{\alpha\beta}\nabla\mu^{\beta\sigma}\\ &= -\sum_\beta\Big(\sum_\alpha \mathbf{M}^{\alpha\beta}\Big)\nabla\mu^{\beta\sigma}\\ &= \mathbf{0},\end{aligned}$$

which is satisfied for each choice of σ provided the term in parenthesis vanishes. Thus, recalling (72.30), we are led to the **mobility constraints**[352]

$$\sum_\alpha \mathbf{M}^{\alpha\beta} = \mathbf{0} \qquad \text{and} \qquad \sum_\beta \mathbf{M}^{\alpha\beta} = \mathbf{0}, \tag{72.31}$$

which must hold for all β and α, respectively.

72.6 Thermodynamic Restrictions

Our next step is to determine restrictions on the constitutive equations (72.24) and (72.25) that ensure satisfaction of the dissipation inequality (72.9). Because of the lattice constraint, thermodynamic arguments involving arbitrary variations of the atomic densities are delicate. In this regard, we show at the end of this section that

(‡) given any admissible density-list $\vec{n}_{R*}$, any scalar a, any two atomic species $\alpha \neq \beta$, and any time τ, there is a time-dependent, admissible density-list $\vec{n}_R(t)$ such that, at τ,

$$\vec{n}_R(\tau) = \vec{n}_{R*}, \qquad \dot{n}_R^\alpha(\tau) = -\dot{n}_R^\beta(\tau) = a, \qquad \dot{n}_R^\gamma(\tau) = 0 \quad \text{for} \quad \gamma \neq \alpha, \beta. \tag{72.32}$$

[352] Cf. equations (8.2) and (8.3) of LARCHÉ & CAHN (1985).

Fix a reference species σ. Recall that, by (72.12), (72.15), and the sentence containing (72.27),

$$\frac{\partial^{(\sigma)}\bar{\psi}_{\rm R}}{\partial n_{\rm R}^{\sigma}} = \frac{\partial \bar{\psi}_{\rm R}^{(\sigma)}}{\partial n_{\rm R}^{\sigma}} = 0,$$

where $\bar{\psi}_{\rm R}^{(\sigma)}$ is the description of $\bar{\psi}_{\rm R}$ with respect to σ.[353] Choose an arbitrary process consistent with the constitutive equations (72.24) and (72.25). Then, by (72.22),

$$\begin{aligned}\dot{\psi}_{\rm R} &= \frac{\partial \bar{\psi}_{\rm R}(\mathbf{C}, \vec{n}_{\rm R})}{\partial \mathbf{C}} : \dot{\mathbf{C}} + \sum_{\alpha} \frac{\partial \bar{\psi}_{\rm R}^{(\sigma)}(\mathbf{C}, \vec{n}_{\rm R})}{\partial n_{\rm R}^{\alpha}} \dot{n}_{\rm R}^{\alpha} \\ &= \frac{\partial \bar{\psi}_{\rm R}(\mathbf{C}, \vec{n}_{\rm R})}{\partial \mathbf{C}} : \dot{\mathbf{C}} + \sum_{\alpha} \frac{\partial^{(\sigma)} \bar{\psi}_{\rm R}(\mathbf{C}, \vec{n}_{\rm R})}{\partial n_{\rm R}^{\alpha}} \dot{n}_{\rm R}^{\alpha}.\end{aligned} \tag{72.33}$$

The requirement that the dissipation inequality (72.23) hold in all such processes leads to the inequality

$$\left(\frac{\partial \bar{\psi}_{\rm R}(\mathbf{C}, \vec{n}_{\rm R})}{\partial \mathbf{C}} - \tfrac{1}{2}\mathbf{T}_{\rm RR}(\mathbf{C}, \vec{n}_{\rm R})\right) : \dot{\mathbf{C}} + \sum_{\alpha} \left(\frac{\partial^{(\sigma)} \bar{\psi}_{\rm R}(\mathbf{C}, \vec{n}_{\rm R})}{\partial n_{\rm R}^{\alpha}} - \hat{\mu}^{\alpha\sigma}(\mathbf{C}, \vec{n}_{\rm R})\right) \dot{n}_{\rm R}^{\alpha}$$
$$- \sum_{\alpha,\beta} \nabla\mu^{\alpha\sigma} \cdot \mathbf{M}^{\alpha\beta}(\mathbf{C}, \vec{n}_{\rm R}) \nabla\mu^{\beta\sigma} \le 0, \tag{72.34}$$

for each choice of the free-index σ. If we momentarily restrict attention to spatially constant processes, then the inequality (72.34) reduces to

$$\left(\frac{\partial \bar{\psi}_{\rm R}(\mathbf{C}, \vec{n}_{\rm R})}{\partial \mathbf{C}} - \tfrac{1}{2}\mathbf{T}_{\rm RR}(\mathbf{C}, \vec{n}_{\rm R})\right) : \dot{\mathbf{C}} + \sum_{\alpha} \left(\frac{\partial^{(\sigma)} \bar{\psi}_{\rm R}(\mathbf{C}, \vec{n}_{\rm R})}{\partial n_{\rm R}^{\alpha}} - \hat{\mu}^{\alpha\sigma}(\mathbf{C}, \vec{n}_{\rm R})\right) \dot{n}_{\rm R}^{\alpha} \le 0.$$

This inequality must hold for all $\mathbf{C}(t)$ and all admissible density lists $\vec{n}_{\rm R}(t)$. Assuming that the atomic densities are independent of time leads to the requirement that

$$\mathbf{T}_{\rm RR}(\mathbf{C}, \vec{n}_{\rm R}) = 2\frac{\partial \bar{\psi}_{\rm R}(\mathbf{C}, \vec{n}_{\rm R})}{\partial \mathbf{C}}. \tag{72.35}$$

Similarly, assuming that the strain is independent of time leads to the requirement that

$$\hat{\mu}^{\alpha\sigma}(\mathbf{C}, \vec{n}_{\rm R}) = \frac{\partial^{(\sigma)} \bar{\psi}_{\rm R}(\mathbf{C}, \vec{n}_{\rm R})}{\partial n_{\rm R}^{\alpha}}. \tag{72.36}$$

Recalling that, by (72.12) and (72.28),

$$\frac{\partial^{(\sigma)} \bar{\psi}_{\rm R}(\mathbf{C}, \vec{n}_{\rm R})}{\partial n_{\rm R}^{\sigma}} = 0 \qquad \text{and} \qquad \hat{\mu}^{\sigma\sigma}(\mathbf{C}, \vec{n}_{\rm R}) = 0,$$

we see that (72.36) holds for each atomic species α and each choice of the free index σ corresponding to the chosen reference species. Next, (72.35) and (72.36) reduce (72.34) to the inequality

$$\sum_{\alpha,\beta} \nabla\mu^{\alpha\sigma} \cdot \mathbf{M}^{\alpha\beta}(\mathbf{C}, \vec{n}_{\rm R}) \nabla\mu^{\beta\sigma} \ge 0,$$

for each choice of the reference species σ. Consistent with this inequality we assume that the matrix of mobilities is positive-definite.

353 Cf. (72.11).

72.7 Verification of (‡)

Note first that a simple choice $\vec{n}_R(t)$ consistent with the lattice constraint and with (72.32) is given by

$$n^\alpha_R(t) = n^\alpha_{*R} + (t-\tau)a, \qquad n^\beta_R(t) = n^\beta_{*R} - (t-\tau)a,$$

and

$$n^\gamma_R(t) = n^\gamma_R \quad \text{for} \quad \gamma \neq \alpha, \beta. \tag{72.37}$$

But this choice does not furnish a solution of our problem, since the densities $n^\alpha_R(t)$ and $n^\beta_R(t)$ may be negative. This is easily remedied: Given any $\epsilon > 0$, we can always find a scalar function $T(t)$ such that $T(\tau) = 0$, $\dot{T}(\tau) = 1$, and $|T(t)| < \epsilon$. The density list $\vec{n}_R(t)$ defined by

$$n^\alpha_R(t) = n^\alpha_{*R} + T(t)a \quad \text{and} \quad n^\beta_R(t) = n^\beta_{*R} - T(t)a,$$

supplemented by (72.37), satisfies (72.32) and will be admissible for all t provided we choose ϵ small enough. This completes the proof.

72.8 Normalization Based on the Elimination of the Lattice Constraint

Because of the lattice constraint (72.2), we may omit the atomic balance for one of the atomic species, say σ, and simply define

$$n^\sigma_R = n^{\text{sites}}_R - \sum_{\alpha\neq\sigma} n^\alpha_R. \tag{72.38}$$

Thus, by the substitutional-flux constraint (72.5),

$$\dot{n}^\sigma_R = -\sum_{\alpha\neq\sigma} \dot{n}^\alpha_R, \qquad \mathbf{h}^\sigma_R = -\sum_{\alpha\neq\sigma} \mathbf{h}^\alpha_R, \qquad \text{and} \qquad h^\sigma_R = -\sum_{\alpha\neq\sigma} h^\alpha_R, \tag{72.39}$$

so that the atomic balance for species σ is satisfied automatically provided the atomic balances for each of the remaining species $\alpha \neq \sigma$ are satisfied.

Without loss in generality, one may therefore employ the following normalization in which a given species σ is used as reference:

- Take the atomic density n^σ_R, the atomic flux $\mathbf{h}^\sigma_R$, and atomic supply h^σ_R to be defined by the lattice constraint and the substitutional-flux constraint via (72.39).
- Omit from consideration the atomic balance for species σ.
- Use as chemical potentials for species α the relative chemical-potentials $\mu^{\alpha\sigma}$.
- Use the free-energy imbalance (72.23) with species σ as reference (since this law is independent of n^σ_R and $\mathbf{h}^\sigma_R$).

EXERCISES

1. For a binary substitutional alloy with densities n^1_R and n^2_R, fluxes $\mathbf{h}^1_R$ and $\mathbf{h}^2_R$, supplies h^1_R and h^2_R, and chemical potentials μ^1 and μ^2, use the definitions

$$n_R = n^1_R = n^{\text{sites}}_R - n^2_R,$$

$$\mathbf{h}_R = \mathbf{h}^1_R = -\mathbf{h}^2_R,$$

$$h_R = h^1_R = -h^2_R,$$

$$\mu = \mu^{12} = \mu^1 - \mu^2$$

to reduce the species mass balances and the free-energy imbalance to the mass balance (65.8) and free-energy imbalance (65.9) for a single (unconstrained) species.

2. Use (72.35) and (72.36) to show that the Gibbs relation

$$\dot{\psi}_R = \tfrac{1}{2}\mathbf{T}_{RR} : \dot{\mathbf{C}} + \sum_\alpha \mu^{\alpha\sigma} \dot{n}^\alpha_R \tag{72.40}$$

and the Maxwell relations

$$\frac{\partial^{(\sigma)}\bar{\mathbf{T}}_{\mathrm{RR}}(\mathbf{C}, \vec{n}_{\mathrm{R}})}{\partial n_{\mathrm{R}}^{\alpha}} = 2\frac{\partial \bar{\mu}^{\alpha\sigma}(\mathbf{C}, \vec{n}_{\mathrm{R}})}{\partial \mathbf{C}}, \tag{72.41}$$

hold for each choice of the reference species σ.

3. Show that for a substitutional alloy the free-energy imbalance (65.13) is invariant under all transformations of the form

$$\mu^{\alpha}(\mathbf{x}, t) \rightarrow \mu^{\alpha}(\mathbf{x}, t) + \lambda(\mathbf{x}, t) \qquad \text{for all species } \alpha$$

and, moreover, use the fundamental lemma of the calculus of variations (page 167) to prove that this invariance is equivalent to the substitutional-flux constraint (72.5).

73 Linearization

Our derivation of the linear theory is based on the following hypotheses in which $\vec{n}_{0R}$ is a prescribed (constant) list of species densities:

C1 the reference configuration is natural for the density list $\vec{n}_{0R}$;
C2 the density list $\vec{n}_R$ is everywhere close to $\vec{n}_{0R}$;
C3 the displacement gradient $\mathbf{H} = \nabla\mathbf{u}$ is everywhere small.

We derive asymptotic forms of the governing equations appropriate to the limit as

$$\epsilon = \sqrt{|\mathbf{H}|^2 + \sum_\alpha \left| \frac{n_R^\alpha - n_{0R}^\alpha}{n_{0R}^\alpha} \right|^2}$$

tends to zero.[354] In this regard, given a function $\Phi(\mathbf{C}, \vec{n}_R)$, we write

$$\Phi_0 = \Phi|_0 \ \text{for}\ \Phi\ \text{evaluated for}\ (\mathbf{H}, \vec{n}_R) = (\mathbf{0}, \vec{n}_{0R}). \tag{73.1}$$

Bearing in mind that $\mathbf{C} = \mathbf{0}$ and $\mathbf{F} = \mathbf{1}$ when $\mathbf{H} = \mathbf{0}$, this notation makes sense for functions $\Phi(\mathbf{C}, \vec{n}_R)$ and $\Phi(\mathbf{F}, \vec{n}_R)$: in the former case, $\Phi_0 = \Phi(\mathbf{0}, \vec{n}_{0R})$; in the latter case, $\Phi_0 = \Phi(\mathbf{1}, \vec{n}_{0R})$.

73.1 Approximate Constitutive Equations for the Stress, Chemical Potentials, and Fluxes

We now determine asymptotic constitutive equations for the stress and chemical potentials appropriate to small departures from a reference configuration that is natural at the density list $\vec{n}_{0R}$.

Expanding $\bar{\mathbf{T}}_{RR}(\mathbf{C}, \vec{n}_R)$ and $\tilde{\mu}^\alpha(\mathbf{C}, \vec{n}_R)$ about $\mathbf{C} = \mathbf{1}$ and $\vec{n}_R = \vec{n}_{0R}$, we obtain

$$\bar{\mathbf{T}}_{RR}(\mathbf{C}, \vec{n}_R) = \bar{\mathbf{T}}_{RR}|_0 + \left.\frac{\partial \bar{\mathbf{T}}_{RR}}{\partial \mathbf{C}}\right|_0 (\mathbf{C} - \mathbf{1}) + \sum_\alpha \left.\frac{\partial \bar{\mathbf{T}}_{RR}}{\partial n_R^\alpha}\right|_0 (n_R^\alpha - n_{0R}^\alpha) + o(\epsilon),$$

$$\tilde{\mu}^\alpha(\mathbf{C}, \vec{n}_R^\alpha) = \tilde{\mu}^\alpha|_0 + \left.\frac{\partial \tilde{\mu}^\alpha}{\partial \mathbf{C}}\right|_0 : (\mathbf{C} - \mathbf{1}) + \sum_\beta \left(\left.\frac{\partial \tilde{\mu}^\alpha}{\partial n_R^\beta}\right|_0 \right) (n_R^\beta - n_{0R}^\beta) + o(\epsilon).$$

[354] This makes precise the sense of the term "small" in C2 and C3.

so that, by (70.21),

$$\left.\frac{\partial \tilde{\mathbf{T}}_{\mathrm{RR}}}{\partial \mathbf{E}}\right|_0 = \tfrac{1}{2}\mathbb{C}_0,$$

$$\left.\frac{\partial \tilde{\mathbf{T}}_{\mathrm{RR}}}{\partial n^\alpha_{\mathrm{R}}}\right|_0 = 2\left.\frac{\partial \tilde{\mu}^\alpha}{\partial \mathbf{C}}\right|_0 = \mathbf{S}^\alpha_0,$$

$$\left.\frac{\partial \tilde{\mu}^\alpha}{\partial n^\beta_{\mathrm{R}}}\right|_0 = \Lambda^{\alpha\beta}_0,$$

and, by (70.20),

$$\tilde{\mathbf{T}}_{\mathrm{RR}}|_0 = \mathbf{0} \qquad \text{and} \qquad \tilde{\mu}^\alpha|_0 = 0.$$

Thus, introducing the shorthand notation

$$\mathbb{C} = \mathbb{C}_0, \qquad \mathbf{S}^\alpha = \mathbf{S}^\alpha_0, \qquad \Lambda^{\alpha\beta} = \Lambda^{\alpha\beta}_0, \tag{73.2}$$

and using the relation

$$\mathbf{E} = \tfrac{1}{2}(\mathbf{C} - \mathbf{1}),$$

we obtain the following estimates for the second Piola stress and the chemical potentials:

$$\begin{aligned}\mathbf{T}_{\mathrm{RR}} &= \mathbb{C}\mathbf{E} + \sum_\alpha (n^\alpha_{\mathrm{R}} - n^\alpha_{0\mathrm{R}})\mathbf{S}^\alpha + \mathrm{o}(\epsilon),\\ \mu^\alpha &= \mathbf{S}^\alpha\colon\mathbf{E} + \sum_\beta \Lambda^{\alpha\beta}(n^\beta_{\mathrm{R}} - n^\beta_{0\mathrm{R}}) + \mathrm{o}(\epsilon),\end{aligned} \tag{73.3}$$

as $\epsilon \to 0$. Further, arguing as in the derivations of (52.26) and (52.27),[355] we find that, as $\epsilon \to 0$,

$$\begin{aligned}\mathbf{T}_{\mathrm{R}} &= \mathbb{C}\mathbf{E} + \sum_\alpha (n^\alpha_{\mathrm{R}} - n^\alpha_{0\mathrm{R}})\mathbf{S}^\alpha + \mathrm{o}(\epsilon),\\ \mathbf{T} &= \mathbb{C}\mathbf{E} + \sum_\alpha (n^\alpha_{\mathrm{R}} - n^\alpha_{0\mathrm{R}})\mathbf{S}^\alpha + \mathrm{o}(\epsilon),\end{aligned} \tag{73.4}$$

so that, as in elasticity theory, to within a small error the Cauchy and Piola stresses are all symmetric and coincident. Similarly, arguing as is the derivation of (52.31), we may show that the free energy function $\tilde{\psi}_{\mathrm{R}}(\mathbf{C}, \vec{n}_{\mathrm{R}})$ admits the estimate[356]

$$\psi_{\mathrm{R}} = \frac{1}{2}\mathbf{E}\colon\mathbb{C}\mathbf{E} + \sum_\alpha (n^\alpha_{\mathrm{R}} - n^\alpha_{0\mathrm{R}})\mathbf{S}^\alpha\colon\mathbf{E} + \sum_{\alpha,\beta} \Lambda^{\alpha\beta}(n^\alpha_{\mathrm{R}} - n^\alpha_{0\mathrm{R}})(n^\beta_{\mathrm{R}} - n^\beta_{0\mathrm{R}}) + \mathrm{o}(\epsilon^2) \tag{73.5}$$

as $\epsilon \to 0$. Moreover, by (18.9), the spatial and material forms of the density and conventional body force are related through

$$\rho = [1 + \mathrm{o}(1)]\rho_{\mathrm{R}} \qquad \text{and} \qquad \mathbf{b}_0 = [1 + \mathrm{o}(1)]\mathbf{b}_{0\mathrm{R}}. \tag{73.6}$$

Next, expanding $\mathbf{M}^{\alpha\beta}(\mathbf{C}, \vec{n}_{\mathrm{R}})$, we obtain

$$\begin{aligned}\mathbf{M}^{\alpha\beta}(\mathbf{C}, \vec{n}_{\mathrm{R}}) &= \mathbf{M}^{\alpha\beta}\big|_0 + \mathrm{o}(1)\\ &= \mathbf{M}^{\alpha\beta}_0 + \mathrm{o}(1),\end{aligned}$$

[355] Cf. (59.3).
[356] Cf. (59.4).

and it follows from (70.21), (71.2), and the relation $\mathbf{E} = \frac{1}{2}(\mathbf{C} - \mathbf{1})$ that

$$\mathbf{h}^{\alpha}_{\mathrm{R}} = -\sum_{\beta} \mathbf{M}^{\alpha\beta}_{0}\Bigg(\sum_{\gamma} \frac{\partial^2 \bar{\psi}}{\partial n^{\beta}_{\mathrm{R}} \partial n^{\gamma}_{\mathrm{R}}}\bigg|_0 \nabla n^{\gamma}_{\mathrm{R}} + \mathbf{S}^{\beta}_{0} : \nabla\mathbf{E}\Bigg) + \mathrm{o}(1).$$

Thus, introducing the shorthand notation

$$\mathbf{M}^{\alpha\beta} = \mathbf{M}^{\alpha\beta}_{0}$$

and drawing on (73.2), we have

$$\mathbf{h}^{\alpha}_{\mathrm{R}} = -\sum_{\beta} \mathbf{M}^{\alpha\beta}\Bigg(\sum_{\gamma} \Lambda^{\beta\gamma} \nabla n^{\gamma}_{\mathrm{R}} + \mathbf{S}^{\beta} : \nabla\mathbf{E}\Bigg) + \mathrm{o}(1). \tag{73.7}$$

By (70.16), the matrix of mobility tensors in positive-semidefinite:

$$\sum_{\alpha,\beta} \mathbf{p}^{\alpha} \cdot \mathbf{M}^{\alpha\beta}\mathbf{p}^{\beta} \geq 0. \tag{73.8}$$

73.2 Basic Equations of the Linear Theory

The linear theory of species diffusion coupled to elasticity is based on approximate equations obtained when the higher-order terms in

$$\mathbf{E} = \tfrac{1}{2}(\mathbf{H} + \mathbf{H}^{\top}) + \mathrm{o}(|\mathbf{H}|^2), \tag{73.9}$$

(73.3), (73.4), (73.5), and (73.7) are neglected.[357] We therefore take $\rho = \rho_{\mathrm{R}}$, $\mathbf{T} = \mathbf{T}_{\mathrm{R}}$, $\mathbf{b}_0 = \mathbf{b}_{\mathrm{R}0}$, and $h^{\alpha} = h^{\alpha}_{\mathrm{R}}$ and base the theory on the **strain-displacement relation**

$$\mathbf{E} = \tfrac{1}{2}(\mathbf{H} + \mathbf{H}^{\top}), \tag{73.10}$$

and the **constitutive equations**

$$\begin{aligned}
\psi_{\mathrm{R}} &= \frac{1}{2}\mathbf{E} : \mathbb{C}\mathbf{E} + \sum_{\alpha}(n^{\alpha}_{\mathrm{R}} - n^{\alpha}_{0\mathrm{R}})\mathbf{S}^{\alpha} : \mathbf{E} + \sum_{\alpha,\beta} \Lambda^{\alpha\beta}(n^{\alpha}_{\mathrm{R}} - n^{\alpha}_{0\mathrm{R}})(n^{\beta}_{\mathrm{R}} - n^{\beta}_{0\mathrm{R}}), \\
\mathbf{T} &= \mathbb{C}\mathbf{E} + \sum_{\alpha}(n^{\alpha}_{\mathrm{R}} - n^{\alpha}_{0\mathrm{R}})\mathbf{S}^{\alpha}, \\
\mu^{\alpha} &= \mathbf{S}^{\alpha} : \mathbf{E} + \sum_{\alpha,\beta} \Lambda^{\alpha\beta}(n^{\beta}_{\mathrm{R}} - n^{\beta}_{0\mathrm{R}}), \\
\mathbf{h}^{\alpha} &= -\sum_{\beta} \mathbf{M}^{\alpha\beta}\Bigg(\sum_{\gamma} \Lambda^{\beta\gamma} \nabla n^{\gamma}_{\mathrm{R}} + \mathbf{S}^{\beta} : \nabla\mathbf{E}\Bigg),
\end{aligned} \tag{73.11}$$

where $\mathbb{C}$, $\mathbf{S}^{\alpha}$, $\Lambda^{\alpha\beta}$, and $\mathbf{M}^{\alpha\beta}$ are, respectively, the elasticity tensor, the chemistry-strain modulus of species α, the chemistry modulus of species α and β, and the mobility of species α with respect to species β.

By (68.7), $\mathbb{C}$ is symmetric and positive-semidefinite; here we assume, in addition, that $\mathbb{C}$ is *positive-definite*:

$$\mathbf{A} : \mathbb{C}\mathbf{A} > 0 \quad \text{for all symmetric tensors } \mathbf{A} \neq \mathbf{0}. \tag{73.12}$$

[357] Precisely, the $\mathrm{o}(\epsilon)$ terms in (73.3) and (73.4) along with the $\mathrm{o}(1)$ terms in (73.19) and (73.6) and the $\mathrm{o}(\epsilon^2)$ terms in (73.5) are neglected.

Similarly, we strengthen the condition (70.24) by requiring that the matrix of entries $\Lambda^{\alpha\beta}$ is *positive-definite*:

$$\sum_{\alpha,\beta} \kappa^\alpha \kappa^\beta \Lambda^{\alpha\beta} > 0 \tag{73.13}$$

for all $\vec{\kappa}$. Further, we strengthen (73.8) by requiring that the matrix of mobilities be positive-definite:

$$\sum_{\alpha,\beta} \mathbf{p}^\alpha \cdot \mathbf{M}^{\alpha\beta}\mathbf{p}^\beta > 0 \quad \text{for all } \vec{\mathbf{p}} \neq \mathbf{0}. \tag{73.14}$$

We recall, also, that $\mathbf{S}^\alpha$ is symmetric for each α.

The basic equations of the linear theory consist of (73.10), (73.11), the local force balance (65.4)

$$\mathrm{Div}(\mathbb{C}\mathbf{E}) + \sum_\alpha \mathbf{S}^\alpha \nabla n_{\mathrm{R}}^\alpha + \mathbf{b}_0 = \mathbf{0} \tag{73.15}$$

and the species balance

$$\dot{n}_{\mathrm{R}}^\alpha = \mathrm{Div}\left[\sum_\beta \mathbf{M}^{\alpha\beta}\left(\sum_\gamma \Lambda^{\beta\gamma}\nabla n_{\mathrm{R}}^\gamma + \mathbf{S}^\beta : \nabla\mathbf{E}\right)\right] + h_{\mathrm{R}}^\alpha \tag{73.16}$$

for each $\alpha = 1, 2, \ldots, N$.

Next, introducing, for each species α, a **chemical diffusivity**

$$\mathbf{D}^{\alpha\beta} = \sum_\gamma \mathbf{M}^{\alpha\gamma}\Lambda^{\gamma\beta}, \tag{73.17}$$

for species α with respect to species β, and an **elastic diffusivity**

$$\mathbf{J}^\alpha = \sum_\beta \mathbf{M}^{\alpha\beta}\mathbf{S}^\beta, \tag{73.18}$$

we may rewrite (73.7) as

$$\mathbf{h}_{\mathrm{R}}^\alpha = -\sum_\beta \mathbf{D}^{\alpha\beta}\nabla n_{\mathrm{R}}^\beta - \mathbf{J}^\alpha : \nabla\mathbf{E} + \mathrm{o}(1). \tag{73.19}$$

In terms of these quantities, the species balance (73.16) becomes

$$\dot{n}_{\mathrm{R}}^\alpha = \sum_\beta \mathbf{D}^{\alpha\beta} : \nabla\nabla n_{\mathrm{R}}^\beta + \mathbf{J}^\alpha : \Delta\mathbf{E} + h_{\mathrm{R}}^\alpha. \tag{73.20}$$

73.3 Isotropic Linear Theory

When one uses $\mathbf{E}$ instead of $\mathbf{C}$ to express the constitutive equations for the theory, the transformation rules (71.4) and (71.5) under a symmetry transformation $\mathbf{Q}$, respectively, become

$$\tilde{\psi}_{\mathrm{R}}(\mathbf{Q}^{\top}\mathbf{E}\mathbf{Q}, \vec{n}_{\mathrm{R}}) = \tilde{\psi}_{\mathrm{R}}(\mathbf{E}, \vec{n}_{\mathrm{R}}) \quad \text{and} \quad \mathbf{Q}^{\top}\tilde{\mathbf{M}}^{\alpha\beta}(\mathbf{E}, \vec{n}_{\mathrm{R}})\mathbf{Q} = \tilde{\mathbf{M}}^{\alpha\beta}(\mathbf{Q}^{\top}\mathbf{E}\mathbf{Q}, \vec{n}_{\mathrm{R}}). \tag{73.21}$$

For the free-energy function $(73.11)_1$, these rules immediately imply that

$$\mathbf{Q}^{\top}(\mathbb{C}\mathbf{E})\mathbf{Q} = \mathbb{C}(\mathbf{Q}^{\top}\mathbf{E}\mathbf{Q}) \qquad \text{and} \qquad \mathbf{Q}^{\top}\mathbf{S}^\alpha\mathbf{Q} = \mathbf{S}^\alpha, \tag{73.22}$$

for a symmetry transformation $\mathbf{Q}$ and all symmetric tensors $\mathbf{E}$. Thus, if the body is isotropic then $\mathbb{C}\mathbf{E}$, $\mathbf{S}^\alpha$, and $\mathbf{M}^{\alpha\beta}$ have the specific forms

$$\left.\begin{aligned}\mathbb{C}\mathbf{E} &= 2\mu\mathbf{E} + \lambda(\operatorname{tr}\mathbf{E})\mathbf{1},\\ \mathbf{S}^\alpha &= s^\alpha\mathbf{1},\\ \mathbf{M}^{\alpha\beta} &= m^{\alpha\beta}\mathbf{1},\end{aligned}\right\} \tag{73.23}$$

with μ and λ **elastic moduli**, s^α the (scalar) **chemistry-strain modulus** for species α, and $m^{\alpha\beta}$ the (scalar) **mobility** of species α with respect to species β.[358]

As before,[359] the requirement (73.12) that $\mathbb{C}$ be positive-definite implies that the elastic moduli μ and λ satisfy

$$\mu > 0 \qquad \text{and} \qquad \kappa = \lambda + \tfrac{2}{3}\mu > 0, \tag{73.24}$$

with κ being the **compressibility**. Additionally, the requirement (73.14) that the matrix of (tensorial) mobilities be positive-definite implies that the matrix

$$\begin{bmatrix} m^{11} & m^{12} & \dots & m^{1N}\\ m^{21} & m^{22} & \dots & m^{2N}\\ \vdots & \vdots & \ddots & \vdots\\ m^{N1} & m^{N2} & \dots & m^{NN}\end{bmatrix} \tag{73.25}$$

is positive-definite.

By (73.23), the defining constitutive equations for flow through an isotropic, linear elastic solid are

$$\left.\begin{aligned}\psi_{\mathrm{R}} &= \mu|\mathbf{E}|^2 + \frac{\lambda}{2}(\operatorname{tr}\mathbf{E})^2 + \sum_\alpha s^\alpha(n^\alpha_{\mathrm{R}} - n^\alpha_{0\mathrm{R}})\operatorname{tr}\mathbf{E}\\ &\qquad + \sum_{\alpha,\beta}\Lambda^{\alpha\beta}(n^\alpha_{\mathrm{R}} - n^\alpha_{0\mathrm{R}})(n^\beta_{\mathrm{R}} - n^\beta_{0\mathrm{R}}),\\ \mathbf{T} &= 2\mu\mathbf{E} + \lambda(\operatorname{tr}\mathbf{E})\mathbf{1} + \sum_\alpha s^\alpha(n^\alpha_{\mathrm{R}} - n^\alpha_{0\mathrm{R}})\mathbf{1},\\ \mu^\alpha &= s^\alpha\operatorname{tr}\mathbf{E} + \sum_{\alpha,\beta}\Lambda^{\alpha\beta}(n^\beta_{\mathrm{R}} - n^\beta_{0\mathrm{R}}),\\ \mathbf{h}^\alpha &= -\sum_\beta m^{\alpha\beta}\Big(\sum_\gamma \Lambda^{\beta\gamma}\nabla n^\gamma_{\mathrm{R}} + s^\beta\nabla(\operatorname{tr}\mathbf{E})\Big).\end{aligned}\right. \tag{73.26}$$

Granted (59.9), this stress-strain relation $(73.26)_2$ may be inverted to give

$$\mathbf{E} = \frac{1}{2\mu}\left(\mathbf{T} - \frac{\lambda}{2\mu + 3\lambda}(\operatorname{tr}\mathbf{T})\mathbf{1}\right) - \sum_\alpha b^\alpha(n^\alpha_{\mathrm{R}} - n^\alpha_{0\mathrm{R}})\mathbf{1}, \tag{73.27}$$

where

$$b^\alpha \overset{\text{def}}{=} -\frac{s^\alpha}{2\mu + 3\lambda} \tag{73.28}$$

[358] Again, these results follow from standard representation theorems for isotropic functions. Cf. GURTIN (1981, §37).

[359] Cf. §59.3.

is the coefficient of solute expansion for species α. Thus, in terms of the compressibility κ,

$$b^\alpha = -3\kappa s^\alpha. \tag{73.29}$$

For an isotropic material, the basic equations (73.15) and (73.16) specialize to

$$\mu\Delta\mathbf{u} + (\lambda+\mu)\nabla\mathrm{div}\,\mathbf{u} + \sum_\alpha s^\alpha \nabla n^\alpha_{\mathrm{R}} = \mathbf{0} \tag{73.30}$$

and

$$\dot{n}^\alpha_{\mathrm{R}} = \sum_\beta m^{\alpha\beta}\Big(\sum_\gamma \Lambda^{\beta\gamma}\Delta n^\gamma_{\mathrm{R}} + s^\beta \Delta\mathrm{div}\,\mathbf{u}\Big). \tag{73.31}$$

EXERCISES

1. Writing $\tilde{\psi}_{\mathrm{R}}(\mathbf{E},\vec{n}_{\mathrm{R}})$ for the right side of $(73.11)_1$, verify that $\mathbf{T}$ and μ^α as determined by $(73.11)_2$ and $(73.11)_3$ are consistent with the thermodynamic relations

$$\mathbf{T} = \frac{\partial\tilde{\psi}_{\mathrm{R}}(\mathbf{E},\vec{n}_{\mathrm{R}})}{\partial\mathbf{E}} \qquad \text{and} \qquad \mu^\alpha = \frac{\partial\tilde{\psi}_{\mathrm{R}}(\mathbf{E},\vec{n}_{\mathrm{R}})}{\partial n^\alpha_{\mathrm{R}}}.$$

2. Show that the relations (73.11) are consistent with the local free-energy imbalance

$$\dot{\psi}_{\mathrm{R}} - \mathbf{T}\colon\dot{\mathbf{E}} - \sum_\alpha(\mu^\alpha\dot{n}^\alpha_{\mathrm{R}} - \mathbf{h}^\alpha_{\mathrm{R}}\dot{\nabla}\mu^\alpha) \le 0. \tag{73.32}$$

3. Drawing on the balances (73.15) and (73.16), show that (73.32) arises as the local consequence of requiring that the free-energy imbalance

$$\overline{\int_{\mathrm{P}}\psi_{\mathrm{R}}\,dv}^{\,\boldsymbol{\cdot}} \le \int_{\partial\mathrm{P}}\mathbf{Tn}\cdot\dot{\mathbf{u}}\,da + \int_{\mathrm{P}}\mathbf{b}_0\cdot\dot{\mathbf{u}}\,dv + \sum_\alpha\Big(-\int_{\partial\mathrm{P}}\mu^\alpha\mathbf{h}^\alpha_{\mathrm{R}}\cdot\mathbf{n}\,da + \int_{\mathrm{P}}\mu^\alpha h^\alpha_{\mathrm{R}}\,dv\Big) \tag{73.33}$$

hold for any subregion P of the body.

4. Writing $\mathbb{K}$ for the compliance tensor (that is, the inverse of the symmetric and positive-definite elasticity tensor $\mathbb{C}$), show that

$$\mathbf{E} = \mathbb{K}\mathbf{T} + \sum_\alpha(n^\alpha_{\mathrm{R}} - n^\alpha_{0\mathrm{R}})\mathbf{G}^\alpha,$$

where

$$\mathbf{G}^\alpha = -\mathbb{K}\mathbf{S}^\alpha \tag{73.34}$$

denotes the **chemistry-stress tensor**, and, further, that the free energy can be expressed as in terms of $\mathbf{T}$ and $\vec{n}_{\mathrm{R}}$ via

$$\begin{aligned}\psi &= \check{\psi}(\mathbf{T},\vec{n}_{\mathrm{R}})\\ &= \tfrac{1}{2}\mathbf{T}\colon\mathbb{K}\mathbf{T} + \sum_{\alpha,\beta}(n^\alpha_{\mathrm{R}} - n^\alpha_{0\mathrm{R}})(n^\beta_{\mathrm{R}} - n^\beta_{0\mathrm{R}})\mathbf{S}^\alpha\colon\mathbf{G}^\beta + \Psi_0(\vec{n}_{\mathrm{R}}).\end{aligned}$$

5. Using (73.34), show that the stress can be expressed in the form

$$\mathbf{T} = \mathbb{C}(\mathbf{E} - \mathbf{E}_{\mathrm{com}}),$$

where

$$\mathbf{E}_{\mathrm{com}} = \sum_\alpha(n^\alpha_{\mathrm{R}} - n^\alpha_{0\mathrm{R}})\mathbf{G}^\alpha$$

is the **compositional-strain tensor**, and derive an analogous relation for $\mathbf{T}$ involving a compositional-stress tensor.

6. Show that the **free enthalpy** defined via $\varphi_R = \psi_R - \mathbf{T}:\mathbf{E}$, with ψ_R as given by $(73.26)_1$, takes the form

$$\begin{aligned}\varphi_R &= \tilde{\varphi}_R(\mathbf{T}, \vec{n}_R) \\ &= -\tfrac{1}{2}\mathbf{T}:\mathbb{K}\mathbf{T} - \sum_\alpha (n_R^\alpha - n_{0R}^\alpha)\mathbf{G}^\alpha : \mathbf{T} + \Psi_0(\vec{n}_R)\end{aligned}$$

and that the strain and the chemical potential of species α are determined thermodynamically by the relations

$$\mathbf{E} = -\frac{\partial \tilde{\varphi}_R(\mathbf{T}, \vec{n}_R)}{\partial \mathbf{T}} \qquad \text{and} \qquad \mu^\alpha = \frac{\partial \tilde{\varphi}_R(\mathbf{T}, \vec{n}_R)}{\partial n_R^\alpha}.$$

7. Show that the flux relation can be expressed alternatively in the form

$$\mathbf{h}^\alpha = -\sum_\beta \mathbf{M}_0^{\alpha\beta}(\vec{n}_R)\left(\sum_\gamma \frac{\partial^2 \Psi_0(\vec{n}_R)}{\partial n_R^\beta \partial n_R^\gamma}\nabla n_R^\gamma - \mathbf{G}^\alpha : \nabla\mathbf{T}\right),$$

which demonstrates that spatial variations of the stress may drive species transport.

8. Show that for an isotropic material the flux relations (73.20) specialize to

$$\mathbf{h}_R^\alpha = -\sum_\beta m^{\alpha\beta}\left(\sum_\gamma \Lambda^{\beta\gamma}\nabla n_R^\gamma + s^\beta \nabla(\operatorname{tr}\mathbf{E})\right) + \mathrm{o}(1).$$

PART XIII

THEORY OF ISOTROPIC PLASTIC SOLIDS UNDERGOING SMALL DEFORMATIONS

The theory of elasticity furnishes a simple and elegant vehicle for illustrating the basic ideas of continuum mechanics. Among its many uses, elasticity theory is widely utilized for modeling the response of rubber-like, elastomeric materials at large strains. However, elasticity theory can be applied to the description of metals only for extremely small strains, typically not exceeding 10^{-3}. Larger deformations in metals lead to flow, permanent set, hysteresis, and other interesting and important phenomena that fall naturally within the purview of plasticity.

74 Some Phenomenological Aspects of the Elastic-Plastic Stress-Strain Response of Polycrystalline Metals

A stress-strain curve obtained from a simple tension test reveals the major features of the elastic-plastic response of a polycrystalline metal. In such an experiment, the length L_0 and cross-sectional area A_0 of a cylindrical specimen are deformed to L and A, respectively. If P denotes the axial force required to affect such a deformation, then the axial *engineering stress* (i.e., the Piola stress) in the specimen is

$$s = \frac{P}{A_0}.$$

In addition, if $\lambda = L/L_0$ denotes the axial stretch, the corresponding axial engineering strain is

$$e = \lambda - 1.$$

Figure 74.1 shows a curve representing engineering stress versus engineering strain for a metallic specimen. The portion OB of the stress-strain curve is essentially linear, and reversing the direction of strain from any point on OB results in a retracing of the forward straining portion of the stress-strain curve; in this range of small strains, the response of the material is typically idealized to be linearly elastic. Beyond the point B, the stress-strain curve deviates from linearity; accordingly, the point B is called the **elastic limit**, and the stress corresponding to the elastic limit is called the **yield strength** of the material. Beyond the elastic limit, the engineering stress increases with increasing strain; hence, the specimen is able to withstand a greater axial load despite a reduction of its cross-sectional area. This phenomenon is known as **strain-hardening**, and the portion of the curve beyond the elastic limit may be referred to as the **hardening curve**. Upon reversing the direction of strain at any stage C beyond the elastic limit B, the stress and strain values do not retrace the forward straining portion of the stress-strain curve; instead, the stress is reduced along an elastic unloading curve CD. That is, beyond the elastic limit, unloading to zero stress reduces the strain by an amount called the **elastic strain** e^e and leaves a permanent **plastic strain** e^p. Another reversal of the strain direction from D (reloading) retraces the unloading curve, and the stress-strain curve approaches the hardening curve at the point C from which the unloading was initiated, and under further loading the stress-strain curve once again follows the hardening curve.

The stress-strain response of a material may also be expressed in terms of the true stress (i.e., the Cauchy stress), defined by

$$\sigma = \frac{P}{A},$$

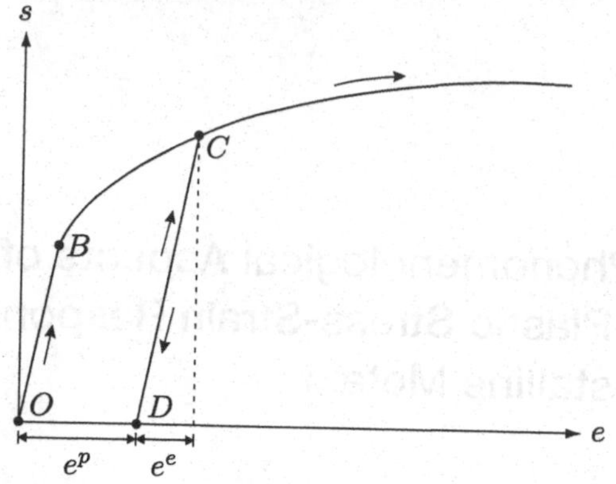

Figure 74.1. Schematic of an engineering stress-strain curve for a metallic material.

and *true strain* (logarithmic strain), defined by

$$\epsilon = \ln \lambda.$$

The true stress at any point in a tension test may be calculated by taking simultaneous measurements of the load P and current cross-sectional area A of the specimen. However, simultaneous measurements of the axial elongation and the diametrical reduction of a specimen are seldom carried out. Instead, use is made of the experimental observation that plastic flow of metals is essentially **incompressible** — volume change in a tension test is associated only with the elastic response of the material. Thus, for a metallic specimen undergoing plastic deformation that is large in comparison to its elastic response, it is reasonable to assume that the volume of the specimen is conserved, so that $AL \approx A_0 L_0$.[360] Hence, for any pair of values (s, e), the corresponding pair (σ, ϵ) may be calculated via the relations

$$\sigma = \frac{P}{A} = \frac{P}{A_0}\frac{L}{L_0} = s(1+e), \qquad \epsilon = \ln(1+e).$$

A true stress-strain curve is contrasted with a corresponding engineering stress-strain curve in Figure 74.2.

When the absolute values of the true stress and true strain obtained from a simple compression test are plotted and compared with corresponding values obtained for a tension test, with both tests conducted at ambient pressures, it is found that for most metallic materials the stress-strain curve obtained from the compression test is nearly coincident with the corresponding tensile stress-strain curve. A similar comparison between tensile and compressive engineering stress versus engineering strain curves does not provide such a nearly coincident response. For this reason, *the true stress-strain curve is believed to represent the intrinsic plastic flow characteristics of a metallic material.*

The stress-strain response of metals under cyclic testing is substantially more complicated than that under monotonic testing. For example, many metals exhibit the *Bauschinger effect*, a phenomenon first observed in metals by Bauschinger (1886), who reported that a metal specimen, after receiving a certain amount of axial extension into the plastic range, showed a decrease in the magnitude of the flow strength upon subsequent compression. A schematic of a reversed-deformation curve is shown in Figure 74.3. The solid line shows a true stress-strain curve in which the specimen was first extended in tension, and then the direction of deformation

[360] Provided the deformation is homogeneous.

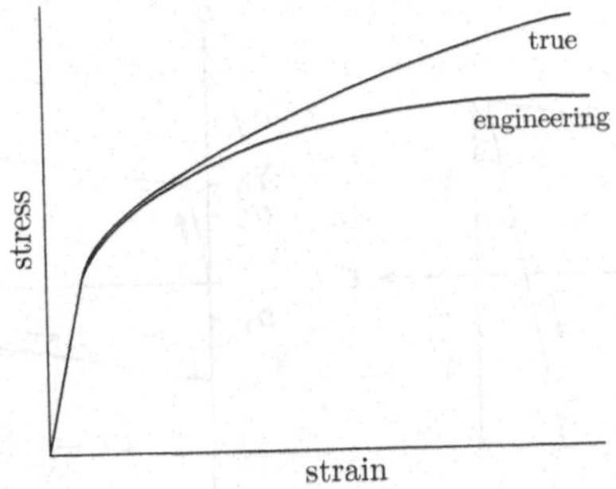

Figure 74.2. Comparison of a true stress-strain curve with the corresponding engineering stress-strain curve.

was reversed. The stress level from which reversed deformation was initiated is denoted by σ_f, and the stress level at which the stress-strain curve in compression begins to deviate from linearity is denoted by $\sigma_r < 0$. The absolute value $|\sigma_r|$ of the elastic limit in compression is smaller than the stress σ_f at which the reversed deformation was initiated.

74.1 Isotropic and Kinematic Strain-Hardening

As is clear from our brief discussion of the phenomenology of the stress-strain response, the deformation of metals beyond the elastic limit is quite complicated. In this subsection, we discuss some idealizations of strain-hardening that are frequently used in theories of plasticity.

A simple idealization of actual material response, referred to as **isotropic hardening**, accounts for strain-hardening but approximates the hardening by a straight line, neglects the Bauschinger effect, and assumes that after reversal of deformation from any level of strain in the plastic regime, the magnitude of the flow stress upon which reverse yielding begins has the same value as the flow stress from which the unloading was initiated. The stress-strain response corresponding to isotropic hardening is shown schematically in Figure 74.4 (a).

Let the stress level from which reversed deformation is initiated be denoted by σ_f, and denote the stress at which the stress-strain curve in compression begins to deviate from linearity by $\sigma_r < 0$. The closed interval $[\sigma_r, \sigma_f]$ is called the **elastic range**, and its end points $\{\sigma_r, \sigma_f\}$ are called the **yield set**. Let Y, with initial value Y_0,

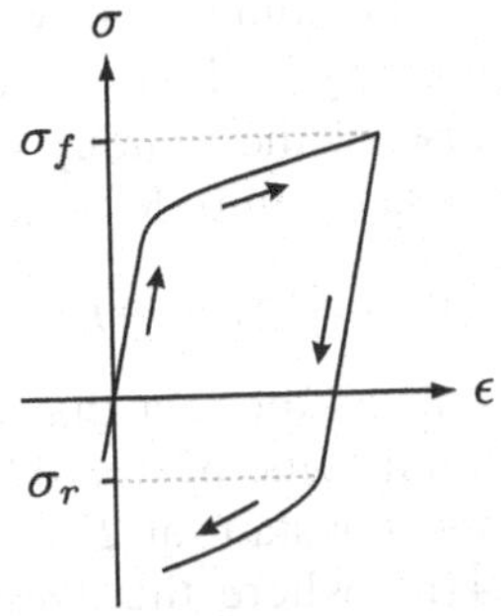

Figure 74.3. A stress-strain diagram obtained from a tension-compression experiment showing the Bauschinger effect.

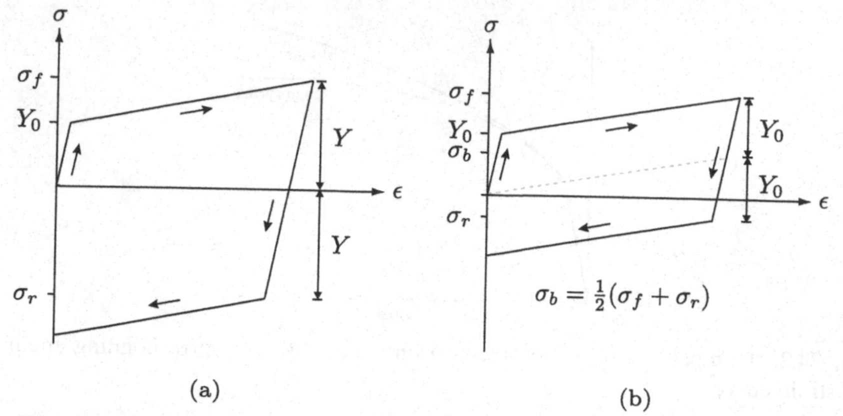

Figure 74.4. Idealized stress-strain response for an elastic-plastic material with (a) linear isotropic hardening; (b) linear kinematic hardening.

denote the **flow resistance** (a material property) of the material. The initial elastic range then has

$$\sigma_r = -Y_0, \qquad \sigma_f = Y_0,$$

and, during subsequent plastic deformation along the hardening curve, the deformation resistance increases linearly from Y_0 to Y due to strain-hardening, and the new elastic range becomes

$$\sigma_r = -Y, \qquad \sigma_f = Y.$$

For such an idealized response of an elastic-plastic material, the magnitude of the stress σ in this elastic range is bounded with the restriction

$$|\sigma| \leq Y,$$

generally referred to as a **yield condition**,[361] with plastic flow possible when $|\sigma| = Y$.

Isotropic strain-hardening is an idealization of the actual hardening behavior of metals; in particular, it does not account for the Bauschinger effect. An alternative simple model for strain-hardening that accounts for this phenomenon is referred to as **kinematic hardening**. The stress-strain response for a material with linear kinematic hardening is shown in Figure 74.4 (b): the initial flow resistance is Y_0; upon deformation in tension into the plastic regime, the stress increases linearly to σ_f, and, upon reversal of deformation, the material starts to plastically flow at a stresslevel $\sigma_r < 0$ and thereafter again continues to harden linearly. The magnitude $|\sigma_r|$ of the stress at which plastic flow recommences upon load reversal is smaller than that, σ_f, at which the reversed loading event was initiated — which embodies the *Bauschinger effect*. The asymmetry in the onset of yield upon reversal of loading in such a model arises because of the buildup of an internal stress called the *backstress* and denoted by σ_b, whose magnitude is equal to

$$\sigma_b = \tfrac{1}{2}(\sigma_f + \sigma_r).$$

The term *kinematic hardening* reflects the fact that the center of the *elastic range* in stress space moves from an initial value of $\sigma = 0$ to a value of $\sigma = \sigma_b$, while the height of the elastic range remains constant at $2Y_0$. This is in contrast to isotropic hardening shown in Figure 74.4 (a), where the *elastic range* in stress space stays

[361] The special case corresponding to no strain-hardening represents an elastic-perfectly plastic material, for which the yield condition becomes $|\sigma| \leq Y_0$, with Y_0 a *constant*.

centered at $\sigma = 0$, while the height of the elastic range expands from $2Y_0$ to $2Y$ as the material strain-hardens. For the material with kinematic hardening described above, the stress obeys the yield condition

$$|\sigma - \sigma_b| \leq Y_0.$$

For many metals, the actual strain-hardening behavior — including the Bauschinger effect — may be approximated by a combination of nonlinear isotropic hardening and nonlinear kinematic hardening.

75 Formulation of the Conventional Theory. Preliminaries

The description of materials, such as metals, that sometimes deform elastically and at other times flow like a fluid requires a constitutive framework far more complicated than those discussed thus far. For that reason, we do not seek the most general theory consistent with the second law in the form of a free-energy imbalance. Instead, bearing in mind the success of the conventional theories of plasticity, we seek to present those theories within a modern framework based on simple and physically meaningful hypotheses; but, because plasticity theory is complex when presented within the context of finite deformations, we begin with the far simpler theory in which the deformations are presumed to be small.

Most metals exhibit rate-independent or nearly rate-independent response, at least at sufficiently low temperatures; for that reason we base our initial discussion of constitutive relations on the notion of rate-independence.

75.1 Basic Equations

We take as our starting point the kinematical assumptions of the linear theory of elasticity as discussed in §52. Many of the equations derived there are independent of constitutive relations and hence applicable to a wide class of materials under the assumption of small deformations. Of particular importance to our discussion of conventional plasticity are the strain-displacement relation (52.32) (supplemented by an equation for the rotation), the momentum balance (52.37) (with stress $\mathbf{T}$ symmetric), and the free-energy imbalance (52.71) modified to account for dissipation:

(i) the *strain-displacement relation*

$$\mathbf{E} = \tfrac{1}{2}(\nabla\mathbf{u} + (\nabla\mathbf{u})^{\top}) \tag{75.1}$$

and the rotation-displacement relation

$$\mathbf{W} = \tfrac{1}{2}(\nabla\mathbf{u} - (\nabla\mathbf{u})^{\top})$$

in which $\mathbf{E}$ is the strain and $\mathbf{W}$ the rotation, so that, trivially, $\nabla\mathbf{u} = \mathbf{E} + \mathbf{W}$;

(ii) the *momentum balances*

$$\rho\ddot{\mathbf{u}} = \mathrm{Div}\,\mathbf{T} + \mathbf{b}_0, \qquad \mathbf{T} = \mathbf{T}^{\top}, \tag{75.2}$$

with $\mathbf{T}$ the stress, ρ the density, and $\mathbf{b}_0$ the conventional body force; the momentum balance $(75.2)_1$ can be written as a force balance

$$\mathrm{Div}\,\mathbf{T} + \mathbf{b} = \mathbf{0}, \tag{75.3}$$

with **b** the total body force

$$\mathbf{b} = \mathbf{b}_0 - \rho\ddot{\mathbf{u}}. \tag{75.4}$$

(iii) the free-energy imbalance

$$\dot{\Psi} - \mathbf{T}:\dot{\mathbf{E}} = -\delta \leq 0, \tag{75.5}$$

with Ψ the free energy and δ the dissipation.

Here and in what follows:

- To avoid cumbersome notation, when discussing small deformations we write Ψ for the free energy measured per unit volume; Ψ should be viewed as ψ_{R}, which for large deformations represents the free energy per unit referential volume.[362] Similarly, δ denotes the *dissipation* measured per unit volume.

75.2 Kinematical Assumptions that Define Plasticity Theory

We begin with some identities associated with deviatoric tensors. Recall that: (i) a tensor **G** is deviatoric if $\mathrm{tr}\,\mathbf{G} = 0$; (ii) for any tensor **B**

$$\mathbf{B}_0 = \mathbf{B} - \tfrac{1}{3}(\mathrm{tr}\,\mathbf{B})\mathbf{1}$$

is the deviatoric part of **B**; (iii) for **G** deviatoric,[363]

$$\mathbf{B}:\mathbf{G} = \mathbf{B}_0:\mathbf{G}; \tag{75.6}$$

(iv) for **G** arbitrary and **H** invertible,[364]

$$(\mathbf{H}\mathbf{G}\mathbf{H}^{-1})_0 = \mathbf{H}\mathbf{G}_0\mathbf{H}^{-1}. \tag{75.7}$$

Underlying most theories of plasticity is a physical picture that associates with a plastic solid a microscopic structure, such as a crystal lattice, that may be stretched and rotated, together with a notion of defects, such as dislocations, capable of flowing through that structure. Here, we mathematize this picture with a kinematical constitutive assumption requiring that the displacement gradient admit a decomposition

$$\nabla\mathbf{u} = \mathbf{H}^e + \mathbf{H}^p, \tag{75.8}$$

in which:

(i) $\mathbf{H}^e$, the **elastic distortion**, represents stretch and rotation of the underlying microscopic structure, and
(ii) $\mathbf{H}^p$, the **plastic distortion**, represents the local deformation of material due to the formation and motion of dislocations through that structure.

We define **elastic** and **plastic strains** $\mathbf{E}^e$ and $\mathbf{E}^p$ through

$$\mathbf{E}^e = \mathrm{sym}\,\mathbf{H}^e \quad\text{and}\quad \mathbf{E}^p = \mathrm{sym}\,\mathbf{H}^p. \tag{75.9}$$

Similarly, $\mathbf{W}^e$ and $\mathbf{W}^p$ are the elastic and plastic rotations

$$\mathbf{W}^e = \mathrm{skw}\,\mathbf{H}^e \quad\text{and}\quad \mathbf{W}^p = \mathrm{skw}\,\mathbf{H}^p, \tag{75.10}$$

so that

$$\mathbf{E} = \mathbf{E}^e + \mathbf{E}^p \quad\text{and}\quad \mathbf{W} = \mathbf{W}^e + \mathbf{W}^p. \tag{75.11}$$

---

362 Theories of small deformations do not distinguish between the observed and referential spaces.
363 Cf. (2.46) and $(2.61)_3$.
364 Cf. (2.78).

Finally, consistent with the general observation that the flow of dislocations does not induce changes in volume,[365] $\mathbf{H}^p$ *is deviatoric*, viz.

$$\operatorname{tr}\mathbf{H}^p = \operatorname{tr}\mathbf{E}^p = 0. \tag{75.12}$$

If, in the free-energy imbalance (75.5), we account explicitly for the elasticity and plasticity of the material via $(75.11)_1$, we then find that

$$\dot{\psi} - \mathbf{T}:\dot{\mathbf{E}}^e - \mathbf{T}:\dot{\mathbf{E}}^p = -\delta \leq 0. \tag{75.13}$$

Further, since $\dot{\mathbf{E}}^p$ is deviatoric, we may conclude that

$$\mathbf{T}:\dot{\mathbf{E}}^p = \mathbf{T}_0:\dot{\mathbf{E}}^p;$$

hence, the free-energy imbalance (75.13) becomes

$$\dot{\psi} - \mathbf{T}:\dot{\mathbf{E}}^e - \mathbf{T}_0:\dot{\mathbf{E}}^p = -\delta \leq 0. \tag{75.14}$$

75.3 Separability Hypothesis

Most theories of plasticity are based on constitutive relations that separate plastic and elastic response, with free energy strictly elastic and dissipation solely plastic. Further, most successful theories are not based on a single constitutive relation for the stress, but instead require that the stress be consistent with two separate constitutive relations:

(i) one elastic, associated with the classical elastic strain energy;
(ii) a second viewed as a constraint on purely elastic response imposed by the plasticity of the material.

This is the approach we shall take.[366]

In accord with the assumption of a free energy that is strictly elastic and dissipation that is solely plastic, we separate the free-energy imbalance (75.14) into an **elastic balance**

$$\dot{\psi} = \mathbf{T}:\dot{\mathbf{E}}^e, \tag{75.15}$$

and a **dissipation inequality**

$$\delta = \mathbf{T}_0:\dot{\mathbf{E}}^p \geq 0 \tag{75.16}$$

characterizing energy dissipated during plastic flow.

In the next few sections, we introduce constitutive assumptions that characterize the elasticity and the plasticity of the material. We assume throughout that the body is homogeneous and isotropic.

75.4 Constitutive Characterization of Elastic Response

We neglect defect energy. Consistent with this, we assume that the free energy and the stress are independent of $\mathbf{E}^p$; specifically, we take as our starting point the free energy (52.60) and stress (52.59) appropriate to an isotropic linearly elastic solid, but we replace the strain $\mathbf{E}$ in these relations with the elastic strain $\mathbf{E}^e$:

$$\begin{aligned}\psi &= \mu|\mathbf{E}^e|^2 + \tfrac{1}{2}\lambda(\operatorname{tr}\mathbf{E}^e)^2,\\ \mathbf{T} &= 2\mu\mathbf{E}^e + \lambda(\operatorname{tr}\mathbf{E}^e)\mathbf{1}.\end{aligned} \tag{75.17}$$

[365] Cf., e.g., Bridgman (1952) and Spitzig, Sober & Richmond (1975).
[366] An alternative approach consistent with classical ideas regarding the formulation of constitutive equations is discussed in §84.

To ensure positive-definiteness of the free energy, we assume that the Lamé moduli μ and λ are consistent with the inequalities $\mu > 0$ and $2\mu + 3\lambda > 0$.[367] By (75.17), the elastic balance (75.15) is satisfied and we have only to consider the dissipation inequality (75.16). To ensure satisfaction of this inequality in all "constitutive processes" requires that additional constitutive restrictions — involving the plastic strain-rate $\dot{\mathbf{E}}^p$ — be placed on the deviatoric stress $\mathbf{T}_0$. In this regard note that, by $(75.17)_2$,

$$\mathbf{T}_0 = 2\mu\mathbf{E}_0^e. \tag{75.18}$$

[367] Cf. (81.54).

76 Formulation of the Mises Theory of Plastic Flow

We begin with a discussion of the classical constitutive theory of Lévy, Mises, Prandtl, and Reuss. It seems best to discuss this theory using conventional notation and Cartesian tensors: σ'_{ij} is the deviatoric stress, ε_{ij} is the ij-th component of total strain, ε^p_{ij} is the ij-th component of the plastic strain. HILL (1950, pp. 38–39), in discussing the foundations of plasticity theory, writes:[368] "Theoretical speculation about the relation between stress and strain [for plastic materials] began in 1870 with Saint-Venant's treatment of plane plastic strain. With great physical insight Saint-Venant proposed that the principal axes of the strain–increment (and not the total strain) coincided with the principal axes of stress. ... A general relationship between the ratios of the components of the strain-increment and the stress ratios was first suggested by Lévy (1871). Lévy's work remained largely unknown outside his own country, and it was not until the same equations were suggested independently by VON MISES in 1913 that they became widely used as the basis of plasticity theory. The Lévy–Mises equations, as they are known, may be expressed ... compactly ... as

$$d\varepsilon_{ij} = \sigma'_{ij}\,d\lambda,$$

where $d\lambda$ is [an arbitrary] scalar factor of proportionality [that generally depends on position and time]. Since LÉVY and von Mises used the total strain–increment, the equations are strictly applicable only to a fictitious material in which the elastic strains are zero. ... The extension of the Lévy–Mises equations to allow for the elastic component of strain was carried out by PRANDTL (1924) for the plane problem, and in complete generality by Reuss (1930). REUSS assumed that

$$d\varepsilon^p_{ij} = \sigma'_{ij}\,d\lambda.\text{"} \tag{76.1}$$

In the Lévy–Mises–Reuss theory, the equations (76.1) are supplemented by a yield condition

$$\sigma'_{ij}\sigma'_{ij} = 2k^2, \tag{76.2}$$

due to VON MISES (1913).[369]

Most modern discussions of plasticity are based on generalizations and structural variations of the theory of Lévy, Mises, Prandtl, and Reuss, and for that reason our treatment of plasticity begins with a discussion of that theory; however,

- *because the presence of the arbitrary increment $d\lambda$ in* (76.1) *is inconsistent with the modern notion of a constitutive equation, the path we take to this theory is slightly different from that familiar to most plasticians.*

To justify our approach, we begin by writing (76.1) and (76.2) in our notation:

$$\dot{\mathbf{E}}^p = \dot{\lambda}\mathbf{T}_0 \qquad \text{and} \qquad |\mathbf{T}_0| = \sqrt{2}k \qquad \text{for} \qquad \dot{\lambda} \neq 0. \tag{76.3}$$

[368] Hill's footnotes, which give complete citations, are omitted. The relevant references for us are LÉVY (1871), VON MISES (1913), and REUSS (1930).

[369] As noted by HILL (1950, p. 20): "Von Mises' criterion was anticipated, to some extent, by HUBER (1904) in a paper in Polish which did not attract general attention until nearly twenty years later." According to HILL (1950), this criterion was also "anticipated by Clerk Maxwell in a letter to W. Thomson dated December 18, 1856." Huber's paper has been recently translated into English and published in the Archives of Mechanics **56** (2004), 171–172.

Combining these equations and using (75.16), we obtain

$$\dot{\lambda} = \frac{|\dot{\mathbf{E}}^p|}{\sqrt{2}k}, \tag{76.4}$$

an equation that allows us to rewrite (76.3) as a single equation

$$\mathbf{T}_0 = \sqrt{2}k\frac{\dot{\mathbf{E}}^p}{|\dot{\mathbf{E}}^p|} \quad \text{for} \quad \dot{\mathbf{E}}^p \neq \mathbf{0} \tag{76.5}$$

(in which the arbitrary field $\dot{\lambda}$ does not appear). We find (76.5) preferable to (76.3) for the following reasons:

- The equation (76.5) encapsulates the yield condition $(76.3)_2$; there is no need to introduce the arbitrary field $\dot{\lambda}$. (Defining $\dot{\lambda}$ via (76.4) reduces (76.5) to $(76.3)_1$.)
- The underlying structure of (76.5) with deviatoric stress as dependent constitutive variable[370] is consistent with modern constitutive theory as presented in this book.[371]

Throughout this book we refer to constitutive relations of the general structure (76.5) as **Mises flow rules**; in so doing, we do not mean to diminish the contributions of St. Venant, Lévy, Prandtl, and Reuss.

76.1 General Constitutive Equations for Plastic Flow

As noted previously, theories of plasticity are typically based on constitutive relations that separate plastic and elastic response: In accord with this, and guided by the dissipation inequality (75.16) and the equations of Lévy, Mises, Prandtl, and Reuss in the form (76.5),[372] we consider a constitutive relation giving the deviatoric stress $\mathbf{T}_0$ when the material is flowing and provided that the plastic strain–rate $\dot{\mathbf{E}}^p$ is known. But such a dependence is not by itself sufficiently robust to model *strain-hardening*, a phenomenon characterized by an increase in flow resistance with plastic deformation. For that reason we begin with a constitutive relation of the form[373]

$$\mathbf{T}_0 = \hat{\mathbf{T}}_0(\dot{\mathbf{E}}^p, S), \tag{76.6}$$

with S a scalar internal variable introduced to characterize *strain-hardening*. We refer to S as the **hardening variable**.[374] Internal variables generally evolve according to differential equations. We follow this tradition and complete the general constitutive theory with the assumption that S evolve according to a differential equation — called the **hardening equation** — of the form

$$\dot{S} = h(\dot{\mathbf{E}}^p, S). \tag{76.7}$$

We refer to $h(\dot{\mathbf{E}}^p, S)$ as the **hardening rate**.

---

370 Cf. Gurtin (2000b, 2003).

371 Truesdell & Noll (1965, p. 56) lay down a principle of determinism for stress which asserts that "the stress in a body is determined by the history of the motion of that body," a postulate they credit to Cauchy.

372 Cf. the two bullets following (76.5). Further, taking deviatoric stress as dependent constitutive variable allows us to introduce and discuss the notion of rate-independence and, in addition, provides a structure that generalizes in a straightforward manner to include constitutive dependencies on plastic–strain gradients and their rates.

373 We develop our theory based on this constitutive assumption, but we show in §77 that, for a conventional rate-independent theory, $\dot{\mathbf{E}}^p$ may be expressed as a function of the total strain rate $\dot{\mathbf{E}}$, the deviatoric stress $\mathbf{T}_0$, and a hardening variable S; while, for a conventional rate-dependent theory, we show in §78.2 that $\dot{\mathbf{E}}^p$, under a suitable invertibility assumption, is determined by $\mathbf{T}_0$ and S.

374 More generally, internal variables are used to describe phenomena at a microstructural level.

Granted sufficient smoothness, the hardening equation when augmented by an initial condition giving S at some initial time, say $t = 0$, has a unique solution $S(t)$ on any time interval on which $\dot{\mathbf{E}}^p(t)$ is known. This observation and (76.6) render precise the well-known view that *the stress* $\mathbf{T}_0$ *depends on the past history of the plastic strain.* While we develop the general theory within a framework based on the abstract notion of a hardening variable as embraced in the differential equation (76.7), a concrete example of such a variable is the **accumulated plastic strain** e^p, an internal variable presumed to evolve according to the differential equation

$$\dot{e}^p = |\dot{\mathbf{E}}^p|, \tag{76.8}$$

a differential equation that we generally supplement with the initial condition

$$e^p(\mathbf{X}, 0) = 0. \tag{76.9}$$

The notion of accumulated plastic strain is due to Hill (1950, p. 30). We show in §76.6 that *for rate-independent materials the accumulated plastic strain represents the most general hardening variable*, so that the generality afforded by (76.7) is illusory.

Remark. *In considering* e^p *as defined by* (76.8) *and* (76.9) *it is tacit that*

$$\mathbf{E}^p(\mathbf{X}, 0) = \mathbf{0}. \tag{76.10}$$

Because most metals exhibit rate-independent or nearly rate-independent response at low temperatures, we begin with a discussion of isotropic, *rate-independent* constitutive relations involving the deviatoric stress and the plastic strain–rate. We later generalize these to account for slightly rate-dependent behavior.

76.2 Rate-Independence

The spatial dependence of fields (described referentially) is irrelevant to our discussion of rate-independence and can safely be suppressed.

A **change in time scale** is a transformation of the form

$$t_* = \tau + \kappa t, \tag{76.11}$$

with $\kappa > 0$ the **rate constant** of the transformation and τ an arbitrarily chosen time. In the present discussion we may, without loss in generality, take $\tau = 0$ and write (76.11) in the form

$$t_* = \kappa t. \tag{76.12}$$

Let Θ denote an arbitrary function of time. In what follows $\dot{\Theta}$ denotes the derivative of the function Θ, no matter the argument:

$$\dot{\Theta}(\bullet) = \frac{d\Theta(\bullet)}{d\bullet}. \tag{76.13}$$

Under the time-scale change (76.12) the function Θ transforms to the function Θ_κ, with

$$\Theta_\kappa(t) = \Theta(\kappa t), \tag{76.14}$$

so that, by the chain-rule,

$$\begin{aligned}\dot{\Theta}_\kappa(t) &= \frac{d}{dt}\Theta(\kappa t)\\ &= \left(\frac{d}{dt^*}\Theta(t^*)\Big|_{t^*=\kappa t}\right)\kappa\\ &= \kappa\dot{\Theta}(\kappa t).\end{aligned} \tag{76.15}$$

Assume that the material is rate-independent, or, more precisely, that the constitutive equations (76.6) and (76.7) are rate-independent.[375] Consider first the constitutive equation (76.6) for the deviatoric stress. By (76.14) and (76.15), under the time-scale change (76.12) this equation has the form

$$\begin{aligned}\mathbf{T}_{0\kappa}(t) &= \hat{\mathbf{T}}_0\big(\dot{\mathbf{E}}^p_\kappa(t), S_\kappa(t)\big)\\ &= \hat{\mathbf{T}}_0\big(\kappa\dot{\mathbf{E}}^p(\kappa t), S(\kappa t)\big),\end{aligned} \tag{76.16}$$

and, for (76.6) to be rate-independent, (76.6) evaluated at time κt must be equal to the right side of (76.16). Thus, omitting the argument κt, we must have

$$\hat{\mathbf{T}}_0(\dot{\mathbf{E}}^p, S) = \hat{\mathbf{T}}_0(\kappa\dot{\mathbf{E}}^p, S) \tag{76.17}$$

and, since the rate constant κ was chosen arbitrarily, (76.17) must hold for *all* rate constants $\kappa > 0$ and all $\dot{\mathbf{E}}^p$ and S.

Choose $\dot{\mathbf{E}}^p \neq \mathbf{0}$ arbitrarily and — bearing in mind that κ is dimensionless — choose a reference flow-rate $d_0 > 0$ and take $\kappa = d_0|\dot{\mathbf{E}}^p|^{-1}$; then, by (76.17), the response function $\hat{\mathbf{T}}_0$ must have the specific form

$$\hat{\mathbf{T}}_0(\dot{\mathbf{E}}^p, S) = \hat{\mathbf{T}}_0\left(d_0\frac{\dot{\mathbf{E}}^p}{|\dot{\mathbf{E}}^p|}, S\right).$$

Thus, redefining $\hat{\mathbf{T}}_0$ to absorb the constant d_0, we see that the most general rate-independent constitutive relation of the form $\mathbf{T}_0 = \hat{\mathbf{T}}_0(\dot{\mathbf{E}}^p, S)$ must have the specific form

$$\mathbf{T}_0 = \hat{\mathbf{T}}_0\left(\frac{\dot{\mathbf{E}}^p}{|\dot{\mathbf{E}}^p|}, S\right). \tag{76.18}$$

Further, since

$$\frac{\dot{\mathbf{E}}^p}{|\dot{\mathbf{E}}^p|} \quad \text{is defined only when} \quad \dot{\mathbf{E}}^p \neq \mathbf{0}, \tag{76.19}$$

the constitutive relation (76.18) *is defined only when* $\dot{\mathbf{E}}^p \neq \mathbf{0}$; that is, (76.18) is defined only when there is **plastic flow**.

In the space of symmetric, deviatoric tensors,

$$\boxed{\mathbf{N}^p \stackrel{\text{def}}{=} \frac{\dot{\mathbf{E}}^p}{|\dot{\mathbf{E}}^p|}} \tag{76.20}$$

375 The precise meaning of rate-independence should be clear from the ensuing discussion. Cf. GURTIN (2000b), who uses (a slightly flawed notion of) rate-independence to reduce constitutive relations for plastic flow in single crystals.

represents the (plastic) **flow direction**. Using the flow direction we can express the general rate-independent constitutive relation (76.18) in the form

$$\boxed{\mathbf{T}_0 = \hat{\mathbf{T}}_0(\mathbf{N}^p, S).} \tag{76.21}$$

We refer to the relation (76.21) as a **flow rule** because *it gives the deviatoric stress* $\mathbf{T}_0$ *whenever the flow direction* $\mathbf{N}^p$ *is known.* To avoid pedantry, let us agree that

- *whenever the flow direction* $\mathbf{N}^p$ *is mentioned — for example in an equation — it is tacit that there is flow*:

$$\dot{\mathbf{E}}^p \neq \mathbf{0}.$$

Consider, next, the evolution equation (76.7). By (76.15), under the time-scale change (76.12) the differential equation (76.7) takes the form

$$\kappa \dot{S} = h(\kappa \dot{\mathbf{E}}^p, S)$$

or, equivalently,

$$\dot{S} = \kappa^{-1} h(\kappa \dot{\mathbf{E}}^p, S).$$

Rate-independence requires that this equation hold for all $\kappa > 0$; if we consider an arbitrary time t and choose $\kappa = d_0|\dot{\mathbf{E}}^p(t)|^{-1}$, we arrive at the relation[376]

$$\dot{S}(t) = d_0^{-1}|\dot{\mathbf{E}}^p(t)|h(d_0\mathbf{N}^p(t), S(t)).$$

Since t was chosen arbitrarily, this relation must hold for all time. Finally, if we redefine $h(\mathbf{N}^p, S)$ to absorb the constant d_0 in $d_0^{-1}h(d_0\mathbf{N}^p, S)$, we arrive at the rate-independent form of the hardening equation:

$$\dot{S} = h(\mathbf{N}^p, S)|\dot{\mathbf{E}}^p|. \tag{76.22}$$

The constitutive theory developed thus far for rate-independent plastic flow consists of two relations:

(i) the *flow rule* (76.21);
(ii) the *hardening equation* (76.22).

76.3 Strict Dissipativity

Assuming that there is plastic flow ($\dot{\mathbf{E}}^p \neq \mathbf{0}$) and that the flow rule (76.21) is satisfied, we can rewrite the dissipation

$$\delta = \mathbf{T}_0 : \dot{\mathbf{E}}^p \geq 0$$

in the form

$$\begin{aligned}\delta &= \hat{\delta}(\dot{\mathbf{E}}^p, S)\\ &= \hat{\mathbf{T}}_0(\mathbf{N}^p, S) : \dot{\mathbf{E}}^p \geq 0.\end{aligned} \tag{76.23}$$

A basic hypothesis — and one tacit throughout this book — is that plastic flow incurs dissipation in the sense that

$$\hat{\delta}(\dot{\mathbf{E}}^p, S) > 0 \quad \text{whenever} \quad \dot{\mathbf{E}}^p \neq \mathbf{0}. \tag{76.24}$$

376 The scalar d_0 is required for dimensional consistency.

This requirement, which is referred to as **strict dissipativity**, has an important consequence; namely, granted (76.21),

$$\mathbf{T}_0 \neq \mathbf{0} \quad \text{whenever} \quad \dot{\mathbf{E}}^p \neq \mathbf{0}. \tag{76.25}$$

Thus, when there is plastic flow, dividing (76.23) by $|\dot{\mathbf{E}}^p|$ and using (76.24) yields

$$Y(\mathbf{N}^p, S) \stackrel{\text{def}}{=} \mathbf{N}^p : \hat{\mathbf{T}}_0(\mathbf{N}^p, S) > 0. \tag{76.26}$$

The quantity $Y(\mathbf{N}^p, S)$ represents the dissipation $\hat{\delta}(\dot{\mathbf{E}}^p, S)$ measured per unit $|\dot{\mathbf{E}}^p|$.

76.4 Formulation of the Mises Flow Equations

We now introduce specific assumptions regarding the constitutive relations (76.21) and (76.22); these assumptions are satisfied by a large class of materials discussed in the literature.

A defining property of a plastic solid is its ability to flow like a fluid. In this regard, note that the constitutive equation for the extra stress $\mathbf{T}_0$ as a function of the stretching $\mathbf{D}$ in an incompressible Newtonian fluid[377] has the simple form $\mathbf{T}_0 = 2\mu\mathbf{D}$ and, hence, trivially satisfies

$$\frac{\mathbf{T}_0}{|\mathbf{T}_0|} = \frac{\mathbf{D}}{|\mathbf{D}|}, \tag{76.27}$$

granted $\mathbf{D} \neq \mathbf{0}$.

The counterpart of (76.27) for a plastic solid is the subject of the following hypothesis requiring that

(CD) (**codirectionality**) *the direction of the deviatoric stress and the direction of the plastic strain-rate coincide:*[378]

$$\frac{\mathbf{T}_0}{|\mathbf{T}_0|} = \frac{\dot{\mathbf{E}}^p}{|\dot{\mathbf{E}}^p|} \equiv \mathbf{N}^p \qquad \text{whenever} \qquad \dot{\mathbf{E}}^p \neq \mathbf{0}. \tag{76.28}$$

Note that this condition is satisfied by the classical Reuss equation (76.3).

The codirectionality hypothesis requires more than the coincidence of the principal directions of the plastic strain-rate and the principal directions of the deviatoric stress (*coaxiality*). Codirectionality requires that the tensors themselves — viewed as vectors in the space of symmetric, deviatoric tensors — point in the same direction. For real materials the extent to which the codirectionality hypothesis (or (76.1)) is satisfied is characterized by comparison with a Lode diagram: for T_i and $\dot{E}_i^p$ the principal stresses and principal plastic strain-rates the Lode diagram is a graph of $\mu = (2\,T_3 - T_1 - T_2)/(T_1 - T_2)$ versus $\nu = (2\,\dot{E}_3^p - \dot{E}_1^p - \dot{E}_2^p)/(\dot{E}_1^p - \dot{E}_2^p)$.[379] The Lode diagram for a material consistent with (76.29) has $\mu = \nu$. Cf. Figure 8 of Hill (1950) for plots of μ versus ν from combined tension-torsion tests of Taylor & Quinney (1931) for copper, aluminum, and mild steel.

Next, by (76.28),

$$\mathbf{T}_0 = |\mathbf{T}_0|\mathbf{N}^p; \tag{76.29}$$

thus, by (76.26),

$$\begin{aligned} |\mathbf{T}_0| &= \mathbf{T}_0 : \mathbf{N}^p \\ &= Y(\mathbf{N}^p, S). \end{aligned} \tag{76.30}$$

377 Cf. §46.2.

378 This form of the codirectionality hypothesis is generally inapplicable for theories that involve a backstress; cf. Exercise 2 on page 441, where it is the dissipative part of the deviatoric stress and the flow direction that are codirectional.

379 Note that μ and ν, as defined here, are not the same as the elastic shear modulus and Poisson's ratio.

An important consequence of (76.29) and (76.30) is that the flow rule $\mathbf{T}_0 = \hat{\mathbf{T}}_0(\mathbf{N}^p, S)$ must have the specific form

$$\mathbf{T}_0 = Y(\mathbf{N}^p, S)\mathbf{N}^p \qquad \text{for } \dot{\mathbf{E}}^p \neq \mathbf{0}. \tag{76.31}$$

We refer to the scalar constitutive variable $Y(\mathbf{N}^p, S)$ as the **flow resistance**. By (76.26) and (76.30),

- *during plastic flow the flow resistance is strictly positive and coincides with the magnitude of the deviatoric stress.*

In view of (76.23), (76.31), and the identity

$$\mathbf{N}^p : \dot{\mathbf{E}}^p = |\dot{\mathbf{E}}^p|, \tag{76.32}$$

we can write the dissipation $\delta = \mathbf{T}_0 : \dot{\mathbf{E}}^p \geq 0$ in the simple form

$$\delta = Y(\mathbf{N}^p, S)|\dot{\mathbf{E}}^p|. \tag{76.33}$$

The codirectionality hypothesis, while central to our definition of a plastic material, is not sufficient to characterize even the simple Mises-type equation (76.5); we require, in addition, that[380]

(SI) (**strong isotropy**) *the flow resistance* $Y(\mathbf{N}^p, S)$ *and the function* $h(\mathbf{N}^p, S)$ *characterizing the hardening equation be* independent *of the flow direction* $\mathbf{N}^p$.

As a consequence of the strict dissipativity requirement (76.24), (CD), and (SI), the constitutive equations (76.22) and (76.31) take a simple form in which $Y = Y(S)$ and $h = h(S)$. Thus, summarizing, we have a result that is basic to much of what follows:

MISES RELATION AND HARDENING EQUATION *Granted rate-independence and strict dissipativity, as well as the codirectionality and strong isotropy hypotheses* (CD) *and* (SI), *the general constitutive relations* (76.6) *and* (76.7) *reduce to the* ***Mises flow equations***

$$\left.\begin{aligned} \mathbf{T}_0 &= Y(S)\mathbf{N}^p \qquad \textit{for } \dot{\mathbf{E}}^p \neq \mathbf{0}, \\ \dot{S} &= h(S)|\dot{\mathbf{E}}^p|. \end{aligned}\right\} \tag{76.34}$$

Conversely,[381] *the relations* (76.34) *with* $Y(S) > 0$ *are rate-independent, strictly dissipative, and consistent with* (CD) *and* (SI). *We refer to* $(76.34)_1$ *as the* **Mises flow rule**.

Note that, by (76.24) the dissipation (76.33) satisfies

$$\left.\begin{aligned} \delta &= Y(S)|\dot{\mathbf{E}}^p| \geq 0, \\ \delta &> 0 \quad \text{if} \quad \dot{\mathbf{E}}^p \neq \mathbf{0}. \end{aligned}\right\} \tag{76.35}$$

An important consequence of the Mises flow rule $(76.34)_1$ is the **yield condition**:[382]

$$|\mathbf{T}_0| = Y(S) \qquad \text{for } \dot{\mathbf{E}}^p \neq \mathbf{0}. \tag{76.36}$$

---

[380] The standard assumption of isotropy for $Y(\mathbf{N}^p, S)$ and $h(\mathbf{N}^p, S)$ is weaker: It requires that these functions depend on $\mathbf{N}^p$ through $\det \mathbf{N}^p$, which is the only nontrivial invariant of $\mathbf{N}^p$; cf. Exercise 3 on page 434.

[381] Cf. Exercise 4 on page 434.

[382] Cf. MISES (1913). In the plasticity literature, the Mises yield condition is often written in terms of an equivalent tensile stress $\bar{\sigma}$ or an equivalent shear stress $\bar{\tau}$, where

$$\bar{\sigma} = \sqrt{\tfrac{3}{2}}|\mathbf{T}_0|, \qquad \bar{\tau} = \tfrac{1}{\sqrt{2}}|\mathbf{T}_0|.$$

Here, to avoid cumbersome factors of $\sqrt{3/2}$ or $1/\sqrt{2}$, we simply use $|\mathbf{T}_0|$.

In contrast, $(76.34)_1$ *asserts nothing about the deviatoric stress* $\mathbf{T}_0$ *when* $\dot{\mathbf{E}}^p = \mathbf{0}$ (and there is no plastic flow). When Y is constant, the material is said to be *perfectly plastic*. We leave it as an exception to show that

- *granted the yield condition* (76.36), *the Mises flow rule* $(76.34)_1$ *for a perfectly plastic material is equivalent to the Reuss equations* (76.1).

The constitutive relations (76.34) do not suffice to characterize rate-independent plastic materials an *additional assumption*, namely the **boundedness hypothesis**[383]

$$\boxed{|\mathbf{T}_0| \le Y(S)} \tag{76.37}$$

— asserting that the norm of the deviatoric stress not exceed the flow resistance — is needed. Note that when $|\mathbf{T}_0| < Y(S)$ the yield condition (76.36) is not satisfied, so that, necessarily, $\dot{\mathbf{E}}^p = \mathbf{0}$; therefore

$$\dot{\mathbf{E}}^p = \mathbf{0} \qquad \text{for } |\mathbf{T}_0| < Y(S). \tag{76.38}$$

The central results of this section, summarized as follows, define the **Mises theory of rate-independent plastic response**:

$$\left.\begin{aligned} \mathbf{T}_0 = Y(S)\mathbf{N}^p \quad \text{for } \dot{\mathbf{E}}^p \neq \mathbf{0} \qquad &\text{(Mises flow rule)},\\ \dot{S} = h(S)|\dot{\mathbf{E}}^p| \qquad &\text{(hardening equation)},\\ |\mathbf{T}_0| \le Y(S) \qquad &\text{(boundedness hypothesis)},\\ \dot{\mathbf{E}}^p = \mathbf{0} \quad \text{for } |\mathbf{T}_0| < Y(S) \qquad &\text{(no-flow condition)}. \end{aligned}\right\} \tag{76.39}$$

The basic nonkinematic constitutive assumptions of the theory thus reduce to (75.17), which characterizes the elastic response of the material, and (76.39), which characterizes plastic flow. The basic elastic "stress-strain" relation $(75.17)_2$ is assumed to hold in all motions of the body, even during plastic flow, but during plastic flow the deviatoric stress $\mathbf{T}_0$ is constrained, via $(76.39)_{2,3}$, by restrictions imposed by the plasticity of the material.

Remark. The relation $(76.39)_1$, which is basic to what follows, is a functional relationship giving $\mathbf{T}_0$ when $\mathbf{N}^p$ and S are known and, hence, rules out the classical Tresca theory, since on each face of the Tresca-hexagon a single flow direction corresponds to a range of values of the deviatoric stress $\mathbf{T}_0$.[384] As Hill (1950, p. 21) remarks: "For most metals the Mises criterion fits the data more closely than Tresca's."

EXERCISES

1. Establish the assertion made in the bullet on page 433.
2. Show that the constitutive relations (76.39) are isotropic.

---

[383] Cf. §79.2, where the boundedness hypothesis is shown to be a consequence of the notion of maximum dissipation.

[384] Cf. Hill (1950, p. 19) and Malvern (1969, p. 348).

3. Use the following steps to give an alternative derivation of the constitutive equations $(76.39)_{1,2}$ based primarily on isotropy:
   (a) Show that, since $\mathbf{N}^p$ is symmetric and deviatoric and has unit magnitude, its principal invariants as defined in (2.142) are
$$I_1(\mathbf{N}^p) = 0, \qquad I_2(\mathbf{N}^p) = -\tfrac{1}{2}, \qquad I_3(\mathbf{N}^p) = \det\mathbf{N}^p. \tag{76.40}$$
   (b) Show that, consequently, the functions $h(\mathbf{N}^p, S)$ and $\hat{\mathbf{T}}_0(\mathbf{N}^p, S)$ are given by
$$\begin{aligned}\hat{\mathbf{T}}_0(\mathbf{N}^p, S) &= \alpha_1(\det\mathbf{N}^p, S)\mathbf{N}^p + \alpha_2(\det\mathbf{N}^p, S)\big[\mathbf{N}^{p2} - \tfrac{1}{3}\mathbf{1}\big]\\ h(\mathbf{N}^p, S) &= h(\det\mathbf{N}^p, S).\end{aligned} \tag{76.41}$$
   (Use the representation theorems for isotropic scalar and tensor functions (113.3) and (113.6), together with the fact that $\mathbf{T}_0$ is deviatoric.)
   (c) Show that if the moduli α_1, α_2, and h are independent of $\det\mathbf{N}^p$ and if[385]
$$\hat{\mathbf{T}}_0(\mathbf{N}^p, S) = -\hat{\mathbf{T}}_0(-\mathbf{N}^p, S), \tag{76.42}$$
   then the constitutive equations (76.41) reduce to $(76.39)_{1,2}$.
4. Show that the Mises flow rule $(76.34)_1$ and the hardening equation $(76.34)_2$ are rate-independent and consistent with (CD) and (SI).

76.5 Initializing the Mises Flow Equations

Granted that the functions $Y(S)$ and $h(S)$ are smooth, the hardening variable S in the Mises flow equations
$$\begin{aligned}\mathbf{T}_0 &= Y(S)\mathbf{N}^p \qquad \text{for } \dot{\mathbf{E}}^p \neq \mathbf{0},\\ \dot{S} &= h(S)|\dot{\mathbf{E}}^p|\end{aligned} \tag{76.43}$$
is determined uniquely when and only when it is known at some initial time, say $t = 0$. Let S_0 denote this initial value,
$$S(\mathbf{X}, 0) = S_0, \tag{76.44}$$
and assume that S_0 is constant.[386] In discussing the initial condition (76.44), we consider two special cases:

(i) equations with $Y(S)$ not identically equal to S;
(ii) equations with $Y(S) \equiv S$, so that the flow resistance is taken as the hardening variable.

76.5.1 Flow Equations With $Y(S)$ not Identically Equal to S

If we introduce a new hardening variable S^* defined by
$$S^* = S - S_0 \tag{76.45}$$
and new functions $Y^*(S^*)$ and $h^*(S^*)$ defined by
$$Y^*(S^*) = Y(S^* + S_0) \qquad \text{and} \qquad h^*(S^*) = h(S^* + S_0), \tag{76.46}$$

[385] And hence that $(76.41)_1$ displays no Bauschinger effect; cf. the discussion in §74.

[386] The choice of a constant initial condition S_0 in (76.43) is equivalent to our initial assumption of homogeneous constitutive relations; were $S_0 = S_0(\mathbf{X})$, then the transformated functions (76.46) would not be homogeneous.

then the Mises flow equations become

$$\left.\begin{aligned} \mathbf{T}_0 &= Y^*(S^*)\mathbf{N}^p \qquad \text{for } \dot{\mathbf{E}}^p \neq \mathbf{0}, \\ \dot{S} &= h^*(S^*)|\dot{\mathbf{E}}^p|, \end{aligned}\right\} \tag{76.47}$$

and the appropriate initial condition is given by

$$S^*(\mathbf{X}, 0) = 0. \tag{76.48}$$

Granted the transformations (76.45) and (76.46), the equations (76.43) and (76.44) are thus equivalent to the equations (76.47) and (76.48). The difference in the forms of these two systems is simply a difference in initial conditions. For specificity we use the system defined by (76.47) and (76.48), so that, with an obvious change in notation, we henceforth work with the Mises flow equations (76.43) subject to the null initial conditions

$$S(\mathbf{X}, 0) = 0. \tag{76.49}$$

We assume that the initial condition (76.49) holds only when the flow resistance is not the hardening variable. The case when the flow resistance is the hardening variable is discussed in the next section.

76.5.2 Theory with Flow Resistance as Hardening Variable

This choice of flow resistance is based on the assumption that[387]

$$Y(S) \equiv S. \tag{76.50}$$

In this case, the hardening variable S is the **flow resistance**, S_0 is the **initial flow resistance**, and the flow equations and initial condition become

$$\left.\begin{aligned} \mathbf{T}_0 &= S\mathbf{N}^p \qquad \text{for } \dot{\mathbf{E}}^p \neq \mathbf{0}, \\ \dot{S} &= h(S)|\dot{\mathbf{E}}^p| \qquad S(\mathbf{X}, 0) = S_0. \end{aligned}\right\} \tag{76.51}$$

EXERCISE

1. Consider the special case in which $Y(S) = cS$, with c constant. Transform the hardening variable via the transformation $S^* = cS$ and use transformations of Y and h analogous to (76.45) and (76.46) to show the transformed Mises flow equations and initial condition are of the same form as (76.51).

76.6 Solving the Hardening Equation. Accumulated Plastic Strain is the Most General Hardening Variable

In this section, we work within the framework, discussed on page 434, in which the flow resistance $Y(S)$ is not the hardening variable S:

$$Y(S) \not\equiv S.$$

We show that[388]

(†) *a theory based on a general hardening variable can be converted, without loss in generality, to a theory in which the hardening variable is the accumulated plastic strain e^p* defined by[389]

$$\dot{e}^p = |\dot{\mathbf{E}}^p|, \qquad e^p(\mathbf{X}, 0) = 0. \tag{76.52}$$

[387] Cf. Brown, Kim & Anand (1989) and Weber & Anand (1990).

[388] Within the present isothermal, rate-independent framework.

[389] Cf. (76.8) and the Remark on page 428.

We accomplish this by showing that the hardening equation equipped with suitable initial conditions can be solved to give a one-to-one correspondence between the hardening variable S and e^p.

Bearing in mind (76.49), the initial-value problem associated with the hardening equation $(76.39)_2$ is given by

$$\dot{S} = h(S)|\dot{\mathbf{E}}^p| \qquad \text{for all time} \geq 0, \qquad S(0) = 0. \tag{76.53}$$

We refer to (76.53) as the hardening problem because it describes the evolution of the hardening variable S *granted a knowledge of the function* $|\dot{\mathbf{E}}^p|$. To solve the hardening problem, we note that the initial-value problem (76.52) for the accumulated plastic strain can be trivially solved to give a function[390]

$$\hat{e}^p(t) = \int_0^t |\dot{\mathbf{E}}^p(\tau)|\, d\tau; \tag{76.54}$$

we then show that the hardening problem (76.53) may be rephrased as a standard initial-value problem

$$\frac{dS}{de^p} = h(S), \qquad e^p \geq 0, \qquad S(0) = 0. \tag{76.55}$$

We assume that the function $h(S)$ is smooth and uniformly bounded for $-\infty < S < \infty$; then (76.55) has a unique solution[391]

$$S = \hat{S}(e^p), \qquad e^p \geq 0. \tag{76.56}$$

We now show that the composite function

$$S(t) = \hat{S}(\hat{e}^p(t)) \tag{76.57}$$

furnishes a solution of (76.53). Since $\hat{e}^p(0) = 0$, $S(0) = 0$, which is $(76.53)_2$. By (76.54),

$$\dot{S} = \frac{d\hat{S}}{de^p}\dot{e}^p = h(S)|\dot{\mathbf{E}}^p|,$$

which is $(76.53)_1$. Thus, (76.57) represents a solution of (76.53).

Points S_{eq} that satisfy

$$h(S_{\text{eq}}) = 0 \tag{76.58}$$

are referred to as **equilibrium points** of the differential equation (76.55) because *a solution of* (76.55) *that reaches* S_{eq} *must remain there*:

$$\text{if } \hat{S}(e_0^p) = S_{\text{eq}}, \text{ then } \hat{S}(e^p) \equiv S_{\text{eq}} \text{ for all } e^p \geq e_0^p; \tag{76.59}$$

indeed, by (76.58), (76.59) represents a solution for $e^p \geq e_0^p$ and a second solution *not consistent with* (76.59) would violate uniqueness.

Because of (76.59), we may, without loss in generality, redefine the domain $\mathcal{H}$ of $h(S)$ such that

$$h(S) \neq 0 \tag{76.60}$$

[390] For convenience we have suppressed the argument **X**.

[391] The assumption of boundedness precludes solutions that become infinite at finite values of e^p.

on the *interior* of $\mathcal{H}$, such that $\mathcal{H}$ contains S_0 in its interior, and such that $\mathcal{H}$ equals $(-\infty, \infty)$, or a semi-infinite interval with one endpoint an equilibrium point, or an interval with both endpoints equilibrium points. Granted this, we may conclude from $(76.55)_1$ that

(‡) *the solution* $S = \hat{S}(e^p)$ *of the initial-value problem* (76.55) *is a strictly monotone function and hence represents a one-to-one correspondence between accumulated plastic strains* e^p *and values* S *of the hardening variable.*

As we shall see, the result (‡) allows us to omit mention of the hardening equation. Specifically, given any function $f(S)$, if we let

$$\hat{f}(e^p) = f(\hat{S}(e^p)),$$

then by (76.26)

$$\hat{Y}(e^p) = Y(\hat{S}(e^p)) > 0, \tag{76.61}$$

and the Mises flow rule $(76.39)_1$ has the form

$$\mathbf{T}_0 = \hat{Y}(e^p)\frac{\dot{\mathbf{E}}^p}{|\dot{\mathbf{E}}^p|} \quad \text{for } \dot{\mathbf{E}}^p \neq \mathbf{0}. \tag{76.62}$$

Further, since

$$\frac{d\hat{S}(e^p)}{de^p} = h(\hat{S}(e^p)) = \hat{h}(e^p), \tag{76.63}$$

it follows that, by (76.109),

$$\begin{aligned}\frac{d\hat{Y}(e^p)}{de^p} &= Y'(\hat{S}(e^p))\frac{d\hat{S}(e^p)}{de^p}\\ &= Y'(S)h(S)\big|_{S=\hat{S}(e^p)}\\ &= \hat{H}(e^p).\end{aligned} \tag{76.64}$$

Thus, by (76.64), *the material hardens or softens according as*

$$\hat{H}(e^p) > 0 \qquad \text{or} \qquad \hat{H}(e^p) < 0. \tag{76.65}$$

The function $\hat{Y}(e^p)$ gives the flow resistance as a function of the accumulated plastic strain and, by (76.65), the material hardens or softens according as the flow resistance increases or decreases with accumulated plastic strain. We refer to $Y(e^p)$ as the *strain-hardening curve*; what makes this formulation of the theory important is that

- the strain isotropic hardening may be specified *a priori*; it does not involve solving a differential equation.

A **hardening transition** is defined by a value e^p_{T} of e^p at which

$$\hat{H}(e^p) = \frac{d\hat{Y}(e^p)}{de^p} \tag{76.66}$$

suffers a change in sign; $e^p = e^p_{\mathrm{T}}$ therefore corresponds to a *transition from hardening to softening* if, near e^p_{T},

$$\begin{aligned}\frac{d\hat{Y}(e^p)}{de^p} &> 0 \qquad \text{for} \qquad e^p < e^p_{\mathrm{T}},\\ \frac{d\hat{Y}(e^p)}{de^p} &< 0 \qquad \text{for} \qquad e^p > e^p_{\mathrm{T}},\end{aligned}$$

and an analogous result holds for a transition from softening to hardening. Thus:

(i) *for a transition at* e^p_{T} *from hardening to softening*

$$\hat{Y}(e^p) \text{ has a strict local maximum at } e^p = e^p_{\mathrm{T}}; \tag{76.67}$$

(ii) *for a transition at* e^p_{T} *from softening to hardening*

$$\hat{Y}(e^p) \text{ has a strict local minimum at } e^p = e^p_{\mathrm{T}}. \tag{76.68}$$

We now summarize the central result of this section:

MISES FLOW EQUATIONS BASED ON ACCUMULATED PLASTIC STRAIN *Granted the initial condition* $S(0) = 0$, *the formulation of rate-independent plasticity based on the general Mises flow equations* (76.39) *may, without loss in generality, be replaced by a formulation in which the hardening variable is the accumulated plastic strain*[392]

$$\begin{aligned} \mathbf{T}_0 &= \hat{Y}(e^p)\frac{\dot{\mathbf{E}}^p}{|\dot{\mathbf{E}}^p|} \qquad \text{for } \dot{\mathbf{E}}^p \neq \mathbf{0}, \\ \dot{e}^p &= |\dot{\mathbf{E}}^p|, \qquad e^p(0) = 0. \end{aligned} \tag{76.69}$$

Remark. Many different function pairs $\big(Y(S), h(S)\big)$ yield the same isotropic hardening $Y(e^p)$ and hence yield theories indistinguishable from one and other. Further, given a strictly positive function $Y(e^p)$, defined for $e^p \geq 0$, (76.69) represents a special case of the general Mises flow equations (76.39) (and initial condition $S(0) = 0$) in which $S = e^p$ and $h(S) = 1$. The physics may dictate a hardening equation that has no simple solution, or one may prefer to work with a hardening variable rather than the accumulated plastic strain e^p, since, by (76.67) and (76.68), the physical nature of such transitions is embodied in the function $\hat{Y}(e^p)$.

Notational agreement Henceforth, we work almost exclusively with the hardening variable $S = e^p$. To avoid excessive notation, we therefore write

$$Y(e^p) = \hat{Y}(e^p) \qquad \text{and} \qquad H(e^p) = \hat{H}(e^p),$$

so that the Mises flow rule has the form

$$\mathbf{T}_0 = Y(e^p)\frac{\dot{\mathbf{E}}^p}{|\dot{\mathbf{E}}^p|}.$$

Combining (76.39) and (76.56), we arrive at the **Mises–Hill equations**:

$$\boxed{\begin{aligned} &\mathbf{T}_0 = Y(e^p)\mathbf{N}^p \quad \text{for } \dot{\mathbf{E}}^p \neq \mathbf{0} \quad \text{(Mises flow rule)}, \\ &\dot{e}^p = |\dot{\mathbf{E}}^p|, \quad e^p(\mathbf{X}, 0) = 0 \quad \text{(hardening equation)}, \\ &|\mathbf{T}_0| \leq Y(e^p) \quad \text{(boundedness hypothesis)}, \\ &\dot{\mathbf{E}}^p = \mathbf{0} \quad \text{for } |\mathbf{T}_0| < Y(e^p) \quad \text{(no-flow condition)}; \end{aligned}} \tag{76.70}$$

these equations define **rate-independent plastic response** when the hardening variable is the accumulated plastic strain.

By (76.35), with S replaced by e^p,

$$\delta = Y(e^p)|\dot{\mathbf{E}}^p| \tag{76.71}$$

represents the dissipation associated with the Mises–Hill equations.

[392] Cf. HILL (1950), p. 30.

EXERCISE

1. The scalar field τ defined by

$$\tau = \mathbf{N}^p : \mathbf{T}_0 \tag{76.72}$$

may be termed the resolved shear because it represents the deviatoric stress $\mathbf{T}_0$ resolved on the direction of plastic flow. Show that the resolved shear obeys the constitutive relation

$$\tau = Y(e^p). \tag{76.73}$$

Show, further, that, conversely, granted the codirectionality hypothesis (CD) on 471, the relations (76.72) and (76.73) imply the Mises flow rule $(76.70)_1$.

76.7 Flow Resistance as Hardening Variable, Revisited

In this case, the Mises flow equations and initial condition have the form (76.51), viz.

$$\begin{aligned} \mathbf{T}_0 &= S\mathbf{N}^p \qquad \text{for } \dot{\mathbf{E}}^p \neq \mathbf{0}, \\ \dot{S} &= h(S)|\dot{\mathbf{E}}^p|, \qquad S(\mathbf{X}, 0) = S_0, \end{aligned} \tag{76.74}$$

with S the **flow resistance** and S_0 the **initial flow resistance**. Here the dissipation (76.35) becomes

$$\delta = S|\dot{\mathbf{E}}^p|, \tag{76.75}$$

while the flow resistance is given by $Y(S) = S$.[393] For these equations $Y'(S) \equiv 1$ and we may conclude from (76.64) that the function h that characterizes the hardening equation obeys

$$h(S) = H(S). \tag{76.76}$$

Thus, trivially, the hardening equation becomes

$$\dot{S} = H(S)|\dot{\mathbf{E}}^p| \tag{76.77}$$

and the steps leading to (76.56) are valid without change; since $Y(S) = S$, (76.56) gives the hardening curve $S = \hat{S}(e^p)$ and, as in the general theory, $\hat{S}(e^p)$ must be strictly monotone in e^p. We therefore have a conclusion that seems basic to a discussion of hardening transitions:

- *the theory with flow resistance as hardening variable cannot describe transitions from hardening to softening or from softening to hardening.*[394]

76.8 Yield Surface. Yield Function. Consistency Condition

Throughout this section, we work within the framework of the Mises–Hill equations (76.70), so that the accumulated plastic strain e^p is the hardening variable. The

393 Cf. (76.50).

394 Softening in metals is often caused by internal damage mechanisms such as the nucleation, growth, and coalescence of microcavities. Within the framework of (76.74), the single internal variable S is insufficient to describe macroscopic strain softening; additional internal variables describing damage are needed to account for hardening-softening transitions.

spherical surface with "radius" $Y(e^p)$ in the space of symmetric deviatoric tensors is called the **yield surface**;[395] a consequence of the yield condition (76.36) is then that

- *plastic flow is possible only when the deviatoric stress* $\mathbf{T}_0$ *lies on the yield surface.*

The closed ball with radius $Y(e^p)$ in the space of symmetric deviatoric tensors is the **elastic range**, because by $(76.70)_4$ plastic flow vanishes in the interior of the elastic range.

The ensuing discussion is simplified somewhat if we introduce a **yield function**[396]

$$f = |\mathbf{T}_0| - Y(e^p), \tag{76.78}$$

which, by $(76.70)_3$, is constrained by

$$-Y(e^p) \le f \le 0. \tag{76.79}$$

The yield condition (76.36) is then equivalent to the requirement that[397]

$$f = 0 \quad \text{for } \dot{\mathbf{E}}^p \ne \mathbf{0}, \tag{76.80}$$

and (76.38) takes the form

$$\dot{\mathbf{E}}^p = \mathbf{0} \qquad \text{for } f < 0. \tag{76.81}$$

The conditions (76.80) and (76.81) taken together imply that $|\dot{\mathbf{E}}^p| f = 0$, a relation known as the *Kuhn–Tucker condition*.[398]

Consider a fixed time t and assume that, at that time, $f(t) = 0$, so that the yield condition is satisfied. Then, by (76.79), $f(t + \tau) \le 0$ for all τ and, consequently, $\dot{f}(t) \le 0$.

Thus,

$$\text{if } f = 0, \text{ then } \dot{f} \le 0. \tag{76.82}$$

Next, if $f(t) = 0$ and $\dot{f}(t) < 0$, then $f(t + \tau) < 0$ for all sufficiently small $\tau > 0$, so that, by (76.81), $\dot{\mathbf{E}}^p(t + \tau) = \mathbf{0}$ for all such τ. Hence $\dot{\mathbf{E}}^p(t) = 0$. Thus

$$\text{if } f = 0 \text{ and } \dot{f} < 0, \text{ then } \dot{\mathbf{E}}^p = \mathbf{0}. \tag{76.83}$$

The equations (76.81) and (76.83) combine to form the **no-flow condition**:

$$\boxed{\dot{\mathbf{E}}^p = \mathbf{0} \text{ if } f < 0 \text{ or if } f = 0 \text{ and } \dot{f} < 0.} \tag{76.84}$$

Next, if $\dot{\mathbf{E}}^p \ne \mathbf{0}$ at a time t, then $\dot{\mathbf{E}}^p \ne \mathbf{0}$ in some neighborhood N of t, so that, by (76.80), $f = 0$ for all times in N; hence, $\dot{f}(t) = 0$. We therefore have the **consistency condition**:

$$\boxed{\text{if } \dot{\mathbf{E}}^p \ne \mathbf{0}, \text{ then } f = 0 \text{ and } \dot{f} = 0.} \tag{76.85}$$

EXERCISES

1. Making reference to (76.35), consider the dissipation δ viewed as a function of $(\dot{\mathbf{E}}^p, e^p)$:

$$\delta(\dot{\mathbf{E}}^p, e^p) = Y(e^p)|\dot{\mathbf{E}}^p|. \tag{76.86}$$

---

395 In this case, trivially, the flow direction is normal to the yield surface and the flow rule is referred to as being associative. Nonassociative flow rules are discussed in §79.3.

396 Considered as a function of $(\mathbf{X}, t)$ rather than $(\mathbf{T}, e^p)$.

397 As is often the case, the plastic strain-rate is discontinuous at the onset of plastic flow; for that reason — at that time — $\dot{\mathbf{E}}^p$ and various related fields should be considered as right-derivatives.

398 Cf., e.g., Simo & Hughes (1998).

Use the identity (3.32) to show that

$$\frac{\partial \delta(\dot{\mathbf{E}}^p, e^p)}{\partial \dot{\mathbf{E}}^p} = Y(e^p)\mathbf{N}^p \tag{76.87}$$

and, hence, that (i) $\partial\delta(\dot{\mathbf{E}}^p)/\partial\dot{\mathbf{E}}^p$ is normal to the yield surface, and (ii) the constitutive relation $(76.39)_1$ is equivalent to

$$\mathbf{T}_0 = \frac{\partial \delta(\dot{\mathbf{E}}^p, e^p)}{\partial \dot{\mathbf{E}}^p}. \tag{76.88}$$

2. Show that

$$\dot{e}^p f = 0 \tag{76.89}$$

and discuss its consequences.

3. This exercise describes a generalization of the theory that accounts for a free energy associated with plastic flow resulting in kinematic hardening and an associated backstress. Specifically, we assume that the free energy has the form

$$\Psi = \Psi^e + \Psi^p$$

with Ψ^e an elastic energy and Ψ^p an energy associated with plastic flow, and we replace the separability hypothesis as defined in (75.15) and (75.16) by an elastic balance

$$\dot{\Psi}^e = \mathbf{T}\colon\dot{\mathbf{E}}^e \tag{76.90}$$

and a plastic free-energy imbalance

$$\delta = -\dot{\Psi}^p + \mathbf{T}_0\colon\dot{\mathbf{E}}^p \geq 0 \tag{76.91}$$

characterizing energy dissipated during plastic flow. As before, we assume that $\mathbf{T}$ and Ψ^e are given by the conventional elastic constitutive relations (75.17) (with Ψ replaced by Ψ^e). Then, (76.90) is satisfied and we are left with the plastic free-energy imbalance (76.91). Within the present framework, the basic ingredient in the choice of constitutive relations describing plastic flow is a dimensionless internal variable $\mathbf{A}$ that, like $\mathbf{E}^p$, is symmetric and deviatoric. The field $\mathbf{A}$ enters the theory through a constitutive equation for the plastic free energy,

$$\Psi^p = \hat{\Psi}^p(\mathbf{A}), \tag{76.92}$$

with $\hat{\Psi}^p$ an isotropic function. The remaining constitutive relations consist of a second relation

$$\mathbf{T}_0 = \hat{\mathbf{T}}_0(\dot{\mathbf{E}}^p, e^p, \mathbf{A}) \tag{76.93}$$

for the deviatoric stress supplemented by an evolution equation for $\mathbf{A}$ of the form

$$\dot{\mathbf{A}} = \dot{\mathbf{E}}^p - \mathbf{G}(\dot{\mathbf{E}}^p, e^p, \mathbf{A}). \tag{76.94}$$

The constitutive response functions $\hat{\Psi}^p$, $\hat{\mathbf{T}}_0$, and $\mathbf{G}$ are presumed to be isotropic.

(a) Assuming that the constitutive relations (76.93) and (76.94) are rate-independent, show that they must have the form

$$\begin{aligned} \mathbf{T}_0 &= \hat{\mathbf{T}}_0(\mathbf{N}^p, e^p, \mathbf{A}), \\ \dot{\mathbf{A}} &= \dot{\mathbf{E}}^p - \mathbf{G}(\mathbf{N}^p, e^p, \mathbf{A})|\dot{\mathbf{E}}^p|. \end{aligned} \tag{76.95}$$

(b) Assume that $\mathbf{G}(\mathbf{N}^p, e^p, \mathbf{A})$ is independent of $\mathbf{N}^p$ and linear in $\mathbf{A}$. Show that, since $\mathbf{G}$ is isotropic, there is a modulus $c(e^p)$ such that

$$\mathbf{G}(\mathbf{N}^p, e^p, \mathbf{A}) = c(e^p)\mathbf{A}.$$

(Recall that $\mathbf{A}$ is symmetric and deviatoric.)

(c) Show that the dissipation has the form

$$\delta = \underbrace{\left[\hat{\mathbf{T}}_0(\mathbf{N}^p, e^p, \mathbf{A}) - \frac{\partial \hat{\Psi}^p}{\partial \mathbf{A}} + c(e^p)\left(\mathbf{A}\colon \frac{\partial \hat{\Psi}^p}{\partial \mathbf{A}}\right)\mathbf{N}^p\right]}_{\mathbf{J}(\mathbf{N}^p, e^p, \mathbf{A})} \colon \dot{\mathbf{E}}^p, \tag{76.96}$$

so that, trivially, the stress admits a decomposition into dissipative and energetic parts as follows:

$$\hat{\mathbf{T}}_0(\mathbf{N}^p, e^p, \mathbf{A}) = \underbrace{\mathbf{J}(\mathbf{N}^p, e^p, \mathbf{A})}_{\text{dissipative}} + \underbrace{\frac{\partial \hat{\Psi}^p}{\partial \mathbf{A}} - c(e^p)\left(\mathbf{A}\colon \frac{\partial \hat{\Psi}^p}{\partial \mathbf{A}}\right)\mathbf{N}^p}_{\text{energetic}}. \tag{76.97}$$

(d) The presence of the term $\partial \hat{\Psi}^p / \partial \mathbf{A}$ in (76.97) generally precludes the codirectionality of $\mathbf{T}_0$ and $\mathbf{N}^p$; in fact, an appropriate codirectionality hypothesis for the present theory is the requirement that the dissipative stress $\mathbf{J}(\mathbf{N}^p, e^p, \mathbf{A})$ and the flow direction $\mathbf{N}^p$ be codirectional:

$$\mathbf{J}(\mathbf{N}^p, e^p, \mathbf{A}) = |\mathbf{J}(\mathbf{N}^p, e^p, \mathbf{A})|\mathbf{N}^p. \tag{76.98}$$

Show that the flow resistance $Y(\cdot)$ defined as in (76.26) should here have the form

$$Y(\mathbf{N}^p, e^p, \mathbf{A}) \stackrel{\text{def}}{=} \mathbf{N}^p \colon \mathbf{J}(\mathbf{N}^p, e^p, \mathbf{A}) \geq 0, \tag{76.99}$$

and show that, consequently,

$$\mathbf{J}(\mathbf{N}^p, e^p, \mathbf{A}) = Y(\mathbf{N}^p, e^p, \mathbf{A})\mathbf{N}^p.$$

(e) Assume that $Y(\mathbf{N}^p, e^p, \mathbf{A})$ is independent of $\mathbf{N}^p$. Assume further that the plastic free-energy (76.92) has the form

$$\Psi^p = \tfrac{1}{2}\xi|\mathbf{A}|^2, \tag{76.100}$$

with $\xi > 0$ constant. Show that the constitutive relation $\mathbf{T}_0 = \hat{\mathbf{T}}_0(\dot{\mathbf{E}}^p, e^p, \mathbf{A})$ must take the form

$$\mathbf{T}_0 - \xi\mathbf{A} = (Y(e^p, \mathbf{A}) - c(e^p)\xi|\mathbf{A}|^2)\mathbf{N}^p, \tag{76.101}$$

while the evolution equation for $\mathbf{A}$ becomes

$$\dot{\mathbf{A}} = \dot{\mathbf{E}}^p - |\dot{\mathbf{E}}^p|c(e^p)\mathbf{A}. \tag{76.102}$$

(f) Assume that the flow resistance has the specific form

$$Y(e^p, \mathbf{A}) = Y(e^p) + c(e^p)\xi|\mathbf{A}|^2 > 0, \tag{76.103}$$

with

$$Y(e^p) > 0$$

a flow resistance. Show that the constitutive relation (76.101) takes the form of modified Mises flow rule

$$\mathbf{T}_0 - \mathbf{T}_{\text{back}} = Y(e^p)\mathbf{N}^p, \qquad \text{for } \dot{\mathbf{E}}^p \neq \mathbf{0}, \tag{76.104}$$

with

$$\mathbf{T}_{\text{back}} \stackrel{\text{def}}{=} \xi\mathbf{A} \tag{76.105}$$

usually referred to as a backstress.[399] Show that plastic flow is possible only if the *yield condition*

$$Y(e^p) = |\mathbf{T}_0 - \mathbf{T}_{\text{back}}| \tag{76.106}$$

is satisfied. Thus, as in §76.8, the yield surface is a spherical surface with "radius" $Y(e^p)$ in the space of symmetric deviatoric tensors, but here the surface is centered at the backstress $\mathbf{T}_{\text{back}}$; as in §76.8, plastic flow is possible only when the deviatoric stress $\mathbf{T}_0$ lies on the yield surface. Finally, show that

$$\mathbf{N}^p = \frac{\mathbf{T}_0 - \mathbf{T}_{\text{back}}}{|\mathbf{T}_0 - \mathbf{T}_{\text{back}}|}.$$

(g) The counterpart of the boundedness inequality $(76.70)_3$ within the present framework is

$$|\mathbf{T}_0 - \mathbf{T}_{\text{back}}| \le Y(e^p) \qquad \text{for } all\ \dot{\mathbf{E}}^p, \tag{76.107}$$

an hypothesis asserting that the deviatoric stress minus the backstress can never exceed the flow resistance in magnitude. Show that $\dot{\mathbf{E}}^p$ vanishes whenever $|\mathbf{T}_0 - \mathbf{T}_{\text{back}}| < Y(e^p)$, an inequality that defines the elastic range in deviatoric-stress space.

76.9 Hardening and Softening

Let

$$Y'(e^p) = \frac{dY(e^p)}{de^p}, \tag{76.108}$$

so that, by $(76.70)_2$,

$$\overline{Y(e^p)}^{\boldsymbol{\cdot}} = Y'(e^p)|\dot{\mathbf{E}}^p|;$$

if we let

$$H(e^p) \stackrel{\text{def}}{=} Y'(e^p), \tag{76.109}$$

then

$$\overline{Y(e^p)}^{\boldsymbol{\cdot}} = H(e^p)|\dot{\mathbf{E}}^p|. \tag{76.110}$$

Thus, if we assume that

$$\dot{\mathbf{E}}^p \neq \mathbf{0},$$

then, by (76.36) with $S = e^p$, $|\mathbf{T}_0| - Y(e^p) = 0$; hence (76.110) yields

$$\boxed{\overline{|\mathbf{T}_0|}^{\boldsymbol{\cdot}} = H(e^p)|\dot{\mathbf{E}}^p|,} \tag{76.111}$$

an important relation giving the rate of change of the magnitude of the deviatoric stress as a function of the magnitude of the plastic strain-rate. Consistent

399 The backstress leads to kinematic hardening; the term $Y(e^p)$—which, by (76.103), is part dissipative and part energetic—leads to isotropic hardening. A detailed discussion of conventional theories of plasticity with combined isotropic and kinematic hardening may be found in LEMAITRE & CHABOCHE (1990).

with standard terminology, we say that the material **hardens** or **softens** according to whether[400]

$$\dot{\overline{|\mathbf{T}_0|}} > 0 \qquad \text{or} \qquad \dot{\overline{|\mathbf{T}_0|}} < 0; \tag{76.112}$$

since[401]

$$\operatorname{sgn} \dot{\overline{|\mathbf{T}_0|}} = \operatorname{sgn} H(e^p), \tag{76.113}$$

(†) *the material hardens or softens according to whether*

$$H(e^p) > 0 \qquad \text{or} \qquad H(e^p) < 0. \tag{76.114}$$

We refer to H as the **hardening modulus.**

[400] Cf. (76.67), (76.68).
[401] For any scalar $y \neq 0$, $\operatorname{sgn} y$ denotes the sign of y defined by $\operatorname{sgn} y = y/|y|$.

77 Inversion of the Mises Flow Rule: $\dot{\mathbf{E}}^p$ in Terms of $\dot{\mathbf{E}}$ and $\mathbf{T}$

We now show that the no-flow and consistency conditions may be used to invert the Mises flow rule to get an equation for $\dot{\mathbf{E}}^p$ in terms of $\dot{\mathbf{E}}$ and $\mathbf{T}$. In deriving the inverted relation the following identity, which follows from (76.29) and (76.20), is basic:

$$\mathbf{N}^p = \frac{\dot{\mathbf{E}}^p}{|\dot{\mathbf{E}}^p|} = \frac{\mathbf{T}_0}{|\mathbf{T}_0|}. \tag{77.1}$$

Assume, unless otherwise specified, that the yield condition is satisfied:[402]

$$f = 0 \qquad (\text{so that } \dot{f} \le 0). \tag{77.2}$$

Then, by (76.110), the definition (77.1) of $\mathbf{N}^p$, the constitutive relation $(75.17)_2$ for $\mathbf{T}_0$, and the decomposition $(75.11)_1$ of $\mathbf{E}$,

$$\begin{aligned}
\dot{f} &= \overline{|\mathbf{T}_0|}^{\boldsymbol{\cdot}} - \overline{Y(e^p)}^{\boldsymbol{\cdot}} \\
&= \frac{\mathbf{T}_0}{|\mathbf{T}_0|} : \dot{\mathbf{T}}_0 - H(e^p)|\dot{\mathbf{E}}^p| \\
&= \mathbf{N}^p : \dot{\mathbf{T}}_0 - H(e^p)|\dot{\mathbf{E}}^p| \\
&= 2\mu \mathbf{N}^p : (\dot{\mathbf{E}}_0 - \dot{\mathbf{E}}^p) - H(e^p)|\dot{\mathbf{E}}^p|,
\end{aligned}$$

or, since $\mathbf{N}^p : \dot{\mathbf{E}}^p = |\dot{\mathbf{E}}^p|$ and $\mathbf{N}^p$ is deviatoric,

$$\dot{f} = 2\mu\, \mathbf{N}^p : \dot{\mathbf{E}} - [2\mu + H(e^p)]|\dot{\mathbf{E}}^p|. \tag{77.3}$$

We henceforth restrict attention to the class of materials for which

$$2\mu + H(e^p) > 0. \tag{77.4}$$

In view of the sentence containing (76.114), the inequality (77.4) always holds for strain-hardening materials, for which $H(e^p) > 0$ for all e^p. In contrast, strain-softening materials satisfy $H(e^p) < 0$, and (77.4) places a restriction on the maximum allowable softening, as it requires that $H(e^p) > -2\mu$ for all e^p. It is convenient to introduce the **stiffness ratio**

$$\beta(e^p) \overset{\text{def}}{=} \frac{2\mu}{2\mu + H(e^p)}. \tag{77.5}$$

[402] Cf. (76.82).

We now deduce conditions that determine whether the material leaves — or remains on — the yield surface when the material point in question is subjected to a loading program characterized by the total strain-rate $\dot{\mathbf{E}}$.

(i) **Elastic unloading** is defined by the condition $\mathbf{N}^p : \dot{\mathbf{E}} < 0$. In this case, since $|\dot{\mathbf{E}}^p| \geq 0$, (77.3) implies that $\dot{f} < 0$, and the no-flow conditions (76.84) imply that $\dot{\mathbf{E}}^p = \mathbf{0}$.

(ii) **Neutral loading** is defined by the condition $\mathbf{N}^p : \dot{\mathbf{E}} = 0$. In this case $|\dot{\mathbf{E}}^p| > 0$ cannot hold, for if it did, then (77.3) would imply that $\dot{f} < 0$, which would violate the no-flow conditions (76.84). Hence, once again, $\dot{\mathbf{E}}^p = \mathbf{0}$.

(iii) **Plastic loading** is defined by the condition $\mathbf{N}^p : \dot{\mathbf{E}} > 0$. In this case, if $|\dot{\mathbf{E}}^p| = 0$, then $\dot{f} > 0$, which violates $\dot{f} \leq 0$. Hence $|\dot{\mathbf{E}}^p| > 0$, and, since the consistency condition (76.85) then requires that $\dot{f} = 0$, (77.3) yields

$$|\dot{\mathbf{E}}^p| = \beta(e^p)\mathbf{N}^p : \dot{\mathbf{E}}. \tag{77.6}$$

Thus, since, by (77.1), $\dot{\mathbf{E}}^p = |\dot{\mathbf{E}}^p|\mathbf{N}^p$,

$$\dot{\mathbf{E}}^p = \beta(e^p)(\mathbf{N}^p : \dot{\mathbf{E}})\mathbf{N}^p \neq \mathbf{0}. \tag{77.7}$$

At this point, it is important to note that

- *in view of* (77.1), *we may consider* $\mathbf{N}^p$ *as defined by* $\mathbf{T}_0/|\mathbf{T}_0|$ *— then* (77.7) *determines* $\dot{\mathbf{E}}^p$ *in terms of* $\dot{\mathbf{E}}$, $\mathbf{T}$, *and the current hardening modulus* $H(e^p)$.

Combining the results of (i)–(iii) with the condition (76.81), we arrive at an equation for the plastic strain-rate $\dot{\mathbf{E}}^p$ that holds for all time, no matter whether $f = 0$ or $f < 0$:

$$\dot{\mathbf{E}}^p = \begin{cases} \mathbf{0} & \text{if } f < 0 \text{ (behavior within the elastic range),} \\ \mathbf{0} & \text{if } f = 0 \text{ and } \mathbf{N}^p : \dot{\mathbf{E}} < 0 \text{ (elastic unloading),} \\ \mathbf{0} & \text{if } f = 0 \text{ and } \mathbf{N}^p : \dot{\mathbf{E}} = 0 \text{ (neutral loading),} \\ \beta(e^p)(\mathbf{N}^p : \dot{\mathbf{E}})\mathbf{N}^p & \text{if } f = 0 \text{ and } \mathbf{N}^p : \dot{\mathbf{E}} > 0 \text{ (plastic loading).} \end{cases} \tag{77.8}$$

The result (77.8) is embodied in the **inverted Mises flow rule**

$$\dot{\mathbf{E}}^p = \chi\beta(e^p)(\mathbf{N}^p : \dot{\mathbf{E}})\mathbf{N}^p, \qquad \text{with} \qquad \mathbf{N}^p = \frac{\mathbf{T}_0}{|\mathbf{T}_0|}, \tag{77.9}$$

where

$$\chi = \begin{cases} 0 & \text{if } f < 0, \text{ or if } f = 0 \text{ and } \mathbf{N}^p : \dot{\mathbf{E}} \leq 0, \\ 1 & \text{if } f = 0 \text{ and } \mathbf{N}^p : \dot{\mathbf{E}} > 0, \end{cases} \tag{77.10}$$

is a **switching parameter**. A central result of the theory may be stated as follows:

EQUIVALENCY THEOREM *Asssume that the boundedness relation*

$$f = |\mathbf{T}_0| - Y(e^p) \leq 0 \tag{77.11}$$

is satisfied. The inverted Mises flow rule, as defined by (77.9) *and* (77.10), *is therefore equivalent to the Mises flow rule* $(76.70)_1$.

The proof of this theorem is given following (77.13).

Note that, given $\mathbf{E}^p$ and $\mathbf{E}$, we can consider the elastic strain as defined by $\mathbf{E}^e = \mathbf{E} - \mathbf{E}^p$ and write the elastic stress-strain relation $(75.17)_2$ in the form

$$\mathbf{T} = 2\mu(\mathbf{E} - \mathbf{E}^p) + \lambda \operatorname{tr} \mathbf{E}. \tag{77.12}$$

Differentiating this relation with respect to time and using the inverted Mises flow rule (77.9), we arrive at an evolution equation for the stress:

$$\dot{\mathbf{T}} = 2\mu\dot{\mathbf{E}} + \lambda(\operatorname{tr}\dot{\mathbf{E}})\mathbf{1} - 2\mu\chi\beta(e^p)(\mathbf{N}^p\!:\!\dot{\mathbf{E}})\mathbf{N}^p. \tag{77.13}$$

Proof of the Equivalency Theorem. The derivation of the inverted Mises flow rule establishes (77.9) and (77.10) as consequences of the Mises flow rule. To prove the converse assertion, assume that (77.9) and (77.10) are satisfied. We must show that if $\dot{\mathbf{E}}^p \neq \mathbf{0}$ then the Mises flow rule

$$\mathbf{T}_0 = Y(e^p)\frac{\dot{\mathbf{E}}^p}{|\dot{\mathbf{E}}^p|}$$

is satisfied. Thus, assume that $\dot{\mathbf{E}}^p \neq \mathbf{0}$ and note that by (77.9) we may ignore the conditions in (77.10) that yield $\chi = 0$. We, therefore, restrict attention to $f = 0$ so that, by (76.26) and the boundedness relation (77.11), $|\mathbf{T}_0| = Y(e^p) > 0$ and $\mathbf{N}^p$ is well defined. Then, (77.9) and (77.10) imply that

$$f = 0 \quad \text{and} \quad \mathbf{N}^p\!:\!\dot{\mathbf{E}} > 0, \quad \text{so that} \quad \chi = 1.$$

Then, by (77.9), bearing in mind that $\beta(e^p) > 0$, there is a scalar $c > 0$ such that $\dot{\mathbf{E}}^p = c\mathbf{T}_0$; thus trivially

$$\mathbf{T}_0 = |\mathbf{T}_0|\frac{\dot{\mathbf{E}}^p}{|\dot{\mathbf{E}}^p|}$$

and, since $f = 0$, (77.11) implies that $|\mathbf{T}_0| = Y(e^p)$, which completes the proof.

The Equivalency Theorem allows us to replace the Mises flow rule by its inverse;[403] within this format the complete set of constitutive equations consist, of:

(i) the elastic stress-strain relation

$$\mathbf{T} = 2\mu(\mathbf{E} - \mathbf{E}^p) + \lambda(\operatorname{tr}\mathbf{E})\mathbf{1}; \tag{77.14}$$

(ii) the plastic boundedness inequality

$$f = |\mathbf{T}_0| - Y(e^p) \leq 0, \tag{77.15}$$

with f the yield function and $Y(e^p)$ the flow resistance;

(iii) a system of evolution equations

$$\boxed{\begin{aligned} \dot{\mathbf{E}}^p &= \chi\beta(e^p)(\mathbf{N}^p\!:\!\dot{\mathbf{E}})\mathbf{N}^p, \quad \mathbf{N}^p = \frac{\mathbf{T}_0}{|\mathbf{T}_0|}, \\ \dot{e}^p &= |\dot{\mathbf{E}}^p|, \qquad e^p(\mathbf{X}, 0) = 0, \end{aligned}} \tag{77.16}$$

for the plastic strain and the accumulated plastic strain, in which

$$\beta(e^p) = \frac{2\mu}{2\mu + H(e^p)}$$

is the stiffness ratio,

$$H(e^p) = Y'(e^p) \tag{77.17}$$

is the hardening modulus, and

$$\chi = \begin{cases} 0 & \text{if } f < 0, \text{ or if } f = 0 \text{ and } \mathbf{N}^p\!:\!\dot{\mathbf{E}} \leq 0, \\ 1 & \text{if } f = 0 \quad \text{and} \quad \mathbf{N}^p\!:\!\dot{\mathbf{E}} > 0, \end{cases} \tag{77.18}$$

is a switching parameter.

---

[403] As we show in §90, the issue is not so simple when the theory is generalized to allow for constitutive dependencies on plastic strain gradients and/or their rates.

Remark. Constitutive equations of the form (77.16) need to be accompanied by initial conditions. Typical initial conditions presume that the body is initially (at time $t = 0$, say) in a **virgin state** in the sense that[404]

$$\mathbf{E}(\mathbf{X}, 0) = \mathbf{E}^p(\mathbf{X}, 0) = \mathbf{0}, \tag{77.19}$$

so that, by $(75.11)_1$, $\mathbf{E}^e(\mathbf{X}, 0) = \mathbf{0}$.

EXERCISE

1. Consider the modified Mises flow equations with backstress as defined by (76.104), (76.105), and (76.102). Define a yield function f through the relation

$$f = |\mathbf{T}_0 - \mathbf{T}_{\text{back}}| - Y(e^p). \tag{77.20}$$

Use arguments of §76.9 and §77 to establish an appropriate analog of (77.9).

[404] The initial condition for the hardening variable e^p is given in $(76.70)_3$.

78 Rate-Dependent Plastic Materials

78.1 Background

Physical considerations of the mechanisms of plastic deformation in metals and experimental observations show that plastic flow is both *temperature- and rate-dependent*, but a standard rule of thumb is that rate-dependence is sufficiently large to merit consideration only for absolute temperatures greater than $0.35\vartheta_{\mathrm{m}}$, where ϑ_m is the melting temperature of the material in degrees absolute. In the temperature range $\vartheta < 0.35\vartheta_{\mathrm{m}}$, plastic stress-strain response is only slightly rate-sensitive, and in this low homologous temperature regime the plastic stress-strain response of metallic materials is generally assumed to be rate-independent,[405] as considered in the previous sections.

Values of the melting temperatures ϑ_{m} and homologous temperatures $0.35\vartheta_m$ for some metals are shown in Table 78.1. At room temperature, the stress-strain response of Ti may be idealized as rate-independent, whereas that of Pb cannot be so idealized.

In what follows, to account for rate-dependence, we discuss a simple but widely used extension of the rate-independent theory discussed previously.

78.2 Materials with Simple Rate-Dependence

We begin with the general constitutive relation (76.6); viz.

$$\mathbf{T}_0 = \hat{\mathbf{T}}_0(\dot{\mathbf{E}}^p, S). \tag{78.1}$$

For convenience, we introduce a **flow-rate**

$$d^p \stackrel{\text{def}}{=} |\dot{\mathbf{E}}^p|, \tag{78.2}$$

so that the flow direction is given by

$$\mathbf{N}^p = \frac{\dot{\mathbf{E}}^p}{d^p}.$$

405 Materials scientists use the term "homologous temperature" to refer to the temperature of a material expressed as some fraction of its melting temperature. The rate sensitivity of the plastic stress-strain response of two materials is (approximately) the same when compared at the same homologous temperature. The rate-sensitivity of the plastic stress-strain response of Ti at 679K is therefore about the same as that of Al at 327K.

Table 78.1. *Melting temperatures* $\vartheta_{\rm m}$ *and homologous temperatures* $0.35\,\vartheta_m$ *for titanium (Ti), iron (Fe), copper (Cu), aluminum (Al), and lead (Pb) in Kelvin with corresponding Celcius values in parentheses.*

Material	Melting temperature $\vartheta_{\rm m}$, K(°C)	Homologous temperature $0.35\,\vartheta_{\rm m}$, K(°C)
Ti	1941 (1668)	679 (406)
Fe	1809 (1536)	633 (360)
Cu	1356 (1083)	452 (201)
Al	933 (600)	327 (54)
Pb	600 (327)	210 (−63)

We assume that the deviatoric stress is well-defined when the flow-rate vanishes; in fact, we assume that

$$\hat{\mathbf{T}}_0(\mathbf{0}, S) = \lim_{\dot{\mathbf{E}}^p \to \mathbf{0}} \hat{\mathbf{T}}_0(\dot{\mathbf{E}}^p, S) = \mathbf{0}, \tag{78.3}$$

so that — in contrast to the rate-independent theory — the deviatoric stress vanishes when the flow rate vanishes.

For $d^p = 0$ we consider the dependence of $\mathbf{T}_0$ on $\dot{\mathbf{E}}^p$ as a dependence on the pair $(d^p, \mathbf{N}^p)$ and, hence, rewrite (78.1) in the form[406]

$$\mathbf{T}_0 = \hat{\mathbf{T}}_0(d^p, \mathbf{N}^p, S). \tag{78.4}$$

We restrict attention to constitutive relations (78.1) that are consistent with natural generalizations of the *codirectionality* and *strong isotropy* restrictions (CD) and (SI) in §76.3. If we write $\bar{Y}(d^p, \mathbf{N}^p, S)$ for the *flow resistance*,

$$\bar{Y}(d^p, \mathbf{N}^p, S) \overset{\text{def}}{=} \mathbf{N}^p : \hat{\mathbf{T}}_0(d^p, \mathbf{N}^p, S), \tag{78.5}$$

then, granted the codirectionality requirement in the form (76.29), (78.1) takes the form

$$\mathbf{T}_0 = \bar{Y}(d^p, \mathbf{N}^p, S)\mathbf{N}^p. \tag{78.6}$$

Further, the natural analog of the strong isotropy hypothesis (SI) applied to (78.6) requires that $\bar{Y}(d^p, \mathbf{N}^p, S)$ be independent of $\mathbf{N}^p$, so that

$$\mathbf{T}_0 = \bar{Y}(d^p, S)\mathbf{N}^p. \tag{78.7}$$

We now describe a type of moderate rate-dependence common to a large class of rate-dependent plastic materials described in the literature. Stated precisely, a material described by (78.1) has **simple rate-dependence** if the flow resistance $Y(d^p, S)$ has the simple form

$$\bar{Y}(d^p, S) = g(d^p)Y(S), \tag{78.8}$$

where, consistent with (78.3),

$$g(0) = 0, \text{ and g is a strictly increasing function of } d^p. \tag{78.9}$$

406 Cf. (76.21).

We refer to $g(d^p)$ as the **rate-sensitivity function.** A material with simple rate-dependence is therefore governed by a constitutive equation of the form[407]

$$\mathbf{T}_0 = \underbrace{g(d^p)}_{\text{rate-dependent}}\,\underbrace{Y(S)\mathbf{N}^p}_{\text{rate-independent}}. \tag{78.10}$$

Note that for $g \equiv 1$, (78.10) reduces to the rate-independent Mises flow rule (76.34).

A simple form of (78.10) — patterned after the rate-independent relation (76.51) (which has flow resistance as hardening variable) — has the form[408]

$$\mathbf{T}_0 = g(d^p)S\mathbf{N}^p. \tag{78.11}$$

An additional property required of $g(d^p)$, and one that characterizes the small rate-dependence of most metals at low temperatures, is that — except for a very small time interval at the onset of loading — $g(d^p) \approx 1$ for a range of flow-rates of interest.

If we take the absolute value of (78.10), we arrive at the useful relation

$$|\mathbf{T}_0| = g(d^p)Y(S). \tag{78.12}$$

Note that, by (78.10), the dissipation (75.16) has the simple form

$$\delta = d^p g(d^p)Y(S) \tag{78.13}$$

and, by (78.9), the dissipation is strictly positive for $d^p \neq 0$ if and only if

$$Y(S) > 0, \tag{78.14}$$

as in the rate-independent theory.[409]

Next, by (78.9) the function $g(d^p)$ is invertible, and the inverse function $f = g^{-1}$ is strictly increasing and hence strictly positive for all nonzero arguments; further $f(0) = 0$. Hence the general relation (78.12) may be inverted to give an expression

$$d^p = g^{-1}\left(\frac{|\mathbf{T}_0|}{Y(S)}\right) \equiv f\left(\frac{|\mathbf{T}_0|}{Y(S)}\right) \tag{78.15}$$

for the flow-rate. Thus, in contrast to the rate-independent theory developed in §76,

- *the plastic strain-rate is nonzero whenever the stress is nonzero: There is no elastic range in which the response of the material is purely elastic, and there are no considerations of a yield condition, a consistency condition, loading/unloading conditions, and so forth.*

As a direct consequence of (78.10)

$$\frac{\mathbf{T}_0}{|\mathbf{T}_0|} = \mathbf{N}^p, \tag{78.16}$$

and by (78.15) the relation (78.10) may be inverted to give a constitutive relation

$$\dot{\mathbf{E}}^p = f\left(\frac{|\mathbf{T}_0|}{Y(S)}\right)\frac{\mathbf{T}_0}{|\mathbf{T}_0|} \tag{78.17}$$

---

407 There are classes of constitutive relations for which (78.3), (78.8), and (78.10) are not satisfied. One such alternative constitutive relation has the form

$$\mathbf{T}_0 = [Y_1 + g(d^p)Y_2(S)]\mathbf{N}^p,$$

with Y_1 a *rate-independent frictional resistance.*

408 Cf., e.g., Brown et al. (1989).

409 Cf. (76.26).

for the plastic strain-rate; in contrast to the rate-independent theory, the basic constitutive relation may thus be inverted to give plastic strain-rate as a function of stress.

To complete the constitutive theory,[410] we consider the general hardening equation (76.7), assuming that — consistent with the strong isotropy hypothesis (SI) on page 432 — $h(\dot{\mathbf{E}}^p, S)$ depends on $\dot{\mathbf{E}}^p = d^p\mathbf{N}^p$ at most through d^p:

$$\dot{S} = h(d^p, S). \tag{78.18}$$

Because the material is rate-dependent, the result expressed in the paragraph containing (76.69) no longer applies. Even so, there may be circumstances of interest in which the choice

$$S = e^p, \tag{78.19}$$

with e^p the accumulated plastic strain, defined by $(76.70)_{2,3}$, may be applicable.

78.3 Power-Law Rate-Dependence

An example of a commonly used rate-sensitivity function is the **power-law function**[411]

$$g(d^p) = \left(\frac{d^p}{d_0}\right)^m, \tag{78.20}$$

where $m > 0$, a constant, is a **rate-sensitivity parameter** and $d_0 > 0$, also a constant, is the reference flow-rate. The power-law function satisfies $g(d^p) \approx 1$ for $d^p \approx d_0$, and (78.20) is therefore intended to model plastic flows with rates close to d_0. In view of (78.20), the relation (78.10) becomes

$$\mathbf{T}_0 = \left(\frac{d^p}{d_0}\right)^m Y(S)\mathbf{N}^p, \tag{78.21}$$

and implies that

$$\ln(|\mathbf{T}_0|) = \ln(Y(S)) + m \ln\left(\frac{d^p}{d_0}\right);$$

thus the rate-sensitivity factor m is the slope of the graph of $\ln(|\mathbf{T}_0|)$ versus $\ln(d^p/d_0)$.

The power-law function allows one to characterize *nearly rate-independent behavior*, for which m is very small. Such rate-dependent models also serve as a *regularization of rate-independent* behavior.[412] Note that, since

$$\lim_{m\to 0}\left(\frac{d^p}{d_0}\right)^m = 1$$

the limit $m \to 0$ in (78.21) corresponds to *rate-independent response* as described by (76.34).

---

[410] The elastic constitutive relations remain as described in §75.4.

[411] This function is widely used because of its utility in characterizing experimental data. Another example is furnished by

$$g(d^p) = \tanh(cd^p), \qquad c > 0,$$

which, unlike the relation (78.20), has a limit as $d^p \to \infty$, and, hence, might be useful when rate-sensitivity is introduced as a means of regularizing the rate-independent theory.

[412] Cf. Footnote 411.

Further, granted the power-law function (78.20), the expressions (78.15) and (78.17) have the specific form

$$d^p = d_0 \left(\frac{|\mathbf{T}_0|}{Y(S)} \right)^{\frac{1}{m}} \tag{78.22}$$

and

$$\dot{\mathbf{E}}^p = d_0 \left(\frac{|\mathbf{T}_0|}{Y(S)} \right)^{\frac{1}{m}} \frac{\mathbf{T}_0}{|\mathbf{T}_0|}, \tag{78.23}$$

when expressed in terms of the inverse of the power-law function (78.15).

An important problem in application of the rate-dependent theory concerns the temporal integration of the rate equations (78.23) and (78.18) for $\mathbf{E}^p$ and S. These equations are typically highly nonlinear, coupled, and mathematically stiff. With reference to the power-law model (78.23) for $\dot{\mathbf{E}}^p$, the stiffness of the equations depends on the strain-rate sensitivity parameter m, and the stiffness increases to infinity as m tends to zero, the rate-independent limit.[413]

[413] For small values of m, special care is required to develop stable constitutive time-integration procedures; cf., e.g., LUSH, WEBER & ANAND (1989).

79 Maximum Dissipation

A standard assumption underlying many theories of plastic flow is that the basic fields evolve to *maximize the dissipation*, an assumption that we use to develop a mathematical framework more general than that of Mises. Specifically, we abandon the codirectionality hypothesis (CD) on page 471 in favor of an hypothesis based on maximum dissipation.[414]

Our starting point is the rate-independent **flow rule**[415]

$$\mathbf{T}_0 = \hat{\mathbf{T}}_0(\mathbf{N}^p, e^p) \tag{79.1}$$

(with accumulated plastic strain e^p as hardening variable) and the consequence (76.26) of strict dissipativity,

$$Y(\mathbf{N}^p, e^p) \overset{\text{def}}{=} \mathbf{N}^p : \hat{\mathbf{T}}_0(\mathbf{N}^p, e^p) > 0. \tag{79.2}$$

79.1 Basic Definitions

The physical fields involved in the basic constitutive relations (79.1) and (79.2) are the deviatoric stress $\mathbf{T}_0$, the plastic strain-rate $\dot{\mathbf{E}}^p$, and the hardening variable $S \equiv e^p$. Because both $\mathbf{T}_0$ and $\dot{\mathbf{E}}^p$ are symmetric and deviatoric, the following definition is useful:

$$\text{SymDev} \overset{\text{def}}{=} \text{the space of all symmetric and deviatoric tensors.} \tag{79.3}$$

Granted that the shear modulus μ is nonzero, the elastic stress-strain relation (75.18) places no restriction on the values of the deviatoric stress $\mathbf{T}_0$. But we know from experience with real materials that *plastic flow severely limits the set of observed values of* $\mathbf{T}_0$ — a set generally referred to as the **elastic range**.

Successful theories of rate-independent plasticity generally begin with the notion of a yield surface — a closed surface in SymDev — and then define the elastic range to be the subset of SymDev enclosed by the yield surface.

414 As noted by Simo & Hughes (1998, p. 98) — and as is clear from the book of Han & Reddy (1999) — the notion of maximum dissipation plays a crucial role in the variational formulation of plasticity [and] "is central in the mathematical formulation of plasticity; see, e.g., Duvaut & Lions (1972), Johnson (1976,1978), Moreau (1976), and the recent account of Temam (1985)." There is a large and growing literature based on maximum dissipation: cf., e.g., Hackl (1997) and Mielke (2003). In §79.3, we relate the notion of maximum dissipation to a notion of material stability.

415 Cf. (76.21).

- Here the notion of *maximum dissipation*[416] provides a framework much richer than that used classically and leads to natural choices for both the elastic range and the yield surface.

The following definition is basic to what follows: a pair $(\mathbf{T}_0, \dot{\mathbf{E}}^p)$ with $\mathbf{T}_0$ a deviatoric stress and $\dot{\mathbf{E}}^p \neq \mathbf{0}$ a plastic strain-rate is **physically attainable** if it is consistent with the flow rule (79.1); in this case, we also refer to $(\mathbf{T}_0, \mathbf{N}^p)$ as **physically attainable.** But — and what is most important —

(‡) *we do not limit our discussion to physically attainable pairs*, even though these are the only pairs that one might see in actual experiments.[417]

Indeed, it is often profitable to allow for "thought experiments" involving fields not necessarily consistent with the underlying equations. Here, since the relevant equation is the flow rule (79.1), we work with pairs $(\mathbf{T}_0, \dot{\mathbf{E}}^p)$ and $(\mathbf{T}_0, \mathbf{N}^p)$ — called **flows** and **normalized flows**, respectively — that *may or may not be physically attainable.*

We now "flesh-out" the constitutive theory using the assumption that the basic fields evolve to *maximize the dissipation.* Roughly speaking, the notion of maximum dissipation is the requirement that in no flow $(\mathbf{T}_0, \dot{\mathbf{E}}^p)$ with $\mathbf{T}_0$ in the elastic range should it be possible to expend power at a rate that is greater than the dissipation $\hat{\delta}(\dot{\mathbf{E}}^p, e^p)$:

$$\mathbf{T}_0 : \dot{\mathbf{E}}^p \leq \hat{\delta}(\dot{\mathbf{E}}^p, e^p). \tag{79.4}$$

By (76.23), (79.4) may be written as

$$\mathbf{T}_0 : \dot{\mathbf{E}}^p \leq \hat{\mathbf{T}}_0(\mathbf{N}^p, e^p) : \dot{\mathbf{E}}^p,$$

and a more elemental form this inequality obtains upon division by $|\dot{\mathbf{E}}^p|$:

$$\mathbf{T}_0 : \mathbf{N}^p \leq \hat{\mathbf{T}}_0(\mathbf{N}^p, e^p) : \mathbf{N}^p. \tag{79.5}$$

Using (79.2) we can write (79.4) in the more transparent form

$$\mathbf{T}_0 : \mathbf{N}^p \leq Y(\mathbf{N}^p, e^p). \tag{79.6}$$

The next definition, which is based on the foregoing discussion, allows us to delineate the class of stresses $\mathbf{T}_0$ that comprise the elastic range. Precisely,

(i) we say that a deviatoric stress $\mathbf{T}_0$ is **admissible in the sense of maximum dissipation** if

$$\mathbf{T}_0 : \mathbf{N}^p \leq Y(\mathbf{N}^p, e^p) \quad \text{for every flow direction } \mathbf{N}^p; \tag{79.7}$$

(ii) and we define the **elastic range** to be the closed set[418]

$$\mathcal{E}(e^p) \stackrel{\text{def}}{=} \text{the set of all } \mathbf{T}_0 \text{ that are admissible in the sense of maximum dissipation.} \tag{79.8}$$

Some terminology is useful. We use the term **flow stress** to denote a stress $\mathbf{T}_0$ such that

$$\mathbf{T}_0 = \hat{\mathbf{T}}_0(\mathbf{N}^p, e^p) \quad \text{for some flow direction } \mathbf{N}^p, \tag{79.9}$$

and we refer to

$$\mathcal{Y}(e^p) \stackrel{\text{def}}{=} \text{the set of all } \mathbf{T}_0 \text{ that are flow stresses} \tag{79.10}$$

---

416 In the plasticity literature this notion is referred to as *maximum plastic dissipation*. Here the underlying dissipation is solely plastic and the adjective "plastic" may safely be omitted.

417 Cf. the paragraph in petite type on page 496.

418 In the literature one often finds the elastic range defined as the interior of $\mathcal{E}(e^p)$.

as the **yield set** — $\mathcal{Y}(e^p)$ consists of all deviatoric stresses that correspond to physically attainable normalized flows $(\mathbf{T}_0, \mathbf{N}^p)$. We are now in a position to derive the Mises flow equations.

79.2 Warm-up: Derivation of the Mises Flow Equations Based on Maximum Dissipation

We here derive the Mises theory within a setting that we view as physically more satisfying than that of §76.4. Our derivation is based on two hypotheses:

(FS) (**flow-stress admissibility**) each flow stress is admissible in the sense of maximum dissipation, so that

$$\mathcal{Y}(e^p) \subset \mathcal{E}(e^p); \tag{79.11}$$

(SI) (**strong isotropy**) $Y(\mathbf{N}^p, e^p)$ is independent of the flow direction $\mathbf{N}^p$; viz.

$$Y(\mathbf{N}^p, e^p) = Y(e^p).$$

Our first step is to use the hypotheses (FS) and (SI) to simplify the general flow rule (79.1). By (FS), (SI), and (79.7), we have the important inequality

$$\mathbf{T}_0 : \mathbf{N}^p \leq Y(e^p) \quad \text{for every flow stress } \mathbf{T}_0 \text{ and every flow direction } \mathbf{N}^p. \tag{79.12}$$

Choose an arbitrary flow direction $\mathbf{N}^p$ and an arbitrary tensor $\mathbf{\Lambda}$ in SymDev such that[419]

$$\mathbf{\Lambda} : \mathbf{N}^p = 0. \tag{79.13}$$

Let $\mathcal{N}$ denote the set of all flow directions:

$$\mathcal{N} \stackrel{\text{def}}{=} \text{the set of all symmetric and deviatoric } unit \text{ tensors } \mathbf{N}^p; \tag{79.14}$$

$\mathcal{N}$ represents the unit sphere in the space SymDev. Then, there is a curve $\tilde{\mathbf{N}}^p(\lambda)$ on $\mathcal{N}$ such that, for some λ_0,[420]

$$\tilde{\mathbf{N}}^p(\lambda_0) = \mathbf{N}^p, \qquad \frac{d\tilde{\mathbf{N}}^p(\lambda)}{d\lambda}\bigg|_{\lambda=\lambda_0} = \mathbf{\Lambda}. \tag{79.15}$$

Further, for $\mathbf{T}_0 = \hat{\mathbf{T}}_0(\mathbf{N}^p, e^p)$, (79.2) and (79.12) imply that $\Phi(\lambda)$ defined by

$$\Phi(\lambda) = Y(e^p) - \mathbf{T}_0 : \tilde{\mathbf{N}}^p(\lambda)$$

satisfies $\Phi(\lambda) \geq 0$ and $\Phi(\lambda_0) = 0$; hence, $\Phi(\lambda)$ has a minimum at $\lambda = \lambda_0$. Consequently,

$$\begin{aligned} \frac{d\Phi(\lambda)}{d\lambda}\bigg|_{\lambda=\lambda_0} &= -\mathbf{T}_0 : \frac{d\tilde{\mathbf{N}}^p(\lambda)}{d\lambda}\bigg|_{\lambda=\lambda_0} \\ &= 0, \end{aligned}$$

so that, by (79.13) and (79.15),

$$\mathbf{T}_0 : \mathbf{\Lambda} = 0 \tag{79.16}$$

for every $\mathbf{\Lambda} \in$ SymDev tangent to $\mathcal{N}$ at $\mathbf{N}^p$; the stress $\mathbf{T}_0$ must therefore be normal to $\mathcal{N}$ at $\mathbf{N}^p$; hence, there is a scalar β such that

$$\mathbf{T}_0 = \beta \mathbf{N}^p,$$

---

[419] Such a $\mathbf{\Lambda}$ is constructed as follows: Let $\mathbf{A}$ be an arbitrary tensor in SymDev and define $\mathbf{\Lambda}$ by $\mathbf{\Lambda} = \mathbf{A} - (\mathbf{A} : \mathbf{N}^p)\mathbf{N}^p$.

[420] Given a surface S, a point $\mathbf{z} \in S$, and a vector $\mathbf{t}$ tangent to S at $\mathbf{z}$, there is always a curve on S through $\mathbf{z}$ whose tangent vector at $\mathbf{z}$ is $\mathbf{t}$.

and, by (79.2), $\beta = Y(e^p)$. Finally, bearing in mind that $\mathbf{N}^p$ is an arbitrary flow direction and that $\mathbf{T}_0$ is given by (79.1), we have the *Mises flow rule*:

$$\mathbf{T}_0 = Y(e^p)\mathbf{N}^p \qquad \text{for } \dot{\mathbf{E}}^p \neq \mathbf{0}, \tag{79.17}$$

where, by (79.2), $Y(e^p) > 0$.

Our next step is to determine the forms of the yield set $\mathcal{Y}(e^p)$ and the elastic range $\mathcal{E}(e^p)$ when all flow stresses have the Mises form (79.17). Since $\mathcal{Y}(e^p)$ is the set of all flow stresses, $\mathbf{T}_0 \in \mathcal{Y}(e^p)$ and (79.17) lead to the **yield condition**

$$|\mathbf{T}_0| = Y(e^p). \tag{79.18}$$

Conversely, given a deviatoric stress $\mathbf{T}_0$ that satisfies the yield condition (79.18), there is a flow direction, namely

$$\mathbf{N}^p = \frac{\mathbf{T}_0}{|\mathbf{T}_0|}, \tag{79.19}$$

such that $\mathbf{T}_0 = Y(e^p)\mathbf{N}^p$. The yield set is therefore given by

$$\mathcal{Y}(e^p) = \text{the set of all } \mathbf{T}_0 \text{ such that } |\mathbf{T}_0| = Y(e^p)$$

and is, hence, a spherical surface with radius $Y(e^p)$. Moreover, it is clear that the flow direction $\mathbf{N}^p$ represents the *outward unit normal* to $\mathcal{Y}(e^p)$.

To determine the elastic range $\mathcal{E}(e^p)$, assume first that $\mathbf{T}_0 \in \mathcal{E}(e^p)$. Then, by (79.8), $\mathbf{T}_0$ must be admissible in the sense of maximum dissipation, so that, by (79.12), given any flow direction $\mathbf{N}^p$,

$$\mathbf{T}_0 : \mathbf{N}^p \leq Y(e^p). \tag{79.20}$$

In particular, the choice (79.19) implies that $\mathbf{T}_0 : \mathbf{N}^p = |\mathbf{T}_0|$; thus, we have the **boundedness inequality**

$$|\mathbf{T}_0| \leq Y(e^p). \tag{79.21}$$

(In contrast to the argument leading to the boundedness hypothesis (76.37), the inequality (79.21) is therefore not a separate hypothesis but follows instead as a consequence of our starting assumptions.) We have therefore shown that if $\mathbf{T}_0 \in \mathcal{E}(e^p)$, then $|\mathbf{T}_0| \leq Y(e^p)$. To see that the elastic range is the set of all $\mathbf{T}_0$ consistent with (79.21) we must show that, conversely, if $\mathbf{T}_0$ obeys (79.21), then $\mathbf{T}_0$ is admissible in the sense of maximum dissipation. Thus, assume that (79.21) is satisfied. Then, given *any* flow direction $\mathbf{N}^p$, we may use the Schwarz inequality and the requirement $|\mathbf{N}^p| = 1$ to conclude that

$$\mathbf{T}_0 : \mathbf{N}^p \leq |\mathbf{T}_0||\mathbf{N}^p| = |\mathbf{T}_0|.$$

Thus, (79.21) implies that $\mathbf{T}_0 : \mathbf{N}^p \leq Y(e^p)$, and, hence, that $\mathbf{T}_0$ is admissible in the sense of maximum dissipation. The **elastic range** is therefore the set

$$\mathcal{E}(e^p) = \text{the set of all } \mathbf{T}_0 \text{ such that } |\mathbf{T}_0| \leq Y(e^p)$$

and, hence, is the closed ball in SymDev of radius $Y(e^p)$. It follows that

$$\mathcal{Y}(e^p) = \partial\mathcal{E}(e^p).$$

Finally, when $|\mathbf{T}_0| < Y(e^p)$ the yield condition (76.36) is not satisfied, so that, necessarily, $\dot{\mathbf{E}}^p = \mathbf{0}$; therefore,

$$\dot{\mathbf{E}}^p = \mathbf{0} \qquad \text{for } |\mathbf{T}_0| < Y(e^p) \tag{79.22}$$

and there is no plastic flow in the interior of the elastic range.

Summarizing, we have established the complete set[421] (76.70) of *Mises–Hill equations* as consequences of (FS) and (SI).

EXERCISE

1. Consider the general case in which the flow resistance $Y(\mathbf{N}^p)$ depends on the flow direction $\mathbf{N}^p$. Retaining the hypothesis (FS), extend the argument leading to (79.17) to establish the general flow rule[422]

$$\mathbf{T}_0 = Y(\mathbf{N}^p)\mathbf{N}^p + \frac{\partial Y(\mathbf{N}^p)}{\partial \mathbf{N}^p}. \qquad (79.23)$$

Interestingly, for this general rule the constitutive function $Y(\mathbf{N}^p)$ for the flow resistance determines the explicit form of the flow rule.

79.3 More General Flow Rules. Drucker's Theorem

In this section, we establish a major result of Drucker (1950, 1952),[423] who gave a formal argument to show that — for a "stable material" — the yield surface must be *convex with outward unit normal the flow direction*. This result is often referred to as *Drucker's theorem*. While our conclusions coincide with those of Drucker, we base our derivation on the notion of maximum dissipation rather than on stability.

Our starting point is the (rate-independent) flow rule (79.1); viz.

$$\mathbf{T}_0 = \hat{\mathbf{T}}_0(\mathbf{N}^p). \qquad (79.24)$$

Here, and throughout this section,

- *we suppress the accumulated plastic strain e^p as an argument.*

The corresponding results should therefore be interpreted as being valid at each fixed value of e^p.

79.3.1 Yield-Set Hypotheses

To begin, we note that, by (79.2) and (79.7), a deviatoric stress $\mathbf{T}_0$ is **admissible in the sense of maximum dissipation** if

$$(\hat{\mathbf{T}}_0(\mathbf{N}^p) - \mathbf{T}_0) : \mathbf{N}^p \geq 0 \quad \text{for every flow direction } \mathbf{N}^p. \qquad (79.25)$$

The yield set $\mathcal{Y}$ defined in (79.10) is the range of the mapping $\hat{\mathbf{T}}_0(\cdot)$; that is, the set of stresses $\hat{\mathbf{T}}_0(\mathbf{N}^p)$ generated as $\mathbf{N}^p$ ranges over the set (79.14) of all flow directions (i.e., the surface of the unit ball in SymDev. Our first step in establishing Drucker's theorem is to introduce hypotheses that render the yield set $\mathcal{Y}$ a *surface*. A basic hypothesis that one might place on the yield set $\mathcal{Y}$ is (FS) on page 456, which by (79.25) is, for our purposes, best expressed as

$$(\hat{\mathbf{T}}_0(\mathbf{N}^p) - \mathbf{T}_0) : \mathbf{N}^p \geq 0 \quad \text{for every flow stress } \mathbf{T}_0 \text{ and flow direction } \mathbf{N}^p. \qquad (79.26)$$

But given any flow stress $\mathbf{T}_0$ there is a flow direction $\bar{\mathbf{N}}^p$ such that $\mathbf{T}_0 = \hat{\mathbf{T}}_0(\bar{\mathbf{N}}^p)$; thus (79.26) is equivalent to the requirement that

$$[\hat{\mathbf{T}}_0(\mathbf{N}^p) - \hat{\mathbf{T}}_0(\bar{\mathbf{N}}^p)] : \mathbf{N}^p \geq 0 \quad \text{for all flow directions } \mathbf{N}^p, \bar{\mathbf{N}}^p. \qquad (79.27)$$

---

[421] The condition $(76.70)_2$ is tacit.

[422] Gurtin & Anand (2005c, eq. (8.12)).

[423] Cf. Drucker (1964) and the references therein. A discussion of Drucker's ideas is given by Malvern (1969, pp. 356–363). See also Il'yushin (1954, 1961), Pipkin & Rivlin (1965), and Lucchesi & Podio-Guidugli (1990).

Actually, we need a hypothesis stronger than (79.27).

(Y1) (**strict flow-stress admissibility**)

$$[\hat{\mathbf{T}}_0(\mathbf{N}^p) - \hat{\mathbf{T}}_0(\bar{\mathbf{N}}^p)] : \mathbf{N}^p > 0 \quad \text{for all flow directions } \mathbf{N}^p \neq \bar{\mathbf{N}}^p. \tag{79.28}$$

This hypothesis has an important consequence. To explore this, we arbitrarily choose flow directions $\mathbf{N}^p$ and $\bar{\mathbf{N}}^p$ obeying

$$\mathbf{N}^p \neq \bar{\mathbf{N}}^p. \tag{79.29}$$

Then (79.27) yields the inequalities

$$[\hat{\mathbf{T}}_0(\mathbf{N}^p) - \hat{\mathbf{T}}_0(\bar{\mathbf{N}}^p)] : \mathbf{N}^p > 0,$$

$$[\hat{\mathbf{T}}_0(\bar{\mathbf{N}}^p) - \hat{\mathbf{T}}_0(\mathbf{N}^p)] : \bar{\mathbf{N}}^p > 0,$$

which when added imply that[424]

$$[\hat{\mathbf{T}}_0(\mathbf{N}^p) - \hat{\mathbf{T}}_0(\bar{\mathbf{N}}^p)] : (\mathbf{N}^p - \bar{\mathbf{N}}^p) > 0. \tag{79.30}$$

Thus, $\mathbf{N}^p \neq \bar{\mathbf{N}}^p$ implies that $\hat{\mathbf{T}}_0(\mathbf{N}^p) \neq \hat{\mathbf{T}}_0(\bar{\mathbf{N}}^p)$, and the function $\hat{\mathbf{T}}_0$ is one-to-one. We therefore have an important result:

- *The function $\hat{\mathbf{T}}_0$, as a mapping of flow directions $\mathbf{N}^p$ onto flow stresses*

$$\mathbf{T}_0 = \hat{\mathbf{T}}_0(\mathbf{N}^p),$$

is a one-to-one mapping of $\mathcal{N}$ onto the yield set $\mathcal{Y}$. Writing $\hat{\mathbf{N}}^p$ for the inverse of $\hat{\mathbf{T}}_0$, we may consider the flow direction $\mathbf{N}^p$ as a function

$$\mathbf{N}^p = \hat{\mathbf{N}}^p(\mathbf{T}_0) \tag{79.31}$$

of the flow stress $\mathbf{T}_0$, in which case we say that $\mathbf{N}^p$ is the ***flow direction corresponding to $\mathbf{T}_0$***.

An important consequence of this bullet is that the assertion (79.25) defining a stress $\mathbf{T}_0$ that is admissible in the sense of maximum dissipation may be written equivalently in the form[425]

$$(\bar{\mathbf{T}}_0 - \mathbf{T}_0) : \hat{\mathbf{N}}^p(\bar{\mathbf{T}}_0) \geq 0 \quad \text{for every flow stress } \bar{\mathbf{T}}_0. \tag{79.32}$$

To verify (79.32), simply appeal to (79.25) with

$$\mathbf{N}^p = \hat{\mathbf{N}}^p(\bar{\mathbf{T}}_0), \quad \text{so that} \quad \bar{\mathbf{T}}_0 = \hat{\mathbf{T}}^p(\mathbf{N}^p). \tag{79.33}$$

The second yield-set hypothesis is a smoothness assumption:

(Y2) (**smoothness hypothesis**) the mapping $\hat{\mathbf{T}}_0$ and its inverse $\hat{\mathbf{N}}^p$ are smooth functions.

We refer to the hypotheses (Y1) and (Y2) as the **yield-set hypotheses**. These hypotheses render the mapping $\hat{\mathbf{T}}_0$ a *diffeomorphism*[426] of the spherical surface $\mathcal{N}$ onto the yield set $\mathcal{Y}$; and, since diffeomorphisms carry smooth surfaces that are closed and bounded onto other such surfaces,

(‡) *$\mathcal{Y}$ is a smooth surface that is closed and bounded.*

---

424 The inequality (79.30) renders the function $\hat{\mathbf{T}}_0$ strictly monotonic.

425 This inequality corresponds to the classical *postulate of maximum dissipation*, which LUBLINER (1990, pp. 115–120) attributes variously to von Mises, Taylor, and Hill.

426 Roughly speaking, a diffeomorphism is a smooth, invertible mapping whose inverse is also smooth.

We henceforth refer to $\mathcal{Y}$ as the **yield surface**.[427]

EXERCISE

1. For the purpose of this exercise, replace (79.28) by (79.7). As is clear from (79.7), a stress $\mathbf{T}_0$ is admissible in the sense of maximum dissipation if and only if its resolved value $\mathbf{T}_0\!:\!\mathbf{N}^p$ with respect to any flow direction $\mathbf{N}^p$ is not greater than the flow resistance[428]
$$Y(\mathbf{N}^p) \stackrel{\text{def}}{=} \mathbf{N}^p\!:\!\hat{\mathbf{T}}_0(\mathbf{N}^p) \geq 0. \tag{79.34}$$
Since the flow direction determines the flow stress through the flow rule $\mathbf{T}_0 = \hat{\mathbf{T}}_0(\mathbf{N}^p)$, one might ask if there is a natural method of determining those flow directions $\mathbf{N}^p$ that correspond to a prescribed flow stress $\mathbf{T}_0$. In fact, one might expect that such flow directions $\mathbf{N}^p$ could be found variationally by the requirement that the corresponding resolved values $\mathbf{T}_0\!:\!\mathbf{N}^p$ be as close as possible to $Y(\mathbf{N}^p)$. Guided by this expectation, consider the **flow-direction problem**: Given $\mathbf{T}_0 \in \mathcal{Y}$, find a flow direction $\mathbf{N}^p$ that minimizes the function
$$\Phi(\mathbf{N}^p) \stackrel{\text{def}}{=} Y(\mathbf{N}^p) - \mathbf{T}_0\!:\!\mathbf{N}^p \geq 0 \tag{79.35}$$
over the set $\mathcal{N}$ of all flow direction $\mathbf{N}^p$. (The inequality follows from (79.34).) Show that if the flow direction problem has a unique solution $\mathbf{N}^p$ for every flow stress $\mathbf{T}_0$, then the bullet on page 459 is satisfied.

79.3.2 Digression: Some Definitions and Results Concerning Convex Surfaces

Let $\mathcal{A}$ be a finite-dimensional vector space with elements $a, q, r, s, \ldots$ and inner product $a \cdot q$. Let Ω be a **closed** region in $\mathcal{A}$ with boundary
$$\mathcal{S} = \partial\Omega \tag{79.36}$$
a smooth connected surface. Given any $q \in \mathcal{S}$:

(i) n_q denotes the *outward unit normal* to $\mathcal{S}$ at q;
(ii) $\mathcal{H}_q$ is the *half space* with $q \in \partial\mathcal{H}_q$ and n_q the outward unit normal to $\partial\mathcal{H}_q$.

The mapping
$$q \mapsto n_q \tag{79.37}$$
that associates with every $q \in \mathcal{S}$ the outward unit normal n_q is called the **Gauss map** of the surface.

The following definitions are central: $\mathcal{S}$ is **convex** if, given any $q \in \mathcal{S}$, the surface $\mathcal{S}$ is contained in $\mathcal{H}_q$; equivalently, $\mathcal{S}$ is convex if
$$(q - a) \cdot n_q \geq 0 \quad \text{for all } q, a \in \mathcal{S}; \tag{79.38}$$
$\mathcal{S}$ is **strictly convex** if
$$(q - a) \cdot n_q > 0 \quad \text{for every } q, a \in \mathcal{S}, a \neq q. \tag{79.39}$$
Important consequences of this definition are that:

(C1) if $\mathcal{S}$ is strictly convex, then the interior of Ω is the set of all $a \in \mathcal{A}$ such that
$$(q - a) \cdot n_q > 0 \quad \text{for all } q \in \mathcal{S}; \tag{79.40}$$
the *exterior* of Ω is the set of all $a \in \mathcal{A}$ such that
$$(q - a) \cdot n_q < 0 \quad \text{for some } q \in \mathcal{S}; \tag{79.41}$$
(C2) $\mathcal{S}$ is strictly convex if and only if the Gauss map is one-to-one.[429]

The definitions and results stated above are basic to §79.3.3; there the role of $\mathcal{A}$ is played by the space SymDev of symmetric, deviatoric tensors.

[427] For most of this book we have not fussed over smoothness hypotheses, but here some care is required because the hypothesis (Y2) has a strong consequence: It implies that the yield surface cannot have corners. Cf. SIMO & HUGHES (1998) and HAN & REDDY (1999) for formulations that allow for corners on the yield surface.

[428] Cf. (79.2).

[429] This assertion is generally false if $\mathcal{S}$ is not smooth.

79.3.3 Drucker's Theorem

First of all we note that, since (79.28) implies (79.26),

$$\mathcal{Y} \subset \mathcal{E}. \tag{79.42}$$

The following definition eases the statement of the next result: We refer to a stress $\mathbf{T}_0$ as **strictly admissible in the sense of maximum dissipation**[430] if

$$(\hat{\mathbf{T}}_0(\mathbf{N}^p) - \mathbf{T}_0) : \mathbf{N}^p > 0 \quad \text{for every flow direction } \mathbf{N}^p; \tag{79.43}$$

or equivalently, in view of the substitutions (79.33),

$$(\bar{\mathbf{T}}_0 - \mathbf{T}_0) : \hat{\mathbf{N}}^p(\bar{\mathbf{T}}_0) > 0 \quad \text{for every flow stress } \bar{\mathbf{T}}_0. \tag{79.44}$$

We are now in a position to establish

DRUCKER'S THEOREM[431] *Assume that the yield-surface hypothesis* (Y1) on page 459 *and the subsequent hypothesis* (Y2) *are satisfied. Then:*

(D1) *given any flow stress* $\mathbf{T}_0$, *the flow direction* $\mathbf{N}^p$ *corresponding to* $\mathbf{T}_0$ *is the* outward unit normal *to* $\mathcal{Y}$ *at* $\mathbf{T}_0$,
(D2) *the yield surface is* strictly convex
(D3) *The elastic range* $\mathcal{E}$ *is the closed, bounded region whose boundary is* $\mathcal{Y}$ *and for which the flow direction is directed outward from* $\mathcal{E}$;
(D4) *The interior of the elastic range is the set of all stresses* $\mathbf{T}_0$ *that are strictly admissible in the sense of maximum dissipation — and*

$$\dot{\mathbf{E}}^p = \mathbf{0} \text{ throughout the interior of the elastic range.} \tag{79.45}$$

PROOF. To establish (D1), choose:

(i) an arbitrary flow stress $\mathbf{T}_0$ on $\mathcal{Y}$ and let $\bar{\mathbf{N}}^p$ denote the corresponding flow direction;
(ii) an arbitrary curve $\tilde{\mathbf{T}}_0(\lambda)$ on $\mathcal{Y}$ through $\mathbf{T}_0$, so that, for some λ_0,

$$\tilde{\mathbf{T}}_0(\lambda_0) = \mathbf{T}_0.$$

Note that an immediate consequence of (79.42) is that *every flow stress* $\mathbf{T}_0$ *is admissible in the sense of maximum dissipation*. Thus, by (79.25),

$$\Phi(\lambda) \stackrel{\text{def}}{=} \hat{\mathbf{T}}_0(\bar{\mathbf{N}}^p) : \bar{\mathbf{N}}^p - \tilde{\mathbf{T}}_0(\lambda) : \bar{\mathbf{N}}^p$$
$$\geq 0 \qquad \text{for all } \lambda,$$

and, since $\Phi(\lambda_0) = 0$, $\Phi(\lambda)$ must have a minimum at $\lambda = \lambda_0$; hence, $d\Phi(\lambda)/d\lambda$ must vanish at $\lambda = \lambda_0$:

$$\bar{\mathbf{N}}^p : \frac{d\tilde{\mathbf{T}}_0(\lambda)}{d\lambda}\bigg|_{\lambda=\lambda_0} = 0. \tag{79.46}$$

Since the curve $\tilde{\mathbf{T}}_0(\lambda)$ on $\mathcal{Y}$ through $\mathbf{T}_0$ was arbitrarily chosen, $\bar{\mathbf{N}}^p$ must be normal to every curve on $\mathcal{Y}$ through $\mathbf{T}_0$. Thus, $\bar{\mathbf{N}}^p$ is normal to the yield surface at $\mathbf{T}_0$. In addition, by (79.34),

$$\mathbf{T}_0 : \bar{\mathbf{N}}^p \geq 0$$

and $\bar{\mathbf{N}}^p$ is outward from $\mathcal{Y}$ at $\mathbf{T}_0$. This establishes (D1).

---

430 Note that flow stresses cannot be strictly admissible in the sense of maximum dissipation because each such stress $\mathbf{T}_0$ satisfies $\mathbf{T}_0 = \hat{\mathbf{T}}_0(\mathbf{N}^p)$ for some flow direction $\mathbf{N}^p$.

431 The results (D1) and (D2) were proposed by DRUCKER (1951) based on an heuristic notion of material stability. In contrast to the present treatment, Drucker allowed for corners on the yield surface. The proof of Drucker's Theorem given below is due to GURTIN.

We turn next to the verification of (D2). In view of (79.28) and the definition (79.37) of the Gauss map,

(‡) *the function* $\hat{\mathbf{N}}^p$ *represents the Gauss map of the yield surface* $\mathcal{Y}$.

The result (C2) on page 460 therefore implies that the yield surface is strictly convex, which is (D2).

Next, in view of (‡), the result (C1) on page 460 implies that the region interior to $\mathcal{Y}$ is the set of all $\mathbf{T}_0$ such that

$$(\bar{\mathbf{T}}_0 - \mathbf{T}_0) : \hat{\mathbf{N}}^p(\bar{\mathbf{T}}_0) > 0 \quad \text{for every flow stress } \bar{\mathbf{T}}_0; \tag{79.47}$$

the region exterior to $\mathcal{Y}$ is the set of all $\mathbf{T}_0$ such that

$$(\bar{\mathbf{T}}_0 - \mathbf{T}_0) : \hat{\mathbf{N}}^p(\bar{\mathbf{T}}_0) < 0 \quad \text{for every flow stress } \bar{\mathbf{T}}_0. \tag{79.48}$$

Since $\mathcal{E}$ is the set of all stresses $\bar{\mathbf{T}}_0$ that are admissible in the sense of maximum dissipation and, hence, all $\bar{\mathbf{T}}_0$ such that (79.32) is satisfied, it follows that the exterior of $\mathcal{E}$ (the set of all tensors in SymDev not consistent with (79.32)) must coincide with the set of stresses $\bar{\mathbf{T}}_0$ consistent with (79.48), a set that defines the region exterior to $\mathcal{Y}$. Thus,

(i) $\mathcal{E}$ is closed with $\mathcal{Y}$ its boundary,

$$\mathcal{Y} = \partial\mathcal{E}; \tag{79.49}$$

(ii) the interior of $\mathcal{E}$ coincides with the set of stresses $\bar{\mathbf{T}}_0$ consistent with (79.47), a set that defines the region interior to $\mathcal{Y}$.

Since (79.47) is equivalent to (79.43), the interior of $\mathcal{E}$ is strictly admissible in the sense of maximum dissipation. The first part of (D4) is therefore valid. Further, since the interior of the elastic range is the region interior to the yield surface $\mathcal{Y}$, and since the flow direction coincides with the outward unit normal to $\mathcal{Y}$, the flow direction is directed outward from $\mathcal{E}$. This result and the assertion in the phrase ending in (79.49) comprise the content of (D3). Finally, to establish the second part of (D4) we have to establish (79.45). Our prooof is by contradiction. Were (79.45) false, there would be a deviatoric stress $\mathbf{T}_0$ in the interior of $\mathcal{E}$ and a plastic strain-rate $\dot{\mathbf{E}}^p \neq \mathbf{0}$ such that the normalized flow $(\mathbf{T}_0, \mathbf{N}^p)$ — with $\mathbf{N}^p = \dot{\mathbf{E}}^p/|\dot{\mathbf{E}}^p|$ — is physically attainable. But this implies that $\mathbf{T}_0$ belongs to the yield surface and hence cannot belong to the interior of $\mathcal{E}$.

This completes the proof of Drucker's Theorem.

79.4 The Conventional Theory of Perfectly Plastic Materials Fits within the Framework Presented Here

Conventional formulations of rate-independent perfectly plastic materials — that is materials without work-hardening[432] — are typically based on

(i) a smooth, strictly convex yield function $f(\mathbf{T}_0)$ defined for all deviatoric stresses $\mathbf{T}_0$; together with
(ii) a flow rule of the form

$$\dot{\mathbf{E}}^p = \lambda \frac{\partial f(\mathbf{T}_0)}{\partial \mathbf{T}_0}, \tag{79.50}$$

with $\lambda > 0$ an arbitrary field.

432 The inclusion of hardening is left as an exercise.

The elastic range $\mathcal{E}$ and the yield surface $\mathcal{Y}$ are then defined by

$$\mathcal{E} \stackrel{\text{def}}{=} \text{the set of all } \mathbf{T}_0 \in \text{SymDev such that } f(\mathbf{T}_0) \le 0 \tag{79.51}$$

and

$$\mathcal{Y} \stackrel{\text{def}}{=} \text{the set of all } \mathbf{T}_0 \text{ such that } f(\mathbf{T}_0) = 0, \tag{79.52}$$

so that $\mathcal{Y} = \partial\mathcal{E}$. Further, the null-stress $\mathbf{0} \in \text{SymDev}$ is assumed to lie in the interior of $\mathcal{E}$.

We now show that this formulation fits within the framework discussed in §79.1 and §79.3. First of all, since $\mathcal{Y}$ is convex, the tensor function

$$\hat{\mathbf{N}}^p(\mathbf{T}_0) \stackrel{\text{def}}{=} \frac{\dfrac{\partial f(\mathbf{T}_0)}{\partial \mathbf{T}_0}}{\left|\dfrac{\partial f(\mathbf{T}_0)}{\partial \mathbf{T}_0}\right|}$$

represents the outward unit normal to the yield surface at $\mathbf{T}_0$. On the other hand, by (79.50), this outward unit normal is also given by $\mathbf{N}^p = \dot{\mathbf{E}}^p/|\dot{\mathbf{E}}^p|$; thus

$$\hat{\mathbf{N}}^p(\mathbf{T}_0) = \mathbf{N}^p. \tag{79.53}$$

Further, because the yield surface $\mathcal{Y}$ is strictly convex, there is a one-to-one correspondence between deviatoric stresses $\mathbf{T}_0$ on $\mathcal{Y}$ and outward unit normals

$$\mathbf{N}^p = \frac{\dot{\mathbf{E}}^p}{|\dot{\mathbf{E}}^p|}$$

to $\mathcal{Y}$ at $\mathbf{T}_0$;[433] hence we can invert the function $\hat{\mathbf{N}}^p(\mathbf{T}_0)$ and arrive at the flow rule in the form

$$\mathbf{T}_0 = \hat{\mathbf{T}}_0(\mathbf{N}^p), \tag{79.54}$$

which is the flow rule (79.1) of the general theory discussed in §79.1 and §79.3,[434] neglecting hardening. Moreover, in view of the discussion in the paragraph containing (79.10), $\mathcal{Y}$ — as defined in (79.52) — can equally well be defined as the set of all flow stresses.[435]

Next, by (79.54), the dissipation during plastic flow is given by

$$\hat{\mathbf{T}}_0(\mathbf{N}^p):\dot{\mathbf{E}}^p \tag{79.55}$$

and we now show that, given $\dot{\mathbf{E}}^p$, this dissipation represents the maximum value of $\mathbf{T}_0:\dot{\mathbf{E}}^p$ over all $\mathbf{T}_0 \in \mathcal{E}$ — or, by(79.51),

$$\text{the minimum value of } -\mathbf{T}_0:\dot{\mathbf{E}}^p \text{ over all } \mathbf{T}_0 \text{ such that } f(\mathbf{T}_0) \le 0.$$

It therefore follows from the theory of constrained minima that this minimum value occurs at a deviatoric stress $\mathbf{T}_0$ consistent with the **Kuhn–Tucker optimality conditions:**[436]

$$\frac{\partial}{\partial \mathbf{T}_0}\big[-\mathbf{T}_0:\dot{\mathbf{E}}^p + \lambda f(\mathbf{T}_0)\big] = \mathbf{0}, \qquad \lambda \ge 0, \qquad \lambda f(\mathbf{T}_0) = 0. \tag{79.56}$$

Thus, $\mathbf{T}_0$ satisfies (79.50) and hence (79.54). We have therefore shown that

$$\mathbf{T}_0:\dot{\mathbf{E}}^p \le \hat{\mathbf{T}}_0(\mathbf{N}^p):\dot{\mathbf{E}}^p,$$

---

433 Cf. (C2) on page 460.

434 The fact that (79.53) and (79.54) are inverses of one and other is simply the bullet on page 459.

435 Cf. (79.10).

436 Cf. Strang (1986, p. 724).

and hence, dividing this inequality by $|\dot{\mathbf{E}}^p|$, that

$$\mathbf{T}_0\!:\!\mathbf{N}^p \le Y(\mathbf{N}^p),$$

with $Y(\mathbf{N}^p)$ defined in (79.2). Thus, in view of (i) and (ii) on page 455, the elastic range $\mathcal{E}$, defined in (79.51), can equally well be defined as the set of all deviatoric stresses $\mathbf{T}_0$ that are admissible in the sense of maximum dissipation, a definition consistent with (79.8).

We have therefore established the following:

CONSISTENCY THEOREM *The conventional theory of perfectly plastic materials, as described in the paragraph containing* (79.51), *is a special case of a theory based on a flow rule of the form* (79.54), *with yield surface the range of the function* $\hat{\mathbf{T}}_0$ *and with elastic range the set of all* $\mathbf{T}_0$ *that are admissible in the sense of maximum dissipation.*

EXERCISES

1. A conventional treatment of hardening, with accumulated plastic strain e^p as hardening variable, starts with a smooth yield function $f(\mathbf{T}_0, e^p) = \bar{f}(\mathbf{T}_0) - F(e^p)$ defined for all $\mathbf{T}_0 \in \text{SymDev}$ and $e^p \ge 0$ with $\bar{f}(\mathbf{T}_0)$ strictly convex in $\mathbf{T}_0$ and $F(e^p) > 0$. Establish analogs of the results of this section for this more general starting point. How would you define a hardening-softening transition within this framework?
2. Show that the flow rule (79.54), derived above, is consistent with (Y1) on page 459.
3. Show that the conventional theory as described in the paragraph containing (79.50) is strictly dissipative in the sense that, whenever $\mathbf{T}_0$ and $\dot{\mathbf{E}}^p \neq \mathbf{0}$ are related through the flow rule (79.50), then

$$\mathbf{T}_0\!:\!\dot{\mathbf{E}}^p > 0.$$

80 Hardening Characterized by a Defect Energy

Strain hardening is generally seen as a byproduct of the formation of defects during plastic flow. In §76 this hardening is viewed as *dissipative*; in fact, as characterized by the Mises–Hill equations

$$\mathbf{T}_0 = \hat{Y}(e^p)\frac{\dot{\mathbf{E}}^p}{|\dot{\mathbf{E}}^p|} \qquad \text{for } \dot{\mathbf{E}}^p \neq \mathbf{0}, \tag{80.1}$$
$$\dot{e}^p = |\dot{\mathbf{E}}^p|, \qquad e^p(0) = 0,$$

with e^p the accumulated plastic strain, and with

$$\delta(\dot{\mathbf{E}}^p, e^p) = Y(e^p)|\dot{\mathbf{E}}^p| \tag{80.2}$$

the associated dissipation.[437] We now show that — surprisingly — this hardening may equally well be regarded as *energetic*.[438]

80.1 Free-Energy Imbalance Revisited

We view the free energy Ψ as a sum

$$\Psi = \Psi^e + \Psi^p,$$

where Ψ^p is a *defect energy*,[439] plastic in nature, associated with the formation of defects, while Ψ^e, the elastic energy, is assumed related to the elastic stress-power through the standard balance[440]

$$\dot{\Psi}^e = \mathbf{T}:\dot{\mathbf{E}}^e.$$

Granted these assumptions, the free-energy imbal,ance (75.14) reduces to an imbalance

$$\dot{\Psi}^p - \mathbf{T}_0:\dot{\mathbf{E}}^p = -\delta \leq 0, \tag{80.3}$$

associated solely with plastic flow.

437 Cf. (76.69) and (76.71).
438 This section is taken from GURTIN & REDDY (2009).
439 More precisely, a defect *free*-energy.
440 Cf. §75.15.

80.2 Constitutive Equations. Flow Rule

We view the accumulated plastic strain e^p — defined by $(80.1)_2$ — as a measure of the defectiveness of the microscopic structure and, based on this, take[441]

$$\Psi^p = \hat{\Psi}^p(e^p) \tag{80.4}$$

as the constitutive relation for the defect energy.[442] It is convenient to introduce a thermodynamic **hardening stress**

$$\hat{g}(e^p) \stackrel{\text{def}}{=} \frac{d\hat{\Psi}^p(e^p)}{de^p} \tag{80.5}$$

and to assume that, consistent with $(80.1)_2$,[443]

$$\hat{g}(0) = 0, \qquad \hat{g}(e^p) > 0 \text{ for } e^p > 0. \tag{80.6}$$

Then, since by (76.20) and (76.52)

$$\dot{e}^p = |\dot{\mathbf{E}}^p| = \mathbf{N}^p : \dot{\mathbf{E}}^p,$$

it follows that

$$\overline{\hat{\Psi}^p(e^p)}^{\cdot} = g(e^p)|\dot{\mathbf{E}}^p| \geq 0. \tag{80.7}$$

The imbalance (80.3) therefore takes the form

$$(\mathbf{T}_0 - \hat{g}(e^p)\mathbf{N}^p) : \dot{\mathbf{E}}^p = \delta \geq 0. \tag{80.8}$$

The quantity

$$\mathbf{T}_0^{\text{dis}} \stackrel{\text{def}}{=} \mathbf{T}_0 - \hat{g}(e^p)\mathbf{N}^p, \tag{80.9}$$

which represents the *dissipative part* of the deviatoric stress $\mathbf{T}_0$, must then be consistent with the *dissipation inequality*

$$\mathbf{T}_0^{\text{dis}} : \dot{\mathbf{E}}^p = \delta \geq 0. \tag{80.10}$$

Further, granted codirectionality, $\mathbf{T}_0^{\text{dis}}$ must be positively proportional to $\mathbf{N}^p$. Thus, if hardening is completely characterized by the hardening stress $\hat{g}(e^p)$, the constitutive relation for $\mathbf{T}_0^{\text{dis}}$ must the simple form

$$\mathbf{T}_0^{\text{dis}} = Y_0\mathbf{N}^p \tag{80.11}$$

in which the flow resistance Y_0 — assumed constant and strictly positive — mimics that of a material without hardening. Combining (80.9) and (80.11) we arrive at the **flow rule**

$$\boxed{\mathbf{T}_0 = [Y_0 + \hat{g}(e^p)]\mathbf{N}^p.} \tag{80.12}$$

Further, the dissipation may be easily calculated from (80.10) and (80.11); the result is

$$\delta(\dot{\mathbf{E}}^p) = Y_0|\dot{\mathbf{E}}^p|. \tag{80.13}$$

Consider, once again, the *Mises–Hill flow rule* $(80.1)_1$, whose derivation restricts attention to *dissipative* inelasticity — there is no defect energy. Interestingly, that

---

441 Cf. Simo & Hughes (1998) and Han & Reddy (1999), and the references therein. These books are based on the use of free energies to describe hardening.

442 Often referred to as the stored energy of cold work.

443 Granted that $e^p = 0$ corresponds to an absence of defects.

flow-rule is *identical* to the *energetically* based flow rule (80.12) provided[444]

$$\boxed{Y(e^p) = Y_0 + \hat{g}(e^p).} \tag{80.14}$$

Thus, while the flow rules $(80.1)_1$ and (80.12) rest upon different physical concepts, the field equations resulting from the two flow rules are *identical*. In comparing the view of $\hat{\Psi}^p(e^p)$ as a free energy with the more conventional view in which there is no perceived need to introduce such an energy, we note that, by (80.7), $\hat{\Psi}^p(e^p)$ *cannot* decrease with time, no matter the loading path. This energy therefore corresponds to loading processes that are *irreversible*. In that sense:

- the defect energy $\hat{\Psi}^p(e^p)$ *mimics dissipative behavior.*

We know of no other example from continuum mechanics where an energetic quantity mimics dissipative behavior.

Moreover, in view of the equivalence of the flow rule (80.12) to the Mises–Hill flow-rule $(80.1)_1$ and of the specific relationship (80.14) between flow resistances, we can define an **effective dissipation** via the relation

$$\boxed{D_{\rm eff}(\dot{\mathbf{E}}^p, e^p) \stackrel{\rm def}{=} [Y_0 + \hat{g}(e^p)]|\dot{\mathbf{E}}^p|.} \tag{80.15}$$

Further, using (80.7) we can express this effective dissipation as the sum of

(i) a term $Y_0|\dot{\mathbf{E}}^p| = \delta(\dot{\mathbf{E}}^p)$, which represents the *actual* dissipation, and
(ii) a term

$$\hat{g}(e^p)|\dot{\mathbf{E}}^p| = \overline{\hat{\Psi}^p(e^p)}^{\,\cdot},$$

which represents the rate of change of a *nonrecoverable* energy.

The central result of this section is summarized as follows:

THEOREM *The theory without a defect energy is equivalent to the theory with a defect energy provided the flow resistance $Y(e^p)$ and the dissipation $\delta(\dot{\mathbf{E}}^p, e^p)$ in the former are replaced by*

$$Y_0 + \hat{g}(e^p) \qquad \text{and} \qquad D_{\rm eff}(\dot{\mathbf{E}}^p, e^p) \tag{80.16}$$

in the latter. Thus whether or not one chooses to use a theory with a defect energy could very well be based on which of these choices is more amenable to analysis and/or computation.[445]

More generally, suppose we are given a flow rule in the Mises–Hill form $(80.1)_1$ with *flow resistance* $Y(e^p)$. Suppose further, that we are given a defect energy $\Psi^p(e^p)$ consistent with (80.5) and (81.31) and a function $Y_{\rm dis}(e^p) > 0$ such that

$$Y(e^p) = Y_{\rm dis}(e^p) + \frac{d\hat{\Psi}^p(e^p)}{de^p}. \tag{80.17}$$

Then, defining $\hat{g}(e^p)$ through (80.5) and arguing as in the steps from (80.4)–(80.13), we find that the dissipative part of the deviatoric stress, the dissipation, and the flow

444 Consistent with this, we assume that the *boundedness inequality* has the form $|\mathbf{T}_0| \le Y_0 + \hat{g}(e^p)$.

445 Existence and uniqueness of solutions to initial/boundary-value problems for the theory *with* a defect energy were established by REDDY (1992). A trivial corollary of our theorem and Reddy's result therefore establishes existence and uniqueness for the theory *without* a defect energy.

rule have the respective forms

$$\mathbf{T}_0^{\mathrm{dis}} = Y_{\mathrm{dis}}(e^p)\mathbf{N}^p, \qquad \delta(\dot{\mathbf{E}}^p, e^p) = Y_{\mathrm{dis}}(e^p)|\dot{\mathbf{E}}^p|,$$
$$\mathbf{T}_0 = \big[Y_{\mathrm{dis}}(e^p) + \hat{g}(e^p)\big]\mathbf{N}^p. \tag{80.18}$$

It therefore makes sense to refer (80.17) as a *partition* of the flow resistance into dissipative and energetic parts. Since each such partition leaves the original flow rule invariant, the field equations are independent of the partition. Thus, interestingly, the mechanical theory provides no information about the partition of the flow resistance into dissipative and energetic parts.

A theory that differentiates between dissipative and energetic flow rules requires a framework that allows for thermal variations, because within such a framework the production of entropy is affected by dissipation, but not by irreversibility induced by a defect energy. Such a theory is given by ROSAKIS, ROSAKIS, RAVICHANDRAN & HODOWANY (2000) — with corresponding experimental results given by HODOWANY, RAVICHANDRAN, ROSAKIS & ROSAKIS (2000).[446] These works establish *separate* and *unique* constitutive relations for the defect energy and the dissipative part of the flow resistance. This issue is discussed in detail in §81.

[446] See HODOWANY, RAVICHANDRAN, ROSAKIS & ROSAKIS (2000) and ROSAKIS, ROSAKIS, RAVICHANDRAN & HODOWANY (2000) for references and a comprehensive overview — dating back to the pioneering experiments of FAREN & TAYLOR (1925) and TAYLOR & QUINNEY (1937) — of the relevant experimental and theoretical literature.

81 The Thermodynamics of Mises–Hill Plasticity

81.1 Background

In §80.2 — which was limited to isothermal conditions — we showed that all Mises–Hill flow rules of the form

$$\mathbf{T}_0 = \underbrace{\left[Y_{\text{dis}}(e^p) + \frac{d\hat{\Psi}^p(e^p)}{de^p} \right]}_{Y(e^p)} \mathbf{N}^p, \tag{81.1}$$

with

$$Y_{\text{dis}}(e^p) \quad \text{and} \quad \frac{d\hat{\Psi}^p(e^p)}{de^p} \tag{81.2}$$

the dissipative and energetic parts of a prescribed flow resistance $Y(e^p)$, result in the same field equations.

A basic issue concerning theories of the type discussed above involves the determination of the fraction β of the plastic stress-power converted to heating. Within such theories the plastic stress-power is given by

$$\mathbf{T}_0 \colon \dot{\mathbf{E}}^p = Y_{\text{dis}}(e^p)\dot{e}^p + \frac{d\hat{\Psi}^p(e^p)}{de^p}\dot{e}^p, \tag{81.3}$$

and the fraction of this stress power converted to heating is the dissipative part $Y_{\text{dis}}(e^p)\dot{e}^p$. Thus, by (81.3),

$$\beta = \frac{Y_{\text{dis}}(e^p)}{Y_{\text{dis}}(e^p) + \frac{d\hat{\Psi}^p(e^p)}{de^p}}. \tag{81.4}$$

But, as noted in the paragraph containing (80.18),

- *the mechanical theory provides no information about the partition of the flow resistance into dissipative and energetic parts.*

Thus (in this purely mechanical theory) β is *indeterminate*, as it may take any value in the interval $[0, 1]$.

To discuss a theory that differentiates between dissipative and energetic flow rules, we work within a framework that allows for *thermal variations*, because within such a framework the energy balance is affected by dissipation, but not by irreversibility induced by a defect energy. Specifically, we work within the framework of continuum *thermodynamics* based on the first two laws of thermodynamics: balance of energy and the Clausius–Duhem inequality.

81.2 Thermodynamics

Our discussion of thermodynamics involves five fields:

ε	the internal energy,
ϑ	the absolute temperature ($\vartheta > 0$),
η	the entropy,
$\mathbf{q}$	the heat flux,
q	the external heat supply.

The first two laws of thermodynamics for a continuum are *balance of energy*

$$\dot{\varepsilon} = \mathbf{T}:\dot{\mathbf{E}} - \operatorname{div}\mathbf{q} + q \tag{81.5}$$

and the *Clausus–Duhem inequality*

$$\dot{\eta} \geq -\operatorname{div}\left(\frac{\mathbf{q}}{\vartheta}\right) + \frac{q}{\vartheta}, \tag{81.6}$$

which describes the *growth of entropy*.[447] These laws, when combined, form a *free-energy imbalance*

$$\dot{\Psi} + \eta\dot{\vartheta} - \mathbf{T}:\dot{\mathbf{E}} + \frac{1}{\vartheta}\mathbf{q}\cdot\nabla\vartheta = -\vartheta\Gamma \leq 0 \tag{81.7}$$

in which

$$\Psi = \varepsilon - \vartheta\eta \tag{81.8}$$

is the *free energy*. Here Γ is the left side of (81.6) minus the right side and hence represents the rate of *entropy production.*

Note that, since $\dot{\mathbf{E}}^p = \dot{e}^p\mathbf{N}^p$, we can write the *plastic stress-power* in the form

$$\mathbf{T}:\dot{\mathbf{E}}^p = \tau\dot{e}^p, \tag{81.9}$$

where

$$\tau \stackrel{\text{def}}{=} \mathbf{N}^p:\mathbf{T}_0, \tag{81.10}$$

the *resolved shear*, represents the deviatoric stress resolved on the direction of plastic flow. Further, using (75.11) and (81.9), we can decompose the stress power $\mathbf{T}:\dot{\mathbf{E}}$ into elastic and plastic power expenditures:

$$\mathbf{T}:\dot{\mathbf{E}} = \mathbf{T}:\dot{\mathbf{E}}^e + \tau\dot{e}^p. \tag{81.11}$$

81.3 Constitutive Equations

We assume that the free energy is the sum

$$\Psi = \Psi^e + \Psi^p \tag{81.12}$$

of a conventional elastic *strain energy* Ψ^e and a *defect energy* Ψ^p, plastic in nature, associated with the formation of defects. We assume that these energies are described by constitutive equations of the form

$$\Psi^e = \hat{\Psi}^e(\mathbf{E}^e, \vartheta), \qquad \Psi^p = \hat{\Psi}^p(e^p, \vartheta); \tag{81.13}$$

and that the elastic energy generates the stress $\mathbf{T}$ through the conventional relation

$$\mathbf{T} = \frac{\partial\hat{\Psi}^e(\mathbf{E}^e, \vartheta)}{\partial\mathbf{E}^e}. \tag{81.14}$$

[447] Cf. (26.8) and (27.13). Because the discussion here is for small deformations, the stress power in (81.5) is $\mathbf{T}:\dot{\mathbf{E}}$ rather than $\mathbf{T}:\mathbf{D}$ and ε and η are measured per unit volume, rather than per unit mass.

By (81.12) and (81.13) the (total) free energy Ψ is given by the auxilliary relation

$$\Psi = \hat{\Psi}_{\rm tot}(\mathbf{E}^e, e^p, \vartheta) \stackrel{\rm def}{=} \hat{\Psi}^e(\mathbf{E}^e, \vartheta) + \hat{\Psi}^p(e^p, \vartheta); \tag{81.15}$$

hence (81.14) implies that

$$\begin{aligned}\dot{\Psi} &= \frac{\partial \hat{\Psi}^e(\mathbf{E}^e, \vartheta)}{\partial \mathbf{E}^e} : \dot{\mathbf{E}}^e + \frac{\partial \hat{\Psi}^p(e^p, \vartheta)}{\partial e^p} \dot{e}^p + \frac{\partial \hat{\Psi}_{\rm tot}(\mathbf{E}^e, e^p, \vartheta)}{\partial \vartheta} \dot{\vartheta} \\ &= \mathbf{T} : \dot{\mathbf{E}}^e + \frac{\partial \hat{\Psi}^p(e^p, \vartheta)}{\partial e^p} \dot{e}^p + \frac{\partial \hat{\Psi}_{\rm tot}(\mathbf{E}^e, e^p, \vartheta)}{\partial \vartheta} \dot{\vartheta}.\end{aligned} \tag{81.16}$$

Our next step is to use the free-energy imbalance (81.7) in conjunction with the Coleman–Noll procedure to develop thermodynamically consistent constitutive equations. Using (81.14) and (81.16) we can express this imbalance in a form,

$$\left[\frac{\partial \hat{\Psi}^p(e^p, \vartheta)}{\partial e^p} - \tau\right] \dot{e}^p + \left[\frac{\partial \hat{\Psi}_{\rm tot}(\mathbf{E}^e, e^p, \vartheta)}{\partial \vartheta} + \eta\right] \dot{\vartheta} + \frac{1}{\vartheta} \mathbf{q} \cdot \nabla\vartheta = -\vartheta\Gamma \le 0, \tag{81.17}$$

that indicates a need for constitutive relations for the fields τ, η, and $\mathbf{q}$. Regarding these fields, we limit our discussion to constitutive equations that are *independent of* $\dot{\vartheta}$, a limitation that renders (81.17) *linear* in $\dot{\vartheta}$ and hence requires that the coeficient of $\dot{\vartheta}$ must vanish (for otherwise $\dot{\vartheta}$ could be chosen to violate (81.17)). The constitutive relation for the entropy must therefore have a classical structure giving the entropy as the negative of the partial derivative of the free energy with respect to temperature, viz.

$$\hat{\eta}(\mathbf{E}^e, e^p, \vartheta) = -\frac{\partial \hat{\Psi}_{\rm tot}(\mathbf{E}^e, e^p, \vartheta)}{\partial \vartheta}. \tag{81.18}$$

This entropy-relation simplifies the inequality (81.17) as follows:

$$\left[\frac{\partial \hat{\Psi}^p(e^p, \vartheta)}{\partial e^p} - \tau\right] \dot{e}^p + \frac{1}{\vartheta} \mathbf{q} \cdot \nabla\vartheta = -\vartheta\Gamma \le 0. \tag{81.19}$$

Most theories of plasticity are *not* based on a *single* constitutive relation for the stress $\mathbf{T}$, but instead require that the stress be consistent with *two separate* relations:[448]

(C1) The first is the constitutive relation (81.14) describing the elastic response of the material.

(C2) The second — a relation for the deviatoric stress $\mathbf{T}_0$ — represents a constraint on the elastic response imposed by the inelasticity of the material. Here, because our goal is a thermodynamics of Mises–Hill materials, we base the relation for $\mathbf{T}_0$ on a *codirectionality constraint* requiring that the flow direction $\mathbf{N}^p$ coincide with the direction of the deviatoric stress:

$$\mathbf{N}^p = \frac{\mathbf{T}_0}{|\mathbf{T}_0|}. \tag{81.20}$$

Trivially, (81.20) implies that $\mathbf{T}_0 = |\mathbf{T}_0|\mathbf{N}^p$, and hence that, by (81.10),

$$\tau = |\mathbf{T}_0|;$$

hence the codirectionality constraint takes the form

$$\mathbf{T}_0 = \tau\mathbf{N}^p. \tag{81.21}$$

---

[448] Cf. §75.3 and §76.4.

Guided by (81.1) and (81.19), we supplement this constraint with a constitutive relation

$$\tau = \hat{Y}(e^p, \vartheta) \tag{81.22}$$

for the resolved shear, thereby arriving at a constitutive relation

$$\mathbf{T}_0 = \hat{Y}(e^p, \vartheta)\mathbf{N}^p \tag{81.23}$$

characterizing plastic flow.

Finally, we consider a constitutive relation for the heat flux in the form

$$\mathbf{q} = \hat{\mathbf{q}}(e^p, \vartheta, \nabla\vartheta). \tag{81.24}$$

Then, by (81.22), the free-energy imbalance (81.19) takes the form

$$\left[\frac{\partial \hat{\Psi}^p(e^p, \vartheta)}{\partial e^p} - \hat{Y}(e^p, \vartheta)\right]\dot{e}^p + \frac{1}{\vartheta}\hat{\mathbf{q}}(e^p, \vartheta, \nabla\vartheta)\cdot\nabla\vartheta = -\vartheta\Gamma \le 0; \tag{81.25}$$

since the mechanical term is independent of $\nabla\vartheta$, while the thermal term is independent of $\dot{e}^p$, the choice $\nabla\vartheta = \mathbf{0}$ yields the *mechanical dissipation inequality*

$$\left[\frac{\partial \hat{\Psi}^p(e^p, \vartheta)}{\partial e^p} - \hat{Y}(e^p, \vartheta)\right]\dot{e}^p = -\vartheta\Gamma_{\mathrm{mec}} \le 0, \tag{81.26}$$

while the choice $\dot{e}^p = 0$ yields the *heat-conduction inequality*

$$\frac{1}{\vartheta}\hat{\mathbf{q}}(e^p, \vartheta, \nabla\vartheta)\cdot\nabla\vartheta = -\vartheta\Gamma_{\mathrm{ther}} \le 0. \tag{81.27}$$

The function $\hat{Y}_{\mathrm{dis}}(e^p, \vartheta)$ defined via the relation

$$\hat{Y}(e^p, \vartheta) = \hat{Y}_{\mathrm{dis}}(e^p, \vartheta) + \frac{\partial \hat{\Psi}^p(e^p, \vartheta)}{\partial e^p} \tag{81.28}$$

represents the dissipative part of the resolved shear, because, by (81.26),

$$\hat{Y}_{\mathrm{dis}}(e^p, \vartheta)\dot{e}^p = \vartheta\Gamma_{\mathrm{mec}} \ge 0. \tag{81.29}$$

This inequality must hold for all $\dot{e}^p \ge 0$;[449] thus

$$\hat{Y}_{\mathrm{dis}}(e^p, \vartheta) \ge 0. \tag{81.30}$$

81.4 Nature of the Defect Energy

The dependence of the defect energy Ψ^p on the accumulated plastic strain e^p has interesting and somewhat unexpected properties, which we now discuss. Since e^p is an increasing function of time with $e^p\big|_{t=0} = 0$,[450] and since we expect the defect energy to increase as the number of defects (as described by e^p) increases, we assume that

$$\hat{\Psi}^p(e^p, \vartheta)\big|_{e^p=0} = 0, \qquad \frac{\partial \hat{\Psi}^p(e^p, \vartheta)}{\partial e^p} \ge 0, \tag{81.31}$$

so that

$$\frac{\partial \hat{\Psi}^p(e^p, \vartheta)}{\partial e^p}\dot{e}^p \ge 0. \tag{81.32}$$

[449] Cf. $(80.1)_2$.
[450] Cf. $(80.1)_{2,3}$.

The defect energy Ψ^p thus increases with time in any process that has $\dot{\vartheta} \equiv 0$; hence Ψ^p is *not recoverable* during isothermal processes. The defect energy therefore mimics dissipative behavior by describing a class of processes that are *irreversible*. In fact, if the defect energy is independent of ϑ, then the defect energy *always* describes irreversible behavior. On the other hand, since ϑ may increase or decrease at will, the defect energy during any process in which e^p is constant *is recoverable*, at least in principle. Thus the individual arguments e^p and ϑ of the defect energy characterize *disparate* physical behaviors — and for an arbitrary process these two behaviors are *coupled*. Further, by (81.22), (81.28), (81.9), (81.29), and (81.32), the plastic stress-power is *nonnegative*,

$$\mathbf{T}_0 : \dot{\mathbf{E}}^p = \hat{Y}(e^p, \vartheta)\dot{e}^p \geq 0 \tag{81.33}$$

$$= \hat{Y}_{\mathrm{dis}}(e^p, \vartheta)\dot{e}^p + \frac{\partial \hat{\Psi}^p(e^p, \vartheta)}{\partial e^p}\dot{e}^p \geq 0. \tag{81.34}$$

Summarizing, a *defect energy dependent on accumulated plastic strain mimics dissipative behavior* in the sense that:

- During isothermal conditions — or when the defect energy is independent of temperature — the defect energy describes processes that are *irreversible*.

- The plastic stress-power $\mathbf{T}_0 : \dot{\mathbf{E}}^p$ is nonnegative.

81.5 The Flow Rule and the Boundedness Inequality

This section follows §76.8. By (81.23) and (81.28), the deviatoric stress must satisfy

$$\mathbf{T}_0 = \underbrace{\left[\hat{Y}_{\mathrm{dis}}(e^p, \vartheta) + \frac{\partial \hat{\Psi}^p(e^p, \vartheta)}{\partial e^p} \right]}_{\hat{Y}(e^p, \vartheta)} \mathbf{N}^p, \tag{81.35}$$

a relation that represents the **flow rule**; consequently, we refer to the function $\hat{Y}(e^p, \vartheta)$ as the *flow resistance*. An important consequence of (81.35) is the *yield condition*

$$|\mathbf{T}_0| = \hat{Y}(e^p, \vartheta) \quad \text{for } \dot{e}^p \neq 0. \tag{81.36}$$

As is well understood by plasticians, the flow rule does not suffice to characterize the rate-independent behavior of plastic materials. An additional assumption, namely the *boundedness hypothesis*

$$|\mathbf{T}_0| \leq \hat{Y}(e^p, \vartheta), \tag{81.37}$$

which asserts that that the norm of the deviatoric stress not exceed the flow resistance, is needed. When $|\mathbf{T}_0| < \hat{Y}(e^p, \vartheta)$, the **yield condition** (81.35) is not satisfied; thus, necessarily, $\dot{e}^p = 0$. Therefore,

$$\dot{e}^p = 0 \quad \text{for} \quad |\mathbf{T}_0| < \hat{Y}(e^p, \vartheta). \tag{81.38}$$

Finally, the set of deviatoric stresses $\mathbf{T}_0$ that satisfy (81.37) represents the *elastic range* corresponding to e^p and ϑ.

81.6 Balance of Energy Revisited

Using (81.8) and (81.11), we can write the energy balance (81.5) in the form

$$\dot{\Psi} + \vartheta\dot{\eta} + \eta\dot{\vartheta} = \mathbf{T} : \dot{\mathbf{E}}^e + \tau\dot{e}^p - \mathrm{div}\,\mathbf{q} + q. \tag{81.39}$$

Hence, appealing to (81.16), (81.18), and (81.23)$_1$, we see that

$$\frac{\partial \hat{\Psi}^p(e^p, \vartheta)}{\partial e^p} \dot{e}^p + \vartheta \dot{\eta} = \hat{Y}(e^p, \vartheta)\dot{e}^p - \mathrm{div}\mathbf{q} + q, \tag{81.40}$$

so that, by (81.28),

$$\vartheta \dot{\eta} = \hat{Y}_{\mathrm{dis}}(e^p, \vartheta)\dot{e}^p - \mathrm{div}\mathbf{q} + q. \tag{81.41}$$

If we restrict attention to classical Fourier conduction in which

$$\mathbf{q} = -k\nabla\vartheta \tag{81.42}$$

with conductivity $k > 0$, and assume that $q = 0$, we find that

$$\vartheta \dot{\eta} = \hat{Y}_{\mathrm{dis}}(e^p, \vartheta)\dot{e}^p + k\Delta\vartheta. \tag{81.43}$$

If we neglect heat conduction, then, by (81.29), (81.43) has the *adiabatic* form

$$\boxed{\vartheta \dot{\eta} = \hat{Y}_{\mathrm{dis}}(e^p, \vartheta)\dot{e}^p \geq 0} \tag{81.44}$$

and is said to describe *adiabatic plastic flow*. Thus, during adiabatic plastic flow the entropy increases with time. On the the other hand, conduction of heat in the body results in a concomitant flow of entropy, and the entropy may increase or decrease.

A result analogous to (81.44) holds when the region of space $\mathcal{B}$ occupied by the body is thermally isolated in the sense that

$$\mathbf{q} \cdot \mathbf{n} = 0 \text{ on } \partial\mathcal{B}, \tag{81.45}$$

with $\mathbf{n}$ the outward unit normal on the boundary $\partial\mathcal{B}$. To establish this result we assume that $q \equiv 0$ and rewrite (81.41) as an *entropy balance*:

$$\dot{\eta} = \vartheta^{-1}\hat{Y}_{\mathrm{dis}}(e^p, \vartheta)\dot{e}^p - \vartheta^{-1}\mathrm{div}\mathbf{q}. \tag{81.46}$$

Using the divergence theorem we find that

$$\int_{\mathcal{B}} \vartheta^{-1}\mathrm{div}\mathbf{q}\, dv = \int_{\mathcal{B}} \mathrm{div}\frac{\mathbf{q}}{\vartheta}\, dv - \int_{\mathcal{B}} \mathbf{q} \cdot \nabla\frac{1}{\vartheta}\, dv = \int_{\partial\mathcal{B}} \frac{\mathbf{q}}{\vartheta} \cdot \mathbf{n}\, da + \int_{\mathcal{B}} \vartheta^{-2}\mathbf{q} \cdot \nabla\vartheta\, dv$$

and hence, by (81.45) and (81.46), that

$$\int_{\mathcal{B}} \dot{\eta}\, dv = \int_{\mathcal{B}} \vartheta^{-1}\hat{Y}_{\mathrm{dis}}(e^p, \vartheta)\dot{e}^p\, dv - \int_{\mathcal{B}} \vartheta^{-2}\mathbf{q} \cdot \nabla\vartheta\, dv. \tag{81.47}$$

Next, appealing to (81.29) and the heat-conduction inequality (81.27), we find that

$$\boxed{\overline{\int_{\mathcal{B}} \eta\, dv}^{\,\boldsymbol{\cdot}} \geq 0,} \tag{81.48}$$

and hence that *the entropy of the body $\mathcal{B}$ increases with time.*

As our last step, we assume, for simplicity, that the entropy $\hat{\eta}(\mathbf{E}^e, e^p, \vartheta)$ as given by (81.18) is independent of the accumulated plastic strain e^p, so that

$$\dot{\eta} = \frac{\partial \hat{\eta}(\mathbf{E}^e, \vartheta)}{\partial \mathbf{E}^e} : \dot{\mathbf{e}}^e + \frac{\partial \hat{\eta}(\mathbf{E}^e, \vartheta)}{\partial \vartheta}\dot{\vartheta},$$

and, by (81.41), we can rewrite (81.33) in a form,

$$\underbrace{\mathbf{T}_0 : \dot{\mathbf{E}}^p}_{\substack{\text{plastic}\\ \text{stress power}}} = \underbrace{\vartheta \frac{\partial \hat{\eta}(\mathbf{E}^e, \vartheta)}{\partial \vartheta}\dot{\vartheta}}_{\text{heating}} + \underbrace{\frac{\partial \hat{\Psi}^p(e^p, \vartheta)}{\partial e^p}\dot{e}^p}_{\substack{\text{rate of isothermal}\\ \text{storage of defects}}} + \underbrace{\vartheta \frac{\partial \hat{\eta}(\mathbf{E}^e, \vartheta)}{\partial \mathbf{E}^e} : \dot{\mathbf{E}}^e}_{\substack{\text{rate of isothermal}\\ \text{changes in entropy}}} + \underbrace{\mathrm{div}\mathbf{q} - q}_{\text{heat flow}}, \tag{81.49}$$

displaying the partition of the plastic stress power into heating and various other physically meaningful quantities. In this regard, a quantity of great interest is the fraction

$$\beta \stackrel{\text{def}}{=} \frac{\vartheta \dfrac{\partial \hat{\eta}(\mathbf{E}^e, \vartheta)}{\partial \vartheta} \dot{\vartheta}}{\mathbf{T}_0 : \dot{\mathbf{E}}^p} \tag{81.50}$$

of the plastic stress-power converted to heating.

81.7 Thermally Simple Materials

With a view toward applications, we now discuss a simple theory of Mises–Hill thermoplasticity based on the following simplifying assumptions:

(i) The defect energy Ψ^p is independent of temperature. Then, by $(81.13)_2$, the constitutive relation for the defect energy has the form

$$\Psi^p = \hat{\Psi}^p(e^p), \tag{81.51}$$

and, by (81.15) and (81.18), the entropy-relation has the form

$$\hat{\eta}(\mathbf{E}^e, \vartheta) = -\frac{\partial \hat{\Psi}^e(\mathbf{E}^e, \vartheta)}{\partial \vartheta}. \tag{81.52}$$

(ii) The constitutive relations for Ψ^e, $\mathbf{T}$, η, and $\mathbf{q}$ are of the standard form that defines conventional, small deformation, thermoelasticity:

$$\Psi^e = \mu |\mathbf{E}^e|^2 + \frac{\lambda}{2}(\operatorname{tr}\mathbf{E}^e)^2 + \xi(\vartheta - \vartheta_0)\operatorname{tr}\mathbf{E}^e - \frac{c}{2\vartheta_0}(\vartheta - \vartheta_0)^2,$$

$$\mathbf{T} = 2\mu \mathbf{E}^e + \lambda(\operatorname{tr}\mathbf{E}^e)\mathbf{1} + \xi(\vartheta - \vartheta_0)\mathbf{1}, \tag{81.53}$$

$$\eta = -\xi \operatorname{tr}\mathbf{E}^e + \frac{c}{\vartheta_0}(\vartheta - \vartheta_0),$$

$$\mathbf{q} = -k\nabla\vartheta.$$

(iii) The heat supply q vanishes.

Here ϑ_0 is a reference temperature, μ is the *shear modulus* (first Lamé modulus), λ is the second Lamé modulus, and we assume that

$$\mu > 0, \qquad 2\mu + 3\lambda > 0, \tag{81.54}$$

so that Ψ^e is a positive-definite function of $\mathbf{E}^e$. Further, $c > 0$ is the specific heat, ξ is a modulus related to the coefficient of thermal expansion α via the relation

$$\xi = -(2\mu + 3\lambda)\alpha, \tag{81.55}$$

which is positive for metals. We assume that the moduli μ, λ, c, ξ, and k are constant. By (81.35) and (81.51), the flow rule becomes

$$\mathbf{T}_0 = \left[\hat{Y}_{\text{dis}}(e^p, \vartheta) + \frac{d\hat{\Psi}^p(e^p)}{de^p} \right] \mathbf{N}^p, \tag{81.56}$$

with $\hat{Y}_{\text{dis}}(e^p, \vartheta) > 0$. On the other hand, $(81.53)_3$ implies that

$$\vartheta \dot{\eta} = -\xi \operatorname{tr}\dot{\mathbf{E}}^e + c\frac{\vartheta}{\vartheta_0}\dot{\vartheta},$$

and, granted the approximation

$$\frac{\vartheta}{\vartheta_0} \approx 1, \tag{81.57}$$

this equation becomes

$$\vartheta\dot{\eta} = -\xi \operatorname{tr}\dot{\mathbf{E}}^e + c\dot{\vartheta}; \tag{81.58}$$

thus (81.43) reduces to a generalized heat equation:

$$\boxed{c\dot{\vartheta} = \xi \operatorname{tr}\dot{\mathbf{E}}^e + \hat{Y}_{\rm dis}(e^p, \vartheta)\dot{e}^p + k\Delta\vartheta.} \tag{81.59}$$

In practice, a simplifying approximation often imposed to facilitate the solution of actual problems, is to neglect the small coupling term

$$\xi \operatorname{tr}\dot{\mathbf{E}}^e$$

in the partial differential equation (81.59). Under this approximation, the generalized heat equation is referred to as being weakly elastic. In addition, if we neglect heat conduction, then we arrive at the *weakly elastic, adiabatic heat equation*

$$\boxed{c\dot{\vartheta} = \hat{Y}_{\rm dis}(e^p, \vartheta)\dot{e}^p.} \tag{81.60}$$

Finally, using $(81.53)_3$ and the appproximation (81.57), we can rewrite (81.50) in the form

$$\boxed{\beta = \frac{c\dot{\vartheta}}{\mathbf{T}_0 : \dot{\mathbf{E}}^p}.} \tag{81.61}$$

and, in addition, this result with (81.34) and (81.60) yield the alternative relation (81.4) for β.

81.8 Determination of the Defect Energy by the Rosakis Brothers, Hodowany, and Ravichandran

As noted in §81.1, the purely mechanical theory says nothing about the partition of the flow resistance into dissipative and energetic parts — all partitions that yield the same flow resistance yield the same flow rule and, hence, the same field equations. To the contrary, in this subsection we follow ROSAKIS, ROSAKIS, RAVICHANDRAN & HODOWANY (2000) and consider a coupled theory based on the flow rule (81.56) in conjunction with the weakly elastic, adiabatic heat equation (81.60).[451]

Specifically, we consider a flow in which $\dot{e}^p > 0$ for all time. Then e^p is a strictly increasing (and, hence, invertible) function of time, and, consequently, we may consider $\vartheta(t)$ to be a function $\bar{\vartheta}(e^p)$, so that, by the chain-rule,

$$\dot{\vartheta}(t) = \frac{d\bar{\vartheta}(e^p)}{de^p}\dot{e}^p,$$

an equation that allows us to write (81.60) in the form

$$c\frac{d\bar{\vartheta}(e^p)}{de^p} = \hat{Y}_{\rm dis}(e^p, \bar{\vartheta}(e^p)) \stackrel{\rm def}{=} \bar{Y}_{\rm dis}(e^p). \tag{81.62}$$

Here, for any function $\hat{\Lambda}(e^p, \vartheta)$, we define

$$\bar{\Lambda}(e^p) = \hat{\Lambda}(e^p, \bar{\vartheta}(e^p)),$$

[451] See also the allied experimental work of HODOWANY, RAVICHANDRAN, ROSAKIS & ROSAKIS (2000).

and hence write (81.28) in the form

$$\bar{Y}(e^p) = \bar{Y}_{\text{dis}}(e^p) + \frac{\partial \hat{\Psi}^p(e^p)}{\partial e^p}. \tag{81.63}$$

Integrating (81.62) we find that

$$c(\bar{\vartheta}(e^p) - \bar{\vartheta}_0) = \int_0^{e^p} \bar{Y}_{\text{dis}}(e)\,de, \qquad \bar{\vartheta}_0 = \bar{\vartheta}(e^p)\big|_{e^p=0}, \tag{81.64}$$

and, bearing in mind (81.64), we can integrate (81.63) to arrive at

$$\hat{\Psi}^p(e^p) = \int_0^{e^p} \bar{Y}(e)\,de - c(\bar{\vartheta}(e^p) - \bar{\vartheta}_0). \tag{81.65}$$

Granted that the functions $\bar{Y}(e^p)$ and $\bar{\vartheta}(e^p)$ describing the flow resistance and temperature can be determined experimentally, then the relations (81.65) and (81.63) allow for the determination of the functions $\hat{\Psi}(e^p)$ and $\bar{Y}_{\text{dis}}(e^p)$ describing the defect energy and the dissipative part of the flow flow resistance $\bar{Y}(e^p)$.

Next, since the fraction β of the plastic stress-power converted to heating is given by

$$\beta = \frac{c\dot{\vartheta}(t)}{\bar{Y}(e^p)\dot{e}^p},$$

and we may use (81.62) and (81.63) to conclude that

$$\beta(e^p) = \frac{\bar{Y}_{\text{dis}}(e^p)}{\bar{Y}_{\text{dis}}(e^p) + \frac{d\hat{\Psi}^p(e^p)}{de^p}}. \tag{81.66}$$

Interestingly, (81.66) is basically a repeat of the result (81.4) for the purely mechanical theory, except that now, because our accounting for thermal effects has rendered the dissipative and energetic parts of the flow resistance unique,

- *β is no longer indeterminate.*

However, since the flow rule (81.63) has the form (81.1), if one were to use (81.63) in a purely mechanical theory, then the resulting flow stress would remain independent of the partition of the flow resistance into dissipative and energetic parts, an observation that would allow one to use a different partition, were it more amenable to analysis or computation.

81.9 Summary of the Basic Equations

The basic equations of the theory of thermally simple materials consist of:

(i) The kinematical equations

$$\mathbf{E} = \text{sym}\,\nabla\mathbf{u} = \mathbf{E}^e + \mathbf{E}^p, \qquad \text{tr}\,\mathbf{E}^p = 0,$$

$$\mathbf{N}^p = \frac{\dot{\mathbf{E}}^p}{|\dot{\mathbf{E}}^p|}, \qquad \dot{e}^p = |\dot{\mathbf{E}}^p|, \qquad e^p(\mathbf{x}, 0) = 0.$$

(ii) The codirectionality constraint and the relation for the resolved-shear

$$\mathbf{N}^p = \frac{\mathbf{T}_0}{|\mathbf{T}_0|}, \qquad \tau = \mathbf{N}^p : \mathbf{T}_0,$$

where $\mathbf{T}_0$ is the deviatoric stress.

(iii) The elasticity equations

$$\begin{aligned} \mathbf{T} &= 2\mu\mathbf{E}^e + \lambda(\operatorname{tr}\mathbf{E}^e)\mathbf{1} + \xi(\vartheta - \vartheta_0)\,\mathbf{1}, \\ \operatorname{div}\mathbf{T} + \mathbf{b}_0 &= \rho\ddot{\mathbf{u}}. \end{aligned} \tag{81.67}$$

(iv) The constitutive relations for the defect stress and for the resolved shear

$$\Psi^p = \hat{\Psi}^p(e^p), \qquad \tau = \hat{Y}(e^p, \vartheta) = \underbrace{\hat{Y}_{\rm dis}(e^p, \vartheta)}_{\text{dissipative}} + \underbrace{\frac{\partial \hat{\Psi}^p(e^p, \vartheta)}{\partial e^p}}_{\text{energetic}}$$

with $\hat{Y}_{\rm dis}(e^p, \vartheta) \geq 0$ and $d\hat{\Psi}^p(e^p)/de^p \geq 0$.

(v) The flow rule

$$\mathbf{T}_0 = \left[\hat{Y}_{\rm dis}(e^p, \vartheta) + \frac{d\hat{\Psi}^p(e^p)}{de^p} \right] \mathbf{N}^p \tag{81.68}$$

together with the restrictions

$$\begin{aligned} |\mathbf{T}_0| = \hat{Y}(e^p, \vartheta) \quad \text{for } \dot{e}^p \neq 0, \qquad & |\mathbf{T}_0| \leq \hat{Y}(e^p, \vartheta), \\ \dot{e}^p = 0 \quad \text{for} \quad |\mathbf{T}_0| < \hat{Y}(e^p, \vartheta). & \end{aligned} \tag{81.69}$$

(vi) The generalized heat equation

$$c\dot{\vartheta} = \xi \operatorname{tr}\dot{\mathbf{E}}^e + \hat{Y}_{\rm dis}(e^p, \vartheta)\dot{e}^p + k\Delta\vartheta. \tag{81.70}$$

These equations must be augmented by mechanical and thermal boundary and initial conditions.

82 Formulation of Initial/Boundary-Value Problems for the Mises Flow Equations as Variational Inequalities

Variational inequalities provide a basic tool for formulating, discussing, and solving the initial/boundary-value problems of rate-independent plasticity.[452] They are, in essence, an analog of the "first variation" in minimization problems, as they relate to the governing equations in the same way as the virtual-power principle relates to corresponding force balances. In addition, variational inequalities constitute a weak formulation of the governing equations — weak in the sense that less smoothness is demanded of the solution, which is required to satisfy a global condition. Further, it is such weak formulations that form the basis of finite-element approximations.

Here we first derive an equivalent formulation of the Mises flow equations in terms of dissipation, and thereafter the full set of governing equations, including the standard force balance, as a global variational inequality.[453]

82.1 Reformulation of the Mises Flow Equations in Terms of Dissipation

Our starting point is the Mises flow rule

$$\mathbf{T}_0 = Y(e^p)\frac{\dot{\mathbf{E}}^p}{|\dot{\mathbf{E}}^p|} \qquad \text{for } \dot{\mathbf{E}}^p \neq \mathbf{0}, \tag{82.1}$$

with $Y(e^p) > 0$ the flow resistance. Associated with this flow rule are the *dissipation*

$$\delta(\dot{\mathbf{E}}^p, e^p) = Y(e^p)|\dot{\mathbf{E}}^p| \tag{82.2}$$

and the *boundedness inequality*

$$|\mathbf{T}_0| \leq Y(e^p), \tag{82.3}$$

an inequality that defines those deviatoric stresses admissible to the theory.[454]

452 The use of variational inequalities as a basis for a mathematical theory of rate-independent hardening plasticity is discussed in detail by HAN & REDDY (1999). Reference should also be made to the monograph of SIMO & HUGHES (1997) and to the early works of MOREAU (1976, 1977). The variational-inequality approach discussed in this section is based on ideas of MARTIN (1981), who introduced an internal-variable formulation in which the flow rule is expressed in terms of dissipation; see also BIRD & MARTIN (1986, 1990), who discuss algorithmic approaches based on MARTIN (1981). Existence and uniqueness of solutions to the corresponding variational inequality — that is, the inequality based on dissipation — were established for the problem with hardening by REDDY (1992); cf. HAN & REDDY (1999).

453 This section is taken from GURTIN & REDDY (2009).

454 Cf. $(76.71)_{1,2}$ and (76.71). Admissibility is in the sense of maximum dissipation; cf. (79.7) and (79.21).

The central ingredient in our reformulation of the Mises flow equations is the

EQUIVALENCY THEOREM *Choose tensors* $\mathbf{T}_0$, $\dot{\mathbf{E}}^p \in \mathrm{SymDev}$ *and a scalar* $e^p \geq 0$. *Then* $\mathbf{T}_0$, $\dot{\mathbf{E}}^p$, *and* e^p *satisfy the flow rule* (82.1) *and the boundedness inequality* (82.3) *if and only if* $\mathbf{T}_0$, $\dot{\mathbf{E}}^p$, *and* e^p *satisfy the* **local inequality**

$$\boxed{\delta(\tilde{\mathbf{E}}^p, e^p) \geq \delta(\dot{\mathbf{E}}^p, e^p) + \mathbf{T}_0 : (\tilde{\mathbf{E}}^p - \dot{\mathbf{E}}^p) \quad \text{for all } \tilde{\mathbf{E}}^p \in \mathrm{SymDev}.} \tag{82.4}$$

PROOF. Throughout this proof the accumulated plastic strain e^p is fixed and, for convenience, suppressed as an argument; that is, we write

$$Y \text{ for } Y(e^p); \quad \delta(\dot{\mathbf{E}}^p) \text{ for } \delta(\dot{\mathbf{E}}^p, e^p); \quad \delta(\tilde{\mathbf{E}}^p) \text{ for } \delta(\tilde{\mathbf{E}}^p, e^p).$$

As our first step we show that if

$$\dot{\mathbf{E}}^p \neq \mathbf{0}, \tag{82.5}$$

then[455]

$$(82.1) \Longleftrightarrow (82.4). \tag{82.6}$$

If (82.1) is satisfied, then, by (82.2),

$$\begin{aligned} Y^{-1}\big[\delta(\tilde{\mathbf{E}}^p) - \delta(\dot{\mathbf{E}}^p) - \mathbf{T}_0 : (\tilde{\mathbf{E}}^p - \dot{\mathbf{E}}^p)\big] &= |\tilde{\mathbf{E}}^p| - |\dot{\mathbf{E}}^p| - \frac{\dot{\mathbf{E}}^p}{|\dot{\mathbf{E}}^p|} : (\tilde{\mathbf{E}}^p - \dot{\mathbf{E}}^p) \\ &= |\tilde{\mathbf{E}}^p| - \frac{\dot{\mathbf{E}}^p}{|\dot{\mathbf{E}}^p|} : \tilde{\mathbf{E}}^p \\ &= |\dot{\mathbf{E}}^p|^{-1}\big(|\dot{\mathbf{E}}^p||\tilde{\mathbf{E}}^p| - \dot{\mathbf{E}}^p : \tilde{\mathbf{E}}^p\big). \end{aligned} \tag{82.7}$$

Further, the Schwarz inequality implies that $\dot{\mathbf{E}}^p : \tilde{\mathbf{E}}^p \leq |\dot{\mathbf{E}}^p||\tilde{\mathbf{E}}^p|$ and, hence, that the last line of (82.7) is nonnegative; hence, (82.4) is satisfied. Thus (82.1) implies (82.4).

Next we prove the converse; that is, granted $\dot{\mathbf{E}}^p \neq \mathbf{0}$, (82.4) implies (82.1). We begin by using (82.2) and its counterpart for $\tilde{\mathbf{E}}^p$ to rewrite the local inequality (82.4) in the form

$$Y\big(|\tilde{\mathbf{E}}^p| - |\dot{\mathbf{E}}^p|\big) \geq \mathbf{T}_0 : (\tilde{\mathbf{E}}^p - \dot{\mathbf{E}}^p) \quad \text{for all } \tilde{\mathbf{E}}^p \in \mathrm{SymDev}. \tag{82.8}$$

Choose $|\tilde{\mathbf{E}}^p| = |\dot{\mathbf{E}}^p|$, divide (82.8) by $|\dot{\mathbf{E}}^p|$, and let

$$\mathbf{N}^p = \frac{\dot{\mathbf{E}}^p}{|\dot{\mathbf{E}}^p|} \quad \text{and} \quad \tilde{\mathbf{N}}^p = \frac{\tilde{\mathbf{E}}^p}{|\tilde{\mathbf{E}}^p|};$$

the result is

$$\mathbf{T}_0 : (\tilde{\mathbf{N}}^p - \mathbf{N}^p) \leq 0. \tag{82.9}$$

Next, choose an arbitrary tensor $\mathbf{\Lambda} \in \mathrm{SymDev}$ such that[456]

$$\mathbf{\Lambda} : \mathbf{N}^p = 0. \tag{82.10}$$

Let $\mathcal{N}$ denote the unit sphere in the space SymDev.[457] There is then a curve $\tilde{\mathbf{N}}^p(\lambda)$ on $\mathcal{N}$ such that, for some λ_0,[458]

$$\tilde{\mathbf{N}}^p(\lambda_0) = \mathbf{N}^p \quad \text{and} \quad \frac{d\tilde{\mathbf{N}}^p(\lambda)}{d\lambda}\bigg|_{\lambda=\lambda_0} = \mathbf{\Lambda} \tag{82.11}$$

---

455 Since (82.1) implies that $|\mathbf{T}_0| = Y$, (82.6) is equivalent to the assertion that (82.1) and (82.3) are together equivalent to (82.4).

456 Cf. Footnote 419.

457 Cf. (79.14).

458 Cf. Footnote 420.

— and such that, by (82.9) and $(82.11)_1$, the function

$$\Phi(\lambda) \stackrel{\text{def}}{=} \mathbf{T}_0 : (\tilde{\mathbf{N}}^p(\lambda) - \mathbf{N}^p) \le 0 \tag{82.12}$$

has a maximum at $\lambda = \lambda_0$. Consequently,

$$\begin{aligned} \frac{d\Phi(\lambda)}{d\lambda}\bigg|_{\lambda=\lambda_0} &= \mathbf{T}_0 : \frac{d\tilde{\mathbf{N}}^p(\lambda)}{d\lambda}\bigg|_{\lambda=\lambda_0} \\ &= 0, \end{aligned}$$

so that, by (82.11),

$$\mathbf{T}_0 : \mathbf{\Lambda} = 0 \tag{82.13}$$

for every $\mathbf{\Lambda} \in$SymDev tangent to $\mathcal{N}$ at $\mathbf{N}^p$. The stress $\mathbf{T}_0$ must therefore be normal to $\mathcal{N}$ at $\mathbf{N}^p$; there is hence a scalar β such that

$$\mathbf{T}_0 = \beta \mathbf{N}^p. \tag{82.14}$$

Next, to show that $\beta = Y$, we first take $\tilde{\mathbf{E}}^p = \mathbf{0}$. Then, by (82.14) and, since

$$\mathbf{N}^p : \dot{\mathbf{E}}^p = |\dot{\mathbf{E}}^p|,$$

(82.8) becomes $Y|\dot{\mathbf{E}}^p| \le \beta|\dot{\mathbf{E}}^p|$; consequently,

$$Y \le \beta. \tag{82.15}$$

On the other hand, assume that $\mathbf{N}^p = \tilde{\mathbf{N}}^p$. Then, $\mathbf{N}^p : \tilde{\mathbf{E}}^p = |\tilde{\mathbf{E}}^p|$ and (82.8) becomes

$$Y(|\tilde{\mathbf{E}}^p| - |\dot{\mathbf{E}}^p|) \ge \beta(|\tilde{\mathbf{E}}^p| - |\dot{\mathbf{E}}^p|),$$

so that $Y \ge \beta$. Thus, $Y = \beta$ and, by (82.14), the flow rule $\mathbf{T}_0 = Y\mathbf{N}^p$ is satisfied. Thus, (82.4) implies (82.1). We have therefore shown that, for $\dot{\mathbf{E}}^p \ne \mathbf{0}$, (82.1) is equivalent to (82.4).

Assume next that

$$\dot{\mathbf{E}}^p = \mathbf{0}, \tag{82.16}$$

in which case — since (82.4) is equivalent to (82.8) — we must show that

$$(82.3) \Longleftrightarrow (82.8). \tag{82.17}$$

Note, first, that by (82.16), (82.8) becomes

$$Y|\tilde{\mathbf{E}}^p| \ge \mathbf{T}_0 : \tilde{\mathbf{E}}^p \quad \text{for all } \tilde{\mathbf{E}}^p \in \text{SymDev}. \tag{82.18}$$

Dividing by $|\tilde{\mathbf{E}}^p|$ and then choosing $\tilde{\mathbf{E}}^p = \mathbf{T}_0$, we find that

$$\begin{aligned} Y &\ge \mathbf{T}_0 : \frac{\mathbf{T}_0}{|\mathbf{T}_0|} \\ &= |\mathbf{T}_0|, \end{aligned}$$

which is (82.3). Thus, (82.8) implies (82.3). Conversely, assume that (82.3) is satisfied. Choose $\tilde{\mathbf{E}}^p \in$ SymDev arbitrarily. Then, (82.3) and the Schwarz inequality imply that

$$\begin{aligned} Y|\tilde{\mathbf{E}}^p| &\ge |\mathbf{T}_0||\tilde{\mathbf{E}}^p| \\ &\ge \mathbf{T}_0 : \tilde{\mathbf{E}}^p, \end{aligned}$$

which is (82.18). This completes the proof of the Equivalency Theorem.

82.2 The Global Variational Inequality

We now turn to a derivation of the global variational inequality appropriate to problems involving the Mises flow equations. Without loss in generality we replace $\mathbf{T}_0$ in (82.4) by $\mathbf{T}$. Assume that the displacement $\mathbf{u}$ and $\dot{\mathbf{E}}^p$, e^p, $\mathbf{T}$, and $\tilde{\mathbf{E}}^p$ are fields on B (with e^p consistent with (82.2) and $\tilde{\mathbf{E}}^p$ an arbitrary tensor in SymDev). Further, let

$$J(\tilde{\mathbf{E}}^p, e^p) = \int_{\mathrm{B}} \delta(\tilde{\mathbf{E}}^p, e^p)\, dv$$

$$= \int_{\mathrm{B}} Y(e^p)\,|\tilde{\mathbf{E}}^p|\, dv, \qquad (82.19)$$

where we have used (82.2). Then — bearing in mind the (linear) elastic stress-strain relation[459]

$$\mathbf{T} = \mathbb{C}(\mathrm{sym}\,\nabla\mathbf{u} - \mathbf{E}^p) \qquad (82.20)$$

with elasticity tensor $\mathbb{C}$ — we integrate (82.4) over B to give

$$J(\tilde{\mathbf{E}}^p, e^p) \geq J(\dot{\mathbf{E}}^p, e^p) + \int_{\mathrm{B}} (\tilde{\mathbf{E}}^p - \dot{\mathbf{E}}^p)\colon \mathbb{C}\mathbf{E}^e\, dv, \qquad (82.21)$$

with

$$\mathbf{E}^e \stackrel{\mathrm{def}}{=} \mathrm{sym}\nabla\mathbf{u} - \mathbf{E}^p \qquad (82.22)$$

the elastic strain.

Consider boundary conditions of the form

$$\mathbf{Tn} = \mathbf{t}_{\mathcal{S}} \quad \text{on } \mathcal{S} \qquad \text{and} \qquad \mathbf{u} = \mathbf{0} \quad \text{on } \partial\mathrm{B} \setminus \mathcal{S}, \qquad (82.23)$$

in which $\mathbf{t}_{\mathcal{S}}$ is a prescribed function on a subsurface $\mathcal{S}$ of ∂B. In the ensuing analysis we restrict attention to virtual fields, termed *kinematically admissible*, that satisfy

$$\tilde{\mathbf{v}} = \mathbf{0} \quad \text{on } \partial\mathrm{B} \setminus \mathcal{S}. \qquad (82.24)$$

Granted the symmetry of $\mathbf{T}$,[460] a consequence of the classical **principle of virtual power** §22.2 is then that the virtual power balance

$$\int_{\mathrm{B}} \mathbf{T}\colon \nabla\tilde{\mathbf{v}}\, dv = \int_{\mathcal{S}} \mathbf{t}_{\mathcal{S}} \cdot \tilde{\mathbf{v}}\, da + \int_{\mathrm{B}} \mathbf{b} \cdot \tilde{\mathbf{v}}\, dv \qquad (82.25)$$

is satisfied for all kinematically admissible virtual fields $\tilde{\mathbf{v}}$ on B if and only if

$$\mathrm{Div}\mathbf{T} + \mathbf{b} = \mathbf{0} \quad \text{in B} \qquad \text{and} \qquad \mathbf{Tn} = \mathbf{t}_{\mathcal{S}} \quad \text{on } \mathcal{S}. \qquad (82.26)$$

Without loss in generality we may replace $\tilde{\mathbf{v}}$ in (82.25) by $\tilde{\mathbf{u}} - \dot{\mathbf{u}}$, with $\tilde{\mathbf{u}}$ kinematically admissible, so that, by (82.20),

$$\int_{\mathcal{S}} \mathbf{t}_{\mathcal{S}} \cdot (\tilde{\mathbf{u}} - \dot{\mathbf{u}})\, da + \int_{\mathrm{B}} \mathbf{b} \cdot (\tilde{\mathbf{u}} - \dot{\mathbf{u}})\, dv + \int_{\mathrm{B}} (\mathrm{sym}\nabla\dot{\mathbf{u}} - \mathrm{sym}\nabla\tilde{\mathbf{u}})\colon \mathbb{C}\mathbf{E}^e\, dv = 0. \qquad (82.27)$$

459 We find it easier to work with the elasticity tensor $\mathbb{C}$ than with its isotropic form.
460 Cf. (82.20).

Finally, if we add the left side of (82.27) to the right side of (82.21) we arrive at a global *variational inequality for Mises materials*:

$$J(\tilde{\mathbf{E}}^p, e^p) \geq J(\dot{\mathbf{E}}^p, e^p) + \int\limits_{\mathrm{B}} (\dot{\mathbf{E}}^e - \tilde{\mathbf{E}}^e) : \mathbb{C}\mathbf{E}^e \, dv$$

$$+ \int\limits_{\mathcal{S}} \mathbf{t}_{\mathcal{S}} \cdot (\tilde{\mathbf{u}} - \dot{\mathbf{u}}) \, da + \int\limits_{\mathrm{B}} \mathbf{b} \cdot (\tilde{\mathbf{u}} - \dot{\mathbf{u}}) \, dv, \qquad (82.28)$$

with

$$\tilde{\mathbf{E}}^e \stackrel{\text{def}}{=} \operatorname{sym}\nabla\tilde{\mathbf{u}} - \tilde{\mathbf{E}}^p \qquad (82.29)$$

the virtual elastic-strain.

The variational inequality (82.28) poses an initial/boundary-value problem to be solved for $\mathbf{u}$ and $\mathbf{E}^p$, with e^p related to $\mathbf{E}^p$ through $(80.1)_2$

EXERCISE

1. Use the fundamental lemma of the calculus of variations (page 167) to show that the virtual power balance (82.25) is satisfied for all kinematically admissible $\tilde{\mathbf{v}}$ if and only if (82.26) are satisfied.

82.3 Alternative Formulation of the Global Variational Inequality When Hardening is Described by a Defect Energy

Tacit in the derivation of (82.28) is the assumption that the flow rule is purely dissipative. But, as noted in the theorem on page 467, this purely dissipative theory is equivalent to the theory with defect energy described in §80 provided $Y(e^p)$ and $\delta(\dot{\mathbf{E}}^p, e^p)$ in the former are replaced by

$$Y_0 + \hat{g}(e^p) \qquad \text{and} \qquad D_{\text{eff}}(\dot{\mathbf{E}}^p, e^p) \qquad (82.30)$$

in the latter, where $\hat{g}(e^p)$, the hardening stress, is defined in terms of the defect energy via (80.5), while

$$D_{\text{eff}}(\dot{\mathbf{E}}^p, e^p) = \big[Y_0 + \hat{g}(e^p)\big]|\dot{\mathbf{E}}^p|$$

is the effective dissipation (80.15). Thus, by (82.19), $J(\tilde{\mathbf{E}}^p, e^p)$ for the theory based on a defect energy has the form

$$J(\tilde{\mathbf{E}}^p, e^p) = \int\limits_{\mathrm{B}} [Y_0 + \hat{g}(e^p)]|\tilde{\mathbf{E}}^p| \, dv \qquad (82.31)$$

and, granted (82.31), the global variational inequality (82.28) remains valid. Further, if we let

$$J_0(\tilde{\mathbf{E}}^p) = \int\limits_{\mathrm{B}} Y_0|\tilde{\mathbf{E}}^p| \, dv, \qquad (82.32)$$

then, by (82.31),

$$J(e^p, \tilde{\mathbf{E}}^p) = J_0(\tilde{\mathbf{E}}^p) + \int\limits_{\mathrm{B}} \hat{g}(e^p)\,|\tilde{\mathbf{E}}^p| \, dv \qquad (82.33)$$

and (82.28) becomes

$$J_0(\tilde{\mathbf{E}}^p) \geq J_0(\dot{\mathbf{E}}^p) + \int\limits_{\mathrm{B}} (\dot{\mathbf{E}}^e - \tilde{\mathbf{E}}^e) : \mathbb{C}\mathbf{E}^e \, dv - \int\limits_{\mathrm{B}} \hat{g}(e^p)(|\tilde{\mathbf{E}}^p| - |\dot{\mathbf{E}}^p|) \, dv$$

$$+ \int\limits_{\mathcal{S}} \mathbf{t}_{\mathcal{S}} \cdot (\tilde{\mathbf{u}} - \dot{\mathbf{u}}) \, da + \int\limits_{\mathrm{B}} \mathbf{b} \cdot (\tilde{\mathbf{u}} - \dot{\mathbf{u}}) \, dv. \qquad (82.34)$$

In this form the terms involving the functional $J_0(\cdot)$ are dissipative, while the term

$$\int\limits_{\mathrm{B}} \hat{g}(e^p)(|\tilde{\mathbf{E}}^p| - |\dot{\mathbf{E}}^p|) \, dv,$$

which represents hardening, is energetic.

PART XIV

SMALL DEFORMATION, ISOTROPIC PLASTICITY BASED ON THE PRINCIPLE OF VIRTUAL POWER

83 Introduction

The classical principle of virtual power[461] represents, for any subregion P and any virtual velocity $\tilde{\mathbf{u}}$, a balance

$$\int_{\partial P} \mathbf{t}(\mathbf{n}) \cdot \tilde{\mathbf{u}}\, da + \int_{P} \mathbf{b} \cdot \tilde{\mathbf{u}}\, dv = \int_{P} \mathbf{T} : \nabla\tilde{\mathbf{u}}\, dv \tag{83.1}$$

between

(i) the power expended *within* P by the stress **T** and
(ii) the power expended *on* P by the traction $\mathbf{t}(\mathbf{n})$ and the body force **b**;

and, what is most important, the balance (83.1) is equivalent to the classical balance

$$\operatorname{Div}\mathbf{T} + \mathbf{b} = \mathbf{0} \tag{83.2}$$

and traction condition $\mathbf{t}(\mathbf{n}) = \mathbf{T}\mathbf{n}$. The chief feature of this classical principle — and one that may be used as a paradigm for the formulation of more general theories — is a physical structure involving

- *the introduction of stresses, body forces, and surface tractions through the manner in which they expend power.*

This observation is nontrivial — it allows one to use the virtual-power principle as a basic tool in determining local force balances (and concomitant surface-traction conditions) *when the forms of the balances are not known a priori.*[462]

461 Formulated as in §22.2, but in notation appropriate to the current discussion. The Remarks on page 167 list important features of the virtual-power principle.
As is clear from §22.2,

the principle of virtual power is independent of constitutive equations; as such this principle is valid for both solids and fluids and, most importantly, for plastic materials.

One often finds this principle formulated *erroneously* in terms of an underlying free energy, an error arising from the observation that — in the absence of dissipation — energy minimization generally results in partial-differential equations equivalent to those resulting from a corresponding virtual-power principle.

462 For example, to formulate the classical Cosserat theory of couple stress one would assume that the power expended within the body is represented, not only by the classical stress power, but, in addition, by a *couple stress* **M** that expends power in concert with the gradient $\nabla\boldsymbol{\omega}$ of the rotation $\boldsymbol{\omega} = \operatorname{Curl}\mathbf{u}$, so that

$$\int_{P} (\mathbf{T} : \nabla\tilde{\mathbf{u}} + \mathbf{M} : \nabla\tilde{\boldsymbol{\omega}})\, dv \tag{$\star$}$$

To fix ideas we begin by discussing conventional plasticity within the virtual-power framework; what is unconventional about our approach is our allowance for separate internal-power expenditures associated with elastic and plastic response. Such a formulation, while consistent with conventional theory, provides a convenient framework for a discussion of material stability. Further, an understanding of the conventional theory within a virtual-power framework provides a foundation upon which to build theories involving the gradient of plastic strain.[463]

represents the internal-power expenditure. This expenditure is balanced by an external-power expenditure more general than the left side of (83.1), and a consequence of the corresponding virtual-power principle is that the classical balance (83.2) and associated traction condition are replaced by the balance

$$\operatorname{Div}\mathbf{T} + \operatorname{Curl}\operatorname{Div}\mathbf{M} + \mathbf{b} = \mathbf{0}$$

and the traction conditions

$$\mathbf{t}_{\partial\mathrm{P}} = \mathbf{Tn} + \operatorname{Div}(\mathbf{Mn}\times) - \mathbf{n}\cdot[\nabla(\mathbf{Mn}\times)]\mathbf{n} + \mathbf{n}\times(\operatorname{Div}\mathbf{M} - 2H\mathbf{Gn}),$$

$$\mathbf{m}_{\partial\mathrm{P}} = \mathbf{n}\times\mathbf{Mn},$$

with H being the mean curvature of $\partial\mathrm{P}$; cf. FRIED & GURTIN (2009, §3.6, eqs. (3.24)–(3.25)), where the couple stress is denoted by $\mathbf{G}$ instead of by $\mathbf{M}$. The complicated nature of these relations contrasted with the simplicity and physical relevance of the power expenditure ($\star$) demonstrates the utility of the principle of virtual power in developing new theories of continuum mechanics.

[463] Cf. Part XV.

84 Conventional Theory Based on the Principle of Virtual Power

The virtual-power formulation is based on the belief that

- *the power expended by each independent "rate-like" kinematical descriptor is expressible in terms of an associated force system consistent with its own balance.*

But the basic "rate-like" descriptors — the velocity $\dot{\mathbf{u}}$ and the elastic and plastic distortions $\dot{\mathbf{H}}^e$ and $\dot{\mathbf{H}}^p$ — are not independent, since, by (75.8)–(75.12), they are constrained by

$$\nabla\dot{\mathbf{u}} = \dot{\mathbf{H}}^e + \dot{\mathbf{H}}^p, \qquad \mathrm{tr}\,\dot{\mathbf{H}}^p = 0, \tag{84.1}$$

and it is not apparent what forms the associated force balances should take. It is in such situations that the virtual-power principle displays its strength, because

- *this principle automatically determines the underlying force balances.*

84.1 General Principle of Virtual Power

Let P denote an arbitrary subregion of the body with $\mathbf{n}$ the outward unit normal on the boundary ∂P of P.

The classical virtual-power principle is discussed in §22.2 and here, as there, the formulation of this principle is based on a balance between the external power $\mathcal{W}(\mathrm{P})$ expended on P and the internal power $\mathcal{I}(\mathrm{P})$ expended within P. But, in contrast to the discussion of §22.2, we replace the classical stress power $\mathbf{T}\colon\nabla\dot{\mathbf{u}}$ by a detailed reckoning that individually characterizes the disparate kinematical processes involved, namely,

- the stretching of the underlying microscopic structure as described by the elastic distortion-rate $\dot{\mathbf{H}}^e$, and
- the flow of dislocations through that structure as described by the plastic strain-rate $\dot{\mathbf{H}}^p$.

Specifically, we allow for power expended internally by

- an **elastic stress** $\mathbf{T}^e$ power conjugate to $\dot{\mathbf{H}}^e$, and
- a **plastic stress** $\mathbf{T}^p$ power conjugate to $\dot{\mathbf{H}}^p$,

and we write the **internal power** in the form

$$\mathcal{I}(\mathrm{P}) = \int\limits_{\mathrm{P}} (\mathbf{T}^e\colon\dot{\mathbf{H}}^e + \mathbf{T}^p\colon\dot{\mathbf{H}}^p)\,dv. \tag{84.2}$$

Here, $\mathbf{T}^e$ and $\mathbf{T}^p$ are defined over the body for all time. Since $\dot{\mathbf{H}}^p$ is deviatoric, we may, without loss in generality, assume that $\mathbf{T}^p$ is deviatoric; viz.

$$\operatorname{tr}\mathbf{T}^p = 0. \tag{84.3}$$

The internal power is balanced by power expended externally by tractions on $\partial\mathrm{P}$ and body forces acting within P. We assume that this external power is applied in part by surface tractions $\mathbf{t}(\mathbf{n})$ (for each unit vector $\mathbf{n}$) acting over $\partial\mathrm{P}$ and by a body force $\mathbf{b}$, presumed to account for inertia, acting over the interior of P. Thus, granted an inertial frame,

$$\mathbf{b} = \mathbf{b}_0 - \rho\ddot{\mathbf{u}}. \tag{84.4}$$

As is standard, we assume that $\mathbf{t}(\mathbf{n})$ and $\mathbf{b}$ are power-conjugate to the velocity $\dot{\mathbf{u}}$, so that

$$\mathcal{W}(\mathrm{P}) = \int\limits_{\partial\mathrm{P}} \mathbf{t}(\mathbf{n})\cdot\dot{\mathbf{u}}\,da + \int\limits_{\mathrm{P}} \mathbf{b}\cdot\dot{\mathbf{u}}\,dv \tag{84.5}$$

represents the power expended by standard surface tractions and body forces. Here $\mathbf{t}(\mathbf{n})$ (for each unit vector $\mathbf{n}$) and $\mathbf{b}$ are defined over the body for all time.[464]

The principle of virtual power takes as its starting point the requirement that the internal and external power expenditures be balanced, an hypothesis that may be motivated by its classical counterpart (22.11). We, therefore, require that

$$\underbrace{\int\limits_{\partial\mathrm{P}} \mathbf{t}(\mathbf{n})\cdot\dot{\mathbf{u}}\,da + \int\limits_{\mathrm{P}} \mathbf{b}\cdot\dot{\mathbf{u}}\,dv}_{\mathcal{W}(\mathrm{P})} = \underbrace{\int\limits_{\mathrm{P}} (\mathbf{T}^e\!:\!\dot{\mathbf{H}}^e + \mathbf{T}^p\!:\!\dot{\mathbf{H}}^p)\,dv}_{\mathcal{I}(\mathrm{P})}, \tag{84.6}$$

a relation we refer to as the **power balance.**

Consider the fields $\dot{\mathbf{u}}$, $\dot{\mathbf{H}}^e$, and $\dot{\mathbf{H}}^p$ — at some arbitrarily chosen but *fixed* time — as "virtual velocities" to be specified in a manner consistent with (84.1); that is, denoting the virtual fields by $\tilde{\mathbf{u}}$, $\tilde{\mathbf{H}}^e$, and $\tilde{\mathbf{H}}^p$ to differentiate them from fields associated with the actual evolution of the body, we require that

$$\nabla\tilde{\mathbf{u}} = \tilde{\mathbf{H}}^e + \tilde{\mathbf{H}}^p \qquad \text{and} \qquad \operatorname{tr}\tilde{\mathbf{H}}^p = 0. \tag{84.7}$$

We define a (generalized) virtual velocity to be a list

$$\mathcal{V} = (\tilde{\mathbf{u}}, \tilde{\mathbf{H}}^e, \tilde{\mathbf{H}}^p) \tag{84.8}$$

of such fields consistent with the constraint (84.7), and we write

$$\left.\begin{aligned} \mathcal{W}(\mathrm{P},\mathcal{V}) &= \int\limits_{\partial\mathrm{P}} \mathbf{t}(\mathbf{n})\cdot\tilde{\mathbf{u}}\,da + \int\limits_{\mathrm{P}} \mathbf{b}\cdot\tilde{\mathbf{u}}\,dv, \\ \mathcal{I}(\mathrm{P},\mathcal{V}) &= \int\limits_{\mathrm{P}} (\mathbf{T}^e\!:\!\tilde{\mathbf{H}}^e + \mathbf{T}^p\!:\!\tilde{\mathbf{H}}^p)\,dv \end{aligned}\right\} \tag{84.9}$$

for the corresponding external and internal expenditures of virtual power. The **principle of virtual power** is then the requirement that, given any subregion P (of the body),

$$\underbrace{\int\limits_{\partial\mathrm{P}} \mathbf{t}(\mathbf{n})\cdot\tilde{\mathbf{u}}\,da + \int\limits_{\mathrm{P}} \mathbf{b}\cdot\tilde{\mathbf{u}}\,dv}_{\mathcal{W}(\mathrm{P},\mathcal{V})} = \underbrace{\int\limits_{\mathrm{P}} (\mathbf{T}^e\!:\!\tilde{\mathbf{H}}^e + \mathbf{T}^p\!:\!\tilde{\mathbf{H}}^p)\,dv}_{\mathcal{I}(\mathrm{P},\mathcal{V})} \tag{84.10}$$

for all virtual velocities $\mathcal{V}$.

---

[464] Cf. (22.11).

Our discussion in §22.2 of the classical virtual-power principle is based on a virtual internal-power expenditure for large deformations of a form

$$\int_{\mathcal{P}_t} \mathbf{T}\colon \operatorname{grad}\tilde{\mathbf{v}}\, dv. \tag{84.11}$$

The argument in §22.2 that renders $\mathbf{T}$ symmetric and converts the integrand of (84.11) to the more standard form $\mathbf{T}\colon\tilde{\mathbf{D}}$ is based on the requirement that (84.11) be frame-indifferent. A similar argument may be applied to the theory currently under consideration, but beginning with a virtual-power framework based on the internal power $\mathcal{I}(\mathrm{P}, \mathcal{V})$ specified in $(84.9)_2$.

Specifically, frame-indifference, as applied to small deformations, requires that the internal power $\mathcal{I}(\mathrm{P}, \mathcal{V})$ be invariant under transformations of the form[465]

$$\tilde{\mathbf{H}}^{e*} = \tilde{\mathbf{H}}^e + \mathbf{\Omega}, \qquad \tilde{\mathbf{H}}^{p*} = \tilde{\mathbf{H}}^p, \tag{84.12}$$

with $\mathbf{\Omega}$ an arbitrary spatially constant skew tensor field. A consequence of this requirement is that

$$\int_{\mathrm{P}} \mathbf{T}^e\colon(\tilde{\mathbf{H}}^e + \mathbf{\Omega})\, dv = \int_{\mathrm{P}} \mathbf{T}^e\colon\tilde{\mathbf{H}}^e\, dv \tag{84.13}$$

for all P and all skew tensors $\mathbf{\Omega}$, so that, necessarily, $\mathbf{T}^e$ *is symmetric*,

$$\mathbf{T}^e = \mathbf{T}^{e\top}, \tag{84.14}$$

and

$$\int_{\mathrm{P}} \mathbf{T}^e\colon\tilde{\mathbf{H}}^e\, dv = \int_{\mathrm{P}} \mathbf{T}^e\colon\tilde{\mathbf{E}}^e\, dv. \tag{84.15}$$

Remark. This argument based on frame-indifference and leading to the symmetry of the stress $\mathbf{T}(=\mathbf{T}^e)$ and hence to the observation that the internal power expenditure has the form

$$\mathcal{I}(\mathrm{P}) = \int_{\mathrm{P}} (\mathbf{T}\colon\dot{\mathbf{E}}^e + \cdots)\, dv \tag{84.16}$$

applies without change in many of the later sections and will not be repeated.

Our next step is to determine the local force balances implied by virtual power. In applying the power balance (84.10), we are at liberty to choose any $\mathcal{V}$ consistent with the constraint (84.7). Consider a virtual velocity $\mathcal{V}$ with $\tilde{\mathbf{u}}$ arbitrary and

$$\tilde{\mathbf{H}}^e = \nabla\tilde{\mathbf{u}}, \tag{84.17}$$

so that, by (84.7),

$$\tilde{\mathbf{H}}^p \equiv \mathbf{0}. \tag{84.18}$$

For this virtual velocity (84.10) reduces to a balance[466]

$$\int_{\partial\mathrm{P}} \mathbf{t}(\mathbf{n})\cdot\tilde{\mathbf{u}}\, da + \int_{\mathrm{P}} \mathbf{b}\cdot\tilde{\mathbf{u}}\, dv = \int_{\mathrm{P}} \mathbf{T}^e\colon\nabla\tilde{\mathbf{u}}\, dv \tag{84.19}$$

involving a single kinematic variable: the velocity $\tilde{\mathbf{u}}$ of points of the body; for that reason we refer to virtual velocities $\mathcal{V}$ consistent with (84.17) as **macroscopic**.

---

465 These transformation laws are best understood when viewed as small-deformation counterparts of corresponding laws appropriate to large deformations. Cf. §91.5; in particular, (91.28).

466 The form of this balance is directly analogous to that of the classical virtual-power balance (22.13).

Next, as a consequence of the divergence theorem

$$\int_{\mathrm{P}} \mathbf{T}^e : \nabla\tilde{\mathbf{u}}\, dv = -\int_{\mathrm{P}} \mathrm{Div}\,\mathbf{T}^e \cdot \tilde{\mathbf{u}}\, dv + \int_{\partial\mathrm{P}} (\mathbf{T}^e\mathbf{n}) \cdot \tilde{\mathbf{u}}\, da$$

and the balance (84.19) becomes

$$\int_{\partial\mathrm{P}} (\mathbf{t}(\mathbf{n}) - \mathbf{T}^e\mathbf{n}) \cdot \tilde{\mathbf{u}}\, da + \int_{\mathrm{P}} (\mathrm{Div}\,\mathbf{T}^e + \mathbf{b}) \cdot \tilde{\mathbf{u}}\, dv = 0.$$

Thus, since both $\tilde{\mathbf{u}}$ and P are arbitrary, we conclude from the fundamental lemma of the calculus of variations (page 167) that the traction condition

$$\mathbf{t}(\mathbf{n}) = \mathbf{T}^e\mathbf{n} \tag{84.20}$$

and local force balance

$$\mathrm{Div}\,\mathbf{T}^e + \mathbf{b} = \mathbf{0} \tag{84.21}$$

are satisfied.

This traction condition and force balance and the symmetry condition (84.14) are the classical conditions satisfied by the (standard) Cauchy stress $\mathbf{T}$, an observation that allows us to write

$$\mathbf{T} \stackrel{\text{def}}{=} \mathbf{T}^e \tag{84.22}$$

and to view (the symmetric tensor) $\mathbf{T}$ as the **macroscopic stress** and (84.21) as the local **macroscopic force balance**. This use of the adjective macroscopic would seem justified by the foregoing analysis, as the derivation of the balance (84.21) involved only the macroscopic virtual velocity $\mathcal{V}$ described via (84.17) and (84.18). In view of (84.4) and (84.22), (84.21) is equivalent to the *momentum balance*

$$\mathrm{Div}\,\mathbf{T} + \mathbf{b}_0 = \rho\ddot{\mathbf{u}}. \tag{84.23}$$

We now show that, unlike conventional theories of plasticity,

- *there is an additional force balance associated with the plastic stress* $\mathbf{T}^p$.

To derive this balance we consider a virtual velocity with $\tilde{\mathbf{H}}^p$ an arbitrary deviatoric tensor field, with $\tilde{\mathbf{H}}^e$ given by

$$\tilde{\mathbf{H}}^e = -\tilde{\mathbf{H}}^p, \tag{84.24}$$

and with

$$\tilde{\mathbf{u}} \equiv \mathbf{0}, \tag{84.25}$$

consistent with (84.7). Virtual velocities $\mathcal{V}$ of this type might be termed **microscopic** because they involve no macroscopic motion of any part of the body: *There are only microscopic motions in which local changes in shape induced by plastic flow are balanced by local stretch and rotation of the material structure.* For such a $\mathcal{V}$, (84.10) reduces to

$$\int_{\mathrm{P}} (\mathbf{T}^p - \mathbf{T}) : \tilde{\mathbf{H}}^p\, dv = 0. \tag{84.26}$$

Next, by (84.3), $\mathbf{T}^p$ is deviatoric and, since $\tilde{\mathbf{H}}^p$ is deviatoric, (75.6) yields $\mathbf{T} : \tilde{\mathbf{H}}^p = \mathbf{T}_0 : \tilde{\mathbf{H}}^p$; thus, since P is arbitrary, (84.26) implies that

$$(\mathbf{T}^p - \mathbf{T}_0) : \tilde{\mathbf{H}}^p = 0. \tag{84.27}$$

If we take $\tilde{\mathbf{H}}^p = \tilde{\mathbf{W}}^p$, an arbitrary skew tensor, then, since $\mathbf{T}$ is symmetric, (84.27) becomes $\mathbf{T}^p : \tilde{\mathbf{W}}^p = 0$ for all skew $\tilde{\mathbf{W}}^p$, so that $\mathbf{T}^p$ is symmetric:

$$\mathbf{T}^p = \mathbf{T}^{p\top}, \tag{84.28}$$

and by (84.3) also deviatoric. Finally, since the deviatoric tensor field $\tilde{\mathbf{H}}^p$ in (84.27) is arbitrary, we are led to the **microscopic force balance**

$$\mathbf{T}_0 = \mathbf{T}^p. \tag{84.29}$$

The appellation microscopic force balance would seem justified by the argument leading to (84.29), an argument involving only microscopic virtual velocities as defined by (84.24) and (84.25). Further, (84.29) might be viewed as a balance between

- *forces described by the macroscopic stress* $\mathbf{T}$ *and associated with the material structure, and*
- *forces described by the microscopic stress* $\mathbf{T}^p$ *and associated with the system of dislocations.*[467]

Finally, we note that — as a consequence of the symmetries (84.14) and (84.28) of the stresses $\mathbf{T}^e$ and $\mathbf{T}^p$ together with the identification $\mathbf{T}^e = \mathbf{T}$ — the power balance (84.6) becomes

$$\underbrace{\int\limits_{\partial P} \mathbf{t}(\mathbf{n}) \cdot \dot{\mathbf{u}}\, da + \int\limits_{P} \mathbf{b} \cdot \dot{\mathbf{u}}\, dv}_{\mathcal{W}(P)} = \underbrace{\int\limits_{P} (\mathbf{T} : \dot{\mathbf{E}}^e + \mathbf{T}^p : \dot{\mathbf{E}}^p)\, dv}_{\mathcal{I}(P)}. \tag{84.30}$$

84.2 Principle of Virtual Power Based on the Codirectionality Constraint

84.2.1 General Principle Based on Codirectionality

We now develop a more restrictive form of the virtual-power principle by requiring — from the outset — that the **codirectionality constraint**[468]

$$\frac{\mathbf{T}_0}{|\mathbf{T}_0|} = \mathbf{N}^p \qquad \text{and} \qquad \mathbf{N}^p \equiv \frac{\dot{\mathbf{E}}^p}{|\dot{\mathbf{E}}^p|} \tag{84.31}$$

be satisfied. Thus, necessarily, $\mathbf{T}(\equiv \mathbf{T}^e)$ *is symmetric*, consistent with frame-indifference.[469]

By $(84.31)_2$ we can rewrite $\dot{\mathbf{E}}^p$ in the form[470]

$$\dot{\mathbf{E}}^p = \dot{e}^p \mathbf{N}^p \qquad \text{and} \qquad \dot{e}^p = |\dot{\mathbf{E}}^p| \geq 0, \tag{84.32}$$

so that

$$\dot{\mathbf{H}}^p = \dot{e}^p \mathbf{N}^p + \dot{\mathbf{W}}^p \qquad \text{and} \qquad \dot{\mathbf{W}}^p = \operatorname{skw} \dot{\mathbf{H}}^p,$$

with $\dot{\mathbf{W}}^p$ the plastic spin. The kinematical constraint (84.1) then takes the form

$$\nabla \dot{\mathbf{u}} = \dot{\mathbf{H}}^e + \dot{e}^p \mathbf{N}^p + \dot{\mathbf{W}}^p, \tag{84.33}$$

---

[467] Cf. §75.3, where we required that the stress $\mathbf{T}$ be consistent with two separate constitutive relations: (i) one elastic, associated with the classical elastic strain energy; and (ii) a second viewed as a constraint on purely elastic response imposed by the plasticity of the material. The microscopic force balance (84.29) demonstrates the physical consistency of the present view with that of §75.3.

[468] Cf. (81.20).

[469] Cf. the Remark on page 491.

[470] Cf. (76.8).

and the definitions

$$\tau^p = \mathbf{N}^p : \mathbf{T}^p \qquad \text{and} \qquad \mathbf{T}^p_{\text{skw}} = \operatorname{skw} \mathbf{T}^p \tag{84.34}$$

applied to (84.6) (with $\mathbf{T}(\equiv \mathbf{T}^e)$ symmetric) yield the power balance

$$\underbrace{\int_{\partial \mathrm{P}} \mathbf{t}(\mathbf{n}) \cdot \dot{\mathbf{u}}\, da + \int_{\mathrm{P}} \mathbf{b} \cdot \dot{\mathbf{u}}\, dv}_{\mathcal{W}(\mathrm{P})} = \underbrace{\int_{\mathrm{P}} (\mathbf{T} : \dot{\mathbf{E}}^e + \tau^p \dot{e}^p + (\operatorname{skw} \mathbf{T}^p) : \dot{\mathbf{W}}^p)\, dv}_{\mathcal{I}(\mathrm{P})}. \tag{84.35}$$

As before we consider the fields $\dot{\mathbf{u}}$, $\dot{\mathbf{H}}^e$, and $\dot{\mathbf{H}}^p$ as "virtual velocities" $\tilde{\mathbf{u}}$, $\tilde{\mathbf{H}}^e$, and $\tilde{\mathbf{H}}^p$ consistent with

$$\nabla \tilde{\mathbf{u}} = \tilde{\mathbf{H}}^e + \tilde{\mathbf{H}}^p, \qquad \operatorname{tr} \tilde{\mathbf{H}}^p = 0. \tag{84.36}$$

But we now assume that the codirectionality constraint $(84.31)_1$ — viewed as a constraint on the flow direction $\mathbf{N}^p$ — is satisfied. Consistent with this we restrict attention to virtual plastic strain-rates

$$\tilde{\mathbf{E}}^p = \tilde{e}^p \mathbf{N}^p, \qquad \tilde{e}^p \geq 0,$$

in which:

(i) the accumulated plastic strain-rate $\tilde{e}^p$, being virtual, is nonnegative but otherwise arbitrary;
(ii) the flow direction $\mathbf{N}^p$ is not arbitrary but instead constrained to satisfy the codirectionality constraint $(84.31)_1$.

Based on this, we define a (generalized) virtual velocity to be a list

$$\mathcal{V} = (\tilde{\mathbf{u}}, \tilde{\mathbf{H}}^e, \tilde{e}^p, \tilde{\mathbf{W}}^p) \tag{84.37}$$

consistent with the kinematical constraint

$$\nabla \tilde{\mathbf{u}} = \tilde{\mathbf{H}}^e + \underbrace{\tilde{e}^p \mathbf{N}^p + \tilde{\mathbf{W}}^p}_{\tilde{\mathbf{H}}^p}, \tag{84.38}$$

and, since $\mathbf{T} : \tilde{\mathbf{E}}^e = \mathbf{T} : \tilde{\mathbf{H}}^e$, we write

$$\begin{aligned} \mathcal{W}(\mathrm{P}, \mathcal{V}) &= \int_{\partial \mathrm{P}} \mathbf{t}(\mathbf{n}) \cdot \tilde{\mathbf{u}}\, da + \int_{\mathrm{P}} \mathbf{b} \cdot \tilde{\mathbf{u}}\, dv, \\ \mathcal{I}(\mathrm{P}, \mathcal{V}) &= \int_{\mathrm{P}} (\mathbf{T} : \tilde{\mathbf{H}}^e + \tau^p \tilde{e}^p + (\operatorname{skw} \mathbf{T}^p) : \tilde{\mathbf{W}}^p)\, dv \end{aligned} \tag{84.39}$$

for the corresponding external and internal expenditures of virtual power. The **principle of virtual power** is then the requirement that, given any subregion P,

$$\int_{\partial \mathrm{P}} \mathbf{t}(\mathbf{n}) \cdot \tilde{\mathbf{u}}\, da + \int_{\mathrm{P}} \mathbf{b} \cdot \tilde{\mathbf{u}}\, dv = \int_{\mathrm{P}} (\mathbf{T} : \tilde{\mathbf{H}}^e + \tau^p \tilde{e}^p + (\operatorname{skw} \mathbf{T}^p) : \tilde{\mathbf{W}}^p)\, dv \tag{84.40}$$

for all virtual velocities $\mathcal{V}$. Further, the argument leading to the local force balance

$$\operatorname{Div} \mathbf{T} + \mathbf{b} = \mathbf{0}$$

remains valid.[471]

---

[471] Cf. (84.21).

Our final step is to derive the microscopic force balance. Here, the microscopic virtual velocity defined by (84.24) and (84.25) has the counterpart

$$\tilde{\mathbf{H}}^e = -(\tilde{e}^p\mathbf{N}^p + \tilde{\mathbf{W}}^p), \qquad \tilde{\mathbf{u}} \equiv \mathbf{0},$$

and (84.40) implies that, since $\mathbf{T}:\tilde{\mathbf{W}}^p = 0$,

$$\int_{\mathrm{P}} [(\tau^p - \mathbf{T}:\mathbf{N}^p)\tilde{e}^p + (\mathrm{skw}\,\mathbf{T}^p):\tilde{\mathbf{W}}^p]\, dv = 0.$$

But both P and the skew field $\mathbf{W}^p$ are arbitrary; thus, necessarily, $\mathrm{skw}\,\mathbf{T}^p = \mathbf{0}$ and

$$(\tau^p - \mathbf{T}:\mathbf{N}^p)\tilde{e}^p = 0.$$

Finally, since $\tilde{e}^p \geq 0$ is arbitrary, and since, by $(84.31)_1$,

$$\mathbf{N}^p:\mathbf{T}_0 = |\mathbf{T}_0|, \tag{84.41}$$

we have the **microscopic force balance**

$$|\mathbf{T}_0| = \tau^p. \tag{84.42}$$

An interesting consequence of the codirectionality constraint is the scalar nature of the balance (84.42). A second (and related) consequence is that only the normal part $\tau^p = \mathbf{N}^p:\mathbf{T}^p$ of $\mathbf{T}^p$ is important; indeed, the skew part of $\mathbf{T}^p$ vanishes and the "tangential part" is indeterminate.

84.2.2 Streamlined Principle Based on Codirectionality

A streamlined version of the virtual-power principle (84.40) may be based on assuming from the outset that the basic kinematical descriptors are the velocity $\dot{\mathbf{u}}$, the elastic strain-rate $\dot{\mathbf{E}}^e$, and the rate $\dot{e}^p$ of the accumulated plastic strain. These descriptors are subject to the kinematical constraint

$$\mathrm{sym}\,\nabla\dot{\mathbf{u}} = \dot{\mathbf{E}}^e + \dot{e}^p\mathbf{N}^p, \tag{84.43}$$

where throughout this subsection $\mathbf{N}^p$ is presumed to be consistent with the codirectionality constraint

$$\frac{\mathbf{T}_0}{|\mathbf{T}_0|} = \mathbf{N}^p \tag{84.44}$$

(so that, necessarily, $|\mathbf{N}^p| = 1$). The plastic strain-rate may then be defined by

$$\dot{\mathbf{E}}^p = \dot{e}^p\mathbf{N}^p.$$

The streamlined principle is based on the conventional form (84.5) of the external power, but replaces the internal power (84.39) with the simple relation

$$\mathcal{I}(\mathrm{P}) = \int_{\mathrm{P}} (\mathbf{T}:\dot{\mathbf{E}}^e + \tau^p\dot{e}^p)\, dv. \tag{84.45}$$

The **streamlined virtual-power principle** is then the requirement that the virtual balance

$$\int_{\partial\mathrm{P}} \mathbf{t}(\mathbf{n})\cdot\tilde{\mathbf{u}}\, da + \int_{\mathrm{P}} \mathbf{b}\cdot\tilde{\mathbf{u}}\, dv = \int_{\mathrm{P}} (\mathbf{T}:\tilde{\mathbf{E}}^e + \tau^p\tilde{e}^p)\, dv \tag{84.46}$$

be satisfied for all subregions P and all virtual velocities $\tilde{\mathbf{u}}$, $\tilde{\mathbf{E}}^e$, and $\tilde{e}^p$ consistent with the kinematical constraint

$$\mathrm{sym}\,\nabla\tilde{\mathbf{u}} = \tilde{\mathbf{E}}^e + \tilde{e}^p\mathbf{N}^p. \tag{84.47}$$

Direct consequences of the streamlined virtual-power principle are the macroscopic and microscopic force balances

$$\operatorname{Div}\mathbf{T} + \mathbf{b} = \mathbf{0} \quad \text{and} \quad |\mathbf{T}_0| = \tau^p. \tag{84.48}$$

EXERCISE

1. Establish the balances (84.48) as consequences of the streamlined virtual-power principle defined in the paragraph containing (84.46).

84.3 Virtual External Forces Associated with Dislocation Flow

We find it convenient to allow also for external *microscopic* forces associated with the flow of dislocations through the body. Because such flows are characterized by the plastic strain-rate $\dot{\mathbf{E}}^p$, we introduce an arbitrary symmetric and deviatoric **external microscopic force** B^p power-conjugate to $\dot{\mathbf{E}}^p$, and, hence, add the term[472]

$$\int\limits_{\mathrm{P}} B^p : \dot{\mathbf{E}}^p \, dv \tag{84.49}$$

to the external power (84.5). The **external power** therefore has the expanded form

$$\mathcal{W}(\mathrm{P}) = \int\limits_{\partial\mathrm{P}} \mathbf{t}(\mathbf{n}) \cdot \dot{\mathbf{u}} \, da + \int\limits_{\mathrm{P}} \mathbf{b} \cdot \dot{\mathbf{u}} \, dv + \int\limits_{\mathrm{P}} B^p : \dot{\mathbf{E}}^p \, dv. \tag{84.50}$$

It would not seem a simple matter to produce the external microscopic force B^p in the laboratory; for that reason B^p should be considered as virtual. Thus one might ask:

(‡) *Why introduce the external microscopic force* B^p?

The answer: The inclusion of B^p allows for physically meaningful "thought experiments" that we use to motivate a precise notion of material stability. Specifically, we consider the external microscopic force B^p as a test force applied by an external agency to evolve the system of dislocations via the power expenditure $B^p : \dot{\mathbf{E}}^p$, an expenditure that — when positive — serves as an indication of the stability of the system with $B^p = \mathbf{0}$. Further, because we account for the power expended by B^p, this force is also useful in determining thermodynamically consistent constitutive relations for $\mathbf{T}^e$ and $\mathbf{T}^p$.[473]

The use of external forces to justify kinematical processes is not new. In fact, the mechanics literature is replete with arguments in which various kinematical fields are independently varied without consideration of forces needed to incur such variations. For example, RICE (1971) argues that: "... one may consider the structural rearrangements and the applied stress or strains as independently prescribable quantities ...." Other workers assume the existence of an "external agency that performs work," but say little about the form of this work or how it enters the theory; for example, DRUCKER (1950) asserts that: "The concept of work-hardening ... can be expressed in terms of the work done by an external agency ...." In contrast, *the theory presented here makes explicit the external forces needed to support the "virtual processes" used, and, in so doing, ensures that these forces, whether virtual or not, enter the theory in a consistent manner.*

The only change resulting from the use of the external power $\mathcal{W}(\mathrm{P})$ in the form (84.50) rather than in the form (84.5) used in §84.1 is that the microscopic force balance (84.29) is replaced by the **virtual microscopic force balance**

$$\mathbf{T}^p - \mathbf{T}_0 = B^p. \tag{84.51}$$

[472] A further discussion of the external microscopic force B^p is given in §84.5.
[473] Cf. the sentence following (85.1).

Further, for the virtual-power principle based on the codirectionality constraint — but with $\mathcal{W}(\mathrm{P})$ defined by (84.50) — the microscopic force balance has the form

$$\tau^p - |\mathbf{T}_0| = b^p, \tag{84.52}$$

with b^p the scalar external force defined by

$$b^p = \mathbf{N}^p : B^p. \tag{84.53}$$

Remark. Throughout this book, we include the external force B^p (or b^p) only when: (a) its use is required to establish a given result; or (b) its use leads to a better understanding of a physical notion or assumption.

EXERCISES

1. Verify (84.51).
2. Establish (84.52) and (84.53).

84.4 Free-Energy Imbalance

Because our discussion of force is here based on a paradigm far different from that introduced previously, and because this paradigm is based on nonclassical notions of stress and, consequently, power, it is necessary to derive anew an appropriate free-energy imbalance. With this as a goal, we note that the general free-energy imbalance introduced in §29.1 here leads to the requirement that.

(‡) for any subregion P of the body, *the temporal increase in free energy of* P *must be balanced by the power expended on* P *minus the dissipation within P, and the dissipation is nonnegative.*

Thus, letting ψ denote the **free energy** and

$$\delta \geq 0$$

the **dissipation**, both measured per unit volume, we conclude from (84.5) that (‡) takes the form of a **free-energy imbalance**

$$\overline{\int_{\mathrm{P}} \psi \, dv} = \underbrace{\int_{\partial \mathrm{P}} \mathbf{t}(\mathbf{n}) \cdot \dot{\mathbf{u}} \, da + \int_{\mathrm{P}} \mathbf{b} \cdot \dot{\mathbf{u}} \, dv}_{\mathcal{W}(\mathrm{P})} - \int_{\mathrm{P}} \delta \, dv. \tag{84.54}$$

Further, appealing to the power balance (84.30), we find that (84.54) becomes

$$\overline{\int_{\mathrm{P}} \psi \, dv} = \int_{\mathrm{P}} (\mathbf{T} : \dot{\mathbf{E}}^e + \mathbf{T}^p : \dot{\mathbf{E}}^p) \, dv - \int_{P} \delta \, dv. \tag{84.55}$$

Thus, since $\overline{\int_P \psi \, dv} = \int_P \dot{\psi} \, dv$ and P is arbitrary, (84.55) has the local form

$$\dot{\psi} - \mathbf{T} : \dot{\mathbf{E}}^e - \mathbf{T}^p : \dot{\mathbf{E}}^p = -\delta \leq 0. \tag{84.56}$$

This inequality represents the local **free-energy imbalance** of the general theory.

For the theory of §84.2, which is based on the codirectionality constraint, (84.32) and (84.34) imply that

$$\mathbf{T}^p : \dot{\mathbf{E}}^p = \tau^p \dot{e}^p$$

and hence that the free-energy imbalance has the form

$$\dot{\psi} - \mathbf{T} : \dot{\mathbf{E}}^e - \tau^p \dot{e}^p = -\delta \leq 0. \tag{84.57}$$

EXERCISE

1. Show that the inclusion of the external force B^p (or b^p) does not alter the conclusions (84.56) and (84.57).

84.5 Discussion of the Virtual-Power Formulation

From a pragmatic point of view the standard and virtual-power formulations of the theory are equivalent, at least in the absence of the external microscopic force B^p; indeed for $B^p = \mathbf{0}$ the microscopic force balance (84.26) becomes

$$\mathbf{T}_0 = \mathbf{T}^p,$$

a result that renders

$$\mathbf{T}^p : \dot{\mathbf{E}}^p = \mathbf{T}_0 : \dot{\mathbf{E}}^p \tag{84.58}$$

and reduces the free-energy imbalance (84.56) to the conventional imbalance (75.14). Even so, we believe that the virtual-power formulation is conceptually deeper than conventional formulations because (84.58) is a "theorem" rather than an assumption, and because:

(i) The virtual-power formulation reinforces the view that the stretching of the underlying microscopic structure and the flow of dislocations through that structure are *disparate* kinematical processes.
(ii) The virtual-power formulation leads to a final free-energy imbalance (84.56) that involves elastic and plastic stresses $\mathbf{T}$ and $\mathbf{T}^p$, *each* available for constitutive prescription; it does not involve subjecting the stress $\mathbf{T}$ to separate elastic and plastic constitutive prescriptions — a procedure inconsistent with what is common in continuum mechanics — nor does it involve an *a priori* separation of the free-energy imbalance into elastic and plastic parts as in (75.15) and (75.16).
(iii) As we shall see, the virtual-power formulation equipped with the external microscopic force B^p leads naturally to a notion of material stability that, interestingly, is identical to that of maximum dissipation.
(iv) The virtual-power formulation allows for a straightforward generalization (§90) of the conventional theory to situations in which *gradients* of the plastic strain and its rate are independent constitutive variables; it is difficult to see how such a generalization could emanate from a conventional formulation of the theory.

85 Basic Constitutive Theory

We neglect defect energy and assume separability of elastic and plastic constitutive response; therefore, guided by (84.56), we consider constitutive relations of the form

$$\begin{gathered}\psi = \hat{\psi}(\mathbf{E}^e), \qquad \mathbf{T} = \hat{\mathbf{T}}(\mathbf{E}^e),\\ \mathbf{T}^p = \hat{\mathbf{T}}^p(\dot{\mathbf{E}}^p, e^p).\end{gathered} \tag{85.1}$$

As in §76, the assumption of rate-independence reduces $(85.1)_3$ to a relation of the form[474]

$$\mathbf{T}^p = \hat{\mathbf{T}}^p(\mathbf{N}^p, e^p), \qquad \mathbf{N}^p = \frac{\dot{\mathbf{E}}^p}{|\dot{\mathbf{E}}^p|}, \tag{85.2}$$

where, as in (76.70), the accumulated plastic strain e^p is introduced to characterize strain-hardening. Importantly, the constitutive relation (85.2) is *undefined* when $\dot{\mathbf{E}}^p = \mathbf{0}$; in this case,

- we consider the plastic stress $\mathbf{T}^p$ as *indeterminate* and the microscopic force balance $\mathbf{T}^p - \mathbf{T}_0 = B^p$ as trivially satisfied.

The presence of the external body force B^p (as well as the conventional body force $\mathbf{b}$) allows us to use the Coleman–Noll procedure[475] to establish constitutive restrictions that are necessary and sufficient that all constitutive processes[476] be consistent with the free-energy imbalance (84.56). Here, guided by §48.2, we assume that the macroscopic and microscopic body forces[477]

$$\mathbf{b}_0 = \rho\ddot{\mathbf{u}} - \operatorname{Div}\mathbf{T} \qquad \text{and} \qquad B^p = \mathbf{T}^p - \mathbf{T}_0$$

are arbitrarily assignable, so that the macroscopic and microscopic balance laws in no way restrict the class of constitutive processes under consideration.

---

474 Cf. (76.20) and the bullet on page 430. Equation (85.2) is the counterpart of the constitutive relation

$$\mathbf{T}_0 = \hat{\mathbf{T}}_0(\mathbf{N}^p, e^p) \tag{$\star\star$}$$

in the conventional formulation; cf. (76.21). In (85.2), the response function is denoted by $\hat{\mathbf{T}}^p$ since its values represent plastic stresses; in contrast, the response function in the conventional relation ($\star\star$) is $\hat{\mathbf{T}}_0$, because its values represent standard deviatoric stresses $\mathbf{T}_0$. In view of the microscopic force balance (84.51), the present theory reduces to the conventional theory if we assume that $B^p \equiv \mathbf{0}$, for then we could simply redefine $\hat{\mathbf{T}}^p$ to be $\hat{\mathbf{T}}_0$, the deviatoric elastic stress (75.18).

475 Discussed in §48.2. We sketch this procedure to demonstrate that its (rigorous) use in plasticity requires the presence of the external microscopic force B^p.

476 I.e., "processes" consistent with the constitutive relations (85.1).

477 Cf. (84.23) and (84.51).

Consider a constitutive process with $\dot{\mathbf{E}}^p \equiv \mathbf{0}$. Then the free-energy imbalance (84.56) reduces to $\dot{\psi} \le \mathbf{T}:\dot{\mathbf{E}}^e$, an inequality satisfied in all such constitutive processes only if $\hat{\mathbf{T}} = \partial\hat{\psi}/\partial\mathbf{E}^e$, a result that reduces (84.56) to $\mathbf{T}^p:\dot{\mathbf{E}}^p \ge 0$. The requirement that (84.56) hold in all constitutive processes therefore yields the following *thermodynamic restrictions*:

$$\hat{\mathbf{T}}(\mathbf{E}^e) = \frac{\partial\hat{\psi}(\mathbf{E}^e)}{\partial\mathbf{E}^e}, \qquad \hat{\mathbf{T}}^p(\mathbf{N}^p, e^p):\dot{\mathbf{E}}^p \ge 0. \tag{85.3}$$

Conversely, processes related through (85.3) are (trivially) consistent with the free-energy imbalance (84.56).

Finally, assuming that the free energy is quadratic and isotropic we arrive at the elastic constitutive equations (75.17).

86 Material Stability and Its Relation to Maximum Dissipation

In this section we develop a notion of stability useful in the derivation of general flow rules.[478] Metaphysical ideas underlying the notion of stability for plastic materials are due to DRUCKER (1950, 1952), who gave formal arguments showing that yield surfaces for stable materials must be convex with outward unit normal the flow direction. This result is often referred to as *Drucker's Theorem.*[479]

While our conclusions coincide with those of Drucker, our notion of stability, introduced by GURTIN (2003), is far different from that of Drucker. Like Drucker we base our discussion on a notion of "work done by an external agency." For Drucker:

(i) The forces applied by the external agency are introduced by incrementing the standard macroscopic forces.
(ii) The corresponding work corresponds to work done in closed, quasi-static, homogeneous cycles involving standard paths in the elastic range and infinitesimal paths on the yield surface.

But, unlike Drucker, our notion of work done — actually power expended — by an external agency is made precise. Here — because we base the theory on the principle of virtual power — we have at our disposal a (virtual) external microscopic force B^p related to the elastic and plastic stresses $\mathbf{T}$ and $\mathbf{T}^p$ via the microscopic force balance[480]

$$\mathbf{T}^p - \mathbf{T}_0 = B^p. \tag{86.1}$$

To facilitate a comparison of our ideas with those of Drucker, we view B^p as a force exerted by an external agency.

The force balance (86.1) is accompanied by an associated power balance

$$(\mathbf{T}^p - \mathbf{T}_0):\dot{\mathbf{E}}^p = B^p:\dot{\mathbf{E}}^p, \tag{86.2}$$

in which $B^p:\dot{\mathbf{E}}^p$ represents power expended by the external agency.[481] To simplify the ensuing discussion we divide the microscopic power balance (86.2) by $|\dot{\mathbf{E}}^p|$

[478] Note the following remark of DRUCKER (1964, p. 239): "Postulating material to be stable in the mechanical sense [discussed here] does not mean that all materials are stable. ... Isothermal mechanical stability of material is a means of classification, it is not a law of nature."

[479] Cf. DRUCKER (1964) and the references therein. A discussion of Drucker's ideas is given by MALVERN (1969, pp. 356–363). A closely related proof of Drucker's theorem is due to IL'YUSHIN (1954, 1961), who bases his discussion on the requirement that the work done around a closed cycle be nonnegative; cf. PIPKIN & RIVLIN (1965) and LUCCHESI & PODIO-GUIDUGLI (1990).

[480] Cf. (84.51).

[481] Cf. (84.49).

(presumed to be strictly positive); the result is a (normalized) virtual microscopic power-balance

$$B^p : \mathbf{N}^p = (\mathbf{T}^p - \mathbf{T}_0) : \mathbf{N}^p \tag{86.3}$$

that is central to what follows.

Our notion of stability is based on the balance (86.1), and for that reason we work with the constitutive relation

$$\mathbf{T}^p = \hat{\mathbf{T}}^p(\mathbf{N}^p, e^p) \tag{86.4}$$

for the plastic stress $\mathbf{T}^p$.[482] In so doing

(i) $\mathbf{T}^p$ *always denotes the* plastic stress; *as such it is related to the flow direction* $\mathbf{N}^p$ *through the constitutive relation* (86.4);
(ii) $\mathbf{T}_0$ *always denotes the* deviatoric macroscopic stress; *as such it is arbitrary.*

As before we use the term *normalized flow* for a pair $(\mathbf{T}_0, \mathbf{N}^p)$ with $\mathbf{T}_0$ a (macroscopic deviatoric) stress and $\mathbf{N}^p$ a flow direction. For $(\mathbf{T}_0, \mathbf{N}^p)$ such a flow and $\mathbf{T}^p$ — given by the constitutive relation (86.4) — the plastic stress corresponding to $\mathbf{N}^p$, the external force B^p computed via (86.1)[483] represents a force exerted by an external agency in support of the normalized flow $(\mathbf{T}_0, \mathbf{N}^p)$.

In this case, we refer to a normalized flow $(\mathbf{T}_0, \mathbf{N}^p)$ as physically attainable if the external force needed to support it vanishes; viz.

$$B^p = \mathbf{0}. \tag{86.5}$$

Thus, by (86.1) and (86.4), $(\mathbf{T}_0, \mathbf{N}^p)$ is physically attainable[484] if and only if $\mathbf{T}_0$ and $\mathbf{N}^p$ are related through the flow rule[485]

$$\mathbf{T}_0 = \hat{\mathbf{T}}^p(\mathbf{N}^p, e^p). \tag{86.6}$$

Hence,

(‡) *physically attainable flows — that is, flows that do not require a virtual external microscopic power-expenditure — are possible when and only when the deviatoric macroscopic stress* $\mathbf{T}_0$ *and the flow direction* $\mathbf{N}^p$ *are related through the flow rule* (86.6).

The only flows that one might see in actual experiments are those that are physically attainable and hence consistent with the flow rule. More generally, the presence of an external agency as represented by the external force B^p allows us to at least contemplate "thought experiments" involving normalized flows $(\mathbf{T}_0, \mathbf{N}^p)$ that are not compatible with the flow rule, an observation essential to our discussion of stability. Specifically, given such a flow $(\mathbf{T}_0, \mathbf{N}^p)$, we consider $B^p : \mathbf{N}^p$ as a *test expenditure of power* applied by the external agency to investigate the stability of the normalized flow $(\mathbf{T}_0, \mathbf{N}^p)$ — and we note that conventional notions of material stability[486] would then imply that

- if

$$B^p : \mathbf{N}^p > 0 \quad \text{for all } \mathbf{N}^p \tag{86.7}$$

---

482 Cf. (85.2).
483 Cf. (84.51).
484 This definition of a physically attainable flow is consistent with that given on page 455.
485 Within this virtual-power-based framework the flow rule is not simply a constitutive assumption (as in most theories of plasticity), but instead it represents the microscopic force balance augmented by a constitutive relation for the plastic stress.
486 Cf. Drucker (1964,§2).

then the external agency expends power for flow in any direction; in this case there is an energetic barrier for flow and the "system" is stable; as we wish to include neutral stability in our notion of stability, we replace (86.7) by

$$B^p : \mathbf{N}^p \geq 0 \quad \text{for all flow directions } \mathbf{N}^p. \tag{86.8}$$

The following definition — based on the foregoing bullet and the power balance (86.3) — is central to what follows:[487] A deviatoric stress $\mathbf{T}_0$ is **materially stable** if

$$(\hat{\mathbf{T}}^p(\mathbf{N}^p, e^p) - \mathbf{T}_0) : \mathbf{N}^p \geq 0 \quad \text{for every flow direction } \mathbf{N}^p. \tag{86.9}$$

Thus $\mathbf{T}_0$ is materially stable if and only if, given any flow direction $\mathbf{N}^p$, plastic flow in the direction of $\mathbf{N}^p$ requires a nonnegative expenditure of power by an external agency.

Within the present framework the expression (76.26) for the flow resistance takes the form[488]

$$Y(\mathbf{N}^p, e^p) = \mathbf{N}^p : \hat{\mathbf{T}}^p(\mathbf{N}^p, e^p). \tag{86.10}$$

A stress $\mathbf{T}_0$ *is thus materially stable if and only if*

$$\mathbf{T}_0 : \mathbf{N}^p \leq Y(\mathbf{N}^p, e^p) \quad \text{for every flow direction } \mathbf{N}^p. \tag{86.11}$$

Comparing this result with (79.7) we arrive at the

EQUIVALENCY THEOREM *A deviatoric stress* $\mathbf{T}_0$ *is materially stable if and only if* $\mathbf{T}_0$ *is admissible in the sense of maximum dissipation. In view of this equivalence, each of the results established in §79 — the section on maximum dissipation — becomes a result based on material stability.*[489]

Remark. We based our proof of Drucker's Theorem (page 461) on the notion of maximum dissipation, chiefly because that notion is used throughout the plasticity literature, and because the notion of material stability as used here requires the use of virtual power in conjunction with ideas that would seem unfamiliar to most plasticians. Having said this, we believe that material stability is the more compelling notion, because it is not simply presented as an "axiom" but is instead grounded on a rigorous argument based on the principle of virtual power and classical notions of stability.

---

487 Cf. GURTIN (2003, p. 65). Note that, while the external microscopic force B^p provides a motivation for the notion of material stability, B^p does not enter the inequality (86.9) upon which this definition is based.

488 We assume that, as in the conventional theory discussed in §76.3, the dissipation is strict.

489 In particular, using this equivalency one easily converts the proof of Drucker's Theorem on page 461 to a proof based on material stability.

PART XV

STRAIN GRADIENT PLASTICITY BASED ON THE PRINCIPLE OF VIRTUAL POWER

87 Introduction

As reviewed by Hutchinson (2000), a number of experimental results including those from nano/micro-indentation, torsion of micron-dimensioned wires, and bending of micron-dimensioned thin films, all show that[490]

- in the approximate size range between 100 nm and 50 μm, the strength of metallic components undergoing inhomogeneous plastic flow is inherently size-dependent, with *smaller components being stronger than larger components.*

Micromechanical studies of dispersion-strengthening in metals by hard particles lead to a similar conclusion:[491]

- For the same volume fraction of particles, the flow strength of the material increases as the average particle size and the average particle spacing decrease.

These experimental observations cannot be captured by conventional theories of plasticity, because such theories do not contain intrinsic material length-scales.

What is, apparently, the earliest attempt at a plasticity theory with a material length-scale is contained in the seminal work of Aifantis who — working within the framework of small deformations — proposed the flow rule[492]

$$\mathbf{T}_0 = (Y(e^p) - \beta \Delta e^p)\mathbf{N}^p, \tag{87.1}$$

obtained by simply adding the term $-\beta \Delta e^p$ to the conventional flow resistance $Y(e^p)$. Here

$$\Delta = \text{Div}\,\nabla$$

is the Laplace operator, e^p is the accumulated plastic strain, and $\beta > 0$ is a material constant. If we let τ denote the **resolved shear** defined by

$$\tau = \mathbf{T}_0 : \mathbf{N}^p, \tag{87.2}$$

then (87.1) implies that

$$\tau = Y(e^p) - \beta \Delta e^p. \tag{87.3}$$

The presence of a flow rule, such as (87.3), in the form of a partial-differential equation would seem indicative of the absence of a basic force balance, a possibility

490 Cf., e.g., Stelmashenko, Walls, Brown & Milman (1993), Ma & Clarke (1995), Fleck, Muller, Ashby & Hutchinson (1994), and Stolken & Evans (1998).

491 Cf., e.g., Ashby (1970) and Lloyd (1994).

492 Cf. Aifantis (1987, eq. (99)). Cf. also Aifantis (1984, eq. (31)) (which pertains to one space-dimension and contains a typographical error) and Mülhaus & Aifantis (1991).

reinforced by the bullet on page 492 and the discussion surrounding (84.29). In fact, the Aifantis flow rule is the precursor to a large class of flow rules for plasticity in the form of partial-differential equations, many of which are based on the principle of virtual power,[493] a paradigm that automatically delivers the missing force balance from assumptions concerning the manner in which power is expended by stresses, surface tractions, and body forces.[494] We refer to (87.3) as a *microscopic force balance*.[495]

Within the present context, the term *gradient theory* connotes a theory involving constitutive dependencies on the gradient of the plastic strain and/or its rate. In this chapter, we discuss two gradient theories:

(I) The *Aifantis theory*. In that theory ∇e^p is the relevant gradient field and a basic assumption is the (conventional) codirectionality hypothesis requiring that the direction of plastic flow coincide with the direction of the deviatoric stress.[496] The Aifantis theory is fairly simple and results in a scalar flow rule amenable to numerical simulations. But this simplicity results in several deficiencies, specifically, the theory (as presented here) seems incapable of:
   (i) accounting for a backstress and hence for the Bauschinger effect;
   (ii) accounting directly for a defect energy dependent on the Burgers vector;[497]
   (iii) characterizing strengthening.[498]

(II) The *Gurtin–Anand theory*.[499] This theory has $\nabla \mathbf{E}^p$ as the relevant gradient field and the resulting flow rule is a tensorial partial-differential equation, rendering the theory far richer than that of Aifantis. On the other hand, the theory is able to account for a backstress, an energy dependent on the Burgers vector, and strengthening.

In discussing these theories we do not find it necessary to introduce the external virtual microscopic force B^p (or b^p).[500] In the same vein, we shall content ourselves with constitutive equations that are sufficient — but generally not necessary — for compatibility with thermodynamics.

---

493 A general survey of the early work in gradient-plasticity is contained in the review of Fleck & Hutchinson (1997). More recent theories — substantially different from one another — are due to Fleck & Hutchinson (2001), Gurtin (2000b, 2002, 2003, 2004), Cermelli & Gurtin (2002), Gurtin & Needleman (2004), Gudmundson (2004), and Gurtin & Anand (2005a,b). This list omits theories not based on the principle of virtual power.

494 Cf. the discussion surrounding the bullet on page 487.

495 Cf. (84.29) and the ensuing discussion.

496 Cf. (81.20).

497 As described by the tensor $\mathbf{G} = \text{Curl}\,\mathbf{H}^p$ defined in (88.5).

498 An increase in the coarse-grain flow resistance with decreasing length-scales.

499 Cf. Gurtin & Anand (2005a).

500 Cf. the Remark on page 497.

88 Kinematics

As a precursor to a discussion of these theories, we introduce two basic kinematical notions: the *Burger vector* and *irrotational plastic flow*.

88.1 Characterization of the Burgers Vector

Plasticity in crystals arises in response to the motion of dislocations, and the dislocation-induced defectiveness of a crystal may be characterized by the **Burgers vector**,[501] a geometric quantity that measures the closure failure of circuits in the atomic lattice. Both dislocations and their accompanying Burgers vector are microscopic quantities: There are no dislocations in a continuum theory. Even so, the microscopic definition of the Burgers vector may be lifted, almost without change, to form a macroscopic kinematical concept appropriate to a continuous body undergoing plastic deformation.

Consider a two-dimensional crystal lattice as displayed schematically in Figure 88.1; in that figure, (a) shows the undeformed defect-free crystal lattice, while (b) shows the deformed lattice with a dislocation at the point marked with the symbol ⊥. Consider a counterclockwise closed circuit $\mathcal{C}$, with starting lattice point S and finishing lattice point F, that lies in the deformed lattice and surrounds the dislocation. Then, because of the presence of the dislocation, $\mathcal{C}_{\mathrm{R}}$, which is $\mathcal{C}$ as viewed in the dislocation-free undeformed crystal, is not closed. The vector $\mathbf{b}$ closing $\mathcal{C}_{\mathrm{R}}$ and directed from the end point F to the starting point S is called the *Burgers vector.*

To derive the macroscopic counterpart of this notion, consider first the decomposition (75.8) of the displacement gradient $\nabla\mathbf{u}$ into elastic and plastic parts $\mathbf{H}^e$ and $\mathbf{H}^p$, viz.

$$\nabla\mathbf{u} = \mathbf{H}^e + \mathbf{H}^p. \tag{88.1}$$

Now, consider a closed curve $\mathcal{C}$ in the body. Assume that $\mathcal{C}$ is the boundary curve of a smooth oriented surface $\mathcal{S}$ in the body, with $\mathbf{e}$ the unit normal field for $\mathcal{S}$. Then by Stokes' theorem $(4.8)_3$ and, since Curl $\nabla\mathbf{u} = \mathbf{0}$,[502]

$$\begin{aligned}\int_{\mathcal{C}} (\nabla\mathbf{u})d\mathbf{X} &= \int_{\mathcal{S}} (\mathrm{Curl}\,\nabla\mathbf{u})^{\top}\mathbf{e}\,da \\ &= \mathbf{0}.\end{aligned} \tag{88.2}$$

[501] According to Teodosiu (1982, p. 101), whose explanation we follow, the correct definition is due to Frank (1951), although the basic idea stems from work of Burgers (1939).

[502] Cf. $(3.25)_6$.

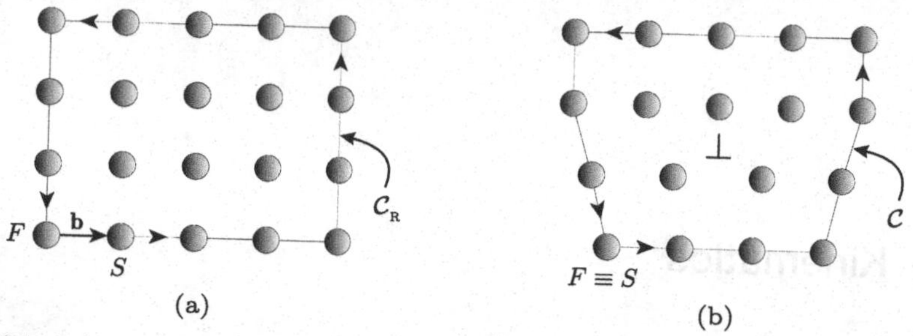

Figure 88.1. Schematic of an edge dislocation in a crystal: (a) the undeformed defect-free crystal lattice; (b) the deformed lattice with an edge dislocation at the point marked ⊥. The *Burgers circuit* $\mathcal{C}$ in the deformed lattice around the edge dislocation is a *closed* circuit that starts and finishes at the lattice point S. The inverse image $\mathcal{C}_R$ of the circuit $\mathcal{C}$ in the undeformed defect-free crystal starts at S and ends at F, and is therefore not closed; the closure-failure of $\mathcal{C}_R$, as described by the vector from F to S, is called the Burgers vector and denoted by $\mathbf{b}$.

Next, because of our assumption that the deformation is small, the deformed and undeformed lattices are geometrically indistinguishable. However, because $\mathbf{H}^p$ represents the distortion of the lattice due to the formation of dislocations, the corresponding integration around $\mathcal{C}$ in the distorted lattice is represented by the integral

$$\begin{aligned}\mathbf{b}(\mathcal{C}) &= \int_{\mathcal{C}} \mathbf{H}^p\,\mathbf{dX}\\ &= \int_{\mathcal{S}} (\mathrm{Curl}\,\mathbf{H}^p)^{\top}\mathbf{e}\,da, \end{aligned} \tag{88.3}$$

an equation that has the component form

$$\begin{aligned} b_j(\mathcal{C}) &= \int_{\mathcal{C}} H^p_{jk}\,dX_k\\ &= \int_{\mathcal{S}} \epsilon_{ipq}\frac{\partial H^p_{jq}}{\partial X_p} e_i\,da. \end{aligned}$$

Because $\mathbf{H}^p$ is not generally the gradient of a vector field, the integral (88.3) does not generally vanish. The vector $\mathbf{b}(\mathcal{C})$, which represents the Burgers vector corresponding to the curve $\mathcal{C}$, is a macroscopic analog of the Burgers vector as defined at the microscopic level.

The local consequence of (88.3) is basic to what follows. We associate the vector measure

$$(\mathrm{Curl}\,\mathbf{H}^p)^{\top}\mathbf{e}\,da \tag{88.4}$$

with the Burgers vector corresponding to the boundary curve of the surface-element $\mathbf{e}\,da$. In this sense, the tensor field[503]

$$\boxed{\mathbf{G} \stackrel{\text{def}}{=} \mathrm{Curl}\,\mathbf{H}^p,} \tag{88.5}$$

[503] The transpose of $-\mathbf{G}$ is often referred to as Nye's tensor, although Nye's (1953) result involves elastic rotations, neglecting elastic strains. The general form (88.5) is apparently due to Kröner (1960), although the counterpart of (88.6) for finite deformations is due to Kondo (1952).

which we refer to as the **Burgers tensor**, provides a local characterization of the Burgers vector. Specifically,

- $\mathbf{G}^{\top}\mathbf{e}$ *represents the Burgers vector, measured per unit area, for infinitesimal closed circuits on any plane* Π *with unit normal* $\mathbf{e}$;

that is, $\mathbf{G}^{\top}\mathbf{e}$ is the local Burgers vector for those dislocation lines piercing Π. On account of (88.1) and because $\mathrm{Curl}\,\nabla\mathbf{u} = \mathbf{0}$, it follows that

$$\mathbf{G} = -\mathrm{Curl}\,\mathbf{H}^{e}, \tag{88.6}$$

and we have a second relation for the Burgers tensor. That there are two relations for the Burgers tensor is of great value. The relation $\mathbf{G} = \mathrm{Curl}\,\mathbf{H}^{p}$ seems most relevant to theories of plasticity involving plastic-strain gradients.[504] On the other hand, in discussing single crystals, materials scientists typically neglect lattice strains, taking $\mathbf{H}^{e} = \mathrm{skw}\,\mathbf{H}^{e} = \mathbf{W}^{e}$, an (infinitesimal) rotation; in this case $\mathbf{G} = -\mathrm{Curl}\,\mathbf{W}^{e}$ may be determined via measurements of lattice rotations.

88.2 Irrotational Plastic Flow

In the classical theory of isotropic plasticity as discussed in §1 the plastic rotation $\mathbf{W}^{p} = \mathrm{skw}\,\mathbf{H}^{p}$ is essentially irrelevant, as it may be absorbed by its elastic counterpart without affecting the resulting field equations. Here, in view of the form (88.5) of the Burgers tensor, the role of the plastic rotation is not so easily dismissed. Indeed, with one exception, any constitutive theory involving the Burgers tensor must necessarily account for the plastic rotation, the exception being when one assumes, from the outset, that the flow is irrotational:[505]

$$\boxed{\mathbf{W}^{p} \equiv \mathbf{0}.} \tag{88.7}$$

In this chapter, we follow this path,[506] so that

$$\mathbf{H}^{p} \equiv \mathbf{E}^{p} \qquad \text{and} \qquad \mathrm{tr}\,\mathbf{E}^{p} \equiv 0, \tag{88.8}$$

and the Burgers tensor is given by

$$\mathbf{G} = \mathrm{Curl}\,\mathbf{E}^{p}. \tag{88.9}$$

---

[504] Cf. §105.

[505] A gradient theory that accounts for plastic rotations is given by Gurtin (2004), but the complicated nature of the resulting nonlocal flow rule would seem to justify the development of a plastically irrotational theory, as would the observation that many of the existing gradient-plasticity theories do not involve plastic rotations; e.g., Fleck & Hutchinson (2001), Gudmundson (2004).

[506] We do not assume that $\mathbf{W}^{p} \equiv \mathbf{0}$ in our discussion of single crystals beginning on page 581.

89 The Gradient Theory of Aifantis

In this section we discuss the Aifantis flow rule

$$\tau = Y(e^p) - \beta \triangle e^p, \tag{89.1}$$

in which $\tau = \mathbf{T}_0 \colon \mathbf{N}^p$ represents the resolved shear (87.2). As noted in the paragraph containing (87.2), Aifantis provides no derivation of (89.1). Granted a knowledge of τ, (89.1) represents a second-order partial-differential equation for the accumulated plastic strain e^p, and one may ask:

- Is (89.1) a constitutive relation, a balance law, or a combination of both?

Because a partial-differential equation such as (89.1) generally requires concomitant boundary conditions, (89.1) cannot be simply constitutive — in fact, it is our view[507] that relations such as (89.1) represent microscopic force balances supplemented by appropriate constitutive equations.

Interestingly, with the exception of GUDMUNDSON (2004) and GURTIN & ANAND (2008), we have not been able to find in the literature any discussion of whether the "nonlocal" term

$$\beta \triangle e^p \tag{89.2}$$

is energetic or dissipative — or even whether or not the underlying theory is consistent with thermodynamics. Aside from the two exceptions mentioned above, we are unaware of any discussion of this issue in the literature, although a common belief seems to be that the nonlocal term (89.2) is dissipative.

In what follows we give a rigorous discussion of the Aifantis theory based on the work of GUDMUNDSON (2004) and GURTIN & ANAND (2009) — these studies are based on the laws of thermodynamics as embodied in the free-energy imbalance and show that, contrary to the common belief, the nonlocal term (89.2) is energetic.

Throughout our discussion of the Aifantis theory we restrict attention to *irrotational plastic flow*:

$$\mathbf{W}^p \equiv \mathbf{0}. \tag{89.3}$$

89.1 The Virtual-Power Principle of Fleck and Hutchinson

Our discussion of the Aifantis theory is based on a virtual-power principle of FLECK & HUTCHINSON (2001),[508] a principle that represents a gradient-theory counterpart

[507] Based on GURTIN (2000b, 2002) and FLECK & HUTCHINSON (2001).

[508] Interestingly, Fleck and Hutchinson did not use their principle to derive the Aifantis flow rule (as did GURTIN & ANAND (2009)), but instead used it — without the aid of thermodynamics and

of the streamlined virtual-power principle discussed in §84.2.2. As in that section, the basic kinematical descriptors are the velocity $\dot{\mathbf{u}}$, the elastic strain-rate $\dot{\mathbf{E}}^e$, and the rate $\dot{e}^p$ of the accumulated plastic strain. Bearing in mind (89.3), these descriptors are subject to the kinematical constraint

$$\operatorname{sym}\nabla\dot{\mathbf{u}} = \dot{\mathbf{E}}^e + \dot{e}^p\mathbf{N}^p \tag{89.4}$$

with flow direction $\mathbf{N}^p$ consistent with the *codirectionality constraint*[509]

$$\frac{\mathbf{T}_0}{|\mathbf{T}_0|} = \mathbf{N}^p, \tag{89.5}$$

so that, necessarily, the conventional stress $\mathbf{T}$ is symmetric.

The streamlined principle discussed in §84.2.2 begins with the expression

$$\int\limits_{\mathrm{P}} (\mathbf{T}\colon\dot{\mathbf{E}}^e + \tau^p\dot{e}^p)\,dv \tag{89.6}$$

for the internal power. Here, because our goal is a theory that accounts explicitly for the gradient of the accumulated plastic strain, we allow also for power expended internally by

- a **microscopic hyperstress** $\boldsymbol{\xi}^p$ power-conjugate to $\nabla\dot{e}^p$

and write the **internal power** in the form

$$\mathcal{I}(\mathrm{P}) = \int\limits_{\mathrm{P}} (\mathbf{T}\colon\dot{\mathbf{E}}^e + \tau^p\dot{e}^p + \boldsymbol{\xi}^p\cdot\nabla\dot{e}^p)\,dv. \tag{89.7}$$

This internal power must be balanced by power expended externally by tractions on $\partial\mathrm{P}$ and body forces acting within P. But the conventional traction $\mathbf{t}(\mathbf{n})$, which gives rise to the conventional expenditure of external power

$$\int\limits_{\partial\mathrm{P}} \mathbf{t}(\mathbf{n})\cdot\dot{\mathbf{u}}\,da + \int\limits_{\mathrm{P}} \mathbf{b}\cdot\dot{\mathbf{u}}\,dv \tag{89.8}$$

is not sufficiently general to accommodate the internal power (89.7); indeed, an additional external power expenditure is needed to balance the internal expenditure $\boldsymbol{\xi}^p\cdot\nabla\dot{e}^p$. In the classical discussion of the principle of virtual power in §22.2 and, in particular, in the argument embodied in (22.14), we see that the internal power $\mathbf{T}\colon\operatorname{grad}\mathbf{v}$ gives rise to the boundary term $\mathbf{Tn}\cdot\mathbf{v}$ and hence to the traction condition $\mathbf{t}(\mathbf{n}) = \mathbf{Tn}$. Arguing by analogy, the gradient term $\boldsymbol{\xi}^p\cdot\nabla\dot{e}^p$ should give rise to a traction term associated with the microscopic hyperstress $\boldsymbol{\xi}^p$. The following identity, which is based on the divergence theorem, helps us to choose an appropriate form for this traction:

$$\int\limits_{\mathrm{P}} \boldsymbol{\xi}^p\cdot\nabla\dot{e}^p\,dv = -\int\limits_{\mathrm{P}} (\operatorname{Div}\boldsymbol{\xi}^p)\dot{e}^p\,dv + \int\limits_{\partial\mathrm{P}} (\boldsymbol{\xi}^p\cdot\mathbf{n})\dot{e}^p\,da. \tag{89.9}$$

Guided by the term $(\boldsymbol{\xi}^p\cdot\mathbf{n})\dot{e}^p$, we assume that power is expended externally by a microscopic hypertraction $\chi(\mathbf{n})$ conjugate to $\dot{e}^p$; we therefore assume that the **external**

---

neglecting inelastic free energy — to develop a flow rule that, in its simplest form, is given by

$$\dot{\tau} = H(\gamma^p)\dot{\gamma}^p - \ell^2\operatorname{Div}(H(\gamma^p)\nabla\dot{\gamma}^p),$$

with ℓ a length scale. Gurtin & Anand (2009) show that this flow rule is consistent with the free-energy imbalance only if $\ell = 0$, thereby obviating the gradient nature of the theory.

509 Cf. (81.20) and (84.31).

power has the form[510]

$$\mathcal{W}(\mathrm{P}) = \int\limits_{\partial \mathrm{P}} \mathbf{t}(\mathbf{n}) \cdot \dot{\mathbf{u}}\, da + \int\limits_{\mathrm{P}} \mathbf{b} \cdot \dot{\mathbf{u}}\, dv + \int\limits_{\partial \mathrm{P}} \chi(\mathbf{n}) \dot{e}^p\, da \tag{89.10}$$

with $\chi(\mathbf{n})$ (for each unit vector $\mathbf{n}$) defined over the body for all time. Our discussion of virtual power is therefore based on the **power balance**

$$\underbrace{\int\limits_{\partial \mathrm{P}} (\mathbf{t}(\mathbf{n})\dot{\mathbf{u}} + \chi(\mathbf{n})\dot{e}^p)\, da + \int\limits_{\mathrm{P}} \mathbf{b} \cdot \dot{\mathbf{u}}\, dv}_{\mathcal{W}(\mathrm{P})} = \underbrace{\int\limits_{\mathrm{P}} (\mathbf{T}\colon \dot{\mathbf{E}}^e + \tau^p \dot{e}^p + \boldsymbol{\xi}^p \cdot \nabla \dot{e}^p)\, dv}_{\mathcal{I}(P)}. \tag{89.11}$$

Bearing in mind (89.4), we use the term virtual velocity for a list

$$\mathcal{V} = (\tilde{\mathbf{u}}, \tilde{\mathbf{E}}^e, \tilde{e}^p)$$

consistent with the kinematical constraint

$$\mathrm{sym}\, \nabla \tilde{\mathbf{u}} = \tilde{\mathbf{E}}^e + \tilde{e}^p \mathbf{N}^p, \tag{89.12}$$

with flow direction $\mathbf{N}^p$ related to $\mathbf{T}_0$ through the codirectionality constraint (89.5). Thus, based on (89.11), we have the **Fleck–Hutchinson principle**[511] requiring that

$$\underbrace{\int\limits_{\partial P} (\mathbf{t}(\mathbf{n})\tilde{\mathbf{u}} + \chi(\mathbf{n})\tilde{e}^p)\, da + \int\limits_{\mathrm{P}} \mathbf{b} \cdot \tilde{\mathbf{u}}\, dv}_{\mathcal{W}(P,\mathcal{V})} = \underbrace{\int\limits_{\mathrm{P}} (\mathbf{T}\colon \tilde{\mathbf{E}}^e + \tau^p \tilde{e}^p + \boldsymbol{\xi}^p \cdot \nabla \tilde{e}^p)\, dv}_{\mathcal{I}(P,\mathcal{V})} \tag{89.13}$$

for any subregion P and any choice of the virtual velocity $\mathcal{V}$.

Guided by the analysis in §84 leading to (84.21), consider a virtual velocity $\mathcal{V}$ with

$$\tilde{\mathbf{E}}^e = \mathrm{sym}\, \nabla \tilde{\mathbf{u}}, \tag{89.14}$$

so that, by (89.12),

$$\tilde{e}^p \equiv 0. \tag{89.15}$$

Then, since $\mathbf{T}$ is symmetric,

$$\mathbf{T}\colon \tilde{\mathbf{E}}^e = \mathbf{T}\colon \nabla \tilde{\mathbf{u}}$$

and (89.13) becomes

$$\int\limits_{\partial \mathrm{P}} \mathbf{t}(\mathbf{n}) \cdot \tilde{\mathbf{u}}\, da + \int\limits_{\mathrm{P}} \mathbf{b} \cdot \tilde{\mathbf{u}}\, dv = \int\limits_{\mathrm{P}} \mathbf{T}\colon \nabla \tilde{\mathbf{u}}\, dv. \tag{89.16}$$

This balance, which is required to hold for all $\tilde{\mathbf{u}}$ and P, coincides with (84.19), and its consequences[512] are the macroscopic traction condition

$$\mathbf{t}(\mathbf{n}) = \mathbf{T}\mathbf{n} \tag{89.17}$$

and the macroscopic force balance

$$\mathrm{div}\mathbf{T} + \mathbf{b} = \mathbf{0}. \tag{89.18}$$

[510] To simplify the presentation we do not include the external microscopic power $b^p \dot{e}^p$; cf. (84.50).
[511] Fleck and Hutchinson actually use the principle of virtual work, which is equivalent.
[512] Cf. (84.20) and (84.21).

To discuss the microscopic counterparts of these results, we define the resolved shear-stress through

$$\tau = \mathbf{T}_0 : \mathbf{N}^p, \tag{89.19}$$

so that, by (89.19) and the codirectionality constraint (89.5),

$$\tau = |\mathbf{T}_0| \qquad \text{and} \qquad \mathbf{T}_0 = \tau \mathbf{N}^p. \tag{89.20}$$

Consider a virtual velocity $\mathcal{V}$ with $\tilde{\mathbf{u}} \equiv \mathbf{0}$, so that (89.13) becomes

$$\int\limits_{\partial \mathrm{P}} \chi(\mathbf{n})\tilde{e}^p \, da = \int\limits_{\mathrm{P}} (\mathbf{T}:\tilde{\mathbf{E}}^e + \tau^p \tilde{e}^p + \boldsymbol{\xi}^p \cdot \nabla \tilde{e}^p)\, dv. \tag{89.21}$$

Further, choose the virtual field $\tilde{e}^p$ arbitrarily and let

$$\tilde{\mathbf{E}}^e = -\tilde{e}^p \mathbf{N}^p;$$

then,

$$\mathbf{T}:\tilde{\mathbf{E}}^e = -\tau \tilde{e}^p \tag{89.22}$$

and the power balance (89.21) yield the microscopic virtual power relation

$$\int\limits_{\partial \mathrm{P}} \chi(\mathbf{n})\tilde{e}^p \, da = \int\limits_{\mathrm{P}} [(\tau^p - \tau)\tilde{e}^p + \boldsymbol{\xi}^p \cdot \nabla \tilde{e}^p]\, dv \tag{89.23}$$

to be satisfied for all $\tilde{e}^p$ and all P. Equivalently, using the divergence theorem, we find that

$$\int\limits_{\partial \mathrm{P}} (\chi(\mathbf{n}) - \boldsymbol{\xi}^p \cdot \mathbf{n})\tilde{e}^p \, da + \int\limits_{\mathrm{P}} (\tau - \tau^p + \mathrm{Div}\,\boldsymbol{\xi}^p)\tilde{e}^p \, dv = 0, \tag{89.24}$$

and a standard argument based on the fundamental lemma of the calculus of variations (page 167) yields the **microscopic traction condition**

$$\chi(\mathbf{n}) = \boldsymbol{\xi}^p \cdot \mathbf{n} \tag{89.25}$$

and the **microscopic force balance**[513]

$$\tau = \tau^p - \mathrm{Div}\,\boldsymbol{\xi}^p. \tag{89.26}$$

89.2 Free-Energy Imbalance

Arguing as in §84.4 and appealing to (89.11), we arrive at the free-energy imbalance

$$\overline{\int\limits_{\mathrm{P}} \Psi \, dv} = \underbrace{\int\limits_{\partial \mathrm{P}} (\mathbf{t}(\mathbf{n})\dot{\mathbf{u}} + \chi(\mathbf{n})\dot{e}^p)\, da + \int\limits_{\mathrm{P}} \mathbf{b}\cdot\dot{\mathbf{u}}\, dv}_{\mathcal{W}(\mathrm{P})} - \int\limits_{\mathrm{P}} \delta \, dv$$

$$= \underbrace{\int\limits_{\mathrm{P}} (\mathbf{T}:\dot{\mathbf{E}}^e + \tau^p \dot{e}^p + \boldsymbol{\xi}^p \cdot \nabla \dot{e}^p)\, dv}_{\mathcal{I}(\mathrm{P})} - \int\limits_{\mathrm{P}} \delta \, dv, \tag{89.27}$$

[513] Fleck & Hutchinson (2001). A microscopic force balance of this form was shown by Gurtin (2000b, eq. (48)) to hold on each slip system of a single crystal; in this case τ represents the resolved shear on that slip system.

with Ψ the free energy and $\delta \geq 0$ the dissipation. Thus, since $\overline{\int_{\mathrm{P}} \Psi \, dv}^{\cdot} = \int_{\mathrm{P}} \dot{\Psi} \, dv$ and P is arbitrary, (89.27) yields the local **free-energy imbalance**[514]

$$\dot{\Psi} - \mathbf{T}:\dot{\mathbf{E}}^e - \tau^p \dot{e}^p - \boldsymbol{\xi}^p \cdot \nabla \dot{e}^p = -\delta \leq 0. \tag{89.28}$$

89.3 Constitutive Equations

The constitutive theory we present is based on a physical picture that associates with an (elastic)-plastic solid a *microscopic structure*, such as a crystal lattice together with a notion of *defects*, such as dislocations, capable of being stored within — and of flowing through — that structure.[515]

By (76.8) and (76.9) the accumulated plastic strain satisfies

$$e^p(0) = 0, \qquad \dot{e}^p(t) \geq 0, \tag{89.29}$$

and, hence, increases with time in any "process." We view e^p as a macroscopic measure of *dislocations* stored in the microscopic structure, and we view its gradient

$$\mathbf{g}^p \stackrel{\text{def}}{=} \nabla e^p \tag{89.30}$$

as a measure of the *inhomogeneity* of the microscopic structure induced by the presence of stored dislocations.

We assume throughout that the free energy admits a decomposition

$$\Psi = \Psi^e + \Psi^p, \tag{89.31}$$

in which Ψ^e represents elastic strain energy, while

$$\Psi^p = \hat{\Psi}^p(e^p, \mathbf{g}^p) \tag{89.32}$$

is a defect energy.

We assume that the *elastic energy* Ψ^e has the standard isotropic form

$$\Psi^e = \mu |\mathbf{E}^e|^2 + \tfrac{1}{2}\lambda (\operatorname{tr} \mathbf{E}^e)^2 \tag{89.33}$$

and generates the stress $\mathbf{T}$ through the classical relation

$$\mathbf{T} = 2\mu \mathbf{E}^e + \lambda (\operatorname{tr} \mathbf{E}^e)\mathbf{1} \tag{89.34}$$

with Lamé moduli μ and λ consistent with the inequalities $\mu > 0$ and $2\mu + 3\lambda > 0$.[516] A consequence of these relations is the standard balance

$$\dot{\Psi}^e = \mathbf{T}:\dot{\mathbf{E}}^e. \tag{89.35}$$

Turning to the defect energy, we introduce energetic microscopic stresses τ^p_{en} and $\boldsymbol{\xi}^p_{\text{en}}$ through the relations

$$\begin{aligned} \tau^p_{\text{en}} &= \hat{\tau}^p_{\text{en}}(e^p, \mathbf{g}^p) = \frac{\partial \hat{\Psi}^p(e^p, \mathbf{g}^p)}{\partial e^p}, \\ \boldsymbol{\xi}^p_{\text{en}} &= \hat{\boldsymbol{\xi}}^p_{\text{en}}(e^p, \mathbf{g}^p) = \frac{\partial \hat{\Psi}^p(e^p, \mathbf{g}^p)}{\partial \mathbf{g}^p}. \end{aligned} \tag{89.36}$$

Then,

$$\dot{\Psi}^p = \tau^p_{\text{en}} \dot{e}^p + \boldsymbol{\xi}^p_{\text{en}} \cdot \dot{\mathbf{g}}^p; \tag{89.37}$$

[514] Cf. (84.57).
[515] Cf. the paragraph containing (75.8).
[516] Cf. (81.54).

the relation (89.37) asserts that temporal changes in the defect energy are balanced by power expended by a microscopic stress $\tau^p_{\rm en}$ conjugate to $\dot{e}^p$ and a microscopic hyperstress $\boldsymbol{\xi}^p_{\rm en}$ conjugate to $\dot{\mathbf{g}}^p$. Equations (89.31), (89.35), and (89.37) imply that

$$\dot{\Psi} = \mathbf{T}\colon\dot{\mathbf{E}}^e + \boldsymbol{\xi}^p_{\rm en}\cdot\dot{\mathbf{g}}^p + \tau^p_{\rm en}\dot{e}^p \tag{89.38}$$

and, hence, that the dissipation δ defined by (89.28) is given by

$$\delta = (\tau^p - \tau^p_{\rm en})\dot{e}^p + (\boldsymbol{\xi}^p - \boldsymbol{\xi}^p_{\rm en})\cdot\dot{\mathbf{g}}^p \geq 0. \tag{89.39}$$

Based on this inequality we refer to the microscopic stresses

$$\tau^p_{\rm dis} = \tau^p - \tau^p_{\rm en} \qquad \text{and} \qquad \boldsymbol{\xi}^p_{\rm dis} = \boldsymbol{\xi}^p - \boldsymbol{\xi}^p_{\rm en} \tag{89.40}$$

as dissipative and rewrite (89.39) in the form

$$\delta = \tau^p_{\rm dis}\dot{e}^p + \boldsymbol{\xi}^p_{\rm dis}\cdot\dot{\mathbf{g}}^p \geq 0. \tag{89.41}$$

We refer to (89.41) as the *plastic-flow inequality.*

Consistent with our choice of independent variables for the free energy (89.32), we consider constitutive equations for the dissipative microstresses of the general form

$$\tau^p_{\rm dis} = \hat{\tau}^p_{\rm dis}(e^p, \mathbf{g}^p), \qquad \boldsymbol{\xi}^p_{\rm dis} = \hat{\boldsymbol{\xi}}^p_{\rm dis}(e^p, \mathbf{g}^p), \tag{89.42}$$

presumed consistent with the plastic-flow inequality (89.41).

89.4 Flow Rules

Within the present framework the flow rule is the microscopic force balance (89.26) augmented by constitutive relations for the microscopic stresses τ^p and $\boldsymbol{\xi}^p$. Using (89.40) we can write this balance in a form

$$\tau = \tau^p_{\rm en} + \tau^p_{\rm dis} - \mathrm{Div}\,(\boldsymbol{\xi}^p_{\rm en} + \boldsymbol{\xi}^p_{\rm dis}), \tag{89.43}$$

so that, by (89.32), (89.36), and (89.42), we have the **general flow rule**

$$\tau = \hat{\tau}^p_{\rm dis}(e^p, \mathbf{g}^p) + \frac{\partial\hat{\Psi}^p(e^p, \mathbf{g}^p)}{\partial e^p} - \mathrm{Div}\left(\hat{\boldsymbol{\xi}}^p_{\rm dis}(e^p, \mathbf{g}^p) + \frac{\partial\hat{\Psi}^p(e^p, \mathbf{g}^p)}{\partial \mathbf{g}^p}\right). \tag{89.44}$$

The Aifantis theory is based on the following additional assumptions:

(i) the defect energy has the form

$$\Psi^p = \tfrac{1}{2}\beta\,|\mathbf{g}^p|^2, \tag{89.45}$$

with $\beta > 0$ constant, so that, by (89.36),

$$\boldsymbol{\xi}^p_{\rm en} = \beta\,\mathbf{g}^p \qquad \text{and} \qquad \tau^p_{\rm en} \equiv 0; \tag{89.46}$$

(ii) the constitutive relations for the dissipative microscopic stresses have the form

$$\tau^p_{\rm dis} = Y(e^p) \qquad \text{and} \qquad \boldsymbol{\xi}^p_{\rm dis} \equiv 0, \tag{89.47}$$

with **coarse-grain flow resistance**[517] $Y(e^p) > 0$.

Since

$$\mathrm{Div}\,\boldsymbol{\xi}^p_{\rm en} = \beta\Delta e^p,$$

[517] I.e., the flow resistance that would obtain in the absence of plastic-strain gradients.

these constitutive relations reduce the general flow rule (89.44) to the **Aifantis flow rule** (87.1); viz.

$$\boxed{\tau = Y(e^p) - \beta\Delta e^p.} \tag{89.48}$$

The coarse-grain flow-resistance $Y(e^p)$ is therefore dissipative; the nonlocal term $\beta\Delta e^p$ is energetic.

EXERCISE

1. Consider a defect energy of the form

$$\Psi^p = \tfrac{1}{2}\beta(e^p)|\mathbf{g}^p|^2,$$
$$\beta(e^p) > 0, \qquad \beta'(e^p) \geq 0. \tag{89.49}$$

(a) Show that

$$\hat{\tau}^p_{\rm en}(e^p, \mathbf{g}^p) = \tfrac{1}{2}\beta'(e^p)\,|\mathbf{g}^p|^2, \tag{89.50}$$

and that

$$\hat{\tau}^p_{\rm en}(e^p, \mathbf{g}^p) \geq 0 \qquad \text{and} \qquad \hat{\tau}^p_{\rm en}(e^p, \mathbf{g}^p)\dot{e}^p \geq 0. \tag{89.51}$$

(This type of behavior is usually described as dissipative; interestingly, here the underlying phenomenon is energetic.)

(b) Show that the defect energy $\Psi^p(t)$ increases during any process that has $\dot{\mathbf{g}}^p \equiv \mathbf{0}$. The defect energy during any such process is therefore nonrecoverable.[518]

(c) Show that

$$\tau^p\dot{e}^p + \boldsymbol{\xi}^p_{\rm dis}\cdot\dot{\mathbf{g}}^p = \delta + \tau^p_{\rm en}\dot{e}^p \tag{89.52}$$

and that[519]

$$\tau^p\dot{e}^p + \boldsymbol{\xi}^p_{\rm dis}\cdot\dot{\mathbf{g}}^p \geq 0. \tag{89.53}$$

(d) Assume that $\tau^p_{\rm dis}$ is independent of $\mathbf{g}^p$ and strictly positive, so that

$$\hat{\tau}^p_{\rm dis}(e^p) = Y(e^p) \tag{89.54}$$

with $Y(e^p) > 0$. Establish the flow rule[520]

$$\tau = Y(e^p) - \tfrac{1}{2}\beta'(e^p)|\mathbf{g}^p|^2 - \beta(e^p)\,\Delta e^p. \tag{89.55}$$

(Note that, since both $\beta'(e^p)$ and $|\mathbf{g}^p|^2$ are nonnegative, the term $\tfrac{1}{2}\beta'(e^p)|\mathbf{g}^p|^2$ characterizes softening.)

89.5 Microscopically Simple Boundary Conditions

Unlike conventional plasticity theories,

- the general flow rule (89.44) and the Aifantis flow rule (89.48) are partial-differential equations and hence require concomitant *boundary conditions.*

---

518 On the other hand, since $|\mathbf{g}^p|$ may increase or decrease at will, the energy in a process during which e^p is constant is recoverable, at least in principle. Thus, the individual arguments e^p and $\mathbf{g}^p$ of the defect energy characterize disparate physical behaviors.

519 One might view this inequality as a basis for a discussion of constitutive equations, but because satisfaction of (89.53) does not imply satisfaction of the plastic-flow imbalance (89.41) and hence does not ensure that the dissipation be nonnegative, such a view would seem to be conceptually flawed.

520 Cf. AIFANTIS (1984), where the gradient terms in the flow rule — unlike those in (89.55) — have moduli that are unrelated.

To discuss such boundary conditions we focus on the boundary ∂B, with outward unit normal $\mathbf{n}$. The external power expended on B is given by (89.10), and, in view of (89.25), the microscopic portion of this power has the form

$$\int_{\partial B} (\boldsymbol{\xi}^p \cdot \mathbf{n})\dot{e}^p \, da, \tag{89.56}$$

with $(\boldsymbol{\xi}^p \cdot \mathbf{n})\dot{e}^p$ the microscopic power expended, per unit area, on ∂B by the material in contact with the body.

We limit our discussion to boundary conditions that result in a *null expenditure of microscopic power* in the sense that

$$(\boldsymbol{\xi}^p \cdot \mathbf{n})\dot{e}^p = 0 \quad \text{on } \partial B. \tag{89.57}$$

Specifically, we consider **microscopically simple boundary conditions** asserting that

$$\dot{e}^p = 0 \ \text{ on } \mathcal{S}_{\rm hard} \qquad \text{and} \qquad \boldsymbol{\xi}^p \cdot \mathbf{n} = 0 \ \text{ on } \mathcal{S}_{\rm free}, \tag{89.58}$$

where $\mathcal{S}_{\rm hard}$ and $\mathcal{S}_{\rm free}$ are complementary subsurfaces[521] of ∂B. We refer to $\mathcal{S}_{\rm hard}$ and $\mathcal{S}_{\rm free}$, respectively, as the **microscopically hard** and the **microscopically free** portions of ∂B. The microscopically hard condition corresponds to a boundary surface that cannot pass dislocations (e.g., a boundary surface that abuts a hard material); the microscopically free condition corresponds to a boundary across which dislocations can flow freely from the body; this condition would seem consistent with the macroscopic condition $\mathbf{T}\mathbf{n} = \mathbf{0}$.

89.6 Variational Formulation of the Flow Rule

The macroscopic balance $\mathrm{div}\mathbf{T} + \mathbf{b} = \mathbf{0}$ and associated traction boundary-conditions can be formulated variationally using the classical principle of virtual power based on (89.16), a formulation central to analysis and computation. The flow rule (89.48) and the microscopically free boundary-condition (89.58) have an analogous variational formulation based on the microscopic virtual-power relation (89.23), applied to B. Because the boundary conditions (89.58) render the power expenditure null on ∂B, we consider (89.23) with the boundary term omitted and with $\tilde{e}^p = V$:

$$\boxed{\int_{B} [(\tau^p - \tau)V + \boldsymbol{\xi}^p \cdot \nabla V] \, dv = 0.} \tag{89.59}$$

We refer to V as a *test field* and assume that V is *kinematically admissible* in the sense that

$$V = 0 \quad \text{on } \mathcal{S}_{\rm hard}. \tag{89.60}$$

In view of (89.9) (with $\dot{e}^p = V$) and (89.60),

$$\int_{B} \boldsymbol{\xi}^p \cdot \nabla V \, dv = -\int_{B} (\mathrm{Div}\,\boldsymbol{\xi}^p) V \, dv + \int_{\mathcal{S}_{\rm free}} (\boldsymbol{\xi}^p \cdot \mathbf{n}) V \, da.$$

[521] I.e, $\partial B = \mathcal{S}_{\rm hard} \cup \mathcal{S}_{\rm free}$ with $\mathcal{S}_{\rm hard} \cap \mathcal{S}_{\rm free}$ a smooth curve.

Thus, (89.59) holds if and only if

$$\int_{\mathcal{S}_{\rm free}} (\boldsymbol{\xi}^p \cdot \mathbf{n}) V\,da + \int_{\rm B} (\tau^p - \tau - {\rm Div}\,\boldsymbol{\xi}^p) V\,dv = 0; \tag{89.61}$$

therefore, appealing to (an obvious analog of) the fundamental lemma of the calculus of variations (page 167), we see that (89.59) holds for all kinematically admissible test fields V if and only if the microscopic force balance (89.26) and the microscopically free boundary-condition $(89.58)_2$ are satisfied. Since the microscopic force balance — when supplemented by the constitutive equations (89.46) and (89.47) — is equivalent to the flow rule (89.48), we are led to the

VARIATIONAL FORMULATION OF THE AIFANTIS FLOW RULE *Assume that the constitutive equations* (89.46) *and* (89.47) *are satisfied. The Aifantis flow rule* (89.48) *on* B *and the microscopically free boundary-condition* $(89.58)_2$ *are therefore together equivalent to the requirement that* (89.59) *be satisfied for all test fields* V.[522]

89.7 Plastic Free-Energy Balance

Assume that the microscopically simple boundary conditions (89.58) are satisfied, so that

$$(\boldsymbol{\xi}^p \cdot \mathbf{n})\dot{e}^p = 0 \text{ on } \partial{\rm B}. \tag{89.62}$$

Then, by (89.37) and the microscopic force balance (89.26),

$$\begin{aligned}\overline{\int_{\rm B} \hat{\Psi}^p(e^p, \mathbf{g}^p)\,dv} &= \int_{\rm B} (\tau^p_{\rm en}\dot{e}^p + \boldsymbol{\xi}^p_{\rm en} \cdot \nabla\dot{e}^p)\,dv \\ &= \int_{\rm B} (\tau^p_{\rm en}\dot{e}^p + \boldsymbol{\xi}^p_{\rm en} \cdot \nabla\dot{e}^p + \underbrace{(\tau - \tau^p + {\rm Div}\,\boldsymbol{\xi}^p)}_{=0} \cdot \dot{e}^p)\,dv \\ &= \int_{\rm B} (\tau^p_{\rm en}\dot{e}^p + \boldsymbol{\xi}^p_{\rm en} \cdot \nabla\dot{e}^p + (\tau - \tau^p) \cdot \dot{e}^p - \boldsymbol{\xi}^p \cdot \nabla\dot{e}^p)\,dv \end{aligned} \tag{89.63}$$

and, by (89.9) and (89.40), the right side of (89.63) becomes

$$\int_{\rm B} [(\boldsymbol{\xi}^p_{\rm en} - \boldsymbol{\xi}^p) \cdot \nabla\dot{e}^p + (\tau^p_{\rm en} - \tau^p) : \dot{e}^p + \tau\dot{e}^p]\,dv = \int_{\rm B} \tau\dot{e}^p\,dv - \int_{\rm B} (\tau^p_{\rm dis}\dot{e}^p + \boldsymbol{\xi}^p_{\rm dis} \cdot \nabla\dot{e}^p)\,dv;$$

thus, we have the **plastic free-energy balance**[523]

$$\overline{\int_{\rm B} \hat{\Psi}^p(e^p, \mathbf{g}^p)\,dv} = \underbrace{\int_{\rm B} \tau\dot{e}^p\,dv}_{\text{plastic working}} - \underbrace{\int_{\rm B} (\tau^p_{\rm dis}\dot{e}^p + \boldsymbol{\xi}^p_{\rm dis} \cdot \nabla\dot{e}^p)dv}_{\text{dissipation}\geq 0}, \tag{89.64}$$

- *the temporal increase in defect energy can never exceed the plastic working.*

[522] Cf. GURTIN & REDDY (2009) for a reformulation — in terms of global variational inequalities — of initial/boundary-value problems associated with the Aifantis theory.

[523] Cf. GURTIN (2003, eq. (9.23); 2004, eq. (9.4)).

89.8 Spatial Oscillations. Shear Bands

Although conventional theories can characterize the onset of localization, they cannot address issues associated with the attendant instabilities, issues such as the wave length of ensuing oscillations and the thickness of shear bands. We now show that the Aifantis flow rule is capable of modeling scale-dependent phenomena such as these.

89.8.1 Oscillations

In this section, we discuss some simple time-independent solutions of the Aifantis flow rule (89.48) considered as a partial differential equation for e^p, granted a knowledge of τ. Specifically, we consider a one-dimensional theory in which the sole nonzero fields are the displacement u_1, the elastic and plastic strains E^e_{12} and E^p_{12}, and the stress T_{12}, with these fields functions of the coordinate

$$x = X_2.$$

In this case the flow direction has values ± 1 and we restrict attention to the positive value 1. Then, $\dot{E}^p_{12} > 0$ and, in view of the Remark on page 428,

$$e^p \equiv E^p_{12},$$

an observation that, because of (76.8) and (76.9), allows us to refer to e^p simply as the **plastic shear**.

We neglect body forces; the force balance (89.18) then implies that

$$\tau \equiv \text{constant} \tag{89.65}$$

and the Aifantis flow rule (89.48) takes the form

$$\tau = Y(e^p) - \beta \frac{d^2 e^p}{dx^2}. \tag{89.66}$$

For the constitutive equation

$$Y(e^p) = Y_0 - \kappa e^p \qquad \text{with} \qquad \kappa > 0, \tag{89.67}$$

which represents the simplest example of *strain-softening*, the flow rule (89.66) yields oscillations; specifically, the differential equation (89.66) has oscillatory solutions

$$e^p = \frac{Y_0 - \tau}{\kappa} + C \exp\left(\frac{\pm ix}{\lambda}\right)$$

with wave length $\lambda = \sqrt{\beta/\kappa}$. Since $\lambda \to 0$ as $\beta \to 0$, the oscillations become finer and finer as $\beta \to 0$, indicating instability. The Aifantis flow rule (89.66) can also produce shear bands, but for this both hardening and softening seem needed, at least in one space dimension.[524]

89.8.2 Single Shear Bands and Periodic Arrays of Shear Bands

We now consider solutions of the differential equation (89.66) for prescribed values of τ, assuming that the dissipative-hardening function $Y(e^p)$ exhibits a hardening-softening transition. Specifically, we are interested in solutions of (89.66) on the interval

$$-\infty < x < \infty;$$

[524] Cf., e.g., Gurtin (2000b, §13.2).

such solutions may be determined via their phase portraits — which are curves of

$$\frac{de^p}{dx} \quad \text{versus} \quad e^p$$

representing solutions of (89.66); such curves are referred to as *orbits*.

Consider a value of τ such that

$$\tau = Y(e^{p-}) = Y(e^{p+}), \tag{89.68}$$

where e^{p-} lies in the hardening interval, while e^{p+} lies in the softening interval. The values e^{p-} and e^{p+} of plastic shear represent *equilibrium points* of (89.66), because, by (89.68),

$$e^p(x) \equiv e^{p-} \qquad \text{and} \qquad e^p(x) \equiv e^{p+}$$

represent solutions of (89.66).

Granted that τ satisfies (89.68), two classes of solutions are interesting and important:

(i) The first is described by an orbit that begins and ends at the equilibrium e^{p-}. (Such orbits are termed homoclinic.) It follows from standard results from the theory of ordinary differential equations that this orbit represents a solution e^p with the following properties:[525]

$$e^p(\pm\infty) = e^{p\pm},$$
$$\frac{de^p(x)}{dx} \to 0 \text{ as } x \to \pm\infty.$$

Moreover e^p is strictly increasing until reaching its maximum value, and then strictly decreasing. This solution represents a *single shear-band*, as it begins and ends at the same value of the plastic shear e^{p-}. Each shear e^{p-} in the hardening interval with $Y(e^{p-}) = Y(e^{p+})$ for some shear e^{p+} in the softening interval corresponds to a single shear-band starting and ending at e^{p-}. To the contrary, no value of shear in the softening interval is the initial and terminal point of a single shear-band.

(ii) The second class of solutions is represented by smooth closed curves (closed orbits) in the phase portrait. Each such orbit has $e^p(x) \geq e^{p-}$ for all x and encloses the equilibrium e^{p+}. These orbits represent *periodic solutions* of the differential equation (89.66). Each orbit that lies close to e^{p-} represents a periodic array of nearly constant plastic shear separated by shear bands. On the other hand, the orbits close to e^{p+} are approximately sinusoidal, more and more so as their maximum distance from e^{p+} approaches zero. Finally, the ratio of the width of the shear bands to the width of the valleys between shear bands tends to zero as their minimum distance ζ from e^{p-} tends to zero; the ratio of shear-band width to valley width can thus have any value, depending on the choice of ζ.

Remark. Letting

$$\bar{Y}(e^p) = \frac{Y(e^p)}{Y(0)}, \qquad \bar{\tau} = \frac{\tau}{Y(0)}, \qquad \bar{x} = \frac{x}{\sqrt{\beta/Y(0)}}$$

[525] Cf., e.g., Hirsh & Smale (1972) or Percival & Richards (1982, §3.2).

we arrive at a dimensionless form of (89.66),

$$\bar{\tau} = \bar{Y}(e^p) + \frac{d^2 e^p}{d\bar{x}^2}, \tag{89.69}$$

with $\sqrt{Y(0)/\beta}$ the relevant length-scale. Working with (89.69), GURTIN (2000b, pp. 1025–1028) gives the phase portrait (his Figure 1) and examples of a single shear-band and a periodic array of shear bands (his Figure 2) for $\bar{Y}$ a cubic of the particular form

$$\bar{Y}(e^p) = 0.028(e^p - 1)(e^p - 3)(e^p - 6) + 1.5, \qquad \bar{\tau} = 1.5.$$

90 The Gradient Theory of Gurtin and Anand

This gradient theory[526] differs in many respects from that of Aifantis discussed in §89, chiefly because the codirectionality constraint (89.5) is not employed. For that reason, the basic kinematical field of the Gurtin–Anand description of flow is $\dot{\mathbf{E}}^p$ rather than $\dot{e}^p$. Further, because a gradient theory is desired, the theory is based on an extension of the conventional virtual-power formulation of §84 to include power expended in concert with $\nabla\dot{\mathbf{E}}^p$. Finally, the gradient theory we develop is *rate-dependent.*

90.1 Third-Order Tensors

We begin with a discussion of third-order tensors, which we view as linear transformations $\mathbb{K}$ that associate with each vector $\mathbf{v}$ a (second-order) tensor

$$\mathbf{A} = \mathbb{K}\mathbf{v} \qquad (A_{ij} = K_{ijk}v_k).$$

The inner product of third-order tensors $\mathbb{K}$ and $\mathbb{A}$ is defined in the natural manner; viz.

$$\mathbb{K}\vdots\mathbb{A} = K_{ijk}A_{ijk};$$

the divergence of a third-order tensor $\mathbb{K}$ is the second-order tensor

$$(\mathrm{Div}\,\mathbb{K})_{ij} = \frac{\partial K_{ijk}}{\partial X_k}; \tag{90.1}$$

the gradient of a second-order tensor $\mathbf{U}$ is the third-order tensor

$$(\nabla\mathbf{U})_{ijk} = \frac{\partial U_{ij}}{\partial X_k}. \tag{90.2}$$

A third-order tensor $\mathbb{K}$ is **symmetric and deviatoric in its first two subscripts** if

$$K_{ijk} = K_{jik}, \qquad K_{ppk} = 0. \tag{90.3}$$

An example of such a tensor, and one that we consider, is the gradient $\nabla\mathbf{U}$ of a symmetric, deviatoric (second-order) tensor $\mathbf{U}$.[527]

Given a third-order tensor $\mathbb{A}$, it is useful to note that the part of $\mathbb{A}$ that is symmetric and deviatoric in its first two subscripts is given by

$$\tfrac{1}{2}(A_{ijk} + A_{jik}) - \tfrac{1}{3}\delta_{ij}A_{rrk}. \tag{90.4}$$

---

[526] Gurtin & Anand (2005a).

[527] Cf. (90.2).

90.2 Virtual-Power Formulation: Macroscopic and Microscopic Force Balances

Let P denote an *arbitrary* subregion of the body with **n** the outward unit normal on the boundary ∂P of P.

Bearing in mind the Remark on page 491 and our assumption that the plastic spin $\mathbf{W}^p$ vanishes, we base our discussion on the "rate-like" descriptors $\dot{\mathbf{u}}$, $\dot{\mathbf{E}}^e$, and $\dot{\mathbf{E}}^p$ — restricted by the constraint

$$\operatorname{sym} \nabla\dot{\mathbf{u}} = \dot{\mathbf{E}}^e + \dot{\mathbf{E}}^p, \qquad \operatorname{tr}\dot{\mathbf{E}}^p = 0. \tag{90.5}$$

Consistent with our choice of descriptors, we begin with the conventional internal power expenditure as expressed on the right side of (84.30):

$$\int_{\mathrm{P}} (\mathbf{T}:\dot{\mathbf{E}}^e + \mathbf{T}^p:\dot{\mathbf{E}}^p)\, dv \tag{90.6}$$

with

$$\mathbf{T} \text{ symmetric} \qquad \text{and} \qquad \mathbf{T}^p \text{ symmetric and deviatoric.} \tag{90.7}$$

But — because our goal is a theory that accounts explicitly for plastic-strain gradients — we allow also for power expended internally by

- a (third-order) **microscopic hyperstress** $\mathbb{K}^p$ power-conjugate to $\nabla\dot{\mathbf{E}}^p$

and, therefore, write the **internal power** in the form

$$\mathcal{I}(\mathrm{P}) = \int_{\mathrm{P}} (\mathbf{T}:\dot{\mathbf{E}}^e + \mathbf{T}^p:\dot{\mathbf{E}}^p + \mathbb{K}^p\vdots\nabla\dot{\mathbf{E}}^p)\, dv, \tag{90.8}$$

or, equivalently,

$$\mathcal{I}(\mathrm{P}) = \int_{\mathrm{P}} \left(T^e_{ij}\dot{E}^e_{ij} + T^p_{ij}\dot{E}^p_{ij} + K^p_{ijk}\frac{\partial \dot{E}^p_{ij}}{\partial X_k}\right) dv.$$

Since $\dot{\mathbf{E}}^p$ is symmetric and deviatoric, we may, without loss in generality, assume that

$$\mathbb{K}^p \text{ is symmetric and deviatoric in its first two subscripts;} \tag{90.9}$$

that is,

$$K^p_{ijk} = K^p_{jik}, \quad K^p_{qqk} = 0. \tag{90.10}$$

Here $\mathbf{T}$, $\mathbf{T}^p$, and $\mathbb{K}^p$ are defined over the body for all time.

The internal power (90.8) must be balanced by power expended externally by tractions on ∂P and body forces acting within P. Arguing as in the paragraph leading to (89.10), we supplement the conventional external power-expenditure (89.8) with a higher-order power expenditure involving a hypertraction $\mathbf{K}(\mathbf{n})$ associated with the hyperstress $\mathbb{K}^p$. The following integral identity guides us in choosing an appropriate form for this hypertraction and concomitant power conjugate:

$$\int_{\mathrm{P}} \mathbb{K}^p\vdots\nabla\dot{\mathbf{E}}^p\, dv = -\int_{\mathrm{P}} \operatorname{Div}\mathbb{K}^p:\dot{\mathbf{E}}^p\, dv + \int_{\partial\mathrm{P}} \mathbb{K}^p\mathbf{n}:\dot{\mathbf{E}}^p\, da. \tag{90.11}$$

The verification of (90.11), which involves the divergence theorem, proceeds as follows:

$$\int_P K^p_{ijk} \frac{\partial \dot{E}^p_{ij}}{\partial X_k}\, dv = \int_P \frac{\partial}{\partial X_k}(\dot{E}^p_{ij} K^p_{ijk})\, dv - \int_P \dot{E}^p_{ij} \frac{\partial K^p_{ijk}}{\partial X_k}\, dv$$
$$= -\int_P \frac{\partial K^p_{ijk}}{\partial X_k} \dot{E}^p_{ij}\, dv + \int_{\partial P} \dot{E}^p_{ij} K^p_{ijk} n_k\, da.$$

Guided by the term $\mathbb{K}^p\mathbf{n}\colon\dot{\mathbf{E}}^p$ in (90.11), we assume that power is expended externally by a microscopic traction $\mathbf{K}(\mathbf{n})$ (with components $K_{ij}(\mathbf{n})$) conjugate to the plastic strain-rate $\dot{\mathbf{E}}^p$ and, therefore, assume that the **external power** has the form[528]

$$\mathcal{W}(P) = \int_{\partial P} \mathbf{t}(\mathbf{n}) \cdot \dot{\mathbf{u}}\, da + \int_P \mathbf{b} \cdot \dot{\mathbf{u}}\, dv + \int_{\partial P} \mathbf{K}(\mathbf{n})\colon \dot{\mathbf{E}}^p\, da, \tag{90.12}$$

with $\mathbf{K}(\mathbf{n})$ (for each unit vector $\mathbf{n}$) defined over the body for all time. Since $\dot{\mathbf{E}}^p$ is symmetric and deviatoric, we assume that $\mathbf{K}(\mathbf{n})$ is symmetric and deviatoric.

The principle of virtual power is based on the **power balance**[529]

$$\underbrace{\int_{\partial P} \mathbf{t}(\mathbf{n}) \cdot \dot{\mathbf{u}}\, da + \int_P \mathbf{b} \cdot \dot{\mathbf{u}}\, dv + \int_{\partial P} \mathbf{K}(\mathbf{n})\colon \dot{\mathbf{E}}^p\, da}_{\mathcal{W}(P)} = \underbrace{\int_P (\mathbf{T}\colon\dot{\mathbf{E}}^e + \mathbf{T}^p\colon\dot{\mathbf{E}}^p + \mathbb{K}^p \vdots \nabla\dot{\mathbf{E}}^p)\, dv}_{\mathcal{I}(P)}. \tag{90.13}$$

We now consider the fields $\dot{\mathbf{u}}$, $\dot{\mathbf{E}}^e$, and $\dot{\mathbf{E}}^p$ as virtual fields $\tilde{\mathbf{u}}$, $\tilde{\mathbf{E}}^e$, and $\tilde{\mathbf{E}}^p$ consistent with the constraint[530]

$$\nabla\tilde{\mathbf{u}} = \tilde{\mathbf{E}}^e + \tilde{\mathbf{E}}^p, \qquad \operatorname{tr}\tilde{\mathbf{E}}^p = 0. \tag{90.14}$$

Then, given a *generalized virtual velocity* $\mathcal{V} = (\tilde{\mathbf{u}}, \tilde{\mathbf{E}}^e, \tilde{\mathbf{E}}^p)$ *consistent with* (90.14), we write

$$\begin{aligned} \mathcal{W}(P, \mathcal{V}) &= \int_{\partial P} (\mathbf{t}(\mathbf{n}) \cdot \tilde{\mathbf{u}} + \mathbf{K}(\mathbf{n})\colon\tilde{\mathbf{E}}^p)\, da + \int_P \mathbf{b} \cdot \tilde{\mathbf{u}}\, dv, \\ \mathcal{I}(P, \mathcal{V}) &= \int_P (\mathbf{T}\colon\tilde{\mathbf{E}}^e + \mathbf{T}^p\colon\tilde{\mathbf{E}}^p + \mathbb{K}^p \vdots \nabla\tilde{\mathbf{E}}^p)\, dv \end{aligned} \tag{90.15}$$

for the corresponding external and internal power expenditures. The **principle of virtual power** is then the requirement that, given any subregion P of the body, the

[528] Cf. the Remark on page 497.
[529] Cf. (84.6).
[530] Cf. (90.5).

expenditures (90.15) are balanced:

$$\underbrace{\int\limits_{\partial P} (\mathbf{t}(\mathbf{n}) \cdot \tilde{\mathbf{u}} + \mathbf{K}(\mathbf{n}) : \tilde{\mathbf{E}}^p) \, da + \int\limits_{P} \mathbf{b} \cdot \tilde{\mathbf{u}} \, dv}_{\mathcal{W}(P,\mathcal{V})} = \underbrace{\int\limits_{P} (\mathbf{T} : \tilde{\mathbf{E}}^e + \mathbf{T}^p : \tilde{\mathbf{E}}^p + \mathbb{K}^p \vdots \nabla \tilde{\mathbf{E}}^p) \, dv}_{\mathcal{I}(P,\mathcal{V})} \tag{90.16}$$

for all virtual velocities $\mathcal{V}$.[531]

Our next step is to determine the macroscopic and microscopic force balances. Regarding the former, we assume that $\tilde{\mathbf{u}}$ is arbitrary and that $\tilde{\mathbf{E}}^e = \operatorname{sym} \nabla \tilde{\mathbf{u}}$ so that, by (90.14), $\tilde{\mathbf{E}}^p \equiv \mathbf{0}$. Then, (90.5) is satisfied, (90.16) gives

$$\int\limits_{\partial P} \mathbf{t}(\mathbf{n}) \cdot \tilde{\mathbf{u}} \, da + \int\limits_{P} \mathbf{b} \cdot \tilde{\mathbf{u}} = \int\limits_{P} \mathbf{T} : \nabla \tilde{\mathbf{u}} \, dv,$$

and steps identical to those leading to (84.20) and (84.21) yield identical results: the classical *traction condition*

$$\mathbf{t}(\mathbf{n}) = \mathbf{T}\mathbf{n} \tag{90.17}$$

and *local macroscopic force balance*

$$\operatorname{Div} \mathbf{T} + \mathbf{b} = \mathbf{0}. \tag{90.18}$$

To derive the microscopic force balance, we consider a virtual velocity with: $\tilde{\mathbf{E}}^p$ an arbitrary symmetric, deviatoric tensor field; $\tilde{\mathbf{E}}^e$ given by

$$\tilde{\mathbf{E}}^e = -\tilde{\mathbf{E}}^p; \tag{90.19}$$

and

$$\tilde{\mathbf{u}} \equiv \mathbf{0}, \tag{90.20}$$

consistent with (90.14). Then, (90.16) reduces to the *microscopic virtual-power relation*

$$\int_{\partial P} \mathbf{K}(\mathbf{n}) : \tilde{\mathbf{E}}^p \, da = \int_{P} [(\mathbf{T}^p - \mathbf{T}) : \tilde{\mathbf{E}}^p + \mathbb{K}^p \vdots \nabla \tilde{\mathbf{E}}^p] \, dv. \tag{90.21}$$

Next, using the identity (90.11) in (90.21) we find that

$$\int_{\partial P} (\mathbf{K}(\mathbf{n}) - \mathbb{K}^p \mathbf{n}) : \tilde{\mathbf{E}}^p \, da = \int_{P} (\mathbf{T}^p - \mathbf{T} - \operatorname{Div} \mathbb{K}^p) : \tilde{\mathbf{E}}^p \, dv \tag{90.22}$$

must hold for all parts P and all symmetric-deviatoric tensor fields $\tilde{\mathbf{E}}^p$. Thus, since the terms in (90.22) multiplying this field are themselves symmetric-deviatoric tensor fields, an analog of the fundamental lemma of the calculus of variations (page 167) yields the *microscopic force balance*

$$\mathbf{T}_0 = \mathbf{T}^p - \operatorname{Div} \mathbb{K}^p \qquad (T_{0ij} = T^p_{ij} - K^p_{ijk,k}), \tag{90.23}$$

and the *microscopic traction condition*

$$\mathbf{K}(\mathbf{n}) = \mathbb{K}^p \mathbf{n} \qquad (K_{ij}(\mathbf{n}) = K^p_{ijk} n_k). \tag{90.24}$$

531 Cf. also GUDMUNDSON (2004, eq. (5)).

90.3 Free-Energy Imbalance

Our derivation of the free-energy imbalance follows that of §84.4 and consequently begins with a counterpart of (84.54); viz.

$$\overline{\int_{\mathrm{P}} \Psi\, dv} = \mathcal{W}(\mathrm{P}) - \int_{\mathrm{P}} \delta\, dv. \tag{90.25}$$

As before $\mathcal{W}(\mathrm{P}) = \mathcal{I}(\mathrm{P})$, but we now use (90.8) to write (90.25) equivalently as

$$\overline{\int_{\mathrm{P}} \Psi\, dv} = \int_{\mathrm{P}} (\mathbf{T}{:}\dot{\mathbf{E}}^e + \mathbf{T}^p{:}\dot{\mathbf{E}}^p + \mathbb{K}^p{\vdots}\nabla\dot{\mathbf{E}}^p)\, dv - \int_{\mathrm{P}} \delta\, dv. \tag{90.26}$$

Thus, since $\overline{\int_{\mathrm{P}} \Psi\, dv} = \int_{\mathrm{P}} \dot{\Psi}\, dv$, (90.26) yields the local **free-energy imbalance**

$$\boxed{\dot{\Psi} - \mathbf{T}{:}\dot{\mathbf{E}}^e - \mathbf{T}^p{:}\dot{\mathbf{E}}^p - \mathbb{K}^p{\vdots}\nabla\dot{\mathbf{E}}^p = -\delta \le 0.} \tag{90.27}$$

90.4 Energetic Constitutive Equations

Our goal is a rate-dependent constitutive theory that allows for dependencies on the gradient $\nabla\dot{\mathbf{E}}^p$ of the plastic strain-rate and on the Burgers tensor $\mathbf{G} = \mathrm{curl}\,\mathbf{E}^p$ but that does not otherwise depart drastically from the more conventional theory developed in §75.

We assume that the free energy Ψ is the sum of a quadratic, isotropic elastic energy of standard form[532] and a **defect energy** $\Psi^p(\mathbf{G})$:

$$\Psi = \underbrace{\mu|\mathbf{E}^e|^2 + \tfrac{1}{2}\lambda(\mathrm{tr}\,\mathbf{E}^e)^2}_{\substack{\text{elastic}\\ \text{energy}}} + \underbrace{\Psi^p(\mathbf{G})}_{\substack{\text{defect}\\ \text{energy}}}. \tag{90.28}$$

Consistent with this, we assume that the elastic stress is given by

$$\mathbf{T} = 2\mu\mathbf{E}^e + \lambda(\mathrm{tr}\,\mathbf{E}^e)\mathbf{1}, \tag{90.29}$$

so that

$$\overline{\mu|\mathbf{E}^e|^2 + \tfrac{1}{2}\lambda(\mathrm{tr}\,\mathbf{E}^e)^2} = \mathbf{T}{:}\dot{\mathbf{E}}^e. \tag{90.30}$$

Next, since $\mathbf{G} = \mathrm{Curl}\,\mathbf{E}^p$,

$$\begin{aligned}\overline{\Psi^p(\mathbf{G})} &= \frac{\partial\Psi^p}{\partial\mathbf{G}}:\dot{\mathbf{G}}\\ &= \frac{\partial\Psi^p}{\partial G_{ij}}\epsilon_{ipq}\frac{\partial\dot{E}^p_{jq}}{\partial X_{\mathrm{P}}},\end{aligned} \tag{90.31}$$

so that, letting $\mathbb{A}$ denote the third-order tensor with components

$$A_{jqp} = \frac{\partial\Psi^p}{\partial G_{ij}}\epsilon_{ipq}, \tag{90.32}$$

we find that

$$\overline{\Psi^p(\mathbf{G})} = \mathbb{A}{\vdots}\nabla\dot{\mathbf{E}}^p. \tag{90.33}$$

532 To ensure positive definiteness of the elastic energy, we assume that the Lamé moduli μ (the shear modulus) and λ satisfy $\mu > 0$ and $2\mu + 3\lambda > 0$; cf. (81.54) and (75.17).

Thus, since $\nabla\dot{\mathbf{E}}^p$ is symmetric and deviatoric in its first two subscripts, it is only that part of $\mathbb{A}$ — namely

$$(\mathbb{K}^p_{\rm en})_{ijk} \stackrel{\rm def}{=} \tfrac{1}{2}(A_{ijk} + A_{jik}) - \tfrac{1}{3}\delta_{ij}A_{rrk} \tag{90.34}$$

— that contributes to temporal changes in the defect energy $\psi^p(\mathbf{G})$.[533] Thus,

$$\mathbb{A}\,\vdots\,\nabla\dot{\mathbf{E}}^p = \mathbb{K}^p_{\rm en}\,\vdots\,\nabla\dot{\mathbf{E}}^p$$

and (90.31) implies that

$$\overline{\psi^p(\mathbf{G})}^{\boldsymbol{\cdot}} = \mathbb{K}^p_{\rm en}\,\vdots\,\nabla\dot{\mathbf{E}}^p. \tag{90.35}$$

The tensor field $\mathbb{K}^p_{\rm en}$ is an *energetic hyperstress* associated with temporal changes in the defect energy. Finally, by (90.28) and (90.30),

$$\dot{\psi} = \mathbf{T}:\dot{\mathbf{E}}^e + \mathbb{K}^p_{\rm en}\,\vdots\,\nabla\dot{\mathbf{E}}^p. \tag{90.36}$$

We assume that the defect energy $\psi^p(\mathbf{G})$ is *quadratic* and *isotropic*. Then $\psi^p(\mathbf{G})$ must reduce to a function of $|\mathrm{skw}\,\mathbf{G}|^2$ and the principal invariants (2.142) of sym $\mathbf{G}$.[534] But by (88.9),

$$\mathrm{tr}\,\mathbf{G} = \epsilon_{ijk}E^p_{ik,j} = 0$$

and $\det\mathbf{G}$ is cubic. Thus, $\psi^p(\mathbf{G})$ must be a function of $|\mathrm{sym}\,\mathbf{G}|^2$ and $|\mathrm{skw}\,\mathbf{G}|^2$ or, equivalently, since[535]

$$|\mathbf{G}|^2 = |\mathrm{sym}\,\mathbf{G}|^2 + |\mathrm{skw}\,\mathbf{G}|^2, \tag{90.37}$$

a function of $|\mathbf{G}|^2$ and $|\mathrm{skw}\,\mathbf{G}|^2$; $\psi^p(\mathbf{G})$ must therefore have the form

$$\psi^p(\mathbf{G}) = \alpha_1|\mathbf{G}|^2 + \alpha_2|\mathbf{G} - \mathbf{G}^\top|^2, \tag{90.38}$$

with α_1 and α_2 scalar constants. Using (90.37), we can rewrite (90.38) as

$$\psi^p(\mathbf{G}) = \tfrac{1}{2}\alpha_1|\mathbf{G} + \mathbf{G}^\top|^2 + (\alpha_2 + \tfrac{1}{2}\alpha_1)|\mathbf{G} - \mathbf{G}^\top|^2; \tag{90.39}$$

thus, since the symmetric and skew parts of $\mathbf{G}$ may be specified independently, $\psi^p(\mathbf{G})$ is positive-definite if and only if

$$\alpha_1 > 0 \qquad \text{and} \qquad 2\alpha_2 + \alpha_1 > 0. \tag{90.40}$$

Granted that $\psi^p(\mathbf{G})$ is *positive-definite*, addition of the two inequalities in (90.40) yields $\alpha_2 + \alpha_1 > 0$; the definitions

$$\lambda_1 = \frac{\alpha_1}{\alpha}, \qquad \lambda_2 = \frac{\alpha_2}{\alpha}, \qquad \alpha = \alpha_1 + \alpha_2$$

therefore allow us to write (90.38) in the form

$$\psi^p(\mathbf{G}) = \alpha\big(\lambda_1|\mathbf{G}|^2 + \lambda_2|\mathbf{G} - \mathbf{G}^\top|^2\big), \tag{90.41}$$

with λ_1 and λ_2 *dimensionless moduli* consistent with

$$\lambda_1 + \lambda_2 = 1. \tag{90.42}$$

Finally, since $\mathbf{E}^p$ is dimensionless and $\mathbf{G}$ carries dimensions of $(\text{length})^{-1}$, it follows from (90.28) and (90.41) that α/μ carries dimensions of $(\text{length})^2$. We can

---

533 Cf. (90.4).

534 Cf. TRUESDELL & NOLL (1965); specifically, the paragraph containing (11.24) with $\mathbf{u}$ the axial vector corresponding to skw $\mathbf{G}$.

535 Cf. (2.58).

therefore define an *energetic length-scale* $L > 0$ through

$$L \stackrel{\text{def}}{=} \sqrt{\frac{2\alpha}{\mu}} \tag{90.43}$$

and write (90.41) in the form

$$\Psi^p(\mathbf{G}) = \tfrac{1}{2}\mu L^2(\lambda_1|\mathbf{G}|^2 + \lambda_2|\mathbf{G} - \mathbf{G}^\top|^2). \tag{90.44}$$

Our next step is to compute the energetic hyperstress $\mathbb{K}^p_{\text{en}}$. By (90.42) and (90.44),

$$\begin{aligned}\frac{\partial \Psi^p(\mathbf{G})}{\partial \mathbf{G}} &= \mu L^2[\lambda_1 \mathbf{G} + \lambda_2(\mathbf{G} - \mathbf{G}^\top)] \\ &= \mu L^2(\mathbf{G} - \lambda_2 \mathbf{G}^\top),\end{aligned} \tag{90.45}$$

and a condition necessary and sufficient that (90.40) hold is that

$$\lambda \stackrel{\text{def}}{=} \lambda_2 \qquad \text{satisfy} \qquad -1 < \lambda < 1.$$

By (90.32), the third-order tensor $\mathbb{A}$ in the expression (90.34) for $\mathbb{K}^p_{\text{en}}$ is given by

$$\begin{aligned}A_{jqp} &= \mu L^2 \epsilon_{ipq}(G_{ij} - \lambda G_{ji}) \\ &= \mu L^2 \epsilon_{ipq}\left(\epsilon_{irs}\frac{\partial E^p_{js}}{\partial X_r} - \lambda \epsilon_{jrs}\frac{\partial E^p_{is}}{\partial X_r}\right) \\ &= \mu L^2\left[\left(\frac{\partial E^p_{jq}}{\partial X_p} - \frac{\partial E^p_{jp}}{\partial X_q}\right) - \lambda \epsilon_{ipq}\epsilon_{jrs}\frac{\partial E^p_{is}}{\partial X_r}\right];\end{aligned} \tag{90.46}$$

(90.4) therefore yields

$$\begin{aligned}(\mathbb{K}^p_{\text{en}})_{jqp} = \mu L^2 \Bigg[&\frac{\partial E^p_{jq}}{\partial X_p} - \tfrac{1}{2}\left(\frac{\partial E^p_{jp}}{\partial X_q} + \frac{\partial E^p_{qp}}{\partial X_j}\right) \\ &+ \tfrac{1}{3}(1+\lambda)\,\delta_{jq}\frac{\partial E^p_{rp}}{\partial X_r} - \tfrac{1}{2}\lambda\,(\epsilon_{ipq}\epsilon_{jrs} + \epsilon_{ipj}\epsilon_{qrs})\frac{\partial E^p_{is}}{\partial X_r}\Bigg].\end{aligned} \tag{90.47}$$

90.5 Dissipative Constitutive Equations

Substituting (90.36) into the local free-energy imbalance (90.27), we obtain

$$(\mathbb{K}^p_{\text{en}} - \mathbb{K}^p) \vdots \nabla\dot{\mathbf{E}}^p - \mathbf{T}^p : \dot{\mathbf{E}}^p \le 0. \tag{90.48}$$

Thus, if we define a *dissipative hyperstress* $\mathbb{K}^p_{\text{dis}}$ through the decomposition

$$\mathbb{K}^p = \mathbb{K}^p_{\text{en}} + \mathbb{K}^p_{\text{dis}}, \tag{90.49}$$

then (90.48) yields the reduced dissipation-inequality

$$\delta \stackrel{\text{def}}{=} \mathbf{T}^p : \dot{\mathbf{E}}^p + \mathbb{K}^p_{\text{dis}} \vdots \nabla\dot{\mathbf{E}}^p \ge 0. \tag{90.50}$$

Note that, since both $\mathbb{K}$ and $\mathbb{K}^p_{\text{en}}$ are symmetric and deviatoric in their first two subscripts, so also is $\mathbb{K}^p_{\text{dis}}$.[536] The relation (90.49) represents a decomposition of the microscopic hyperstress $\mathbb{K}^p$ into energetic and dissipative hyperstresses $\mathbb{K}^p_{\text{en}}$ and $\mathbb{K}^p_{\text{dis}}$.

[536] Cf. (90.10) and (90.34).

Our next step is to develop rate-dependent constitutive relations for $\mathbf{T}^p$ and $\mathbb{K}^p_{\text{dis}}$ consistent with the reduced dissipation inequality (90.50). We consider first the plastic stress $\mathbf{T}^p$. Guided by the conventional relations (78.10) and (78.19), we introduce

(i) a *generalized flow-rate*

$$d^p \overset{\text{def}}{=} \sqrt{|\dot{\mathbf{E}}^p|^2 + \ell^2\,|\nabla\dot{\mathbf{E}}^p|^2} \tag{90.51}$$

with $\ell > 0$ a constant *dissipative length-scale*;

(ii) a *generalized accumulated plastic-strain* $\dot{E}^p$ defined by

$$\dot{E}^p = d^p, \qquad E^p(\mathbf{X}, 0) = 0; \tag{90.52}$$

(iii) a *flow resistance* $Y(E^p) > 0$;

(iv) a *rate-sensitivity function* $g(d^p)$ consistent with[537]

$$g(0) = 0, \qquad g(d^p) \text{ is a strictly increasing function of } d^p; \tag{90.53}$$

and we lay down a constitutive relation for $\mathbf{T}^p$ of the form[538]

$$\mathbf{T}^p = g(d^p)Y(E^p)\frac{\dot{\mathbf{E}}^p}{d^p}. \tag{90.54}$$

Deciding on a constitutive relation for the dissipative hyperstress $\mathbb{K}^p_{\text{dis}}$ requires some thought. As noted at the start of §90.4, our goal is a constitutive theory that does not depart drastically from the conventional Mises theory of §75 and §78. As formulated there, the Mises theory is based on a codirectionality hypothesis requiring that the direction of plastic flow coincide with the direction of deviatoric stress. Here, we generalize that formulation. We introduce a generalized plastic strain-rate

$$\dot{\mathbb{E}}^p \overset{\text{def}}{=} (\dot{\mathbf{E}}^p, \ell\nabla\dot{\mathbf{E}}^p),$$

a generalized flow direction

$$\mathbb{N}^p \overset{\text{def}}{=} \frac{\dot{\mathbb{E}}^p}{d^p},$$

and a generalized plastic stress

$$\mathbb{T}^p \overset{\text{def}}{=} (\mathbf{T}^p, \ell^{-1}\mathbb{K}^p_{\text{dis}}),$$

definitions which imply that

$$\mathbb{T}^p \bullet \dot{\mathbb{E}}^p = \mathbf{T}^p : \dot{\mathbf{E}}^p + \mathbb{K}^p_{\text{dis}} \,\vdots\, \nabla\dot{\mathbf{E}}^p$$

and hence convert the reduced dissipation inequality (90.50) to an inequality of the form

$$\delta = \mathbb{T}^p \bullet \dot{\mathbb{E}}^p \geq 0, \tag{90.55}$$

---

537 Cf. (78.9).

538 For $g(d^p) = 1$ and $l = 0$ (so that $d^p = |\dot{\mathbf{E}}^p|$), (90.54) reduces to the conventional flow rule $(76.69)_1$. Since the theory is rate-dependent, one might replace $Y(E^p)$ by a flow resistance in the form of an internal variable S defined by a hardening equation and concomitant initial condition,

$$\dot{S} = h(d^p, S), \qquad S(\mathbf{X}, 0) = S_0,$$

with $S_0 > 0$, a constant, the (initial) flow strength; cf. (78.18).

where $\bullet$ denotes a generalized inner product.[539] Important here is the observation that the reduced dissipation inequality (90.55) as expressed in terms of the fields $\mathbb{T}^p$ and $\dot{\mathbb{E}}^p$ has the conventional structure[540]

$$\delta = \mathbf{T}_0 : \dot{\mathbf{E}}^p \geq 0.$$

We base the constitutive relation for $\mathbb{K}^p_{\text{dis}}$ on a **codirectionality hypothesis** requiring that the stress $\mathbb{T}^p$ point in the direction $\mathbb{N}^p$; specifically, we assume that

$$\mathbb{T}^p = \phi \mathbb{N}^p, \tag{90.56}$$

with ϕ a scalar function (to be specified), and then note that (90.56) is consistent with (90.54) if and only if

$$\phi = g(d^p)Y(E^p). \tag{90.57}$$

Thus we are led naturally to a constitutive relation for the dissipative hyperstress; viz.

$$\mathbb{K}^p_{\text{dis}} = \ell^2 g(d^p)Y(E^p)\frac{\nabla\dot{\mathbf{E}}^p}{d^p}. \tag{90.58}$$

The dissipative constitutive equations of the gradient theory therefore take the form[541]

$$\left.\begin{aligned} \mathbf{T}^p &= g(d^p)Y(E^p)\frac{\dot{\mathbf{E}}^p}{d^p}, \\ \mathbb{K}^p_{\text{dis}} &= \ell^2 g(d^p)Y(E^p)\frac{\nabla\dot{\mathbf{E}}^p}{d^p}. \end{aligned}\right\} \tag{90.59}$$

The relations (90.59) represent the complete set of *dissipative* constitutive relations; interestingly, these equations render the dissipation (90.50) of the simple form

$$\delta = g(d^p)Y(E^p)d^p. \tag{90.60}$$

90.6 Flow Rule

The microscopic force balance augmented by the constitutive equations for $\mathbf{T}^p$, $\mathbb{K}^p_{\text{dis}}$, and $\mathbb{K}^p_{\text{en}}$ forms the flow rule. In view of (90.49), we may rewrite the microscopic force balance (90.23) in the form

$$\mathbf{T}_0 + \operatorname{Div}\mathbb{K}^p_{\text{en}} = \mathbf{T}^p - \operatorname{Div}\mathbb{K}^p_{\text{dis}}, \tag{90.61}$$

where we have placed the term $\operatorname{Div}\mathbb{K}^p_{\text{en}}$ on the left, since, being energetic, its negative represents a *backstress* $\mathbf{T}_{\text{back}}$:

$$\mathbf{T}_{\text{back}} = -\operatorname{Div}\mathbb{K}^p_{\text{en}}. \tag{90.62}$$

[539] The underlying vector space consists of pairs $(\mathbf{A}, \mathbb{B})$ with $\mathbf{A}$ a symmetric, deviatoric tensor and $\mathbb{B}$ a third-order tensor symmetric and deviatoric in its first two subscripts — and, given another such pair $(\mathbf{C}, \mathbb{D})$,

$$(\mathbf{A}, \mathbb{B}) \bullet (\mathbf{C}, \mathbb{D}) \stackrel{\text{def}}{=} \mathbf{A}:\mathbf{C} + \mathbb{D} \vdots \mathbb{B}.$$

[540] Cf. (75.16).

[541] Constitutive relations of this more or less basic structure were proposed by Gurtin (2000b, §14) for single crystals and Gudmundson (2004) for isotropic materials. Cf. Footnote 642.

In fact, a consequence of (90.47) is that

$$\begin{aligned}(\mathrm{Div}\,\mathbb{K}^p_{\mathrm{en}})_{jq} &= \frac{\partial(\mathbb{K}^p_{\mathrm{en}})_{jqp}}{\partial X_p} \\ &= \mu L^2 \frac{\partial}{\partial X_p}\left[\frac{\partial^2 E^p_{jq}}{\partial X_p} - \tfrac{1}{2}\left(\frac{\partial E^p_{jp}}{\partial X_q} + \frac{\partial E^p_{qp}}{\partial X_j}\right)\right. \\ &\quad \left. + \tfrac{1}{3}(1+\lambda)\,\delta_{jq}\frac{\partial E^p_{rp}}{\partial X_r} - \tfrac{1}{2}\lambda\,(\epsilon_{ipq}\epsilon_{jrs} + \epsilon_{ipj}\epsilon_{qrs})\frac{\partial E^p_{is}}{\partial X_r}\right],\end{aligned} \tag{90.63}$$

so that, for

$$(\Delta \mathbf{E}^p)_{ij} = \frac{\partial^2 E^p_{ij}}{\partial X_k \partial X_k},$$

it follows that the **backstress** (90.62) is given by

$$\boxed{\mathbf{T}_{\mathrm{back}} = -\mu L^2(\Delta\mathbf{E}^p - \mathrm{sym}\,(\nabla \mathrm{Div}\,\mathbf{E}^p) + \tfrac{1}{3}(1+\lambda)(\mathrm{Div}\,\mathrm{Div}\,\mathbf{E}^p)\mathbf{1} - \lambda \mathrm{Curl}\,\mathrm{Curl}\,\mathbf{E}^p).} \tag{90.64}$$

Finally, (90.54), (90.58), (90.62), and (90.64), when substituted into (90.61), yield the **flow rule**

$$\boxed{\mathbf{T}_0 - \mathbf{T}_{\mathrm{back}} = \underbrace{Y(E^p)g(d^p)\frac{\dot{\mathbf{E}}^p}{d^p} - \ell^2 \mathrm{Div}\left(Y(E^p)g(d^p)\frac{\nabla\dot{\mathbf{E}}^p}{d^p}\right)}_{\text{dissipative hardening}},} \tag{90.65}$$

which is the central result of the gradient theory.

Given the deviatoric stress $\mathbf{T}_0$, (90.65) represents a *second-order partial-differential equation* for the plastic strain $\mathbf{E}^p$. Thus, unlike conventional plasticity theories,

- the flow rule (90.65) is *nonlocal* and needs to be augmented by appropriate *boundary conditions.*

EXERCISE

1. Establish (90.63) and (90.64).

90.7 Microscopically Simple Boundary Conditions

We focus on the boundary ∂B, with outward unit normal $\mathbf{n}$. The external power expended on B is given by (90.12), and, in view of (90.24), the microscopic portion of this power has the form

$$\int_{\partial \mathrm{B}} \mathbb{K}^p\mathbf{n} : \dot{\mathbf{E}}^p \, da, \tag{90.66}$$

with $\mathbb{K}^p\mathbf{n} : \dot{\mathbf{E}}^p$ the microscopic power expended, per unit area, on ∂B by the material in contact with the body.

We limit our discussion to boundary conditions that result in a null expenditure of microscopic power in the sense that $\mathbb{K}^p\mathbf{n} : \dot{\mathbf{E}}^p = 0$ on ∂B. Specifically, we consider

microscopically simple boundary conditions asserting that

$$\dot{\mathbf{E}}^p = \mathbf{0} \text{ on } \mathcal{S}_{\text{hard}} \qquad \text{and} \qquad \mathbb{K}^p\mathbf{n} = \mathbf{0} \text{ on } \mathcal{S}_{\text{free}}, \tag{90.67}$$

where $\mathcal{S}_{\text{hard}}$ and $\mathcal{S}_{\text{free}}$ are complementary subsurfaces of $\partial\mathrm{B}$, respectively referred to as the **microscopically hard** and the **microscopically free** portions of $\partial\mathrm{B}$.[542] The interpretation of the conditions (90.67) is completely analogous to that of the simple conditions (89.58) of the Aifantis theory. Specifically, the microscopically hard condition corresponds to a boundary surface that cannot pass dislocations (for example, a boundary surface that abuts a hard material); the microscopically free condition corresponds to a boundary across which dislocations can flow freely from the body and would seem consistent with the macroscopic condition $\mathbf{Tn} = \mathbf{0}$.

90.8 Variational Formulation of the Flow Rule

Our next step is to establish a variational formulation of the flow rule based on the microscopically simple boundary conditions (90.67). We begin with the microscopic virtual power relation (90.21) applied to B.[543] Because the boundary conditions (90.67) render the power expenditure null on $\partial\mathrm{B}$, we consider (90.21) with the boundary term omitted and with $\tilde{\mathbf{E}}^p = \mathbf{V}$:

$$\boxed{\int_{\mathrm{B}} [(\mathbf{T}^p - \mathbf{T}_0) : \mathbf{V} + \mathbb{K}^p \vdots \nabla\mathbf{V}]\, dv = 0.} \tag{90.69}$$

We refer to $\mathbf{V}$ as a *test field* and assume that $\mathbf{V}$ is **kinematically admissible** in the sense that $\mathbf{V}$ is symmetric and deviatoric, and that

$$\mathbf{V} = \mathbf{0} \text{ on } \mathcal{S}_{\text{hard}}. \tag{90.70}$$

In view of (90.11) (with $\dot{\mathbf{E}}^p = \mathbf{V}$) and (90.70),

$$\int_{\mathrm{B}} \mathbb{K}^p \vdots \nabla\mathbf{V}\, dv = -\int_{\mathrm{B}} \mathbf{V} : \operatorname{Div}\mathbb{K}^p\, dv + \int_{\mathcal{S}_{\text{free}}} (\mathbb{K}^p\mathbf{n}) : \mathbf{V}\, da.$$

Thus, (90.69) holds if and only if

$$\int_{\mathcal{S}_{\text{free}}} (\mathbb{K}^p\mathbf{n}) : \mathbf{V}\, da + \int_{\mathrm{B}} (\mathbf{T}^p - \mathbf{T}_0 - \operatorname{Div}\mathbb{K}^p) : \mathbf{V}\, dv = 0; \tag{90.71}$$

therefore, appealing to (an obvious analog of) the fundamental lemma of the calculus of variations (page 167), we see that (90.69) holds for all kinematically admissible test fields $\mathbf{V}$ if and only if the microscopic force balance (90.23) and the microscopically free boundary-condition $(90.67)_1$ are satisfied. Since the microscopic force balance, when supplemented by the constitutive equations (90.54) and (90.58), is equivalent to the flow rule (90.65), we are led to the[544]

---

542 For the special case in which we neglect gradient dissipation via the assumption $\mathbb{K}^p_{\text{dis}} \equiv \mathbf{0}$, the boundary conditions (90.67) need to be replaced by the weaker conditions

$$\dot{\mathbf{E}}^p(\mathbf{n}\times) = \mathbf{0} \text{ on } \mathcal{S}_{\text{hard}} \qquad \text{and} \qquad \mathbb{K}^p(\mathbf{n}\times) = \mathbf{0} \text{ on } \mathcal{S}_{\text{free}}, \tag{90.68}$$

with $(\mathbf{n}\times)_{ij} = \epsilon_{ikj}n_k$; cf. Gurtin & Needleman (2004).

543 Cf. the first paragraph of §89.6.

544 Cf. Reddy, Ebobisse, & McBride (2008), who establish the well posedness of boundary-value problems associated with the theory of Gurtin & Anand (2005).

VARIATIONAL FORMULATION OF THE FLOW RULE *Assume that the constitutive equations* (90.59) *are satisfied. The flow rule* (90.65) *on B and the microscopically free boundary-condition* $(90.67)_2$ *are therefore together equivalent to the requirement that* (90.69) *be satisfied for all test fields* $\mathbf{V}$.

90.9 Plastic Free-Energy Balance. Flow-Induced Strengthening

Assume that the microscopically simple boundary conditions (90.67) are satisfied, so that

$$\mathbb{K}^p\mathbf{n}:\dot{\mathbf{E}}^p = 0 \text{ on } \partial\mathrm{B}. \tag{90.72}$$

Then, by (90.35) and the microscopic force balance (90.23),

$$\begin{aligned}\overline{\int_{\mathrm{B}} \Psi^p(\mathbf{G})\,dv}^{\boldsymbol{\cdot}} &= \int_{\mathrm{B}} \mathbb{K}^p_{\mathrm{en}} \vdots \nabla\dot{\mathbf{E}}^p\,dv \\ &= \int_{\mathrm{B}} (\mathbb{K}^p_{\mathrm{en}} \vdots \nabla\dot{\mathbf{E}}^p + \underbrace{(\mathbf{T}_0 - \mathbf{T}^p + \mathrm{Div}\,\mathbb{K}^p)}_{=\mathbf{0}} : \dot{\mathbf{E}}^p)\,dv\end{aligned} \tag{90.73}$$

and, by (90.11) and (90.49), the right side of (90.73) becomes

$$\int_{\mathrm{B}} [(\mathbb{K}^p_{\mathrm{en}} - \mathbb{K}^p) \vdots \nabla\dot{\mathbf{E}}^p + (\mathbf{T}_0 - \mathbf{T}^p):\dot{\mathbf{E}}^p]\,dv = \int_{\mathrm{B}} [(\mathbf{T}_0 - \mathbf{T}^p):\dot{\mathbf{E}}^p - \mathbb{K}^p_{\mathrm{dis}} \vdots \nabla\dot{\mathbf{E}}^p]\,dv;$$

thus we have the **plastic free-energy balance**[545]

$$\overline{\int_{\mathrm{B}} \Psi^p(\mathbf{G})\,dv}^{\boldsymbol{\cdot}} = \underbrace{\int_{\mathrm{B}} \mathbf{T}_0:\dot{\mathbf{E}}^p\,dv}_{\text{plastic working}} - \underbrace{\int_{\mathrm{B}} (\mathbf{T}^p:\dot{\mathbf{E}}^p + \mathbb{K}^p_{\mathrm{dis}} \vdots \nabla\dot{\mathbf{E}}^p)\,dv}_{\text{dissipation} \geq 0}. \tag{90.74}$$

Thus, since the dissipation is nonnegative, *the temporal increase in defect energy can never exceed the plastic working.*

The balance (90.74) is independent of the particular (thermodynamically consistent) constitutive relations for $\mathbf{T}^p$ and $\mathbb{K}^p_{\mathrm{dis}}$; if, in particular, the constitutive equations (90.54) and (90.58) are used, then (90.74) takes the form[546]

$$\overline{\int_{\mathrm{B}} \Psi^p(\mathbf{G})\,dv}^{\boldsymbol{\cdot}} = \int_{\mathrm{B}} \mathbf{T}_0:\dot{\mathbf{E}}^p\,dv - \int_{\mathrm{B}} g(d^p)Y(E^p)\sqrt{|\dot{\mathbf{E}}^p|^2 + \ell^2|\nabla\dot{\mathbf{E}}^p|^2}\,dv. \tag{90.75}$$

Since the rate-dependence of most metals at room temperature is very small, insight may be gained by choosing g to be the *power-law function* (78.20) and discussing the *rate-independent limit* $m \to 0$, so that, in effect, $g(d^p) \equiv 1$. Here, we wish to focus on the microscopic stresses $\mathbf{T}^p$ and $\mathbb{K}^p_{\mathrm{dis}}$ and, in particular, on the consequences of their dependence on the strain-rate gradient $\nabla\dot{\mathbf{E}}^p$. To isolate the effects of $\nabla\dot{\mathbf{E}}^p$, *we neglect the defect energy as well as hardening due to the generalized accumulated plastic strain* with the assumptions:

(A1) $L = 0$, so that $\Psi^p(\mathbf{G}) \equiv 0$ and, hence, $\mathbb{K}^p_{\mathrm{en}} \equiv \mathbf{0}$;
(A2) $Y(E^p) \equiv Y_0$, a constant.

[545] Cf. (90.50) and GURTIN (2003, eq. (9.23); 2004, eq. (9.4)).
[546] Cf. (90.51) and (90.60).

In addition, we assume that:

(A3) the material is rate-independent ($g(d^p) \equiv 1$);
(A4) at no time is the plastic strain-rate spatially constant;
(A5) the boundary conditions are microscopically simple in the sense of (90.67).

Then, by (90.75) with $g(d^p) \equiv 1$, we find, with the aid of (A1)–(A3), that

$$\int_B \mathbf{T}_0 : \dot{\mathbf{E}}^p \, dv = Y_0 \int_B \sqrt{|\dot{\mathbf{E}}^p|^2 + \ell^2 |\nabla \dot{\mathbf{E}}^p|^2} \, dv. \tag{90.76}$$

Further, fixing the time and letting

$$\max_B |\mathbf{T}_0| = \text{ maximum value of } |\mathbf{T}_0| \text{ over the body,}$$

we may use (A4) (and the fact that $\dot{\mathbf{E}}^p$ is deviatoric) to conclude that

$$\int_B \mathbf{T}_0 : \dot{\mathbf{E}}^p \, dv \leq \max_B |\mathbf{T}_0| \int_B |\dot{\mathbf{E}}^p| \, dv \qquad \text{and} \qquad \sqrt{|\dot{\mathbf{E}}^p|^2 + \ell^2 \, |\nabla \dot{\mathbf{E}}^p|^2} > |\dot{\mathbf{E}}^p|.$$

Thus (90.76) yields the inequality

$$\max_B |\mathbf{T}_0| > Y_0. \tag{90.77}$$

A consequence of this inequality is that, given any time, there is a nontrivial subregion of the body that is strengthened by flow; that is, a subregion on which the magnitude $|\mathbf{T}_0|$ of the flow stress is strictly greater than the flow strength Y_0.[547]

EXERCISE

1. Show that if the boundary is microscopically free, then at each time some nontrivial part of the body must be weakened by flow.

90.10 Rate-Independent Theory

The theory discussed thus far has a rate-independent counterart that follows upon taking

$$g(d^p) \equiv 1 \tag{90.78}$$

in the dissipative constitutive equations (90.59):

$$\mathbf{T}^p = Y(E^p)\frac{\dot{\mathbf{E}}^p}{d^p},$$
$$\mathbb{K}^p_{\text{dis}} = \ell^2 Y(E^p)\frac{\nabla \dot{\mathbf{E}}^p}{d^p}. \tag{90.79}$$

For this rate-independent theory there is **yield condition**, which follows from (90.56), (90.57), and (90.78):

$$|\mathbb{T}^p| = Y(E^p), \tag{90.80}$$

[547] Actually one can prove a stronger result, also based on (A1)–(A5): If the boundary conditions are microscopically simple, then at each time some nontrivial part of the body must be strengthened by flow. This result is independent of the shape of the body or of the particular macroscopic boundary conditions under consideration.

Related discussions of strengthening based on dissipative-hardening of the form discussed here are given by Fredriksson & Gudmundson (2005), Anand, Gurtin, Lele & Gething (2005), Gurtin & Anand (2005a). Cf. also Ohno & Okumura (2007), who show that the self-energy (105.65) leads to strengthening.

or, equivalently, and plastic flow (defined by $d^p \neq 0$) is possible when and only when the yield condition is satisfied. Finally, by (90.65) and (90.78), the **flow rule** has the form

$$\mathbf{T}_0 - \mathbf{T}_{\text{back}} = Y(E^p)\frac{\dot{\mathbf{E}}^p}{d^p} - \ell^2 \text{Div}\left(Y(E^p)\frac{\nabla\dot{\mathbf{E}}^p}{d^p}\right) \tag{90.81}$$

with backstress $\mathbf{T}_{\text{back}}$ given by (90.64).

PART XVI

LARGE-DEFORMATION THEORY OF ISOTROPIC PLASTIC SOLIDS

While small-deformation plasticity theories are widely used in the analysis and design of metal structures, such theories fall short in providing an adequate basis for design against plastic-buckling and other structural instabilities — situations in which, although the strains may be small, the rotations are often large. Also, a proper analysis of the stress and strain states associated with tips of cracks in structural components requires a theory of finite plasticity, especially under conditions in which the size of the plastic zone at the crack tip is large relative to the remaining characteristic dimensions of the body. Further, and perhaps most importantly, products made from ductile metals are often subjected to processing operations such as forging, rolling, extrusion, and drawing, as well as finishing operations such as machining, and large plastic deformations are ubiquitous to such manufacturing processes; for that reason, large-deformation theories of plasticity form the *basis* for numerically based computational methods for the design and analysis of processing operations.

Finally, since even metals can undergo large elastic dilational changes under high pressures, such as under high-velocity impact, it is important to formulate the theory within a thermodynamically consistent frame-indifferent description that allows both elastic and plastic deformations to be large. Here we develop such a theory.

91 Kinematics

91.1 The Kröner Decomposition

As in its small-deformation counterpart, the framework of large-deformation plasticity associates with an (elastic)-plastic solid a microscopic structure, such as a crystal lattice, that may be stretched and rotated, together with a notion of defects, such as dislocations, capable of flowing through that structure. But in contrast to its small-deformation counterpart, the rich kinematical framework of the finite theory allows for a deep characterization of that structure.

As we are working within the framework of large deformations, we replace the additive decomposition (75.8) of the displacement gradient **H** by a multiplicative decomposition

$$\boxed{\mathbf{F} = \mathbf{F}^e\mathbf{F}^p, \qquad F_{ij} = F^e_{ik}F^p_{kj}} \tag{91.1}$$

of the deformation gradient $\mathbf{F}(\mathbf{X})$, in which:

(i) $\mathbf{F}^e(\mathbf{X})$, the **elastic distortion**, represents the local deformation of material in an infinitesimal neighborhood of $\mathbf{X}$ due to stretch and rotation of the microscopic structure;
(ii) $\mathbf{F}^p(\mathbf{X})$, the **plastic distortion**, represents the local deformation of material $\mathbf{X}$ in an infinitesimal neighborhood due to the flow of defects through that microscopic structure.

We refer to (91.1) as the **Kröner decomposition.**[548]

Consistent with our stipulation that[549]

$$J = \det \mathbf{F} > 0,$$

we assume that

$$\det \mathbf{F}^e > 0 \qquad \text{and} \qquad \det \mathbf{F}^p > 0, \tag{91.2}$$

so that both $\mathbf{F}^e$ and $\mathbf{F}^p$ are invertible.

In discussing the Kröner decomposition, it is important to fully understand the differences between the tensor fields $\mathbf{F}$, $\mathbf{F}^e$, and $\mathbf{F}^p$. First of all, while $\mathbf{F} = \nabla\boldsymbol{\chi}$ is the gradient of a point field, in general there is no point field $\boldsymbol{\chi}^p$ such that

548 Introduced by Kröner (1960) within a purely kinematical context. Cf. Bilby, Bullough, & Smith (1955) and Bilby (1960) for the special case in which $\mathbf{F}^e$ is an infinitesimal rotation. Somewhat later and apparently independently, Lee & Liu (1967) and Lee (1969) introduced (91.1) in an attempt to develop a complete dynamical theory.

549 Cf. (6.2).

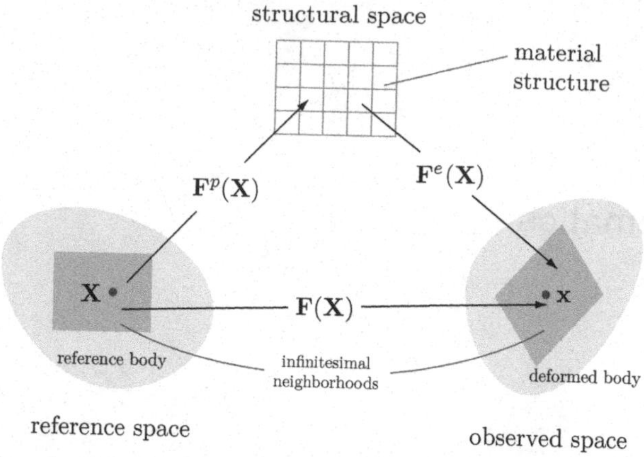

Figure 91.1. Schematic of the Kröner decomposition. The grey squares denote infinitesimal neighborhoods of the points $\mathbf{X}$ and $\mathbf{x} = \chi(\mathbf{X})$. The arrows are meant to indicate the mapping properties of the linear transformations $\mathbf{F}$, $\mathbf{F}^p$, and $\mathbf{F}^e$.

$\mathbf{F}^p = \nabla\chi^p$, nor is there a point field χ^e such that $\mathbf{F}^e = \nabla\chi^e$. Thus we can at most describe the physical nature of the tensor fields $\mathbf{F}^e$ and $\mathbf{F}^p$ through their pointwise mapping properties as linear transformations. With this in mind, consider the formal relation

$$d\mathbf{x} = \mathbf{F}(\mathbf{X}, t)d\mathbf{X}. \tag{91.3}$$

As discussed in the paragraph containing (6.7), (91.3) represents a mapping of an infinitesimal neighborhood of $\mathbf{X}$ in the undeformed body to an infinitesimal neighborhood of $\mathbf{x} = \chi_t(\mathbf{X})$ in the deformed body, and characterizes the tensor field $\mathbf{F}$, at each $\mathbf{X}$, as a linear transformation of material vectors to spatial vectors. As is clear from (91.3) and the Kröner decomposition (91.1), suppressing the argument $(\mathbf{X}, t)$,

$$d\mathbf{x} = \mathbf{F}^e\mathbf{F}^p d\mathbf{X}. \tag{91.4}$$

For want of a better notation, let $d\mathbf{l}$ denote $\mathbf{F}^p d\mathbf{X}$,[550]

$$d\mathbf{l} = \mathbf{F}^p d\mathbf{X},$$

so that, by (91.4),

$$d\mathbf{x} = \mathbf{F}^e d\mathbf{l}.$$

The output of the linear transformation $\mathbf{F}^p$ must therefore coincide with the input of the linear transformation $\mathbf{F}^e$; that is,

$$\text{the range of } \mathbf{F}^p = \text{the domain of } \mathbf{F}^e. \tag{91.5}$$

We refer to this common space as the **structural space** and to vectors in this space as **structural vectors**.[551] Thus, $\mathbf{F}^p$ and $\mathbf{F}^e$ have the following mapping properties (Figure 91.1):

(P1) $\mathbf{F}^p$ *maps material vectors to structural vectors*;
(P2) $\mathbf{F}^e$ *maps structural vectors to spatial vectors.*

Recall our agreement in §12.1 to refer to a tensor field $\mathbf{G}$ as a *spatial tensor field* if $\mathbf{G}$ maps spatial vectors to spatial vectors, a *material tensor field* if $\mathbf{G}$ maps material

550 We do not mean to infer from this that there is a vector $\mathbf{l}$ with differential $d\mathbf{l}$.
551 The structural space is commonly referred to as the intermediate or relaxed configuration.

vectors to material vectors; in the same vein,

(‡) we refer to $\mathbf{G}$ as a **structural tensor field** if $\mathbf{G}$ *maps structural vectors to structural vectors.*

91.2 Digression: Single Crystals

The physical nature of the structural space is most easily comprehended in the special case of a single crystal, for then the structural space houses the undistorted crystal lattice (Figure 91.1). Some readers may disagree with this: In the literature one sometimes finds the assertion, either verbally or implicitly via a figure, that the undistorted lattice resides in the reference space (and is hence material). We believe this to be a misconception. Physically, because a flow of dislocations involves a flow of material relative to the undistorted lattice, lattice vectors cannot be material. A more rigorous argument proceeds as follows. Assume for the moment that, in accord with an approximative assumption made by materials scientists, the material is rigid-plastic, so that $\mathbf{F}^e = \mathbf{R}^e$, a rotation. Most workers would agree with the assertion that, for such a material, a lattice vector as viewed in the deformed body is simply the undistorted lattice rotated by $\mathbf{R}^e$.[552] Granted this, then, since as noted in (P2), $\mathbf{R}^e$ maps (i.e., rotates) a vector $\mathbf{s}$ in the structural space to a vector $\mathbf{R}^e\mathbf{s}$ in the observed space, the lattice must reside in the structural space. Mathematics then dictates that lattice vectors be pulled back to the reference space via the transformation $\mathbf{F}^{p-1}$ and, hence, the image of the lattice as it appears in the reference space would be distorted and transient. Were a lattice vector $\mathbf{s}$ undistorted in the reference, then it would appear in the deformed body as $\mathbf{F}\mathbf{s}$, rather than as $\mathbf{R}^e\mathbf{s}$, which contradicts the underlying mathematical structure. On the other hand, if $\mathbf{s}$ appears in the reference space as $\mathbf{F}^{p-1}\mathbf{s}$, then it would appear in the observed space as

$$\mathbf{F}\mathbf{F}^{p-1}\mathbf{s} = \mathbf{R}^e\mathbf{s},$$

as expected.

A further justification of the assumption that the undistorted lattice live in the structural space, as implied by Figure 91.1, is the discussion in Footnote 688.

91.3 Elastic and Plastic Stretching and Spin. Plastic Incompressibility

The velocity gradient

$$\mathbf{L} = \operatorname{grad} \dot{\chi}$$

is related to the deformation gradient $\mathbf{F}$ through the identity[553]

$$\mathbf{L} = \dot{\mathbf{F}}\mathbf{F}^{-1},$$

and we may use the Kröner decomposition (91.1) to relate $\mathbf{L}$ to $\mathbf{F}^p$ and $\mathbf{F}^e$. By (91.1),

$$\dot{\mathbf{F}} = \dot{\mathbf{F}}^e\mathbf{F}^p + \mathbf{F}^p\dot{\mathbf{F}}^p, \qquad \mathbf{F}^{-1} = \mathbf{F}^{p-1}\mathbf{F}^{e-1}, \tag{91.6}$$

and, therefore,

$$\begin{aligned}\mathbf{L} &= (\dot{\mathbf{F}}^e\mathbf{F}^p + \mathbf{F}^e\dot{\mathbf{F}}^p)(\mathbf{F}^{p-1}\mathbf{F}^{e-1}) \\ &= \dot{\mathbf{F}}^e\mathbf{F}^{e-1} + \mathbf{F}^e(\dot{\mathbf{F}}^p\mathbf{F}^{p-1})\mathbf{F}^{e-1}.\end{aligned}$$

552 This is confirmed by experiment. Indeed, a typical method of measuring lattice rotations is to view the lattice in the deformed body, where it appears essentially undistorted, at least locally. Were this not true a standard method of measuring lattice rotations via orientation-imaging microscopy (OIM) measurements would not be possible; cf., e.g., Sun, Adams, Shett, Saigal & King (1998).

553 Cf. (9.12).

Thus, defining **elastic** and **plastic distortion-rate tensors** $\mathbf{L}^e$ and $\mathbf{L}^p$ through the relations

$$\mathbf{L}^e = \dot{\mathbf{F}}^e\mathbf{F}^{e-1} \qquad \text{and} \qquad \mathbf{L}^p = \dot{\mathbf{F}}^p\mathbf{F}^{p-1}, \tag{91.7}$$

we have the decomposition

$$\mathbf{L} = \mathbf{L}^e + \mathbf{F}^e\mathbf{L}^p\mathbf{F}^{e-1}. \tag{91.8}$$

Guided by (11.2), we define the **elastic stretching** $\mathbf{D}^e$ and the **elastic spin** $\mathbf{W}^e$ through the relations

$$\left.\begin{aligned}\mathbf{D}^e &= \tfrac{1}{2}(\mathbf{L}^e + \mathbf{L}^{e\top}),\\ \mathbf{W}^e &= \tfrac{1}{2}(\mathbf{L}^e - \mathbf{L}^{e\top});\end{aligned}\right\} \tag{91.9}$$

similarly, we define the **plastic stretching** $\mathbf{D}^p$ and the **plastic spin** $\mathbf{W}^p$ through

$$\left.\begin{aligned}\mathbf{D}^p &= \tfrac{1}{2}(\mathbf{L}^p + \mathbf{L}^{p\top}),\\ \mathbf{W}^p &= \tfrac{1}{2}(\mathbf{L}^p - \mathbf{L}^{p\top}).\end{aligned}\right\} \tag{91.10}$$

As in the small-deformation theory, we assume that plastic flow does not induce changes in volume: Consistent with this we assume that $\mathbf{L}^p$ and (hence) $\mathbf{D}^p$ are *deviatoric*, viz. $\operatorname{tr}\mathbf{L}^p = \operatorname{tr}\mathbf{D}^p = 0$.[554] Hence, modulo a change in the choice of reference space, we may assume that

$$\det\mathbf{F}^p \equiv 1. \tag{91.11}$$

Then, since by (91.1), $J = \det\mathbf{F} = (\det\mathbf{F}^e)(\det\mathbf{F}^p)$, it follows that

$$\begin{aligned} J &= \det\mathbf{F}\\ &= \det\mathbf{F}^e, \end{aligned} \tag{91.12}$$

and, hence, that[555]

$$\dot{J} = J\operatorname{tr}\mathbf{D}^e. \tag{91.13}$$

A consequence of (P1) and (P2) on page 542, (91.7), (91.9), and (91.10) is that

(P3) $\mathbf{L}^e$ *and* $\mathbf{D}^e$ *are spatial tensor fields*;
(P4) $\mathbf{L}^p$, $\mathbf{D}^p$, *and* $\mathbf{W}^p$ *are structural tensor fields.*

91.4 Elastic and Plastic Polar Decompositions

As in §7.1, our definition of the elastic stretch and rotation tensors is based on the right and left polar decompositions:[556]

$$\mathbf{F}^e = \mathbf{R}^e\mathbf{U}^e = \mathbf{V}^e\mathbf{R}^e. \tag{91.14}$$

Here $\mathbf{R}^e$ is the **elastic rotation**, while $\mathbf{U}^e$ and $\mathbf{V}^e$ are the right and left **elastic stretch tensors**,[557] so that, as in (7.2) and (7.3),

$$\left.\begin{aligned}\mathbf{U}^e &= \sqrt{\mathbf{F}^{e\top}\mathbf{F}^e},\\ \mathbf{V}^e &= \sqrt{\mathbf{F}^e\mathbf{F}^{e\top}},\end{aligned}\right\} \tag{91.15}$$

---

[554] Cf. $(53.2)_2$ and (53.3).

[555] Cf. (9.15).

[556] The corresponding right and left polar decompositions of $\mathbf{F}^p$ are defined analogously.

[557] The stretch tensors are therefore symmetric and positive-definite.

and the right and left **elastic Cauchy–Green tensors** $\mathbf{C}^e$ and $\mathbf{B}^e$ are defined by

$$\begin{aligned}\mathbf{C}^e &= \mathbf{U}^{e2} = \mathbf{F}^{e\top}\mathbf{F}^e,\\ \mathbf{B}^e &= \mathbf{V}^{e2} = \mathbf{F}^e\mathbf{F}^{e\top}.\end{aligned}\tag{91.16}$$

A useful quantity is the elastic Green–St. Venant strain

$$\mathbf{E}^e = \tfrac{1}{2}(\mathbf{C}^e - \mathbf{1}) \tag{91.17}$$

$$= \tfrac{1}{2}(\mathbf{U}^{e2} - \mathbf{1}) \tag{91.18}$$

$$= \tfrac{1}{2}(\mathbf{F}^{e\top}\mathbf{F}^e - \mathbf{1}), \tag{91.19}$$

which we henceforth refer to as the **elastic strain**. Differentiating (91.19) with respect to time results in the following expression for the *elastic strain-rate*:

$$\dot{\mathbf{E}}^e = \tfrac{1}{2}\dot{\mathbf{C}}^e \tag{91.20}$$

$$\begin{aligned}&= \tfrac{1}{2}(\mathbf{F}^{e\top}\dot{\mathbf{F}}^e + \dot{\mathbf{F}}^{e\top}\mathbf{F}^e)\\ &= \operatorname{sym}\big(\mathbf{F}^{e\top}\dot{\mathbf{F}}^e\big).\end{aligned}\tag{91.21}$$

A consequence of (P2) on page 542, (91.14), (91.16), and (91.17) is that:

(P5) $\mathbf{U}^e$, $\mathbf{C}^e$, *and* $\mathbf{E}^e$ *are structural tensor fields*;
(P6) $\mathbf{R}^e$ *maps structural vectors to spatial vectors.*

Next, by (91.8) and (91.16),

$$\mathbf{F}^{e\top}\mathbf{L}\mathbf{F}^e = \mathbf{F}^{e\top}\mathbf{L}^e\mathbf{F}^e + \mathbf{C}^e\mathbf{L}^p. \tag{91.22}$$

Since $\mathbf{D} = \operatorname{sym}\mathbf{L}$ and $\mathbf{D}^e = \operatorname{sym}\mathbf{L}^e$, the identity (2.32) implies that

$$\operatorname{sym}(\mathbf{F}^{e\top}\mathbf{L}\mathbf{F}^e) = \mathbf{F}^{e\top}\mathbf{D}\mathbf{F}^e \qquad \text{and} \qquad \operatorname{sym}(\mathbf{F}^{e\top}\mathbf{L}^e\mathbf{F}^e) = \mathbf{F}^{e\top}\mathbf{D}^e\mathbf{F}^e;$$

therefore, taking the symmetric part of (91.22) we conclude that

$$\mathbf{F}^{e\top}\mathbf{D}\mathbf{F}^e = \mathbf{F}^{e\top}\mathbf{D}^e\mathbf{F}^e + \operatorname{sym}(\mathbf{C}^e\mathbf{L}^p). \tag{91.23}$$

Next, the computation (54.11) applied with $\mathbf{F}$, $\mathbf{C}$, $\mathbf{L}$, and $\mathbf{D}$ replaced by $\mathbf{F}^e$, $\mathbf{C}^e$, $\mathbf{L}^e$, and $\mathbf{D}^e$ yields the important identity

$$2\mathbf{F}^{e\top}\mathbf{D}^e\mathbf{F}^e = \dot{\mathbf{C}}^e, \tag{91.24}$$

or, equivalently, by (91.20),

$$\mathbf{F}^{e\top}\mathbf{D}^e\mathbf{F}^e = \dot{\mathbf{E}}^e. \tag{91.25}$$

EXERCISES

1. Using (P1) and (P2), establish (P3)–(P6).
2. Establish (91.24) and (91.26).
3. Show that

$$\mathbf{T}\!:\!\mathbf{D} = \mathbf{T}\!:\!\mathbf{D}^e + \mathbf{T}\!:\!(\mathbf{F}^e\mathbf{L}^p\mathbf{F}^{e-1}).$$

4. Writing $\mathbf{F}^p = \mathbf{R}^p\mathbf{U}^p$ for the right polar decomposition of $\mathbf{F}^p$, show that

$$\begin{aligned}\mathbf{W}^e &= \dot{\mathbf{R}}^e\mathbf{R}^{e\top} + \mathbf{R}^e\big[\operatorname{skw}(\dot{\mathbf{U}}^e\mathbf{U}^{e-1})\big]\mathbf{R}^{e\top},\\ \mathbf{W}^p &= \dot{\mathbf{R}}^p\mathbf{R}^{p\top} + \mathbf{R}^p\big[\operatorname{skw}(\dot{\mathbf{U}}^p\mathbf{U}^{p-1})\big]\mathbf{R}^{p\top}.\end{aligned}\tag{91.26}$$

Thus temporal changes in the plastic stretch $\mathbf{U}^p$ can induce plastic spin. (Hint: Consider the argument leading to (11.8).)

5. Show that if the principal directions of $\mathbf{U}^p$ are independent of time when expressed materially, then

$$\mathbf{W}^p = \dot{\mathbf{R}}^p\mathbf{R}^{p\top}.$$

(A similar result is associated with the right elastic stretch tensor.)

91.5 Change in Frame Revisited in View of the Kröner Decomposition

As described in §20.1, a change of frame is, at each time, a rotation and translation of the *observed space* (the space through which the body moves); it does not affect the reference space,[558] nor does it affect the structural space; thus,

(‡) *material vectors and structural vectors are invariant under changes in frame,*

an assertion that should be at least intuitively clear from Figure 91.1.

As noted in the bullet on page 147, because observers view only the deformed body, tensor fields that map material vectors to material vectors are invariant under changes in frame.[559] In view of (‡), the exact same argument yields the following result:

(†) *tensor fields*

(a) *that map material vectors to material vectors, or*
(b) *that map material vectors to structural vectors, or*
(c) *that map structural vectors to material vectors, or*
(d) *that map structural vectors to structural vectors,*
are invariant under changes in frame.

Moreover, by (‡) on page 543, we see that, arguing as in the bullet on page 147,

(‡) *structural tensor fields are invariant under changes in frame.*

Next, recall the transformation law $(20.16)_1$:

$$\mathbf{F}^* = \mathbf{QF}. \tag{91.27}$$

By (91.1) and (91.27),

$$(\mathbf{F}^e\mathbf{F}^p)^* = \mathbf{Q}(\mathbf{F}^e\mathbf{F}^p).$$

On the other hand, by (P2) and (b) of (†), $\mathbf{F}^{p*} = \mathbf{F}^p$, so that

$$\begin{aligned}(\mathbf{F}^e\mathbf{F}^p)^* &= \mathbf{F}^{e*}\mathbf{F}^{p*}\\ &= \mathbf{F}^{e*}\mathbf{F}^p;\end{aligned}$$

hence,

$$\mathbf{QF}^e\mathbf{F}^p = \mathbf{F}^{e*}\mathbf{F}^p.$$

Thus,

$$\mathbf{F}^{e*} = \mathbf{QF}^e \qquad \text{and} \qquad \mathbf{F}^p \text{ is invariant.} \tag{91.28}$$

Similarly, appealing to (P4) and (‡),

$$\mathbf{L}^p,\, \mathbf{D}^p, \text{ and } \mathbf{W}^p \text{ are invariant.} \tag{91.29}$$

[558] Cf. the bullet on page 147
[559] Cf. the argument preceding that bullet.

Next, by (91.14) and (91.28),

$$\begin{aligned}\mathbf{F}^{e*} &= \mathbf{R}^{e*}\mathbf{U}^{e*} = \mathbf{Q}\mathbf{F}^{e} = \underline{\mathbf{Q}\mathbf{R}^{e}}\mathbf{U}^{e},\\ \mathbf{F}^{e*} &= \mathbf{V}^{e*}\mathbf{R}^{e*} = \mathbf{Q}\mathbf{F}^{e} = \underline{\mathbf{Q}\mathbf{V}^{e}\mathbf{Q}^{\top}}\,\mathbf{Q}\mathbf{R}^{e},\end{aligned} \tag{91.30}$$

and we may conclude from the uniqueness of the polar decomposition that $\mathbf{R}^{e*} = \mathbf{Q}\mathbf{R}^{e}$ and

$$\mathbf{U}^{e} \text{ and (hence) } \mathbf{E}^{e} \text{ are invariant} \quad \text{and} \quad \mathbf{V}^{e*} = \mathbf{Q}\mathbf{V}^{e}\mathbf{Q}^{\top}, \tag{91.31}$$

so that, by (91.16),

$$\mathbf{C}^{e} \text{ is invariant} \quad \text{and} \quad \mathbf{B}^{e*} = \mathbf{Q}\mathbf{B}^{e}\mathbf{Q}^{\top}. \tag{91.32}$$

Turning to the tensor field $\mathbf{L}^{e} = \dot{\mathbf{F}}^{e}\mathbf{F}^{e-1}$, we see that (91.30) yields

$$\begin{aligned}\mathbf{L}^{e*} &= \overline{\mathbf{F}^{e*}}^{\boldsymbol{\cdot}}(\mathbf{F}^{e*})^{-1}\\ &= (\mathbf{Q}\dot{\mathbf{F}}^{e} + \dot{\mathbf{Q}}\mathbf{F}^{e})\mathbf{F}^{e-1}\mathbf{Q}^{\top}\\ &= \mathbf{Q}\dot{\mathbf{F}}^{e}\mathbf{F}^{e-1}\mathbf{Q}^{\top} + \dot{\mathbf{Q}}\mathbf{Q}^{\top}\\ &= \mathbf{Q}\mathbf{L}^{e}\mathbf{Q}^{\top} + \boldsymbol{\Omega},\end{aligned} \tag{91.33}$$

with $\boldsymbol{\Omega}$ the frame-spin (20.2).

EXERCISES

1. Establish the invariance of $\mathbf{U}^{e}$ and $\mathbf{C}^{e}$ using their mapping properties.
2. Show that $\mathbf{D}^{e}$ is frame-indifferent.

92 Virtual-Power Formulation of the Standard and Microscopic Force Balances

We here develop a large-deformation counterpart[560] of the virtual-power formulation of §84.

92.1 Internal and External Expenditures of Power

We assume that at some arbitrarily chosen but fixed time the fields χ, $\mathbf{F}^e$, and (hence) $\mathbf{F}^p$ are known and

(‡) *we denote by $\mathcal{P}_t$ an arbitrary subregion of the deformed body, at that time, and by* $\mathbf{n}$ *the outward unit normal on $\partial\mathcal{P}_t$.*

The basic "rate-like" descriptors for a body undergoing large deformations are the velocity $\mathbf{v}$ and the elastic and plastic distortion-rate tensors $\mathbf{L}^e$ and $\mathbf{L}^p$ as constrained by (91.8), or, equivalently, since $\mathbf{L} = \operatorname{grad}\mathbf{v}$,

$$\operatorname{grad}\mathbf{v} = \mathbf{L}^e + \mathbf{F}^e\mathbf{L}^p\mathbf{F}^{e-1}. \tag{92.1}$$

The formulation of the principle of virtual power is based on a balance between the external power $\mathcal{W}(\mathcal{P}_t)$ expended *on* $\mathcal{P}_t$ and the internal power $\mathcal{I}(\mathcal{P}_t)$ expended *within* $\mathcal{P}_t$.[561] Consider first the internal power. As in §84.1 we replace the classical stress power $\mathbf{T}\colon\operatorname{grad}\mathbf{v}$ by a more detailed reckoning that *individually* characterizes:

(i) the stretching and spinning of the underlying microscopic structure as described by the elastic distortion-rate $\mathbf{L}^e$, and
(ii) the flow of defects through that structure as described by the plastic distortion-rate $\mathbf{L}^p$.

We therefore allow for power expended internally by:

- an *elastic stress* $\mathbf{S}^e$ power conjugate to $\mathbf{L}^e$; and
- a *plastic stress* $\mathbf{T}^p$ power conjugate to $\mathbf{L}^p$,

so that $\mathbf{S}^e\colon\mathbf{L}^e$ and $\mathbf{T}^p\colon\mathbf{L}^p$ are the relevant stress powers; since $\mathbf{L}^p$ is deviatoric, we assume that

$$\mathbf{T}^p \text{ is deviatoric.} \tag{92.2}$$

Because $\mathbf{L}^e$ is a spatial tensor field, we view $\mathbf{S}^e$ as a spatial tensor field, and because $\mathbf{L}^p$ is a *structural* tensor field, we view $\mathbf{T}^p$ as a *structural* tensor field.[562] Therefore,

560 Cf. Gurtin & Anand (2005c).
561 Cf. §84.1.
562 Cf. (P3) and (P4) on page 544.

for consistency, we view the elastic stress power as measured per unit volume in the deformed body, the plastic stress power as measured per unit volume in the structural space. Thus, since

$$\mathbf{S}^e : \mathbf{L}^e \qquad \text{and} \qquad J^{-1}\mathbf{T}^p : \mathbf{L}^p \tag{92.3}$$

then represent the elastic and plastic stress powers, measured per unit volume in the deformed body, we assume that the internal power has the form

$$\mathcal{I}(\mathcal{P}_t) = \int_{\mathcal{P}_t} (\mathbf{S}^e : \mathbf{L}^e + J^{-1}\mathbf{T}^p : \mathbf{L}^p)\, dv. \tag{92.4}$$

Turning to the external power $\mathcal{W}(\mathcal{P}_t)$, we note that, by (22.11), the conventional form of this power is given by

$$\int_{\partial\mathcal{P}_t} \mathbf{t}(\mathbf{n}) \cdot \mathbf{v}\, da + \int_{\mathcal{P}_t} \mathbf{b} \cdot \mathbf{v}\, dv, \tag{92.5}$$

with body force $\mathbf{b}$ presumed to account for inertia; that is, granted the underlying frame is inertial,

$$\mathbf{b} = \mathbf{b}_0 - \rho\dot{\mathbf{v}}, \tag{92.6}$$

with $\mathbf{b}_0$ the conventional body force.[563]

Guided by the discussion in §84.1, we allow also for an arbitrary **external microscopic force** B^p power-conjugate to $\mathbf{L}^p$ with

$$B^p \text{ deviatoric} \tag{92.7}$$

and, hence, add the term

$$\int_{\mathcal{P}_t} J^{-1} B^p : \mathbf{L}^p\, dv$$

to the conventional external power (92.5).[564] We therefore assume that the external power has the form

$$\mathcal{W}(\mathcal{P}_t) = \int_{\partial\mathcal{P}_t} \mathbf{t}(\mathbf{n}) \cdot \mathbf{v}\, da + \int_{\mathcal{P}_t} \mathbf{b} \cdot \mathbf{v}\, dv + \int_{\mathcal{P}_t} J^{-1} B^p : \mathbf{L}^p\, dv. \tag{92.8}$$

92.2 Principle of Virtual Power

Consider the velocity $\mathbf{v}$ and the elastic and plastic distortion-rates $\mathbf{L}^e$ and $\mathbf{L}^p$ as virtual velocities that may be specified independently in a manner consistent with (92.1); that is, denoting the virtual fields by $\tilde{\mathbf{v}}$, $\tilde{\mathbf{L}}^e$, and $\tilde{\mathbf{L}}^p$ to differentiate them from fields associated with the actual evolution of the body, we require that

$$\operatorname{grad}\tilde{\mathbf{v}} = \tilde{\mathbf{L}}^e + \mathbf{F}^e\tilde{\mathbf{L}}^p\mathbf{F}^{e-1} \tag{92.9}$$

and refer to the list

$$\mathcal{V} = (\tilde{\mathbf{v}}, \tilde{\mathbf{L}}^e, \tilde{\mathbf{L}}^p)$$

563 Cf. (19.15).
564 Cf. (92.3).

as a (generalized) *virtual velocity*. Further, writing

$$\mathcal{I}(\mathcal{P}_t, \mathcal{V}) = \int\limits_{\mathcal{P}_t} (\mathbf{S}^e : \tilde{\mathbf{L}}^e + J^{-1}\mathbf{T}^p : \tilde{\mathbf{L}}^p)\, dv,$$

and

$$\mathcal{W}(\mathcal{P}_t, \mathcal{V}) = \int\limits_{\partial\mathcal{P}_t} \mathbf{t}(\mathbf{n}) \cdot \tilde{\mathbf{v}}\, da + \int\limits_{\mathcal{P}_t} (\mathbf{b} \cdot \tilde{\mathbf{v}} + J^{-1}\mathbb{B}^p : \tilde{\mathbf{L}}^p)\, dv$$

for the corresponding internal and external expenditures of *virtual power*, the **principle of virtual power** is the requirement that, given any subregion $\mathcal{P}_t$ of the deformed body,

$$\underbrace{\int\limits_{\partial\mathcal{P}_t} \mathbf{t}(\mathbf{n}) \cdot \tilde{\mathbf{v}}\, da + \int\limits_{\mathcal{P}_t} (\mathbf{b} \cdot \tilde{\mathbf{v}} + J^{-1}\mathbb{B}^p : \tilde{\mathbf{L}}^p)\, dv}_{\mathcal{W}(\mathcal{P}_t,\mathcal{V})} = \underbrace{\int\limits_{\mathcal{P}_t} (\mathbf{S}^e : \tilde{\mathbf{L}}^e + J^{-1}\mathbf{T}^p : \tilde{\mathbf{L}}^p)\, dv}_{\mathcal{I}(\mathcal{P}_t,\mathcal{V})} \tag{92.10}$$

for all virtual velocities $\mathcal{V}$.

92.2.1 Consequences of Frame-Indifference

We assume that the internal power $\mathcal{I}(\mathcal{P}_t, \mathcal{V})$ is invariant under a change in frame and that the virtual fields transform in a manner identical to their nonvirtual counterparts. Given a change in frame, if $\mathcal{P}_t^*$ and $\mathcal{I}^*(\mathcal{P}_t^*, \mathcal{V}^*)$ represent the region and the internal power in the new frame, invariance of the internal power then requires that

$$\mathcal{I}(\mathcal{P}_t, \mathcal{V}) = \mathcal{I}^*(\mathcal{P}_t^*, \mathcal{V}^*), \tag{92.11}$$

where $\mathcal{V}^*$ is the generalized virtual velocity in the new frame. Further, by (91.33),

$$\tilde{\mathbf{L}}^{e*} = \mathbf{Q}\tilde{\mathbf{L}}^e\mathbf{Q}^{\top} + \mathbf{\Omega}, \tag{92.12}$$

where $\mathbf{L}^{e*}$ is the elastic distortion rate in the new frame, and where $\mathbf{Q}$ is the frame-rotation and $\mathbf{\Omega}$ the frame-spin. Further, since $\mathbf{T}^p$ is a structural tensor field and since by (P4) $\tilde{\mathbf{L}}^p$ is also, (‡) on page 546 implies that

$$\mathbf{T}^p \text{ and } \tilde{\mathbf{L}}^p \text{ are invariant} \tag{92.13}$$

and, hence, that the plastic stress power $J^{-1}\mathbf{T}^p : \tilde{\mathbf{L}}^p$ is invariant under a change in frame. Thus, by (92.11) and the relation represented by the right side of (92.10)

$$\int\limits_{\mathcal{P}_t} \mathbf{S}^e : \tilde{\mathbf{L}}^e\, dv = \int\limits_{\mathcal{P}_t^*} \mathbf{S}^{e*} : \tilde{\mathbf{L}}^{e*}\, dv, \tag{92.14}$$

where $\mathbf{S}^{e*}$ is the stress $\mathbf{S}^e$ in the new frame. The relation (92.14) is identical in form to (22.18) in our discussion of the virtual-power principle within a classical framework: $\mathbf{S}^e$ and $\mathbf{L}^e$ in (92.14) play the roles of the Cauchy stress $\mathbf{T}$ and the velocity gradient $\mathbf{L}$ in (22.18). Further, the transformation law (92.12) for $\mathbf{L}^e$ is identical to the law (22.17) for $\mathbf{L}$. Thus the argument starting from (22.18) in §22.2 and resulting in the frame-indifference and symmetry of the Cauchy stress $\mathbf{T}$ as expressed in (22.21) and (22.22) applies without change within the present framework; here, it leads to the conclusion that the elastic stress $\mathbf{S}^e$ is *frame-indifferent*,

$$\mathbf{S}^{e*} = \mathbf{Q}\mathbf{S}^e\mathbf{Q}^{\top}, \tag{92.15}$$

and *symmetric*,

$$\mathbf{S}^e = \mathbf{S}^{e\top}. \tag{92.16}$$

92.2.2 Macroscopic Force Balance

Consider a **macroscopic** virtual velocity $\mathcal{V}$ for which $\tilde{\mathbf{v}}$ is arbitrary and

$$\tilde{\mathbf{L}}^e = \operatorname{grad}\tilde{\mathbf{v}}. \tag{92.17}$$

In this case, (92.9) implies that

$$\tilde{\mathbf{L}}^p = \mathbf{0} \tag{92.18}$$

and (92.10) reduces to

$$\int_{\partial\mathcal{P}_t} \mathbf{t}(\mathbf{n})\cdot\tilde{\mathbf{v}}\,da + \int_{\mathcal{P}_t} \mathbf{b}\cdot\tilde{\mathbf{v}}\,dv = \int_{\mathcal{P}_t} \mathbf{S}^e : \operatorname{grad}\tilde{\mathbf{v}}\,dv. \tag{92.19}$$

Further, by the divergence theorem

$$\int_{\mathcal{P}_t} \mathbf{S}^e : \operatorname{grad}\tilde{\mathbf{v}}\,dv = -\int_{\mathcal{P}_t} \operatorname{div}\mathbf{S}^e\cdot\tilde{\mathbf{v}}\,dv + \int_{\partial\mathcal{P}_t} (\mathbf{S}^e\mathbf{n})\cdot\tilde{\mathbf{v}}\,da$$

and (92.19) becomes

$$\int_{\partial\mathcal{P}_t} (\mathbf{t}(\mathbf{n}) - \mathbf{S}^e\mathbf{n})\cdot\tilde{\mathbf{v}}\,da + \int_{\mathcal{P}_t} (\operatorname{div}\mathbf{S}^e + \mathbf{b})\cdot\tilde{\mathbf{v}}\,dv = 0. \tag{92.20}$$

Since (92.20) must hold for all $\mathcal{P}_t$ and all $\tilde{\mathbf{v}}$, an argument identical to that given in the paragraph containing (22.15) yields the *traction condition*

$$\mathbf{t}(\mathbf{n}) = \mathbf{S}^e\mathbf{n}, \tag{92.21}$$

and the *local force balance*

$$\operatorname{div}\mathbf{S}^e + \mathbf{b} = \mathbf{0}. \tag{92.22}$$

This traction condition and force balance and the symmetry and frame-indifference of $\mathbf{S}^e$ are classical conditions satisfied by the *Cauchy stress* $\mathbf{T}$, an observation that allows us to write

$$\mathbf{T} \stackrel{\text{def}}{=} \mathbf{S}^e \tag{92.23}$$

and to view

$$\mathbf{T} = \mathbf{T}^{\top} \tag{92.24}$$

as the *macroscopic stress* and (92.22) as the local **macroscopic force balance.** Granted that we are working in an inertial frame, so that (92.6) is satisfied, (92.22) reduces to the local balance law for linear momentum:

$$\rho\dot{\mathbf{v}} = \operatorname{div}\mathbf{T} + \mathbf{b}_0, \tag{92.25}$$

with $\mathbf{b}_0$ the conventional body force.

92.2.3 Microscopic Force Balance

Assume that $\mathcal{V}$ is **microscopic** in the sense that

$$\tilde{\mathbf{L}}^e = -\mathbf{F}^e\tilde{\mathbf{L}}^p\mathbf{F}^{e-1} \tag{92.26}$$

and

$$\tilde{\mathbf{v}} \equiv \mathbf{0}, \tag{92.27}$$

consistent with (92.9). Then, by (92.10) and (92.23),

$$\int_{\mathcal{P}_t} (\mathbf{T}:\tilde{\mathbf{L}}^e + J^{-1}\mathbf{T}^p:\tilde{\mathbf{L}}^p - J^{-1}B^p:\tilde{\mathbf{L}}^p)\,dv = 0$$

for all $\mathcal{P}_t$, so that, by (92.26), the requirement that $\tilde{\mathbf{L}}^p$ is deviatoric and (75.6), and the identity (75.7),

$$\begin{aligned} J^{-1}(\mathbf{T}^p - B^p):\tilde{\mathbf{L}}^p &= -\mathbf{T}:\tilde{\mathbf{L}}^e \\ &= \mathbf{T}:(\mathbf{F}^e\tilde{\mathbf{L}}^p\mathbf{F}^{e-1}) \\ &= (\mathbf{F}^{e\top}\mathbf{T}\mathbf{F}^{e-\top}):\tilde{\mathbf{L}}^p \\ &= (\mathbf{F}^{e\top}\mathbf{T}\mathbf{F}^{e-\top})_0:\tilde{\mathbf{L}}^p \\ &= (\mathbf{F}^{e\top}\mathbf{T}_0\mathbf{F}^{e-\top}):\tilde{\mathbf{L}}^p. \end{aligned}$$

Thus, since $\tilde{\mathbf{L}}^p$ is arbitrary, we have the **microscopic force balance**

$$\boxed{J\mathbf{F}^{e\top}\mathbf{T}_0\mathbf{F}^{e-\top} = \mathbf{T}^p - B^p.} \tag{92.28}$$

93 Free-Energy Imbalance

93.1 Free-Energy Imbalance Expressed in Terms of the Cauchy Stress

The general free-energy imbalance introduced in §29.1 here leads to the requirement that

(‡) *for any spatial region $\mathcal{P}_t$ convecting with the body, the temporal increase in free-energy of $\mathcal{P}_t$ be less than or equal to the power expended on $\mathcal{P}_t$ minus the dissipation within $\mathcal{P}_t$.*

Let φ denote the *free-energy* and $\delta \geq 0$ the *dissipation*, with φ and δ measured per unit volume in the structural space, so that, by (15.6) and since $J = \det \mathbf{F} = \det \mathbf{F}^e$,

$$\int_{\mathcal{P}_t} \varphi J^{-1}\, dv \qquad \text{and} \qquad \int_{\mathcal{P}_t} \delta J^{-1}\, dv,$$

respectively, represent the free energy of — and the dissipation within — $\mathcal{P}_t$. The free-energy imbalance (‡), stated precisely, is then the assertion that

$$\overline{\int_{\mathcal{P}_t} \varphi J^{-1}\, dv}^{\,\boldsymbol{\cdot}} - \mathcal{W}(\mathcal{P}_t) = -\int_{\mathcal{P}_t} \delta J^{-1}\, dv \leq 0. \tag{93.1}$$

Since $\mathcal{W}(\mathcal{P}_t) = \mathcal{I}(\mathcal{P}_t)$, (92.4) and (92.23) imply that

$$\overline{\int_{\mathcal{P}_t} \varphi J^{-1}\, dv}^{\,\boldsymbol{\cdot}} - \int_{\mathcal{P}_t} (\mathbf{T}\colon\mathbf{D}^e + J^{-1}\mathbf{T}^p\colon\mathbf{L}^p)\, dv \leq 0. \tag{93.2}$$

Further, $\mathcal{P}_t$ convects with the body and hence, by (5.10), there is a (fixed) *material region* P such that $\mathcal{P}_t = \boldsymbol{\chi}_t(\mathrm{P})$ for all t; therefore, by (15.6),

$$\begin{aligned}\overline{\int_{\mathcal{P}_t} \varphi J^{-1}\, dv}^{\,\boldsymbol{\cdot}} &= \overline{\int_{\mathrm{P}} \varphi\, dv_{\mathrm{R}}}^{\,\boldsymbol{\cdot}} \\ &= \int_{\mathrm{P}} \dot{\varphi}\, dv_{\mathrm{R}} \\ &= \int_{\mathcal{P}_t} \dot{\varphi} J^{-1}\, dv, \end{aligned} \tag{93.3}$$

and it follows that

$$\int_{\mathcal{P}_t} (J^{-1}\dot{\varphi} - \mathbf{T}\colon\mathbf{D}^e - J^{-1}\mathbf{T}^p\colon\mathbf{L}^p)\,dv = -\int_{\mathcal{P}_t} \delta J^{-1}\,dv \le 0.$$

Thus, since $\mathcal{P}_t$ was arbitrarily chosen,

$$J^{-1}\dot{\varphi} - \mathbf{T}\colon\mathbf{D}^e - J^{-1}\mathbf{T}^p\colon\mathbf{L}^p = -J^{-1}\delta \le 0,$$

and we have the local free-energy imbalance

$$\dot{\varphi} - J\mathbf{T}\colon\mathbf{D}^e - \mathbf{T}^p\colon\mathbf{L}^p = -\delta \le 0. \tag{93.4}$$

94 Two New Stresses

In this section we introduce two new stresses derived from the Cauchy stress $\mathbf{T}$; these stresses allow us to express the free-energy imbalance and the microscopic force balance in forms more amenable to applications.

94.1 The Second Piola Elastic-Stress $\mathbf{T}^e$

The term

$$\mathbf{T}:\mathbf{D}^e = \mathbf{T}:\mathbf{L}^e,$$

which represents the elastic stress-power, is most conveniently expressed in terms of the elastic-strain-rate $\dot{\mathbf{E}}^e$.[565] To accomplish this we note that, since $\mathbf{T}$ is symmetric, so also is $\mathbf{F}^{e-1}\mathbf{T}\mathbf{F}^{e-\top}$ and, therefore, by $(91.7)_1$ and (91.21),

$$\begin{aligned}\mathbf{T}:\mathbf{L}^e &= \mathbf{T}:(\dot{\mathbf{F}}^e\mathbf{F}^{e-1})\\ &= (\mathbf{T}\mathbf{F}^{e-\top}):\dot{\mathbf{F}}^e\\ &= (\mathbf{F}^{e-1}\mathbf{T}\mathbf{F}^{e-\top}):(\mathbf{F}^{e\top}\dot{\mathbf{F}}^e)\\ &= J^{-1}\,\mathbf{T}^e:\dot{\mathbf{E}}^e \end{aligned} \tag{94.1}$$

with

$$\boxed{\mathbf{T}^e \stackrel{\text{def}}{=} J\,\mathbf{F}^{e-1}\mathbf{T}\mathbf{F}^{e-\top}.} \tag{94.2}$$

The stress $\mathbf{T}^e$ is a counterpart of the standard second Piola stress $\mathbf{T}_{RR}$ discussed in §25.1 in the sense that $\mathbf{T}^e$ is computed using $\mathbf{F}^e$ in place of $\mathbf{F}$. We refer to $\mathbf{T}^e$ as the **second Piola elastic-stress**. A consequence of the symmetry of $\mathbf{T}$ is that

$$\mathbf{T}^e \text{ is symmetric.} \tag{94.3}$$

Note that (94.2) may be inverted to give an expression for $\mathbf{T}$ as a function of $\mathbf{T}^e$:

$$\mathbf{T} = J^{-1}\mathbf{F}^e\mathbf{T}^e\mathbf{F}^{e\top}. \tag{94.4}$$

Finally, the definition (94.2) allows us to rewrite (93.4) in the form

$$\boxed{\dot{\varphi} - \mathbf{T}^e:\dot{\mathbf{E}}^e - \mathbf{T}^p:\mathbf{L}^p = -\delta \le 0.} \tag{94.5}$$

[565] Cf. (91.20).

This local **free-energy imbalance** is central to the development of a suitable constitutive theory.

As is clear from (94.2), the input space for $\mathbf{T}^e$ is the same as that for $\mathbf{F}^{e-\top}$, which, by (P2) on page 542, is the structural space. Similarly, the output space for $\mathbf{T}^e$ is the same as that for $\mathbf{F}^{e-1}$, which is again the structural space. $\mathbf{T}^e$ therefore maps structural vectors to structural vectors. Thus, by (92.2),

(P7) *$\mathbf{T}^p$ and $\mathbf{T}^e$ are structural tensor fields.*

Therefore, by (‡) on page 546,

$$\mathbf{T}^e \text{ and } \mathbf{T}^p \text{ are invariant under changes in frame.} \tag{94.6}$$

94.2 The Mandel Stress $\mathbf{M}^e$

Elasticity and plastic flow interact through the microscopic force balance[566]

$$\underline{J\mathbf{F}^{e\top}\mathbf{T}_0\mathbf{F}^{e-\top}} = \mathbf{T}^p - B^p. \tag{94.7}$$

Focusing on the underlined term, which represents the elastic contribution to this balance, we note that, by (75.7), $(91.16)_1$, and (94.2),[567]

$$\begin{aligned} J\mathbf{F}^{e\top}\mathbf{T}_0\mathbf{F}^{e-\top} &= J\left(\mathbf{F}^{e\top}\mathbf{T}\mathbf{F}^{e-\top}\right)_0 \\ &= J\,\mathrm{dev}\left[(\mathbf{F}^{e\top}\mathbf{F}^e)\mathbf{F}^{e-1}\mathbf{T}\mathbf{F}^{e-\top}\right] \\ &= (\mathbf{C}^e\mathbf{T}^e)_0. \end{aligned}$$

Important to a discussion of plasticity within the framework of large deformations is the **Mandel stress**[568]

$$\boxed{\mathbf{M}^e \stackrel{\text{def}}{=} \mathbf{C}^e\mathbf{T}^e;} \tag{94.8}$$

using this stress we can rewrite the microscopic force balance (94.7) in the simple form

$$\mathbf{M}^e_0 = \mathbf{T}^p - B^p. \tag{94.9}$$

Next, by (91.16), (94.4), and (94.8), since $\mathbf{T}$ is symmetric,

$$\begin{aligned} \mathbf{T} &= J^{-1}\mathbf{F}^e\mathbf{C}^{e-1}\mathbf{M}^e\mathbf{F}^{e\top} \\ &= J^{-1}\mathbf{F}^e\mathbf{F}^{e-1}\mathbf{F}^{e-\top}\mathbf{M}^e\mathbf{F}^{e\top} \\ &= J^{-1}\mathbf{F}^e\mathbf{M}^{e\top}\mathbf{F}^{e-1}; \end{aligned} \tag{94.10}$$

the Cauchy and Mandel stresses are therefore related by

$$\mathbf{T} = J^{-1}\mathbf{F}^e\mathbf{M}^{e\top}\mathbf{F}^{e-1}, \tag{94.11}$$

or, equivalently,

$$\boxed{\mathbf{M}^e = J\mathbf{F}^{e\top}\mathbf{T}\mathbf{F}^{e-\top}.} \tag{94.12}$$

566 Cf. (92.28).
567 Recall that $\mathrm{dev}\mathbf{A} = \mathbf{A}_0$ represents the deviatoric part of $\mathbf{A}$; cf. (2.46).
568 Mandel (1973).

95 Constitutive Theory

In this section, we develop a constitutive theory appropriate to rate-independent Mises plasticity.

95.1 General Separable Constitutive Theory

We neglect defect energy and restrict attention to constitutive relations that separate elastic and plastic constitutive response; therefore, guided by (94.5), we consider constitutive relations of the form

$$\left.\begin{aligned} \varphi &= \hat{\varphi}(\mathbf{E}^e), \\ \mathbf{T}^e &= \hat{\mathbf{T}}^e(\mathbf{E}^e), \\ \mathbf{T}^p &= \hat{\mathbf{T}}^p(\mathbf{L}^p, e^p), \end{aligned}\right\} \tag{95.1}$$

with e^p, the accumulated plastic strain, consistent with the **hardening equation**[569]

$$\dot{e}^p = |\mathbf{D}^p|, \tag{95.2}$$

and, hence, assumed to range over the interval $0 \le e^p < \infty$. Note that, by (91.29) and (91.31), $\mathbf{E}^e$ and $\mathbf{L}^p$ are invariant under changes in frame, and, by (94.6) so also are $\mathbf{T}^e$ and $\mathbf{T}^p$. Thus, since e^p is a scalar field and, hence, frame-indifferent,

- *the constitutive equations* (95.1) *and* (95.2) *are frame-indifferent.*

The application of the Coleman–Noll procedure in a framework that allows for both elastic and plastic response is more complicated than its application in §48.2, where the material is elastic. By a **constitutive process** we mean a pair $(\chi, \mathbf{F}^p)$ of *fields* χ and $\mathbf{F}^p$ — with χ a motion and $\mathbf{F}^p$ a plastic-distortion tensor (so that $\det \mathbf{F}^p = 1$) — together fields $\mathbf{E}^e$, $\mathbf{L}^p$, e^p, φ, $\mathbf{T}^e$, and $\mathbf{T}^p$, where

(i) $\mathbf{L}^p$ and $\mathbf{E}^e$ are defined by

$$\mathbf{L}^p = \dot{\mathbf{F}}^p\mathbf{F}^{p-1}, \qquad \mathbf{E}^e = \tfrac{1}{2}(\mathbf{F}^{e\top}\mathbf{F}^e - \mathbf{1}), \qquad \mathbf{F}^e = \mathbf{F}\mathbf{F}^{p-1}, \qquad \mathbf{F} = \nabla\chi;$$

(ii) e^p is *any* solution of (95.2);
(iii) φ, $\mathbf{T}^e$, and $\mathbf{T}^p$ are determined by the fields $\mathbf{E}^e$, $\mathbf{L}^p$, and e^p through the constitutive equations (95.1).

Remark. It is important to note that — because (ii) does *not* prescribe initial conditions for the hardening equation (95.2) — the fields χ and $\mathbf{F}^p$ do not uniquely

[569] Cf. (76.8).

determine a constitutive process. In this regard, we view the prescription of initial conditions for e^p as part of the specification of an initial state for the body and, hence, *not part of the basic constitutive theory*.[570]

Given a constitutive process, (94.4) determines the stress $\mathbf{T}$, and the momentum balance (92.25) and the microscopic force balance (92.28) provide explicit relations

$$\begin{aligned} \mathbf{b}_0 &= \rho\dot{\mathbf{v}} - \mathrm{div}\mathbf{T}, \\ B^p &= \mathbf{T}^p - J\mathbf{F}^{e\top}\mathbf{T}_0\mathbf{F}^{e-\top} \end{aligned} \tag{95.3}$$

for the body force $\mathbf{b}$ and the microscopic body force B^p needed to **support** the process. We assume that these body forces are arbitrarily assignable; the force balances thus in no way restrict the class of constitutive processes the material may undergo. But unless the constitutive equations are suitably restricted, not all constitutive processes will be compatible with the free-energy imbalance (94.5).

To determine such restrictions we first note that, given an arbitrary constitutive process, (95.1) implies that

$$\dot{\varphi} = \frac{\partial\hat{\varphi}(\mathbf{E}^e)}{\partial\mathbf{E}^e} : \dot{\mathbf{E}}^e, \tag{95.4}$$

and, hence, that the free-energy imbalance (94.5) reduces to the inequality

$$\left(\hat{\mathbf{T}}^e(\mathbf{E}^e) - \frac{\partial\hat{\varphi}(\mathbf{E}^e)}{\partial\mathbf{E}^e}\right) : \dot{\mathbf{E}}^e + \hat{\mathbf{T}}^p(\mathbf{L}^p, e^p) : \mathbf{L}^p \geq 0. \tag{95.5}$$

We now determine constitutive restrictions that ensue from the requirement that (95.5) hold in all constitutive processes. A central step in accomplishing this is to prove that:

(I) it is possible to find a constitutive process such that

$$\mathbf{L}^p \equiv \mathbf{0} \tag{95.6}$$

and such that $\mathbf{E}^e$ and $\dot{\mathbf{E}}^e$ have arbitrarily prescribed values at some point and time;

(II) it is possible to find a constitutive process such that

$$\mathbf{F}^e \equiv \mathbf{1} \quad (\text{so that } \dot{\mathbf{E}}^e \equiv \mathbf{0}) \tag{95.7}$$

and such that $\mathbf{L}^p$ and e^p have arbitrarily prescribed values — with $\mathbf{L}^p$ deviatoric and $e^p \geq 0$ — at some point and time.

The verification of (I) and (II) is given at the end of the section.

Assume that (I) and (II) are satisfied with $(\mathbf{X}_0, t_0)$ the point at which the relevant fields have arbitrarily prescribed values. Then by (95.6) the inequality (95.5) (at $(\mathbf{X}_0, t_0)$) reduces to

$$\underbrace{\left(\frac{\partial\hat{\varphi}(\mathbf{E}^e)}{\partial\mathbf{E}^e} - \hat{\mathbf{T}}^e(\mathbf{E}^e)\right)}_{\mathbf{\Phi}(\mathbf{E}^e)} : \dot{\mathbf{E}}^e \leq 0, \tag{95.8}$$

and the coefficient $\mathbf{\Phi}(\mathbf{E}^e)$ of $\dot{\mathbf{E}}^e$ in (95.8) must vanish (at $(\mathbf{X}_0, t_0)$), for otherwise $\dot{\mathbf{E}}^e$ may be chosen to violate (95.8). Thus since $\mathbf{E}^e$ at $(\mathbf{X}_0, t_0)$ is arbitrary, we must have

$$\hat{\mathbf{T}}^e(\mathbf{E}^e) = \frac{\partial\hat{\varphi}(\mathbf{E}^e)}{\partial\mathbf{E}^e}. \tag{95.9}$$

[570] Cf. our discussion of fluids in §42 — in particular, the sentence following (C3) on page 245.

Further, an immediate consequence of (95.9) is that, given any constitutive process, the inequality (95.5) reduces to

$$\hat{\mathbf{T}}^p(\mathbf{L}^p, e^p):\mathbf{L}^p \geq 0 \tag{95.10}$$

and (II) ensures that this inequality need hold for all deviatoric $\mathbf{L}^p$ and all $e^p \geq 0$.

Conversely, if the constitutive restrictions (95.9) and (95.10) are satisfied, then the free-energy imbalance (95.5) is satisfied in all constitutive processes. We, therefore, have the following result:

THERMODYNAMIC RESTRICTIONS *The following conditions are both necessary and sufficient that every constitutive process satisfy the free-energy imbalance:*

(i) *the free energy determines the second Piola elastic-stress through the* **stress relation**

$$\hat{\mathbf{T}}^e(\mathbf{E}^e) = \frac{\partial\hat{\varphi}(\mathbf{E}^e)}{\partial\mathbf{E}^e}; \tag{95.11}$$

(ii) *the plastic stress satisfies the* **reduced dissipation inequality**

$$\hat{\mathbf{T}}^p(\mathbf{L}^p, e^p):\mathbf{L}^p \geq 0. \tag{95.12}$$

The left side of (95.12) represents the dissipation as a function

$$\delta(\mathbf{L}^p, e^p) = \hat{\mathbf{T}}^p(\mathbf{L}^p, e^p):\mathbf{L}^p \geq 0. \tag{95.13}$$

Verification of (I) and (II)

Choose a material point $\mathbf{X}_0$ and a time t_0. We first discuss the construction of motions corresponding to the constitutive processes of (I) and (II). To find a motion χ and a plastic distortion $\mathbf{F}^p$ that have the properties specified in (I), we note that, since $\mathbf{F} = \mathbf{F}^e\mathbf{F}^p$, if we let $\mathbf{F}^p \equiv \mathbf{1}$ (so that $\mathbf{L}^p \equiv \mathbf{0}$), then the desired motion would have $\mathbf{F} \equiv \mathbf{F}^e$, so that, necessarily, $\mathbf{L} \equiv \mathbf{L}^e$, $\mathbf{E} \equiv \mathbf{E}^e$, and so forth. On the other hand, regarding (II), the hypothesis $\mathbf{F}^e \equiv \mathbf{1}$ implies that the desired motion must have $\mathbf{F} \equiv \mathbf{F}^p$, and, hence, that $\det\mathbf{F} \equiv 1$ and $\mathbf{L} \equiv \mathbf{L}^p$.

For both (I) and (II) we use the motion defined on all space by (14.5); viz.

$$\chi(\mathbf{X}, t) = \mathbf{x}_0 + e^{(t-t_0)\mathbf{L}}\mathbf{F}_0(\mathbf{X} - \mathbf{X}_0), \qquad -\infty < t < \infty, \tag{95.14}$$

with $\mathbf{L}$ and $\mathbf{F}_0$ constant, and with $\det\mathbf{F}_0 > 0$. In addition, for (II) given any choice of the constant deviatoric tensor $\mathbf{L}$ (and hence $\mathbf{D}$) and any choice of $e^p_0 \geq 0$, a constant, we define the accumulated plastic strain $e^p(t)$ (independent of $\mathbf{X}$) by

$$e^p(t) = e^p_0 + (t - t_0)|\mathbf{D}|,$$

a choice that trivially satisfies the hardening equation (95.2).

To complete the verification of (I) and (II) we must show that

- for some choice of the constant $\mathbf{L}$ the motion χ has the properties specified in (I), and for another choice χ and e^p have the properties specified in (II).

Consider (I). We assume that $\mathbf{L} \equiv \mathbf{D}$ is *symmetric*. We have to show that the Green–St. Venant strain $\mathbf{E} \equiv \mathbf{E}^e$ corresponding to χ and its rate $\dot{\mathbf{E}} \equiv \dot{\mathbf{E}}^e$ may be arbitrarily specified at $(\mathbf{X}_0, t_0)$. But since both $\mathbf{F}_0$ and $\mathbf{D}$ are arbitrary, this is a direct consequence of the identity $\mathbf{E} = \frac{1}{2}(\mathbf{C} - \mathbf{1})$ and the argument specified in the paragraph containing (48.21).

Regarding (II), we choose a deviatoric $\mathbf{L}$, and we let $\mathbf{F}_0 = \mathbf{1}$, so that, by (14.4), $\det\mathbf{F}(t) \equiv \det\mathbf{F}^p(t) \equiv 1$. Then χ has the desired properties, as does e^p, since $e^p_0 \geq 0$ was arbitrarily chosen.

This completes the verification of (I) and (II).

95.2 Structural Frame-Indifference and the Characterization of Polycrystalline Materials Without Texture

Based on the great success of conventional frame-indifference, GREEN & NAGHDI (1971) introduced the notion of a *change in frame of the structural space.*[571]

[571] Cf. also CASEY & NAGHDI (1980).

Consistent with Figure 91.1, this notion leads to transformation laws of the form

$$_*\mathbf{F}^p = \mathbf{Q}\mathbf{F}^p \qquad \text{and} \qquad _*\mathbf{F}^e = \mathbf{F}^e\mathbf{Q}^\top, \tag{95.15}$$

in which $\mathbf{Q}$ is an arbitrary *time-dependent* rotation of the structural space and, for any field Θ, $_*\Theta$ denotes Θ as seen in the new frame.[572] Unfortunately, Green and Naghdi viewed structural frame-indifference as a *general principle*; that is, a principle that stands at a level equivalent to that of conventional frame-indifference. This view has been refuted by many workers,[573] and as a consequence references to structural frame-indifference have almost disappeared from the literature. While we agree with the view that

- *structural frame-indifference is not a general principle,*

this hypothesis does represent an important facet of the behavior of a large class of polycrystalline materials based on the Kröner decomposition. Indeed,

- *for polycrystalline materials without texture the structural space would seem to be associated with a collection of randomly oriented lattices, and hence the evolution of dislocations through that space should be independent of the frame with respect to which this evolution is measured.*[574]

Consequences of $(91.7)_2$, (91.19), and (95.15) are the *transformation laws*

$$\left.\begin{aligned} _*\mathbf{E}^e &= \mathbf{Q}\mathbf{E}^e\mathbf{Q}^\top, \\ _*\mathbf{L}^p &= \mathbf{Q}\mathbf{L}^p\mathbf{Q}^\top + \mathbf{\Lambda}, \\ _*\mathbf{D}^p &= \mathbf{Q}\mathbf{D}^p\mathbf{Q}^\top, \end{aligned}\right\} \tag{95.16}$$

with

$$\mathbf{\Lambda} \stackrel{\text{def}}{=} \dot{\mathbf{Q}}\mathbf{Q}^\top \tag{95.17}$$

the corresponding frame-spin, a *skew tensor*. By (95.16), mimicking the conventional definition of a frame-indifferent field, we may say that $\mathbf{E}^e$ and $\mathbf{D}^p$ are *structurally frame-indifferent*, while $\mathbf{L}^p$ is not.

The requirement that the material defined via the constitutive equations (95.1) be independent of the structural frame is somewhat delicate, because we do not yet know how the elastic and plastic stresses $\mathbf{T}^e$ and $\mathbf{T}^p$ transform. On the other hand, since the free energy φ and the dissipation δ are *scalars* and hence necessarily invariant under structural frame changes, we may use $(95.1)_1$, (95.13), and (95.16) to phrase the requirement of invariance under changes in structural frame as follows:

(SFI) *the constitutive relations* $(95.1)_1$ *and* (95.13) *describing the free energy and the dissipation must satisfy*

$$\begin{aligned} \hat{\varphi}(\mathbf{E}^e) &= \hat{\varphi}(\mathbf{Q}\mathbf{E}^e\mathbf{Q}^\top), \\ \delta(\mathbf{L}^p, e^p) &= \delta(\mathbf{Q}\mathbf{L}^p\mathbf{Q}^\top + \mathbf{\Lambda}, e^p), \end{aligned} \tag{95.18}$$

for all rotations $\mathbf{Q}$ *and all skew tensors* $\mathbf{\Lambda}$.

[572] In this section we restrict attention to a prescribed point of the body so that, without loss in generality, the orthogonal transformations involved may be considered as being independent of the material point $\mathbf{X}$. However, the discussion given here is valid without change if these trasformations are allowed to depend on $\mathbf{X}$, because nowhere in this section are these transformations differentiated with respect to $\mathbf{X}$.

[573] Cf., e.g., Dashner (1986) and the references therein. In fact, invariance under $(95.15)_2$ renders the elastic response isotropic and hence not generic.

[574] Cf. Footnote 3 of Gurtin (2003).

Consider the free energy. By (95.18), $\hat{\varphi}(\mathbf{E}^e)$ must be an isotropic function of $\mathbf{E}^e$ and a standard argument based on the relation (95.11) for the elastic stress yields the same conclusion for $\hat{\mathbf{T}}^e(\mathbf{E}^e)$. A condition both necessary and sufficient that the elastic relations be invariant under changes in structural frame is thus that *the constitutive relations* $(95.1)_{1,2}$ *governing elastic response be isotropic*. We henceforth assume that this condition is satisfied.

Next, for ${}_*\mathbf{T}^p = \hat{\mathbf{T}}^p({}_*\mathbf{L}^p, e^p)$, (95.13) and $(95.18)_2$ imply that

$$\mathbf{T}^p : \mathbf{L}^p = {}_*\mathbf{T}^p : (\mathbf{Q}\mathbf{L}^p\mathbf{Q}^\top + \mathbf{\Lambda}) \tag{95.19}$$

for all rotations $\mathbf{Q}$ and all skew tensors $\mathbf{\Lambda}$. Taking $\mathbf{\Lambda} = \mathbf{0}$ we find that

$$\begin{aligned} \mathbf{T}^p : \mathbf{L}^p &= {}_*\mathbf{T}^p : (\mathbf{Q}\mathbf{L}^p\mathbf{Q}^\top) \\ &= (\mathbf{Q}^\top {}_*\mathbf{T}^p\mathbf{Q}) : \mathbf{L}^p \end{aligned}$$

and, since this relation must hold for all $\mathbf{L}^p$, we must have

$${}_*\mathbf{T}^p = \mathbf{Q}\mathbf{T}^p\mathbf{Q}^\top; \tag{95.20}$$

$\mathbf{T}^p$ is therefore *structurally frame-indifferent.* On the other hand, for $\mathbf{Q} = \mathbf{1}$, so that $\mathbf{T}^p = {}_*\mathbf{T}^p$, we conclude from (95.19) that $\mathbf{T}^p : \mathbf{\Lambda} = 0$ (for every skew tensor $\mathbf{\Lambda}$); hence $\mathbf{T}^p$ is *symmetric*. We have therefore shown that

$$\mathbf{T}^p \text{ is symmetric and structurally frame-indifferent.} \tag{95.21}$$

An important consequence of this result is that the dissipation (95.13) must have the more conventional form

$$\mathbf{T}^p : \mathbf{D}^p \geq 0. \tag{95.22}$$

Next, by $(95.16)_2$ and (95.20) the constitutive response function $\hat{\mathbf{T}}^p(\mathbf{L}^p, e^p)$ must transform according to

$$\mathbf{Q}\hat{\mathbf{T}}^p(\mathbf{L}^p, e^p)\mathbf{Q}^\top = \hat{\mathbf{T}}^p(\mathbf{Q}\mathbf{L}^p\mathbf{Q}^\top + \mathbf{\Lambda});$$

if we take $\mathbf{Q} = \mathbf{1}$ we find that

$$\hat{\mathbf{T}}^p(\mathbf{L}^p, e^p) = \hat{\mathbf{T}}^p(\mathbf{L}^p + \mathbf{\Lambda}, e^p)$$

for very skew tensor $\mathbf{\Lambda}$. Thus, taking $\mathbf{\Lambda} = -\mathbf{W}^p$ we conclude that, in view of (95.21) and since, by $(95.16)_3$, $\mathbf{D}^p$ is structurally frame-indifferent, the constitutive relation $(95.1)_3$ must reduce to an *isotropic* relation

$$\boxed{\mathbf{T}^p = \hat{\mathbf{T}}^p(\mathbf{D}^p, e^p),} \tag{95.23}$$

with $\mathbf{T}^p$ symmetric. Thus,

- *the plastic stress cannot depend on the plastic spin.*

Further, using (95.22) and (95.23), we can rewrite the dissipation (95.13) as a function

$$\delta(\mathbf{D}^p, e^p) = \hat{\mathbf{T}}^p(\mathbf{D}^p, e^p) : \mathbf{D}^p \geq 0 \tag{95.24}$$

of $\mathbf{D}^p$ and e^p.

The main results of this section may be summarized as follows:

CONSEQUENCES OF STRUCTURAL FRAME-INDIFFERENCE:

(i) *The constitutive relations* $(95.1)_{1,2}$ *governing elastic response are isotropic.*
(ii) *The plastic stress* $\mathbf{T}^p$ *is symmetric and structurally frame-indifferent, and can depend on* $\mathbf{L}^p$ *at most isotropically through* $\mathbf{D}^p$ *— thus, importantly,* $\mathbf{T}^p$ *is independent of the plastic spin.*
(iii) *The dissipation* δ *is independent of the plastic spin.*

EXERCISE

1. Establish the following **transformation laws** appropriate to a *change in structural frame*:

$$\left.\begin{aligned} {}_*\mathbf{C}^e &= \mathbf{Q}\mathbf{C}^e\mathbf{Q}^\top, \\ {}_*\mathbf{E}^e &= \mathbf{Q}\mathbf{E}^e\mathbf{Q}^\top, \\ {}_*\mathbf{D}^p &= \mathbf{Q}\mathbf{D}^p\mathbf{Q}^\top, \end{aligned}\right\} \tag{95.25}$$

and

$$\left.\begin{aligned} {}_*\mathbf{R}^p &= \mathbf{Q}\mathbf{R}^p, \\ {}_*\mathbf{L}^p &= \mathbf{Q}\mathbf{L}^p\mathbf{Q}^\top + \mathbf{\Lambda}, \\ {}_*\mathbf{W}^p &= \mathbf{Q}\mathbf{W}^p\mathbf{Q}^\top + \mathbf{\Lambda}, \end{aligned}\right\} \tag{95.26}$$

with $\mathbf{\Lambda}$ the corresponding frame-spin (95.17). (The result (95.25) asserts that the fields $\mathbf{C}^e$, $\mathbf{E}^e$, and $\mathbf{D}^p$ are structurally frame-indifferent, which is not surprising: As is clear from (P4) on page 544 and (P5) on page 545, $\mathbf{C}^e$, $\mathbf{E}^e$, *and* $\mathbf{D}^p$ *are structural tensor fields*, and (95.25) are natural transformation laws for tensors with this mapping property.)

2. Show, as a consequence of (95.11), that $\varphi(\mathbf{E}^e)$ an isotropic function of $\mathbf{E}^e$ implies that $\hat{\mathbf{T}}^e(\mathbf{E}^e)$ is an isotropic function of $\mathbf{E}^e$.

95.3 Interaction of Elasticity and Plastic Flow

The response of most metals is typically associated with *small* elastic strains.[575] Thus, bearing in mind (i) on page 562, we consider elastic constitutive relations of the form[576]

$$\begin{aligned} \varphi &= \mu|\mathbf{E}^e|^2 + \tfrac{1}{2}\lambda|\mathrm{tr}\,\mathbf{E}^e|^2, \\ \mathbf{T}^e &= 2\mu\mathbf{E}^e + \lambda(\mathrm{tr}\,\mathbf{E}^e)\mathbf{1}. \end{aligned} \tag{95.27}$$

Even though the purported application of (95.27) is to small elastic strains, the relations (95.27) are frame-indifferent and consistent with the free-energy imbalance — these relations are therefore theoretically valid constitutive equations, independent of the size of the deformation. But, as is well known, (95.27) characterize observed behavior only for $\mathbf{E}^e$ small.

The identity $\mathbf{C}^e - \mathbf{1} = 2\mathbf{E}^e$ allows us to rewrite $(95.27)_2$ in the form

$$\mathbf{T}^e = \mu(\mathbf{C}^e - \mathbf{1}) + \tfrac{1}{2}\lambda(\mathrm{tr}\,\mathbf{C}^e - 3)\mathbf{1}, \tag{95.28}$$

[575] Except under conditions involving high-velocity impact, where elastic volume changes can become large.

[576] The stress-strain relation $(95.27)_2$ follows from a standard representation theorem provided in §112.7; the verification of $(95.27)_1$, granted (95.9) and $(95.27)_2$, is left as an exercise.

and, since by (94.8) $\mathbf{M}^e = \mathbf{C}^e\mathbf{T}^e$,

$$\mathbf{M}^e = \mu(\mathbf{C}^{e2} - \mathbf{C}^e) + \tfrac{1}{2}\lambda(\operatorname{tr}\mathbf{C}^e - 3)\mathbf{C}^e. \tag{95.29}$$

Thus, $\mathbf{M}^e$ is *symmetric*,

$$\mathbf{M}^e = \mathbf{M}^{e\top}. \tag{95.30}$$

Further, $\mathbf{M}^e$ is an isotropic function of $\mathbf{C}^e$ (an assertion whose proof we leave as an exercise); hence, the commutation property (ii) of Appendix 113.2 yields the conclusion that

$$\mathbf{M}^e\mathbf{C}^e = \mathbf{C}^e\mathbf{M}^e; \tag{95.31}$$

thus since $\mathbf{C}^e$ and $\mathbf{M}^e$ are symmetric, so also is $\mathbf{C}^e\mathbf{M}^e$.

EXERCISES

1. Show that $\mathbf{M}^e$ is an isotropic function of $\mathbf{E}^e$.
2. Show that

$$\mathbf{U}^e\mathbf{M}^e = \mathbf{M}^e\mathbf{U}^e. \tag{95.32}$$

   (Hint: show that $\mathbf{M}^e$ is an isotropic as a function of $\mathbf{U}^e$ and use the commutation property (ii) of Appendix 113.2.)
3. Use the polar decomposition $\mathbf{F}^e = \mathbf{R}^e\mathbf{U}^e$ and (95.32) to show that

$$\mathbf{T} = J^{-1}\mathbf{R}^e\mathbf{M}^e\mathbf{R}^{e\top}. \tag{95.33}$$

95.4 Consequences of Rate-Independence

Our discussion of rate-independence follows §76.2: We suppress the argument $\mathbf{X}$, invoke the notational agreements spelled out in (76.13) and (76.14), and use the identity[577]

$$\dot{\Theta}_\kappa(t) = \kappa\dot{\Theta}(\kappa t) \tag{95.34}$$

appropriate to a change in time-scale (76.12) with rate constant $\kappa > 0$. Then, $\dot{\mathbf{F}}^p_\kappa(t) = \kappa\dot{\mathbf{F}}^p(\kappa t)$ and, since $\mathbf{L}^p = \dot{\mathbf{F}}^p\mathbf{F}^{p-1}$,

$$\begin{aligned}\mathbf{L}^p_\kappa(t) &= \kappa\dot{\mathbf{F}}^p(\kappa t)\mathbf{F}^{p-1}(\kappa t)\\ &= \kappa\mathbf{L}^p(\kappa t).\end{aligned} \tag{95.35}$$

The symmetric part of $\mathbf{L}^p$ is the plastic stretching $\mathbf{D}^p$; thus,

$$\mathbf{D}^p_\kappa(t) = \kappa\mathbf{D}^p(\kappa t). \tag{95.36}$$

We now assume that the constitutive relation (95.23) for the plastic stress,

$$\mathbf{T}^p = \hat{\mathbf{T}}^p(\mathbf{D}^p, e^p), \tag{95.37}$$

is *rate-independent.* Our discussion of the consequences of this assumption is *identical* to a corresponding discussion in §76 and leads to a constitutive relation of the form (76.21), but with the replacements $\mathbf{T}_0 \to \mathbf{T}^p$ and $\dot{\mathbf{E}}^p \to \mathbf{D}^p$:

$$\mathbf{T}^p = \hat{\mathbf{T}}^p(\mathbf{N}^p, e^p), \tag{95.38}$$

[577] Cf. (76.15).

with $\mathbf{N}^p$, the **flow direction**, here defined by

$$\boxed{\mathbf{N}^p = \frac{\mathbf{D}^p}{|\mathbf{D}^p|}.} \tag{95.39}$$

EXERCISE

1. Derive (95.38) without referring to §76. Show all steps.

95.5 Derivation of the Mises Flow Equations Based on Maximum-Dissipation

For the most part, our discussion follows §79, but, because we now have at our disposal the virtual external body force $\mathcal{B}^p$, it begins with the microscopic force-balance (94.9) supplemented by the constitutive relation (95.38) for the plastic stress $\mathbf{T}^p$; viz.

$$\mathbf{M}_0^e = \hat{\mathbf{T}}^p(\mathbf{N}^p, e^p) - \mathcal{B}^p. \tag{95.40}$$

We use the term **normalized flow** for a pair $(\mathbf{M}_0^e, \mathbf{N}^p)$ with $\mathbf{M}_0^e$ a deviatoric Mandel stress and $\mathbf{N}^p$ a flow direction, and we refer to $(\mathbf{M}_0^e, \mathbf{N}^p)$ as **physically attainable** if[578]

$$\mathcal{B}^p \equiv \mathbf{0}. \tag{95.41}$$

Then, by (95.40), $(\mathbf{M}_0^e, \mathbf{N}^p)$ is physically attainable if and only if $\mathbf{M}_0^e$ and $\mathbf{N}^p$ are related through the **flow rule**[579]

$$\mathbf{M}_0^e = \hat{\mathbf{T}}^p(\mathbf{N}^p, e^p). \tag{95.42}$$

Consistent with this, we use the term **flow stress** for a deviatoric Mandel stress $\mathbf{M}_0^e$ such that

$$\mathbf{M}_0^e = \hat{\mathbf{T}}^p(\mathbf{N}^p, e^p) \quad \text{for some flow direction } \mathbf{N}^p, \tag{95.43}$$

and we refer to

$$\mathcal{Y}(e^p) \overset{\text{def}}{=} \text{the set of all } \textit{flow stresses } \mathbf{M}_0^e \tag{95.44}$$

as the *yield set*.

Within the present framework, the expression (76.26) for the *flow resistance* takes the form

$$Y(\mathbf{N}^p, e^p) = \mathbf{N}^p : \hat{\mathbf{T}}^p(\mathbf{N}^p, e^p) > 0, \tag{95.45}$$

where here, as in §79, we have assumed that *the dissipation is strict.*

Consonant with (79.7) and (79.8), we say that a deviatoric Mandel stress $\mathbf{M}_0^e$ is **admissible in the sense of maximum dissipation** if

$$\mathbf{M}_0^e : \mathbf{N}^p \le Y(\mathbf{N}^p, e^p) \quad \text{for every flow direction } \mathbf{N}^p, \tag{95.46}$$

and we refer to the set

$$\mathcal{E}(e^p) \overset{\text{def}}{=} \text{the set of all } \mathbf{M}_0^e \text{ that are admissible in the sense of maximum dissipation} \tag{95.47}$$

as the *elastic range*.

---

[578] Here, as compared to §79, the presence of the external body force $\mathcal{B}^p$ allows us to consider flows that are not physically attainable; cf. our discussion in the paragraph containing (‡) on page 455.

[579] Cf. Footnote 485.

Our derivation of the Mises flow equations is based on the following two hypotheses:

(FS) (**flow-stress admissibility**) each flow stress is admissible in the sense of maximum dissipation, so that

$$\mathcal{Y}(e^p) \subset \mathcal{E}(e^p); \tag{95.48}$$

(SI) (**strong isotropy**) $Y(\mathbf{N}^p, e^p)$ is independent of the flow direction $\mathbf{N}^p$; viz.

$$Y(\mathbf{N}^p, e^p) = Y(e^p).$$

By (FS), (SI), and (95.46), we have the important inequality

$$\mathbf{M}_0^e : \mathbf{N}^p \le Y(e^p) \quad \text{for every flow stress } \mathbf{M}_0^e \text{ and every flow direction } \mathbf{N}^p. \tag{95.49}$$

Then, if we adjoin to the hardening equation (95.2) the null initial-condition $e^p(\mathbf{X}, 0) = 0$, recall that[580]

$$\text{SymDev} = \text{the set of all symmetric and deviatoric tensors},$$

and argue as in §79, we are led to the following result:

MAXIMUM DISSIPATION AND THE MISES FLOW EQUATIONS FOR LARGE DEFORMATIONS *Assume that the hypotheses* (SI) *and* (FS) *are satisfied. The flow rule* (95.42) *and the hardening equation* (95.2), *with* $e^p(0) = 0$, then *take the form*

$$\boxed{\begin{aligned} \mathbf{M}_0^e &= Y(e^p)\mathbf{N}^p \quad \text{for} \quad \mathbf{D}^p \neq \mathbf{0}, \\ \dot{e}^p &= |\mathbf{D}^p|, \quad e^p(\mathbf{X}, 0) = 0. \end{aligned}} \tag{95.50}$$

Further, the yield set $\mathcal{Y}(e^p)$ *is a spherical surface in* SymDev *of radius*

$$Y(e^p) > 0 \tag{95.51}$$

and center at **0**, *and the elastic range* $\mathcal{E}(e^p)$ *is the closed ball in* SymDev *whose boundary is* $\mathcal{Y}(e^p)$. *The requirement that the deviatoric Mandel stress be confined to the elastic range is thus equivalent to the* **boundedness inequality**[581]

$$|\mathbf{M}_0^e| \le Y(e^p). \tag{95.52}$$

We refer to (95.50) *as the* **Mises flow equations** *and to* $(95.50)_1$ *as the* **Mises flow rule**.

EXERCISE

1. Verify the Mises flow equations (95.50) directly without referring to §79.

---

580 Cf. (79.3).

581 In contrast to the conventional theory introduced in §75, this inequality is not a separate hypothesis.

96 Summary of the Basic Equations. Remarks

- *For the remainder of* this part *we assume that*

$$B^p \equiv \mathbf{0}$$

and, consequently, confine our attention to constitutive processes that are physically attainable.

The *basic equations of the theory*, as derived thus far, are: the Kröner decomposition and the plastic incompressibility condition,[582]

$$\mathbf{F} = \nabla\chi = \mathbf{F}^e\mathbf{F}^p, \qquad \det\mathbf{F}^p = 1; \tag{96.1}$$

the kinematical relations[583]

$$\begin{gathered}\mathbf{E}^e = \tfrac{1}{2}(\mathbf{C}^e - \mathbf{1}), \qquad \mathbf{C}^e = \mathbf{F}^{e\top}\mathbf{F}^e,\\ \mathbf{L}^p = \dot{\mathbf{F}}^p\mathbf{F}^{p-1}, \qquad \mathbf{D}^p = \operatorname{sym}\mathbf{L}^p, \qquad \mathbf{N}^p = \frac{\mathbf{D}^p}{|\mathbf{D}^p|};\end{gathered} \tag{96.2}$$

the elastic stress-strain relation[584]

$$\mathbf{T}^e = 2\mu\mathbf{E}^e + \lambda(\operatorname{tr}\mathbf{E}^e)\mathbf{1}; \tag{96.3}$$

the **Mises–Hill equations**[585]

$$\begin{gathered}\mathbf{M}^e_0 = Y(e^p)\mathbf{N}^p \quad \text{for } \mathbf{D}^p \neq \mathbf{0}, \qquad \dot{e}^p = |\mathbf{D}^p| \ (e^p(\mathbf{X},0) = 0),\\ |\mathbf{M}^e_0| \le Y(e^p), \qquad \mathbf{D}^p = \mathbf{0} \quad \text{for } |\mathbf{M}^e_0| < Y(e^p),\end{gathered} \tag{96.4}$$

with

$$\mathbf{M}^e = \mathbf{C}^e\mathbf{T}^e \tag{96.5}$$

the Mandel stress;[586] the balance law for linear momentum[587]

$$\operatorname{div}\mathbf{T} + \mathbf{b}_0 = \rho\dot{\mathbf{v}}, \qquad \mathbf{T} = J^{-1}\mathbf{F}^e\mathbf{T}^e\mathbf{F}^{e\top}. \tag{96.6}$$

Remark. At this point the plastic flow equations are not in a form most suitable for the solution of initial/boundary-value problems.[588]

582 Cf. (91.1) and (91.11).
583 Cf. $(91.7)_2$, $(91.16)_1$, and (91.17).
584 Cf. $(95.27)_2$.
585 Cf. (95.50) and (95.52); $(96.4)_4$ follows from the tacit requirement that the flow be physically attainable.
586 Cf. (94.8).
587 Cf. (92.25) and (94.4).
588 Cf. §99.4.

97 Plastic Irrotationality: The Condition $\mathbf{W}^p \equiv \mathbf{0}$

We now show that — within the present framework for isotropic plasticity and without loss in generality — *we may assume that the plastic spin vanishes.*[589] Specifically, our goal is to establish the following result:

IRROTATIONALITY THEOREM *If the initial/boundary-value problem consisting of the basic equations specified in* §96 *together with concomitant boundary and initial conditions has a solution, then every* spatially inhomogeneous *frame-change* $\mathbf{Q}(\mathbf{X}, t)$ *for the structural space also yields a solution. Moreover, there is always a (possibly spatially inhomogeneous) frame-change that renders the transformed solution as one without plastic spin. Thus we may assume, without loss in generality, that* $\mathbf{W}^p = \mathbf{0}$.

Based on this result, we restrict attention to **irrotational plastic flow**; that is, to flow for which

$$\boxed{\mathbf{W}^p = \mathbf{0},} \tag{97.1}$$

so that, by $(91.7)_2$,

$$\mathbf{D}^p = \dot{\mathbf{F}}^p\mathbf{F}^{p-1}. \tag{97.2}$$

Further, $(91.26)_2$ and (97.1) imply that

$$\dot{\mathbf{R}}^p = -\mathbf{R}^p \,\mathrm{skw}\,(\dot{\mathbf{U}}^p\mathbf{U}^{p-1});$$

thus, even though the plastic spin vanishes, the plastic rotation is generally nontrivial. Note also that, by (97.1), (91.23) takes the form

$$\mathbf{F}^{e\top}\mathbf{D}\mathbf{F}^e = \mathbf{F}^{e\top}\mathbf{D}^e\mathbf{F}^e + \mathrm{sym}\,(\mathbf{C}^e\mathbf{D}^p). \tag{97.3}$$

Proof of the Irrotationality Theorem

The proof is based on our discussion of structural frame-indifference in §95.2. Consider an arbitrary rotation field $\mathbf{Q}(\mathbf{X}, t)$, viewed as a (time-dependent) change in structural frame.[590] For any field Θ let ${}_*\Theta$ denote the field Θ as seen in the new frame. The elastic and plastic distortions then transform according to (95.15) under the frame change $\mathbf{Q}(\mathbf{X}, t)$, viz.

$$\begin{aligned} {}_*\mathbf{F}^p &= \mathbf{Q}\mathbf{F}^p, \\ {}_*\mathbf{F}^e &= \mathbf{F}^e\mathbf{Q}^\top. \end{aligned} \tag{97.4}$$

Consistent with terminology used in §95.2, we say that a field Θ is *structurally invariant* if

$${}_*\Theta = \Theta,$$

[589] GURTIN & ANAND (2005c).

[590] Cf. Footnote 572.

and that a tensor field $\mathbf{\Lambda}$ is *structurally frame-indifferent* if

$$_*\mathbf{\Lambda} = \mathbf{Q}\mathbf{\Lambda}\mathbf{Q}^\top. \tag{97.5}$$

Further, we say that a constitutive relation $\Phi = \hat{\Phi}(\Lambda)$ is *structurally frame-indifferent* if[591]

$$\Phi = \hat{\Phi}(\Lambda) \quad \text{implies that} \quad {}_*\Phi = \hat{\Phi}({}_*\Lambda). \tag{97.6}$$

By (91.1), $\mathbf{F}$ is invariant under (97.4):

$$_*\mathbf{F} = \mathbf{F}, \tag{97.7}$$

so that, modulo an inconsequential time-dependent rigid *displacement* of the body, the motion χ is structurally invariant:

$$_*\chi = \chi. \tag{97.8}$$

Further consequences of (97.4) are the transformation laws (95.25) and (95.26) in which $\mathbf{\Lambda} = \dot{\mathbf{Q}}\mathbf{Q}^\top$ is the corresponding frame spin (95.17).

Next, by (96.3), (95.25), and (96.5), the elastic and Mandel stresses are structurally frame-indifferent,

$$\begin{aligned} {}_*\mathbf{T}^e &= \mathbf{Q}\mathbf{T}^e\mathbf{Q}^\top, \\ {}_*\mathbf{M}^e &= \mathbf{Q}\mathbf{M}^e\mathbf{Q}^\top; \end{aligned} \tag{97.9}$$

conversely, granted the transformation laws (97.4), the elastic stress-strain relation (96.3) and the Mises flow equations and boundedness inequality expressed in (96.4) are structurally frame-indifferent. Thus, under the transformation law (97.9),

(‡) *the elastic stress-strain relation and the Mises flow equations and boundedness inequality are structurally frame-indifferent, while the Kröner decomposition renders* $\mathbf{F}$ *structurally invariant.*

Further, by (94.11), (97.7), and (97.9), the Cauchy stress $\mathbf{T}$ is invariant:

$$_*\mathbf{T} = \mathbf{T}. \tag{97.10}$$

Assume that we are given a motion χ of the body consistent with the basic equations (96.1)–(96.6). Assume that the spin $\mathbf{W}^p$ corresponding to this solution is nonzero. Consider a (for now arbitrary) *time-dependent* rotation field $\mathbf{Q}(\mathbf{X}, t)$, viewed as a spatially inhomogeneous frame-change for the structural space, so that the basic transformation law (97.4) for $\mathbf{F}^e$ and $\mathbf{F}^p$ is satisfied. Then, by (95.25), $\mathbf{C}^e$, $\mathbf{E}^e$, $\mathbf{D}^p$, and $\mathbf{N}^p$ are structurally frame-indifferent, and if we define ${}_*\mathbf{T}^e$, ${}_*\mathbf{M}^e$, and ${}_*\mathbf{T}^p$ through the transformation laws (97.9), then the transformed fields satisfy (96.1)–(96.6)$_1$ with ${}_*\mathbf{F} = \mathbf{F}$ and ${}_*\mathbf{T} = \mathbf{T}$ (granted that ${}_*\mathbf{T}$ is defined in terms of the transformed fields via (94.11)). Thus the transformed fields satisfy all of the basic equations. Moreover, if the original solution satisfied initial and boundary conditions involving fields unrelated to the structural space, as is standard, then the transformed fields satisfy the same initial/boundary-value problem as the original fields.

Consider now the transformed spin ${}_*\mathbf{W}^p$, which, by (95.26)$_3$, satisfies

$$_*\mathbf{W}^p = \mathbf{Q}\mathbf{W}^p\mathbf{Q}^\top + \dot{\mathbf{Q}}\mathbf{Q}^\top. \tag{97.11}$$

Up to this point $\mathbf{Q}(\mathbf{X}, t)$ has been an arbitrary rotation field. We now seek to find a particular rotation field $\mathbf{Q}(\mathbf{X}, t)$ such that

$$\dot{\mathbf{Q}} = -\mathbf{Q}\mathbf{W}^p, \tag{97.12}$$

for then (97.11) would yield

$$_*\mathbf{W}^p = \mathbf{0}. \tag{97.13}$$

Since any solution of (97.12) renders ${}_*\mathbf{W}^p = \mathbf{0}$, we have only to find a solution $\mathbf{Q}$ of (97.12) with $\mathbf{Q}$ a rotation field. We now show that the unique solution $\mathbf{Q}$ consistent with the initial condition

$$\mathbf{Q}(\mathbf{X}, 0) = \mathbf{1} \tag{97.14}$$

is a rotation field. To see this we differentiate $\mathbf{Q}^\top\mathbf{Q}$:

$$\begin{aligned} \overline{\mathbf{Q}^\top\mathbf{Q}}^{\boldsymbol{\cdot}} &= \dot{\mathbf{Q}}^\top\mathbf{Q} + \mathbf{Q}^\top\dot{\mathbf{Q}} \\ &= (\mathbf{Q}^\top\dot{\mathbf{Q}})^\top + \mathbf{Q}^\top\dot{\mathbf{Q}} \\ &= -(\mathbf{W}^p)^\top - \mathbf{W}^p \\ &= 0. \end{aligned}$$

On the other hand, by (97.14),

$$(\mathbf{Q}^\top\mathbf{Q})(\mathbf{X}, 0) = \mathbf{1}. \tag{97.15}$$

Thus $\mathbf{Q}^\top\mathbf{Q} \equiv \mathbf{1}$ and $\mathbf{Q}$ is a rotation field. The transformed solution therefore has vanishing plastic spin.

591 Cf. (36.1).

98 Yield Surface. Yield Function. Consistency Condition

As noted in the paragraph containing (95.52), the **yield surface** is the spherical surface with radius $Y(e^p) > 0$ in the space SymDev of symmetric, deviatoric tensors and the **elastic range** is the closed ball with radius $Y(e^p)$, and a consequence of the yield condition is that *plastic flow is possible only when the deviatoric Mandel stress* $\mathbf{M}_0^e$ *lies on the yield surface.*

Immediate consequences of the Mises flow rule $(95.50)_1$ and the boundedness inequality (95.52) are the **yield condition**

$$|\mathbf{M}_0^e| = Y(e^p) \qquad \text{for } \mathbf{D}^p \neq \mathbf{0} \tag{98.1}$$

and the condition

$$\mathbf{D}^p = \mathbf{0} \qquad \text{for } |\mathbf{M}_0^e| < Y(e^p). \tag{98.2}$$

As in our discussion of small deformations we introduce a **yield function**[592]

$$f = |\mathbf{M}_0^e| - Y(e^p), \tag{98.3}$$

which by (95.52) is constrained by

$$-Y(e^p) \leq f \leq 0. \tag{98.4}$$

The yield condition (98.1) is then equivalent to the requirement that

$$f = 0 \qquad \text{for } \mathbf{D}^p \neq \mathbf{0}, \tag{98.5}$$

and (98.2) takes the form

$$\mathbf{D}^p = \mathbf{0} \qquad \text{for } f < 0. \tag{98.6}$$

Further, as argued in the case of the small-deformation rate-independent theory in § 76.8, the yield function f obeys the following additional restriction:

$$\text{if } f = 0 \text{ then } \dot{f} \leq 0; \tag{98.7}$$

this leads to the **no-flow condition**

$$\boxed{\mathbf{D}^p = 0 \text{ if } f < 0 \text{ or if } f = 0 \text{ and } \dot{f} < 0} \tag{98.8}$$

and the **consistency condition**

$$\boxed{\text{if } \mathbf{D}^p \neq 0, \text{ then } f = 0 \text{ and } \dot{f} = 0.} \tag{98.9}$$

[592] Considered as a function of $(\mathbf{X}, t)$ rather than $(\mathbf{M}^e, e^p)$.

Next, let

$$Y'(e^p) = \frac{dY(e^p)}{de^p}, \tag{98.10}$$

so that, by $(96.4)_2$,

$$\overline{Y(e^p)}^{\cdot} = Y'(e^p)|\mathbf{D}^p|;$$

then, letting

$$H(e^p) \stackrel{\text{def}}{=} Y'(e^p), \tag{98.11}$$

we arrive at

$$\overline{Y(e^p)}^{\cdot} = H(e^p)|\mathbf{D}^p|. \tag{98.12}$$

Thus, if we assume that $\mathbf{D}^p \neq \mathbf{0}$, then, by (98.1), $|\mathbf{M}_0^e| - Y(e^p) = 0$; hence, (98.12) yields

$$\boxed{\overline{|\mathbf{M}_0^e|}^{\cdot} = H(e^p)|\mathbf{D}^p|,} \tag{98.13}$$

which is the large-deformation counterpart of (76.111).

EXERCISE

1. Establish the no-flow condition (98.8) and the consistency condition (98.9).

99 $|\mathbf{D}^p|$ in Terms of $\dot{\mathbf{E}}$ and $\mathbf{M}^e$

99.1 Some Important Identities

First of all, a consequence of the flow rule $(95.50)_1$ is that

$$\mathbf{N}^p = \frac{\mathbf{M}_0^e}{|\mathbf{M}_0^e|}, \tag{99.1}$$

so that the deviatoric Mandel tensor "points" in the direction of plastic flow. Next, the following *commutativity identities* are useful:

$$\mathbf{C}^e\mathbf{M}^e = \mathbf{M}^e\mathbf{C}^e, \qquad \mathbf{C}^e\mathbf{N}^p = \mathbf{N}^p\mathbf{C}^e, \qquad \mathbf{T}^e\mathbf{N}^p = \mathbf{N}^p\mathbf{T}^e. \tag{99.2}$$

The first of these identities is (95.31). The second follows from the first and (99.1) (since a tensor commutes with $\mathbf{G}$ if and only if it commutes with $\mathbf{G}_0$). The third follows from the second and (95.28).

By (12.6), the material time-derivative of the Green–St. Venant strain

$$\mathbf{E} = \tfrac{1}{2}(\mathbf{F}^{\top}\mathbf{F} - \mathbf{1}) \tag{99.3}$$

is related to the stretching $\mathbf{D}$ by

$$\dot{\mathbf{E}} = \mathbf{F}^{\top}\mathbf{D}\mathbf{F},$$

and, in view of the decomposition $\mathbf{F} = \mathbf{F}^e\mathbf{F}^p$, yields the relation

$$\mathbf{F}^{e\top}\mathbf{D}\mathbf{F}^e = \mathbf{F}^{p-\top}\dot{\mathbf{E}}\mathbf{F}^{p-1}. \tag{99.4}$$

99.2 Conditions that Describe Loading and Unloading

Assume unless otherwise specified that[593]

$$f = 0 \qquad (\text{so that } \dot{f} \le 0). \tag{99.5}$$

Then, by (3.2), (98.3), (98.12), and (99.1),

$$\begin{aligned}\dot{f} &= \overline{|\mathbf{M}_0^e|}^{\boldsymbol{\cdot}} - \overline{Y(e^p)}^{\boldsymbol{\cdot}}\\ &= \frac{\mathbf{M}_0^e}{|\mathbf{M}_0^e|} : \dot{\mathbf{M}}_0^e - H(e^p)|\mathbf{D}^p|\\ &= \mathbf{N}^p : \dot{\mathbf{M}}_0^e - H(e^p)|\mathbf{D}^p|.\end{aligned} \tag{99.6}$$

[593] Cf. (98.7).

Our next step is to compute the term $\mathbf{N}^p:\dot{\mathbf{M}}_0^e$. Clearly,

$$\mathbf{N}^p:\dot{\mathbf{M}}_0^e = \mathbf{N}^p:\dot{\mathbf{M}}^e,$$

because $\mathbf{N}^p$ is deviatoric. By (95.29),

$$\dot{\mathbf{M}}^e = \mu(\mathbf{C}^e\dot{\mathbf{C}}^e + \dot{\mathbf{C}}^e\mathbf{C}^e - \dot{\mathbf{C}}^e) + \tfrac{1}{2}\lambda[(\operatorname{tr}\mathbf{C}^e - 3)\dot{\mathbf{C}}^e + (\mathbf{1}:\dot{\mathbf{C}}^e)\mathbf{C}^e]. \tag{99.7}$$

Since $\mathbf{C}^e\dot{\mathbf{C}}^e + \dot{\mathbf{C}}^e\mathbf{C}^e$ is symmetric, as is $\mathbf{N}^p$,

$$\begin{aligned}(\mathbf{C}^e\dot{\mathbf{C}}^e + \dot{\mathbf{C}}^e\mathbf{C}^e):\mathbf{N}^p &= 2(\mathbf{C}^e\dot{\mathbf{C}}^e):\mathbf{N}^p \\ &= 2(C^e_{ik}\dot{C}^e_{kl})N^p_{il} \\ &= 2(C^e_{ki}N^p_{il})\dot{C}^e_{kl} \\ &= 2(\mathbf{C}^e\mathbf{N}^p):\dot{\mathbf{C}}^e.\end{aligned}$$

Also,

$$(\mathbf{1}:\dot{\mathbf{C}}^e)(\mathbf{C}^e:\mathbf{N}^p) = \big[(\mathbf{C}^e:\mathbf{N}^p)\mathbf{1}\big]:\dot{\mathbf{C}}^e.$$

Thus, by (99.7),

$$\mathbf{N}^p:\dot{\mathbf{M}}_0^e = \mathbf{A}:\dot{\mathbf{C}}^e, \tag{99.8}$$

where

$$\boxed{\mathbf{A} = \mu(2\mathbf{C}^e - \mathbf{1})\mathbf{N}^p + \tfrac{1}{2}\lambda[(\operatorname{tr}\mathbf{C}^e - 3)\mathbf{N}^p + (\mathbf{C}^e:\mathbf{N}^p)\mathbf{1}].} \tag{99.9}$$

We refer to $\mathbf{A}$ as the **normal Mandel elasticity tensor**, because it relates the normal Mandel stress rate $\mathbf{N}^p:\dot{\mathbf{M}}_0^e$ to the rate $\dot{\mathbf{C}}^e$ of the elastic Cauchy–Green tensor $\mathbf{C}^e$. By (99.8), (99.6) may be written as

$$\dot{f} = \mathbf{A}:\dot{\mathbf{C}}^e - H(e^p)|\mathbf{D}^p|. \tag{99.10}$$

Next, since $\mathbf{D}^p = |\mathbf{D}^p|\mathbf{N}^p$, we may use $(99.2)_2$ and (99.4) to rewrite (97.3) in the form

$$\mathbf{F}^{p-\top}\dot{\mathbf{E}}\mathbf{F}^{p-1} = \mathbf{F}^{e\top}\mathbf{D}^e\mathbf{F}^e + |\mathbf{D}^p|\mathbf{C}^e\mathbf{N}^p. \tag{99.11}$$

Further, by (91.24),

$$\mathbf{A}:\dot{\mathbf{C}}^e = 2\mathbf{A}:(\mathbf{F}^{e\top}\mathbf{D}^e\mathbf{F}^e), \tag{99.12}$$

and, if we use (99.11) to eliminate the term $\mathbf{F}^{e\top}\mathbf{D}^e\mathbf{F}^e$ in (99.12) and invoke (2.52), we find that

$$\begin{aligned}\mathbf{A}:\dot{\mathbf{C}}^e &= 2\mathbf{A}:\big(\mathbf{F}^{p-\top}\dot{\mathbf{E}}\mathbf{F}^{p-1} - |\mathbf{D}^p|\mathbf{C}^e\mathbf{N}^p\big) \\ &= 2\mathbf{A}:(\mathbf{F}^{p-\top}\dot{\mathbf{E}}\mathbf{F}^{p-1}) - 2|\mathbf{D}^p|\mathbf{A}:(\mathbf{C}^e\mathbf{N}^p) \\ &= 2(\mathbf{F}^{p-1}\mathbf{A}\mathbf{F}^{p-\top}):\dot{\mathbf{E}} - 2|\mathbf{D}^p|\mathbf{A}:(\mathbf{C}^e\mathbf{N}^p).\end{aligned} \tag{99.13}$$

The tensor field

$$\underline{\mathbf{A}} \stackrel{\text{def}}{=} \mathbf{F}^{p-1}\mathbf{A}\mathbf{F}^{p-\top} \tag{99.14}$$

represents the tensor $\mathbf{A}$ pulled back contravariantly from the structural space to the reference space.[594] Using (99.14), we can write (99.13) in the form

$$\mathbf{A}:\dot{\mathbf{C}}^e = 2\underline{\mathbf{A}}:\dot{\mathbf{E}} - 2\mathbf{A}:(\mathbf{C}^e\mathbf{N}^p)|\mathbf{D}^p|$$

[594] Cf. (12.2) with $\mathbf{F}$ replaced by $\mathbf{F}^p$.

and (99.10) becomes

$$\dot{f} = 2\underline{\mathbf{A}}:\dot{\mathbf{E}} - \underbrace{[2\mathbf{A}:(\mathbf{C}^e\mathbf{N}^p) + H(e^p)]}_{G}|\mathbf{D}^p|. \tag{99.15}$$

We henceforth restrict attention to materials for which

$$G \stackrel{\text{def}}{=} 2\mathbf{A}:(\mathbf{C}^e\mathbf{N}^p) + H(e^p) > 0. \tag{99.16}$$

Remark. The restriction (99.16) may seem devoid of physical meaning, but it is not. The elastic strains are typically small; thus since $\mathbf{C}^e - \mathbf{1} = 2\mathbf{E}^e$ and $\mathbf{F}^e = \mathbf{R}^e\mathbf{U}^e$,

$$\mathbf{C}^e \approx \mathbf{1}, \qquad \operatorname{tr}\mathbf{C}^e \approx 3, \qquad \mathbf{C}^e:\mathbf{N}^p \approx 0, \qquad \mathbf{F}^e \approx \mathbf{R}^e, \tag{99.17}$$

and hence, by (99.9),

$$\mathbf{A} \approx \mu\mathbf{N}^p, \qquad G \approx 2\mu + H(e^p), \qquad \underline{\mathbf{A}} \approx \mu\mathbf{F}^{p-1}\mathbf{N}^p\mathbf{F}^{p-\top}. \tag{99.18}$$

The approximation $(99.18)_2$ should justify the restriction (99.16). Further, *G is approximately the modulus* (77.4) *of the small-deformation theory* and, hence, the discussion following (77.4) is also appropriate here.

We now derive conditions that determine whether the material leaves — or remains on — the yield surface when the material point in question is subjected to a loading program characterized by the strain tensor $\mathbf{E}$.

(i) **Elastic unloading** is defined by the condition $\underline{\mathbf{A}}:\dot{\mathbf{E}} < 0$. In this case, since $|\mathbf{D}^p| \geq 0$, (99.15) implies that $\dot{f} < 0$, and from the no-flow conditions (98.8) we conclude that $\mathbf{D}^p = \mathbf{0}$.

(ii) **Neutral loading** is defined by the condition $\underline{\mathbf{A}}:\dot{\mathbf{E}} = 0$. In this case $|\mathbf{D}^p| > 0$ cannot hold, for if it did, then (99.15) would imply that $\dot{f} < 0$, which would violate (98.8). Hence $|\mathbf{D}^p| = 0$, so that, once again, $\mathbf{D}^p = \mathbf{0}$.

(iii) **Plastic loading** is defined by the condition $\underline{\mathbf{A}}:\dot{\mathbf{E}} > 0$. In this case, if $|\mathbf{D}^p| = 0$ then $\dot{f} > 0$, which violates $\dot{f} \leq 0$. Hence $|\mathbf{D}^p| > 0$, and, since the consistency condition (98.9) then requires that $\dot{f} = 0$, (99.15) yields

$$|\mathbf{D}^p| = 2G^{-1}\underline{\mathbf{A}}:\dot{\mathbf{E}} \neq 0. \tag{99.19}$$

Thus, since $\mathbf{D}^p = |\mathbf{D}^p|\mathbf{N}^p$,

$$\mathbf{D}^p = 2G^{-1}(\underline{\mathbf{A}}:\dot{\mathbf{E}})\mathbf{N}^p \neq \mathbf{0}. \tag{99.20}$$

Hence, by (95.29), (99.9), and (99.14), if $\mathbf{N}^p$ is considered as $\mathbf{M}_0^e/|\mathbf{M}_0^e|$,[595] then (99.20) is determined by $\dot{\mathbf{E}}$, $\mathbf{C}^e$, and the current hardening modulus $H(e^p)$.

EXERCISE

1. Give arguments in support of the approximations (99.17).

[595] Cf. (99.1).

99.3 The Inverted Flow Rule

Combining the results of this discussion with the condition (98.8), we arrive at an equation for $|\mathbf{D}^p|$ that holds for all time, no matter whether $f = 0$ or $f < 0$:

$$\mathbf{D}^p = \begin{cases} \mathbf{0} & \text{if } f < 0 \quad \text{(behavior within the elastic range)}, \\ \mathbf{0} & \text{if } f = 0 \text{ and } \underline{\mathbf{A}}:\dot{\mathbf{E}} < 0 \quad \text{(elastic unloading)}, \\ \mathbf{0} & \text{if } f = 0 \text{ and } \underline{\mathbf{A}}:\dot{\mathbf{E}} = 0 \quad \text{(neutral loading)}, \\ 2G^{-1}(\underline{\mathbf{A}}:\dot{\mathbf{E}})\mathbf{N}^p & \text{if } f = 0 \text{ and } \underline{\mathbf{A}}:\dot{\mathbf{E}} > 0 \quad \text{(plastic loading)}. \end{cases} \tag{99.21}$$

The result (99.21) is embodied in the **inverted flow rule**

$$\mathbf{D}^p = 2\chi\, G^{-1}(\underline{\mathbf{A}}:\dot{\mathbf{E}})\mathbf{N}^p, \qquad \mathbf{N}^p = \frac{\mathbf{M}_0^e}{|\mathbf{M}_0^e|}, \tag{99.22}$$

where

$$\chi = \begin{cases} 0 & \text{if } f < 0, \text{ or if } f = 0 \text{ and } \underline{\mathbf{A}}:\dot{\mathbf{E}} \le 0 \\ 1 & \text{if } f = 0 \text{ and } \underline{\mathbf{A}}:\dot{\mathbf{E}} > 0 \end{cases} \tag{99.23}$$

is a **switching parameter**. Then, as in the small deformation theory (page 446), we have the following result whose proof parallels that given on page 447.

EQUIVALENCY THEOREM *Assume that the boundedness relation*

$$f = |\mathbf{M}_0^e| - Y(e^p) \le 0 \tag{99.24}$$

is satisfied. Then the inverted flow rule, as defined by (99.22) *and* (99.23), *is equivalent to the flow rule* $(95.50)_1$.

99.4 Equivalent Formulation of the Constitutive Equations and Plastic Mises Flow Equations Based on the Inverted Flow Rule

The Equivalency Theorem allows us to consider a complete set of plastic Mises flow equations based on the inverted flow rule:

(i) the Kröner decomposition[596]

$$\mathbf{F} = \mathbf{F}^e\mathbf{F}^p, \qquad \det\mathbf{F}^p = 1, \tag{99.25}$$

in which $\mathbf{F}$ is the deformation gradient, while $\mathbf{F}^e$ and $\mathbf{F}^p$ are the elastic and plastic distortions;

(ii) the elastic stress-strain relation

$$\mathbf{M}^e = \mu(\mathbf{C}^{e2} - \mathbf{C}^e) + \tfrac{1}{2}\lambda(\operatorname{tr}\mathbf{C}^e - 3)\mathbf{C}^e, \tag{99.26}$$

with $\mathbf{M}^e$ the Mandel stress and

$$\mathbf{C}^e = \mathbf{F}^{e\top}\mathbf{F}^e \tag{99.27}$$

the right elastic Cauchy–Green tensor;

(iii) the plastic boundedness inequality

$$f = |\mathbf{M}_0^e| - Y(e^p) \le 0, \tag{99.28}$$

where f is the yield function with $Y(e^p) > 0$ the flow resistance a function of a hardening variable e^p;

[596] Viewed as a kinematic constitutive equation.

(iv) a system of evolution equations

$$\boxed{\begin{gathered}\dot{\mathbf{F}}^p = \mathbf{D}^p\mathbf{F}^p, \quad \mathbf{D}^p = 2\chi\, G^{-1}(\underline{\mathbf{A}}:\dot{\mathbf{E}})\mathbf{N}^p, \qquad \mathbf{N}^p = \frac{\mathbf{M}_0^e}{|\mathbf{M}_0^e|}, \\ \dot{e}^p = |\mathbf{D}^p|, \quad e^p(\mathbf{X},0) = 0\end{gathered}} \tag{99.29}$$

for $\mathbf{F}^p$, $\mathbf{D}^p$, and the hardening variable e^p, in which

$$\mathbf{E} = \tfrac{1}{2}(\mathbf{F}^{\top}\mathbf{F} - \mathbf{1}) \tag{99.30}$$

is the Green–St. Venant strain, G is the modulus

$$G = 2\mathbf{A}:(\mathbf{C}^e\mathbf{N}^p) + H(e^p) > 0, \tag{99.31}$$

χ is the switching parameter

$$\chi = \begin{cases} 0 & \text{if } f < 0, \text{ or if } f = 0 \text{ and } \underline{\mathbf{A}}:\dot{\mathbf{E}} \le 0 \\ 1 & \text{if } f = 0 \quad \text{and} \quad \underline{\mathbf{A}}:\dot{\mathbf{E}} > 0, \end{cases} \tag{99.32}$$

the tensor field

$$\underline{\mathbf{A}} \overset{\text{def}}{=} \mathbf{F}^{p-1}\mathbf{A}\mathbf{F}^{p-\top} \tag{99.33}$$

is the normal Mandel elasticity tensor

$$\mathbf{A} = \mu(2\mathbf{C}^e - \mathbf{1})\mathbf{N}^p + \tfrac{1}{2}\lambda\big[(\operatorname{tr}\mathbf{C}^e - 3)\mathbf{N}^p + (\mathbf{C}^e:\mathbf{N}^p)\mathbf{1}\big] \tag{99.34}$$

pulled back contravariantly to the reference space, and $H(e^p)$ is the hardening modulus

$$H(e^p) = Y'(e^p) = \frac{dY(e^p)}{de^p}. \tag{99.35}$$

Remark. Constitutive equations of the form (99.29) need to be accompanied by initial conditions. Typical initial conditions presume that the body is initially (at time $t = 0$, say) in a **virgin state** in the sense that

$$\mathbf{F}(\mathbf{X},0) = \mathbf{F}^p(\mathbf{X},0) = \mathbf{1}, \tag{99.36}$$

so that, by (99.25) and (99.30), $\mathbf{F}^e(\mathbf{X},0) = \mathbf{1}$ and $\mathbf{E}(\mathbf{X},0) = \mathbf{0}$.

100 Evolution Equation for the Second Piola Stress

By (94.4) and $(99.25)_1$, the second Piola stress (25.2) may be expressed in terms of $\mathbf{T}^e$ and $\mathbf{F}^p$,

$$\mathbf{T}_{\mathrm{RR}} = \mathbf{F}^{p-1}\mathbf{T}^e\mathbf{F}^{p-\top}, \tag{100.1}$$

a relation that with (94.8) yields the expression

$$\mathbf{M}^e = \mathbf{C}^e\mathbf{F}^p\mathbf{T}_{\mathrm{RR}}\mathbf{F}^{p\top} \tag{100.2}$$

for the Mandel stress. By (100.1),

$$\begin{aligned}\dot{\mathbf{T}}_{\mathrm{RR}} &= \dot{\mathbf{F}}^{p-1}\mathbf{T}^e\mathbf{F}^{p-\top} + \mathbf{F}^{p-1}\dot{\mathbf{T}}^e\mathbf{F}^{p-\top} + \mathbf{F}^{p-1}\mathbf{T}^e\dot{\mathbf{F}}^{p-\top}\\ &= \mathbf{F}^{p-1}[\mathbf{F}^p\dot{\mathbf{F}}^{p-1}\mathbf{T}^e + \dot{\mathbf{T}}^e + \mathbf{T}^e\dot{\mathbf{F}}^{p-\top}\mathbf{F}^{p\top}]\mathbf{F}^{p-\top}\\ &= \mathbf{F}^{p-1}[-\mathbf{D}^p\mathbf{T}^e + \dot{\mathbf{T}}^e - \mathbf{T}^e\mathbf{D}^p]\mathbf{F}^{p-\top},\end{aligned} \tag{100.3}$$

where the last step uses (97.2). Further, from $(99.2)_3$ we obtain $\mathbf{D}^p\mathbf{T}^e = \mathbf{T}^e\mathbf{D}^p$ and (100.3) becomes

$$\mathbf{F}^p\dot{\mathbf{T}}_{\mathrm{RR}}\mathbf{F}^{p\top} = \dot{\mathbf{T}}^e - 2\mathbf{T}^e\mathbf{D}^p, \tag{100.4}$$

or equivalently, since $\mathbf{D}^p = |\mathbf{D}^p|\mathbf{N}^p$,

$$\mathbf{F}^p\dot{\mathbf{T}}_{\mathrm{RR}}\mathbf{F}^{p\top} = \dot{\mathbf{T}}^e - 2|\mathbf{D}^p|\mathbf{T}^e\mathbf{N}^p. \tag{100.5}$$

On the other hand, (91.25) and (99.11) yield

$$\dot{\mathbf{E}}^e = \mathbf{F}^{e\top}\mathbf{D}\mathbf{F}^e - |\mathbf{D}^p|\mathbf{C}^e\mathbf{N}^p. \tag{100.6}$$

It is convenient to write $\mathbb{C}$ for the fourth-order elasticity tensor defined by

$$\mathbb{C}\mathbf{G} = 2\mu\mathbf{G} + \lambda(\mathrm{tr}\,\mathbf{G})\mathbf{1} \tag{100.7}$$

for every symmetric tensor $\mathbf{G}$, so that, by $(95.27)_2$,

$$\mathbf{T}^e = \mathbb{C}[\mathbf{E}^e] = 2\mu\mathbf{E}^e + \lambda(\mathrm{tr}\,\mathbf{E}^e)\mathbf{1}. \tag{100.8}$$

If we substitute (100.6) into (100.8) and combine the resulting equation with (100.5), we find that

$$\mathbf{F}^p\dot{\mathbf{T}}_{\mathrm{RR}}\mathbf{F}^{p\top} = \mathbb{C}[\mathbf{F}^{e\top}\mathbf{D}\mathbf{F}^e] - |\mathbf{D}^p|(\mathbb{C}[\mathbf{C}^e\mathbf{N}^p] + 2\mathbf{T}^e\mathbf{N}^p), \tag{100.9}$$

and we can write (100.9) in the form

$$\dot{\mathbf{T}}_{\mathrm{RR}} = \underbrace{\mathbf{F}^{p-1}\mathbb{C}[\mathbf{F}^{p-\top}\dot{\mathbf{E}}\mathbf{F}^{p-1}]\mathbf{F}^{p-\top}}_{\mathbf{Z}_1} - |\mathbf{D}^p|\mathbf{F}^{p-1}(\underbrace{\mathbb{C}[\mathbf{C}^e\mathbf{N}^p] + 2(\mathbf{T}^e\mathbf{N}^p)}_{\mathbf{Z}_2})\mathbf{F}^{p-\top}. \tag{100.10}$$

By (100.7)

$$\begin{aligned}\mathbf{Z}_1 &= 2\mu\mathbf{F}^{p-1}\mathbf{F}^{p-\top}\dot{\mathbf{E}}\mathbf{F}^{p-1}\mathbf{F}^{p-\top} + \lambda\,\mathrm{tr}\,(\mathbf{F}^{p-\top}\dot{\mathbf{E}}\mathbf{F}^{p-1})\mathbf{F}^{p-1}\mathbf{F}^{p-\top}\\ &= 2\mu\mathbf{C}^{p-1}\dot{\mathbf{E}}\mathbf{C}^{p-1} + \lambda\,\mathrm{tr}\,(\mathbf{F}^{p-1}\mathbf{F}^{p-\top}\dot{\mathbf{E}})\mathbf{F}^{p-1}\mathbf{F}^{p-\top}\\ &= 2\mu\mathbf{C}^{p-1}\dot{\mathbf{E}}\mathbf{C}^{p-1} + \lambda\,\mathrm{tr}\,(\mathbf{C}^{p-1}\dot{\mathbf{E}})\mathbf{C}^{p-1}\\ &= 2\mu\mathbf{C}^{p-1}\dot{\mathbf{E}}\mathbf{C}^{p-1} + \lambda\,(\dot{\mathbf{E}}:\mathbf{C}^{p-1})\mathbf{C}^{p-1},\end{aligned} \tag{100.11}$$

where

$$\mathbf{C}^p = \mathbf{F}^{p\top}\mathbf{F}^p \tag{100.12}$$

is the right plastic Cauchy–Green tensor. Similarly

$$\begin{aligned}\mathbf{Z}_2 &= 2\mu\mathbf{C}^e\mathbf{N}^p + \lambda(\mathbf{C}^e:\mathbf{N}^p)\mathbf{1} + 2\mu(\mathbf{C}^e - \mathbf{1})\mathbf{N}^p + \lambda(\mathrm{tr}\,\mathbf{C}^e - 3)\mathbf{N}^p\\ &= 2\mu(2\mathbf{C}^e - \mathbf{1})\mathbf{N}^p + \lambda\big[(\mathbf{C}^e:\mathbf{N}^p)\mathbf{1} + (\mathrm{tr}\,\mathbf{C}^e - 3)\mathbf{N}^p\big]\\ &= 2\mathbf{A}.\end{aligned} \tag{100.13}$$

Thus (100.10) reduces to

$$\dot{\mathbf{T}}_{\mathrm{RR}} = 2\mu\mathbf{C}^{p-1}\dot{\mathbf{E}}\mathbf{C}^{p-1} + \lambda\,(\dot{\mathbf{E}}:\mathbf{C}^{p-1})\mathbf{C}^{p-1} - 2|\mathbf{D}^p|\underline{\mathbf{A}}, \tag{100.14}$$

or, equivalently, by (77.10),

$$\boxed{\dot{\mathbf{T}}_{\mathrm{RR}} = 2\mu\mathbf{C}^{p-1}\dot{\mathbf{E}}\mathbf{C}^{p-1} + \lambda\,(\dot{\mathbf{E}}:\mathbf{C}^{p-1})\mathbf{C}^{p-1} - 2\chi\,G^{-1}(\underline{\mathbf{A}}:\dot{\mathbf{E}})\underline{\mathbf{A}}.} \tag{100.15}$$

EXERCISES

1. Use (25.11) and (100.14) to establish the following evolution equation for the Cauchy stress:

$$J\overset{\diamond}{\mathbf{T}} + \dot{J}\mathbf{T} = 2\mu\mathbf{C}^e\mathbf{D}\mathbf{C}^e + \lambda(\mathbf{D}:\mathbf{C}^e)\mathbf{B}^e - 2|\mathbf{D}^p|\bar{\mathbf{A}}. \tag{100.16}$$

   Here,

$$\overset{\diamond}{\mathbf{T}} = \dot{\mathbf{T}} - \mathbf{L}\mathbf{T} - \mathbf{T}\mathbf{L}^{\top}$$

   is the contravariant rate of $\mathbf{T}$ as defined in (20.26),

$$\bar{\mathbf{A}} = \mathbf{F}^e\mathbf{A}\mathbf{F}^{e\top} \tag{100.17}$$

   is $\mathbf{A}$ pushed forward contravariantly to the observed space, and $\mathbf{B}^e$ is given by $(91.16)_2$.
2. Show that, granted the small elastic-strain approximations

$$\mathbf{B}^e \approx \mathbf{C}^e \approx \mathbf{1}, \quad \mathrm{tr}\,\mathbf{B}^e \approx \mathrm{tr}\,\mathbf{C}^e \approx 3, \quad \mathbf{C}^e:\mathbf{N}^p \approx 0, \quad \mathbf{F}^e \approx \mathbf{R}^e, \tag{100.18}$$

the constitutive equations for large-deformation rate-independent isotropic plasticity may be written as the following evolution equations for $\mathbf{T}$ and e^p:[597]

$$\begin{aligned}\overset{\diamond}{\mathbf{T}} &= 2\mu\mathbf{D} + \lambda(\operatorname{tr}\mathbf{D})\mathbf{1} - 2\mu\,|\mathbf{D}^p|\,\bar{\mathbf{N}}^p,\\ \dot{e}^p &= |\mathbf{D}^p|, \quad e^p(\mathbf{X},0)=0,\end{aligned} \tag{100.19}$$

with

$$\bar{\mathbf{N}}^p = \frac{\mathbf{T}_0}{|\mathbf{T}_0|}, \qquad |\mathbf{D}^p| = \chi\,\beta\,\bar{\mathbf{N}}^p\!:\!\mathbf{D},$$

and

$$f = |\mathbf{T}_0| - Y(e^p) \le 0,$$

$$\chi = \begin{cases} 0 & \text{if } f<0, \text{ or if } f=0 \text{ and } \bar{\mathbf{N}}^p\!:\!\mathbf{D}\le 0,\\ 1 & \text{if } f=0 \ \text{ and } \ \bar{\mathbf{N}}^p\!:\!\mathbf{D}>0,\end{cases}$$

$$\beta = \frac{2\mu}{2\mu + H(e^p)} > 0, \qquad H(e^p) = Y'(e^p) = \frac{dY(e^p)}{de^p}.$$

[597] In the plasticity literature, the stress rate $\overset{\diamond}{\mathbf{T}}$ is often replaced by the corotational stress rate $\overset{\circ}{\mathbf{T}} = \dot{\mathbf{T}} - \mathbf{W}\mathbf{T} + \mathbf{T}\mathbf{W}$, a replacement that seems inconsistent with the approximation (100.18) of small elastic strains. However, note that the final stress rate appearing in (100.16) and thereby in (100.19) depends on the initial choice of elastic strain-measure and corresponding stress-measure used in the elastic constitutive equations $(95.1)_{1,2}$ and their linear version (95.27).

101 Rate-Dependent Plastic Materials

101.1 Rate-Dependent Flow Rule

The generalization of the theory to allow for rate-dependence follows as in §78; in fact, the discussion of that section follows verbatim the discussion of §78, provided we make the replacements

$$\mathbf{T}_0 \to \mathbf{M}_0^e \qquad \text{and} \qquad \dot{\mathbf{E}}^p \to \mathbf{D}^p. \tag{101.1}$$

Thus, in particular, the generalized **flow rate** is here given by

$$d^p = |\mathbf{D}^p| \tag{101.2}$$

and the **rate-dependent flow rule** and **hardening equation** have the form

$$\boxed{\begin{gathered} \mathbf{M}_0^e = g(d^p)Y(S)\mathbf{N}^p, \qquad \mathbf{N}^p = \frac{\mathbf{D}^p}{d^p}, \\ \dot{S} = h(d^p, S), \end{gathered}} \tag{101.3}$$

The rate-sensitivity function $g(d^p)$ is assumed to satisfy $g(0) = 0$ and is assumed further to be a strictly monotonically increasing function of d^p, and, hence, invertible.

101.2 Inversion of the Rate-Dependent Flow Rule

If we take the absolute value of (101.3), we arrive at a relation,

$$\frac{|\mathbf{M}_0^e|}{Y(S)} = g(d^p), \tag{101.4}$$

which may be inverted to give an expression

$$\begin{aligned} d^p &= g^{-1}\left(\frac{|\mathbf{M}_0^e|}{Y(S)}\right) \\ &\equiv f\left(\frac{|\mathbf{M}_0^e|}{Y(S)}\right) \end{aligned} \tag{101.5}$$

for the flow rate. Since g is monotonically increasing, $f = g^{-1}$ is monotonically increasing and, hence, strictly positive for all nonzero values of its argument d^p. For the special case in which the rate-sensitivity function is given in the power-law form (78.20) — with reference flow rate $d_0 > 0$ and rate-sensitivity parameter $m > 0$ as before, but with d^p now equal to $|\mathbf{D}^p|$ — the expression (101.5) has the specific form

$$d^p = d_0 \left(\frac{|\mathbf{M}_0^e|}{Y(S)} \right)^{\frac{1}{m}}. \tag{101.6}$$

As a direct consequence of the rate-dependent flow rule $(101.3)_1$,

$$\frac{\mathbf{M}_0^e}{|\mathbf{M}_0^e|} = \mathbf{N}^p \tag{101.7}$$

with the direction of the plastic strain-rate. Using this relation and (101.5), the rate-dependent flow rule $(101.3)_1$ may be inverted to give

$$\boxed{\mathbf{D}^p = f\left(\frac{|\mathbf{M}_0^e|}{Y(S)} \right) \frac{\mathbf{M}_0^e}{|\mathbf{M}_0^e|}.} \tag{101.8}$$

Thus,

- *the plastic stretching is nonzero whenever the stress is nonzero: There is no elastic range in which the response of the material is purely elastic, and there are no considerations of a yield condition, a consistency condition, loading/unloading conditions, and so forth.*

  Finally, we note that (101.8) takes the form

$$\mathbf{D}^p = d_0 \left(\frac{|\mathbf{M}_0^e|}{Y(S)} \right)^{\frac{1}{m}} \frac{\mathbf{M}_0^e}{|\mathbf{M}_0^e|} \tag{101.9}$$

when expressed in terms of the inverse of the power-law function (78.20).

101.3 Summary of the Complete Constitutive Theory

The complete set of constitutive equations consists of:

(i) the Kröner decomposition[598]

$$\mathbf{F} = \mathbf{F}^e\mathbf{F}^p, \qquad \det \mathbf{F}^p = 1, \tag{101.10}$$

in which $\mathbf{F}$ is the deformation gradient, while $\mathbf{F}^e$ and $\mathbf{F}^p$ are the elastic and plastic distortions;

(ii) the elastic stress-strain relation

$$\mathbf{M}^e = \mu(\mathbf{C}^{e2} - \mathbf{C}^e) + \tfrac{1}{2}\lambda(\mathrm{tr}\,\mathbf{C}^e - 3)\mathbf{C}^e, \tag{101.11}$$

with $\mathbf{M}^e$ the Mandel stress and

$$\mathbf{C}^e = \mathbf{F}^{e\top}\mathbf{F}^e \tag{101.12}$$

the right elastic Cauchy–Green tensor;

[598] Viewed as a kinematical constitutive equation.

(iii) a system of evolution equations

$$\boxed{\begin{aligned}\dot{\mathbf{F}}^p &= \mathbf{D}^p\mathbf{F}^p,\\ \mathbf{D}^p &= f\left(\frac{|\mathbf{M}_0^e|}{Y(S)}\right)\frac{\mathbf{M}_0^e}{|\mathbf{M}_0^e|},\\ \dot{S} &= h(|\mathbf{D}^p|, S),\end{aligned}} \tag{101.13}$$

for $\mathbf{F}^p$, $\mathbf{D}^p$, and the hardening variable S, where $Y(S) > 0$ is the flow resistance, and the function f vanishes when its argument vanishes and is monotonically increasing for all positive values of its argument.

PART XVII

THEORY OF SINGLE CRYSTALS UNDERGOING SMALL DEFORMATIONS

102.1 Introduction

Metals are most often encountered in the form of polycrystalline aggregates, composed of grains separated by grain boundaries, with the grain interiors having a structure close to that of a single crystal (Figure 102.1). At low (< 0.35) homologous temperatures[599] the macroscopic inelastic response of most polycrystalline metallic materials with grain sizes larger than about 100 nm is primarily due to the inelastic response of the interiors of the single crystals, and the boundaries of the crystals may be assumed to be perfectly bonded.

Figure 102.1. Photomicrograph of a metal. At the microstructural scale most metals are an aggregate of a large number of single crystals. The single crystals are called grains, and these are separated by grain boundaries.

In this section, we develop a rate-dependent theory of plasticity for single crystals undergoing small deformations. The deformation of a single crystal is generally presumed to result from two independent microscopic mechanisms:

(i) a local elastic deformation of the lattice;
(ii) a local plastic deformation that does not disturb the geometry of the lattice.

[599] Cf. Footnote 405.

Plastic deformation in the individual crystals (grains) generally occurs via the motion of dislocations on crystallographic slip planes in crystallographic slip directions; this microscopic motion results in macroscopic shearing of the slip planes in the slip directions; such shears are generally referred to as slips. Figure 102.2 shows a two-dimensional schematic of the motion of an edge dislocation on a slip plane in a cubic crystal.

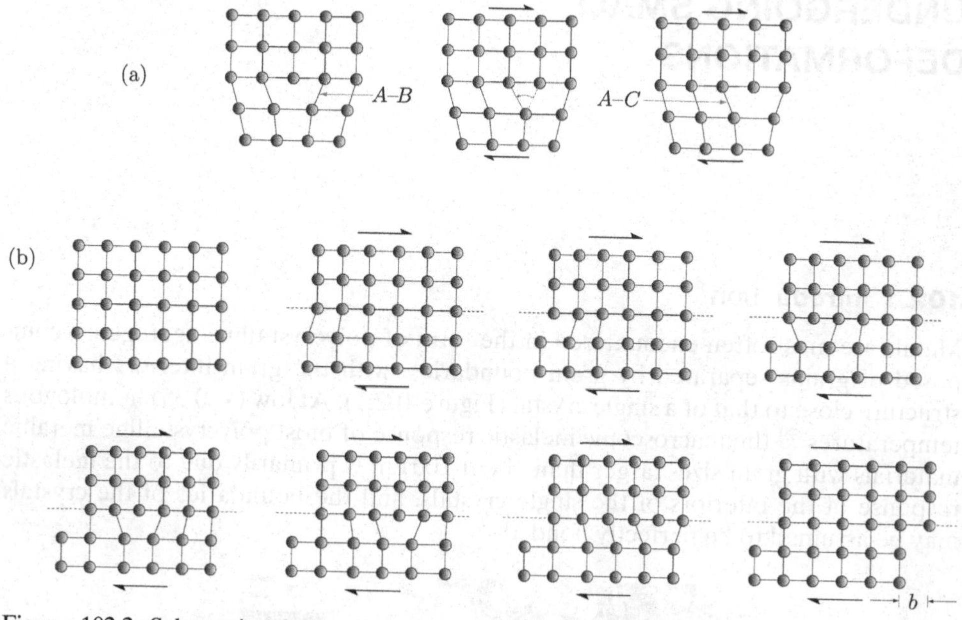

Figure 102.2. Schematic of a motion of an edge dislocation in a crystal under an applied shear stress. (a) An atomic bond A–B in the core of the dislocation breaks and forms a new bond A–C, which allows the dislocation to move. (b) Sequence showing the introduction of a dislocation from the left, its glide through the crystal on the slip plane in the slip direction, and its expulsion at the right to produce a slip step. This process causes the upper part of the crystal to slip by a distance b relative to the lower part.

The most common crystal structures in metals are:

(i) face-centered cubic (fcc); for example, Al, Cu, Ni, Ag, γ-Fe;
(ii) body-centered cubic (bcc); for example, Ta, V, Mo, Cr, α-Fe;
(iii) hexagonal close-packed (hcp); for example, Ti, Mg, Zn, Cd.

Schematics of these structures are shown in Figure 102.3.

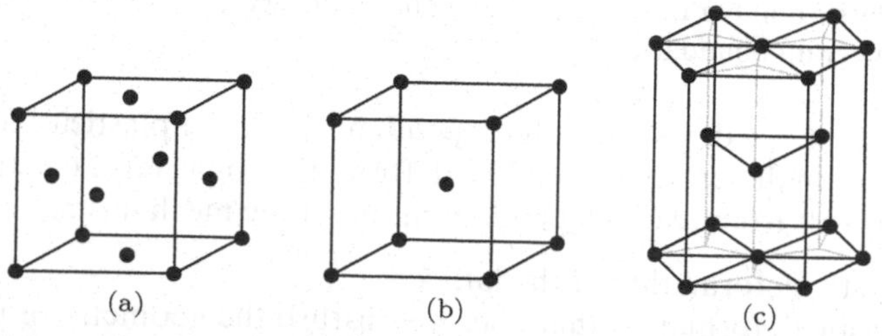

Figure 102.3. Schematics of common crystal structures: (a) face-centered cubic (fcc); (b) body-centered cubic (bcc); (c) hexagonal close-packed (hcp).

Stated precisely, plastic deformation occurs by slip in preferred slip directions

$$\mathbf{s}^\alpha, \qquad \alpha = 1, 2, \ldots, N,$$

on preferred slip planes identified by their normals

$$\mathbf{m}^\alpha, \qquad \alpha = 1, 2, \ldots, N,$$

where $\mathbf{s}^\alpha$ and $\mathbf{m}^\alpha$ are constant orthonormal lattice vectors. The pairs $(\mathbf{s}^\alpha, \mathbf{m}^\alpha)$, $\alpha = 1, 2, \ldots, N$, are then referred to as **slip systems**. The slip planes in a crystal are most often those planes with the highest density of atoms, and the slip directions in these slip planes are the directions in which the atoms are most closely packed.[600]

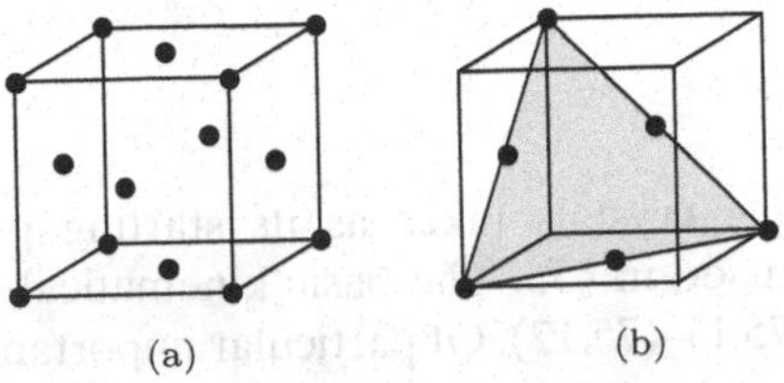

Figure 102.4. Unit cell of an fcc crystal, depicting the lattice sites and a slip plane (shaded). (a) The unit cell and atomic sites. (b) A triangular portion of the slip plane. The corresponding slip directions are parallel to the sides of the triangle.

The foundations of single-crystal plasticity may be traced to papers by Taylor (1938), Mandel (1965, 1972), Hill (1965), Teodosiu (1970), Rice (1971), Teodosiu & Sidoroff (1976), Asaro (1983), Asaro & Needleman (1985), and Kalidindi, Bronkhorst & Anand (1992). Our development of the theory follows Gurtin (2002), whose approach is based on the principle of virtual power.

[600] Cf. Figure 102.4.

102 Basic Single-Crystal Kinematics

The kinematics of single crystals takes as its starting point the kinematics of small deformation discussed in §75. The basic kinematical relations are the small-deformation relations (75.1)–(75.12). Of particular importance is the decomposition

$$\nabla\mathbf{u} = \mathbf{H}^e + \mathbf{H}^p \tag{102.1}$$

in which

(i) $\mathbf{H}^e$, the **elastic distortion**, represents stretch and rotation of the underlying microscopic structure, here a *lattice*;
(ii) $\mathbf{H}^p$, the **plastic distortion**, represents the local deformation of material due to the flow of dislocations through the lattice.

Single-crystal plasticity is based an hypothesis that makes precise an idealized view of the motion of dislocations in crystalline materials. To better understand the physical nature of this hypothesis recall that — within the framework of small deformations — simple shear is a deformation whose displacement gradient has the form

$$\nabla\mathbf{u} = \gamma\,\mathbf{s}\otimes\mathbf{m} \tag{102.2}$$

with $\mathbf{s}$ and $\mathbf{m}$ constant orthonormal unit vectors. Here $\mathbf{m}$ is normal to the planes being sheared, $\mathbf{s}$ is the direction of the shear, and γ is the scalar shear strain.[601]

The single-crystal hypothesis is based on the physical assumption that the motion of dislocations takes place on prescribed slip systems $\alpha = 1, 2, \ldots, N$; the presumption that plastic flow takes place through slip then manifests itself in the requirement that the plastic distortion $\mathbf{H}^p$ be the sum of simple shears

$$\gamma^\alpha\,\mathbf{s}^\alpha\otimes\mathbf{m}^\alpha$$

on the individual slip systems α. Thus, explicitly, the **single-crystal hypothesis** is the requirement that $\mathbf{H}^p$ be governed by **slips** γ^α on the individual slip systems via the relation

$$\mathbf{H}^p = \sum_\alpha \gamma^\alpha\,\mathbf{s}^\alpha\otimes\mathbf{m}^\alpha, \tag{102.3}$$

where, for each slip system α, $\mathbf{s}^\alpha$ is the **slip direction** and $\mathbf{m}^\alpha$ is the associated **slip-plane normal** and where $\mathbf{s}^\alpha$ and $\mathbf{m}^\alpha$ are constant orthonormal lattice vectors

$$\mathbf{s}^\alpha\cdot\mathbf{m}^\alpha = 0, \qquad |\mathbf{s}^\alpha| = |\mathbf{m}^\alpha| = 1. \tag{102.4}$$

[601] Cf. §51.

Here and in what follows:

- Lowercase Greek superscripts $\alpha, \beta, \ldots$ denote slip-system labels and as such range over the integers $1, 2, \ldots, N$.
- *We do not use summation convention for Greek superscripts.*
- We use the shorthand

$$\sum_{\alpha} = \sum_{\alpha=1}^{N}.$$

The tensor

$$\mathbb{S}^{\alpha} = \mathbf{s}^{\alpha} \otimes \mathbf{m}^{\alpha} \tag{102.5}$$

is referred to as the **Schmid tensor** for system α. By (102.3) and (102.4),

$$\operatorname{tr} \mathbb{S}^{\alpha} = 0 \qquad \text{and} \qquad \operatorname{tr} \mathbf{H}^{p} = 0, \tag{102.6}$$

consistent with an assumption of plastic incompressibility.[602] A consequence of (102.1), (102.3), and (102.5) is that

$$\nabla \dot{\mathbf{u}} = \dot{\mathbf{H}}^{e} + \sum_{\alpha} \dot{\gamma}^{\alpha} \mathbb{S}^{\alpha}. \tag{102.7}$$

[602] Cf. (75.12).

103 The Burgers Vector and the Flow of Screw and Edge Dislocations

This section is based on the discussion — beginning on page 509 — of the Burgers vector and the associated Burgers tensor

$$\mathbf{G} = \operatorname{Curl} \mathbf{H}^p, \tag{103.1}$$

where $\mathbf{G}^{\top}\mathbf{e}$ represents the Burgers vector, measured per unit area, for infinitesimal closed circuits on any plane Π with unit normal $\mathbf{e}$.

103.1 Decomposition of the Burgers Tensor G into Distributions of Edge and Screw Dislocations

Thus far, we have made no use of the single-crystal hypothesis (102.3); the theory is much richer when we do so. In view of (102.3) and (103.1),

$$\mathbf{G} = \sum_{\alpha} \operatorname{Curl} (\gamma^\alpha \mathbf{s}^\alpha \otimes \mathbf{m}^\alpha),$$

and, since

$$\begin{aligned}
[\operatorname{Curl} (\gamma^\alpha \mathbf{s}^\alpha \otimes \mathbf{m}^\alpha)]_{ij} &= \varepsilon_{irq} \frac{\partial \gamma^\alpha}{\partial x_r} s_j^\alpha m_q^\alpha \\
&= [(\nabla \gamma^\alpha \times \mathbf{m}^\alpha) \otimes \mathbf{s}^\alpha]_{ij},
\end{aligned}$$

we have the useful identity

$$\boxed{\mathbf{G} = \sum_{\alpha} (\nabla \gamma^\alpha \times \mathbf{m}^\alpha) \otimes \mathbf{s}^\alpha.} \tag{103.2}$$

Let

$$\Pi^\alpha \text{ denote } \textbf{slip plane } \alpha. \tag{103.3}$$

Then, for any α, the vectors $\mathbf{s}^\alpha$ and

$$\mathbf{l}^\alpha \stackrel{\text{def}}{=} \mathbf{m}^\alpha \times \mathbf{s}^\alpha \tag{103.4}$$

form an orthonormal basis for Π^α. Since the vector $\nabla\gamma^\alpha \times \mathbf{m}^\alpha$ in (103.2) is orthogonal to $\mathbf{m}^\alpha$, it can be expanded in terms of $\mathbf{s}^\alpha$ and $\mathbf{l}^\alpha$,

$$\begin{aligned}\nabla\gamma^\alpha \times \mathbf{m}^\alpha &= [\mathbf{l}^\alpha \cdot (\nabla\gamma^\alpha \times \mathbf{m}^\alpha)]\,\mathbf{l}^\alpha + [\mathbf{s}^\alpha \cdot (\nabla\gamma^\alpha \times \mathbf{m}^\alpha)]\,\mathbf{s}^\alpha \\ &= [(\mathbf{m}^\alpha \times \mathbf{l}^\alpha)\cdot\nabla\gamma^\alpha]\,\mathbf{l}^\alpha + [(\mathbf{m}^\alpha \times \mathbf{s}^\alpha)\cdot\nabla\gamma^\alpha]\,\mathbf{s}^\alpha \\ &= (-\mathbf{s}^\alpha\cdot\nabla\gamma^\alpha)\mathbf{l}^\alpha + (\mathbf{l}^\alpha\cdot\nabla\gamma^\alpha)\mathbf{s}^\alpha;\end{aligned}$$

hence we can write (103.2) in the form[603]

$$\mathbf{G} = \sum_\alpha [(\mathbf{l}^\alpha\cdot\nabla\gamma^\alpha)\mathbf{s}^\alpha\otimes\mathbf{s}^\alpha - (\mathbf{s}^\alpha\cdot\nabla\gamma^\alpha)\mathbf{l}^\alpha\otimes\mathbf{s}^\alpha]. \tag{103.5}$$

Within a continuum theory the geometric features of edge and screw dislocations are characterized by dyads of the form

$$\mathbf{l}\otimes\mathbf{s}\ \ldots\ \begin{cases}\mathbf{l}\perp\mathbf{s} & \text{edge,}\\ \mathbf{l}=\mathbf{s} & \text{screw,}\end{cases} \tag{103.6}$$

where $\mathbf{l}$ and $\mathbf{s}$ are *unit vectors*, with $\mathbf{s}$ the **Burgers direction** and $\mathbf{l}$ the **line direction.**[604] A tensor of the form

$$\rho\,\mathbf{l}\otimes\mathbf{s} \tag{103.7}$$

is then viewed as a *distribution of dislocations* with **density** ρ. Central to the current discussion is the observation that[605] the canonical dislocations for slip on system α are *screw dislocations* with both Burgers and line directions equal to $\mathbf{s}^\alpha$ and *edge dislocations* with Burgers direction $\mathbf{s}^\alpha$ and line direction $\mathbf{l}^\alpha = \mathbf{m}^\alpha \times \mathbf{s}^\alpha$. The *canonical dislocation dyads* for slip on α are therefore the *edge and screw dyads*

$$\mathbf{l}^\alpha\otimes\mathbf{s}^\alpha\ \text{(edge)} \qquad \text{and} \qquad \mathbf{s}^\alpha\otimes\mathbf{s}^\alpha\ \text{(screw)}. \tag{103.8}$$

Introducing the symbols "$\vdash$" and "$\odot$" for edge and screw dislocations, we can thus rewrite (103.5) in the form[606]

$$\boxed{\begin{gathered}\mathbf{G} = \sum_\alpha (\rho_\vdash^\alpha\,\mathbf{l}^\alpha\otimes\mathbf{s}^\alpha + \rho_\odot^\alpha\,\mathbf{s}^\alpha\otimes\mathbf{s}^\alpha),\\ \rho_\vdash^\alpha \stackrel{\text{def}}{=} -\mathbf{s}^\alpha\cdot\nabla\gamma^\alpha, \qquad \rho_\odot^\alpha \stackrel{\text{def}}{=} \mathbf{l}^\alpha\cdot\nabla\gamma^\alpha.\end{gathered}} \tag{103.9}$$

In view of the sentences containing (103.7) and (103.8), the tensor fields

$$\rho_\vdash^\alpha\,\mathbf{l}^\alpha\otimes\mathbf{s}^\alpha \qquad \text{and} \qquad \rho_\odot^\alpha\,\mathbf{s}^\alpha\otimes\mathbf{s}^\alpha$$

represent respective distributions of edge and screw dislocations on slip system α; for that reason, we refer to $\rho_\vdash^\alpha$ and $\rho_\odot^\alpha$ as **edge** and **screw dislocation densities**. Thus, appealing to (103.9) we see that

- **G** *can be decomposed into distributions of edge and screw dislocations on the individual slip systems.*

Note that the densities $\rho_\vdash^\alpha$ and $\rho_\odot^\alpha$ carry units of $(\text{length})^{-1}$ and that they may be positive or negative.

---

603 Fleck, Muller, Ashby & Hutchinson (1994).
604 Nye (1953).
605 Cf. Kubin, Canova, Condat, Devincre, Pontikis & Brechet (1992), Sun, Adams, Shet, Saigal & King (1998), Sun, Adams & King (2000), and Arsenlis & Parks (1999).
606 Arsenlis & Parks (1999).

The **glide directions** for dislocations of a given type lie in the slip plane and are orthogonal to the line direction; thus, by (103.4),

$$\left.\begin{aligned} \mathbf{l}^\alpha &= \mathbf{m}^\alpha \times \mathbf{s}^\alpha \quad \text{is the glide direction for screw dislocations on } \alpha;\\ -\mathbf{s}^\alpha &= \mathbf{m}^\alpha \times \mathbf{l}^\alpha \quad \text{is the glide direction for edge dislocations on } \alpha. \end{aligned}\right\} \tag{103.10}$$

Each of the densities $(103.9)_{2,3}$ therefore represents a directional derivative of the slip in the glide direction. Thus, in accord with experience, a pile-up of, say, screw dislocations as characterized by a positive value of the directional derivative $(103.9)_2$ for $\rho^\alpha_\odot > 0$ (or a negative value for $\rho^\alpha_\odot < 0$) results in an increasing absolute density of screw dislocations, and similarly for edge dislocations. *The screw and edge dislocation-densities* $\rho^\alpha_\odot$ *and* $\rho^\alpha_\vdash$ *thus characterize the pile-up of screw and edge dislocations on slip plane* α.

EXERCISE

1. Show that

$$\operatorname{tr}\mathbf{G} = \sum_\alpha \rho^\alpha_\odot \tag{103.11}$$

and that the axial vector of the skew part of **G** is[607]

$$-\tfrac{1}{2}\textstyle\sum_\alpha \rho^\alpha_\vdash \mathbf{m}^\alpha. \tag{103.12}$$

The trace of **G** is therefore completely determined by the screw densities and the skew part of **G** is completely determined by the edge densities.

103.2 Dislocation Balances

Interestingly, the edge and screw dislocation distributions obey more or less standard balance laws relating the density rates to associated fluxes. To derive these balance laws, we introduce *edge and screw dislocation fluxes* by

$$\mathbf{q}^\alpha_\vdash = \dot{\gamma}^\alpha \mathbf{s}^\alpha \qquad \text{and} \qquad \mathbf{q}^\alpha_\odot = -\dot{\gamma}^\alpha \mathbf{l}^\alpha, \tag{103.13}$$

so that, consistent with experience, the dislocation fluxes are parallel to their associated glide directions (103.10). By $(103.9)_{2,3}$

$$\dot{\rho}^\alpha_\vdash = -\mathbf{s}^\alpha \cdot \nabla\dot{\gamma}^\alpha = -\operatorname{Div}(\dot{\gamma}^\alpha \mathbf{s}^\alpha) \qquad \text{and} \qquad \dot{\rho}^\alpha_\odot = \mathbf{l}^\alpha \cdot \nabla\dot{\gamma}^\alpha = \operatorname{Div}(\dot{\gamma}^\alpha \mathbf{l}^\alpha),$$

equations that when combined with (103.13) yield balance laws for distributions of edge and screw dislocations:

$$\dot{\rho}^\alpha_\vdash = -\operatorname{Div}\mathbf{q}^\alpha_\vdash \qquad \text{and} \qquad \dot{\rho}^\alpha_\odot = -\operatorname{Div}\mathbf{q}^\alpha_\odot. \tag{103.14}$$

103.3 The Tangential Gradient ∇^α on the Slip Plane Π^α

Recall our agreement that Π^α represents slip plane α. By $(2.6)_2$, the tensor

$$\mathbf{P}^\alpha \stackrel{\text{def}}{=} \mathbf{1} - \mathbf{m}^\alpha \otimes \mathbf{m}^\alpha \tag{103.15}$$

represents the **projection** onto Π^α: given any vector **v**

$$\mathbf{P}^\alpha\mathbf{v} = \mathbf{v} - (\mathbf{m}^\alpha \cdot \mathbf{v})\mathbf{v},$$

so that $\mathbf{P}^\alpha$ associates with any vector **v** its component $\mathbf{P}^\alpha\mathbf{v}$ *tangent to* Π^α.

---

[607] Cf. (2.40).

Since $\{\mathbf{m}^\alpha, \mathbf{s}^\alpha, \mathbf{l}^\alpha\}$ represents an orthonormal basis for the space $\mathcal{V}$ of all vectors, (2.7) implies that

$$\mathbf{s}^\alpha \otimes \mathbf{s}^\alpha + \mathbf{l}^\alpha \otimes \mathbf{l}^\alpha + \mathbf{m}^\alpha \otimes \mathbf{m}^\alpha = \mathbf{1}, \tag{103.16}$$

and, hence, by (103.15) that

$$\mathbf{P}^\alpha = \mathbf{s}^\alpha \otimes \mathbf{s}^\alpha + \mathbf{l}^\alpha \otimes \mathbf{l}^\alpha. \tag{103.17}$$

Further,

$$\begin{aligned}\mathbf{P}^\alpha\mathbf{P}^\alpha &= (\mathbf{1} - \mathbf{m}^\alpha \otimes \mathbf{m}^\alpha)(\mathbf{1} - \mathbf{m}^\alpha \otimes \mathbf{m}^\alpha)\\ &= \mathbf{1} - \mathbf{m}^\alpha \otimes \mathbf{m}^\alpha,\end{aligned}$$

so that

$$\mathbf{P}^\alpha\mathbf{P}^\alpha = \mathbf{P}^\alpha. \tag{103.18}$$

Given any field φ,

$$\nabla^\alpha\varphi \overset{\text{def}}{=} \mathbf{P}^\alpha\nabla\varphi \tag{103.19}$$

is the **tangential gradient of** φ **on** Π^α. By (103.17),

$$\nabla^\alpha\varphi = (\mathbf{s}^\alpha \cdot \nabla\varphi)\mathbf{s}^\alpha + (\mathbf{l}^\alpha \cdot \nabla\varphi)\mathbf{l}^\alpha, \tag{103.20}$$

and we may conclude from $(103.9)_{2,3}$ that

$$\nabla^\alpha\gamma^\alpha = -\rho^\alpha_\vdash\mathbf{s}^\alpha + \rho^\alpha_\odot\mathbf{l}^\alpha; \tag{103.21}$$

the dislocation densities $\rho^\alpha_\vdash$ and $\rho^\alpha_\odot$ therefore represent components of the tangential slip gradient $\nabla^\alpha\gamma^\alpha$ relative to the basis $\{-\mathbf{s}^\alpha, \mathbf{l}^\alpha\}$ for the slip plane Π^α. The field

$$\rho^\alpha_{\text{net}} \overset{\text{def}}{=} \sqrt{|\rho^\alpha_\vdash|^2 + |\rho^\alpha_\odot|^2} \tag{103.22}$$

represents the **net dislocation density** on Π^α. An interesting and important consequence of (103.21) is then that

$$\rho^\alpha_{\text{net}} = |\nabla^\alpha\gamma^\alpha|. \tag{103.23}$$

Terminology: Our use of the terms Burgers vector and dislocation density are **continuum mechanical** in origin and differ from these terms as used by materials scientists. As noted in the paragraph containing (103.1), the term Burgers vector signifies a vector of a given length, measured per unit area, and the notion of a dislocation density (such as that given by (103.23)) also represents a length measured per unit area. Thus in continuum mechanics both the Burgers vector and the dislocation densities carry the dimension *length*$^{-1}$. In **materials science** the Burgers vector — shown schematically in Figure 88.1 — is the vector that represents the closure failure of a Burgers circuit around a *single* dislocation in a crystal lattice; it's magnitude, denoted by b, is *also* referred to as the Burgers vector, a definition that typically renders b an interatomic spacing. To emphasize this difference in notation, we henceforth refer to b as the *material Burgers vector*. In materials science dislocation densities are measured in dislocations per unit area and hence carry the dimension *length*$^{-2}$. Each continuum-mechanical density, say ρ_{cm}, can be converted to a materials-science density, say ρ_{ms}, via the transformation

$$\rho_{\text{ms}} = b^{-1}\rho_{\text{cm}}. \tag{103.24}$$

Throughout, unless stated to the contrary, *we use the terminology of continuum mechanics*, so that each density is a "ρ_{cm}". As a consequence of the transformation

(103.24), the net dislocation density (103.23) would have the following form as a *materials science density*:[608]

$$(\rho^\alpha_{\text{net}})_{\text{ms}} = b^{-1}|\nabla^\alpha \gamma^\alpha|. \tag{103.25}$$

EXERCISE

1. Consider the dislocation concentrations defined by

$$c^\alpha_\vdash = \frac{\rho^\alpha_\vdash}{\rho^\alpha_{\text{net}}} \qquad \text{and} \qquad c^\alpha_\odot = \frac{\rho^\alpha_\odot}{\rho^\alpha_{\text{net}}}. \tag{103.26}$$

Derive the identities

$$\overline{\rho^\alpha_{\text{net}}}^{\boldsymbol{\cdot}} = c^\alpha_\vdash \dot{\rho}^\alpha_\vdash + c^\alpha_\odot \dot{\rho}^\alpha_\odot \tag{103.27}$$

and

$$\overline{\rho^\alpha_{\text{net}}}^{\boldsymbol{\cdot}} = -\mathrm{Div}\,\mathbf{q}^\alpha_{\text{net}} + \sigma^\alpha_{\text{net}}, \tag{103.28}$$

where

$$\mathbf{q}^\alpha_{\text{net}} = c^\alpha_\vdash \mathbf{q}^\alpha_\vdash + c^\alpha_\odot \dot{\mathbf{q}}^\alpha_\odot \qquad \text{and} \qquad \sigma^\alpha_{\text{net}} = \nabla c^\alpha_\vdash \cdot \mathbf{q}^\alpha_\vdash + \nabla c^\alpha_\odot \cdot \mathbf{q}^\alpha_\odot$$

represent the net dislocation flux and the net dislocation supply on Π^α, respectively. What is the physical significance of these identities?

[608] This relation bears comparison with a well-known relation due to Ashby (1970); cf. Fleck, Muller, Ashby & Hutchinson (1994), Ohashi (1997).

104 Conventional Theory of Single-Crystals

104.1 Virtual-Power Formulation of the Standard and Microscopic Force Balances

For a single crystal, the basic independent kinematical quantities characterizing plastic flow are the slips

$$\gamma^1, \gamma^2, \ldots, \gamma^N; \tag{104.1}$$

therefore, the basic "rate-like" descriptors for a single-crystal undergoing small deformations are the velocity $\dot{\mathbf{u}}$, the elastic distortion-rate $\dot{\mathbf{H}}^e$, and the slip rates $\dot{\gamma}^1, \dot{\gamma}^2, \ldots, \dot{\gamma}^N$; these fields are constrained by (102.7).

As in §84, the formulation of the principle of virtual power for a single crystal is based on a balance between the external power $\mathcal{W}(\mathrm{P})$ expended *on* P and the internal power $\mathcal{I}(\mathrm{P})$ expended *within* P. As before, to describe the internal power we replace the classical stress power $\mathbf{T}\colon\operatorname{grad}\dot{\mathbf{u}}$ by a more detailed reckoning that *individually* characterizes:

- the stretching and spinning of the underlying material structure as described by the lattice distortion-rate $\dot{\mathbf{H}}^e$ and
- dislocation-induced slip as described by the slip rates $\dot{\gamma}^1, \dot{\gamma}^2, \ldots, \dot{\gamma}^N$.

Specifically, we allow for power expended internally by:

- an **elastic-stress** $\mathbf{T}^e$ power-conjugate to $\dot{\mathbf{H}}^e$, and
- a scalar **internal microscopic force** π^α power-conjugate to $\dot{\gamma}^\alpha$ (for each slip system α)

and, therefore, write the **internal power** in the form

$$\mathcal{I}(\mathrm{P}) = \int\limits_{\mathrm{P}} \mathbf{T}^e\colon\dot{\mathbf{H}}^e\,dv + \sum_\alpha \int\limits_{\mathrm{P}} \pi^\alpha\dot{\gamma}^\alpha\,dv. \tag{104.2}$$

We assume that the **external power** has the standard form (84.5); viz.

$$\mathcal{W}(\mathrm{P}) = \int\limits_{\partial\mathrm{P}} \mathbf{t}(\mathbf{n})\cdot\dot{\mathbf{u}}\,da + \int\limits_{\mathrm{P}} \mathbf{b}\cdot\dot{\mathbf{u}}\,dv. \tag{104.3}$$

Here, for convenience, we do not include a scalar external virtual microscopic force b^α power conjugate to $\dot{\gamma}^\alpha$ (for each slip system α). Such an external force b^α would be a counterpart of the external microscopic force B^p of the isotropic theory[609] and would result in an external-power expenditure of the form

$$\sum_\alpha \int_P b^\alpha \dot{\gamma}^\alpha \, dv. \tag{104.4}$$

Slip rates are directly related to dislocation flow; the external force b^α therefore represents an external virtual force associated with the flow of dislocations on slip plane α.

The principle of virtual power is based on a view of the velocity $\dot{\mathbf{u}}$, the elastic distortion-rate $\dot{\mathbf{H}}^e$, and the slip rates $\dot{\gamma}^1, \dot{\gamma}^2, \ldots, \dot{\gamma}^N$ as virtual velocities to be specified independently in a manner consistent

with (102.7); that is, denoting the virtual velocities by $\tilde{\mathbf{u}}$, $\tilde{\mathbf{H}}^e$, and $\tilde{\gamma}^1, \tilde{\gamma}^2, \ldots \tilde{\gamma}^N$, we require that

$$\nabla\tilde{\mathbf{u}} = \tilde{\mathbf{H}}^e + \sum_\alpha \tilde{\gamma}^\alpha \mathbb{S}^\alpha. \tag{104.5}$$

Thus if we define a (generalized) *virtual velocity* to be a list

$$\mathcal{V} = (\tilde{\mathbf{u}}, \tilde{\mathbf{H}}^e, \tilde{\gamma}^1, \tilde{\gamma}^2, \ldots, \tilde{\gamma}^N) \tag{104.6}$$

and write

$$\left.\begin{aligned} \mathcal{W}(\mathrm{P}, \mathcal{V}) &= \int_{\partial \mathrm{P}} \mathbf{t}(\mathbf{n}) \cdot \tilde{\mathbf{u}} \, da + \int_{\mathrm{P}} \mathbf{b} \cdot \tilde{\mathbf{u}} \, dv, \\ \mathcal{I}(\mathrm{P}, \mathcal{V}) &= \int_{\mathrm{P}} \mathbf{T}^e \cdot \tilde{\mathbf{H}}^e \, dv + \sum_\alpha \int_{\mathrm{P}} \pi^\alpha \tilde{\gamma}^\alpha \, dv, \end{aligned}\right\} \tag{104.7}$$

for the corresponding external and internal expenditures of *virtual power*, then the **principle of virtual power** is the requirement that, given any subregion P of the body,

$$\underbrace{\int_{\partial \mathrm{P}} \mathbf{t}(\mathbf{n}) \cdot \tilde{\mathbf{u}} \, da + \int_{\mathrm{P}} \mathbf{b} \cdot \tilde{\mathbf{u}} \, dv}_{\mathcal{W}(\mathrm{P}, \mathcal{V})} = \underbrace{\int_{\mathrm{P}} \mathbf{T}^e : \tilde{\mathbf{H}}^e \, dv + \sum_\alpha \int_{\mathrm{P}} \pi^\alpha \tilde{\gamma}^\alpha \, dv}_{\mathcal{I}(\mathrm{P}, \mathcal{V})} \tag{104.8}$$

for all virtual velocities $\mathcal{V}$.

Arguing as in the steps leading to (84.14), we then note that under a change in frame the transformation laws (84.12) should be replaced by

$$\tilde{\mathbf{H}}^{e*} = \tilde{\mathbf{H}}^e + \mathbf{W}, \qquad \tilde{\gamma}^\alpha \text{ is invariant (for each } \alpha), \tag{104.9}$$

with $\mathbf{W}$ an arbitrary spatially constant skew tensor field. Under this replacement, frame-indifference once again leads to (84.13) and hence to the conclusion that, as before,

$$\mathbf{T}^e \text{ is symmetric.} \tag{104.10}$$

Consider a virtual velocity $\mathcal{V}$ with $\tilde{\mathbf{u}}$ arbitrary,

$$\tilde{\mathbf{H}}^e = \nabla\tilde{\mathbf{u}}, \tag{104.11}$$

and

$$\tilde{\gamma}^\alpha \equiv \mathbf{0} \tag{104.12}$$

[609] Cf. the paragraph containing (‡) on page 496. Here such external forces would add nothing of import.

for all α, so that (104.5) is satisfied and (104.8) reduces to the classical virtual-power balance (84.19); viz.

$$\int_{\partial P} \mathbf{t}(\mathbf{n}) \cdot \tilde{\mathbf{u}}\, da + \int_{P} \mathbf{b} \cdot \tilde{\mathbf{u}}\, dv = \int_{P} \mathbf{T}^e : \nabla\tilde{\mathbf{u}}\, dv. \tag{104.13}$$

The steps leading from (84.19) to the equations (84.20)–(84.23) are then valid without change; thus, referring to

$$\mathbf{T} \overset{\text{def}}{=} \mathbf{T}^e \tag{104.14}$$

as the **macroscopic stress**, we are led to the traction condition

$$\mathbf{t}(\mathbf{n}) = \mathbf{T}\mathbf{n} \tag{104.15}$$

and the local force balance

$$\operatorname{Div}\mathbf{T} + \mathbf{b} = \mathbf{0}, \tag{104.16}$$

a relation that, by (84.4) and (84.22), is equivalent to the momentum balance

$$\operatorname{Div}\mathbf{T} + \mathbf{b}_0 = \rho\ddot{\mathbf{u}}. \tag{104.17}$$

Basic to what follows is the **resolved shear stress** τ^α defined by

$$\begin{aligned} \tau^\alpha &\overset{\text{def}}{=} \mathbb{S}^\alpha : \mathbf{T} \\ &= \mathbf{s}^\alpha \cdot \mathbf{T}\mathbf{m}^\alpha; \end{aligned} \tag{104.18}$$

τ^α represents the macroscopic stress $\mathbf{T}$ *resolved on slip system* α.

Consider a generalized virtual velocity $\mathcal{V}$ with

$$\tilde{\mathbf{u}} \equiv \mathbf{0} \qquad \text{and} \qquad \tilde{\mathbf{H}}^e = -\sum_\alpha \tilde{\gamma}^\alpha \mathbb{S}^\alpha, \tag{104.19}$$

with $\tilde{\gamma}^1, \tilde{\gamma}^2, \ldots, \tilde{\gamma}^N$ arbitrary. Then the constraint (104.5) is satisfied and (104.8) and (104.14) imply that

$$\int_P \mathbf{T} : \tilde{\mathbf{H}}^e\, dv + \sum_\alpha \int_P \pi^\alpha \tilde{\gamma}^\alpha\, dv = 0. \tag{104.20}$$

Further, by (104.18) and $(104.19)_2$,

$$\begin{aligned} \mathbf{T} : \tilde{\mathbf{H}}^e &= -\sum_\alpha \tilde{\gamma}^\alpha\, \mathbf{T} : \mathbb{S}^\alpha \\ &= -\sum_\alpha \tau^\alpha \tilde{\gamma}^\alpha \end{aligned} \tag{104.21}$$

and (104.20) becomes

$$\sum_\alpha \int_P (\pi^\alpha - \tau^\alpha)\tilde{\gamma}^\alpha\, dv = 0. \tag{104.22}$$

Since (104.22) is to be satisfied for all $\tilde{\gamma}^\alpha$ and all P, we must have the **microscopic force balance**

$$\boxed{\tau^\alpha = \pi^\alpha} \tag{104.23}$$

for each slip system α.

The microscopic force balance arises as a consequence of the arbitrary nature of the slip rates and for that reason might be viewed as a force balance for the system of dislocations on slip system α. Specifically, one might view:[610]

- π^α as representing internal forces on slip system α associated with the creation, annihilation, motion, and general interaction of dislocations,
- τ^α as representing the force exerted by the lattice on the system of dislocations on slip system α.

Note that, by (104.16) and (104.18), the macroscopic stress $\mathbf{T}$ plays two roles: as the standard stress in the macroscopic balance and as the coupling term between the microscopic and macroscopic force systems via its resolved values on the individual slip systems.

Finally, because $\mathbf{T} = \mathbf{T}^e$ is symmetric,

$$\mathbf{T}:\mathbf{H}^e = \mathbf{T}:\mathbf{E}^e,$$

with

$$\mathbf{E}^e = \operatorname{sym}\mathbf{H}^e \tag{104.24}$$

the elastic strain. Thus, by (104.13) and (104.14), the virtual-power balance (104.8), when applied to the actual fields within the body, yields the power balance

$$\underbrace{\int\limits_{\partial \mathrm{P}} \mathbf{Tn}\cdot\dot{\mathbf{u}}\,da + \int\limits_{\mathrm{P}} \mathbf{b}\cdot\dot{\mathbf{u}}\,dv}_{\mathcal{W}(\mathrm{P})} = \underbrace{\int\limits_{\mathrm{P}} \mathbf{T}:\dot{\mathbf{E}}^e\,dv + \sum_\alpha \int\limits_{\mathrm{P}} \pi^\alpha\dot{\gamma}^\alpha\,dv}_{\mathcal{I}(\mathrm{P})}. \tag{104.25}$$

104.2 Free-Energy Imbalance

Letting Ψ denote the *free energy* and $\delta \geq 0$ the *dissipation*, each measured per unit volume, the general free-energy imbalance introduced in §84.4 leads to the *free-energy imbalance*

$$\overline{\int\limits_{\mathrm{P}} \Psi\,dv}^{\boldsymbol{\cdot}} = \mathcal{W}(\mathrm{P}) - \int\limits_{\mathrm{P}} \delta\,dv, \tag{104.26}$$

or equivalently, by (104.25),

$$\overline{\int\limits_{\mathrm{P}} \Psi\,dv}^{\boldsymbol{\cdot}} = \int\limits_{\mathrm{P}} \mathbf{T}:\dot{\mathbf{E}}^e\,dv + \sum_\alpha \int\limits_{\mathrm{P}} \pi^\alpha\dot{\gamma}^\alpha\,dv - \int\limits_{\mathrm{P}} \delta\,dv. \tag{104.27}$$

Thus, since P is arbitrary, we have the local **free-energy imbalance**

$$\dot{\Psi} - \mathbf{T}:\dot{\mathbf{E}}^e - \sum_\alpha \pi^\alpha\dot{\gamma}^\alpha = -\delta \leq 0. \tag{104.28}$$

104.3 General Separable Constitutive Theory

We neglect defect energy and restrict attention to constitutive assumptions that separate plastic and elastic response; in particular, guided by (104.28), we characterize

[610] Were we to account for scalar external forces b^α via the power expenditure (104.4), then the microscopic force balance (104.23) would have the form $\pi^\alpha - \tau^\alpha = b^\alpha$ and b^α might be viewed as an external force on the system of dislocations on α.

elastic response by the equations

$$\Psi = \hat{\Psi}(\mathbf{E}^e),$$
$$\mathbf{T} = \hat{\mathbf{T}}(\mathbf{E}^e). \qquad (104.29)$$

We consider *rate-dependent* constitutive relations for plastic flow giving the internal microscopic forces π^α as functions of the slip rates $\dot{\gamma}^1, \dot{\gamma}^2, \ldots, \dot{\gamma}^N$ and *hardening variables* $S^1, S^2, \ldots, S^N$. Precisely, introducing lists

$$\vec{\nu} \stackrel{\text{def}}{=} (\dot{\gamma}^1, \dot{\gamma}^2, \ldots, \dot{\gamma}^N) \qquad \text{and} \qquad \vec{S} \stackrel{\text{def}}{=} (S^1, S^2, \ldots, S^N), \qquad (104.30)$$

we characterize plastic response by relations that, for each slip system α, consist of a constitutive equation for the internal microscopic force π^α supplemented by a differential equation for the hardening variable S^α:

$$\pi^\alpha = \hat{\pi}^\alpha(\vec{\nu}, \vec{S}),$$
$$\dot{S}^\alpha = h^\alpha(\vec{\nu}, \vec{S}). \qquad (104.31)$$

The following **thermodynamic restrictions** render the constitutive equations (104.29) and (104.31) consistent with the free-energy imbalance (104.28):

- *The free energy determines the elastic stress through the* **elastic-stress relation**

$$\hat{\mathbf{T}}(\mathbf{E}^e) = \frac{\partial \hat{\Psi}(\mathbf{E}^e)}{\partial \mathbf{E}^e}. \qquad (104.32)$$

- *The internal microscopic forces must satisfy the* **reduced dissipation inequality**

$$\sum_\alpha \hat{\pi}^\alpha(\vec{\nu}, \vec{S})\dot{\gamma}^\alpha \geq 0. \qquad (104.33)$$

We restrict attention to **strict dissipation** with the assumption that

$$\sum_\alpha \hat{\pi}^\alpha(\vec{\nu}, \vec{S})\dot{\gamma}^\alpha > 0 \qquad (104.34)$$

if at least one slip rate $\dot{\gamma}^\beta$ is nonzero.

104.4 Linear Elastic Stress-Strain Law

As we are working within the framework of small deformations, we restrict attention to a quadratic free energy and a linear stress-strain relation and therefore begin with elastic constitutive equations in the form

$$\Psi = \tfrac{1}{2}\mathbf{E}^e : \mathbb{C}\mathbf{E}^e, \qquad \Psi = \tfrac{1}{2}C_{ijkl}E^e_{ij}E^e_{kl},$$
$$\mathbf{T} = \mathbb{C}\mathbf{E}^e, \qquad T_{ij} = C_{ijkl}E^e_{kl}. \qquad (104.35)$$

We assume that the elasticity tensor $\mathbb{C}$ has the properties listed on page 300. If, in particular, the material has cubic symmetry, then the free energy and the stress may be expressed as in §52.4.2.

104.5 Constitutive Equations for Flow with Simple Rate-Dependence

Turning to the constitutive relations (104.31) that govern plastic flow, we assume that

$$\pi^\alpha = S^\alpha \hat{\pi}(\dot{\gamma}^\alpha) \qquad (104.36)$$

for each α, and that

$$S^\alpha > 0 \quad \text{for each } \alpha.$$

Thus

(i) coupling between the individual slip systems is solely through the hardening equations $(104.31)_2$;
(ii) the hardening variable S^α represents the **flow resistance** of slip system α;[611]
(iii) the response of any two slip systems is the same if their flow resistances as well as their slip rates coincide.

Choose a slip system α and consider the following choice for the slip rates:

$$\dot{\gamma}^\beta = 0 \quad \text{for } \beta \neq \alpha,$$

a kinematics referred to as **single slip**. Since $S^\alpha > 0$, this choice for $\vec{\nu}$ and the strict dissipation assumption (104.34) applied to (104.36) yields the conclusion that, for

$$\nu = \dot{\gamma}^\alpha,$$

we must have

$$\hat{\pi}(\nu)\nu > 0$$

for $\nu \neq 0$. Thus

$$\operatorname{sgn}\hat{\pi}(\nu) = \operatorname{sgn}\nu, \tag{104.37}$$

where, for $\nu \neq 0$,

$$\operatorname{sgn}\nu = \frac{\nu}{|\nu|}.$$

Consequently,

$$\hat{\pi}(\nu) = -\hat{\pi}(-\nu), \tag{104.38}$$

so that, granted $\hat{\pi}(\nu)$ is continuous,[612]

$$\hat{\pi}(0) = 0. \tag{104.39}$$

The result (104.38) has an important consequence:[613]

$$\hat{\pi}(\nu) = \hat{\pi}(|\nu|)\operatorname{sgn}\nu. \tag{104.40}$$

To verify (104.40) we simply note that:

(i) (104.40) is satisfied for $\nu > 0$;
(ii) for $\nu < 0$,

$$\begin{aligned}\hat{\pi}(-\nu) &= -\hat{\pi}(-\nu)\\ &= -\hat{\pi}(|\nu|)\\ &= \hat{\pi}(|\nu|)\operatorname{sgn}\nu.\end{aligned}$$

Consistent with the notation used for rate-dependent isotropic materials in §78, we write

$$g(|\nu|) = \hat{\pi}(|\nu|), \tag{104.41}$$

611 Cf. §76.5.2.
612 And hence well defined at $\nu = 0$, a property that would not hold were the material *rate-independent*.
613 By (104.39) $\hat{\pi}(\nu)$ is well defined at $\nu = 0$; in fact $\hat{\pi}(0) = 0$.

so that (104.40) becomes

$$\hat{\pi}(\nu) = g(|\nu|)\,\mathrm{sgn}\,\nu. \tag{104.42}$$

We refer to g as the **rate-sensitivity function**. By (104.38) and (104.39),

$$g(0) = 0, \qquad g(|\nu|) > 0 \quad \text{for } |\nu| > 0. \tag{104.43}$$

Next, by (104.42) the constitutive relations $(104.31)_1$ for the internal microscopic stresses take the simple form

$$\pi^\alpha = S^\alpha g(|\dot{\gamma}^\alpha|)\,\mathrm{sgn}\,\dot{\gamma}^\alpha. \tag{104.44}$$

An immediate consequence of (104.44) is that the dissipation $\delta = \sum_\alpha \pi^\alpha \dot{\gamma}^\alpha$ takes the form

$$\delta = \sum_\alpha S^\alpha g(|\dot{\gamma}^\alpha|)\,|\dot{\gamma}^\alpha|. \tag{104.45}$$

In view of (104.44) and the microscopic force balances (104.23), the constitutive relations (104.44) become relations for the resolved shear stresses

$$\tau^\alpha = S^\alpha g(|\dot{\gamma}^\alpha|)\,\mathrm{sgn}\,\dot{\gamma}^\alpha \tag{104.46}$$

for each slip system α. This relation represents a **flow rule** for slip system α.

We consider next the hardening equations $(104.31)_2$; viz.

$$\dot{S}^\alpha = h^\alpha(\vec{\nu}, \vec{S}). \tag{104.47}$$

We begin with two assumptions:

(H1) The hardening equations are rate-independent.
(H2) The hardening equations are independent of the signs of the slip rates.

These assumptions are consistent with what is often — but not always — assumed in the literature.[614]

To determine the consequences of (H1) and (H2) we let

$$\nu^\alpha \stackrel{\text{def}}{=} \dot{\gamma}^\alpha \quad \text{and} \quad a^\alpha \stackrel{\text{def}}{=} |\nu^\alpha| \tag{104.48}$$

for each α and rewrite the hardening equations in the form

$$\dot{S}^\alpha = h^\alpha(\nu^1, \nu^2, \dots, \nu^N, \vec{S}). \tag{104.49}$$

Then the requirement that the hardening equations be independent of the signs of the slip rates implies that

$$h^\alpha(\nu^1, \nu^2, \dots, \nu^N, \vec{S}) = h^\alpha(a^1, a^2, \dots, a^N, \vec{S}).$$

Let

$$h^{\alpha\beta}(a^1, a^2, \dots, a^N, \vec{S}) \stackrel{\text{def}}{=} \frac{\partial h^\alpha(a^1, a^2, \dots, a^N, \vec{S})}{\partial a^\beta}. \tag{104.50}$$

By (95.34) and (104.48), for each α, the quantities ν^α, a^α, and $\dot{S}^\alpha$ transform as

$$\nu^\alpha_\kappa = \kappa\nu^\alpha, \qquad a^\alpha_\kappa = \kappa a^\alpha, \qquad \text{and} \qquad \dot{\overline{S^\alpha_\kappa}} = \kappa\dot{S}^\alpha$$

under a change in time-scale with rate constant $\kappa > 0$. Thus, under such a change, the hardening equations (104.49) transform according to

$$\kappa\dot{S}^\alpha = h^\alpha(\kappa a^1, \kappa a^2, \dots, \kappa a^N, \vec{S}),$$

614 These assumptions rule out: (a) strain-rate history effects as observed in strain-rate increment and decrement experiments (cf., e.g., Kocks 1987; Balasubramanian & Anand 2002); and (b) Bauschinger-type effects under cyclic loading conditions (cf., e.g., Cailletaud 1992).

and if these equations are to be rate-independent we must have

$$\kappa h^{\alpha}(a^1, a^2, \ldots, a^N, \vec{S}) = h^{\alpha}(\kappa a^1, \kappa a^2, \ldots, \kappa a^N, \vec{S}).$$

Differentiating this equation with respect to κ using the chain-rule and (104.50) we therefore find, after evaluating the result at $\kappa = 0$, that

$$h^{\alpha}(a^1, a^2, \ldots, a^N, \vec{S}) = \sum_{\beta} h^{\alpha\beta}(0, 0, \ldots, 0, \vec{S}) a^{\beta}.$$

Thus, for

$$h^{\alpha\beta}(\vec{S}) \stackrel{\text{def}}{=} h^{\alpha\beta}(0, 0, \ldots, 0, \vec{S})$$

the hardening equation takes the simple form[615]

$$\dot{S}^{\alpha} = \sum_{\beta} h^{\alpha\beta}(\vec{S}) |\dot{\gamma}^{\beta}|. \tag{104.51}$$

We have therefore shown that (104.51) are the most general hardening equations of the form (104.47) that are rate-independent and independent of the signs of the slip rates.

Summarizing, the main results of this section are the flow relations (104.46) and (104.51); viz.

$$\tau^{\alpha} = S^{\alpha} g(|\nu^{\alpha}|) \operatorname{sgn} \nu^{\alpha},$$

$$\dot{S}^{\alpha} = \sum_{\beta} h^{\alpha\beta}(\vec{S}) |\dot{\gamma}^{\beta}| \tag{104.52}$$

for each slip system α.

The equations (104.52) relate the macroscopic response as characterized by the resolved shear stresses τ^{α} and the slip rates $\dot{\gamma}^{\alpha}$. Immediate consequences of $(104.52)_1$ are that:

(i) for each slip system α, the sign of $\dot{\gamma}^{\alpha}$ is determined by the sign of τ^{α},

$$\operatorname{sgn} \dot{\gamma}^{\alpha} = \operatorname{sgn} \tau^{\alpha}, \tag{104.53}$$

(ii) the flow resistance stress τ^{α} satisfies

$$|\tau^{\alpha}| = S^{\alpha} g(|\dot{\gamma}^{\alpha}|). \tag{104.54}$$

Assume that

$$g \text{ is a monotonically increasing function of its argument.} \tag{104.55}$$

Then, g is invertible and writing $f = g^{-1}$ for the corresponding inverse function, (104.43) yields

$$f(0) = 0, \qquad f(z) > 0 \quad \text{for } z > 0, \tag{104.56}$$

and

$$f \text{ is a monotonically increasing function for all positive values of its argument.} \tag{104.57}$$

Thus, inverting (104.54) we find that

$$|\nu^{\alpha}| = f\Big(\frac{|\tau^{\alpha}|}{S^{\alpha}}\Big), \tag{104.58}$$

[615] This derivation of the hardening equations is due to GURTIN (2000b, p. 1006).

and $(104.52)_1$ may be solved for the slip rates

$$\dot{\gamma}^\alpha = f\Big(\frac{|\tau^\alpha|}{S^\alpha}\Big)\,\mathrm{sgn}\,\tau^\alpha. \tag{104.59}$$

104.6 Power-Law Rate Dependence

An example of a commonly used rate-sensitivity function is the **power-law function**

$$g(|\dot{\gamma}^\alpha|) = \left(\frac{|\dot{\gamma}^\alpha|}{d_0}\right)^m, \tag{104.60}$$

where $m > 0$, a constant, is a **rate-sensitivity parameter** and $d_0 > 0$, also a constant, is a reference flow-rate.[616] The power-law function satisfies $g(|\dot{\gamma}^\alpha|) \approx 1$ for $|\dot{\gamma}^\alpha| \approx d_0$, (104.60) is therefore intended to model plastic flows with rates close to d_0. The power-law function allows one to characterize *nearly rate-independent behavior* for which m is very small. Such rate-dependent models also serve as an important *regularization of rate-independent* crystal plasticity response.[617] Granted the power-law function (104.60), the expressions (104.54) and (104.59) have the specific forms

$$|\dot{\gamma}^\alpha| = d_0\left(\frac{|\tau^\alpha|}{S^\alpha}\right)^{\frac{1}{m}} \tag{104.61}$$

and

$$\dot{\gamma}^\alpha = d_0\left(\frac{|\tau^\alpha|}{S^\alpha}\right)^{\frac{1}{m}}\mathrm{sgn}\,\tau^\alpha. \tag{104.62}$$

104.7 Self-Hardening, Latent-Hardening

Recall that for any slip system α, Π^α represents slip plane α.[618] We say that two slip systems α and β are **coplanar** if $\Pi^\alpha = \Pi^\beta$ — that is, if either of the following (equivalent) conditions is satisfied:

$$\mathbf{m}^\beta \times \mathbf{m}^\alpha = \mathbf{0}, \qquad \mathbf{m}^\alpha = \pm\mathbf{m}^\beta. \tag{104.63}$$

The moduli $h^{\alpha\beta}$ for α and β coplanar are referred to as self-hardening moduli, while the moduli for α and β noncoplanar are termed latent-hardening moduli. We refer to the quantities

$$\chi^{\alpha\beta} \stackrel{\text{def}}{=} \begin{cases} 1 & \text{for } \alpha \text{ and } \beta \text{ coplanar slip systems} \\ 0 & \text{otherwise} \end{cases} \tag{104.64}$$

as **coplanarity moduli**, as they differentiate between coplanar and noncoplanar slip systems. Trivially,

$$1 - \chi^{\alpha\beta} = \begin{cases} 0 & \text{for } \alpha \text{ and } \beta \text{ coplanar slip systems} \\ 1 & \text{otherwise.} \end{cases}$$

---

616 Cf. §78.3.
617 Cf. Footnote 411.
618 Cf. (103.3).

The hardening moduli are often presumed to have the specific form[619]

$$h^{\alpha\beta}(\vec{S}) = \underbrace{\chi^{\alpha\beta}h(S^\beta)}_{\text{self-hardening}} + \underbrace{q(1-\chi^{\alpha\beta})h(S^\beta)}_{\text{latent-hardening}}, \tag{104.65}$$

where $h \geq 0$ is a self-hardening function and $q > 0$ is the interaction constant (the ratio of the self-hardening rate to the latent-hardening rate), and, granted (104.65), the hardening equations $(104.52)_2$ become

$$\dot{S}^\alpha = \sum_\beta [\chi^{\alpha\beta} + q(1-\chi^{\alpha\beta})]h(S^\beta)|\nu^\beta|. \tag{104.66}$$

If we use the term individual slip plane to connote the collection of all slip planes coplanar to a given slip plane, then

(i) *self-hardening* characterizes hardening on an individual slip plane due to slip on that slip plane — that is, slip on all slip systems coplanar to the given slip plane;
(ii) *latent-hardening* characterizes hardening on an individual slip plane due to slip on all other individual slip planes.

Bronkhorst, Kalidindi & Anand (1992), in their discussion of fcc crystals, propose a self-hardening function of the form

$$h(S) = \begin{cases} h_0\left(1 - \dfrac{S}{S^*}\right)^a & \text{for } S_0 \leq S \leq S^*, \\ 0 & \text{for } S \geq S^*, \end{cases} \tag{104.67}$$

where S^*, a, and h_0 are constant moduli with $S^* > S_0$, $a \geq 1$, and $h_0 > 0$. The hardening function h defined via (104.67) is strictly decreasing for $S_0 \leq S \leq S^*$ and vanishes for $S \geq S^*$.[620]

Finally, we consider initial conditions for the hardening equations in the form

$$S^\alpha(\mathbf{X}, 0) = S_0, \tag{104.68}$$

with $S_0 > 0$, a constant, the *initial slip-resistance*.[621]

104.8 Summary of the Constitutive Theory

The constitutive equations that characterize a single-crystal undergoing small deformations consist of:

(i) the decomposition

$$\nabla\mathbf{u} = \mathbf{H}^e + \mathbf{H}^p, \qquad \mathrm{tr}\,\mathbf{H}^p = 0, \tag{104.69}$$

with $\mathbf{H}^e$ and $\mathbf{H}^p$ the elastic and plastic distortions,
(ii) the elastic stress-strain relation

$$\mathbf{T} = \mathbb{C}\mathbf{E}^e \tag{104.70}$$

in which

$$\mathbf{E}^e = \mathrm{sym}\,\nabla\mathbf{u} \tag{104.71}$$

is the elastic strain and $\mathbf{T}$ is the macroscopic stress; the elasticity tensor $\mathbb{C}$ is assumed to have the properties listed on page 300,

[619] Cf., e.g., Hutchinson (1970), Asaro (1983), and Peirce, Asaro & Needleman (1983).
[620] For a detailed discussion of latent-hardening and other self-hardening rules; cf. Bassani & Wu (1991) and Bassani (1994).
[621] Cf. Footnote 386.

(iii) the relation

$$\mathbf{H}^p = \sum_\alpha \gamma^\alpha \, \mathbf{s}^\alpha \otimes \mathbf{m}^\alpha, \tag{104.72}$$

for the plastic distortion, where $(\mathbf{s}^\alpha, \mathbf{m}^\alpha)$ denotes slip system α with slip direction $\mathbf{s}^\alpha$ and slip plane normal $\mathbf{m}^\alpha$ constant vectors subject to

$$\mathbf{s}^\alpha \cdot \mathbf{m}^\alpha = 0, \qquad |\mathbf{s}^\alpha| = |\mathbf{m}^\alpha| = 1, \tag{104.73}$$

and where γ^α represents the slip on Π^α,

(iv) the (inverted) constitutive relations[622]

$$\dot{\gamma}^\alpha = f\Big(\frac{|\tau^\alpha|}{S^\alpha}\Big)\operatorname{sgn}\tau^\alpha \tag{104.74}$$

for the individual slip systems α, where

$$\tau^\alpha = \mathbf{s}^\alpha \cdot \mathbf{T}\mathbf{m}^\alpha \tag{104.75}$$

is the resolved shear stress, $S^\alpha > 0$ represents a flow resistance, and the inverted rate-sensitivity function f satisfies (104.56) and (104.57); the special case of power-law rate-dependency corresponds to

$$\dot{\gamma}^\alpha = d_0\left(\frac{|\tau^\alpha|}{S^\alpha}\right)^{\frac{1}{m}} \operatorname{sgn}\tau^\alpha, \tag{104.76}$$

with $d_0 > 0$ a reference slip rate, and $m > 0$ a rate-sensitivity parameter.

(v) the hardening equations and initial condition

$$\dot{S}^\alpha = \sum_\beta h^{\alpha\beta}(\vec{S})|\dot{\gamma}^\beta|, \qquad S^\alpha(\mathbf{X}, 0) = S_0 \ (\text{a constant} > 0), \tag{104.77}$$

where $h^{\alpha\beta}$ are hardening moduli; a widely used special form for the hardening moduli is

$$h^{\alpha\beta}(\vec{S}) = \underbrace{\chi^{\alpha\beta} h(S^\beta)}_{\text{self-hardening}} + \underbrace{q(1-\chi^{\alpha\beta}) h(S^\beta)}_{\text{latent-hardening}}, \tag{104.78}$$

where

$$\chi^{\alpha\beta} = \begin{cases} 1 & \text{if } \mathbf{m}^\beta \times \mathbf{m}^\alpha = \mathbf{0} \\ 0 & \text{otherwise} \end{cases} \tag{104.79}$$

are coplanarity moduli, $h \geq 0$ is a self-hardening function, and $q > 0$ is an interaction constant; a common form for the self-hardening function is

$$h(S) = \begin{cases} h_0\left(1 - \dfrac{S}{S^*}\right)^a & \text{for } S_0 \leq S \leq S^*, \\ 0 & \text{for } S \geq S^*, \end{cases} \tag{104.80}$$

where S^*, a, and h_0 are constant moduli with $S^* > S_0$, $a \geq 1$, and $h_0 > 0$.

[622] Each such relation is actually a combination of a constitutive relation for π^α and the microscopic force balance $\pi^\alpha = \tau^\alpha$.

105 Single-Crystal Plasticity at Small Length-Scales: A Small-Deformation Gradient Theory

In this section we develop a gradient theory of single-crystals building on the conventional theory developed in §104;[623] specifically, we generalize the virtual-power formulation described in §104.1 by allowing for *internal power expenditures* associated with *slip-rate gradients* $\nabla\dot{\gamma}^{\alpha}$.[624]

105.1 Virtual-Power Formulation of the Standard and Microscopic Force Balances of the Gradient Theory

As in section §104.1, the basic "rate-like" descriptors are the velocity $\dot{\mathbf{u}}$, the elastic distortion $\mathbf{H}^e$, and the slip rates $\dot{\gamma}^1, \dot{\gamma}^2, \ldots, \dot{\gamma}^N$, and these fields are constrained by (102.7); viz.

$$\nabla\dot{\mathbf{u}} = \dot{\mathbf{H}}^e + \sum_{\alpha} \dot{\gamma}^{\alpha}\mathbb{S}^{\alpha}, \tag{105.1}$$

with

$$\mathbb{S}^{\alpha} = \mathbf{s}^{\alpha} \otimes \mathbf{m}^{\alpha} \tag{105.2}$$

the Schmid tensor for slip system α.[625]

We follow §104.1 in allowing for power expended internally by

- an **elastic-stress** $\mathbf{T}^e$ power conjugate to $\dot{\mathbf{H}}^e$, and
- a scalar **internal microscopic force** π^{α} power conjugate to $\dot{\gamma}^{\alpha}$ (for each slip system α);

but, in addition, we now allow slip-rate gradients $\nabla\dot{\gamma}^{\alpha}$ to affect the internal power via the introduction of

- a vector **microscopic stress** $\boldsymbol{\xi}^{\alpha}$ power-conjugate to the slip-rate gradient $\nabla\dot{\gamma}^{\alpha}$ (for each slip system α).

We therefore write the **internal power** in the form

$$\mathcal{I}(\mathrm{P}) = \int_{\mathrm{P}} \mathbf{T}^e : \dot{\mathbf{H}}^e \, dv + \sum_{\alpha} \int_{\mathrm{P}} (\pi^{\alpha}\dot{\gamma}^{\alpha} + \boldsymbol{\xi}^{\alpha} \cdot \nabla\dot{\gamma}^{\alpha}) \, dv, \tag{105.3}$$

with P an arbitrary subregion of the body.

Turning to the external power, we allow for the standard power expenditures $\mathbf{t}(\mathbf{n}) \cdot \dot{\mathbf{u}}$ and $\mathbf{b} \cdot \dot{\mathbf{u}}$ by *surface tractions* and (conventional and inertial) *body forces*.

[623] Cf. also the isotropic gradient theories discussed in §89–§90.
[624] This section follows Gurtin, Anand & Lele (2007).
[625] Cf. (102.5).

Further, arguing as in the paragraph containing (90.11), we expect that the gradient terms $\boldsymbol{\xi}^\alpha \cdot \nabla\dot{\gamma}^\alpha$ should give rise to traction terms associated with the microscopic stresses $\boldsymbol{\xi}^\alpha$. Indeed, guided by the term $(\boldsymbol{\xi}^\alpha \cdot \mathbf{n})\dot{\gamma}^\alpha$ in the integral identity

$$\int\limits_{\mathrm{P}} \boldsymbol{\xi}^\alpha \cdot \nabla\dot{\gamma}^\alpha \, dv = \int\limits_{\partial\mathrm{P}} (\boldsymbol{\xi}^\alpha \cdot \mathbf{n})\dot{\gamma}^\alpha \, da - \int\limits_{\mathrm{P}} \dot{\gamma}^\alpha \mathrm{Div}\,\boldsymbol{\xi}^\alpha \, dv, \tag{105.4}$$

we assume that power is expended externally by

- a scalar **microscopic traction** $\Xi^\alpha(\mathbf{n})$ power-conjugate to $\dot{\gamma}^\alpha$

for each slip system α. We therefore assume that the **external power** has the form

$$\mathcal{W}(\mathrm{P}) = \int\limits_{\partial\mathrm{P}} \mathbf{t}(\mathbf{n}) \cdot \dot{\mathbf{u}} \, da + \int\limits_{\mathrm{P}} \mathbf{b} \cdot \dot{\mathbf{u}} \, dv + \sum_\alpha \int\limits_{\partial\mathrm{P}} \Xi^\alpha(\mathbf{n})\dot{\gamma}^\alpha \, da. \tag{105.5}$$

Here, to simplify the presentation, we do not include external microscopic forces.[626]

As before, we consider (generalized) virtual velocities

$$\mathcal{V} = (\tilde{\mathbf{u}}, \tilde{\mathbf{H}}^e, \tilde{\gamma}^1, \tilde{\gamma}^2, \ldots, \tilde{\gamma}^N)$$

consistent with the constraint

$$\nabla\tilde{\mathbf{u}} = \tilde{\mathbf{H}}^e + \sum_\alpha \tilde{\nu}^\alpha \mathbb{S}^\alpha, \tag{105.6}$$

a consideration that leads to the following expressions

$$\left.\begin{aligned} \mathcal{W}(\mathrm{P}, \mathcal{V}) &= \int\limits_{\partial\mathrm{P}} \mathbf{t}(\mathbf{n}) \cdot \tilde{\mathbf{u}} \, da + \int\limits_{\mathrm{P}} \mathbf{b} \cdot \tilde{\mathbf{u}} \, dv + \sum_\alpha \int\limits_{\partial\mathrm{P}} \Xi^\alpha(\mathbf{n})\, \tilde{\gamma}^\alpha \, da \\ \mathcal{I}(\mathrm{P}, \mathcal{V}) &= \int\limits_{\mathrm{P}} \mathbf{T}^e : \tilde{\mathbf{H}}^e \, dv + \sum_\alpha \int\limits_{\mathrm{P}} (\pi^\alpha \tilde{\gamma}^\alpha + \boldsymbol{\xi}^\alpha \cdot \nabla\tilde{\gamma}^\alpha) \, dv, \end{aligned}\right\} \tag{105.7}$$

for the external and internal expenditures of virtual power. The **principle of virtual power** is then the requirement that, given any subregion P of the body,

$$\underbrace{\int\limits_{\partial\mathrm{P}} \mathbf{t}(\mathbf{n}) \cdot \tilde{\mathbf{u}} \, da + \int\limits_{\mathrm{P}} \mathbf{b} \cdot \tilde{\mathbf{u}} \, dv + \sum_\alpha \int\limits_{\partial\mathrm{P}} \Xi^\alpha(\mathbf{n})\, \tilde{\gamma}^\alpha \, da}_{\mathcal{W}(\mathrm{P},\mathcal{V})}$$

$$= \underbrace{\int\limits_{\mathrm{P}} \mathbf{T}^e : \tilde{\mathbf{H}}^e \, dv + \sum_\alpha \int\limits_{\mathrm{P}} (\pi^\alpha \tilde{\gamma}^\alpha + \boldsymbol{\xi}^\alpha \cdot \nabla\tilde{\gamma}^\alpha) \, dv}_{\mathcal{I}(\mathrm{P},\mathcal{V})} \tag{105.8}$$

for all virtual velocities $\mathcal{V}$.

Next, the argument given in the paragraphs containing (104.10) and (104.16) leads to the symmetry of $\mathbf{T}^e$ and to the **traction condition**

$$\mathbf{t}(\mathbf{n}) = \mathbf{T}\mathbf{n} \tag{105.9}$$

and the **macroscopic force balance**

$$\mathrm{Div}\,\mathbf{T} + \mathbf{b} = \mathbf{0}, \tag{105.10}$$

626 Cf. the paragraph containing (104.3).

with macroscopic stress

$$\mathbf{T} \stackrel{\text{def}}{=} \mathbf{T}^e. \tag{105.11}$$

In our discussion of the conventional theory in §104.1, the arbitrary nature of the virtual slip rates leads to a system of microscopic force balances — but these balances are here in the form of partial differential equations (that need concomitant traction conditions). As in §104.1 our derivation of these balances is based on the use of a virtual velocity $\mathcal{V}$ consistent with

$$\tilde{\mathbf{u}} \equiv \mathbf{0} \qquad \text{and} \qquad \tilde{\mathbf{H}}^e = -\sum_\alpha \tilde{\gamma}^\alpha \mathbb{S}^\alpha, \tag{105.12}$$

with $\tilde{\gamma}^1, \tilde{\gamma}^2, \ldots, \tilde{\gamma}^N$ arbitrary.[627] For this choice of $\mathcal{V}$ the constraint (104.5) is satisfied and, for τ^α the resolved shear stress defined by (104.18), the relations (104.21) and $(105.12)_1$ reduce (105.8) to the **microscopic virtual-power relation**

$$\sum_\alpha \int_{\partial \mathrm{P}} \Xi^\alpha(\mathbf{n})\tilde{\gamma}^\alpha \, da = \sum_\alpha \int_{\mathrm{P}} [(\pi^\alpha - \tau^\alpha)\tilde{\gamma}^\alpha + \boldsymbol{\xi}^\alpha \cdot \nabla\tilde{\gamma}^\alpha] \, dv. \tag{105.13}$$

Appealing to the identity (105.4), but with the slip rates replaced by their virtual counterparts, we thus find that

$$\sum_\alpha \left(\int_{\partial \mathrm{P}} (\Xi^\alpha(\mathbf{n}) - \boldsymbol{\xi}^\alpha \cdot \mathbf{n})\tilde{\gamma}^\alpha \, da + \int_{\mathrm{P}} (\mathrm{Div}\,\boldsymbol{\xi}^\alpha + \tau^\alpha - \pi^\alpha)\tilde{\gamma}^\alpha \, dv \right) = 0. \tag{105.14}$$

Since the virtual slip rates are arbitrary, (105.14) must be satisfied for all $\tilde{\gamma}^1, \tilde{\gamma}^2, \ldots, \tilde{\gamma}^N$ and all P; a vectorial version of fundamental lemma of the calculus of variations (page 167) therefore yields the **microscopic force balance**[628]

$$\boxed{\tau^\alpha = \pi^\alpha - \mathrm{Div}\,\boldsymbol{\xi}^\alpha} \tag{105.15}$$

and the **microscopic traction condition**

$$\boxed{\Xi^\alpha(\mathbf{n}) = \boldsymbol{\xi}^\alpha \cdot \mathbf{n}} \tag{105.16}$$

for each slip system α.

Arguing as on page 596, the microscopic force balance (105.15) arises as a consequence of the arbitrary nature of the virtual slip rates and hence might be viewed as a force balance for the system of dislocations on slip system α. Specifically, one might view π^α as representing *internal* forces on the αth slip system associated with the creation, annihilation, and general interaction of dislocations; τ^α as representing the force exerted by the lattice on the system of dislocations on α; and, by virtue of the traction condition (105.16),

- $\boldsymbol{\xi}^\alpha \cdot \mathbf{n}$ as representing forces on slip system α associated with the flow of dislocations across surfaces with normal $\mathbf{n}$.

Consistent with this, in §105.3 we shall relate the microscopic stresses $\boldsymbol{\xi}^\alpha$ to classical Peach–Koehler forces.[629]

---

[627] Cf. (104.19).

[628] Gurtin (2000b, eq. (48); 2002, eq. (5.14)).

[629] Cf., e.g., Teodosiu (1982, p. 191) and Maugin (1993, p. 23) for discussions of classical Peach–Koehler forces.

105.2 Free-Energy Imbalance

Our derivation of the free-energy imbalance follows §104.2, the sole difference being the gradient term $\boldsymbol{\xi}^\alpha \cdot \nabla\dot{\gamma}^\alpha$ in the internal power (105.3); this leads to the free-energy imbalance

$$\overline{\int_P \Psi\, dv} - \int_P \mathbf{T}:\dot{\mathbf{E}}^e\, dv - \sum_\alpha \int_P (\pi^\alpha\dot{\gamma}^\alpha + \boldsymbol{\xi}^\alpha\cdot\nabla\dot{\gamma}^\alpha)\, dv = -\int_P \delta\, dv \le 0, \qquad (105.17)$$

and hence to the local **free-energy imbalance**[630]

$$\boxed{\dot{\Psi} - \mathbf{T}:\dot{\mathbf{E}}^e - \sum_\alpha(\boldsymbol{\xi}^\alpha\cdot\nabla\dot{\gamma}^\alpha + \pi^\alpha\dot{\gamma}^\alpha) = -\delta \ge 0.} \qquad (105.18)$$

The inequality (105.18) is central to the development of a suitable constitutive theory.

105.3 Energetic Constitutive Equations. Peach–Koehler Forces

Our general goal is a constitutive theory that allows for dependencies:

(i) on slip-rate gradients; and
(ii) on screw and edge dislocation densities,

but that does not otherwise depart drastically from the conventional theory developed in §104.3. In the same vein, we content ourselves with constitutive equations that are sufficient — but generally not neceessary — for compatibility with thermodynamics.[631]

We seek a theory that allows for a free energy dependent on the dislocation densities

$$\rho_\vdash^\alpha = -\mathbf{s}^\alpha\cdot\nabla\gamma^\alpha, \qquad \rho_\odot^\alpha = \mathbf{l}^\alpha\cdot\nabla\gamma^\alpha, \qquad (105.19)$$

$\alpha = 1, 2, \dots N$.[632] Specifically, we write

$$\vec{\rho} = (\rho_\vdash^1, \rho_\vdash^2, \dots, \rho_\vdash^N, \rho_\odot^1, \rho_\odot^2, \dots, \rho_\odot^N) \qquad (105.20)$$

for the list of dislocation densities and consider a constitutive relation for the free energy of the form

$$\Psi = \tfrac{1}{2}\mathbf{E}^e:\mathbb{C}\mathbf{E}^e + \Psi^p(\vec{\rho}\,), \qquad (105.21)$$

with Ψ^p a **defect energy**.[633]

We therefore begin with an elastic energy (strain energy) in its classical form with elasticity tensor $\mathbb{C}$ symmetric and positive-definite. Consistent with this, we

630 Gurtin (2000b, eq. (53); 2002, eq. (6.6)).

631 Cf. §104.3.

632 Cf. (103.9).

633 Gurtin (2006, p. 1884) writes: "While it is tempting to consider the dislocation densities (105.19) as appropriate constitutive variables for the characterization of free energy, the description of these densities in terms of slip gradients should give one pause, as the prevailing view among experts is that slip and slip gradients are not suitable *constitutive* variables. One can, of course, rewrite $(103.9)_1$ and (105.19) as rate equations by simply differentiating with respect to t and then argue that the densities are internal variables, but such an argument seems far too facile to justify this use of dislocation densities; for that reason, we turn to the large deformation theory to help settle this issue. . . ." In fact, we do just that in §107.

assume that the macroscopic stress $\mathbf{T}$ is given by the standard relation[634]

$$\mathbf{T} = \mathbb{C}\mathbf{E}^e; \tag{105.22}$$

then,

$$\tfrac{1}{2}\overline{\mathbf{E}^e\!:\mathbb{C}\mathbf{E}^e}^{\boldsymbol{\cdot}} = \mathbf{T}\!:\dot{\mathbf{E}}^e \tag{105.23}$$

and the local free-energy imbalance (105.18) becomes

$$\overline{\Psi^p(\vec{\rho}\,)}^{\boldsymbol{\cdot}} - \sum_\alpha (\boldsymbol{\xi}^\alpha\cdot\nabla\dot{\gamma}^\alpha + \pi^\alpha\dot{\gamma}^\alpha) = -\delta \le 0. \tag{105.24}$$

Central to the theory are the **energetic defect forces** defined by[635]

$$f_\vdash^\alpha(\vec{\rho}\,) = \frac{\partial\Psi^p(\vec{\rho}\,)}{\partial\rho_\vdash^\alpha} \qquad \text{and} \qquad f_\odot^\alpha(\vec{\rho}\,) = \frac{\partial\Psi^p(\vec{\rho}\,)}{\partial\rho_\odot^\alpha}. \tag{105.25}$$

By $(103.9)_{2,3}$

$$\overline{\Psi^p(\vec{\rho}\,)}^{\boldsymbol{\cdot}} = \sum_\alpha \big(f_\vdash^\alpha\,\dot{\rho}_\vdash^\alpha + f_\odot^\alpha\,\dot{\rho}_\odot^\alpha\big) \tag{105.26}$$

$$= \sum_\alpha \big(-f_\vdash^\alpha\mathbf{s}^\alpha + f_\odot^\alpha\mathbf{l}^\alpha\big)\cdot\nabla\dot{\gamma}^\alpha; \tag{105.27}$$

we refer to

$$\boldsymbol{\xi}_{\rm en}^\alpha \overset{\rm def}{=} -f_\vdash^\alpha(\vec{\rho}\,)\mathbf{s}^\alpha + f_\odot^\alpha(\vec{\rho}\,)\mathbf{l}^\alpha \tag{105.28}$$

as the **energetic microscopic stress** for slip system α. Note that $\boldsymbol{\xi}_{\rm en}^\alpha$ is tangent to slip plane α, because $\mathbf{s}^\alpha$ and $\mathbf{l}^\alpha$ lie on this plane.

The classical Peach–Koehler force is the configurational force on a dislocation loop in a linear elastic body.[636] In contrast, the present theory is viscoplastic with dislocations distributed continuously over the body via the density fields $\rho_\vdash^\alpha$ and $\rho_\odot^\alpha$; even so, one might expect there to be a counterpart of the Peach–Koehler force within the present theory.

For each α, we continue to let[637]

$$\Pi^\alpha \text{ denote } \textbf{slip plane } \alpha. \tag{105.29}$$

For a distribution of *pure* dislocations with line direction $\mathbf{l}$ evolving on Π^α, a distributed Peach–Koehler force should be tangent to Π^α and perpendicular to the line direction $\mathbf{l}$. Such a force should therefore have the form

$$\varphi(\mathbf{m}^\alpha\times\mathbf{l}), \tag{105.30}$$

with φ a scalar field. We refer to (105.30) as a distributed **Peach–Koehler force** with density φ.

In view of our agreement that[638]

$$\mathbf{l}^\alpha = \mathbf{m}^\alpha\times\mathbf{s}^\alpha,$$

the energetic microscopic stress (105.28) can be written alternatively as

$$\boldsymbol{\xi}_{\rm en}^\alpha = f_\vdash^\alpha(\mathbf{m}^\alpha\times\mathbf{l}^\alpha) + f_\odot^\alpha(\mathbf{m}^\alpha\times\mathbf{s}^\alpha). \tag{105.31}$$

634 Cf. §104.4.
635 Cf. Gurtin (2006) and Gurtin, Anand & Lele (2007).
636 Cf. Footnote 629.
637 Cf. (103.3).
638 Cf. (103.4).

The microscopic forces $f_{\vdash}^{\alpha}(\mathbf{m}^{\alpha} \times \mathbf{l}^{\alpha})$ and $f_{\odot}^{\alpha}(\mathbf{m}^{\alpha} \times \mathbf{s}^{\alpha})$ each have the form (105.30) and, accordingly, have the following physical interpretations:[639]

$$\underbrace{f_{\vdash}^{\alpha}(\mathbf{m}^{\alpha} \times \mathbf{l}^{\alpha})}_{\substack{\text{distributed Peach–Koehler} \\ \text{force on edge dislocations}}} \qquad \text{and} \qquad \underbrace{f_{\odot}^{\alpha}(\mathbf{m}^{\alpha} \times \mathbf{s}^{\alpha})}_{\substack{\text{distributed Peach–Koehler} \\ \text{force on screw dislocations}}}. \tag{105.32}$$

The energetic defect forces $f_{\vdash}^{\alpha}$ and $f_{\odot}^{\alpha}$ therefore represent densities of distributed Peach–Koehler forces,

- *an observation that allows us to view the energetic microscopic stresses $\boldsymbol{\xi}_{en}^{\alpha}$ as counterparts of Peach–Koehler forces.*

105.4 Constitutive Equations that Account for Dissipation

By (105.27) and (105.28)

$$\overline{\Psi^{P}(\vec{\rho}\,)} = \sum_{\alpha} \boldsymbol{\xi}_{\text{en}}^{\alpha} \cdot \nabla\dot{\gamma}^{\alpha} \tag{105.33}$$

and (105.24) becomes

$$\delta = \sum_{\alpha} [(\boldsymbol{\xi}^{\alpha} - \boldsymbol{\xi}_{\text{en}}^{\alpha}) \cdot \nabla\dot{\gamma}^{\alpha} + \pi^{\alpha}\dot{\gamma}^{\alpha}] \geq 0. \tag{105.34}$$

Thus, if we define **dissipative microscopic stresses** $\boldsymbol{\xi}_{\text{dis}}^{\alpha}$ through the relations

$$\boldsymbol{\xi}_{\text{dis}}^{\alpha} = \boldsymbol{\xi}^{\alpha} - \boldsymbol{\xi}_{\text{en}}^{\alpha}, \tag{105.35}$$

then (105.34) takes the form of a *reduced dissipation inequality*

$$\delta = \sum_{\alpha} (\pi^{\alpha}\dot{\gamma}^{\alpha} + \boldsymbol{\xi}_{\text{dis}}^{\alpha} \cdot \nabla\dot{\gamma}^{\alpha}) \geq 0. \tag{105.36}$$

Our discussion of dissipative constitutive relations is based on this inequality.

Our choice of constitutive relations for the internal microscopic forces π^{α} and the dissipative microscopic stresses $\boldsymbol{\xi}_{\text{dis}}^{\alpha}$ is guided by:

(i) the reduced dissipation inequality (105.36), which suggests a dependence of π^{α} on $\dot{\gamma}^{\alpha}$ and $\boldsymbol{\xi}_{\text{dis}}^{\alpha}$ on $\nabla\dot{\gamma}^{\alpha}$;

(ii) the tacit assumption that the microscopic stress $\boldsymbol{\xi}_{\text{dis}}^{\alpha}$ characterizes dissipative microscopic forces associated with the evolution of dislocations on the slip plane Π^{α}; because the motion of such dislocations is tangent to Π^{α}, we require that $\boldsymbol{\xi}_{\text{dis}}^{\alpha}$ also be tangential to Π^{α};

(iii) our wish to have the dissipation δ and the constitutive relations for the internal microscopic forces π^{α} of a form similar in structure to their conventional counterparts (104.44) and (104.45).

Specifically, we introduce two quantities — an effective flow rate

$$d^{\alpha} \stackrel{\text{def}}{=} \sqrt{|\dot{\gamma}^{\alpha}|^2 + \ell^2|\nabla^{\alpha}\dot{\gamma}^{\alpha}|^2} \tag{105.37}$$

[639] A similar result was established by GURTIN (2002, p. 22) for a defect energy dependent on the Burgers tensor $\mathbf{G}$, rather than on dislocation densities, but the argument used was convoluted. In contrast, the argument leading to (105.32) is closer to the underlying physics — and simpler.

with ℓ a *dissipative length-scale* and $\nabla^\alpha \dot{\gamma}^\alpha = \mathbf{P}^\alpha \nabla \dot{\gamma}^\alpha$ the tangential gradient of $\dot{\gamma}^\alpha$ on Π^α,[640] and a (dimensionless) rate-sensitivity function g consistent with[641]

$$g(0) = 0, \qquad g(d^\alpha) > 0 \text{ for } d^\alpha \neq 0$$

— and we consider constitutive equations for π^α and $\boldsymbol{\xi}^\alpha$ in the form[642]

$$\begin{aligned} \pi^\alpha &= S^\alpha g(d^\alpha)\frac{\dot{\gamma}^\alpha}{d^\alpha}, \\ \boldsymbol{\xi}^\alpha_{\rm dis} &= S^\alpha g(d^\alpha)\ell^2\frac{\nabla^\alpha \dot{\gamma}^\alpha}{d^\alpha}. \end{aligned} \tag{105.38}$$

The relations (105.28), (105.35), and (105.38)₂ combine to form a constitutive equation

$$\boldsymbol{\xi}^\alpha = -f^\alpha_{\vdash}\, \mathbf{s}^\alpha + f^\alpha_{\odot}\, \mathbf{l}^\alpha + S^\alpha g(d^\alpha)\ell^2 \frac{\nabla^\alpha \dot{\gamma}^\alpha}{d^\alpha} \tag{105.39}$$

for the microscopic stress $\boldsymbol{\xi}^\alpha$.

By (103.18) given any scalar field φ

$$\begin{aligned} \nabla\varphi \cdot \nabla^\alpha \varphi &= \nabla\varphi \cdot (\mathbf{P}^\alpha \nabla\varphi) \\ &= \nabla\varphi \cdot (\mathbf{P}^\alpha \mathbf{P}^\alpha \nabla\varphi) \\ &= (\mathbf{P}^\alpha \nabla\varphi) \cdot (\mathbf{P}^\alpha \nabla\varphi) \\ &= |\nabla^\alpha \varphi|^2, \end{aligned} \tag{105.40}$$

since $\mathbf{P}^\alpha$ is symmetric. Thus, $\nabla \dot{\gamma}^\alpha \cdot \nabla^\alpha \dot{\gamma}^\alpha = |\nabla^\alpha \dot{\gamma}^\alpha|^2$ and (105.38) renders the dissipation (105.36) of the simple form

$$\delta = \sum_\alpha S^\alpha g(d^\alpha) d^\alpha. \tag{105.41}$$

The dependence of $\boldsymbol{\xi}^\alpha_{\rm dis}$ on the *tangential gradient* $\nabla^\alpha \dot{\gamma}^\alpha$ renders the constitutive relation for $\boldsymbol{\xi}^\alpha_{\rm dis}$ consistent with (ii). Regarding (iii), the relations (105.38)₁ for π^α and (105.41) for δ differ from the conventional relations (104.44)₁ and (104.45) only through the replacement of $|\dot{\gamma}^\alpha|$ in the conventional relations by the effective flow rate d^α.

Our next step in the prescription of dissipative constitutive equations is the specification of hardening equations for the evolution of the slip resistances S^α. Based on the success of the conventional hardening equations (104.52)₂, we consider hardening equations of the form

$$\dot{S}^\alpha = \sum_\alpha h^{\alpha\beta}(\vec{S}) d^\beta \tag{105.42}$$

with hardening moduli (104.65); viz.

$$h^{\alpha\beta}(\vec{S}) = \underbrace{\chi^{\alpha\beta} h(S^\beta)}_{\text{self-hardening}} + \underbrace{q(1-\chi^{\alpha\beta}) h(S^\beta)}_{\text{latent-hardening}}. \tag{105.43}$$

640 Cf. (103.19).

641 Cf. (104.41). For example, one might take $g(d^\alpha) = (d^\alpha/d_0)^m$; cf. the discussion following (78.20).

642 The equations (105.38) were proposed by Gurtin (2000b, §15). The structure of these equations bears some comparison with equations introduced to characterize strengthening in isotropic plastic materials; cf. FREDRIKSSON & GUDMUNDSON (2005) and GURTIN & ANAND (2005a,b); cf. Footnote 547.

Here, $\chi^{\alpha\beta}$ are the coplanarity moduli (104.64), $h \geq 0$ is the self-hardening function, and $q > 0$ is the interaction constant. We are, therefore, led to the hardening equations[643]

$$\dot{S}^\alpha = \sum_\beta [\chi^{\alpha\beta} + q(1 - \chi^{\alpha\beta})] h(S^\beta)\, d^\beta. \tag{105.44}$$

EXERCISES

1. Develop the dissipative constitutive relations (105.38) using the formulation introduced in the paragraph containing (90.55). That is, introduce, for each slip system α, a generalized slip-rate

$$\dot{\Gamma}^\alpha \stackrel{\text{def}}{=} (\dot{\gamma}^\alpha, \ell \nabla \dot{\gamma}^\alpha)$$

a generalized microscopic stress

$$\Theta^\alpha \stackrel{\text{def}}{=} (\pi^\alpha, \ell^{-1} \boldsymbol{\xi}^\alpha_{\text{dis}}),$$

and so forth. Further, lay down the codirectionality hypothesis

$$\Theta^\alpha = \phi \frac{\dot{\Gamma}^\alpha}{|\dot{\Gamma}^\alpha|},$$

with ϕ a scalar function, to be specified, etc.

2. It is commonly held that dislocations impinging transversely on a slip plane — traditionally called forest dislocations — give rise to a form of hardening referred to as forest-hardening.[644] Consistent with this, CERMELLI & GURTIN (2001), who work within the context of large deformations, show that the field $\mathbf{m}^\alpha \cdot \mathbf{G}\mathbf{m}^\alpha$ characterizes the *distortion* of slip plane Π^α, and — because this field accounts for the normal component of the Burgers vector of dislocations impinging transversally on Π^α — suggest that this field might be useful as a constitutive quantity related to forest-hardening. In light of this discussion, one might also consider hardening moduli dependent on the forest-hardening measure $\zeta^\alpha \geq 0$ for each slip system α by[645]

$$\zeta^\alpha = |\mathbf{m}^\alpha \cdot \mathbf{G}\mathbf{m}^\alpha|. \tag{105.45}$$

Derive the following relations, which express ζ^α in terms of slip gradients and in terms of dislocation densities:

$$\zeta^\alpha = \left| \sum_\beta (\mathbf{s}^\beta \cdot \mathbf{m}^\alpha)(\mathbf{m}^\beta \times \mathbf{m}^\alpha) \cdot \nabla \gamma^\beta \right| \tag{105.46}$$

$$= \left| \sum_\beta (\mathbf{m}^\alpha \cdot \mathbf{s}^\beta)\big[(\mathbf{m}^\alpha \cdot \mathbf{l}^\beta)\rho^\beta_\vdash + (\mathbf{m}^\alpha \cdot \mathbf{s}^\beta)\rho^\beta_\odot\big] \right|. \tag{105.47}$$

Since $\mathbf{m}^\alpha \cdot \mathbf{s}^\beta = 0$ and $\mathbf{m}^\alpha \cdot \mathbf{l}^\beta = 0$ when the slip systems α and β are coplanar,[646] ζ^α is independent of slip gradients (or dislocation densities) associated with slip systems whose slip planes are coplanar to α.

3. Show that

$$|\mathbf{s}^\beta \cdot \mathbf{m}^\alpha| \leq 1, \qquad |(\mathbf{m}^\beta \times \mathbf{m}^\alpha) \cdot \nabla \gamma^\beta| = |(\mathbf{m}^\beta \times \nabla \gamma^\beta) \cdot \mathbf{m}^\alpha| \leq |\nabla^\beta \gamma^\beta|,$$

---

643 Cf. (104.66).

644 Cf., e.g., KUHLMANN-WILSDORF (1989).

645 ACHARYA, BASSANI & BEAUDOIN (2003), propose $|\mathbf{G}^\top \mathbf{m}^\alpha|$ as an alternate measure of "forest dislocations."

646 Cf. (104.63).

and use these identities to establish the following bound for ζ^α:

$$\zeta^\alpha \le \sum_\beta (1-\chi^{\alpha\beta})|\nabla^\beta \gamma^\beta| = \sum_\beta (1-\chi^{\alpha\beta})\sqrt{(\rho_\vdash^\beta)^2 + (\rho_\odot^\beta)^2}; \tag{105.48}$$

ζ^α is therefore bounded by the sum of the absolute values of the tangential gradients over all slip systems not coplanar with Π^α.

105.5 Viscoplastic Flow Rule

The decomposition $\boldsymbol{\xi}^\alpha = \boldsymbol{\xi}^\alpha_{\rm en} + \boldsymbol{\xi}^\alpha_{\rm dis}$ allows us to write the microscopic force balance (105.15) in the form

$$\tau^\alpha + \mathrm{Div}\,\boldsymbol{\xi}^\alpha_{\rm en} = \pi^\alpha - \mathrm{Div}\,\boldsymbol{\xi}^\alpha_{\rm dis}, \tag{105.49}$$

where we have written the term $\mathrm{Div}\,\boldsymbol{\xi}^\alpha_{\rm en}$ on the left, since, being energetic, its negative represents a backstress. When augmented by the constitutive equations (105.28) and (105.38) the balance (105.49) becomes the **flow rule** for slip system α:[647]

$$\tau^\alpha \underbrace{-(-1)\mathrm{Div}(-f_\vdash^\alpha \mathbf{s}^\alpha + f_\odot^\alpha \mathbf{l}^\alpha)}_{\text{energetic backstress}} = \underbrace{S^\alpha g(d^\alpha)\frac{\dot\gamma^\alpha}{d^\alpha} - \ell^2\,\mathrm{Div}\left(S^\alpha g(d^\alpha)\frac{\nabla^\alpha \dot\gamma^\alpha}{d^\alpha}\right)}_{\text{dissipative-hardening}}. \tag{105.50}$$

By (103.9) and (105.25), the defect forces $f_\odot^\alpha$ and $f_\vdash^\alpha$ depend on slip gradients; the flow rule (105.50) therefore relates the resolved stresses to first and second gradients of slip and slip-rate. Moreover, given the resolved stresses τ^α, (105.50) represents a system of partial differential equations for the slips. Thus, unlike the flow rules (104.46) of the conventional theory, the flow rules (105.50) are *nonlocal* and hence require concomitant *boundary conditions*.

To express the flow rules in terms of the slip-gradients, we define *energetic-interaction moduli* by

$$\left.\begin{aligned} D_{\vdash\vdash}^{\alpha\beta} &= D_{\vdash\vdash}^{\beta\alpha} = \frac{\partial f_\vdash^\alpha}{\partial \rho_\vdash^\beta},\\ D_{\odot\odot}^{\alpha\beta} &= D_{\odot\odot}^{\beta\alpha} = \frac{\partial f_\odot^\alpha}{\partial \rho_\odot^\beta},\\ D_{\vdash\odot}^{\alpha\beta} &= D_{\odot\vdash}^{\beta\alpha} = \frac{\partial f_\vdash^\alpha}{\partial \rho_\odot^\beta} = \frac{\partial f_\odot^\beta}{\partial \rho_\vdash^\alpha}, \end{aligned}\right\} \tag{105.51}$$

so that, for example, $D_{\vdash\odot}^{\alpha\beta} = D_{\vdash\odot}^{\alpha\beta}(\vec{\rho}\,)$ represents the energetic interaction between screw dislocations on β and edge dislocations on slip system α; or, more specifically, $D_{\vdash\odot}^{\alpha\beta}$ represents a change in the Peach–Koehler force density for edge dislocations on

[647] Cf. Gurtin (2000b, eq. (164a); 2002, eq. (7.18)); the former has no backstress, the latter has $\boldsymbol{\xi}^\alpha_{\rm dis} = \mathbf{0}$ and defect energy a function of $\mathbf{G}$ (rather than dislocation densities); the complete relation (105.49) is due to Gurtin, Anand, & Lele (2007, eq. (8.2)).

slip system α due to a change in the screw-dislocation density on β.[648] By (105.51),

$$\nabla f_{\vdash}^{\alpha} = \sum_{\beta}(D_{\vdash\vdash}^{\alpha\beta}\nabla\rho_{\vdash}^{\beta} + D_{\vdash\odot}^{\alpha\beta}\nabla\rho_{\odot}^{\beta}),$$

$$\nabla f_{\odot}^{\alpha} = \sum_{\beta}(D_{\odot\vdash}^{\alpha\beta}\nabla\rho_{\vdash}^{\beta} + D_{\odot\odot}^{\alpha\beta}\nabla\rho_{\odot}^{\beta}),$$

and (105.28) yields

$$\begin{aligned}\mathrm{Div}\,\boldsymbol{\xi}_{\mathrm{en}}^{\alpha} &= -\mathbf{s}^{\alpha}\cdot\nabla f_{\vdash}^{\alpha} + \mathbf{l}^{\alpha}\cdot\nabla f_{\odot}^{\alpha} \\ &= \sum_{\beta}[-\mathbf{s}^{\alpha}\cdot(D_{\vdash\vdash}^{\alpha\beta}\nabla\rho_{\vdash}^{\beta} + D_{\vdash\odot}^{\alpha\beta}\nabla\rho_{\odot}^{\beta}) + \mathbf{l}^{\alpha}\cdot(D_{\odot\vdash}^{\alpha\beta}\nabla\rho_{\vdash}^{\beta} + D_{\odot\odot}^{\alpha\beta}\nabla\rho_{\odot}^{\beta})].\end{aligned}$$

But, in view of (103.9):

$$\nabla\rho_{\vdash}^{\beta} = -(\nabla\nabla\gamma^{\beta})\mathbf{s}^{\beta}, \qquad \nabla\rho_{\odot}^{\beta} = (\nabla\nabla\gamma^{\beta})\,\mathbf{l}^{\beta},$$

and, therefore,

$$\begin{aligned}\mathrm{Div}\,\boldsymbol{\xi}_{\mathrm{en}}^{\alpha} = \sum_{\beta}[\mathbf{s}^{\alpha}\cdot(D_{\vdash\vdash}^{\alpha\beta}(\nabla\nabla\gamma^{\beta})\mathbf{s}^{\beta} - D_{\vdash\odot}^{\alpha\beta}(\nabla\nabla\gamma^{\beta})\mathbf{l}^{\beta} \\ + \mathbf{l}^{\alpha}\cdot(-D_{\odot\vdash}^{\alpha\beta}(\nabla\nabla\gamma^{\beta})\mathbf{s}^{\beta} + D_{\odot\odot}^{\alpha\beta}(\nabla\nabla\gamma^{\beta})\mathbf{l}^{\beta})].\end{aligned} \tag{105.52}$$

Thus, if we define energetic-interaction tensors $\mathbf{A}^{\alpha\beta}$ by

$$\mathbf{A}^{\alpha\beta} = D_{\vdash\vdash}^{\alpha\beta}\,\mathbf{s}^{\alpha}\otimes\mathbf{s}^{\beta} - D_{\vdash\odot}^{\alpha\beta}\,\mathbf{s}^{\alpha}\otimes\mathbf{l}^{\beta} - D_{\odot\vdash}^{\alpha\beta}\,\mathbf{l}^{\alpha}\otimes\mathbf{s}^{\beta} + D_{\odot\odot}^{\alpha\beta}\,\mathbf{l}^{\alpha}\otimes\mathbf{l}^{\beta}, \tag{105.53}$$

then

$$\mathrm{Div}\,\boldsymbol{\xi}_{\mathrm{en}}^{\alpha} = \sum_{\beta}\mathbf{A}^{\alpha\beta}:\nabla\nabla\gamma^{\beta} \tag{105.54}$$

and, since the term $\mathrm{Div}(\cdots)$ on the left side of (105.50) is $\mathrm{Div}\,\boldsymbol{\xi}_{\mathrm{en}}^{\alpha}$, we can write this flow rule in the form

$$\boxed{\underbrace{\tau^{\alpha} - (-1)\sum_{\beta}\mathbf{A}^{\alpha\beta}:\nabla\nabla\gamma^{\beta}}_{\text{energetic backstress}} = \underbrace{S^{\alpha}g(d^{\alpha})\frac{\dot{\gamma}^{\alpha}}{d^{\alpha}} - \ell^{2}\mathrm{Div}\left(S^{\alpha}g(d^{\alpha})\frac{\nabla^{\alpha}\dot{\gamma}^{\alpha}}{d^{\alpha}}\right)}_{\text{dissipative-hardening}}.} \tag{105.55}$$

Note that the tensors $\mathbf{A}^{\alpha\beta}$ depend on $\vec{\rho}$ and hence slip gradients, since the moduli $D_{\vdash\vdash}^{\alpha\beta}, \ldots$ depend on $\vec{\rho}$. Note also that (105.55) is valid for any choice of the defect energy.

A simple defect energy has free energy $\Psi^{p}(\vec{\rho})$ uncoupled and quadratic in the net dislocation densities[649]

$$\rho_{\mathrm{net}}^{\alpha} = \sqrt{|\rho_{\vdash}^{\alpha}|^{2} + |\rho_{\odot}^{\alpha}|^{2}} \tag{105.56}$$

and, hence, of the form

$$\Psi^{p}(\vec{\rho}) = \tfrac{1}{2}S_{0}L^{2}\sum_{\alpha}|\rho_{\mathrm{net}}^{\alpha}|^{2}, \tag{105.57}$$

648 Cf. the paragraph containing (105.32).

649 Cf., e.g., Ohno & Okumara (2007) who considered a free energy of the form $\Psi^{p}(\vec{\rho}) = \mathrm{const}\times\sum_{\alpha}\rho_{\mathrm{net}}^{\alpha}$.

with L an energetic length-scale and S_0 the initial slip resistance. An interesting consequence of (103.23) is that this energy has the alternative form

$$\Psi^p(\vec{\rho}) = \tfrac{1}{2} S_0 L^2 \sum_{\alpha} |\nabla^\alpha \gamma^\alpha|^2. \tag{105.58}$$

Granted (105.57), the defect forces (105.25) and energetic microscopic stress (105.28) become

$$\begin{aligned} f_{\vdash}^{\alpha} &= S_0 L^2 \rho_{\vdash}^{\alpha}, \qquad f_{\odot}^{\alpha} = S_0 L^2 \rho_{\odot}^{\alpha}, \\ \boldsymbol{\xi}_{\text{en}}^{\alpha} &= S_0 L^2(-\rho_{\vdash}^{\alpha}\, \mathbf{s}^\alpha + \rho_{\odot}^{\alpha}\, \mathbf{l}^\alpha), \end{aligned} \tag{105.59}$$

results that lead to the following expression for the (slip system α) backstress:

$$-S_0 L^2 \operatorname{Div}(-\rho_{\vdash}^{\alpha} \mathbf{s}^\alpha + \rho_{\odot}^{\alpha}\, \mathbf{l}^\alpha).$$

Further, for this simple energy the only nonzero energetic interaction moduli (105.51) are

$$D_{\vdash\vdash}^{\alpha\alpha} = D_{\odot\odot}^{\alpha\alpha} = S_0 L^2, \qquad \alpha = 1, 2, \ldots, N,$$

and the only nonzero components of the interaction tensor (105.53) are

$$\mathbf{A}^{\alpha\alpha} = S_0 L^2(\mathbf{s}^\alpha \otimes \mathbf{s}^\alpha + \mathbf{l}^\alpha \otimes \mathbf{l}^\alpha), \qquad \alpha = 1, 2, \ldots, N. \tag{105.60}$$

Thus, the flow rule (105.55) for slip system α becomes[650]

$$\boxed{\tau^\alpha - \underbrace{(-1) S_0 L^2\, \Delta^\alpha \gamma^\alpha}_{\text{energetic backstress}} = \underbrace{S^\alpha g(d^\alpha)\frac{\dot{\gamma}^\alpha}{d^\alpha} - \ell^2 \operatorname{Div}\left(S^\alpha g(d^\alpha) \frac{\nabla^\alpha \dot{\gamma}^\alpha}{d^\alpha}\right)}_{\text{dissipative-hardening}},} \tag{105.61}$$

with Δ^α the Laplace operator on slip plane α, defined by

$$\begin{aligned} \Delta^\alpha \varphi &= \operatorname{Div} \nabla^\alpha \varphi \\ &= \mathbf{s}^\alpha \cdot (\nabla\nabla\varphi)\mathbf{s}^\alpha + \mathbf{l}^\alpha \cdot (\nabla\nabla\varphi)\mathbf{l}^\alpha. \end{aligned} \tag{105.62}$$

Unlike (105.55), the backstress involves no coupling between slip systems.

EXERCISES

1. A simple quadratic defect energy with coupling has the form

$$\Psi^p(\vec{\rho}) = \tfrac{1}{2} S_0 \left(\sum_{\alpha} L^2 (\rho_{\text{net}}^{\alpha})^2 + C^2 \sum_{\substack{\alpha,\beta \\ \alpha\neq\beta}} \rho_{\text{net}}^{\alpha} \rho_{\text{net}}^{\beta} \right). \tag{105.63}$$

[650] Computations of Gurtin, Anand & Lele (2007) based on this flow rule show that the theory presented here, with $\ell \neq 0$, leads to strengthening; that is, to an increase in the initial yield stress. Interestingly, Ohno & Okumura (2007) show that the energy (105.65), attributed to the self-energy of the net dislocation density, also leads to strengthening. As is clear from the studies of Fredriksson & Gudmundson (2005) and Anand, Gurtin, Lele & Gething (2005), such an increase is also a consequence of dissipative-hardening when slip-rate gradients enter the constitutive relations for the microscopic stresses $\boldsymbol{\xi}_{\text{dis}}^{\alpha}$, as in (105.38). Since the strengthening introduced by Ohno and Okumura most certainly leads to kinematic-hardening, while dissipative-hardening involves no backstress, it might be possible to experimentally ascertain whether one or both of the hardening mechanisms is the root cause of strengthening.

Show that the energetic defect forces (105.25) are given by

$$f_{\vdash}^{\alpha} = S_0\Big(L^2\rho_{\vdash}^{\alpha} + \tfrac{1}{2}C^2 c_{\vdash}^{\alpha}\sum_{\substack{\beta\\ \beta\neq\alpha}}\rho_{\rm net}^{\beta}\Big),$$

$$f_{\odot}^{\alpha} = S_0\Big(L^2\rho_{\odot}^{\alpha} + \tfrac{1}{2}C^2 c_{\odot}^{\alpha}\sum_{\substack{\beta\\ \beta\neq\alpha}}\rho_{\rm net}^{\beta}\Big), \tag{105.64}$$

where $c_{\vdash}^{\alpha}$ and $c_{\odot}^{\alpha}$ are the dislocation concentrations defined by

$$c_{\vdash}^{\alpha} = \frac{\rho_{\vdash}^{\alpha}}{\rho_{\rm net}^{\alpha}} \quad\text{and}\quad c_{\odot}^{\alpha} = \frac{\rho_{\odot}^{\alpha}}{\rho_{\rm net}^{\alpha}}$$

and C is a length-scale associated with the energetic coupling of slip systems.

2. The defect energy

$$\Psi^{\rm p}(\vec{\rho}) = S_0 L^2(1+r)^{-1}\sum_{\alpha}(\rho_{\rm net}^{\alpha})^{1+r} \tag{105.65}$$

with $r=0$ was introduced by Ohno & Okumura (2007) to account for the *self-energy* of geometrically necessary dislocations.[651] Show that this energy is associated with the defect forces

$$f_{\vdash}^{\alpha} = S_0 L^2(\rho_{\rm net}^{\alpha})^r\frac{\rho_{\vdash}^{\alpha}}{\rho_{\rm net}^{\alpha}} \quad\text{and}\quad f_{\odot}^{\alpha} = S_0 L^2(\rho_{\rm net}^{\alpha})^r\frac{\rho_{\odot}^{\alpha}}{\rho_{\rm net}^{\alpha}}. \tag{105.66}$$

105.6 Microscopically Simple Boundary Conditions

Each of the flow rules discussed in §105.5 involves second slip-rate gradients and hence represents a system of partial-differential equations for the slips, given the resolved shears τ^{α}. The flow rules are therefore *nonlocal* and require associated boundary conditions.

With this in mind, we focus on the boundary $\partial\mathrm{B}$. The external power expended on B is given by (105.5), and, in view of (105.16), the microscopic portion of this power has the form

$$\sum_{\alpha}\int_{\partial\mathrm{B}}(\boldsymbol{\xi}^{\alpha}\cdot\mathbf{n})\dot{\gamma}^{\alpha}\,da; \tag{105.67}$$

(105.67) represents power expended by the material in contact with the body and suggests that the requisite boundary conditions should involve the tractions $\boldsymbol{\xi}^{\alpha}\cdot\mathbf{n}$ and the slip rates $\dot{\gamma}^{\alpha}$. We restrict attention to boundary conditions that result in a *null expenditure of microscopic power* in the sense that

$$(\boldsymbol{\xi}^{\alpha}\cdot\mathbf{n})\dot{\gamma}^{\alpha} = 0 \text{ on } \partial\mathrm{B} \tag{105.68}$$

for all α.[652] Specifically, we consider **microscopically simple** boundary conditions asserting that

$$\dot{\gamma}^{\alpha} = 0 \text{ on } \mathcal{S}_{\rm hard} \quad\text{and}\quad \boldsymbol{\xi}^{\alpha}\cdot\mathbf{n} = 0 \text{ on } \mathcal{S}_{\rm free} \tag{105.69}$$

for all α, where $\mathcal{S}_{\rm hard}$ and $\mathcal{S}_{\rm free}$ are *complementary* subsurfaces of $\partial\mathrm{B}$ respectively referred to as the **microscopically hard** and the **microscopically free** portions of $\partial\mathrm{B}$.[653]

651 The term involving $r > 0$, r small, represents a regularization introduced to ensure that the defect forces (105.25) are defined when $\rho_{\rm net}^{\alpha} = 0$; cf. Ohno & Okumura (2007, eq. (52)).

652 Cf. Gurtin (2000b, eq. (137); 2002, eqs. (9.1) and (9.4)).

653 As Gurtin & Needleman (2005) show, the issue of boundary conditions is delicate when: (i) the defect energy depends on the Burgers tensor **G**; (ii) the theory does *not* include constitutive

The microscopically hard condition corresponds to a boundary surface that cannot pass dislocations (for example, a boundary surface that abuts a hard material); the microscopically free condition corresponds to a boundary across which dislocations can flow freely from the body and would seem consistent with the *macroscopic* condition $\mathbf{Tn} = \mathbf{0}$.

105.7 Variational Formulation of the Flow Rule

The flow rules and the microscopically free boundary-conditions have a variational formulation based on the microscopic virtual-power relation (105.13).[654] To see this, assume that, at some arbitrarily chosen fixed time under consideration, the fields $\mathbf{u}$ and $\mathbf{E}^e$ are known, and let $\mathcal{S}_{\rm hard}$ and $\mathcal{S}_{\rm free}$ be *complementary* subsurfaces of $\partial\mathrm{B}$. Then, given any slip system α, if

(i) $\boldsymbol{\xi}^\alpha \cdot \mathbf{n} = 0$ on $\mathcal{S}_{\rm free}$,
(ii) $\varphi \equiv \tilde{\gamma}^\alpha$ is the only nonzero virtual slip-rate field, and
(iii) $\varphi = 0$ on $\mathcal{S}_{\rm hard}$,

then, by (i)–(iii) and (105.16),

$$\int_{\partial\mathrm{B}} \Xi^\alpha(\mathbf{n})\varphi\, da = 0 \tag{105.70}$$

and (105.13), with P = B, reduces to[655]

$$\boxed{\int_{\mathrm{B}} [(\pi^\alpha - \tau^\alpha)\varphi + \boldsymbol{\xi}^\alpha \cdot \nabla\varphi]\, dv = 0.} \tag{105.71}$$

The foregoing steps were used only to derive the virtual-power relation (105.71): We no longer require that (i)–(iii) be satisfied. We refer to φ in (105.71) as a *test field* and assume that φ is **kinematically admissible** in the sense that

$$\varphi = 0 \quad \text{on } \mathcal{S}_{\rm hard}. \tag{105.72}$$

Then the identity (105.4) (with $\dot{\gamma}^\alpha = \varphi$) implies that

$$\int_{\mathrm{B}} \boldsymbol{\xi}^\alpha \cdot \nabla\varphi\, dv = \int_{\mathcal{S}_{\rm free}} (\boldsymbol{\xi}^\alpha \cdot \mathbf{n})\varphi\, da - \int_{\mathrm{B}} \varphi\, \mathrm{Div}\,\boldsymbol{\xi}^\alpha\, dv \tag{105.73}$$

and, hence, that (105.71) is equivalent to

$$\int_{\mathcal{S}_{\rm free}} (\boldsymbol{\xi}^\alpha \cdot \mathbf{n})\varphi\, da + \int_{\mathrm{B}} (\pi^\alpha - \tau^\alpha - \mathrm{Div}\,\boldsymbol{\xi}^\alpha)\varphi\, dv = 0. \tag{105.74}$$

Moreover, invoking the fundamental lemma of the calculus of variations (page 167), we see that (105.74) holds for all kinematically admissible test fields φ if and only if $\boldsymbol{\xi}^\alpha \cdot \mathbf{n} = 0$ on $\mathcal{S}_{\rm free}$ and the microscopic force balance (105.15) is satisfied in B. Since this force balance — supplemented by the constitutive relations $(105.38)_1$ and

dependencies on *slip-rate* gradients. Here, neither (i) or (ii) is applicable. Even so, GURTIN (2008) conjectured that, granted (ii), the dependence of the defect energy on dislocation densities precludes the problems encountered by Gurtin and Needleman; and that a similar conjecture applies to the small deformation theory of GURTIN, ANAND & LELE (2007).

654 Cf. the first paragraph of §89.6.

655 Cf. GURTIN (2000b, eq. (159); 2002, eq. (10.1)).

(105.39) for π^α and $\boldsymbol{\xi}^\alpha$ — is equivalent to the flow rule (105.50) for α, we have the following result:

VARIATIONAL FORMULATION OF THE FLOW RULE *Suppose that the constitutive relations*

$$\pi^\alpha = S^\alpha g(d^\alpha)\frac{\dot{\gamma}^\alpha}{d^\alpha},$$
$$\boldsymbol{\xi}^\alpha = -f_\vdash^\alpha \mathbf{s}^\alpha + f_\odot^\alpha \mathbf{l}^\alpha + S^\alpha g(d^\alpha)\ell^2\frac{\nabla^\alpha \dot{\gamma}^\alpha}{d^\alpha}, \tag{105.75}$$

are satisfied. The flow rule (105.50) *in* B *and the boundary condition*

$$\boldsymbol{\xi}^\alpha \cdot \mathbf{n} = 0 \quad \textit{on } \mathcal{S}_{free} \tag{105.76}$$

are then together equivalent to the requirement that (105.71) *hold for all kinematically admissible test fields* φ.

This global variational statement of the flow rule should provide a useful basis for computations; in a numerical scheme such as the finite element method, (105.71) augmented by (105.75) would, for each α, reduce to a system of nonlinear algebraic equations for $\dot{\gamma}^\alpha$, granted a knowledge of the "current state" of the system.

105.8 Plastic Free-Energy Balance

As a result of the constitutive relations, the global free-energy imbalance (105.17) has a plastic counterpart, which we now derive.

Assume that the microscopically simple boundary conditions (105.69) are satisfied, so that (105.68) holds and (105.4) takes the form

$$\int_B \boldsymbol{\xi}^\alpha \cdot \nabla \dot{\gamma}^\alpha \, dv = -\int_B \dot{\gamma}^\alpha \, \mathrm{Div}\, \boldsymbol{\xi}^\alpha \, dv.$$

The relation (105.33) for $\overline{\Psi^p(\vec{\rho})}$, the relation $\boldsymbol{\xi}^\alpha_{\rm dis} = \boldsymbol{\xi}^\alpha - \boldsymbol{\xi}^\alpha_{\rm en}$, and the microscopic force balance (105.15) therefore imply that

$$\begin{aligned}
\overline{\int_B \Psi^p(\vec{\rho})\, dv} &= \sum_\alpha \int_B \boldsymbol{\xi}^\alpha_{\rm en} \cdot \nabla \dot{\gamma}^\alpha \, dv \\
&= \sum_\alpha \int_B [\boldsymbol{\xi}^\alpha_{\rm en} \cdot \nabla \dot{\gamma}^\alpha + \underbrace{(\tau^\alpha - \pi^\alpha + \mathrm{Div}\, \boldsymbol{\xi}^\alpha)}_{=0}\dot{\gamma}^\alpha]\, dv \\
&= \sum_\alpha \int_B [(\boldsymbol{\xi}^\alpha_{\rm en} - \boldsymbol{\xi}^\alpha)\cdot \nabla \dot{\gamma}^\alpha + (\tau^\alpha - \pi^\alpha)\dot{\gamma}^\alpha]\, dv \\
&= \sum_\alpha \int_B [(\tau^\alpha - \pi^\alpha)\dot{\gamma}^\alpha - \boldsymbol{\xi}^\alpha_{\rm dis} \cdot \nabla \dot{\gamma}^\alpha]\, dv.
\end{aligned}$$

We, therefore, have the **plastic free-energy balance**

$$\boxed{\overline{\int_B \Psi^p(\vec{\rho})\, dv} = \underbrace{\sum_\alpha \int_B \tau^\alpha \dot{\gamma}^\alpha \, dv}_{\text{plastic working}} - \underbrace{\sum_\alpha \int_B (\pi^\alpha \dot{\gamma}^\alpha + \boldsymbol{\xi}^\alpha_{\rm dis} \cdot \nabla \dot{\gamma}^\alpha)\, dv}_{\text{dissipation} \geq 0}.} \tag{105.77}$$

Thus, *the temporal increase in defect energy can never exceed the plastic working.* Further, if the defect energy vanishes, then — as in conventional theories of plasticity — the plastic working is balanced by the dissipation.

105.9 Some Remarks

The present theory with the energy (105.58) compares well to the discrete dislocation calculations of NICOLA, VAN DER GIESSEN & GURTIN (2005). But — as with most gradient theories — the constitutive length scale L that characterizes the energy (105.58) is *ad hoc*, chosen to match the discrete dislocation simulations. Interesingly, the study of OHNO & OKUMURA (2007), which we now discuss, provides a notable exception to this situation!

Ohno and Okumura show that the theory presented here — when endowed with the Ohno–Okumura self energy given in Footnote 649 — compares well to a large class of experiments on single crystals of Al, Ni, Cu, Fe, and steel at submicron to several micron length scales. But what is most important, Ohno and Okumura do not introduce an *ad hoc* length scale; their energy — patterned after a standard formula for the energy per unit length of a single dislocation in an isotropic crystal[656] — has a built-in gradient length-scale[657] based on the material Burgers vector b.[658]

An alternative to the theory presented here and one based on a defect energy

$$\Psi = \hat{\Psi}^p(\mathbf{G}) \tag{105.78}$$

dependent on the Burgers tensor $\mathbf{G}$ was proposed by GURTIN (2002).[659] Because $\mathbf{G}$ can be decomposed into distributions of edge and screw dislocation via the decomposition (103.9), we can consider $\mathbf{G}$ as a function

$$\mathbf{G} = \hat{\mathbf{G}}(\vec{\rho}\,) \tag{105.79}$$

of the list $\vec{\rho}$ of disclocation densities and convert the energy (105.78) to an energy dependent on dislocation densities. But, as noted by ARSENLIS & PARKS (1999), the function (105.79) is *not* one-to-one and hence cannot be inverted to give the dislocation densities as functions of $\mathbf{G}$.[660] Thus, the energy (105.78) cannot generally be converted to an energy dependent on $\vec{\rho}$. Summarizing,

- *an energetic dependence on* $\mathbf{G}$ *is not equivalent to an energetic dependence on* $\vec{\rho}$.

On a more pragmatic note, an advantage of an energy dependent on $\vec{\rho}$ is that an energetic constitutive equation in the form

$$\Psi^p = \hat{\Psi}^p(\vec{\rho}\,) \tag{105.80}$$

is — for fcc and bcc crystals — automatically consistent with the symmetry of the underlying crystal, but an energy of the form (105.78) is not — a separate analysis is needed to ensure that the function $\hat{\Psi}^p(\mathbf{G})$ be invariant under the symmetry group of such crystals.

Gradient theories of the type considered here extend naturally to situations involving grain boundaries. Indeed, for $\mathcal{S}$ a subsurface of $\partial\mathrm{B}$, the presence of

[656] Cf. HIRTH AND LOTHE (1982, eq. (3.52)).

[657] As noted by GURTIN (2009), the length scale in the Ohno–Okumura theory has the precise form $\mu b/S$, with S the underlying slip resistance.

[658] Cf. the paragraph labelled "terminology" on page 591.

[659] Cf. also GURTIN & NEEDLEMAN (2005).

[660] In fact, ARSENLIS & PARKS (1999) show that certain combinations of densities correspond to $\mathbf{G} = \mathbf{0}$ — so that the absence of a Burgers vector does not generally imply the absence of continuous distributions of dislocations as described by the dislocation densities.

microscopic stresses $\boldsymbol{\xi}^\alpha$ results in a power expenditure

$$\sum_\alpha \int_S (\boldsymbol{\xi}^\alpha \cdot \mathbf{n})\dot{\gamma}^\alpha \, da \tag{105.81}$$

by the material in contact with the body — when this material is that of another grain, then (105.81) and its analog for the other grain leads to an extension of the principle of virtual power that can be used to develop force balances and a free-energy imbalance for the grain boundary. For small deformations, such a procedure was used as a basis for a treatment of grain boundaries by CERMELLI & GURTIN (2002), GURTIN & NEEDLEMAN (2005), and GURTIN (2008a).

PART XVIII

SINGLE CRYSTALS UNDERGOING LARGE DEFORMATIONS

This chapter presents a counterpart — for large deformations — of our treatment of single crystals undergoing small deformations.[661] While much of what we discuss is similar to corresponding material in §101.3, we present almost all arguments in full, not only for completeness, but also because they require notions intrinsic to large deformations. Further, as in §101.3, we base the theory on the principle of virtual power, but here, unlike there, we develop the force balances for the conventional and gradient theories together.

We also discuss the Taylor model. This model, which is based on single-crystal equations developed in this section, is often used to characterize the formation of "texture" in polycrystals.[662]

[661] Cf. §101.3. In this regard, the introductory material on pages 581–585 is applicable here and might be reviewed by those readers not familiar with the materials-science aspects of single crystals.

[662] Cf. §111.

106 Basic Single-Crystal Kinematics

The kinematics of single crystals undergoing large deformations takes as its starting point the kinematics discussed in §91 and, consequently, begins with the Kröner decomposition

$$\mathbf{F} = \mathbf{F}^e\mathbf{F}^p, \tag{106.1}$$

together with the assumption of plastic incompressibility:

$$\det \mathbf{F}^p = 1. \tag{106.2}$$

Here, in agreement with standard terminology[663] we refer to the structural space as the **lattice space** or, more simply, as the **lattice**,[664] and we refer to vectors in that space as **lattice vectors**. Thus $\mathbf{F}^p$ *maps material vectors to lattice vectors*; $\mathbf{F}^e$ *maps lattice vectors to spatial vectors*; a **lattice tensor** is a tensor that maps lattice vectors to lattice vectors (Figure 106.1).

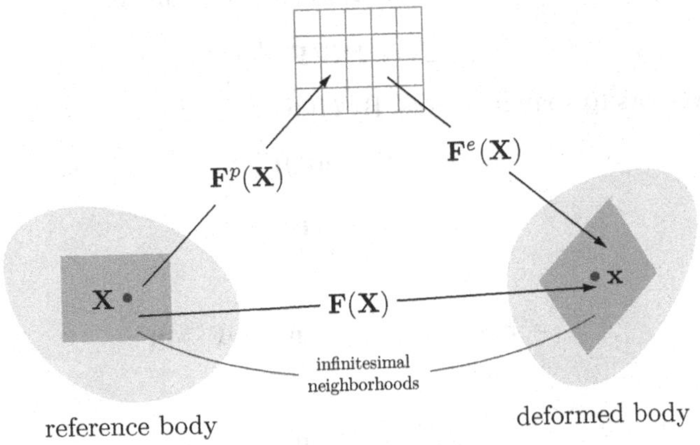

Figure 106.1. Schematic of the Kröner decomposition showing the lattice.

The elastic and plastic rotations and right and left stretch tensors and the elastic and plastic distortion-rate tensors and corresponding spin and stretching tensors are as defined in §91; here we recall only the relations (91.7) and (91.8) for the

663 Cf. §91.2.

664 For us, the term lattice always connotes the *undistorted* lattice.

distortion-rate tensors, as these are needed in the ensuing discussion:

$$\mathbf{L}^e = \dot{\mathbf{F}}^e\mathbf{F}^{e-1}, \qquad \mathbf{L}^p = \dot{\mathbf{F}}^p\mathbf{F}^{p-1},$$
$$\mathbf{L} = \mathbf{L}^e + \mathbf{F}^e\mathbf{L}^p\mathbf{F}^{e-1}. \tag{106.3}$$

Slip systems are as defined in the paragraph containing (102.4); what is important within the present context of large deformations is that, for $\alpha = 1, 2, \ldots, N$,[665]

(‡) the slip direction $\mathbf{s}^\alpha$ and slip-plane normal $\mathbf{m}^\alpha$ are constant *lattice vectors*[666]

consistent with (102.4); viz.

$$\mathbf{s}^\alpha \cdot \mathbf{m}^\alpha = 0, \qquad |\mathbf{s}^\alpha| = |\mathbf{m}^\alpha| = 1. \tag{106.4}$$

Within the framework of large deformations,

- *slips* (as fields on the individual slip systems) *are not well-defined quantities: What are well defined are slip rates.*

For that reason, the **single-crystal hypothesis** is framed in terms of the plastic distortion-rate $\mathbf{L}^p$; specifically, this hypothesis requires that $\mathbf{L}^p$ be governed by **slip rates**[667] ν^α on the individual slip systems via the relation[668]

$$\mathbf{L}^p = \sum_\alpha \nu^\alpha\, \mathbb{S}^\alpha, \tag{106.5}$$

in which, as before,

$$\mathbb{S}^\alpha = \mathbf{s}^\alpha \otimes \mathbf{m}^\alpha \tag{106.6}$$

is the **Schmid tensor** for slip system α. By (106.6), $\mathbb{S}^\alpha$ is a mapping of lattice vectors to lattice vectors and, hence, a *lattice tensor.*

In view of $(106.3)_3$

$$\mathbf{L} = \mathbf{L}^e + \sum_\alpha \nu^\alpha\big(\mathbf{F}^e\mathbb{S}^\alpha\mathbf{F}^{e-1}\big). \tag{106.7}$$

Since $\mathbf{F}^e$ maps lattice vectors to spatial vectors,[669] the tensor

$$\bar{\mathbb{S}}^\alpha \stackrel{\text{def}}{=} \mathbf{F}^e\mathbb{S}^\alpha\mathbf{F}^{e-1} \tag{106.8}$$

maps spatial vectors to spatial vectors. By (106.6)

$$\bar{\mathbb{S}}^\alpha = \mathbf{F}^e(\mathbf{s}^\alpha \otimes \mathbf{m}^\alpha)\mathbf{F}^{e-1}$$
$$= (\mathbf{F}^e\mathbf{s}^\alpha) \otimes (\mathbf{F}^{e-\top}\mathbf{m}^\alpha);$$

thus, letting

$$\bar{\mathbf{s}}^\alpha = \mathbf{F}^e\mathbf{s}^\alpha \qquad \text{and} \qquad \bar{\mathbf{m}}^\alpha = \mathbf{F}^{e-\top}\mathbf{m}^\alpha, \tag{106.9}$$

we see that

$$\bar{\mathbb{S}}^\alpha = \bar{\mathbf{s}}^\alpha \otimes \bar{\mathbf{m}}^\alpha. \tag{106.10}$$

---

665 We continue to use the notational conventions specified in the bullets on page 587.

666 Cf. §91.2, Figure 106.1.

667 It is more common to denote the slip rates by $\dot{\gamma}^\alpha$. We refrain from using this notation because the "dot" has a precise meaning as a material time-derivative, and in the large-deformation theory ν^α is *not* the material time-derivative of a physically meaningful quantity.

668 Cf., e.g., Teodosiu (1970), Rice (1971), Asaro (1983), Havner (1992), Bassani (1994). Cf., also, Deseri & Owen (2002), who discuss this hypothesis within the context of their theory of invertible structured deformations.

669 Cf. Figure 106.1.

The fields $\bar{\mathbf{s}}^\alpha$ and $\bar{\mathbf{m}}^\alpha$ represent the slip direction $\mathbf{s}^\alpha$ and the slip-plane normal $\mathbf{m}^\alpha$ pushed forward to the observed space, with $\mathbf{s}^\alpha$ as a tangent vector and $\mathbf{m}^\alpha$ as a normal vector.[670] Further, since $\mathbf{L} = \operatorname{grad}\mathbf{v}$, (106.7) takes the form

$$\operatorname{grad}\mathbf{v} = \mathbf{L}^e + \sum_\alpha \nu^\alpha\, \bar{\mathbb{S}}^\alpha. \tag{106.11}$$

EXERCISE

1. Show that, for each α, $\mathbb{S}^\alpha$ and $\bar{\mathbb{S}}^\alpha$ are *deviatoric*,

$$\operatorname{tr}\mathbb{S}^\alpha = \operatorname{tr}\bar{\mathbb{S}}^\alpha = 0, \tag{106.12}$$

so that, by (106.5),

$$\operatorname{tr}\mathbf{L}^p = 0,$$

consistent with (106.2).

[670] Cf. (6.15) and (8.4).

107 The Burgers Vector and the Flow of Screw and Edge Dislocations

The macroscopic notion of a Burgers vector is much richer when discussed within the framework of large deformations,[671] but such a discussion requires mathematical machinery more sophisticated than that used in §103, where the deformation is assumed small.[672]

107.1 Transformation of Vector Area Measures Between the Reference, Observed, and Lattice Spaces

Let S_R denote an arbitrary oriented material surface with $S = \chi_t(S_R)$ the corresponding deformed surface, and let $\mathbf{n}_R$ and $\mathbf{n}$ denote the unit normal fields for S_R and S, so that

$$\mathbf{n} = \frac{\mathbf{F}^{-\top}\mathbf{n}_R}{|\mathbf{F}^{-\top}\mathbf{n}_R|}. \tag{107.1}$$

Then writing $\mathbf{n}_R\, da_R$ and $\mathbf{n}\, da$ for the (vector) area measures on S_R and S, we have the transformation rules[673]

$$\int_{S_R} \mathbf{A}\mathbf{n}_R\, da_R = \int_{S} J^{-1}\mathbf{A}\mathbf{F}^{\top}\mathbf{n}\, da \qquad \text{and} \qquad \int_{S} \mathbf{A}\mathbf{n}\, da = \int_{S_R} J\mathbf{A}\mathbf{F}^{-\top}\mathbf{n}_R\, da_R, \tag{107.2}$$

with the material description of $\mathbf{A}$ used in the integrals over S_R and the spatial description used in the integrals over S.[674] These rules are expressed succinctly by the formal relations[675]

$$\mathbf{n}_R\, da_R = J^{-1}\mathbf{F}^{\top}\mathbf{n}\, da \qquad \text{and} \qquad \mathbf{n}\, da = J\mathbf{F}^{-\top}\mathbf{n}_R\, da_R. \tag{107.3}$$

We now establish counterparts of (107.3) for the transformation of vector measures between the reference space and the *lattice* and between the *lattice* and the observed space. First of all, note that by (106.1) and (107.1)

$$\begin{aligned}\mathbf{F}^{p-\top}\mathbf{n}_R &= \mathbf{F}^{e\top}\mathbf{F}^{-\top}\mathbf{n}_R\\ &= |\mathbf{F}^{-\top}\mathbf{n}_R|\mathbf{F}^{e\top}\mathbf{n}.\end{aligned} \tag{107.4}$$

[671] In fact, the approximation of small deformations obscures much of the physics underlying the notion of a Burgers vector.

[672] This section follows GURTIN (2006).

[673] Cf. $(15.8)_2$.

[674] Cf. §9.

[675] Cf. (15.5).

Since $\mathbf{F}^{p-\top}$ maps material vectors to lattice vectors, the quantity $\mathbf{F}^{p-\top}\mathbf{n}_{\rm R}$ is a lattice vector and, by (107.4),

$$\frac{\mathbf{F}^{p-\top}\mathbf{n}_{\rm R}}{|\mathbf{F}^{p-\top}\mathbf{n}_{\rm R}|} = \frac{\mathbf{F}^{e\top}\mathbf{n}}{|\mathbf{F}^{e\top}\mathbf{n}|}.$$

It is therefore meaningful to let $\mathbf{n}_\#$ denote the unit *lattice vector*

$$\mathbf{n}_\# = \frac{\mathbf{F}^{p-\top}\mathbf{n}_{\rm R}}{|\mathbf{F}^{p-\top}\mathbf{n}_{\rm R}|} = \frac{\mathbf{F}^{e\top}\mathbf{n}}{|\mathbf{F}^{e\top}\mathbf{n}|}; \tag{107.5}$$

$\mathbf{n}_\#$ is the unit normal $\mathbf{n}_{\rm R}$ pushed forward (as a unit normal) to the lattice, or, equivalently, the unit normal $\mathbf{n}$ pulled back to the lattice.

Integration is meaningless in the lattice, because the lattice is not a point space, but the vector area measure $\mathbf{n}_\#\,da_\#$ in the lattice (with unit normal $\mathbf{n}_\#$ and area measure $da_\#$) defined formally by

$$\mathbf{n}_\#\,da_\# = \mathbf{F}^{p-\top}\mathbf{n}_{\rm R}\,da_{\rm R} \qquad \text{or, equivalently, by} \qquad \mathbf{n}_\# da_\# = J^{-1}\mathbf{F}^{e\top}\mathbf{n}\,da \tag{107.6}$$

has meaning because $\mathbf{n}_{\rm R}\,da_{\rm R}$ and $\mathbf{n}\,da$ are *local*. Further, (107.6) may be formally inverted to give

$$\mathbf{n}_{\rm R}\,da_{\rm R} = \mathbf{F}^{p\top}\mathbf{n}_\#\,da_\# \qquad \text{and} \qquad \mathbf{n}\,da = J\mathbf{F}^{e-\top}\mathbf{n}_\#\,da_\#. \tag{107.7}$$

107.2 Characterization of the Burgers Vector

Let $\partial S_{\rm R}$ denote the boundary curve of a smooth material surface $S_{\rm R}$ in the *reference body*; then, by Stokes' theorem,[676]

$$\begin{aligned} \mathbf{b}^p(\partial S_{\rm R}) &\stackrel{\text{def}}{=} \int\limits_{\partial S_{\rm R}} \mathbf{F}^p d\mathbf{X} \\ &= \int\limits_{S_{\rm R}} (\mathrm{Curl}\,\mathbf{F}^p)^\top \mathbf{n}_{\rm R}\,da_{\rm R}. \end{aligned} \tag{107.8}$$

Since $\mathbf{F}^p d\mathbf{X}$ lies in the *lattice*, so also does $(\mathrm{Curl}\,\mathbf{F}^p)^\top\mathbf{n}_{\rm R}$; hence, one might associate $(\mathrm{Curl}\,\mathbf{F}^p)^\top\mathbf{n}_{\rm R}$ with the Burgers vector associated with the boundary curve of a surface-element with normal $\mathbf{n}_{\rm R}$. But that would be incorrect. The surface element $\mathbf{n}_{\rm R}\,da_{\rm R}$ lies in the *reference space* rather than in the lattice, so that $(\mathrm{Curl}\,\mathbf{F}^p)^\top\mathbf{n}_{\rm R}$ is a *lattice vector* measured per unit area in the *reference body* — a result that contradicts the conventional notion of a Burgers vector, which asserts that

- the Burgers vector is a vector in the *lattice* measured per unit area in the *lattice.*

This defect is easily rectified. By $(107.7)_1$, formally,

$$(\mathrm{Curl}\,\mathbf{F}^p)^\top\mathbf{n}_{\rm R}\,da_{\rm R} = (\mathrm{Curl}\,\mathbf{F}^p)^\top\mathbf{F}^{p\top}\mathbf{n}_\# da_\#, \tag{107.9}$$

with $\mathbf{n}_\# da_\#$ a surface element in the lattice, so that $(\mathbf{F}^p\mathrm{Curl}\,\mathbf{F}^p)^\top\mathbf{n}_\# da_\#$ might be viewed as the *local Burgers vector* corresponding to the "boundary curve" of the surface element $\mathbf{n}_\#\,da_\#$ in the *lattice*. Thus, for

$$\mathbf{G}^p \stackrel{\text{def}}{=} \mathbf{F}^p\mathrm{Curl}\,\mathbf{F}^p, \tag{107.10}$$

$\mathbf{G}^{p\top}\mathbf{n}_\#$ provides a measure of the (local) Burgers vector in the lattice — per unit lattice area — for the plane $\Pi^\#$ with unit normal $\mathbf{n}_\#$. Moreover, as is clear from the foregoing derivation: *$\mathbf{G}^p$ is a lattice tensor as it maps lattice vectors to lattice vectors.*

[676] Cf. $(4.8)_3$.

On the other hand, let S be a smooth surface in the *deformed body* and let

$$\mathbf{b}^e(\partial S) \stackrel{\text{def}}{=} \int_{\partial S} \mathbf{F}^{e-1} d\mathbf{x}$$

$$= \int_S (\text{curl}\,\mathbf{F}^{e-1})^\top \mathbf{n}\, da. \tag{107.11}$$

Arguing as in the steps leading to (107.10), we may use $(107.7)_2$, again formally, to conclude that

$$(\text{curl}\,\mathbf{F}^{e-1})^\top \mathbf{n}\, da = J(\text{curl}\,\mathbf{F}^{e-1})^\top \mathbf{F}^{e-\top} \mathbf{n}_\# \, da_\#.$$

Thus, by (107.11), for

$$\mathbf{G}^e \stackrel{\text{def}}{=} J\mathbf{F}^{e-1} \text{curl}\,\mathbf{F}^{e-1}, \tag{107.12}$$

$\mathbf{G}^{e\top}\mathbf{n}_\#$ also provides a measure of the Burgers vector, per unit lattice area, for the plane $\Pi^\#$ with unit normal $\mathbf{n}_\#$, and $\mathbf{G}^e$, like $\mathbf{G}^p$, is a lattice tensor. The fields $\mathbf{G}^{p\top}\mathbf{n}_\#$ and $\mathbf{G}^{e\top}\mathbf{n}_\#$ purportedly characterize the same Burgers vector. To reconcile this, note that

$$\int_{\partial S_R} \mathbf{F}^p d\mathbf{X} = \int_{\partial S_R} \underbrace{\mathbf{F}^p\mathbf{F}^{-1}}_{\mathbf{F}^{e-1}} \mathbf{F} d\mathbf{X}$$

$$= \int_{\partial S} \mathbf{F}^{e-1} d\mathbf{x},$$

so that

$$\mathbf{b}^e(\partial S) = \mathbf{b}^p(\partial S_R).$$

Therefore, by $(107.2)_2$,

$$\int_{S_R} (\text{Curl}\,\mathbf{F}^p)^\top \mathbf{n}_R \, da_R = \int_S (\text{curl}\,\mathbf{F}^{e-1})^\top \mathbf{n}\, da$$

$$= \int_{S_R} J(\text{curl}\,\mathbf{F}^{e-1})^\top \mathbf{F}^{-\top} \mathbf{n}_R \, da_R,$$

and, since S_R is arbitrary,

$$(\text{Curl}\,\mathbf{F}^p)^\top = J(\text{curl}\,\mathbf{F}^{e-1})^\top \mathbf{F}^{-\top}. \tag{107.13}$$

Finally, a consequence of the decomposition $\mathbf{F} = \mathbf{F}^e\mathbf{F}^p$ is the identity $\mathbf{F}^{-\top} = \mathbf{F}^{e-\top}\mathbf{F}^{p-\top}$; postmultiplying (107.13) by $\mathbf{F}^{p\top}$, we thus arrive at the conclusion that

$$(\text{Curl}\,\mathbf{F}^p)^\top \mathbf{F}^{p\top} = J(\text{curl}\,\mathbf{F}^{e-1})^\top \mathbf{F}^{e-\top}$$

and, hence, that

$$\mathbf{F}^p \text{Curl}\,\mathbf{F}^p = J\mathbf{F}^{e-1} \text{curl}\,\mathbf{F}^{e-1}.$$

We may therefore conclude from (107.10) and (107.12) that *the tensor fields* $\mathbf{G}^p$ *and* $\mathbf{G}^e$ *coincide*. We refer to the lattice tensor $\mathbf{G}$ defined by $\mathbf{G} = \mathbf{G}^p = \mathbf{G}^e$ and, hence, by

$$\boxed{\begin{aligned}\mathbf{G} &= \mathbf{F}^p \mathrm{Curl}\,\mathbf{F}^p \\ &= J\mathbf{F}^{e-1} \mathrm{curl}\,\mathbf{F}^{e-1}\end{aligned}} \tag{107.14}$$

as the **Burgers tensor.**[677]

- *Given any oriented plane* $\Pi^\#$ *in the lattice, with* $\mathbf{n}_\#$ *its unit normal,* $\mathbf{G}^\top\mathbf{n}_\#$ *represents the Burgers vector — as a vector in the lattice measured per unit lattice area — for infinitesimal circuits on* $\Pi^\#$. *In terms more suggestive than precise,* $\mathbf{G}^\top\mathbf{n}_\#$ *represents the Burgers vector, per unit area, for those dislocation lines piercing* $\Pi^\#$.

That there are two relations for the Burgers tensor is of great value. The relation $\mathbf{G} = \mathbf{F}^p\,\mathrm{Curl}\,\mathbf{F}^p$ seems most relevant to theories of plasticity involving plastic-strain gradients.[678] On the other hand, in discussing single crystals, materials scientists typically neglect lattice strains, taking $\mathbf{F}^e = \mathbf{R}^e$, with $\mathbf{R}^e$ a rotation, in which case $\mathbf{G} = \mathbf{R}^{e\top}\mathrm{curl}\,\mathbf{R}^{e\top}$ and $\mathbf{G}$ may be determined via measurements of lattice rotations.

107.3 The Plastically Convected Rate of G

Our next step is to establish a time derivative of $\mathbf{G}$ following the flow of dislocations through the lattice. In view of $(106.3)_2$ and $(107.14)_1$,

$$\begin{aligned}\dot{\mathbf{G}} &= \dot{\mathbf{F}}^p \mathrm{Curl}\,\mathbf{F}^p + \underbrace{\mathbf{F}^p \mathrm{Curl}\,\dot{\mathbf{F}}^p}_{\mathbf{G}^*} \\ &= \mathbf{L}^p\mathbf{G} + \mathbf{G}^*\end{aligned} \tag{107.15}$$

and, by $(106.3)_2$,

$$\frac{\partial \dot{F}^p_{js}}{\partial X_r} = L^p_{jq}\frac{\partial F^p_{qs}}{\partial X_r} + \frac{\partial L^p_{jq}}{\partial X_r}F^p_{qs};$$

therefore

$$\begin{aligned}G^*_{ij} &= F^p_{im}\varepsilon_{mrs}\frac{\partial \dot{F}^p_{js}}{\partial X_r} \\ &= F^p_{im}\varepsilon_{mrs}\frac{\partial F^p_{qs}}{\partial X_r}L^p_{jq} + F^p_{im}\varepsilon_{mrs}F^p_{qs}\frac{\partial L^p_{jq}}{\partial X_r} \\ &= G_{iq}L^p_{jq} + F^p_{im}\varepsilon_{mrs}F^p_{qs}\frac{\partial L^p_{jq}}{\partial X_r}\end{aligned}$$

and (107.15) yields

$$\dot{G}_{ij} - L^p_{im}G_{mj} - G_{iq}L^p_{jq} = \varepsilon_{mrs}F^p_{im}F^p_{qs}\frac{\partial L^p_{jq}}{\partial X_r}. \tag{107.16}$$

We refer to the left side of (107.16), namely to

$$\boxed{\overset{\diamond}{\mathbf{G}} \stackrel{\text{def}}{=} \dot{\mathbf{G}} - \mathbf{L}^p\mathbf{G} - \mathbf{G}\mathbf{L}^{p\top},} \tag{107.17}$$

---

677 Cf. Cermelli & Gurtin (2001).
678 Cf., e.g., Gurtin (2003).

as the **plastically convected rate of G**;[679] that is,

- as the convected rate of **G** following the flow of dislocations through the lattice as characterized by the tensor field $\mathbf{L}^p$.

Then, by (107.16) and (107.17),

$$\overset{\diamond}{G}_{ij} = \varepsilon_{mrs} F^p_{im} F^p_{qs} \frac{\partial L^p_{jq}}{\partial X_r} \tag{107.18}$$

(where $\overset{\diamond}{G}_{ij}$ denotes the ij-th component of $\overset{\diamond}{\mathbf{G}}$) and using the identities[680]

$$F^p_{mr} F^{p-1}_{bm} = \delta_{br}, \qquad \varepsilon_{mrs} F^p_{im} F^p_{qs} F^p_{ar} = \varepsilon_{iaq} \tag{107.19}$$

and, since the body under consideration is a single crystal, by (107.18) and (107.19), that

$$\begin{aligned}
\overset{\diamond}{G}_{ij} &= \varepsilon_{mrs} F^p_{im} F^p_{qs} \frac{\partial L^p_{jq}}{\partial X_b} \delta_{br} \\
&= \varepsilon_{mrs} F^p_{im} F^p_{qs} \frac{\partial L^p_{jq}}{\partial X_b} F^{p-1}_{ba} F^p_{ar} \\
&= \underbrace{\varepsilon_{mrs} F^p_{im} F^p_{qs} F^p_{ar}}_{\varepsilon_{iaq}} \frac{\partial L^p_{jq}}{\partial X_b} F^{p-1}_{ba} \\
&= \varepsilon_{iaq} F^{p-1}_{ba} \frac{\partial L^p_{jq}}{\partial X_b};
\end{aligned}$$

thus, by $(106.5)_2$,

$$\overset{\diamond}{G}_{ij} = \varepsilon_{iaq} \sum_\alpha s^\alpha_j m^\alpha_q \underline{F^{p-1}_{ba} \frac{\partial \nu^\alpha}{\partial X_b}}, \tag{107.20}$$

where the underlined term represents the component form of $\mathbf{F}^{p-\top}\nabla\nu^\alpha$ and, since $\nabla\nu^\alpha = \mathbf{F}^\top \operatorname{grad}\nu^\alpha$ and $\mathbf{F} = \mathbf{F}^e\mathbf{F}^p$, this term can equally well be expressed in terms of $\mathbf{F}^e$:

$$\begin{aligned}
\mathbf{F}^{p-\top}\nabla\nu^\alpha &= \mathbf{F}^{e\top} \operatorname{grad}\nu^\alpha \\
&\overset{\text{def}}{=} \nabla^{\#}\nu^\alpha.
\end{aligned} \tag{107.21}$$

The relation (107.20) together with the definition (107.21) of $\nabla^{\#}\nu^\alpha$ yields the central result of this subsection:[681]

$$\boxed{\overset{\diamond}{\mathbf{G}} = \sum_\alpha (\nabla^{\#}\nu^\alpha \times \mathbf{m}^\alpha) \otimes \mathbf{s}^\alpha.} \tag{107.22}$$

The identity (107.21) asserts that $\mathbf{F}^{p-\top}\nabla\nu^\alpha$, the material gradient of ν^α *pushed forward* from the reference space to the lattice, is equal to $\mathbf{F}^{e\top}\operatorname{grad}\nu^\alpha$, the spatial

[679] In view of (20.26), the plastically convected rate is the contravariant (i.e., Oldroyd) rate with **L** replaced by $\mathbf{L}^p$. Our use of the symbol $\diamond$ to denote the plastically convected rate should not cause confusion because the contravariant rate is not used in the present context.

[680] $(107.19)_2$ follows from (2.84), since $\det\mathbf{F}^p = 1$.

[681] Cf. Cermelli & Gurtin (2001, eq. (11.11)) and Gurtin (2006, eq. (4.7)). The relation (107.22) should be compared to its small-deformation counterpart (103.2).

gradient of ν^α *pulled back* from the observed space to the lattice; thus $\nabla^\# \nu^\alpha$ might be viewed as *the gradient of* ν^α *in the lattice.*

EXERCISE

1. Establish the following identity for the plastically convected rate of $\mathbf{G}$:

$$\overset{\diamond}{\mathbf{G}} = \mathbf{F}^p \overline{(\mathbf{F}^{p-1}\mathbf{G}\mathbf{F}^{p-\top})}^{\boldsymbol{\cdot}} \mathbf{F}^{p\top}. \tag{107.23}$$

107.4 Densities of Screw and Edge Dislocations

We emulate the discussion of dislocation densities in §103.1.

Since the vector $\nabla^\# \nu^\alpha \times \mathbf{m}^\alpha$ is orthogonal to $\mathbf{m}^\alpha$, it may be expanded in terms of $\mathbf{s}^\alpha$ and the lattice vector

$$\mathbf{l}^\alpha = \mathbf{m}^\alpha \times \mathbf{s}^\alpha \tag{107.24}$$

as follows:[682]

$$\nabla^\# \nu^\alpha \times \mathbf{m}^\alpha = \underbrace{\mathbf{s}^\alpha \cdot (\nabla^\# \nu^\alpha \times \mathbf{m}^\alpha)}_{\mathbf{l}^\alpha \cdot \nabla^\# \nu^\alpha} \mathbf{s}^\alpha + \underbrace{\mathbf{l}^\alpha \cdot (\nabla^\# \nu^\alpha \times \mathbf{m}^\alpha)}_{-\mathbf{s}^\alpha \cdot \nabla^\# \nu^\alpha} \mathbf{l}^\alpha. \tag{107.25}$$

Based on this expression we introduce **screw** and **edge dislocation-densities** $\rho_\odot^\alpha$ and $\rho_\vdash^\alpha$ defined formally as solutions of the differential equations

$$\begin{aligned} \dot{\rho}_\odot^\alpha &= \mathbf{l}^\alpha \cdot \nabla^\# \nu^\alpha, \\ \dot{\rho}_\vdash^\alpha &= -\mathbf{s}^\alpha \cdot \nabla^\# \nu^\alpha, \end{aligned} \tag{107.26}$$

subject to initial conditions at, say, $t = 0$.[683] The dislocation densities $\rho_\odot^\alpha$ and $\rho_\vdash^\alpha$ therefore represent *internal variables* with evolution governed by (107.26).

In view of (107.25) and (107.29), the expression (107.22) reduces to an important decomposition for the plastically convected rate of $\mathbf{G}$:[684]

$$\boxed{\begin{gathered} \overset{\diamond}{\mathbf{G}} = \sum_\alpha (\dot{\rho}_\odot^\alpha \, \mathbf{s}^\alpha \otimes \mathbf{s}^\alpha + \dot{\rho}_\vdash^\alpha \, \mathbf{l}^\alpha \otimes \mathbf{s}^\alpha), \\ \dot{\rho}_\vdash^\alpha = -\mathbf{s}^\alpha \cdot \nabla^\# \nu^\alpha, \qquad \dot{\rho}_\odot^\alpha = \mathbf{l}^\alpha \cdot \nabla^\# \nu^\alpha. \end{gathered}} \tag{107.27}$$

We view the tensor fields[685]

$$\rho_\odot^\alpha \, \mathbf{s}^\alpha \otimes \mathbf{s}^\alpha \qquad \text{and} \qquad \rho_\vdash^\alpha \, \mathbf{l}^\alpha \otimes \mathbf{s}^\alpha \tag{107.28}$$

as *macroscopic distributions of screw and edge dislocations* on slip system α with *densities* $\rho_\odot^\alpha$ and $\rho_\vdash^\alpha$; granted this view, $(107.27)_1$ asserts that

- *temporal changes in* $\mathbf{G}$ *— as characterized by its plastically convected rate* $\overset{\diamond}{\mathbf{G}}$ *— may be decomposed into temporal changes in distributions of screw and edge dislocations on the individual slip systems.*

---

682 Cf. (107.21).

683 E.g., if the body is initially in a *virgin state* in the sense that $\mathbf{F}(\mathbf{X}, 0) = \mathbf{F}^p(\mathbf{X}, 0) = \mathbf{1}$, then we might assume that $\rho_\odot^\alpha(\mathbf{X}, 0) = 0$ and $\rho_\vdash^\alpha(\mathbf{X}, 0) = 0$.

684 Gurtin (2006, eq. (17.27)). Cf. (103.9).

685 Cf. the paragraph containing (103.8).

By (107.21), the differential equations (107.26) for the screw and edge dislocation densities can be expressed as

$$\begin{aligned}\dot{\rho}^\alpha_\odot &= \mathbf{l}^\alpha \cdot \mathbf{F}^{p-\top}\nabla\nu^\alpha = \mathbf{l}^\alpha \cdot \mathbf{F}^{e\top}\mathrm{grad}\,\nu^\alpha,\\ \dot{\rho}^\alpha_\vdash &= -\mathbf{s}^\alpha \cdot \mathbf{F}^{p-\top}\nabla\nu^\alpha = -\mathbf{s}^\alpha \cdot \mathbf{F}^{e\top}\mathrm{grad}\,\nu^\alpha.\end{aligned} \tag{107.29}$$

Consider the vector pairs defined by

$$\mathbf{l}^\alpha_\mathrm{R} = \mathbf{F}^{p-1}\mathbf{l}^\alpha, \quad \mathbf{s}^\alpha_\mathrm{R} = \mathbf{F}^{p-1}\mathbf{s}^\alpha \qquad \text{and} \qquad \bar{\mathbf{l}}^\alpha = \mathbf{F}^e\mathbf{l}^\alpha, \quad \bar{\mathbf{s}}^\alpha = \mathbf{F}^e\mathbf{s}^\alpha; \tag{107.30}$$

the first pair represents the lattice vectors $\mathbf{l}^\alpha$ and $\mathbf{s}^\alpha$ pulled back to the reference space, the second represents $\mathbf{l}^\alpha$ and $\mathbf{s}^\alpha$ pushed forward to the observed space; using these pairs we can rewrite the differential equations (107.29) in the referential form

$$\dot{\rho}^\alpha_\odot = \mathbf{l}^\alpha_\mathrm{R} \cdot \nabla\nu^\alpha, \qquad \dot{\rho}^\alpha_\vdash = -\mathbf{s}^\alpha_\mathrm{R} \cdot \nabla\nu^\alpha \tag{107.31}$$

or in the equivalent spatial form

$$\dot{\rho}^\alpha_\odot = \bar{\mathbf{l}}^\alpha \cdot \mathrm{grad}\,\nu^\alpha, \qquad \dot{\rho}^\alpha_\vdash = -\bar{\mathbf{s}}^\alpha \cdot \mathrm{grad}\,\nu^\alpha. \tag{107.32}$$

Arguing as in the paragraph containing (103.10), we conclude that:

(i) $\mathbf{l}^\alpha = \mathbf{m}^\alpha \times \mathbf{s}^\alpha$ and $-\mathbf{s}^\alpha = \mathbf{m}^\alpha \times \mathbf{l}^\alpha$ are the respective glide directions for screw and edge dislocations on α;

(ii) $\rho^\alpha_\odot$ and $\rho^\alpha_\vdash$ characterize the pile-up of screw and edge dislocations on slip plane α.

Since slip results from the flow of dislocations, the vector fields

$$\mathbf{q}^\alpha_\odot \stackrel{\text{def}}{=} -\nu^\alpha\mathbf{l}^\alpha \qquad \text{and} \qquad \mathbf{q}^\alpha_\vdash \stackrel{\text{def}}{=} \nu^\alpha\mathbf{s}^\alpha \tag{107.33}$$

represent respective *fluxes of screw and edge dislocations.* Using (107.30) these fluxes may be pulled back to the reference space and they may also be pushed forward to the observed space; the results are

$$\mathbf{q}^\alpha_{\mathrm{R}\odot} = -\nu^\alpha\mathbf{l}^\alpha_\mathrm{R}, \quad \mathbf{q}^\alpha_{\mathrm{R}\vdash} = \nu^\alpha\mathbf{s}^\alpha_\mathrm{R}, \qquad \text{and} \qquad \bar{\mathbf{q}}^\alpha_\odot = -\nu^\alpha\bar{\mathbf{l}}^\alpha, \quad \bar{\mathbf{q}}^\alpha_\vdash = \nu^\alpha\bar{\mathbf{s}}^\alpha. \tag{107.34}$$

Thus, by (107.31), (107.32), and (107.34),

$$\begin{aligned}\dot{\rho}^\alpha_\odot &= -\mathrm{Div}\,\mathbf{q}^\alpha_{\mathrm{R}\odot} - \nu^\alpha\mathrm{Div}\,\mathbf{l}^\alpha_\mathrm{R}, & \dot{\rho}^\alpha_\vdash &= -\mathrm{Div}\,\mathbf{q}^\alpha_{\mathrm{R}\vdash} + \nu^\alpha\mathrm{Div}\,\mathbf{l}^\alpha_\mathrm{R},\\ \dot{\rho}^\alpha_\odot &= -\mathrm{div}\bar{\mathbf{q}}^\alpha_\odot - \nu^\alpha\mathrm{div}\bar{\mathbf{l}}^\alpha, & \dot{\rho}^\alpha_\vdash &= -\mathrm{div}\bar{\mathbf{q}}^\alpha_\vdash + \nu^\alpha\mathrm{div}\bar{\mathbf{l}}^\alpha,\end{aligned}$$

and introducing referential and spatial screw and edge *supplies* defined by

$$\begin{aligned}\sigma^\alpha_{\mathrm{R}\odot} &= -\nu^\alpha\mathrm{Div}\,\mathbf{l}^\alpha_\mathrm{R}, & \sigma^\alpha_{\mathrm{R}\vdash} &= \nu^\alpha\mathrm{Div}\,\mathbf{s}^\alpha_\mathrm{R},\\ \bar{\sigma}^\alpha_\odot &= -\nu^\alpha\mathrm{div}\,\bar{\mathbf{l}}^\alpha, & \bar{\sigma}^\alpha_\vdash &= \nu^\alpha\mathrm{div}\,\bar{\mathbf{s}}^\alpha,\end{aligned} \tag{107.35}$$

we have the **referential dislocation balances**

$$\begin{aligned}\dot{\rho}^\alpha_\odot &= -\mathrm{Div}\,\mathbf{q}^\alpha_{\mathrm{R}\odot} + \sigma^\alpha_{\mathrm{R}\odot},\\ \dot{\rho}^\alpha_\vdash &= -\mathrm{Div}\,\mathbf{q}^\alpha_{\mathrm{R}\vdash} + \sigma^\alpha_{\mathrm{R}\vdash},\end{aligned} \tag{107.36}$$

and the *equivalent* **spatial dislocation balances**

$$\begin{aligned}\dot{\rho}^\alpha_\odot &= -\mathrm{div}\bar{\mathbf{q}}^\alpha_\odot + \bar{\sigma}^\alpha_\odot,\\ \dot{\rho}^\alpha_\vdash &= -\mathrm{div}\bar{\mathbf{q}}^\alpha_\vdash + \bar{\sigma}^\alpha_\vdash.\end{aligned} \tag{107.37}$$

107.5 Comparison of Small- and Large-Deformation Results Concerning Dislocation Densities

If we compare the small-deformation equations[686]

$$\dot{\mathbf{G}} = \sum_{\alpha} \left(\dot{\rho}_{\vdash}^{\alpha} \mathbf{l}^{\alpha} \otimes \mathbf{s}^{\alpha} + \dot{\rho}_{\odot}^{\alpha} \mathbf{s}^{\alpha} \otimes \mathbf{s}^{\alpha} \right),$$
$$\dot{\rho}_{\vdash}^{\alpha} = -\mathbf{s}^{\alpha} \cdot \nabla \dot{\gamma}^{\alpha}, \qquad \dot{\rho}_{\odot}^{\alpha} = \mathbf{l}^{\alpha} \cdot \nabla \dot{\gamma}^{\alpha} \tag{107.38}$$

to their large-deformation counterparts[687]

$$\overset{\diamond}{\mathbf{G}} = \sum_{\alpha} \left(\dot{\rho}_{\odot}^{\alpha} \mathbf{s}^{\alpha} \otimes \mathbf{s}^{\alpha} + \dot{\rho}_{\vdash}^{\alpha} \mathbf{l}^{\alpha} \otimes \mathbf{s}^{\alpha} \right),$$
$$\dot{\rho}_{\vdash}^{\alpha} = -\mathbf{s}^{\alpha} \cdot \nabla^{\#} \nu^{\alpha}, \qquad \dot{\rho}_{\odot}^{\alpha} = \mathbf{l}^{\alpha} \cdot \nabla^{\#} \nu^{\alpha}, \tag{107.39}$$

we note their strong similarity:[688] the large-deformation equations (107.39) "map to" the small-deformation equations (107.38) via the "transformations"

$$\overset{\diamond}{\mathbf{G}} \to \dot{\mathbf{G}}, \qquad \nu^{\alpha} \to \dot{\gamma}^{\alpha}, \qquad \nabla \to \nabla^{\#},$$

or, equivalently — if in place of the density-rate equations $(107.39)_{2,3}$, we use the referential forms of these equations expressed in (107.31) — via the "transformations"

$$\overset{\diamond}{\mathbf{G}} \to \dot{\mathbf{G}}, \qquad \nu^{\alpha} \to \dot{\gamma}^{\alpha}, \qquad \mathbf{l}_{\mathrm{R}}^{\alpha} \to \mathbf{l}^{\alpha}, \qquad \mathbf{s}_{\mathrm{R}}^{\alpha} \to \mathbf{s}^{\alpha}.$$

Also interesting is a comparison of the dislocation balances in the small- and large-deformation theories. The dislocation balances (103.14) under small deformations have no supply terms. In contrast, the referential and spatial balances (107.36) and (107.37) exhibit the supply terms (107.35). In particular, the referential screw and edge supplies $\sigma_{\mathrm{R}\odot}^{\alpha}$ and $\sigma_{\mathrm{R}\vdash}^{\alpha}$ arise as a consequence of *the distortion of the lattice* as characterized by the fields

$$\mathrm{Div}\,\mathbf{l}_{\mathrm{R}}^{\alpha} \quad \text{and} \quad \mathrm{Div}\,\mathbf{s}_{\mathrm{R}}^{\alpha}; \tag{107.40}$$

thus, by $(107.30)_{1,2}$, the supplies are functions of $\nabla \mathbf{F}^{p}$ and would hence be most important in regions in which $\mathbf{F}^{p}$ suffers large spatial variations. Analogous assertions apply to the spatial supplies $(107.35)_{3,4}$.

Finally, as the slip-rates ν^{α} approach zero, the equations (107.39) of the large-deformation theory are asymptotic to the equations (107.38) of the small-deformation theory.[689]

---

686 Cf. (103.9).
687 Cf. (107.27).
688 The extent of this similarity is impressive. In this regard we note that the validity of the large-deformation relations (107.39) hinges on the requirement that the undistorted lattice be positioned between the reference and observed spaces as shown in Figure 106.1. Cf. §91.2.
689 GURTIN (2006, §9).

108 Virtual-Power Formulation of the Standard and Microscopic Force Balances

For a single-crystal, the kinematical quantities characterizing plastic flow are the slip-rates

$$\nu^1, \nu^2, \ldots, \nu^N. \tag{108.1}$$

The basic "rate-like" descriptors for a single-crystal undergoing large deformations are therefore the velocity $\mathbf{v}$, the elastic distortion-rate $\mathbf{L}^e$, and the slip rates (108.1), with these fields constrained by (106.11).

As in §92, we assume that, at some arbitrarily chosen but fixed time, the fields χ and $\mathbf{F}^e$ are known and

- *we denote by $\mathcal{P}_t$ an arbitrary subregion of the deformed body at that time and by* $\mathbf{n}$ *the outward unit normal on $\partial\mathcal{P}_t$.*

108.1 Internal and External Expenditures of Power

The formulation of the principle of virtual power for a single-crystal is based on a balance between the external power $\mathcal{W}(\mathcal{P}_t)$ expended on $\mathcal{P}_t$ and the internal power $\mathcal{I}(\mathcal{P}_t)$ expended within $\mathcal{P}_t$. As before,[690] to describe the internal power we replace the classical stress power $\mathbf{T}:\operatorname{grad}\mathbf{v}$ by a more detailed reckoning that *individually* characterizes

- the stretching and spinning of the underlying material structure as described by the lattice distortion-rate $\mathbf{L}^e$; and
- dislocation-induced slip as described by the slip rates $\nu^1, \nu^2, \ldots \nu^N$

via

- an **elastic-stress**[691] $\mathbf{T}$ power-conjugate to $\mathbf{L}^e$,

and, for each slip system α,

- a scalar **internal microscopic force** π^α power-conjugate to ν^α; and
- a vector **microscopic stress** $\boldsymbol{\xi}^\alpha$ power-conjugate to $\operatorname{grad}\nu^\alpha$.

[690] Cf. the paragraph containing (104.2).

[691] In contrast to §92.1, we write the stress power associated with $\mathbf{L}^e$ in the form $\mathbf{T}:\mathbf{L}^e$, rather than $\mathbf{S}^e:\mathbf{L}^e$, a change that is motivated by (92.23) and justified in the first paragraph of §108.3.

We, therefore, write the **internal power** in the form

$$\mathcal{I}(\mathcal{P}_t) = \int\limits_{\mathcal{P}_t} \mathbf{T}\colon\mathbf{L}^e\,dv + \sum_\alpha \int\limits_{\mathcal{P}_t} (\pi^\alpha \nu^\alpha + \boldsymbol{\xi}^\alpha \cdot \operatorname{grad}\nu^\alpha)\,dv. \tag{108.2}$$

Turning to the external power, we allow for the standard power expenditures $\mathbf{t}(\mathbf{n})\cdot\mathbf{v}$ by surface tractions and $\mathbf{b}\cdot\mathbf{v}$ by (conventional and inertial) body forces. Further, we expect that the gradient term $\boldsymbol{\xi}^\alpha \cdot \operatorname{grad}\nu^\alpha$ should give rise to a traction associated with the microscopic stress $\boldsymbol{\xi}^\alpha$ — and we use the underlined term in the identity

$$\int\limits_{\mathcal{P}_t} \boldsymbol{\xi}^\alpha \cdot \operatorname{grad}\nu^\alpha\,dv = \int\limits_{\partial\mathcal{P}_t} \underline{(\boldsymbol{\xi}^\alpha\cdot\mathbf{n})\nu^\alpha\,da} - \int\limits_{\mathcal{P}_t} \nu^\alpha \operatorname{div}\boldsymbol{\xi}^\alpha\,dv. \tag{108.3}$$

Specifically, we assume that

- power is expended externally by a scalar *microscopic traction* $\Xi^\alpha(\mathbf{n})$ power-conjugate to ν^α.

In addition, we allow for

- a (generally virtual) scalar **external microscopic force** b^α power-conjugate to ν^α

and therefore assume that the **external power** has the form

$$\mathcal{W}(\mathcal{P}_t) = \int\limits_{\partial\mathcal{P}_t} \mathbf{t}(\mathbf{n})\cdot\mathbf{v}\,da + \int\limits_{\mathcal{P}_t} \mathbf{b}\cdot\mathbf{v}\,dv + \sum_\alpha \int\limits_{\partial\mathcal{P}_t} \Xi^\alpha(\mathbf{n})\nu^\alpha\,da + \sum_\alpha \int\limits_{\mathcal{P}_t} b^\alpha \nu^\alpha\,dv. \tag{108.4}$$

Assume that, at some arbitrarily chosen time, the fields $\boldsymbol{\chi}$, $\mathbf{F}^e$, and $\mathbb{S}^1, \mathbb{S}^2, \dots, \mathbb{S}^N$ are known, and consider the velocity $\mathbf{v}$, the elastic distortion-rate $\mathbf{L}^e$, and the slip rates $\nu^1, \nu^2, \dots, \nu^N$ as virtual velocities to be specified independently in a manner consistent with (106.7); that is, denoting the virtual fields by $\tilde{\mathbf{v}}$, $\tilde{\mathbf{L}}^e$, and $\tilde{\nu}^1, \tilde{\nu}^2, \dots, \tilde{\nu}^N$, we require that

$$\operatorname{grad}\tilde{\mathbf{v}} = \tilde{\mathbf{L}}^e + \sum_\alpha \tilde{\nu}^\alpha\,\bar{\mathbb{S}}^\alpha. \tag{108.5}$$

If we define a (generalized) *virtual velocity* to be a list

$$\mathcal{V} = (\tilde{\mathbf{v}}, \tilde{\mathbf{L}}^e, \tilde{\nu}^1, \tilde{\nu}^2, \dots, \tilde{\nu}^N) \tag{108.6}$$

and write $\mathcal{I}(\mathcal{P}_t, \mathcal{V})$ and $\mathcal{W}(\mathcal{P}_t, \mathcal{V})$ for the corresponding internal and external expenditures of *virtual power*, then the *principle of virtual power* is the requirement that the virtual power balance

$$\underbrace{\int\limits_{\partial\mathcal{P}_t} \mathbf{t}(\mathbf{n})\cdot\tilde{\mathbf{v}}\,da + \int\limits_{\mathcal{P}_t} \mathbf{b}\cdot\tilde{\mathbf{v}}\,dv + \sum_\alpha \int\limits_{\partial\mathcal{P}_t} \Xi^\alpha(\mathbf{n})\,\tilde{\nu}^\alpha\,da + \sum_\alpha \int\limits_{\mathcal{P}_t} b^\alpha \nu^\alpha\,dv}_{\mathcal{W}(\mathcal{P}_t,\mathcal{V})}$$

$$= \underbrace{\int\limits_{\mathcal{P}_t} \mathbf{T}\colon\tilde{\mathbf{L}}^e\,dv + \sum_\alpha \int\limits_{\mathcal{P}_t} (\pi^\alpha \tilde{\nu}^\alpha + \boldsymbol{\xi}^\alpha \cdot \operatorname{grad}\tilde{\nu}^\alpha)\,dv}_{\mathcal{I}(\mathcal{P}_t,\mathcal{V})} \tag{108.7}$$

be satisfied for any subregion $\mathcal{P}_t$ of the deformed body and any virtual velocity $\mathcal{V}$.

108.2 Consequences of Frame-Indifference

Arguing as in §92.2.1, we assume that under a change in frame the internal power $\mathcal{I}(\mathcal{P}_t, \mathcal{V})$ is invariant, and that the virtual fields transform in a manner identical to their nonvirtual counterparts. Then, (91.33) implies that the elastic distortion-rate transforms according to

$$\tilde{\mathbf{L}}^{e*} = \mathbf{Q}\tilde{\mathbf{L}}^{e}\mathbf{Q}^{\top} + \mathbf{\Omega},$$

with $\mathbf{Q}$ the frame-rotation and $\mathbf{\Omega}$ the frame-spin. Further, for each α,

$$\pi^{\alpha} \text{ and } \tilde{\nu}^{\alpha} \text{ are invariant} \tag{108.8}$$

(as they are scalar fields), but because "grad" represents the gradient in the deformed body, the transformation law for grad $\tilde{\nu}^{\alpha}$ has the form

$$\operatorname{grad}^{*}\tilde{\nu}^{\alpha} = \mathbf{Q}\operatorname{grad}\tilde{\nu}^{\alpha}.$$

Next, by (108.8), writing $\mathcal{P}_t^*$ and $\mathcal{I}^*(\mathcal{P}_t^*)$ for the region and the internal power in the new frame, we see that

$$\mathcal{I}^*(\mathcal{P}_t^*) = \int\limits_{\mathcal{P}_t^*} \mathbf{T}^*:\tilde{\mathbf{L}}^{e*} + \sum_{\alpha} \int\limits_{\mathcal{P}_t^*} (\pi^{\alpha}\tilde{\nu}^{\alpha} + \boldsymbol{\xi}^{\alpha*}\cdot \operatorname{grad}^*\tilde{\nu}^{\alpha})\, dv, \tag{108.9}$$

where $\mathbf{T}^*$ and $\boldsymbol{\xi}^{\alpha*}$ are the stresses $\mathbf{T}$ and $\boldsymbol{\xi}^{\alpha}$ in the new frame. Since $\mathcal{P}_t^*$ is $\mathcal{P}_t$ transformed *rigidly*, we may replace the region of integration $\mathcal{P}_t^*$ in (108.9) by $\mathcal{P}_t$. Thus, the requirement that $\mathcal{I}^*(\mathcal{P}_t^*) = \mathcal{I}(\mathcal{P}_t)$ implies that

$$\begin{aligned}\int\limits_{\mathcal{P}_t}\Big(\mathbf{T}:\tilde{\mathbf{L}}^{e} + \sum_{\alpha}\boldsymbol{\xi}^{\alpha}\cdot\operatorname{grad}\tilde{\nu}^{\alpha}\Big)\,dv &= \int\limits_{\mathcal{P}_t}\Big(\mathbf{T}^*:\tilde{\mathbf{L}}^{e*} + \sum_{\alpha}\boldsymbol{\xi}^{\alpha*}\cdot\operatorname{grad}^*\tilde{\nu}^{\alpha}\Big)\,dv \\ &= \int\limits_{\mathcal{P}_t}\Big(\mathbf{T}^*:(\mathbf{Q}\tilde{\mathbf{L}}^{e}\mathbf{Q}^{\top} + \mathbf{\Omega}) + \sum_{\alpha}\boldsymbol{\xi}^{\alpha*}\cdot(\mathbf{Q}\operatorname{grad}\tilde{\nu}^{\alpha})\Big)\,dv,\end{aligned}$$

and, hence, that, since $\mathcal{P}_t$ is arbitrary,

$$\mathbf{T}:\tilde{\mathbf{L}}^{e} + \sum_{\alpha}\boldsymbol{\xi}^{\alpha}\cdot\operatorname{grad}\tilde{\nu}^{\alpha} = \mathbf{T}^*:(\mathbf{Q}\tilde{\mathbf{L}}^{e}\mathbf{Q}^{\top} + \mathbf{\Omega}) + \sum_{\alpha}\boldsymbol{\xi}^{\alpha*}\cdot(\mathbf{Q}\operatorname{grad}\tilde{\nu}^{\alpha}). \tag{108.10}$$

Since the virtual slip-rates $\tilde{\nu}^{\alpha}$ are arbitrary, we may set them to zero; thus

$$\mathbf{T}:\tilde{\mathbf{L}}^{e} = \mathbf{T}^*:(\mathbf{Q}\tilde{\mathbf{L}}^{e}\mathbf{Q}^{\top} + \mathbf{\Omega}),$$

and, since the orthogonal tensor $\mathbf{Q}$, and the skew tensor $\mathbf{\Omega}$ are arbitrary, it follows that

$$\mathbf{T}^*:\mathbf{\Omega} = 0 \qquad \text{and} \qquad (\mathbf{T} - \mathbf{Q}^{\top}\mathbf{T}^*\mathbf{Q}):\tilde{\mathbf{L}}^{e} = 0;$$

but $\tilde{\mathbf{L}}^{e}$ is also arbitrary; therefore $\mathbf{T}^*$ and, hence, $\mathbf{T}$ are symmetric and

$$\mathbf{T}^* = \mathbf{Q}\mathbf{T}\mathbf{Q}^{\top}.$$

Thus,

$$\mathbf{T} \text{ is symmetric and frame-indifferent} \tag{108.11}$$

and (108.10) becomes

$$\begin{aligned}\sum_{\alpha}\boldsymbol{\xi}^{\alpha}\cdot\operatorname{grad}\tilde{\nu}^{\alpha} &= \sum_{\alpha}\boldsymbol{\xi}^{\alpha*}\cdot(\mathbf{Q}\operatorname{grad}\tilde{\nu}^{\alpha}) \\ &= \sum_{\alpha}(\mathbf{Q}^{\top}\boldsymbol{\xi}^{\alpha*})\cdot\operatorname{grad}\tilde{\nu}^{\alpha}.\end{aligned} \tag{108.12}$$

Assume that $\tilde{\nu}^\alpha$ is the sole nonzero virtual slip-rate, let $\boldsymbol{\phi}$ be an arbitrary spatial vector, and let

$$\tilde{\nu}^\alpha(\mathbf{x}) = \boldsymbol{\phi} \cdot (\mathbf{x} - \mathbf{o});$$

then $\operatorname{grad} \tilde{\nu}^\alpha = \boldsymbol{\phi}$ and (108.12) implies that

$$(\boldsymbol{\xi}^\alpha - \mathbf{Q}^\top \boldsymbol{\xi}^{\alpha *}) \cdot \boldsymbol{\phi} = 0.$$

Thus, since both $\boldsymbol{\phi}$ and α are arbitrary,

$$\boldsymbol{\xi}^{\alpha *} = \mathbf{Q}\boldsymbol{\xi}^\alpha,$$

and

$$\boldsymbol{\xi}^\alpha \text{ is frame-indifferent for all } \alpha. \tag{108.13}$$

108.3 Macroscopic and Microscopic Force Balances

Consider a virtual velocity $\mathcal{V}$ for which $\tilde{\mathbf{v}}$ is arbitrary, $\tilde{\mathbf{L}}^e = \operatorname{grad} \tilde{\mathbf{v}}$, and $\tilde{\nu^\alpha} = 0$ for each α, so that the constraint (108.5) is satisfied. Then, (108.7) reduces to (92.19) (with $\mathbf{S}^e$ replaced by $\mathbf{T}$) and we conclude from the ensuing argument (which leads to (92.21) and (92.22)) that $\mathbf{T}$ satisfies the *local force-balance*

$$\operatorname{div}\mathbf{T} + \mathbf{b} = \mathbf{0} \tag{108.14}$$

and traction condition

$$\mathbf{t}(\mathbf{n}) = \mathbf{T}\mathbf{n}$$

and, hence, may be viewed as the Cauchy stress. Further, granted that we are working in an inertial frame, so that (92.6) is satisfied, (108.14) reduces to the balance law for linear momentum:

$$\rho\dot{\mathbf{v}} = \operatorname{div}\mathbf{T} + \mathbf{b}_0, \tag{108.15}$$

with $\mathbf{b}_0$ the conventional body force.

To derive the microscopic force balances we introduce, for each slip system α, the *resolved shear* τ^α defined by

$$\begin{aligned}\tau^\alpha &\stackrel{\text{def}}{=} \bar{\mathbb{S}}^\alpha : \mathbf{T} \\ &= \bar{\mathbf{s}}^\alpha \cdot \mathbf{T}\bar{\mathbf{m}}^\alpha,\end{aligned} \tag{108.16}$$

so that τ^α represents the Cauchy stress $\mathbf{T}$ resolved on the deformed αth slip system. Assume that the macroscopic virtual velocity vanishes,

$$\tilde{\mathbf{v}} \equiv \mathbf{0},$$

so that, by (108.5),

$$\tilde{\mathbf{L}}^e = -\sum_\alpha \tilde{\nu}^\alpha \bar{\mathbb{S}}^\alpha$$

with $\tilde{\nu}^1, \tilde{\nu}^2, \ldots, \tilde{\nu}^N$ arbitrary. Thus, by (108.16),

$$\begin{aligned}\mathbf{T} : \tilde{\mathbf{L}}^e &= -\sum_\alpha \tilde{\nu^\alpha}\, \mathbf{T} : \bar{\mathbb{S}}^\alpha \\ &= -\sum_\alpha \tau^\alpha \tilde{\nu^\alpha},\end{aligned}$$

and (108.7) reduces to the *microscopic virtual-power relation*

$$\sum_{\alpha} \int_{\partial \mathcal{P}_t} \Xi^{\alpha}(\mathbf{n})\tilde{\nu}^{\alpha}\, da + \sum_{\alpha} \int_{\mathcal{P}_t} b^{\alpha} \nu^{\alpha}\, dv = \sum_{\alpha} \int_{\mathcal{P}_t} [(\pi^{\alpha} - \tau^{\alpha})\tilde{\nu}^{\alpha} + \boldsymbol{\xi}^{\alpha} \cdot \operatorname{grad} \tilde{\nu}^{\alpha}]\, dv. \tag{108.17}$$

Appealing to the virtual counterpart of the identity (108.3), we thus find that

$$\sum_{\alpha} \left(\int_{\partial \mathcal{P}_t} (\Xi^{\alpha}(\mathbf{n}) - \boldsymbol{\xi}^{\alpha} \cdot \mathbf{n})\tilde{\nu}^{\alpha}\, da + \int_{\mathcal{P}_t} (\operatorname{div} \boldsymbol{\xi}^{\alpha} + \tau^{\alpha} - \pi^{\alpha} + b^{\alpha})\tilde{\nu}^{\alpha}\, dv \right) = 0. \tag{108.18}$$

Since the virtual slip rates $\tilde{\nu}$ are arbitrary, (108.18) must be satisfied for all virtual slip rates and all $\mathcal{P}_t$; thus a vectorial version of fundamental lemma of the calculus of variations (page 167) yields the **microscopic force balance**[692]

$$\operatorname{div} \boldsymbol{\xi}^{\alpha} + \tau^{\alpha} - \pi^{\alpha} = -b^{\alpha} \tag{108.19}$$

and the **microscopic traction condition**

$$\Xi^{\alpha}(\mathbf{n}) = \boldsymbol{\xi}^{\alpha} \cdot \mathbf{n}$$

for each slip system α.

[692] GURTIN (2000b, eqt. (48); 2002, eqt. (5.14)).

109 Free-Energy Imbalance

Arguing as in §93, we consider an arbitrary spatial region $\mathcal{P}_t$ convecting with the body, and we let φ denote the *free energy* and $\delta \geq 0$ the *dissipation*, with φ measured per unit volume in the lattice, but with δ measured per unit volume in the observed space.[693] Then, by (15.6) and since $J = \det \mathbf{F}^e$,

$$\int\limits_{\mathcal{P}_t} \varphi J^{-1}\, dv \qquad \text{and} \qquad \int\limits_{\mathcal{P}_t} \delta\, dv,$$

respectively, represent the free energy of — and the dissipation within — $\mathcal{P}_t$. The free-energy imbalance for $\mathcal{P}_t$ is then the assertion that[694]

$$\overline{\int\limits_{\mathcal{P}_t} \varphi J^{-1}\, dv}^{\,\boldsymbol{\cdot}} - \mathcal{W}(\mathcal{P}_t) = -\int\limits_{\mathcal{P}_t} \delta\, dv \leq 0. \tag{109.1}$$

Therefore, since, by (108.7), $\mathcal{W}(\mathcal{P}_t) = \mathcal{I}(\mathcal{P}_t)$, (108.7) implies that

$$\overline{\int\limits_{\mathcal{P}_t} \varphi J^{-1}\, dv}^{\,\boldsymbol{\cdot}} - \int\limits_{\mathcal{P}_t} \mathbf{T} \cdot \mathbf{L}^e\, dv - \sum_\alpha \int\limits_{\mathcal{P}_t} (\pi^\alpha \nu^\alpha + \boldsymbol{\xi}^\alpha \cdot \operatorname{grad} \nu^\alpha)\, dv \leq 0. \tag{109.2}$$

Further, since $\mathcal{P}_t$ convects with the body, (93.3) is satisfied; viz.

$$\overline{\int\limits_{\mathcal{P}_t} \varphi J^{-1}\, dv}^{\,\boldsymbol{\cdot}} = \int\limits_{\mathcal{P}_t} \dot{\varphi} J^{-1}\, dv. \tag{109.3}$$

Thus,

$$\int\limits_{\mathcal{P}_t} \Big(J^{-1} \dot{\varphi} - \mathbf{T} : \mathbf{L}^e\, dv - \sum_\alpha (\pi^\alpha \nu^\alpha + \boldsymbol{\xi}^\alpha \cdot \operatorname{grad} \nu^\alpha)\Big)\, dv \leq 0$$

and, since $\mathcal{P}_t$ was arbitrarily chosen,

$$J^{-1}\dot{\varphi} - \mathbf{T} : \mathbf{L}^e - \sum_\alpha (\boldsymbol{\xi}^\alpha \cdot \operatorname{grad} \nu^\alpha + \pi^\alpha \nu^\alpha) = -\delta \leq 0. \tag{109.4}$$

---

693 In contrast to §93, where δ is measured per unit volume in the structural space.

694 Cf. (93.1).

The term $\mathbf{T}:\mathbf{L}^e$, which represents the elastic stress-power, is most conveniently expressed in terms of the elastic strain-rate $\dot{\mathbf{E}}^e$ via (94.1) and (94.2); viz.

$$\mathbf{T}:\mathbf{L}^e = J^{-1}\,\mathbf{T}^e:\dot{\mathbf{E}}^e \tag{109.5}$$

with

$$\mathbf{T}^e = J\mathbf{F}^{e-1}\mathbf{T}\mathbf{F}^{e-\top}. \tag{109.6}$$

As noted following (94.2), $\mathbf{T}^e$ is a second Piola stress computed using $\mathbf{F}^e$ in place of $\mathbf{F}$. The definition (109.6) allows us to rewrite the imbalance (109.4) in the form

$$\boxed{J^{-1}\,\dot{\varphi} - J^{-1}\,\mathbf{T}^e:\dot{\mathbf{E}}^e - \sum_{\alpha}(\boldsymbol{\xi}^{\alpha}\cdot \operatorname{grad}\nu^{\alpha} + \pi^{\alpha}\nu^{\alpha}) = -\delta \le 0.} \tag{109.7}$$

110 Conventional Theory

Within the present framework the conventional theory is based on the assumption that

$$\boldsymbol{\xi}^\alpha \equiv \mathbf{0} \quad \text{for all } \alpha,$$

so that the microscopic force balance (108.19) and the free-energy imbalance (109.7) become

$$\tau^\alpha - \pi^\alpha = -b^\alpha \tag{110.1}$$

(for all α) and

$$J^{-1}\dot{\varphi} - J^{-1}\mathbf{T}^e\colon\dot{\mathbf{E}}^e - \sum_\alpha \pi^\alpha \nu^\alpha = -\delta \le 0. \tag{110.2}$$

110.1 Constitutive Relations

For convenience we introduce lists

$$\vec{\nu} \overset{\text{def}}{=} (\nu^1, \nu^2, \ldots, \nu^N) \qquad \text{and} \qquad \vec{S} \overset{\text{def}}{=} (S^1, S^2, \ldots, S^N) \tag{110.3}$$

of slip-rates and hardening variables, with

$$S^\alpha > 0 \quad \text{for each } \alpha.$$

As in §104.3, we neglect defect energy and restrict attention to constitutive assumptions that separate plastic and elastic response. Hence, guided by (110.2), we characterize elastic response by the standard equations

$$\begin{aligned} \varphi &= \hat{\varphi}(\mathbf{E}^e), \\ \mathbf{T}^e &= \hat{\mathbf{T}}^e(\mathbf{E}^e), \end{aligned} \tag{110.4}$$

and plastic response by relations that, for each slip system α, consist of a constitutive equation for the internal microscopic force π^α supplemented by a hardening equation

$$\begin{aligned} \pi^\alpha &= \bar{\pi}^\alpha(\vec{\nu}, \vec{S}), \\ \dot{S}^\alpha &= h^\alpha(\vec{\nu}, \vec{S}). \end{aligned} \tag{110.5}$$

Arguing as in the steps leading to (‡) on page 546, we see that $\mathbf{E}^e$ and $\mathbf{T}^e$ are frame-indifferent and — because they are scalar fields — ν^α, π^α, and S^α are invariant. Thus,

- *the constitutive equations* (110.4) *and* (110.5) *are frame-indifferent.*

It eases notation if we introduce a list

$$\vec{\pi} \stackrel{\text{def}}{=} (\pi^1, \pi^2, \dots, \pi^N) \tag{110.6}$$

of internal microscopic forces. By a **constitutive process**, we mean a pair $(\chi, \vec{\nu})$ of *fields* χ and $\vec{\nu}$ — with χ a motion and $\vec{\nu}$ a list of slip rates — together with fields $\mathbf{E}^e$, $\vec{S}$, φ, $\mathbf{T}^e$, and $\vec{\pi}$, where

(i) $\mathbf{E}^e$ is given by

$$\mathbf{E}^e = \tfrac{1}{2}(\mathbf{F}^{e\top}\mathbf{F}^e - \mathbf{1}), \qquad \mathbf{F}^e = \mathbf{F}\mathbf{F}^{p-1}, \qquad \mathbf{F} = \nabla\chi,$$

and where $\mathbf{F}^p$ is *any* solution of the differential equation[695]

$$\dot{\mathbf{F}}^p = \Big(\sum_\alpha \nu^\alpha\, \mathbf{s}^\alpha \otimes \mathbf{m}^\alpha\Big)\mathbf{F}^p, \tag{110.7}$$

with $\det \mathbf{F}^p \equiv 1$;

(ii) $\vec{S}$ is any solution of the system $(110.5)_2$ of hardening equations;

(iii) φ, $\mathbf{T}^e$, and $\vec{\pi}$ are determined by the fields $\mathbf{E}^e$, $\vec{\nu}$, and $\vec{S}$ through the constitutive equations (110.4) and $(110.5)_1$.

Given a constitutive process, (109.6) determines the stress $\mathbf{T}$, and the standard and microscopic force balances (108.15) and (110.1) provide explicit relations

$$\mathbf{b}_0 = \rho\dot{\mathbf{v}} - \operatorname{div}\mathbf{T} \qquad \text{and} \qquad b^\alpha = \pi^\alpha - \tau^\alpha, \tag{110.8}$$

$\alpha = 1, 2, \dots, N$, for the conventional body force $\mathbf{b}_0$ and the external microscopic forces b^α needed to *support* the process. As before[696] we assume that these forces are arbitrarily assignable and refer to a constitutive process as **physically attainable** if

$$b^\alpha = 0 \qquad \text{for all } \alpha, \tag{110.9}$$

so that external microscopic power expenditures are not needed to support the process.[697]

To determine restrictions imposed by the free-energy imbalance, we use the constitutive equations (110.4) and $(110.5)_1$ to write the free-energy imbalance (109.7) in the form

$$\delta = J^{-1}\bigg(\hat{\mathbf{T}}^e(\mathbf{E}^e) - \frac{\partial\hat{\varphi}(\mathbf{E}^e)}{\partial\mathbf{E}^e}\bigg) : \dot{\mathbf{E}}^e + \sum_\alpha \hat{\pi}^\alpha(\vec{\nu}, \vec{S})\nu^\alpha \geq 0. \tag{110.10}$$

Our derivation of thermodynamic restrictions that ensue from this inequality is based on the following two assertions:[698]

(I) It is possible to find a constitutive process such that

$$\nu^1 \equiv \nu^2 \equiv \cdots \equiv \nu^N \equiv 0 \tag{110.11}$$

and such that $\mathbf{E}^e$ and $\dot{\mathbf{E}}^e$ have arbitrarily prescribed values at some point and time.

(II) It is possible to find a constitutive process such that

$$\mathbf{F}^e \equiv \mathbf{1} \tag{110.12}$$

[695] Cf. the remark on page 558.
[696] E.g., as in the paragraph containing (44.6).
[697] Cf. the sentence containing (86.5).
[698] Which are direct counterparts of (I) and (II) on page 558.

and such that the slip rates and hardening variables have arbitrarily prescribed values at some point and time.

Then, arguing as in the steps leading to (95.9) and (95.10) — but using (I) and (II) above — we see that (95.9) remains valid, while (95.10) is replaced by the requirement that

$$\delta = \sum_{\alpha} \bar{\pi}^{\alpha}(\vec{\nu}, \vec{S})\nu^{\alpha} \geq 0. \tag{110.13}$$

Conversely, granted (95.10) and (110.13), the free-energy imbalance (110.10) is satisfied in all constitutive processes. We therefore have the following result:

THERMODYNAMIC RESTRICTIONS *Conditions both necessary and sufficient that each constitutive process satisfies the free-energy imbalance are that*

(i) *the free energy determines the elastic stress through the* **stress relation**

$$\hat{\mathbf{T}}^e(\mathbf{E}^e) = \frac{\partial\hat{\varphi}(\mathbf{E}^e)}{\partial\mathbf{E}^e}; \tag{110.14}$$

(ii) *the internal microscopic stresses must satisfy the* **reduced dissipation inequality** (110.13).

We assume, henceforth, that the internal microscopic stresses are **strictly dissipative** in the sense that

$$\sum_{\alpha} \bar{\pi}^{\alpha}(\vec{\nu}, \vec{S})\nu^{\alpha} > 0 \tag{110.15}$$

whenever not all slip rates vanish.

110.2 Simplified Constitutive Theory

We restrict attention to a quadratic free energy and linear stress-strain relation[699] and therefore begin with elastic constitutive equations

$$\begin{aligned} \varphi &= \tfrac{1}{2}\mathbf{E}^e : \mathbb{C}\mathbf{E}^e, \\ \mathbf{T}^e &= \mathbb{C}\mathbf{E}^e. \end{aligned} \tag{110.16}$$

We assume that the elasticity tensor $\mathbb{C}$ has the properties listed on page 300.

Consider next the constitutive equations (110.5) introduced to describe flow. These equations and the reduced dissipation inequality (110.13) are of the same form as the constitutive equations (104.31) and dissipation inequality (104.33) of the small deformation theory, the sole difference being that $\vec{\nu}$ and ν^{α} in that theory are given by

$$\vec{\nu} = (\dot{\gamma}^1, \dot{\gamma}^2, \ldots, \dot{\gamma}^N) \qquad \text{and} \qquad \nu^{\alpha} = \dot{\gamma}^{\alpha}. \tag{110.17}$$

Further, by (110.17) the constitutive assumptions and analysis of §104.5 when applied here starting from (110.5) result in the relations[700]

$$\begin{aligned} \pi^{\alpha} &= S^{\alpha} g(|\nu^{\alpha}|)\,\mathrm{sgn}\,\nu^{\alpha}, \\ \dot{S}^{\alpha} &= \sum_{\beta} h^{\alpha\beta}(\vec{S})|\nu^{\beta}|, \end{aligned} \tag{110.18}$$

699 Cf. the discussion following (95.27).

700 The specific hardening moduli specified in (104.65) and (104.67) are also used within the present context of large deformations.

to be satisfied for each slip system α. Here g is a *rate-sensitivity function* consistent with

$$g(0) = 0, \qquad g(\nu) > 0 \quad \text{for } \nu > 0. \tag{110.19}$$

If, as in §104.7, we consider hardening moduli of the form (104.65), with $h \geq 0$ a self-hardening function and $q > 0$ an interaction constant, then the hardening equations $(110.18)_2$ take the form (104.66); viz.

$$\dot{S}^\alpha = \sum_\beta [\chi^{\alpha\beta} + q(1 - \chi^{\alpha\beta})]h(S^\beta)|\nu^\beta|. \tag{110.20}$$

If we restrict attention to constitutive processes that are physically attainable, then $(110.8)_2$ with $b^\alpha \equiv 0$ implies that $\tau^\alpha = \pi^\alpha$ for all α; $(110.18)_1$, therefore, yields a relation

$$\tau^\alpha = S^\alpha g(|\nu^\alpha|) \tag{110.21}$$

for the resolved shear stress. Finally, by (110.17), the discussion from (104.53) to (104.68) with $\dot{\gamma}^\alpha$ replaced by ν^α holds equally well here and we are led to the constitutive theory summarized in the next section.

The constitutive equations described above need to be accompanied by concomitant *initial conditions*. Typical initial conditions presume that the body is initially (at time $t = 0$, say) in a *virgin state* in the sense that

$$\mathbf{F}(\mathbf{X}, 0) = \mathbf{F}^p(\mathbf{X}, 0) = \mathbf{1}, \qquad S^\alpha(\mathbf{X}, 0) = S_0, \tag{110.22}$$

with $S_0 > 0$ the *initial slip resistance.*

110.3 Summary of Basic Equations

The constitutive equations that characterize a single-crystal undergoing large deformations consist of:

(i) the Kröner decomposition of the deformation gradient $\mathbf{F}$,

$$\mathbf{F} = \mathbf{F}^e\mathbf{F}^p, \qquad \det \mathbf{F}^p = 1, \tag{110.23}$$

in which $\mathbf{F}^e$ and $\mathbf{F}^p$ are the elastic and plastic distortions;

(ii) the elastic stress-strain relation

$$\mathbf{T}^e = \mathbb{C}\mathbf{E}^e, \tag{110.24}$$

in which

$$\mathbf{E}^e = \tfrac{1}{2}(\mathbf{F}^{e\top}\mathbf{F}^e - \mathbf{1}) \tag{110.25}$$

is the elastic strain, $\mathbf{T}^e$ is given by

$$\mathbf{T}^e = J\mathbf{F}^{e-1}\mathbf{T}\mathbf{F}^{e-\top}, \tag{110.26}$$

with $\mathbf{T}$ the Cauchy stress, and the elasticity tensor $\mathbb{C}$ is assumed to have the properties listed on page 300;

(iii) a differential equation

$$\dot{\mathbf{F}}^p = \mathbf{L}^p\mathbf{F}^p \qquad \text{with} \qquad \mathbf{L}^p = \sum_{\alpha=1}^{N} \nu^\alpha\, \mathbf{s}^\alpha \otimes \mathbf{m}^\alpha \tag{110.27}$$

that describes the evolution of the plastic distortion $\mathbf{F}^p$, where $\mathbf{s}^\alpha$ and $\mathbf{m}^\alpha$, with

$$\mathbf{s}^\alpha \cdot \mathbf{m}^\alpha = 0, \qquad |\mathbf{s}^\alpha| = |\mathbf{m}^\alpha| = 1, \tag{110.28}$$

denote the slip direction and slip plane normal for slip system α, and where ν^α represents the corresponding slip rate;

(iv) the (inverted) constitutive relation[701]

$$\nu^\alpha = f\Big(\frac{|\tau^\alpha|}{S^\alpha}\Big)\operatorname{sgn}\tau^\alpha, \tag{110.29}$$

where τ^α, defined by

$$\begin{gathered}\tau^\alpha = \bar{\mathbf{s}}^\alpha \cdot \mathbf{T}\bar{\mathbf{m}}^\alpha,\\ \bar{\mathbf{s}}^\alpha = \mathbf{F}^e\mathbf{s}^\alpha, \qquad \bar{\mathbf{m}}^\alpha = \mathbf{F}^{e-\top}\mathbf{m}^\alpha,\end{gathered} \tag{110.30}$$

represents the Cauchy stress $\mathbf{T}$ resolved on the deformed αth slip system,[702] $S^\alpha > 0$ represents the flow resistance on α, and f is an inverted rate-sensitivity function consistent with (99.5) and (104.57); the special case of a power-law rate-dependency corresponds to

$$\nu^\alpha = d_0\left(\frac{|\tau^\alpha|}{S^\alpha}\right)^{\frac{1}{m}}\operatorname{sgn}\tau^\alpha, \tag{110.31}$$

with $d_0 > 0$ a reference slip rate and $m > 0$ a rate-sensitivity parameter;

(v) the hardening equations

$$\dot{S}^\alpha = \sum_\beta [\chi^{\alpha\beta} + q(1-\chi^{\alpha\beta})]h(S^\beta)|\nu^\beta| \tag{110.32}$$

with $h(\cdot) \geq 0$ a self-hardening function and $q > 0$ an interaction constant.

[701] Cf. Footnote 622.

[702] $\bar{\mathbf{s}}^\alpha$ and $\bar{\mathbf{m}}^\alpha$ represent the slip direction and the slip plane normal pushed forward to the observed space.

111 Taylor's Model of Polycrystal

Anisotropy in the plastic response of polycrystalline metals undergoing large deformations is most often a consequence of *reorientation* of the deformed lattices of the individual grains, a process generally referred to as *crystallographic texturing*. In this section, we discuss a well-known theory of crystallographic texturing.[703] This theory, which is based on ideas of TAYLOR (1938), has, as central ingredients

(i) a theory for single crystals undergoing large deformations;
(ii) a method of homogenizing the response of the individual material points of a polycrystal.

111.1 Kinematics of a Taylor Polycrystal

A classical homogenization method for high-symmetry polycrystals[704] posits a prescribed set $\mathcal{T}$ of *reference grains*, labelled

$$g = 1, 2, \ldots, G, \tag{111.1}$$

such that

(T1) the complete set $\mathcal{T}$ of grains is assumed to be active at each material point $\mathbf{X}$ of the polycrystal;
(T2) *all grains* in $\mathcal{T}$ share the (macroscopic) deformation gradient $\mathbf{F}$ at $\mathbf{X}$;
(T3) for each grain g, $\mathbf{Q}_g$, an orthogonal tensor, represents the *misorientation* of g relative to, say, grain 1 (so that $\mathbf{Q}_1 = \mathbf{1}$).

The central idea underlying these abstract requirements is that, although the specific granular structure (grain boundaries, junctions, etc.) is not visible at the macroscopic level, we can account for microscopic elastic and plastic distortions of each grain g as chacterized by elastic and plastic distortion tensors $\mathbf{F}^e_g$ and $\mathbf{F}^p_g$. A consequence of (T2) is then that $\mathbf{F}^e_g$ and $\mathbf{F}^p_g$ are related to the deformation gradient $\mathbf{F}$ through the *Kröner decompositions*

$$\boxed{\mathbf{F} = \mathbf{F}^e_g\mathbf{F}^p_g \qquad \text{for each grain } g,} \tag{111.2}$$

[703] Cf., e.g., ASARO & NEEDLEMAN (1985), ANAND, BALASUBRAMANIAN & KOTHARI (1997), DAWSON & MARIN (1998), MIEHE, SCHRODER & SCHOTTE (1999), and the references therein. As is clear from these and related references, the theory can predict macroscopic anisotropic stress-strain response, shape changes, and the evolution of crystallographic texture under complex deformations.
[704] E.g., fcc and bcc polycrystals.

or, less succintly,

$$\mathbf{F} = \mathbf{F}_1^e\mathbf{F}_1^p = \mathbf{F}_2^e\mathbf{F}_2^p = \cdots = \mathbf{F}_G^e\mathbf{F}_G^p.$$

As in (106.2), we assume that

$$\det \mathbf{F}_g^p = 1 \qquad \text{for all grains } g. \tag{111.3}$$

Thus, and what is most important, each grain g is equipped with its peculiar Kröner decomposition *and lattice*. In this regard, Figure 106.1 would apply for each grain g with $\mathbf{F}^e$ and $\mathbf{F}^p$ replaced by $\mathbf{F}_g^e$ and $\mathbf{F}_g^p$, and with "lattice" replaced by "lattice for grain g."

Slip systems are as defined in the paragraphs containing (102.4) and (106.4); but within the present context of Taylor polycrystals each grain g is associated with slip systems $\alpha = 1, 2, \ldots, N$. Thus, for each grain g and slip system α the slip direction $\mathbf{s}_g^\alpha$ and slip-plane normal $\mathbf{m}_g^\alpha$ are constant *lattice vectors* (in the undistorted lattice for grain g) with

$$\mathbf{s}_g^\alpha \cdot \mathbf{m}_g^\alpha = 0, \qquad |\mathbf{s}_g^\alpha| = |\mathbf{m}_g^\alpha| = 1; \tag{111.4}$$

and with

$$\mathbb{S}_g^\alpha = \mathbf{s}_g^\alpha \otimes \mathbf{m}_g^\alpha \tag{111.5}$$

the *Schmid tensor* for slip system α in grain g. Then, by (T3), the Schmid tensors of any grain g are related to those of grain 1 via the relations

$$\mathbb{S}_g^\alpha = \mathbf{Q}_g\mathbb{S}_1^\alpha\mathbf{Q}_g^\top \tag{111.6}$$

or, equivalently,

$$\mathbf{s}_g^\alpha = \mathbf{Q}_g\mathbf{s}_1^\alpha, \qquad \mathbf{m}_g^\alpha = \mathbf{Q}_g\mathbf{m}_1^\alpha.$$

Further, fixing attention on a given grain g, we let

$$\bar{\mathbb{S}}_g^\alpha \stackrel{\text{def}}{=} \mathbf{F}_g^e\mathbb{S}_g^\alpha\mathbf{F}_g^{e-1}; \tag{111.7}$$

so that

$$\bar{\mathbf{s}}_g^\alpha = \mathbf{F}_g^e\mathbf{s}_g^\alpha \qquad \text{and} \qquad \bar{\mathbf{m}}_g^\alpha = \mathbf{F}_g^{e-\top}\mathbf{m}_g^\alpha. \tag{111.8}$$

Thus,

$$\bar{\mathbb{S}}_g^\alpha = \bar{\mathbf{s}}_g^\alpha \otimes \bar{\mathbf{m}}_g^\alpha. \tag{111.9}$$

The fields $\bar{\mathbf{s}}_g^\alpha$ and $\bar{\mathbf{m}}_g^\alpha$ represent the slip direction $\mathbf{s}_g^\alpha$ and the slip-plane normal $\mathbf{m}_g^\alpha$ pushed forward (from the lattice for g) to the observed space (deformed body), $\mathbf{s}_g^\alpha$ as a tangent vector, $\mathbf{m}_g^\alpha$ as a normal.[705]

Remark. Consider a given polycrystal and assume that the Schmid tensors $\mathbb{S}_g^\alpha$ (as they appear in the undistorted lattices of the individual grains) are *distributed randomly* with respect to the grains $g = 1, 2, \ldots, G$. Assume that the polycrystal is *elastically undistorted* at time $t = 0$ in the sense that

$$\mathbf{F}_g^e(\mathbf{X}, 0) = \mathbf{1} \quad \text{for all grains } g \text{ and all } \mathbf{X}.$$

Then, by (111.7),

$$\bar{\mathbb{S}}_g^\alpha(\mathbf{X}, 0) \equiv \mathbb{S}_g^\alpha \quad \text{for all grains } g \text{ and all } \mathbf{X}. \tag{111.10}$$

[705] Cf. the paragraph containing (106.8).

Thus, initially, the observed (deformed) crystal is essentially isotropic. On the other hand, as the deformation progresses the deformed Schmid tensors are given by

$$\bar{\mathbb{S}}^\alpha_g(\mathbf{X},t) = \mathbf{F}^e_g(\mathbf{X},t)\mathbb{S}^\alpha_g\mathbf{F}^{e-1}_g(\mathbf{X},t) \quad \text{for all grains } g \text{ and all } \mathbf{X}. \tag{111.11}$$

Variations — with $\mathbf{X}$ — of the (deformed) Schmid tensors

$$\bar{\mathbb{S}}^\alpha_1(\mathbf{X},t),\ \bar{\mathbb{S}}^\alpha_2(\mathbf{X},t),\ldots,\ \bar{\mathbb{S}}^\alpha_G(\mathbf{X},t)$$

represent an example of crystallographic texturing at time t.

Given any grain g, the elastic and plastic rotations, the right and left stretch tensors, and the elastic and plastic distortion-rate tensors and corresponding spin and stretching tensors are as defined in §91, but now these fields depend on the grain g in question. Here, we recall only the relations (91.19) and (91.21) for the elastic strain and its rate, and the relations (91.7) and (91.8) for the distortion-rate tensors; viz.

$$\mathbf{E}^e_g = \tfrac{1}{2}(\mathbf{F}^{e\top}_g\mathbf{F}^e_g - \mathbf{1}), \qquad \dot{\mathbf{E}}^e_g = \operatorname{sym}(\mathbf{F}^{e\top}_g\dot{\mathbf{F}}^e_g) \tag{111.12}$$

and

$$\begin{aligned} \mathbf{L}^e_g = \dot{\mathbf{F}}^e_g\mathbf{F}^{e-1}_g, \qquad \mathbf{L}^p_g = \dot{\mathbf{F}}^p_g\mathbf{F}^{p-1}_g, \\ \mathbf{L} = \mathbf{L}^e_g + \mathbf{F}^e_g\mathbf{L}^p_g\mathbf{F}^{e-1}_g \end{aligned} \tag{111.13}$$

for each grain g. An analog of the single-crystal hypothesis (106.5) then requires that $\mathbf{L}^p_g$ be governed by *slip rates* ν^α_g on the individual slip systems of the individual grains g via the relation

$$\mathbf{L}^p_g = \sum_\alpha \nu^\alpha_g\,\mathbb{S}^\alpha_g. \tag{111.14}$$

Then, by $(106.3)_3$,

$$\mathbf{L} = \mathbf{L}^e_g + \sum_\alpha \nu^\alpha(\mathbf{F}^e_g\mathbb{S}^\alpha\mathbf{F}^{e-1}_g) \qquad \text{for each grain } g, \tag{111.15}$$

and, since $\mathbf{L} = \operatorname{grad}\mathbf{v}$, we may conclude from (111.9) that

$$\operatorname{grad}\mathbf{v} = \mathbf{L}^e_g + \sum_\alpha \nu^\alpha_g\,\bar{\mathbb{S}}^\alpha_g \qquad \text{for each grain } g. \tag{111.16}$$

111.2 Principle of Virtual Power

As in §108, which we follow, we assume that, at some arbitrarily chosen but fixed time the fields χ and $\mathbf{F}^e_1, \mathbf{F}^e_2, \ldots, \mathbf{F}^e_G$ are known, and we denote by $\mathcal{P}_t$ an arbitrary subregion of the deformed body at that time, and by $\mathbf{n}$ the outward unit normal on $\partial\mathcal{P}_t$.

For each grain g, we allow for power expended internally by

- an elastic-stress $\mathbf{S}^e_g$ power-conjugate to $\mathbf{L}^e_g$; and
- a scalar internal microscopic force π^α_g, for each slip system α, power-conjugate to ν^α_g.

Thus, granted the external power is conventional,[706] we write the internal and external power-expenditures $\mathcal{I}(\mathcal{P}_t)$ and $\mathcal{W}(\mathcal{P}_t)$ in the form

$$\mathcal{I}(\mathcal{P}_t) = \int\limits_{\mathcal{P}_t} \sum_g \mathbf{S}^e_g : \mathbf{L}^e_g \, dv + \int\limits_{\mathcal{P}_t} \sum_{g,\alpha} \pi^\alpha_g \nu^\alpha_g \, dv,$$
$$\mathcal{W}(\mathcal{P}_t) = \int\limits_{\partial\mathcal{P}_t} \mathbf{t}(\mathbf{n}) \cdot \mathbf{v} \, da + \int\limits_{\mathcal{P}_t} \mathbf{b} \cdot \mathbf{v} \, dv. \tag{111.17}$$

The virtual counterpart of the constraint (111.16) is

$$\operatorname{grad} \tilde{\mathbf{v}} = \tilde{\mathbf{L}}^e_g + \sum_\alpha \tilde{\nu}^\alpha_g \, \mathbb{S}^\alpha_g. \tag{111.18}$$

Thus, if we define a (generalized) *virtual velocity* to be a list

$$\mathcal{V} = (\tilde{\mathbf{v}}, \tilde{\mathbf{L}}^e_g, \tilde{\nu}^\alpha_g \mid \; g = 1, 2, \ldots, G; \; \alpha = 1, 2, \ldots, N), \tag{111.19}$$

then the *principle of virtual power* is the requirement that, given any $\mathcal{P}_t$,

$$\underbrace{\int\limits_{\mathcal{P}_t} \mathbf{t}(\mathbf{n}) \cdot \tilde{\mathbf{v}} \, da + \int\limits_{\mathcal{P}_t} \mathbf{b} \cdot \tilde{\mathbf{v}} \, dv}_{\mathcal{W}(\mathcal{P}_t, \mathcal{V})} = \underbrace{\int\limits_{\mathcal{P}_t} \sum_g \mathbf{S}^e_g \cdot \tilde{\mathbf{L}}^e_g \, dv + \int\limits_{\mathcal{P}_t} \sum_{g,\alpha} \pi^\alpha_g \tilde{\nu}^\alpha_g \, dv}_{\mathcal{I}(\mathcal{P}_t, \mathcal{V})} \tag{111.20}$$

for all virtual velocities $\mathcal{V}$. A direct analog of the argument leading to (108.11) then yields the conclusion that, for each grain g,

$$\mathbf{S}^e_g \text{ is symmetric and frame-indifferent.} \tag{111.21}$$

With a view toward deriving the macroscopic force balance, we define a *macroscopic* virtual velocity $\mathcal{V}$ as follows: $\tilde{\mathbf{v}}$ is chosen arbitrarily; $\tilde{\mathbf{L}}^e_g$ for *each* g is defined by

$$\tilde{\mathbf{L}}^e_g = \operatorname{grad} \tilde{\mathbf{v}}; \tag{111.22}$$

the slip-rates are assumed to vanish,

$$\tilde{\nu}^\alpha_g = 0 \tag{111.23}$$

for all g and α. The virtual velocity $\mathcal{V}$ defined in this manner is consistent with the constraint (111.18) and, by (111.22),

$$\sum_g \mathbf{S}^e_g : \tilde{\mathbf{L}}^e_g = \Big(\sum_g \mathbf{S}^e_g \Big) : \operatorname{grad} \tilde{\mathbf{v}}$$

and (111.20) reduces to

$$\int\limits_{\mathcal{P}_t} \mathbf{t}(\mathbf{n}) \cdot \tilde{\mathbf{v}} \, da + \int\limits_{\mathcal{P}_t} \mathbf{b} \cdot \tilde{\mathbf{v}} \, dv = \int\limits_{\mathcal{P}_t} \Big(\sum_g \mathbf{S}^e_g \Big) : \operatorname{grad} \tilde{\mathbf{v}} \, dv. \tag{111.24}$$

Thus, if we let

$$\boxed{\mathbf{T} \stackrel{\text{def}}{=} \sum_g \mathbf{S}^e_g,} \tag{111.25}$$

then (111.24) reduces to (92.19) (with $\mathbf{S}^e$ replaced by $\mathbf{T}$) and we conclude from the ensuing argument (which leads to (92.21) and (92.22)) that $\mathbf{T}$ satisfies the (standard)

706 For convenience, we do not allow for a scalar external microscopic forces b^α_g.

local force balance

$$\operatorname{div}\mathbf{T} + \mathbf{b} = \mathbf{0} \tag{111.26}$$

and traction condition

$$\mathbf{t}(\mathbf{n}) = \mathbf{T}\mathbf{n}$$

and hence should be viewed as the *Cauchy stress*. The conclusion that

- *the Cauchy stress* $\mathbf{T}$ *is the sum of the elastic stresses* $\mathbf{S}^e_g$ *in the individual grains* g

is a central result of the Taylor model.

We turn next to the derivation of the microscopic force balance. The *resolved shear stress* τ^α_g within any grain g is defined by

$$\begin{aligned}\tau^\alpha_g &= \bar{\mathbb{S}}^\alpha_g\colon \mathbf{S}^e_g \\ &= \bar{\mathbf{s}}^\alpha \cdot \mathbf{S}^e_g \bar{\mathbf{m}}^\alpha,\end{aligned} \tag{111.27}$$

and hence represents the elastic-stress $\mathbf{S}^e_g$ resolved on the *deformed* α-th slip system of grain g.

Consider a *microscopic* virtual velocity $\mathcal{V}$; that is, a $\mathcal{V}$ whose corresponding macroscopic velocity vanishes,

$$\tilde{\mathbf{v}} \equiv \mathbf{0}. \tag{111.28}$$

For such a $\mathcal{V}$, (111.20) implies that

$$\int\limits_{\mathcal{P}_t} \sum_g \mathbf{S}^e_g\colon \tilde{\mathbf{L}}^e_g \, dv + \int\limits_{\mathcal{P}_t} \sum_{g,\alpha} \pi^\alpha_g \tilde{\nu}^\alpha_g \, dv = 0,$$

and, since $\mathcal{P}_t$ is arbitrary,

$$\sum_g \mathbf{S}^e_g\colon \tilde{\mathbf{L}}^e_g + \sum_{g,\alpha} \pi^\alpha_g \tilde{\nu}^\alpha_g = 0. \tag{111.29}$$

Next, the constraint (111.18) requires that

$$\tilde{\mathbf{L}}^e_g = -\sum_\alpha \tilde{\nu}^\alpha_g \bar{\mathbb{S}}^\alpha_g \qquad \text{for all grains } g,$$

with $\tilde{\nu}^1, \tilde{\nu}^2, \ldots \tilde{\nu}^N$ arbitrary. Thus,

$$\sum_g \mathbf{S}^e_g\colon \mathbf{L}^e_g = -\sum_{g,\alpha} \tau^\alpha_g \tilde{\nu}^\alpha_g$$

and, by (111.29),

$$\sum_{g,\alpha} (\pi^\alpha_g - \tau^\alpha_g)\tilde{\nu}^\alpha_g = 0. \tag{111.30}$$

Since (111.30) is to be satisfied for all virtual slip-rates $\tilde{\nu}^\alpha_g$, we must have the *microscopic force balance*

$$\pi^\alpha_g - \tau^\alpha_g = 0 \tag{111.31}$$

for all grains g and slip systems α.

EXERCISE

1. Give a detailed verification of (111.21).

111.3 Free-Energy Imbalance

Consider an arbitrary spatial region $\mathcal{P}_t$ convecting with the body, and let φ denote the *free energy* measured per unit volume in the lattice. Then, arguing as in §93, we are led to the free-energy imbalance[707]

$$\overline{\int_{\mathcal{P}_t} \varphi J^{-1}\,dv} - \mathcal{W}(\mathcal{P}_t) \leq 0, \tag{111.32}$$

and, since $\mathcal{W}(\mathcal{P}_t) = \mathcal{I}(\mathcal{P}_t)$, $(111.17)_1$ and (111.32) imply that

$$\overline{\int_{\mathcal{P}_t} \varphi J^{-1}\,dv} - \int_{\mathcal{P}_t} \sum_g \mathbf{S}^e_g : \mathbf{L}^e_g\,dv - \int_{\mathcal{P}_t} \sum_{g,\alpha} \pi^\alpha_g \nu^\alpha_g\,dv \leq 0, \tag{111.33}$$

and, hence, that

$$J^{-1}\dot{\varphi} - \sum_g \mathbf{S}^e_g : \mathbf{L}^e_g - \sum_{g,\alpha} \pi^\alpha_g \nu^\alpha_g \leq 0. \tag{111.34}$$

The term $\mathbf{S}^e_g : \mathbf{L}^e_g$, which represents the elastic stress-power, is most conveniently expressed in terms of the elastic strain-rate $\dot{\mathbf{E}}^e_g$ via (94.1) and (94.2); viz.

$$\mathbf{S}^e_g : \mathbf{L}^e_g = J^{-1}\,\mathbf{T}^e_g : \dot{\mathbf{E}}^e_g \tag{111.35}$$

with

$$\mathbf{T}^e_g = J\,\mathbf{F}^{e-1}_g \mathbf{S}^e_g \mathbf{F}^{e-\top}_g; \tag{111.36}$$

$\mathbf{T}^e_g$ represents a second Piola stress computed using $\mathbf{F}^e_g$ in place of $\mathbf{F}$. The definition (111.36) allows us to rewrite the imbalance (111.34) in the form

$$\boxed{J^{-1}\dot{\varphi} - J^{-1}\sum_g \mathbf{T}^e_g : \dot{\mathbf{E}}^e_g - \sum_{g,\alpha} \pi^\alpha_g \nu^\alpha_g \leq 0.} \tag{111.37}$$

111.4 Constitutive Relations

As constitutive relations we take counterparts of the relations summarized in §110.3. The free energy is given by

$$\varphi = \tfrac{1}{2}\sum_g \mathbf{E}^e_g : \mathbb{C}_g \mathbf{E}^e_g, \tag{111.38}$$

and the corresponding elastic stress-strain relations for each grain g are

$$\mathbf{S}^e_g = \mathbb{C}_g \mathbf{E}^e_g. \tag{111.39}$$

Here, $\mathbb{C}_g$ is the elastic tensor for grain g.

We consider dissipative constitutive equations that when expressed with slip-rates as dependent variables have the power-law form

$$\nu^\alpha_g = d_0 \left(\frac{|\tau^\alpha_g|}{S^\alpha}\right)^{\frac{1}{m}} \operatorname{sgn}\tau^\alpha_g \tag{111.40}$$

for each grain g and slip system α. Here $d_0 > 0$ is a reference slip rate, $m > 0$ is a rate-sensitivity parameter, and the slip resistances S^α_g are presumed to satisfy the

[707] Bear in mind (111.3). Here $J = \det\mathbf{F}$.

hardening equations

$$\dot{S}_g^\alpha = \sum_\beta [\chi^{\alpha\beta} + q(1 - \chi_g^{\alpha\beta})]h(S_g^\beta)|\nu_g^\beta|, \tag{111.41}$$

with $h \geq 0$ a self-hardening function and $q > 0$ an interaction constant.

EXERCISE

1. Show that, granted the microscopic force balance (111.31), the constitutive relations (111.38)–(111.41) are consistent with the free-energy imbalance (111.37).

112 Single-Crystal Plasticity at Small Length Scales: A Large-Deformation Gradient Theory

The microscopic force balance and the free-energy imbalance appropriate to a gradient theory are (108.19) and (109.7); to complete the gradient theory we have only to develop appropriate constitutive equations, a project simplified by the presence of an analogous discussion, appropriate to small deformations, in §§105.2–105.4. Here, we refrain from continually referring back to these sections, instead — because some portions of the analysis specifically related to large deformations are delicate — we present a complete discussion.[708]

112.1 Energetic Constitutive Equations. Peach–Koehler Forces

Writing

$$\vec{\rho} = (\rho^1_\vdash, \rho^2_\vdash, \ldots, \rho^N_\vdash, \rho^1_\odot, \rho^2_\odot, \ldots, \rho^N_\odot) \tag{112.1}$$

for the list of dislocation densities and restricting attention to situations in which the elastic strains are small, we assume that the free energy is given by a standard elastic strain-energy augmented by a **defect energy** $\Psi^p(\vec{\rho}\,)$:

$$\varphi = \tfrac{1}{2}\mathbf{E}^e{:}\,\mathbb{C}\mathbf{E}^e + \Psi^p(\vec{\rho}\,), \tag{112.2}$$

with elasticity tensor $\mathbb{C}$ symmetric and positive-definite. We assume further that the stress $\mathbf{T}^e$ is given by the standard stress-strain relation

$$\mathbf{T}^e = \mathbb{C}\mathbf{E}^e; \tag{112.3}$$

then

$$\tfrac{1}{2}\overline{\mathbf{E}^e{:}\,\mathbb{C}\mathbf{E}^e}^{\boldsymbol{\cdot}} = \mathbf{T}^e : \dot{\mathbf{E}}^e \tag{112.4}$$

and the local free-energy imbalance (109.7) becomes

$$J^{-1}\,\overline{\Psi^p(\vec{\rho}\,)}^{\boldsymbol{\cdot}} - \sum_\alpha (\boldsymbol{\xi}^\alpha \cdot \operatorname{grad} \nu^\alpha + \pi^\alpha \nu^\alpha) \le 0. \tag{112.5}$$

Central to the theory are the energetic *defect forces* defined by

$$f^\alpha_\vdash(\vec{\rho}\,) = J^{-1}\frac{\partial \Psi^p(\vec{\rho}\,)}{\partial \rho^\alpha_\vdash} \qquad \text{and} \qquad f^\alpha_\odot(\vec{\rho}\,) = J^{-1}\frac{\partial \Psi^p(\vec{\rho}\,)}{\partial \rho^\alpha_\odot}. \tag{112.6}$$

708 This section follows Gurtin (2008b).

By (107.32),

$$J^{-1}\overline{\Psi^p(\vec{\rho})} = \sum_\alpha \left(f^\alpha_\vdash \,\dot{\rho}^\alpha_\vdash + f^\alpha_\odot \,\dot{\rho}^\alpha_\odot\right) \tag{112.7}$$

$$= \sum_\alpha \left(-\, f^\alpha_\vdash \,\bar{\mathbf{s}}^\alpha + f^\alpha_\odot \,\bar{\mathbf{l}}^\alpha\right)\cdot \operatorname{grad}\nu^\alpha \tag{112.8}$$

with $\bar{\mathbf{s}}^\alpha$ and $\bar{\mathbf{l}}^\alpha$, defined in (107.30), the lattice vectors $\mathbf{s}^\alpha$ and $\mathbf{l}^\alpha$ pushed forward to the observed space. We refer to

$$\boldsymbol{\xi}^\alpha_{\rm en} \overset{\rm def}{=} -\, f^\alpha_\vdash(\vec{\rho})\,\bar{\mathbf{s}}^\alpha + f^\alpha_\odot(\vec{\rho})\,\bar{\mathbf{l}}^\alpha \tag{112.9}$$

as the energetic microscopic stress for slip system α. Note that $\boldsymbol{\xi}^\alpha_{\rm en}$ *is tangent to slip plane* Π^α — because $\mathbf{s}^\alpha$ and $\mathbf{l}^\alpha$ lie on that plane.

Recall the relation[709]

$$\mathbf{l}^\alpha = \mathbf{m}^\alpha \times \mathbf{s}^\alpha.$$

With a view toward showing that the energetic microscopic stress $\boldsymbol{\xi}^\alpha_{\rm en}$ may be viewed as a combination of distributed **Peach–Koehler forces**, we use (107.30) to pull $\boldsymbol{\xi}^\alpha_{\rm en}$ from the deformed body back to the lattice:

$$\begin{aligned}\boldsymbol{\xi}^{\alpha\#}_{\rm en} &\overset{\rm def}{=} \mathbf{F}^{e-1}\boldsymbol{\xi}^\alpha_{\rm en}\\ &= -\, f^\alpha_\vdash\,\mathbf{s}^\alpha + f^\alpha_\odot\,\mathbf{l}^\alpha \qquad (112.10)\\ &= f^\alpha_\vdash\,(\mathbf{m}^\alpha\times\mathbf{l}^\alpha) + f^\alpha_\odot\,(\mathbf{m}^\alpha\times\mathbf{s}^\alpha). \qquad (112.11)\end{aligned}$$

Arguing as in the paragraph containing (105.32) we therefore see that the microscopic forces $f^\alpha_\vdash\,(\mathbf{m}^\alpha\times\mathbf{l}^\alpha)$ and $f^\alpha_\odot\,(\mathbf{m}^\alpha\times\mathbf{s}^\alpha)$ have the form (105.30), an observation that would seem to justify the following physical interpretations:[710]

$$\underbrace{f^\alpha_\vdash\,(\mathbf{m}^\alpha\times\mathbf{l}^\alpha)}_{\substack{\text{distributed Peach–Koehler}\\ \text{force on edge dislocations}}} \qquad \text{and} \qquad \underbrace{f^\alpha_\odot\,(\mathbf{m}^\alpha\times\mathbf{s}^\alpha).}_{\substack{\text{distributed Peach–Koehler}\\ \text{force on screw dislocations}}} \tag{112.12}$$

The defect forces $f^\alpha_\vdash$ and $f^\alpha_\odot$ therefore represent densities of distributed Peach–Koehler forces and based on this *we view the energetic microscopic stresses* $\boldsymbol{\xi}^{\alpha\#}_{en}$ *as counterparts of Peach–Koehler forces.*

112.2 Dissipative Constitutive Equations that Account for Slip-Rate Gradients

By (112.8) and (112.9),

$$J^{-1}\overline{\Psi^p(\vec{\rho})} = \sum_\alpha \boldsymbol{\xi}^\alpha_{\rm en}\cdot \operatorname{grad}\nu^\alpha \tag{112.13}$$

and (112.5) becomes

$$\sum_\alpha [(\boldsymbol{\xi}^\alpha - \boldsymbol{\xi}^\alpha_{\rm en})\cdot \operatorname{grad}\nu^\alpha + \pi^\alpha\nu^\alpha] \ge 0. \tag{112.14}$$

Thus if we define **dissipative microscopic stresses** $\boldsymbol{\xi}^\alpha_{\rm dis}$ through the relations

$$\boldsymbol{\xi}^\alpha_{\rm dis} = \boldsymbol{\xi}^\alpha - \boldsymbol{\xi}^\alpha_{\rm en}, \tag{112.15}$$

then (112.14) takes the form

$$\sum_\alpha (\pi^\alpha\nu^\alpha + \boldsymbol{\xi}^\alpha_{\rm dis}\cdot \operatorname{grad}\nu^\alpha) \ge 0. \tag{112.16}$$

709 Cf. (103.4).

710 Cf. Footnote 639.

Next, let ∇^α denote the *tangential gradient* of φ on the deformed αth slip plane, so that, given any scalar field φ,

$$\nabla^\alpha \varphi = \operatorname{grad}\varphi - (\bar{\mathbf{m}}^\alpha \cdot \operatorname{grad}\varphi)\bar{\mathbf{m}}^\alpha. \tag{112.17}$$

Underlying the present theory is the tacit assumption that

- *the microscopic stresses $\boldsymbol{\xi}^\alpha_{dis}$ characterize dissipative microscopic forces associated with the motion of dislocations on Π^α.*

Because such dislocations migrate *tangentially* on Π^α, we require that $\boldsymbol{\xi}^\alpha_{\rm dis}$ be *tangential* to the deformed α-th slip plane.[711] Granted this, we may, without loss in generality, replace the slip rate gradients $\operatorname{grad}\nu^\alpha$ in (112.16) by the corresponding *tangential gradients* $\nabla^\alpha \nu^\alpha$; the result is a reduced dissipation inequality

$$\sum_\alpha (\pi^\alpha \nu^\alpha + \boldsymbol{\xi}^\alpha_{\rm dis} \cdot \nabla^\alpha \nu^\alpha) \geq 0 \tag{112.18}$$

basic to our discussion of dissipative constitutive relations.

Our next step is to lay down dissipative constitutive relations for π^α and $\boldsymbol{\xi}^\alpha_{\rm dis}$. In deciding on a constitutive relation for π^α we are guided by the relation (110.18) and by the success of the conventional theory. Specifically, we consider a constitutive relation for π^α that differs from (110.18) only through the replacement of $|\nu^\alpha|$ by an **effective flow rate**

$$d^\alpha \stackrel{\rm def}{=} \sqrt{|\nu^\alpha|^2 + \ell^2 |\nabla^\alpha \nu^\alpha|^2}, \tag{112.19}$$

with ℓ a *dissipative length-scale.* This replacement leads us to introduce a (dimensionless) *rate-sensitivity function*[712] g such that

$$g(0) = 0, \qquad g(d^\alpha) > 0 \ \text{for} \ d^\alpha \neq 0,$$

together with a constitutive equation for π^α of a form

$$\pi^\alpha = S^\alpha g(d^\alpha)\frac{\nu^\alpha}{d^\alpha} \tag{112.20}$$

that bears comparison with the conventional relation (110.18).

We next lay down constitutive relations for the dissipative microscopic stresses $\boldsymbol{\xi}^\alpha_{\rm dis}$. Here, the reduced dissipation inequality (112.18) and a desire for mathematical simplicity suggest a constitutive equation of the *form* (112.20), but with ν^α replaced by $\ell^2 \nabla^\alpha \nu^\alpha$; viz.[713]

$$\boldsymbol{\xi}^\alpha_{\rm dis} = S^\alpha g(d^\alpha)\ell^2 \frac{\nabla^\alpha \nu^\alpha}{d^\alpha}. \tag{112.21}$$

This relation, (112.9), and (112.15) combine to form a constitutive equation for the microscopic stress $\boldsymbol{\xi}^\alpha$:

$$\boldsymbol{\xi}^\alpha = -f^\alpha_{\vdash}\, \bar{\mathbf{s}}^\alpha + f^\alpha_{\circ}\, \bar{\mathbf{l}}^\alpha + S^\alpha g(d^\alpha)\ell^2 \frac{\nabla^\alpha \nu^\alpha}{d^\alpha}. \tag{112.22}$$

Since

$$\operatorname{grad}\nu^\alpha \cdot \nabla^\alpha \nu^\alpha = |\nabla^\alpha \nu^\alpha|^2, \tag{112.23}$$

---

711 Some (unrelated) remarks: (i) by (108.2) and (112.21), $\boldsymbol{\xi}^\alpha$ and $\boldsymbol{\xi}^\alpha_{\rm dis}$ are vectors in the observed space (deformed configuration); (ii) note that (112.9) renders the energetic microstress $\boldsymbol{\xi}^\alpha_{\rm en}$ tangential to the deformed α-th slip plane.

712 For example, one might take $g(d^\alpha) = (d^\alpha/d_0)^m$.

713 Cf. Footnote 642.

the dissipation (112.18) has a simple form

$$\delta = \sum_{\alpha} S^{\alpha} g(d^{\alpha}) d^{\alpha} \tag{112.24}$$

that is strictly analogous to the conventional dissipation

$$\sum_{\alpha} \pi^{\alpha} \nu^{\alpha} = \sum_{\alpha} S^{\alpha} g(|\nu^{\alpha}|)|\nu^{\alpha}|$$

associated with (110.18).

Finally, we assume that the *slip resistances* $S^{\alpha} > 0$, $\alpha = 1, 2, \ldots, N$, are consistent with hardening equations[714]

$$\begin{aligned} \dot{S}^{\alpha} &= \sum_{\alpha} h^{\alpha\beta}(\vec{S}) d^{\beta}, \\ h^{\alpha\beta}(\vec{S}) &= \chi^{\alpha\beta} h(S^{\beta}) + (1 - \chi^{\alpha\beta}) q h(S^{\beta}). \end{aligned} \tag{112.25}$$

112.3 Viscoplastic Flow Rule

The decomposition $\boldsymbol{\xi}^{\alpha} = \boldsymbol{\xi}^{\alpha}_{\mathrm{en}} + \boldsymbol{\xi}^{\alpha}_{\mathrm{dis}}$ allows us to write the microscopic force-balance[715] (108.19) in the form

$$\tau^{\alpha} + \operatorname{div} \boldsymbol{\xi}^{\alpha}_{\mathrm{en}} = \pi^{\alpha} - \operatorname{div} \boldsymbol{\xi}^{\alpha}_{\mathrm{dis}}, \tag{112.26}$$

where we have written the term $\operatorname{div}\boldsymbol{\xi}^{\alpha}_{\mathrm{en}}$ on the left, since, being energetic, its negative represents a backstress. When augmented by the constitutive equations (112.9) and (112.21) the balance (112.26) becomes the **flow rule** for slip system α.[716]

In §105.5 — which was appropriate to small deformations — the counterpart, (105.55), of this flow rule is expressed in terms of second slip gradients. This is not possible within the framework of large deformations[717]

$$\tau^{\alpha} - \underbrace{\operatorname{div}\big(f^{\alpha}_{\vdash}(\vec{\rho})\bar{\mathbf{s}}^{\alpha} - f^{\alpha}_{\odot}(\vec{\rho})\bar{\mathbf{l}}^{\alpha}\big)}_{\text{energetic backstress}} = \underbrace{S^{\alpha} g(d^{\alpha}) \frac{\nu^{\alpha}}{d^{\alpha}} - \ell^2 \operatorname{div}\Big(S^{\alpha} g(d^{\alpha}) \frac{\nabla^{\alpha} \nu^{\alpha}}{d^{\alpha}}\Big)}_{\text{dissipative-hardening}}. \tag{112.27}$$

By (107.32) the fields S^{α}, $\rho^{\alpha}_{\vdash}$, and $\rho^{\alpha}_{\odot}$ should be viewed as *internal-state variables* with evolution governed by the differential equations

$$\begin{aligned} \dot{\rho}^{\alpha}_{\odot} &= \bar{\mathbf{l}}^{\alpha} \cdot \operatorname{grad} \nu^{\alpha}, \qquad \dot{\rho}^{\alpha}_{\vdash} = -\bar{\mathbf{s}}^{\alpha} \cdot \operatorname{grad} \nu^{\alpha}, \\ \dot{S}^{\alpha} &= \sum_{\beta} h^{\alpha\beta}(\vec{S}) d^{\beta}. \end{aligned} \tag{112.28}$$

As such, these equations should be supplemented by initial conditions for S^{α}, $\rho^{\alpha}_{\vdash}$, and $\rho^{\alpha}_{\odot}$. If we assume that *the body is initially in a virgin state*, then appropriate *initial conditions* would be

$$S^{\alpha}(\mathbf{x}, 0) = S_0 \qquad \text{and} \qquad \rho^{\alpha}_{\vdash}(\mathbf{x}, 0) = \rho^{\alpha}_{\odot}(\mathbf{x}, 0) = 0 \tag{112.29}$$

for each slip system α, with $S_0 > 0$, a constant, the *initial slip resistance*.[718]

---

714 Cf. (104.65).

715 For physically attainable processes $b^{\alpha} = 0$.

716 Cf. Gurtin (2000b, eq. (164a)), which has no backstress, and Gurtin (2002, eq. (7.18)) in which $\Psi = \Psi^{p}(\mathbf{G})$ and $\boldsymbol{\xi}^{\alpha}_{\mathrm{dis}} \equiv \mathbf{0}$.

717 Cf. the bullet on page 624.

718 Cf. Footnote 683.

Consider the simple defect energy (105.57), which is uncoupled and quadratic in the net dislocation densities (105.56); viz.

$$\Psi^P(\vec{\rho}) = \tfrac{1}{2} S_0 L^2 \sum_\alpha |\rho^\alpha_{\text{net}}|^2, \qquad \rho^\alpha_{\text{net}} = \sqrt{|\rho^\alpha_\vdash|^2 + |\rho^\alpha_\odot|^2}, \tag{112.30}$$

with L an energetic length-scale and S_0 the initial slip resistance. In this case, if we write

$$\bar{\rho}^\alpha_\vdash = J^{-1}\rho^\alpha_\vdash \qquad \text{and} \qquad \bar{\rho}^\alpha_\odot = J^{-1}\rho^\alpha_\odot \tag{112.31}$$

for the dislocation densities measured per unit volume in the deformed body, then the defect forces (112.6) and energetic microscopic stress (112.9) are given by

$$\begin{gathered} f^\alpha_\vdash = S_0 L^2 \bar{\rho}^\alpha_\vdash, \qquad f^\alpha_\odot = S_0 L^2 \bar{\rho}^\alpha_\odot, \\ \boldsymbol{\xi}^\alpha_{\text{en}} = S_0 L^2 (\bar{\rho}^\alpha_\odot \, \bar{\mathbf{l}}^\alpha - \bar{\rho}^\alpha_\vdash \, \bar{\mathbf{s}}^\alpha), \end{gathered} \tag{112.32}$$

and the flow rule for slip system α becomes[719]

$$\boxed{\tau^\alpha - S_0 L^2 \operatorname{div}(\bar{\rho}^\alpha_\vdash \, \bar{\mathbf{s}}^\alpha - \bar{\rho}^\alpha_\odot \, \bar{\mathbf{l}}^\alpha) = S^\alpha g(d^\alpha)\frac{\nu^\alpha}{d^\alpha} - \ell^2 \operatorname{div}\Big(S^\alpha g(d^\alpha)\frac{\nabla^\alpha \nu^\alpha}{d^\alpha}\Big).} \tag{112.33}$$

112.4 Microscopically Simple Boundary Conditions

Let $\mathcal{B}$ denote the deformed body at an arbitrarily chosen time.[720] The presence of microscopic stresses results in an expenditure of power

$$\int_{\partial\mathcal{B}} (\boldsymbol{\xi}^\alpha \cdot \mathbf{n})\nu^\alpha \, da \tag{112.34}$$

by the material in contact with the body, and this necessitates a consideration of boundary conditions on $\partial\mathcal{B}$ involving the microscopic tractions $\boldsymbol{\xi}^\alpha \cdot \mathbf{n}$ and the slip rates ν^α. We restrict attention to boundary conditions that result in a *null expenditure of microscopic power* in the sense that

$$(\boldsymbol{\xi}^\alpha \cdot \mathbf{n})\nu^\alpha = 0 \text{ on } \partial\mathcal{B} \tag{112.35}$$

for all α.[721] Specifically, we consider *microscopically simple boundary-conditions* asserting that

$$\nu^\alpha = 0 \text{ on } \mathcal{S}_{\text{hard}} \qquad \text{and} \qquad \boldsymbol{\xi}^\alpha \cdot \mathbf{n} = 0 \text{ on } \mathcal{S}_{\text{free}} \tag{112.36}$$

for all α, where $\mathcal{S}_{\text{hard}}$ and $\mathcal{S}_{\text{free}}$ are *complementary* subsurfaces of $\partial\mathcal{B}$ respectively referred to as the *microscopically hard* and the *microscopically free* portions of $\partial\mathcal{B}$.

The microscopically hard condition corresponds to a boundary surface that cannot pass dislocations (e.g., a boundary surface that abuts a hard material); the microscopically free condition corresponds to a boundary across which dislocations can flow freely from the body.

---

719 Cf. (105.61).

720 This subsection follows §105.6.

721 Cf. GURTIN (2000b, eq. (137); 2002, eqs. (9.1) and (9.4)).

112.5 Variational Formulation

Assume that,[722] at some arbitrarily chosen fixed time under consideration, the fields χ and $\mathbf{F}^e$ are known, and let $\mathcal{S}_{\text{hard}}$ and $\mathcal{S}_{\text{free}}$ be complementary subsurfaces of the boundary $\partial\mathcal{B}$ of the deformed body. Then, given any slip system α, if

(i) $\boldsymbol{\xi}^\alpha \cdot \mathbf{n} = 0$ on $\mathcal{S}_{\text{free}}$,
(ii) $\phi \equiv \tilde{\nu}^\alpha$ is the only nonzero virtual slip-rate field, and
(iii) $\phi = 0$ on $\partial\mathcal{S}_{\text{hard}}$,

then (108.17) with $\mathcal{P}_t = \mathcal{B}$ reduces to[723]

$$\int_{\mathcal{B}} [(\pi^\alpha - \tau^\alpha)\phi + \boldsymbol{\xi}^\alpha \cdot \operatorname{grad}\phi]\, dv = 0. \tag{112.37}$$

We refer to ϕ as a *test field* and assume that ϕ is **kinematically admissible** in the sense that

$$\phi = 0 \quad \text{on } \mathcal{S}_{\text{hard}}. \tag{112.38}$$

Then, using (108.3) with $\tilde{\gamma}^\alpha = \phi$ and (112.38), we conclude that (112.37) is equivalent to

$$\int_{\mathcal{S}_{\text{free}}} (\boldsymbol{\xi}^\alpha \cdot \mathbf{n})\phi\, da + \int_{\mathcal{B}} (\pi^\alpha - \tau^\alpha - \operatorname{div}\boldsymbol{\xi}^\alpha)\phi\, dv = 0. \tag{112.39}$$

Moreover, invoking the fundamental lemma of the calculus of variations,[724] we see that (112.39) holds for all kinematically admissible test fields ϕ if and only if $\boldsymbol{\xi}^\alpha \cdot \mathbf{n} = 0$ on $\mathcal{S}_{\text{free}}$ and the microscopic force balance (108.19) is satisfied in $\mathcal{B}$. Since this force balance — supplemented by the constitutive relations $(112.21)_1$ and (112.22) for π^α and $\boldsymbol{\xi}^\alpha$ — is equivalent to the flow rule (112.33) for system α, we have the following result:

VARIATIONAL FORMULATION OF THE FLOW RULE *Suppose that the constitutive relations*

$$\begin{aligned} \pi^\alpha &= S^\alpha g(d^\alpha)\frac{\nu^\alpha}{d^\alpha}, \\ \boldsymbol{\xi}^\alpha &= -f_{\vdash}^\alpha\, \bar{\mathbf{s}}^\alpha + f_{\odot}^\alpha\, \bar{\mathbf{l}}^\alpha + S^\alpha g(d^\alpha)\ell^2 \frac{\nabla^\alpha \nu^\alpha}{d^\alpha}, \end{aligned} \tag{112.40}$$

are satisfied. The flow rule (112.33) *in $\mathcal{B}$ and the boundary condition*

$$\boldsymbol{\xi}^\alpha \cdot \mathbf{n} = 0 \quad \textit{on } \mathcal{S}_{\textit{free}} \tag{112.41}$$

are then together equivalent to the requirement that (112.39) *hold for all kinematically admissible test fields ϕ.*

As noted in the last paragraph of §105.7, this weak statement of the flow rule should provide a useful basis for computations: In a numerical scheme such as the finite-element method, (112.37) would, for each α, reduce to a system of nonlinear algebraic equations for the slip rate ν^α, granted a knowledge of the "current state" of the system.

[722] This subsection follows §105.7.
[723] Cf. GURTIN (2000b, eq. (159); 2002, eq. (10.1)).
[724] Cf. page 167.

112.6 Plastic Free-Energy Balance

By (112.13),

$$\overline{\int_{\mathcal{P}_t} J^{-1}\Psi^p(\vec{\rho})\,dv} = \sum_\alpha \int_{\mathcal{P}_t} \boldsymbol{\xi}^\alpha_{\rm en}\cdot \operatorname{grad}\nu^\alpha\,dv; \tag{112.42}$$

also, using (108.3), (108.19), and (112.15) we can rewrite the right side of (112.43) (for each α) as follows

$$\begin{aligned}\int_{\mathcal{P}_t} \boldsymbol{\xi}^\alpha_{\rm en}\cdot\operatorname{grad}\nu^\alpha\,dv &= \int_{\mathcal{P}_t}[\boldsymbol{\xi}^\alpha_{\rm en}\cdot\operatorname{grad}\nu^\alpha + \underbrace{(\tau^\alpha-\pi^\alpha+\operatorname{div}\boldsymbol{\xi}^\alpha)}_{=0}\nu^\alpha]\,dv\\ &= \int_{\mathcal{P}_t}[(\tau^\alpha-\pi^\alpha)\nu^\alpha - \boldsymbol{\xi}^\alpha_{\rm dis}\cdot\operatorname{grad}\nu^\alpha]\,dv + \int_{\partial\mathcal{P}_t}(\boldsymbol{\xi}^\alpha\cdot\mathbf{n})\nu^\alpha\,da.\end{aligned} \tag{112.43}$$

The identities (112.42) and (112.43) yield the *plastic free-energy balance*

$$\overline{\int_{\mathcal{P}_t} J^{-1}\Psi^p(\vec{\rho})\,dv}$$

$$= \underbrace{\sum_\alpha\int_{\mathcal{P}_t}\tau^\alpha\nu^\alpha\,dv}_{\text{plastic working}} + \underbrace{\sum_\alpha\int_{\partial\mathcal{P}_t}(\boldsymbol{\xi}^\alpha\cdot\mathbf{n})\nu^\alpha\,da}_{\text{microscopic power}} - \underbrace{\sum_\alpha\int_{\mathcal{P}_t}(\pi^\alpha\nu^\alpha + \boldsymbol{\xi}^\alpha_{\rm dis}\cdot\operatorname{grad}\nu^\alpha)\,dv}_{\text{dissipation}\ \geq 0}. \tag{112.44}$$

The term in (112.44) labeled "microscopic power" represents the power expended on $\mathcal{P}_t$ by the microscopic tractions $\boldsymbol{\xi}^\alpha\cdot\mathbf{n}$ acting over $\partial\mathcal{P}_t$. By (112.15) the microscopic stress admits a decomposition $\boldsymbol{\xi}^\alpha = \boldsymbol{\xi}^\alpha_{\rm en} + \boldsymbol{\xi}^\alpha_{\rm dis}$ into energetic and dissipative parts, and, using (112.9), we can write the energetic part of the microscopic power in the form

$$\int_{\partial\mathcal{P}_t}(\boldsymbol{\xi}^\alpha_{\rm en}\cdot\mathbf{n})\nu^\alpha\,da = \int_{\partial\mathcal{P}_t}[-f^\alpha_\vdash\nu^\alpha(\bar{\mathbf{s}}^\alpha\cdot\mathbf{n}) + f^\alpha_\odot\nu^\alpha(\bar{\mathbf{l}}^\alpha\cdot\mathbf{n})]\,da. \tag{112.45}$$

The right side of (112.45) represents power expended across $\partial\mathcal{P}_t$ by the normal components of the Peach–Koehler forces $-f^\alpha_\vdash\mathbf{s}^\alpha$ and $f^\alpha_\odot\mathbf{l}^\alpha$ associated with edge and screw dislocations:[725]

Remark. An interesting interpretation of the relation (112.45) pertains to the *transport of dislocations* as described by their fluxes $\bar{\mathbf{q}}^\alpha_\vdash = \nu^\alpha\bar{\mathbf{s}}^\alpha$ and $\bar{\mathbf{q}}^\alpha_\odot = -\nu^\alpha\bar{\mathbf{l}}^\alpha$.[726] Bearing in mind that the defect forces $f^\alpha_\vdash$ and $f^\alpha_\odot$ defined in (112.6) would — when interpreted within the framework of dislocation transport — be considered as **chemical potentials**[727]

$$\mu^\alpha_\vdash = J^{-1}\frac{\partial\Psi^p}{\partial\rho^\alpha_\vdash}(=f^\alpha_\vdash) \qquad\text{and}\qquad \mu^\alpha_\odot = J^{-1}\frac{\partial\Psi^p}{\partial\rho^\alpha_\odot}(=f^\alpha_\odot)$$

corresponding to energetic changes resulting from changes in edge and screw densities. With this interpretation, $\mu^\alpha_\vdash\bar{\mathbf{q}}^\alpha_\vdash$ and $\mu^\alpha_\odot\bar{\mathbf{q}}^\alpha_\odot$ represent fluxes of energy associated

725 Cf. the paragraph containing (105.32).
726 Cf. (107.34).
727 Cf. §62 or Fried & Gurtin (1999, §2.3.1; 2004, §4A).

with flows of edge and screw dislocations as described by the fluxes $\bar{\mathbf{q}}_{\vdash}^{\alpha}$ and $\bar{\mathbf{q}}_{\odot}^{\alpha}$; thus using (107.34) we may rewrite (112.45) as follows:

$$\int\limits_{\partial\mathcal{P}_t} (\boldsymbol{\xi}_{\rm en}^{\alpha} \cdot \mathbf{n})\nu^{\alpha}\, da = -\int\limits_{\partial\mathcal{P}_t} [\mu_{\vdash}^{\alpha}(\bar{\mathbf{q}}_{\vdash}^{\alpha} \cdot \mathbf{n}) + \mu_{\odot}^{\alpha}(\bar{\mathbf{q}}_{\odot}^{\alpha} \cdot \mathbf{n})]\, da; \tag{112.46}$$

the right side of this relation represents *energy carried into $\mathcal{P}_t$ across $\partial\mathcal{P}_t$ by the flow of dislocations.*

112.7 Some Remarks

Not all facets of the large-deformation theory of this section mirror facets of the small-deformation theory of §105. In particular,

- the slips γ^{α} in the small-deformation theory are well-defined, but in the large-deformation theory, slips — as fields on the individual slip systems — are not well-defined quantities: What are well-defined are the slip-rates ν^{α}.[728]

This difference leads to an important difference in the relations satisfied by the dislocation densities of the two theories. In the small-deformation theory, these densities are given by[729]

$$\rho_{\vdash}^{\alpha} = -\mathbf{s}^{\alpha} \cdot \operatorname{grad}\gamma^{\alpha} \qquad \text{and} \qquad \rho_{\odot}^{\alpha} = \mathbf{l}^{\alpha} \cdot \operatorname{grad}\gamma^{\alpha}. \tag{112.47}$$

But, within the large-deformation theory, the dislocation densities are *internal-state variables* that evolve according to the differential equations[730]

$$\dot{\rho}_{\vdash}^{\alpha} = -\bar{\mathbf{s}}^{\alpha} \cdot \operatorname{grad}\nu^{\alpha} \qquad \text{and} \qquad \dot{\rho}_{\odot}^{\alpha} = \bar{\mathbf{l}}^{\alpha} \cdot \operatorname{grad}\nu^{\alpha}. \tag{112.48}$$

Further, the flow rule (112.33) has an energetic backstress of the form

$$\operatorname{div}\big(f_{\vdash}^{\alpha}(\vec{\rho})\bar{\mathbf{s}}^{\alpha} - f_{\odot}^{\alpha}(\vec{\rho})\bar{\mathbf{l}}^{\alpha}\big)$$

and is hence dependent on gradients of dislocation densities. Because the dislocation densities are *internal-state variables* with evolution governed by (112.48) (and suitable initial conditions), the backstress depends on *histories of second slip-rate gradients*; this — an essential feature of the theory[731] — should be compared with the flow rule (105.61) of the corresponding small-deformation theory in which the backstress depends on *current* values of second slip-gradients (not slip-*rate* gradients!).

An essential feature of the theory is the manner in which it qualitatively mimics certain characteristics of dislocations:[732]

- The plastically convected rate $(107.27)_1$ of the Burgers tensor $\mathbf{G}$ is found to be the sum of rates of continuous distributions of screw and edge dislocations on the individual slip systems, distributions that have the canonical forms for slip in a single-crystal. These distributions are automatically consistent with balances in a form

$$\{\text{density rate}\} = -\{\text{divergence of a flux}\} + \{\text{supply}\}$$

728 Cf. the bullet on page 624.
729 Cf. $(103.9)_{2,3}$.
730 Cf. (107.32).
731 We are unaware of other single-crystal theories with this property.
732 The results of the first two bullets, taken from Gurtin (2006), are essential to the complete picture given below.

standard in theories of transport.[733] Moreover the fluxes are in glide directions; each of the density-rates is given, modulo sign, as a directional derivative of the corresponding slip-rate in the glide direction.

- A free energy dependent on dislocation densities leads to thermodynamically conjugate microscopic forces (stresses) parallel to glide directions of the corresponding edge and screw distributions, which is a central feature of Peach–Koehler forces.[734]

---

733 Cf. (107.37).

734 Cf. the bullet on page 609.

APPENDIX

113 Isotropic Functions

In this section we state several important representation theorems for isotropic functions; that is, functions invariant under the proper orthogonal group

$$\text{Orth}^+ = \text{the group of all rotations.} \tag{113.1}$$

Let $\mathcal{A}$ be a subset of the set of all tensors. We say that $\mathcal{A}$ is **invariant under** Orth^+ if, given any $\mathbf{A}$ in $\mathcal{A}$, the tensor $\mathbf{QAQ}^\top$ belongs to $\mathcal{A}$ for all rotations $\mathbf{Q}$. Sets with this property are:

$$\text{Lin}^+ = \text{the set of all tensors with strictly positive determinant,}$$
$$\text{Sym} = \text{the set of all symmetric tensors,}$$
$$\text{Psym} = \text{the set of all symmetric and } \textit{positive-definite} \text{ tensors,}$$
$$\text{Dsym} = \text{the set of all symmetric and } \textit{deviatoric} \text{ tensors.}$$

To help fix notation, we begin by listing some constitutive equations of the general type under consideration.

(i) The elastic constitutive relations[735]

$$\psi = \bar{\psi}(\mathbf{C}),$$
$$\mathbf{T}_{\mathrm{RR}} = \bar{\mathbf{T}}_{\mathrm{RR}}(\mathbf{C}).$$

Here, the free energy ψ is a scalar, the right Cauchy–Green tensor $\mathbf{C}$ is symmetric and positive-definite, and the second Piola stress $\mathbf{T}_{\mathrm{RR}}$ is symmetric: the constitutive response functions $\bar{\psi}$ and $\bar{\mathbf{T}}_{\mathrm{RR}}$ are therefore mappings[736]

$$\bar{\psi} : \text{Psym} \to \mathbb{R} \quad (\mathbb{R} = \text{the set of real numbers}),$$
$$\bar{\mathbf{T}}_{\mathrm{RR}} : \text{Psym} \to \text{Sym}.$$

(ii) The constitutive equation

$$\mathbf{T} = \hat{\mathbf{T}}(\upsilon, \mathbf{D})$$

of a compressible, viscous fluid.[737] Here, because the Cauchy stress $\mathbf{T}$ and the stretching $\mathbf{D}$ are symmetric tensors, while υ is a scalar and hence irrelevant to

---

[735] Cf. (50.18).

[736] The notation

$$f : A \to H$$

is shorthand for the statement: f is a function (or mapping) that associates with each element a in a set A an element $h = f(a)$ in the set H.

[737] Cf. $(45.7)_2$.

a discussion of material symmetry, we may consider the constitutive response function $\hat{\mathbf{T}}$ as a mapping

$$\hat{\mathbf{T}} : \text{Sym} \to \text{Sym}.$$

Note that the domain of $\bar{\mathbf{T}}_{\text{RR}}$ is Psym, while the domain of $\mathbf{T}$ is Sym. To avoid treating such functions individually, we work with a generic subset $\mathcal{A}$ of the set of all tensors assuming only that

- $\mathcal{A}$ is invariant under Orth$^+$.

113.1 Isotropic Scalar Functions

A scalar function

$$\hat{g} : \mathcal{A} \to \mathbb{R},$$

is **isotropic** if, for each $\mathbf{A}$ in $\mathcal{A}$,

$$\hat{g}(\mathbf{A}) = \hat{g}(\mathbf{Q}\mathbf{A}\mathbf{Q}^{\top})$$

for all rotations $\mathbf{Q}$. Examples of isotropic scalar functions are:

(i) tr *and* det *considered as scalar functions on* $\mathcal{A}$;
(ii) *the principal invariants* $I_1(\mathbf{A})$, $I_2(\mathbf{A})$, *and* $I_3(\mathbf{A})$, *defined in* (2.142), *considered as functions of* $\mathbf{A}$ *in* $\mathcal{A}$.

For convenience we write $\mathcal{I}_{\mathbf{A}}$ for the list of principal invariants:

$$\mathcal{I}_{\mathbf{A}} \stackrel{\text{def}}{=} (I_1(\mathbf{A}), I_2(\mathbf{A}), I_3(\mathbf{A})). \tag{113.2}$$

Representation Theorem for Isotropic Scalar Functions *A scalar function*

$$\hat{g} : \mathcal{A} \to \mathbb{R} \qquad (\mathcal{A} \subset \text{Sym})$$

is isotropic if and only if there is a scalar function $\tilde{g}$ *such that*

$$\hat{g}(\mathbf{A}) = \tilde{g}(\mathcal{I}_{\mathbf{A}}) \tag{113.3}$$

for every $\mathbf{A}$ *in* $\mathcal{A}$.[738]

113.2 Isotropic Tensor Functions

A tensor function

$$\hat{\mathbf{G}} : \mathcal{A} \to \text{Sym}$$

is **isotropic** if, for each $\mathbf{A}$ in $\mathcal{A}$,

$$\mathbf{Q}\hat{\mathbf{G}}(\mathbf{A})\mathbf{Q}^{\top} = \hat{\mathbf{G}}(\mathbf{Q}\mathbf{A}\mathbf{Q}^{\top})$$

for all rotations $\mathbf{Q}$.

Properties of Isotropic Tensor Functions *Let*

$$\hat{\mathbf{G}} : \mathcal{A} \to \text{Sym} \qquad (\mathcal{A} \subset \text{Sym})$$

be an isotropic tensor function. Then $\hat{\mathbf{G}}$ *has the following properties:*

(i) (Transfer Property) *Given any* $\mathbf{A}$ *in* $\mathcal{A}$, *every eigenvector of* $\mathbf{A}$ *is an eigenvector of* $\hat{\mathbf{G}}(\mathbf{A})$.

738 Cf., e.g., Truesdell & Noll (1965, p. 28); Gurtin (1981, p. 230). The domain of $\tilde{g}$ is the set of all triplets $\mathcal{I}_{\mathbf{A}}$ with $\mathbf{A}$ in $\mathcal{A}$.

(ii) (Commutation Property) *For every* **A** *in* $\mathcal{A}$,

$$\hat{\mathbf{G}}(\mathbf{A})\mathbf{A} = \mathbf{A}\hat{\mathbf{G}}(\mathbf{A}). \tag{113.4}$$

We shall only prove (ii).[739] Choose **A** in $\mathcal{A}$ and let

$$\mathbf{G} = \hat{\mathbf{G}}(\mathbf{A}).$$

By hypothesis, both **A** and **G** are symmetric. Thus, in particular, **A** has a spectral decomposition

$$\mathbf{A} = \sum_{i=1}^{3} a_i \mathbf{e}_i \otimes \mathbf{e}_i,$$

where, for each i, a_i is an eigenvalue of **A** and $\mathbf{e}_i$ is a corresponding eigenvector; further, $\{\mathbf{e}_i\}$ is an orthonormal basis, so that[740]

$$\mathbf{e}_i \cdot \mathbf{e}_j = \delta_{ij}. \tag{113.5}$$

Next, in view of the transfer property (i), each of the eigenvectors $\mathbf{e}_i$ of **A** must be an eigenvector of **G**; hence **G** must have a spectral decomposition of the form

$$\mathbf{B} = \sum_{i=1}^{3} b_i \mathbf{e}_i \otimes \mathbf{e}_i.$$

Thus, by (2.28) and (113.5),

$$\begin{aligned}
\mathbf{BA} &= \left(\sum_{i=1}^{3} b_i \mathbf{e}_i \otimes \mathbf{e}_i\right)\left(\sum_{j=1}^{3} a_j \mathbf{e}_j \otimes \mathbf{e}_j\right) \\
&= \sum_{i,j=1}^{3} b_i a_j \underbrace{(\mathbf{e}_i \otimes \mathbf{e}_i)(\mathbf{e}_j \otimes \mathbf{e}_j)}_{=\delta_{ij}(\mathbf{e}_i \otimes \mathbf{e}_j)} \\
&= \sum_{i=1}^{3} b_i a_i \mathbf{e}_i \otimes \mathbf{e}_i \\
&= \sum_{i=1}^{3} a_i b_i \mathbf{e}_i \otimes \mathbf{e}_i \\
&= \mathbf{AB},
\end{aligned}$$

which is the desired result.

Representation Theorem for Isotropic Tensor Functions *A tensor function*

$$\hat{\mathbf{G}} : \mathcal{A} \to \mathrm{Sym} \qquad (\mathcal{A} \subset \mathrm{Sym})$$

is isotropic if and only if there are scalar functions β_1, β_2, *and* β_3 *such that*

$$\hat{\mathbf{G}}(\mathbf{A}) = \beta_1(\mathcal{I}_{\mathbf{A}})\mathbf{1} + \beta_2(\mathcal{I}_{\mathbf{A}})\mathbf{A} + \beta_3(\mathcal{I}_{\mathbf{A}})\mathbf{A}^2 \tag{113.6}$$

for every **A** *in* $\mathcal{A}$.[741]

---

739 For a proof of (i), cf., e.g., Truesdell & Noll (1965, p. 32); Gurtin (1981, p. 231).

740 Cf. (2.112).

741 Cf., e.g., Truesdell & Noll (1965, p. 32); Gurtin (1981, p. 233).

Assume now that the domain $\mathcal{A}$ of $\hat{\mathbf{G}}$ is Psym. Then, for any $\mathbf{A}$ in Psym, $\mathbf{A}$ is invertible, so that, by the Cayley–Hamilton equation (2.144) with $\mathbf{S} = \mathbf{A}$,

$$\mathbf{A}^2 = I_1(\mathbf{A})\mathbf{A} - I_2(\mathbf{A}) + I_3(\mathbf{A})\mathbf{A}^{-1},$$

and we may rewrite the representation (113.6) in the form

$$\hat{\mathbf{G}}(\mathbf{A}) = \alpha_1(\mathcal{I}_{\mathbf{A}})\mathbf{1} + \alpha_2(\mathcal{I}_{\mathbf{A}})\mathbf{A} + \alpha_3(\mathcal{I}_{\mathbf{A}})\mathbf{A}^{-1}. \tag{113.7}$$

We have therefore established the "only if" assertion in the following result: A tensor function

$$\hat{\mathbf{G}} : \mathcal{A} \to \text{Sym} \qquad (\mathcal{A} \subset \text{Psym})$$

is isotropic if and only if there are scalar functions α_1, α_2, and α_3 such that

$$\hat{\mathbf{G}}(\mathbf{A}) = \beta_1(\mathcal{I}_{\mathbf{A}})\mathbf{1} + \beta_2(\mathcal{I}_{\mathbf{A}})\mathbf{A} + \beta_3(\mathcal{I}_{\mathbf{A}})\mathbf{A}^{-1} \tag{113.8}$$

for every $\mathbf{A}$ in $\mathcal{A}$.[742]

113.3 Isotropic Linear Tensor Functions

When an isotropic tensor function is linear its representation is quite simple:

REPRESENTATION THEOREMS FOR ISOTROPIC LINEAR TENSOR FUNCTIONS[743]

(i) *A tensor function*

$$\hat{\mathbf{G}} : \text{Sym} \to \text{Sym}$$

is isotropic if and only if there are scalars μ *and* λ *such that*

$$\hat{\mathbf{G}}(\mathbf{A}) = 2\mu\mathbf{A} + \lambda(\operatorname{tr}\mathbf{A})\mathbf{1} \tag{113.9}$$

for every $\mathbf{A}$ *in* Sym.

(ii) *A tensor function*

$$\hat{\mathbf{G}} : \text{Dsym} \to \text{Sym}$$

is isotropic if and only if there is a scalar μ *such that*

$$\hat{\mathbf{G}}(\mathbf{A}) = 2\mu\mathbf{A} \tag{113.10}$$

for every $\mathbf{A}$ *in* Dsym.

EXERCISES

1. Show that the sets Lin^+, Sym, Psym, and Dsym are invariant under Orth^+.
2. Prove that the $\operatorname{tr}\mathbf{A}$, $\det\mathbf{A}$, $I_1(\mathbf{A})$, $I_2(\mathbf{A})$, and $I_3(\mathbf{A})$ considered as scalar functions of $\mathbf{A}$ on $\mathcal{A}$ are isotropic.
3. Show that the mapping $\hat{\mathbf{G}} : \text{Lin}^+ \to \text{Lin}^+$ defined by
$$\hat{\mathbf{G}}(\mathbf{A}) = \mathbf{A}^{-1}$$
is isotropic.
4. Show that the mapping $\hat{\mathbf{G}} : \text{Sym} \to \text{Sym}$ defined by
$$\hat{\mathbf{G}}(\mathbf{A}) = \mathbf{A}^n \qquad (n \geq 1, \text{ an integer})$$
is isotropic.

[742] Cf., e.g., TRUESDELL & NOLL (1965, p. 33) and GURTIN (1981, p. 235).
[743] Cf., e.g., GURTIN (1981, pp. 235–236).

114 The Exponential of a Tensor

Let $\mathbf{A}$ be a tensor and consider the initial-value problem

$$\left.\begin{aligned}\dot{\mathbf{Z}}(t) &= \mathbf{A}\mathbf{Z}(t),\\ \mathbf{Z}(0) &= \mathbf{1},\end{aligned}\right\} \tag{114.1}$$

for a tensor function $\mathbf{Z}(t)$. The existence theorem for ordinary differential equations tells us that this problem has *exactly one solution* $\mathbf{Z}(t)$, $-\infty < t < \infty$, which we write in the form

$$\mathbf{Z}(t) = \mathrm{e}^{t\mathbf{A}}. \tag{114.2}$$

Two important properties of the function $\mathrm{e}^{t\mathbf{A}}$ defined in this manner are:

(i) *given any tensor* $\mathbf{A}$, *the tensor* $\mathrm{e}^{t\mathbf{A}}$ *is invertible for* $-\infty < t < \infty$. *In fact,* $\det(\mathrm{e}^{t\mathbf{A}})$ *is strictly positive and satisfies*

$$\det(\mathrm{e}^{t\mathbf{A}}) = \mathrm{e}^{t(\mathrm{tr}\mathbf{A})}; \tag{114.3}$$

(ii) *let* $\mathbf{\Omega}$ *be a skew tensor. Then* $\mathrm{e}^{t\mathbf{\Omega}}$ *is a rotation for* $-\infty < t < \infty$.

The verification of (i), which is technical, may be found in Gurtin (1981, §36). To prove (ii), let $\mathbf{\Omega}$ be a *skew tensor* and let

$$\mathbf{Z}(t) = \mathrm{e}^{t\mathbf{\Omega}}, \qquad \mathbf{Y}(t) = \mathbf{Z}(t)\mathbf{Z}^{\top}(t). \tag{114.4}$$

Then, by $(114.1)_1$, since $\mathbf{\Omega} = -\mathbf{\Omega}^{\top}$,

$$\begin{aligned}\dot{\mathbf{Y}} &= \dot{\mathbf{Z}}\mathbf{Z}^{\top} + \mathbf{Z}\dot{\mathbf{Z}}^{\top}\\ &= \mathbf{\Omega}\mathbf{Z}\mathbf{Z}^{\top} - \mathbf{Z}\mathbf{Z}^{\top}\mathbf{\Omega}\\ &= \mathbf{\Omega}\mathbf{Y} - \mathbf{Y}\mathbf{\Omega}.\end{aligned}$$

Thus $\mathbf{Y}$ satisfies

$$\left.\begin{aligned}\dot{\mathbf{Y}}(t) &= \mathbf{\Omega}\mathbf{Y}(t) - \mathbf{Y}(t)\mathbf{\Omega},\\ \mathbf{Y}(0) &= \mathbf{1}.\end{aligned}\right\} \tag{114.5}$$

This initial-value problem has a unique solution; by substitution, we see that $\mathbf{Y}(t) \equiv \mathbf{1}$ is that solution. Thus

$$\mathbf{Z}(t)\mathbf{Z}^{\top}(t) \equiv \mathbf{1}$$

and $\mathbf{Z}(t)$ is a rotation. We have therefore established (ii).

Remarks.

(a) The definition of the exponential of a tensor via a solution of an ordinary differential equation is consistent with the series representation[744]

$$e^{\mathbf{A}} = \sum_{n=0}^{\infty} \frac{1}{n!}\mathbf{A}^n.$$

(b) An immediate consequence of (ii) and $(114.1)_2$ is that the rotation

$$\mathbf{Q}(t) = e^{t\boldsymbol{\Omega}}$$

satisfies

$$\mathbf{Q}(0) = \mathbf{1} \qquad \text{and} \qquad \dot{\mathbf{Q}}(0) = \boldsymbol{\Omega}. \tag{114.6}$$

Thus, given an arbitrary skew tensor $\boldsymbol{\Omega}$, it is always possible to find a rotation $\mathbf{Q}(t)$ such that, at some time t_0, $\mathbf{Q}(t_0) = \mathbf{1}$, $\dot{\mathbf{Q}}(t_0) = \boldsymbol{\Omega}$.

744 Cf., e.g., Hirsh & Smale (1974, Chapter 5, §3).

References

ACHARYA, A., BASSANI, J.L., & BEAUDOIN, A., 2003. Geometrically necessary dislocations, hardening, and a simple theory of crystal plasticity. *Scripta Materialia* **48**, 167–172.

ÅGREN, J., 1982. Diffusion in phases with several components and sublattices. *Journal of the Physics and Chemistry of Solids* **43**, 421–430.

AIFANTIS, E.C., 1984. On the microstructural origin of certain inelastic models. *ASME Journal of Engineering Materials and Technology* **106**, 326–330.

AIFANTIS, E.C., 1987. The physics of plastic deformation. *International Journal of Plasticity* **3**, 211–247.

AIFANTIS, E.C., 1999. Strain gradient interpretation of size effects. *International Journal of Fracture* **95**, 299–314.

ANAND, L., 1979. On H. Hencky's approximate strain-energy function for moderate deformations. *Journal of Applied Mechanics* **46**, 78–82.

ANAND, L., 1986. Moderate deformations in extension-torsion of incompressible isotropic elastic materials. *Journal of the Mechanics and Physics of Solids* **34**, 293–304.

ANAND, L., 1996. A constitutive model for compressible elastomeric solids. *Computational Mechanics* **18**, 339–355.

ANAND, L., BALASUBRAMANIAN, S., & KOTHARI, M., 1997. Constitutive modeling of polycrystalline aggregates at large strains. In: TEODOSIU, C. (Ed.), *Large plastic deformation of crystalline aggregated*, CISM Courses and Lectures No. 376., SpringerWien, New York, 109–172.

ANAND, L., GURTIN, M.E., LELE, S.P., & GETHING, C., 2005. A one-dimensional theory of strain-gradient plasticity: formulation, analysis, numerical results. *Journal of the Mechanics and Physics of Solids* **53**, 1789–1826.

ARSENLIS, A., & PARKS, D.M., 1999. Crystallographic aspects of geometrically-necessary and statistically-stored dislocation density. *Acta Materialia* **47**, 1597–1611.

ASARO, R.J., 1983. Micromechanics of crystals and polycrystals. *Advances in Applied Mechanics* **23**, 1–115.

ASARO, R.J., & NEEDLEMAN, A., 1985. Texture development and strain hardening in rate dependent polycrystals. *Acta Metallurgica* **33**, 923–953.

ASHBY, M.F., 1970. The deformation of plastically non-homogeneous alloys. *Philosophical Magazine* **22**, 399–424.

BALASUBRAMANIAN, S., & ANAND, L., 2002. Elasto-viscoplastic constitutive equations for polycrystalline fcc materials at low homologous temperatures. *Journal of the Mechanics and Physics of Solids* **50**, 101–126.

BASSANI, J.L., & WU, T.Y., 1991. Latent hardening in single crystals–II, analytical characterization and predictions. *Proceedings of the Royal Society of London A* **435**, 21–41.

BASSANI, J.L., 1994. Plastic flow of crystals. *Advances in Applied Mechanics* **30**, 191–258.

BAUSCHINGER, J., 1886. Über die Veränderung der Position der Elastizitätsgrenze des Eisens und Stahls durch Strecken und Quetschen und durch Erwärmen und Abkühlen und durch oftmals wiederholte Beanspruchungen. *Mitteilung aus dem Mechanisch-technischen Laboratorium der Königlichen polytechnischen Hochschule in München* **13**, 1–115.

BILBY, B.A., BULLOUGH, R., & SMITH, E., 1955. Continuous distributions of dislocations: a new application of the methods of non-Riemannian geometry. *Proceedings of the Royal Society of London A* **231**, 263–273.

BILBY, B.A., 1960. Continuous distributions of dislocations. In: SNEDDON, I.N., & HILL, R. (Eds.), *Progress in Solid Mechanics*, 329–395.

BIRD, W.W., & MARTIN, J.B., 1986. A secant approximation for holonomic elastic-plastic incremental analysis with a von Mises yield condition. *Engineering Computations* **3**, 192–201.

BIRD, W.W., & MARTIN, J.B., 1990. Consistent predictors and the solution of the piecewise holonomic incremental problem in elasto-plasticity. *Engineering Structures* **12**, 9–14.

BOWEN, F.M., & WANG, C.C., 1976. *Introduction to Vectors and Tensors, A: Linear and Multilinear Algebra*. Plenum, New York.

BRAND, L., 1947. *Vector and tensor analysis*. Wiley, New York.

BRIDGMAN, P.W., 1952. *Studies in Large Plastic Flow and Fracture with Special Emphasis on the Effects of Hydrostatic Pressure*. McGraw-Hill, New York.

BRONKHORST, C.A., KALIDINDI, S.R., & ANAND, L., 1992. Polycrystalline plasticity and the evolution of crystallographic texture in fcc metals. *Philosophical Transactions of the Royal Society of London A* **341**, 443–477.

BROWN, S., KIM, K., & ANAND, L., 1989. An internal variable constitutive model for hot working of metals. *International Journal of Plasticity* **5**, 95–130.

BURGERS, J.M., 1939. Some considerations of the field of stress connected with dislocations in a regular crystal lattice. *Koninklijke Nederlandse Akademie van Wetenschappen* **42**, 293–325 (Part 1); 378–399 (Part 2).

BURNETT, D., 1936. The distribution of molecular velocities and the mean motion in a non-uniform gas. *Proceedings of the London Mathematical Society* **40**, 382–435.

BUTTKE, T.F., 1993. Velocity methods: Lagrangian numerical methods which preserve the Hamiltonian structure of incompressible fluid flow. In: BEALE, J.T., COTTET, G.H., & HUBERSON, S. (Eds.), *Vortex Flows and Related Numerical Methods*, Kluwer, Dordrecht.

CAHN, J.W., & LARCHÉ, F.C., 1983. An invariant formulation of multicomponent diffusion in crystals. *Scripta Metallurgica* **17**, 927–932.

CAILLETAUD, G., 1992. A micromechanical approach to inelastic behavior of metals. *International Journal of Plasticity* **8**, 55–73.

CALLEN, H.B., 1960. *Thermodynamics*. Wiley, New York.

CARR, N.L., KOBAYASHI, R., & BURROWS, D.B., 1954. Viscosity of hydrocarbon gases under pressure. *Petroleum Transactions of the American Institute of Mining and Metallurgical Engineers* **201**, 264–272.

CASEY, J., & NAGHDI, P.M., 1980. A remark on the use of the decomposition $\mathbf{F} = \mathbf{F}^e\mathbf{F}^p$ in plasticity. *Journal of Applied Mechanics* **47**, 672–675.

CERMELLI, P., FRIED, E., & SELLERS, S., 2001. Configurational stress, yield, and flow in rate-independent plasticity. *Proceedings of the Royal Society of London A* **457**, 1447–1467.

CERMELLI, P., GURTIN, M.E., 2001. On the characterization of geometrically necessary dislocations in finite plasticity. *Journal of the Mechanics and Physics of Solids* **49**, 1539–1568.

CERMELLI, P., & GURTIN, M.E., 2002. Geometrically necessary dislocations in viscoplastic single crystals and bicrystals undergoing small deformations. *International Journal of Solids and Structures* **39**, 6281–6309.

CHADWICK, P., 1976. *Continuum mechanics: concise theory and problems*. Allen & Unwin, London.

COLEMAN, B.D., & NOLL, W., 1963. The thermodynamics of elastic materials with heat conduction and viscosity. *Archive for Rational Mechanics and Analysis* **13**, 167–178.

DASHNER, P.A., 1986. Invariance considerations in large strain elasto-plasticity. *Journal of Applied Mechanics* **53**, 55–60.

DAVÌ, F., & GURTIN, M.E., 1990. On the motion of a phase interface by surface diffusion. *Zeitschrift für angewandte Mathematik und Physik* **41**, 782–811.

DESERI, L., & OWEN, D.R., 2002. Invertible structured deformations and the geometry of multiple slip in single crystals. *International Journal of Plasticity* **18**, 833–849.

DAWSON, P.R., & MARIN, E.B., 1998. Computational mechanics for metal deformation processes using polycrystal plasticity. *Advances in Applied Mechanics* **34**, 77–169.

DRUCKER, D.C., 1950. Some implications of work-hardening and ideal plasticity. *Quarterly of Applied Mathematics* **7**, 411–418.

DRUCKER, D. C., 1951. A more fundamental approach to plastic stress strain relations. In: STERNBERG, E. (Ed.), *Proceedings of the US National Congress of Applied Mechanics*, Ann Arbor, Edwards Brothers Inc., 279–291.

DRUCKER, D.C., 1964. On the postulate of stability of material in the mechanics of continua. *Journal de Mechanique* **3**, 235–250.

DUNN, E., & FOSDICK, R., 1974. Thermodynamics, stability, and boundedness of fluids of complexity 2 and fluids of second grade. *Archive for Mechanics and Analysis* **56**, 191–252.

DUVAUT, G. & LIONS, J.L., 1972. *Les Inequations en Mechanique et en Physique*. Dunod, Paris.

E, W., & LIU, J.G., 2003. Gauge methods for viscous incompressible flows. *Communications in Mathematical Sciences* **1**, 317–332.

ECKART, C., 1940. The thermodynamics of irreversible processes, II. Fluid mixtures. *Physical Review* **58**, 269–275.

EDELEN, D.G., & MCLENNAN, J.A., 1973. Material indifference: A principle or a convenience. *International Journal of Engineering Science* **11**, 813–817.

FARREN, W.S., & TAYLOR, G.I., 1925. The heat development during plastic extension of metals. *Proceedings of the Royal Society of London A* **107**, 422–451.

FLECK, N.A., MULLER, G.M., ASHBY, M.F., & HUTCHINSON, J.W., 1994. Strain gradient plasticity: theory and experiment. *Acta Metallurgica et Materialia* **42**, 475–487.

FLECK, N.A., & HUTCHINSON, J.W., 1997. Strain gradient plasticity. *Advances in Applied Mechanics* **33**, 295–361.

FLECK, N.A., & HUTCHINSON, J.W., 2001. A reformulation of strain gradient plasticity. *Journal of the Mechanics and Physics of Solids* **49**, 2245–2271.

FLORY, P.J., 1944. Network structure and the elastic properties of vulcanized rubber. *Chemical Reviews* **35**, 51–75.

Foias, C., Holm, D.D., & Titi, E.S., 2001. The Navier–Stokes-alpha model of fluid turbulence. *Physica D* **152**, 505–519.

Frazer, R.A., Duncan, W.J., & Collar, A.R., 1938. *Elementary Matrices and Some Applications to Dynamics and Differential Equations*. Cambridge University Press, Cambridge.

Frank, F.C., 1951. Crystal dislocations. Elementary concepts and definitions. *Philosophical Magazine* **42**, 809–819.

Fredriksson, P., & Gudmundson, P., 2005. Size dependent yield strength of thin films. *International Journal of Plasticity* **21**, 1834–1854.

Fried, E., & Gurtin, M.E., 1999. Coherent solid-state phase transitions with atomic diffusion: a thermomechanical treatment. *Journal of Statistical Physics* **95**, 1361–1427.

Fried, E., & Gurtin, M.E., 2004. A unified treatment of evolving interfaces accounting for deformation and atomic transport with emphasis on grain-boundaries and epitaxy. *Advances in Applied Mechanics* **40**, 1–177.

Fried, E., & Gurtin, M.E., 2009. Gradient nanoscale polycrystalline elasticity: intergrain interactions and triple-junction conditions. *Journal of the Mechanics and Physics of Solids* (submitted).

Fried, E., & Sellers, S., 2000. Theory for atomic diffusion on fixed and deformable crystal lattices. *Journal of Elasticity* **59**, 67–81.

Gantmacher, F.R., 1959. *Matrix Theory*, Volume I. Chelsea, New York.

Gatski, T.B., & Wallin, S., 2004. Extending the weak-equilibrium condition for algebraic Reynolds stress models to rotating and curved flows, *Journal of Fluid Mechanics* **518**, 147–155.

Gent, A.N., 1996. A new constitutive relation for rubber. *Rubber Chemistry and Technology* **69**, 59–61.

Green, A.E., & Naghdi, P.M., 1971. Some remarks on elastic-plastic deformation at finite strain. *International Journal of Engineering Science* **9**, 1219–1229.

Gough, J., 1805. A description of a property of Caoutchouc or indian rubber; with some reflections on the case of the elasticity of this substance. *Memoirs of the Literary and Philosophical Society of Manchester* **1**, 288–295.

de Groot, D.R., & Mazur, P., 1962. *Thermodynamics of Irreversible Processes*. North-Holland, Amsterdam.

Gudmundson, P., 2004. A unified treatment of strain gradient plasticity. *Journal of the Mechanics and Physics of Solids* **52**, 1379–1406.

van der Gulik, P.S., 1997. Viscosity of carbon dioxide in the liquid phase. *Physica A* **238**, 81–112.

Gurtin, M.E., 1981. *An Introduction to Continuum Mechanics*. Academic Press, New York.

Gurtin, M.E., 1991. On thermomechanical laws for the motion of a phase interface. *Zeitschrift für angewandte Mathematik und Physik* **42**, 370–388.

Gurtin, M.E., 2000a. *Configurational Forces as Basic Concepts of Continuum Physics*. Springer, New York.

Gurtin, M.E., 2000b. On the plasticity of single crystals: free energy, microforces, plastic strain gradients. *Journal of the Mechanics and Physics of Solids* **48**, 989–1036.

Gurtin, M.E., 2002. A gradient theory of single-crystal viscoplasticity that accounts for geometrically necessary dislocations. *Journal of the Mechanics and Physics of Solids* **50**, 5–32.

Gurtin, M.E., 2003. On a framework for small-deformation viscoplasticity: free energy, microforces, strain gradients. *International Journal of Plasticity* **19**, 47–90.

GURTIN, M.E., 2004. A gradient theory of small-deformation isotropic plasticity that accounts for the Burgers vector and for dissipation due to plastic spin. *Journal of the Mechanics and Physics of Solids* **52**, 2545–2568.

GURTIN, M.E., 2006. The Burgers vector and the flow of screw and edge dislocations in finite-deformation single-crystal plasticity. *Journal of the Mechanics and Physics of Solids* **54**, 1882–1898.

GURTIN, M.E., 2008a. A theory of grain boundaries that accounts automatically for grain misorientation and grain-boundary orientation. *Journal of the Mechanics and Physics of Solids* **56**, 640–662.

GURTIN, M.E., 2008b. A finite-deformation, gradient theory of single-crystal plasticity with free energy dependent on densities of geometrically necessary dislocations. *International Journal of Plasticity* **24**, 702–725.

GURTIN, M.E., 2010. A finite-deformation, gradient theory of single-crystal plasticity with free energy dependent on the accumulation of geometrically necessary dislocation. *International Journal of Plasticity*, in preparation.

GURTIN, M.E., & ANAND, L., 2005a. A theory of strain-gradient plasticity for isotropic, plastically irrotational materials. Part I: Small deformations. *Journal of the Mechanics and Physics of Solids* **53**, 1624–1649.

GURTIN, M.E., & ANAND, L., 2005b. A theory of strain-gradient plasticity for isotropic, plastically irrotational materials. Part II: Finite deformations. *International Journal of Plasticity* **21**, 2297–2318.

GURTIN, M.E., & ANAND, L., 2005c. The decomposition $\mathbf{F} = \mathbf{F}^e\mathbf{F}^p$, material symmetry, and plastic irrotationality for solids that are isotropic-viscoplastic or amorphous. *International Journal of Plasticity* **21**, 1686–1719.

GURTIN, M.E., ANAND, L., & LELE, S.P., 2007. Gradient single-crystal plasticity with free energy dependent on dislocation densities. *Journal of the Mechanics and Physics of Solids* **55**, 1853–1878.

GURTIN, M.E., & ANAND, L., 2009. Thermodynamics applied to gradient theories involving the accumulated plastic strain: the theories of Aifantis and Fleck & Hutchinson and their generalization. *Journal of the Mechanics and Physics of Solids* **57**, 405–421.

GURTIN, M.E., & NEEDLEMAN, A., 2005. Boundary conditions in small-deformation, single-crystal plasticity that account for the Burgers vector. *Journal of the Mechanics and Physics of Solids* **53**, 1–31.

GURTIN, M.E., & REDDY, B.D, 2009. Alternative formulations of isotropic hardening for Mises materials, and associated variational inequalities. *Continuum Mechanics and Thermodynamics* **21**, 237–250.

GURTIN, M.E., & STRUTHERS, A., 1990. Multiphase thermomechanics with interfacial structure 3. Evolving phase boundaries in the presence of bulk deformation. *Archive for Rational Mechanics and Analysis* **112**, 97–160.

GURTIN, M.E., & VARGAS, A.S., 1971. On the classical theory of reacting fluid mixtures. *Archive for Rational Mechanics and Analysis* **43**, 179–197.

GURTIN, M.E., & WILLIAMS, W.O., 1966. On the Clausius–Duhem inequality. *Zeitschrift für angewandte Mathematik und Physik* **17**, 626–633.

GURTIN, M.E., & WILLIAMS, W.O., 1967. An axiomatic foundation for continuum thermodynamics. *Archive for Rational Mechanics and Analysis* **26**, 83–117.

GUTH, E., & MARK, J.E., 1934. Zur innermoleckularen Statistik, insbesondere bei Kettenmolekülen I. *Monatshefte für Chemie* **65**, 93–121.

HARTMAN, P., 1964. *Ordinary Differential Equations*. Wiley, New York.

HALMOS, P.R., 1950. *Measure Theory*. Van Nostrand, New York.

HALMOS, P.R., 1958. *Finite-Dimensional Vector Spaces*. Van Nostrand, Princeton.

HACKL, K., 1997. Generalized standard media and variational principles in classical and finite strain elastoplasticity. *Journal of the Mechanics and Physics of Solids* **45**, 667–688.

HAN, W., & REDDY, B.D., 1999. *Plasticity: Mathematical Theory and Numerical Analysis*. Springer, New York.

HAVNER, K.S., 1992. *Finite Plastic Deformation of Crystalline Solids*. Cambridge University Press, Cambridge.

HENCKY, H., 1928. Über die form des elastizitätsgesetzes bei ideal elastischen stoffen. *Zeitschrift für Technische Physik* **9**, 215–223.

HENCKY, H., 1933. The elastic behavior of vulcanised rubber. *Rubber Chemistry and Technology* **6**, 217–224.

HILL, R., 1950. *The Mathematical Theory of Plasticity*. Oxford University Press, New York.

HILL, R., 1965. Continuum micro-mechanics of elastoplastic polycrystals. *Journal of the Mechanics and Physics of Solids* **13**, 89–101.

HILL, R., 1968. On constitutive inequalities for simple materials. *Journal of the Mechanics and Physics of Solids* **16**, 229–242.

HIRSCH, M.W., & SMALE, S., 1974. *Differential Equations, Dynamical Systems, and Linear Algebra*. Academic Press, New York.

HIRTH, J.P., & LOTHE, J., 1992. *Theroy of Dislocations, 2nd Edition*. Wiley, New York.

HODOWANY, J., RAVICHANDRAN, G., ROSAKIS, A.J. & ROSAKIS, P., 2000. Partition of plastic work into heat and sored energy in metals. *Journal of Experimental Mechanics* **40**, 113–123.

HOLZAPFEL, G.A., 2000. *Nonlinear Solid Mechanics*. Wiley, Chichester.

HUBER, M.T., 1904. Specific work of strain as a measure of material effort (in Polish). *Czasopismo Techniczne* **22**, Lwów, Organ Towarzystwa Politechicznego, we Lwowie. English Translation published in *Archives of Mechanics* **56** (2004), 171–172.

HUTCHINSON, J.W., 1970. Elastic-plastic behavior of polycrystalline metals and composites. *Proceedings of the Royal Society of London A* **319**, 247–272.

HUTCHINSON, J.W., 2000. Plasticity at the micron scale. *International Journal of Solids and Structures* **37**, 225–238.

IL'YUSHIN, A.A., 1954. On the relation between stress and small deformations in the mechanics of continuous media (in Russian). *Prikladnaya Matematika I Mekhanika* **18**, 641–666.

IL'YUSHIN, A.A., 1961. On the postulate of plasticity. *Prikladnaya Matematika I Mekhanika* **25**, 503–507.

JAMMER, M., 1957. *Concepts of Force; a Study in the Foundations of Dynamics*. Harvard University Press, Cambridge.

JAUMANN, G., 1911. Geschlossenes System physikalischer und chemischer Differentialgesetze. *Kaiserliche Akademie der Wisenschaften in Wien Mathematisch-Naturwissenschaftliche Klasse* **120**, 385–530.

JEFFREY, A., 2002. *Advanced engineering mathematics*. Harcourt Academic Press, San Diego.

JOHNSON, C., 1976a. Existency theorems for plasticity problems. *Journal de Mathematiques Pures et Appliques* **55**, 431–444.

JOHNSON, C., 1976b. On finite element methods for plasticity problems. *Numerische Mathematik* **26**, 79–84.

JOHNSON, C., 1978. On plasticity with hardening. *Journal of Applied Mathematical Analysis* **62**, 325–336.

JOULE, J.P., 1859. On some thermodynamic properties of solids. *Philosophical Transactions of the Royal Society of London A* **149**, 91–131.

KALIDINDI, S.R., BRONKHORST, C.A., & ANAND, L., 1992. Crystallographic texture evolution in bulk deformation processing of fcc metals. *Journal of the Mechanics and Physics of Solids* **40**, 536–569.

KARIM, S.M., & ROSENHEAD, L., 1952. The second coefficient of viscosity of liquids and gases. *Reviews of Modern Physics* **24**, 108–116.

KELLOGG, O.D., 1953. *Foundations of Potential Theory*. Dover, New York.

KOCKS, U.F., 1987. Constitutive behavior based on crystal plasticity. In: MILLER, A.K. (Ed.), *Constitutive Equations for Creep and Plasticity*, Elsevier Applied Science, 1–88.

KONDO, K., 1952. On the geometrical and physical foundations of the theory of yielding. *Proceedings Japan National Congress of Applied Mechanics* **2**, 41–47.

KRÖNER, E., 1960. Allgemeine kontinuumstheorie der versetzungen und eigenspannungen. *Archive for Rational Mechanics and Analysis* **4**, 273–334.

KUBIN, L.P., CANOVA, G., CONDAT, M., DEVINCRE, B., PONTIKIS, V., & BRECHET, Y., 1992. Dislocation microstructures and plastic flow: a 3D simulation. *Solid State Phenomena* **23**, 455–472.

KUHLMANN-WILSDORF, D., 1989. Theory of plastic deformation: properties of low energy dislocation structures. *Materials Science and Engineering A* **113**, 1–41.

KUHN, W., 1934. Über die Gestalt fadenförmiger Moleküle in Lösungen. *Kolloid Zeitschrift* **68**, 2–15.

LARCHÉ, F.C., & CAHN, J.W., 1973. A linear theory of thermochemical equilibrium of solids under stress. *Acta Metallurgica* **21**, 1051–1063.

LARCHÉ, F.C., & CAHN, J.W., 1985. The interactions of composition and stress in crystalline solids. *Acta Metallurgica* **33**, 331–357.

LEE, E.H., & LIU, D.T., 1967. Finite-strain elastic-plastic theory with application to plane-wave analysis. *Journal of Applied Physics* **38**, 19–27.

LEE, E.H., 1969. Elastic-plastic deformation at finite strains. *ASME Journal of Applied Mechanics* **36**, 1–6.

LEMAITRE, J., & CHABOCHE, J.L., 1990. *Mechanics of Solid Materials*. Cambridge University Press, Cambridge.

LÉVY, M., 1871. Extrait du Mémoire sur les équations generales des mouvements intérieurs des corps solides ductiles au delà des limites òu l'élasticite pourrait les ramener à leur premier état. *Journal de Mathématique Pures et Appliquées, Série II* **16**, 369–372.

LLOYD, D.J., 1994. Particle reinforced aluminum and magnesium matrix composites. *International Metallurgical Reviews* **39**, 1–23.

LOHR, E., 1917. Entropieprinzip und geschlossenes Gleichungssystem. *Denkschriften der Kaiserlichen Akademie der Wissenschaften, Mathematisch-Naturwissenschaftliche Klasse* **93**, 339–421.

LUBLINER, J., 1990. *Plasticity Theory*. MacMillan, New York.

LUCCHESI, M., & PODIO-GUIDUGLI, P., 1990. Equivalent dissipation postulates in classical plasticity. *Meccanica* **25**, 26–31.

LUMLEY, J.L., 1970. Toward a turbulent constitutive relation. *Journal of Fluid Mechanics* **41**, 413–434.

LUMLEY, J.L., 1983. Turbulence modeling. *Journal of Applied Mechanics* **50**, 1097–1103.

LUSH, A.M., WEBER, G., & ANAND, L., 1989. An implicit time-integration procedure for a set of internal variable constitutive equations for isotropic elasto-viscoplasticity. *International Journal of Plasticity* **5**, 521–529.

MA, Q., & CLARKE, D.R., 1995. Size dependent hardness of silver single crystals. *Journal of Materials Research* **10**, 853–863.

Maddocks, J. H., & Pego, R.L., 1995. An unconstrained Hamiltonian formulation for incompressible fluid flow. *Communications in Mathematical Physics* **170**, 207–217.

Malvern, L.E., 1969. *Introduction to the Mechanics of a Continuous Medium*, Prentice-Hall, Englewood Cliffs, NJ.

Mandel, J., 1965. Généralization de la théorie de plasticité de W. T. Koiter. *International Journal of Solids and Structures* **1**, 273–295.

Mandel, J., 1972. *Plasticité Classique et Viscoplasticité*, CISM Courses and Lectures No. 97. Springer-Verlag, Berlin.

Mandel, J., 1973. Thermodynamics and plasticity. In: Delgado Domingas, J. J., Nina, M. N.R., & Whitelaw, J.H. (Eds.), *Proceedings of The International Symposium on Foundations of Continuum Thermodynamics*, Halsted Press, New York, 283–304.

Martin, J.B., 1981. An internal variable approach to the formulation of finite element problems in plasticity. In: Hult, J., & Lemaitre, J. (Eds.), *Physical Nonlinearities in Structural Analysis*, Springer-Verlag, Berlin, 165–176.

Maugin, G., 1993. *Material Inhomogeneities in Elasticity*. Chapman Hall, London.

Meixner, J., & Reik, H.G., 1959. Die Thermodynamik der irreversiblen Prozesse. In: Flügge, S. (Ed.), *Handbuch der Physik* III/2, Springer-Verlag, Berlin.

Miehe, C., Schroder, J., & Schotte, J., (1999). Computational homogenization analysis in finite plasticity simulation of texture development in polycrystalline materials. *Computer Methods in Applied Mechanics and Engineering* **171**, 387–418.

Mielke, A., 2003. Energetic formulation of multiplicative elasto-plasticity using dissipation distances. *Continuum Mechanics and Thermodynamics* **15**, 351–382.

Moreau, J.J., 1976. Application of convex analysis to the treatment of elastoplastic systems. In: Germain, P., & Nayroles, B. (Eds.), *Applications of Methods of Functional Analysis to Problems in Mechanics*, Springer-Verlag, Berlin.

Moreau, J.J., 1977. Evolution problem associated with a moving convex set in a Hilbert space. *Journal of Differential Equations* **26**, 347–374.

Mooney, M., 1940. A theory of large elastic deformation. *Journal of Applied Physics* **11**, 582–592.

Mühlhaus, H.B., & Aifantis, E.C., 1991. A variational principle for gradient plasticity. *International Journal of Solids Structures* **28**, 845–857.

Müller, I., 1967. On the entropy inequality. *Archive for Rational Mechanics and Analysis* **26**, 118–141.

Müller, I., 1968. A thermodynamic theory of mixtures of fluids. *Archive for Rational Mechanics and Analysis* **28**, 1–39.

Müller, I., 1972. On the frame-dependence of stress and heat flux. *Archive for Rational Mechanics and Analysis* **45**, 241–250.

Murdoch, A.I., 2003. Objectivity in classical continuum physics: a rationale for discarding the ‘principle of invariance under superposed rigid body motions’ in favour of purely objective considerations. *Continuum Mechanics and Thermodynamics* **15**, 309–320.

Murdoch, A.I., 2006. Some primitive concepts in continuum mechanics regarded in terms of objective space-time molecular averaging: The key rôle played by inertial observers. *Journal of Elasticity* **84**, 69–97.

Nickerson, H.K., Spencer, D.C., & Steenrod, N.E., 1959. *Advanced Calculus*. Van Nostrand, Princeton, NJ.

Nicola, L., Van der Giessen, E., & Gurtin, M.E., 2005. Effect of defect energy on strain-gradient predictions of confined single-crystal plasticity. *Journal of the Mechanics and Physics of Solids* **53**, 1280–1294.

NOLL, W., 1955. On the continuity of the solid and fluid states. *Journal of Rational Mechanics and Analysis* **4**, 3–81.

NOLL, W., 1995. On material frame-indifference. Center for Nonlinear Analysis Research Report No. 95-NA-022. Carnegie Mellon University, Pittsburgh.

NOLL, W., 1963. Le mećanique classical, Baseé sur un Axiome d'objectiveté. In: *La Méthode Axiomatic dans les Mécanique Classiques et Nouvelles* (Colloque International, Paris, 1959), Gauthiers-Villars, 47–56.

NYE, J.F., 1953. Some geometrical relations in dislocated solids. *Acta Metallurgica* **1**, 153–162.

OHASHI, T., 1997. Finite element analysis of plastic slip and evolution of geometrically necessary dislocations in fcc crystals. *Philosophical Magazine Letters* **75**, 51–57.

OHNO, N., & OKUMURA, D., 2007. Higher-order stress and grain size effects due to self-energy of geometrically necessary dislocations. *Journal of the Mechanics and Physics of Solids* **55**, 1879–1898.

OHNO, N., OKUMURA, D., & SHIBATA, T., 2008. Grain-size dependent yield behavior under loading, unloading, and reverse loading. *International Journal of Modern Physics B* 22, 5937–5942.

ORTIZ, M., & REPETTO, O., 1999. Nonconvex energy minimization and dislocation structures in ductile single crystals. *Journal of the Mechanics and Physics of Solids* **36**, 286–351.

PENROSE, R., 1989. *The Emperor's New Mind: Concerning Computers, Minds, and the Laws of Physics*. Oxford University Press, Oxford.

PEIRCE, D., ASARO, R.D., & NEEDLEMAN, A., 1983. Material rate dependence and localized deformation in crystalline solids. *Acta Metallurgica* **31**, 1951–1976.

PERCIVAL, I., & RICHARDS, D., 1982. *Introduction to Dynamics*. Cambridge University Press, Cambridge.

PIERCE, C.S., 1934. *Collected papers* **5**, HARTSHORNE, C. & WEISS, P. (Eds.), Harvard University Press, Cambridge.

PIPKIN, J., & RIVILIN, R., 1965. Mechanics of rate-independent materials. *Zeitschrift für Angewandte Mathematik und Physik* **16**, 313–326.

PODIO-GUIDUGLI, P., 1997. Inertia and invariance. *Annali di Matematica Pura e Applicata* (IV) **172**, 103–124.

PODIO-GUIDUGLI, P., 2000. *A Primer in Elasticity*. Kluwer, Dordrecht. (Reprinted from the *Journal of Elasticity* **58** (2000), 1–104.)

PODIO-GUIDUGLI, P., 2006. Models of phase segregation and diffusion of atomic species on a lattice. *Richerce di Matematica* **55**, 105–118.

PRANDTL, L., 1924. Spannungsverteilung in plastischen Körpern. In: *Proceedings of the Ist International Congress on Applied Mechanics*, Delft, 43–54.

REDDY, B.D., 1992. Existence of solutions to a quasistatic problem in elastoplasticity. In: BANDLE, C. et al. (Eds.), *Progress in Partial Differential Equations: Calculus of Variations, Applications*, Pitman Research Notes in Mathematics **267**, Longman, London, 233–259.

REDDY, B.D., 1997. *Introductory Functional Analysis*. Springer-Verlag, New York.

REDDY, B.D., EBOBISSE, F., & MCBRIDE, A., 2008. Well-posedness of a model of strain gradient plasticity for plastically irrotational materials. *International Journal of Plasticity* **24**, 55–73.

REUSS, A., 1930. Berückistigung der elastischen Formänderungen in der Plastizitätstheorie. *Zeitsschrift für angewandte Mathematik und Mechanik*. **10**, 266–274.

RICE, J.R., 1971. Inelastic constitutive relations for solids: an internal-variable theory and its applications to metal plasticity. *Journal of the Mechanics and Physics of Solids* **19**, 443–455.

Rosakis, P., Rosakis, A.J., Ravichandran, G. & Hodowany, J.A., 2000. A thermodynamic internal variable model for the partition of plastic work into heat and stored energy in metals. *Journal of Mechanics and Physics of Solids* **48**, 581–607.

Russo, G., & Smereka, P., 1999. Impulse formulation of the Euler equations: general properties and numerical methods. *Journal of Fluid Mechanics* **391**, 189–209.

Simo, J.C., & Hughes, T.J.R., 1998. *Computational Inelasticity.* Springer, New York.

Sirotin, Yu.I., & Shaskolskaya, M.P., 1982. *Fundamentals of Crystal Physics*, Mir, Moscow.

Spitzig, W.A., Sober, R.J., & Richmond, O., 1975. Pressure-dependence of yielding and associated volume expansion in tempered martensite. *Acta Metallurgica* **23**, 885–893.

Stelmashenko, N.A., Walls, M.G., Brown, L.M., & Milman, Y.V., 1993. Microindentations on W and Mo oriented single crystals: an STM study. *Acta Metallurgica et Materialia* **41**, 2855–2865.

Stephenson, R.A., 1980. On the uniqueness of the square-root of a symmetric, positive-definite tensor. *Journal of Elasticity* **10**, 213–214.

Stokes, G.G., 1845. On the theories of the internal friction of fluids in motion, and of the equilibrium and motion of elastic solids. *Transactions of the Cambridge Philosophical Society* **8**, 287–319.

Stolken, J.S., & Evans, A.G., 1998. A microbend test method for measuring the plasticity length scale. *Acta Materialia* **46**, 5109–5115.

Stewart, F.M., 1963. *Introduction to Linear Algebra.* Van Nostrand, Princeton, NJ.

Strang, G., 1986. *Introduction to Applied Mathematics.* Wellesley-Cambridge Press, Wellesley, MA.

Sun, S., Adams, B.L., Shet, C.Q., Saigal, S., & King, W., 1998. Mesoscale investigation of the deformation field of an aluminum bicrystal. *Scripta Materialia* **39**, 501–508.

Sun, S., Adams, B.L. & King, W., 2000. Observations of lattice curvature near the interface of a deformed bicrystal. *Philosophical Magazine A* **80**, 9–25.

Taylor, G.I., 1938. Plastic strain in metals. *Journal of the Institute of Metals* **62**, 307–324.

Taylor, G.I., & Quinney, H., 1931. The plastic distortion of metals. *Philosophical Transactions of the Royal Society of London A* **230**, 323–362.

Taylor, G.I., & Quinney, H., 1937. The latent heat remaining in a metal after cold working. *Proceedings of the Royal Society of London A* **163**, 157–181.

Temam, R., 1985. *Mathematical Problems in Plasticity*, Gauthier-Villars, Paris (Translation of original 1983 French edition).

Teodosiu, C., 1970. A dynamic theory of dislocations and its applications to the theory of the elastic-plastic continuum. In: Simmons, J.A. (Ed.), *Proceedings of the Conference on Fundamental Aspects of Dislocation Theory.* U.S. National Bureau of Standards, 837–876.

Teodosiu, C., & Sidoroff, F., 1976. A theory of finite elastoviscoplasticity of single crystals. *International Journal of Engineering Science* **14**, 165–176.

Teodosiu, C., 1982. *Elastic Models of Crystal Defects.* Springer-Verlag, Berlin.

Ting, T.C.T., 1996. *Anisotropic Elasticity: Theory and Applications.* Oxford: Oxford University Press.

Tisza, L., 1942. Supersonic absorption and Stokes' viscosity relation. *Physical Review* **61**, 531–536.

THORNE, K.S., 1994. *Black Holes and Time Warps, Einstein's Outrageous Legacy.* W. W. Norton and Co., New York.

TRELOAR, L.R.G., 1943. The elasticity of a network of long-chain molecules. I. *Transactions of the Faraday Society* **39**, 36–41.

TRUESDELL, C., 1954. The present status of the controversy regarding the bulk viscosity of fluids. *Proceedings of the Royal Society London A* **226**, 59–65.

TRUESDELL, C., 1966. *The Elements of Continuum Mechanics.* Springer, New York.

TRUESDELL, C., 1966. *Six Lectures on Modern Natural Philosophy.* Springer-Verlag, Berlin.

TRUESDELL, C., 1969. *Rational Thermodynamics, A Course of Lectures on Selected Topics.* McGraw-Hill, New York.

TRUESDELL, C., 1976. Correction of two errors in the kinetic theory of gases which have been used to cast unfounded doubt upon the principle of material frame-indifference. *Meccanica* **11**, 196–199.

TRUESDELL, C.A., 1991. *A First Course in Rational Continuum Mechanics.* Second Edition. Academic Press, New York.

TRUESDELL, C., & NOLL, W., 1965. *The Nonlinear Field Theories of Mechanics.* In FLÜGGE, S. (Ed.), *Handbuch der Physik* III/3. Springer, Berlin.

TRUESDELL, C., & TOUPIN, R., 1960. *The Classical Field Theories.* In: FLÜGGE, S. (Ed.), *Handbuch der Physik* III/1, Springer, Berlin.

VON MISES, R., 2004, 1913. Mechanik der festen Körper im plastisch-deformablen Zustand. *Nachrichten der königlichen Gesellschaft der Wissenschaften zu Göttingen, Mathematisch-Physikalische Klasse*, 582–592.

WALL, F.T., 1942. Statistical thermodynamics of rubber. *Journal of Chemical Physics* **10**, 132–134.

WANG, C.C., 1969. Appendix 7. In: *Rational Thermodynamics, A Course of Lectures on Selected Topics*, McGraw-Hill, New York.

WEBER, G., & ANAND, L., 1990. Finite deformation constitutive equations and a time integration procedure for isotropic, hyperelastic-viscoplastic solids. *Computer Methods in Applied Mechanics and Engineering* **79**, 173–202.

WEIS, J., & HUTTER, K., 2003. On Euclidean invariance of algebraic Reynolds stress models in turbulence, *Journal of Fluid Mechanics* **476**, 63–68.

Index

D

Index

Index

N

Index

Printed in the United States
By Bookmasters